BASIC ORGANIC
STEREOCHEMISTRY

BASIC ORGANIC STEREOCHEMISTRY

ERNEST L. ELIEL
The University of North Carolina at Chapel Hill
Chapel Hill, North Carolina

SAMUEL H. WILEN
(deceased) formerly of
The City College of the City University of New York
New York, New York

MICHAEL P. DOYLE
University of Arizona
Tucson, Arizona

A JOHN WILEY & SONS, INC., PUBLICATION
New York • Chichester • Weinheim • Brisbane • Singapore • Toronto

Library of Congress Cataloging-in-Publication Data is available.

Eliel, Ernest Ludwig, 1921-
    Basic organic stereochemistry / Ernest L. Eliel, Samuel H. Wilen, Michael P. Doyle.
      p. cm.
    Includes bibliographical references and index.
    ISBN 0-471-37499-7 (pbk.)
      1. Stereochemistry. I. Wilen, Samuel H. II. Doyle, Michael P. III. Title.

QD481.E515 2001
547'.1223—dc21   2001017847

10 9 8 7 6 5 4 3

CONTENTS

Preface   **xiii**

1 Introduction   **1**

   1-1. Scope, 1
   1-2. History, 1
   1-3. Polarimetry and Optical Rotation, 5
         References, 7

2 Structure   **8**

   2-1. Meaning, Factorization, Internal Coordinates, Isomers, 8
   2-2. Constitution, 11
   2-3. Configuration, 13
   2-4. Conformation, 15
   2-5. Determination of Structure, 17
   2-6. A Priori Calculation of Structure, 20
   2-7. Molecular Models, 25
         References, 26

3 Stereoisomers   **30**

   3-1. Nature of Stereoisomers, 30
         a. General, 30
         b. Barriers Between Stereoisomers and Residual Stereoisomers, 34
   3-2. Enantiomers, 36
   3-3. Diastereomers, 39
         a. General Cases, 39

                                                                              v

       b. Degenerate Cases, 42
      References, 44

4 Symmetry   **45**

    4-1. Introduction, 45
    4-2. Symmetry Elements, 45
    4-3. Symmetry Operators and Symmetry Point Groups, 48
       a. Point Groups Containing Chiral Molecules, 49
       b. Point Groups Containing Only Achiral Molecules, 51
    4-4. Averaged Symmetry, 58
    4-5. Symmetry and Molecular Properties, 59
       a. Rotation of Polarized Light, 59
       b. Dipole Moment, 61
       c. Symmetry Number, 62
      References, 63

5 Configuration   **65**

    5-1. Definitions: Relative and Absolute Configuration, 65
    5-2. Absolute Configuration and Notation, 67
    5-3. Determination of Absolute Configuration, 75
       a. Bijvoet Method, 75
       b. Theoretical Approaches, 77
       c. Modification of Crystal Morphology in the Presence of Additives, 77
    5-4. Relative Configuration and Notation, 79
    5-5. Determination of Relative Configuration of Saturated Aliphatic Compounds, 84
       a. X-Ray Structure Analysis, 85
       b. Chemical Interconversion Not Affecting Bonds to the Stereogenic Atom, 86
       c. Methods Based on Symmetry Considerations, 86
       d. Correlation Via Compounds with Chiral Centers of Two Types, 89
       e. The Method of Quasi-Racemates, 90
       f. Chemical Correlations Affecting Bonds to a Chiral Atom in a "Known" Way (For an overview, see ref. 32.), 90
       g. Correlation by Stereoselective Synthesis of "Known" Stereochemical Course, 95
       h. Chiroptical, Spectroscopic, and Other Physical Methods, 98
    5-6. Conclusion: Network Arguments, 98
      References, 98

6 Properties of Stereoisomers and Stereoisomer Discrimination   **102**

    6-1. Introduction, 102
    6-2. Stereoisomer Discrimination, 102
    6-3. The Nature of Racemates, 106

6-4. Properties of Racemates and of Their Enantiomer Components, 108
    a. Introduction, 108
    b. Optical Activity, 109
    c. Crystal Shape, 109
    d. Density and Racemate Type, 110
    e. Melting Point, 111
    f. Solubility, 115
    g. Vapor Pressure, 119
    h. Infrared Spectra, 120
    i. Electronic Spectra, 121
    j. Nuclear Magnetic Resonance Spectra, 122
    k. X-Ray Spectra, 123
    l. Liquid State and Interfacial Properties, 124
    m. Chromatography, 128
    n. Mass Spectrometry, 129
    o. Interaction with Other Chiral Substances, 130
    p. Biological Properties, 132
    q. Origins of Enantiomeric Homogeneity in Nature, 138
6-5. Determination of Enantiomer and Diastereomer Composition, 142
    a. Introduction, 142
    b. Chiroptical Methods, 145
    c. NMR Methods Based on Diastereotopicity, 147
    d. Chromatographic and Related Separation Methods Based on Diastereomeric Interactions, 160
    e. Kinetic Methods, 176
    f. Miscellaneous Methods, 178
    References, 180

7 Separation of Stereoisomers, Resolution, and Racemization   **197**

7-1. Introduction, 197
7-2. Separation of Enantiomers by Crystallization, 198
    a. Crystal Picking and Triage, 198
    b. Conglomerates, 198
    c. Preferential Crystallization, 201
    d. Asymmetric Transformation of Racemates and Total Spontaneous Resolution, 204
7-3. Chemical Separation of Enantiomers via Diastereomers, 209
    a. Formation and Separation of Diastereomers; Resolving Agents, 209
    b. Resolution Principles and Practice, 227
    c. Separation Via Complexes and Inclusion Compounds, 231
    d. Chromatographic Resolution, 236
    e. Asymmetric Transformations of Diastereomers, 240
    f. General Methods for the Separation of Diastereomers, 246
7-4. Enantiomeric Enrichment and Resolution Strategy, 253

7-5. Kinetic Resolution, 257
    a. Theory and Stoichiometric and Abiotic Catalytic Kinetic
       Resolution, 258
    b. Enzymatic Resolution, 268
7-6. Miscellaneous Separation Methods, 274
7-7. Racemization, 277
    a. Racemization Processes, 278
    b. Racemization of Amino Acids, 284
    References, 287

**8 Heterotopic Ligands and Faces: Prostereoisomerism and
Prochirality**   **303**

8-1. Introduction and Terminology, 303
8-2. Significance and History, 305
8-3. Homotopic and Heterotopic Ligands and Faces, 307
    a. Homotopic Ligands and Faces, 307
    b. Enantiotopic Ligands and Faces, 310
    c. Diastereotopic Ligands and Faces, 312
    d. Concepts and Nomenclature, 315
8-4. Heterotopicity and Nuclear Magnetic Resonance, 318
    a. General Principles. Anisochrony, 318
    b. NMR in Assignment of Configuration and of Descriptors of
       Prostereoisomerism, 320
    c. Origin of Anisochrony, 323
    d. Conformationally Mobile Systems, 325
8-5. Heterotopic Ligands and Faces in Enzyme-Catalyzed Reactions, 329
    a. Heterotopicity and Stereoelective Synthesis, 329
    b. Heterotopicity and Enzyme-Catalyzed Reactions, 330
    References, 335

9 Stereochemistry of Alkenes   **339**

9-1. Structure of Alkenes and Nature of cis–trans Isomerism, 339
    a. General, 339
    b. Nomenclature, 340
    c. Cumulenes, 342
    d. Alkenes with Low Rotational Barriers and Nonplanar Alkenes, 342
    e. The $C{=}N$ and $N{=}N$ Double Bonds, 346
9-2. Determination of Configuration of cis–trans Isomers, 348
    a. Chemical Methods, 348
    b. Physical Methods, 353
9-3. Interconversion of cis–trans Isomers: Position of Equilibrium and
    Methods of Isomerization, 362
    a. Position of cis–trans Equilibria, 362
    b. Methods of Equilibration, 366

c. Directed cis–trans Interconversion, 368
References, 371

10 Conformation of Acyclic Molecules   **376**

10-1. Conformation of Ethane, Butane, and Other Simple Saturated Acyclic
Molecules, 376
a. Alkanes, 376
b. Saturated Acyclic Molecules with Polar Substituents or Chains
and the Anomeric Effect, 383
10-2. Conformation of Unsaturated Acyclic and Miscellaneous
Compounds, 388
a. Unsaturated Acyclic Compounds, 388
b. Alkylbenzenes, 395
c. Miscellaneous Compounds, 397
10-3. Physical and Spectral Properties of Diastereomers and Conformers, 398
a. General, 398
b. Dipole Moments, 399
c. Infrared Spectra, 400
d. NMR Spectroscopy, 401
10-4. Conformation and Reactivity: The Winstein–Holness Equation and the
Curtin–Hammett Principle, 407
References, 415

11 Configuration and Conformation of Cyclic Molecules   **421**

11-1. Stereoisomerism and Configurational Nomenclature of Ring
Compounds, 421
11-2. Determination of Configuration of Substituted Ring Compounds, 423
a. Introduction, 423
b. Symmetry-Based Methods, 424
c. Methods Based on Physical and Chemical Properties, 425
d. Correlation Methods, 427
11-3. Stability of Cyclic Molecules, 429
a. Strain, 429
b. Ease of Cyclization as a Function of Ring Size, 432
c. Ease of Ring Closure as a Function of the Ring Atoms and
Substituents: The Thorpe–Ingold Effect, 433
d. Baldwin's Rules, 434
11-4. Conformational Aspects of the Chemistry of Six-Membered Ring
Compounds, 436
a. Cyclohexane, 436
b. Monosubstituted Cyclohexanes, 439
c. Disubstituted and Polysubstituted Cyclohexanes, 447
d. Conformation and Physical Properties in Cyclohexane
Derivatives, 453

e. Conformation and Reactivity in Cyclohexanes, 457

f. *sp²* Hybridized Cyclohexyl Systems, 463

g. Six-Membered Saturated Heterocycles, 472

11-5. Chemistry of Ring Compounds Other than Six-Membered Ones, 480

a. Three-Membered Rings, 480

b. Four-Membered Rings, 481

c. Five-Membered Rings, 482

d. Rings Larger Than Six-Membered, 485

11-6. Stereochemistry of Fused, Bridged, and Caged Ring Systems, 491

a. Fused Rings, 492

b. Bridged Rings, 501

c. Propellanes, 505

d. Catenanes, Rotaxanes, Knots, and Möbius Strips, 505

e. Cubane, Tetrahedrane, Dodecahedrane, Adamantane, and Buckminsterfullerene, 513

References, 517

12 Chiroptical Properties   **534**

12-1. Introduction, 534

12-2. Optical Activity and Anisotropic Refraction, 535

a. Origin and Theory, 535

b. Optical Rotatory Dispersion, 541

12-3. Circular Dichroism and Anisotropic Absorption, 544

12-4. Applications of Optical Rotatory Dispersion and Circular Dichroism, 548

a. Determination of Configuration and Conformation: Theory, 548

b. Classification of Chromophores, 550

c. Sector and Helicity Rules, 553

d. Exciton Chirality, 567

e. Other Applications: Induced ORD and CD, 570

f. Circular Dichroism of Chiral Polymers, 576

12-5. Applications of Optical Activity, 585

a. Polarimetry, 585

b. Empirical Rules and Correlations: Calculation of Optical Rotation, 593

12-6. Vibrational Optical Activity, 597

References, 598

13 Chirality in Molecules Devoid of Chiral Centers   **608**

13-1. Introduction and Nomenclature, 608

13-2. Allenes, 611

a. Historical Overview and Natural Occurrence, 611

b. Synthesis of Optically Active Allenes, 612

c. Determination of Configuration and Enantiomeric Purity of Allenes, 613

d. Cyclic Allenes, Cumulenes, and Ketene Imines, 616

13.3. Alkylidenecycloalkanes, 617

13-4. Spiranes, 620

13-5. Biphenyls and Atropisomerism, 622

a. Introduction, 622

b. Biphenyls and Other Atropisomers of the sp^2–sp^2 Single-Bond Type, 623

c. Atropisomerism About sp^2–sp^3 Single Bonds, 629

d. Atropisomerism About sp^3–sp^3 Bonds, 630

13-6. Molecular Propellers, 632

13-7. Helicenes, 636

13-8. Molecules with Planar Chirality, 638

a. Introduction, 638

b. Cyclophanes, 639

c. *trans*-Cycloalkenes, 640

d. Metallocenes and Related Compounds, 642

References, 642

Index   **649**

PREFACE

In 1994 John Wiley & Sons, Inc. published *Stereochemistry of Organic Compounds* co-authored by one of us (E.L.E) with the late Samuel H. Wilen, with a chapter on stereoselective synthesis by Lewis N. Mander. This comprehensive volume was well received in spite its length of 1267 pages. The book serves well as a reference text and a refresher resource for established chemical scientists, and also as a reliable source of stereochemical information for graduate students, but seems too extensive to be used in a semester course. Yet stereochemical thinking is increasing in importance not only in the core area of chemistry, but also in the more recently developed areas of molecular science, such as in materials chemical science and chemical biology. Thus a systematic training in the often-perplexing area of stereochemistry is more than ever necessary, both at the advanced undergraduate and the at the beginning graduate level.

The present volume represents a substantially abbreviated version of the 1994 book, to nearly half of its original length, making it more suitable for teaching the subject at the college level. To achieve the reduction, we have had to omit the chapter on stereoselective synthesis by Professor Mander, not only because of its original size, but also because that particular area of stereochemistry has so grown in the last six years that an adequate chapter would no longer fit into the confines of a general stereochemistry text. (A number of books in this area have appeared since 1994, including a three-volume series dating from 1999.) We have also omitted the glossary since a glossary has now been issued by IUPAC (see Ref. 1 in Chapter 2); moreover, pertinent terms can be traced through the index. We have also omitted most of the original text that was set in smaller font, and we have substantially condensed all other material. We hope and trust, however, that the essential tenets of stereochemistry have been preserved.

Most of the citations have been taken from the 1994 book, although in a few areas of rapid development they have been supplemented with more recent references. We acknowledge that with the availability of such a large body of references, we have succumbed to the temptation to include more citations than is usual in a textbook; in doing so we have given preference to review articles and books. The references may be of use when the teacher wishes to assign a detailed study of a particular area to individual students, or when the topic forms the basis of a separate seminar.

If, when using this book in the classroom, a choice of topics has to be made, we recommend Chapters 1–6, 8–10, and selective material from Chapter 11. The students can read Chapter 7 independently and Chapters 12 and 13 are of a more advanced nature. This book may serve a specific course on stereochemistry or may assist in the development of major components in an even broader subject sequence. The text will be a valued resource in enabling students to understand critical elements in stereochemical concepts, ideas, analysis, techniques, terminology, and history.

Ernest L. Eliel
Michael P. Doyle

BASIC ORGANIC
STEREOCHEMISTRY

1

INTRODUCTION

1-1. SCOPE

Stereochemistry (from the Greek *stereos*, meaning solid) refers to chemistry in three dimensions. Since most molecules are three-dimensional (3D), stereochemistry, in fact, pervades all of chemistry. It is not so much a branch of the subject as a point of view, and whether one chooses to take this point of view in any given situation depends on the problem one wants to solve and on the tools one has available to solve it.

There is little question that, today, the third dimension has become all-important in the understanding of problems not only in organic, but in physical, inorganic, and analytical chemistry as well as biochemistry, so that no chemist can afford to be without a reasonably detailed knowledge of the subject.

It has become customary to factorize stereochemistry into its static and dynamic aspects. *Static stereochemistry* (perhaps better called stereochemistry of molecules) deals with the counting of stereoisomers, with their structure (i.e., molecular architecture), with their energy, and with their physical and most of their spectral properties. *Dynamic stereochemistry* (or stereochemistry of reactions) deals with the stereochemical requirements and the stereochemical outcome of chemical reactions, including interconversion of conformational isomers or topomers (cf. Chapter 2); this topic is deeply interwoven with the study and understanding of reaction mechanisms. Like most categorizations, this one is not truly dichotomous and some subjects fall in between; for example, quantum mechanical treatments of stereochemistry may deal with either its structural or its mechanistic aspects; spectroscopic measurements may fathom reaction rate as well as molecular structure.

1-2. HISTORY

Historically, the origins of stereochemistry stem from the discovery of plane-polarized light by the French physicist Malus[1] in 1809. In 1812 another French scientist, Biot,[2] following an earlier observation of his colleague Arago,[3] discovered that a quartz plate, cut at right angles to its crystal axis, rotates the plane of polarized light through an angle proportional to the thickness of the plate; this constitutes the phenomenon of optical rotation. Some quartz crystals turn the plane of polarization to the right, while others turn it to the left. Three years later, Biot[4] extended these observations to organic substances, both liquids (such as turpentine) and solutions of solids (such as sucrose, camphor, and tartaric acid). Biot recognized the difference between the rotation produced by quartz and that produced by the organic substances he studied: The former is a property of the crystal; it is observed only in the solid state and depends on the direction in which the crystal is viewed, whereas the latter is a property of the individual molecules and may therefore be observed not only in the solid, but in the liquid and gaseous states as well as in solution.

With respect to the question of the cause of optical rotation, the French mineralogist Haüy[5] had already noticed in 1801 that quartz crystals exhibit the phenomenon of hemihedrism. Hemihedrism (cf. Section 6-4.c) implies inter alia that certain facets of the crystal are so disposed as to produce nonsuperposable species (Fig. 1.1, **A** and **B**), which are related as an object to its mirror image. (Such mirror-image crystals are called "enantiomorphous," from the Greek *enantios* meaning opposite and *morphe* meaning form.) In 1822, Sir John Herschel[6] observed that there was a relation between hemihedrism and optical rotation: All the quartz crystals having the odd facets inclined in one direction rotate the plane of polarized light in one and the same sense, whereas the enantiomorphous crystals rotate polarized light in the opposite sense.

It was, however, left to the genius of Louis Pasteur to extend this correlation from the realm of crystals, such as quartz, which rotate polarized light only in the solid state, to the realm of molecules, which rotate both as the solid and in solution. [The naturally occurring *dextro*-tartaric acid, henceforth denoted as (+)-tartaric acid, rotates the plane of polarized light to the right; see Section 1–3.] In 1848 Pasteur[7] had succeeded in separating crystals of the sodium ammonium salts of (+)- and (−)-tartaric acid from the racemic (nonrotating) mixture in the following way. When the salt of the mixed (racemic) acid which is found in wine caskets, was crystallized by slow evaporation of its aqueous solution, large crystals formed, which, to Pasteur's surprise and delight, displayed hemihedric facets similar to those found in quartz (Fig. 1.1). By looking at these crystals with a lens, Pasteur was able to separate the two types (with their

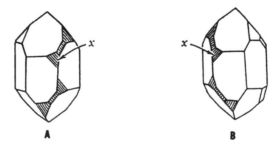

Figure 1.1. Hemihedrism of quartz crystals. [Reprinted with permission from Fieser, L. F. and Fieser, M. *Organic Chemistry*, 3rd ed., Heath, Lexington, MA, 1956.]

dissymmetric facets inclined to the right or left) by means of a pair of tweezers. When he then separately redissolved the two kinds of crystals, he found that one solution rotated polarized light to the right [the crystals being identical with those of the salt of the natural (+)-acid], whereas the other rotated to the left. [(−)-Tartaric acid had never been encountered up to that time.]

Pasteur[8] soon came to realize the analogy between crystals and molecules: In both cases the power to rotate polarized light was caused by dissymmetry, that is, the nonidentity of the crystal or molecule with its mirror image, expressed in the case of the ammonium sodium tartrate crystals by the presence of the hemihedric faces. Similarly, Pasteur postulated, the molecular structures of (+)- and (−)-tartaric acids must be related as an object to its mirror image; he pictured them as nonsuperposable helices of opposite sense. The two acids are thus enantiomorphous at the molecular level; we call them enantiomers. [The ending -mer (as in isomer, polymer, and oligomer, from the Greek meros meaning part) usually refers to a molecular species.]

In 1874 van't Hoff[9] in Utrecht, The Netherlands and Le Bel[10] in Paris, France independently and almost simultaneously proposed the structural base for enantiomerism in a substance of the type Cabcd: the four substituents are arranged tetrahedrally around the central carbon atom to which they are linked. van't Hoff, who had worked with Kekulé and whose views were based on structural theory, specified the 3D arrangement quite precisely: The four linkages to a carbon atom point toward the corners of a regular tetrahedron (Fig. 1.2) and two nonsuperposable arrangements (enantiomers) are thus possible.

We call the model corresponding to a given enantiomer (e.g., Fig. 1.2, **A**) and the molecule that it represents "chiral" (meaning handed, from Greek cheir, hand) because, like hands, the molecules are not superposable with their mirror images. The term chiral was first used by Lord Kelvin[11] in 1893, was rediscovered by Whyte[12] and was firmly reintroduced into the stereochemical literature by Mislow[13] and by Cahn, Ingold, and Prelog,[14] who define a model as chiral when it has no element of symmetry (plane, center, alternating axis; cf. Chapter 4) except at most an axis of rotation. A certain amount of confusion or ambiguity has arisen in the use of the term. When a *molecule* is chiral, it must be either "right-handed" or "left-handed." But if a *substance*

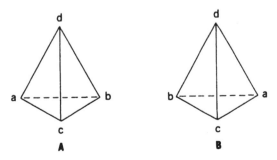

Figure 1.2. Tetrahedral carbon.

or sample is said to be chiral, this merely means that it is made up of chiral molecules; it does not necessarily imply that all the constituent molecules have the same "sense of chirality" (*R* or *S*, or *M* or *P*; cf. Chapter 5). We may distinguish two extreme situations (plus an infinite number of intermediate ones): (a) The sample is made up of molecules that all have the same sense of chirality (homochiral molecules). In that case the sample is said to be chiral and "nonracemic." This serves to distinguish this case from the opposite situation, where (b) the sample is made up of equal (or very nearly equal) numbers of molecules of opposite sense of chirality (heterochiral molecules), in which case the sample is chiral but racemic. Thus the statement that a macroscopic sample (as distinct from an individual molecule) is chiral is ambiguous and therefore sometimes insufficient; it may need to be further stated if the sample is racemic or nonracemic. Lack of precision on this point has led to some confusion, for example, in the titles of articles where the synthesis of a chiral natural product is claimed, but it is not clear whether the investigator simply wishes to draw attention to the chirality of the pertinent structure or whether the product has actually been synthesized as a single enantiomer (i.e., an assembly of homochiral molecules, which should not, however, be called a homochiral sample).

The situation is even slightly more complex than so far implied. There is little ambiguity about the meaning of "chiral, racemic": Chiral, racemic means that (within the limits of normal stochastic fluctuations) the sample is made up of equal numbers of molecules of opposite sense of chirality. But in a "chiral, nonracemic" sample there can be some molecules of a sense of chirality opposite to that of the majority; that is, the sample may not be enantiomerically pure (or *enantiopure*). Experimental tests as to whether a sample is enantiopure or merely *enantioenriched* will be discussed in Section 6-5.

It immediately follows from van't Hoff's hypothesis that in an alkene, where the tetrahedra are linked along one edge (Fig. 1.3) cis–trans isomerism is possible (see Chapter 9) and already in 1875 van't Hoff[9b] predicted the stereoisomerism of allenes, not actually observed in the laboratory until 1935 (cf. Chapter 13).

The hypothesis of van't Hoff and Le Bel has stood with but minor modifications until today. Both the visualization of molecules by X-ray and electron diffraction and the interpretation of vibrational [infrared (IR) and Raman] spectra have confirmed that carbon is, indeed, tetrahedral. Quantum mechanical calculations[15,16] concur in predicting a much lower energy for tetrahedral methane than for (hypothetical) methane of *planar geometry*.

van't Hoff[9a] had already pointed out that if CX_2Y_2 were planar (or, for that matter, square pyramidal), two isomers should exist but only one is found. For a detailed discussion see Wheland.[17]

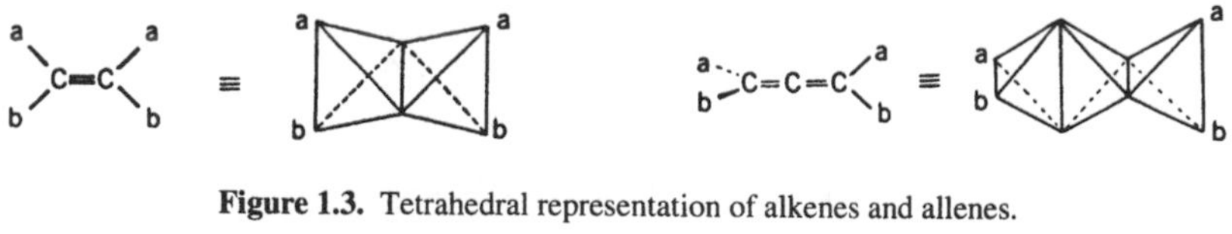

Figure 1.3. Tetrahedral representation of alkenes and allenes.

1-3. POLARIMETRY AND OPTICAL ROTATION

It was mentioned in Section 1-2 that the discoveries of polarized light and optical rotation led to the concept of molecular chirality, which, in turn, is basic to the field of stereochemistry. Polarized light and optical rotation are therefore usually given considerable play in elementary treatments of stereochemistry. In the present text we take the view that the central theme of stereochemistry is molecular architecture, notably including chirality, and the resultant fits (as of a right hand with a right glove or of an enzyme with its natural substrate) or misfits (as of a right hand with a left glove or of an enzyme with the enantiomer of its natural substrate). In this theme, polarimetry and optical rotation are important as diagnostic tools for chirality but not central to its existence. We shall therefore treat polarimetry only briefly at this point, assuming that the nature of polarized light and the workings of a polarimeter are already familiar to the reader.

The observed angle of rotation of the plane of polarization by an optically active liquid, solution, or (more rarely) gas or solid is usually denoted by the symbol α. The angle may be either positive (+) or negative (−) depending on whether the rotation is clockwise, that is, to the right (*dextro*) or counter-clockwise, that is, to the left (*levo*) as seen by an observer *toward whom* the beam of polarized light travels. (This is opposite from the direction of rotation viewed *along* the light beam.) It may be noted that no immediate distinction can be made between rotations of $\alpha \pm 180\, n°$ (n = integer), for if the plane of polarization is rotated in the field of the polarimeter by $\pm 180°$, the new plane will coincide with the old one. In fact α, as measured, is always reported as being between −90° and +90°. Thus, for example, no difference appears between rotations of +50°, +230°, +410°, or −130°. To make the distinction. one must measure the rotation at least at one other concentration. Since optical rotation is proportional to concentration (see below), if solutions of the above rotations were diluted to one-tenth of their original concentrations, their rotations would become +5°, +23°, +41°, and −13°, values that are all clearly distinct. Readings taken at two different concentrations almost always determine α unequivocally. An alternative for solutions and the method of choice for pure liquids is to measure the rotation in a shorter tube. In the above cases, if a tube of a quarter of the original length [e.g., 0.25 decimeters (dm) instead of 1 dm] is used, the rotations as recorded become +12.5°, +57.5°, −77.5° (equivalent to +102.5°), and −32.5°, again all clearly distinguishable. [Note that halving the tube length (e.g., from 1 to 0.5 dm) would have left the ambiguity between the first and third observation (+25° vs. +205° = 180° + 25°) and between the fourth and second (−65° and +115° = 180° − 65°).]

Biot discovered that the observed rotation is proportional to the length ℓ of the cell or tube containing the optically active liquid or solution and the concentration c (or density in the case of a pure liquid): $\alpha = [\alpha] \cdot c \cdot \ell$ (Biot's law). The value of the proportionality constant $[\alpha]$ depends on the units chosen; in polarimetry it is customary to express ℓ in decimeters, because the cells are usually 0.25, 0.5, 1, or 2 dm in length, and c in grams per milliliter (g mL^{-1}) or (and this is preferred for solutions) in g100 mL^{-1}. Thus

$$[\alpha] = \frac{\alpha}{\ell(\text{dm})\, c(\text{g mL}^{-1})} = \frac{100\,\alpha}{\ell(\text{dm})\, c'(\text{g 100mL}^{-1})} \qquad (1.1)$$

The value of $[\alpha]$, the *specific rotation*, depends on wavelength and temperature, which are usually indicated as subscripts and superscripts, respectively; thus $[\alpha]_D^{25}$ denotes the specific rotation for light of the wavelength of the sodium D-line (589 nm) at 25°C. In addition, $[\alpha]$ also depends on the solvent and to some extent to the concentration (in a fashion not taken into account by the concentration term in Biot's law), which must thus also be specified. This is usually done by adding such information in parentheses, thus $[\alpha]_{546}^{20} - 10.8 \pm 0.1$ (c 5.77, 95% ethanol) denotes the specific rotation at 20°C for light of wavelength 546 nm in 95% ethanol solution at a concentration of 5.77 g 100 mL^{-1}. The importance of solvent and concentration is occasioned by association phenomena, which will be discussed in more detail in Chapter 12.

The dimensions of $[\alpha]$ are deg cm^{2}g^{-1} not degrees.[18] In this book $[\alpha]$ will always be given without the units (understood to be 10^{-1} deg cm^2 g^{-1}) and (in contrast to the observed rotation α) will *not* be given in degrees.

For pure liquids, since the density is fixed at a given temperature, one may simply state the observed rotation, along with the cell length, such as $\alpha_D^{25} + 44°$ (neat, $\ell = 1$ dm), the word "neat" (or sometimes "homog" for homogeneous) denoting that the rotation refers to the undiluted (pure) liquid. However, even here it is preferable to give the specific rotation. Thus if the density of the liquid in question at 25°C is 1.1, the specific rotation is $[\alpha]_D^{25} + 40$ (neat) (44/1 × 1.1).

Since optical rotation is proportional to the number of molecules encountered by the beam of polarized light, if two substances have unequal molecular weights but are alike with respect to their power of rotating polarized light, the substance of smaller molecular weight will have the larger specific rotation simply by virtue of having more molecules per unit weight. In order to compensate for this effect and to put rotation on a per-mole basis, one defines the term "molar rotation" as the product of specific rotation and molecular weight divided by 100. [The divisor serves to keep the numerical value of molar rotation on the same approximate scale as that of specific rotation. For a substance of molecular weight (MW) 100, molar and specific rotation are the same.] Thus, denoting molar rotation by $[\Phi]$,

$$[\Phi] = \frac{[\alpha] \cdot \text{MW}}{100} = \frac{\alpha}{\ell(\text{dm}) \cdot c''(\text{mol 100 mL}^{-1})} \qquad (1.2)$$

The choice of solvent particularly affects the rotation of polar compounds because of its intervention in solvation and association phenomena (cf. Chapter 12). Substantial changes of specific rotation with solvent are not uncommon; reversals of sign are less frequent but have been explicitly reported in a number of instances (Chapter 12). A pH dependence of rotation is also common in the case of acids and bases and reversals are recorded, for example, for (S)-(+)-lactic acid, dextrorotatory in water, whose sodium salt is levorotatory[19] and for L-leucine, which is levorotatory in water but dextrorotatory in aqueous hydrochloric acid.[20]

An even more remarkable change in rotation, from positive to negative, is seen in 2-methyl-2-ethylsuccinic acid[21] as its solution in chloroform (containing 0.7% ethanol) is diluted, with a reversal of sign (corresponding to null rotation) occurring at a concentration of 6.3%. The phenomenon (presumably due to association) is confined to solvents of low polarity ($CHCl_3$ or CH_2Cl_2); no reversal is seen in alcohol solvents, pyridine, diglyme, or acetonitrile. 2-Methyl-2-ethylsuccinic acid is also a case where the presence of one enantiomer affects the rotation of the other beyond the obvious way of partially canceling it.[22,23] These points will be returned to later (Chapter 12).

REFERENCES

1. Malus, E. L. *Mem. Soc. d'Arceuil* **1809**, *2*, 143.

2. Biot, J. B. *Mem. Cl. Sci. Math. Phys. Inst. Imp. Fr.* **1812**, *13*, 1.

3. Arago, D. F., *Mem. Cl. Sci. Math. Phys. Inst. Imp. Fr.* **1811**, *12*, 93, 115.

4. Biot, J. B. *Bull. Soc. Philomath. Paris* **1815**, 190.

5. Haüy, R. J. *Traite de Mineralogie,* Chez Louis, Paris, 1801.

6. Herschel, J. F. W. *Trans. Cambridge Philos. Soc.* **1822**, *1*, 43.

7. Pasteur, L. *Ann. Chim. Phys.,* **1848**, *24* [3], 442.

8. Pasteur, L. Two lectures delivered before the Société Chimique de France, Jan. 20 and Feb. 3, 1860. English transl. Richardson, G. M., ed. *The Foundations of Stereochemistry,* American Book Co., New York, 1901 or Alembic Club Reprint, No. 14.

9. (a) van't Hoff, J. H. *Arch Neerl. Sci. Exacles Nat.* **1874**, *9*, 445; see also *Bull. Soc. Chim. Fr.* **1875**, *23* [2], 295. English transl. Richardson, G. M. (see ref. 8) and Bentley, O. T., eds., *Classics in the Theory of Chemical Combination*, Classics of Science, Vol. 1, Dover Publications, New York, 1963; reprinted by Krieger, Malabar, FL, 1981. (b) van't Hoff. J. H. *La Chimie dans l'Espace,* Bazendijk, Rotterdam, The Netherlands, 1875.

10. Le Bel, J. A. *Bull. Soc. Chim. Fr.* **1874**, *22* [2], 337; see also Richardson, G. M. (ref 8).

11. Kelvin, Lord. *The Second Robert Boyle Lecture* in *J. Oxford Univ. Junior Scientific Club* **1894**, [18], 25.

12. Whyte, L. L. *Nature,* **1957**, *180*, 513; *ibid.* **1958**, *182*, 198.

13. Mislow, K. *Introduction to Stereochemistry,* Benjamin, New York, 1965, p. 52.

14. Cahn, R. S., Ingold, Sir C., and Prelog, V. *Angew Chem. Int. Ed. Engl.* **1966**, *5*, 385.

15. Monkhorst, H. J. *Chem. Commun.* **1968**, 1111; Hoffman, R., Alder, R. W., and Wilcox, C. F. *J. Am. Chem. Soc.* **1970**, *92*, 4992.

16. D. Röttiger and G. Erker, *Angew. Chem. Int. Ed. Engl.* **1997**, *36*, 812.

17. Wheland G. A. *Advanced Organic Chemistry,* 3rd ed., Wiley, New York, 1960.

18. Snatzke, G. Personal communication to S. H. Wilen, 1989/90.

19. Borsook H., Huffman, H. M., and Liu, Y.-P. *J. Biol. Chem.* **1933**, *102*, 449.

20. Stoddard M. P. and Dunn, M. S. *J. Biol. Chem.* **1942**, *142*, 329.

21. Krow, G. and Hill, R. K. *Chem. Commun.* **1968**, 430.

22. Horeau, A. *Tetrahedron Lett.* **1969**, 3121.

23. Horeau, A. and Guetté, J. P. *Tetrahedron* **1974**, *30*, 1923.

2

STRUCTURE

2-1. MEANING, FACTORIZATION, INTERNAL COORDINATES, ISOMERS

We have seen (Section 1-1) that static stereochemistry deals with the shape of molecules (molecular architecture or molecular structure). The nomenclature rules of the International Union of Pure and Applied Chemistry (IUPAC)[1] do not provide an unequivocal definition of "structure"; we shall use the term in the sense of the crystallographer as denoting the position in space of all the atoms constituting a molecule. Molecular structure may thus be defined in terms of the Cartesian coordinates of the atoms, or the oblique coordinates that crystallographers often use for crystals belonging to the monoclinic, triclinic, and trigonal hexagonal systems.

For many purposes (cf. Section 2-6), it is more convenient to employ "internal coordinates": bond lengths (or distances) r, bond angles θ, and torsion angles ω or τ. Since the absolute position and orientation of a molecule are of no structural significance, only relative positions of atoms within the molecule need to be specified. For a diatomic molecule A—B (Fig. 2.1), structure is thus completely defined by the nature of the nuclei A and B and the bond distance r between their centers. For a triatomic molecule ABC (Fig. 2.1), besides the nature of A, B, and C, their connectivity (i.e., which atom is connected to which) and the bond lengths A—B (r_1) and B—C (r_2), we must specify the bond angle θ. With a tetratomic molecule ABCD the situation is slightly more complicated: In addition to the nature and connectivity of the atoms and the bond distances r_1, r_2, and r_3 and bond angles θ_1 and θ_2, one must now specify the torsion angle ω (Fig. 2.1) in order to fix the position of all four atoms A, B, C, and D. The torsion angle ω is defined as the angle between the planes ABC and BCD (Fig. 2.2); this angle has sign as well as magnitude. If the turn from the ABC

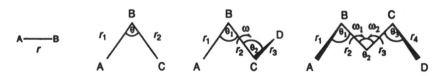

Figure 2.1. Internal coordinates.

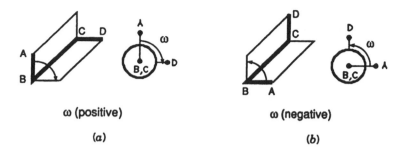

Figure 2.2. Torsion angle.

to the BCD plane (front to back) is clockwise (Fig. 2.2), ω is positive, if it is counterclockwise (Fig. 2.2), ω is negative. In determining the sign of ω it is immaterial whether one views the ABCD array from AB looking toward CD or from CD looking toward AB.

For each additional atom (e.g., E in ABCD—E) three more coordinates need to be specified: the bond length D—E, the bond angle θ(CDE), and the torsion angle ω(BCDE). Thus the total number of independent coordinates for a nonlinear n-atomic molecule is $3n - 6$; for a chain of n atoms these may be taken as $n - 1$ bond distances (the first atom defines no such distance), $n - 2$ bond angles (not defined for the first two atoms), and $n - 3$ torsion angles (defined only for atoms after the third). Figure 2.1 (right) displays an example (5 atoms, 9 coordinates). (The situation is more complicated for branched molecules, rings, or molecules with three or more collinear atoms.)

Whereas torsion angles change widely from one molecule to another and even bond angles can vary appreciably from their standard magnitudes (e.g., the normal CCC angle of 112° in propane is reduced to 88° in cyclobutane), bond distances are usually quite constant (see, however, Sections 2-2 and 2-7). Standard bond distances[2] for common bonds to carbon are given in Table 2.1.

Although structure may thus be described completely by a system of coordinates (Cartesian, internal, or other type), we shall find it convenient, from the conceptual point of view, to subdivide (factorize) structure into a constitutional and stereochemical part. The stereochemical aspect may be further factorized [if with some difficulty (see below)] into configuration and conformation. Structure thus embraces constitution (connectivity), configuration, and conformation (see p. 10).

TABLE 2.1. Standard Bond Distances

Units	C–C	C=C	C≡C	C–H	C–F	C–Cl	C–Br	C–N	C–O	C=O	C–S
Å	1.53	1.34	1.20	1.09	1.38	1.77	1.94	1.47	1.43	1.22	1.82
pm	153	134	120	109	138	177	194	147	143	122	182

By "isomers" we mean substances that have the same composition and molecular weight but differ in properties. On the molecular level, such substances have the same number and kind of atoms, but differ in structure. Again, the structural difference may be in constitution (constitutional isomers) or stereochemical arrangement (stereoisomers). Stereoisomers, in turn, may differ (see Sections 2-3 and 2-4) in configuration (configurational isomers) or, having the same configuration, may differ in conformation (conformational isomers or conformers). A different subdivision of stereoisomers will be considered in Chapter 3.

In a crystal the positions of the molecules and their constituent atoms are generally well defined. Disregarding the generally small atomic motions, molecules in crystals may thus be considered rigid objects with well-defined shape, symmetry, and dimensions and may be represented appropriately by rigid models. However, this rigidity does not persist in the fluid (liquid, solution, or gaseous) state. Molecular vibrations become larger, leading to less well defined bond distances and bond angles, and in some cases, as in the inversion or "umbrella motion" of tertiary amines (NRR'R″) or rotations around single bonds (torsions), for example, around the central bond in $ClH_2C–CH_2Cl$, may actually lead to different structures. Such changes, which frequently occur rapidly at room temperature, put in question what is actually meant by a "molecule." Commonly, the term has been defined as "the smallest entity of a chemical species." It is implicit in this definition that the entities are all the same, but this may not be true. Thus the species chlorocyclohexane (Fig. 2.3) is well known to consist of molecules with equatorial and others with axial chlorine atoms. In the IR spectrum of the species, the two types of molecules can clearly be discerned, for example, by their different C–Cl stretching frequencies. Yet in most respects, for example, in distillation, chromatography, or chemical reaction, chlorocyclohexane appears to be a homogeneous substance. Even that appearance depends on the temperature of observation. At room temperature chlorocyclohexane displays a simple NMR spectrum; thus there are only four different ^{13}C signals. But at −100°C, well below the "coalescence temperature" (cf. Section 8-4.d), separate spectra of the two structures shown in Figure 2.3 are seen and the number of carbon signals doubles to eight. Thus chlorocyclohexane is clearly made up of two kinds of molecules, but whether these molecules can be discerned separately is a matter of both the technique of observation (isolation, NMR, or IR) and the temperature: At −150°C, the two different chlorocyclohexane isomers can actually be isolated[3] (cf. Section 11-4).

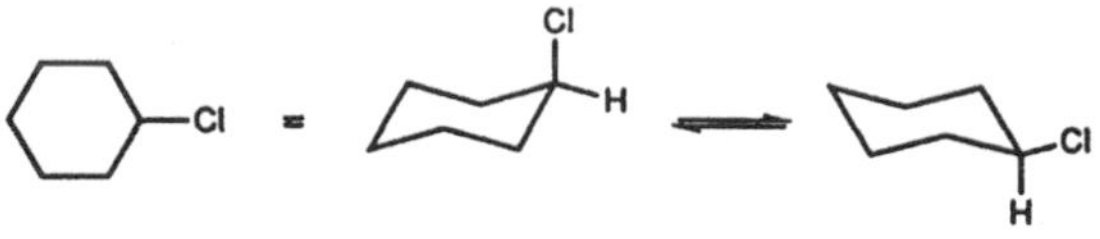

Figure 2.3. Chlorocyclohexane.

2-2. CONSTITUTION

The term "constitution" connotes the number, kind, and connectivity of the atoms in a molecule. Constitution may be represented by a two-dimensional (2D) graph in which the atoms linked to each other are connected by a bond (single, double, or triple). An alternative representation is an n^2 matrix (for a molecule with n atoms) in which the elements are zero for nonbonded atoms, or 1, 2, or 3, respectively, for atoms linked by single, double, or triple bonds. Thus the isomeric molecules ethyl alcohol (ethanol) and dimethyl ether (oxybismethane) (both C_2H_6O) may be represented by the graphical formulas **A** and **B**[4,5] or by the matrices **C** and **D** below[6]

$$
\begin{array}{c}
\quad\overset{\displaystyle H_4}{\underset{}{|}} \quad \overset{\displaystyle H_2}{\underset{}{|}} \\
H_5\!-\!\underset{\underset{\displaystyle H_6}{|}}{C_2}\!-\!\underset{\underset{\displaystyle H_3}{|}}{C_1}\!-\!O\!-\!H_1
\end{array}
\qquad \text{and} \qquad
\begin{array}{c}
\quad\overset{\displaystyle H_3}{\underset{}{|}} \qquad \overset{\displaystyle H_4}{\underset{}{|}} \\
H_2\!-\!\underset{\underset{\displaystyle H_1}{|}}{C_1}\!-\!O\!-\!\underset{\underset{\displaystyle H_6}{|}}{C_2}\!-\!H_5
\end{array}
$$

$$\textbf{A} \qquad\qquad\qquad\qquad\qquad \textbf{B}$$

	C_1	C_2	O	H_1	H_2	H_3	H_4	H_5	H_6
C_1		1	1	0	1	1	0	0	0
C_2	1		0	0	0	0	1	1	1
O	1	0		1	0	0	0	0	0
H_1	0	0	1		0	0	0	0	0
H_2	1	0	0	0		0	0	0	0
H_3	1	0	0	0	0		0	0	0
H_4	0	1	0	0	0	0		0	0
H_5	0	1	0	0	0	0	0		0
H_6	0	1	0	0	0	0	0	0	

$$\textbf{C}$$

	C_1	C_2	O	H_1	H_2	H_3	H_4	H_5	H_6
C_1		0	1	1	1	1	0	0	0
C_2	0		1	0	0	0	1	1	1
O	1	1		0	0	0	0	0	0
H_1	1	0	0		0	0	0	0	0
H_2	1	0	0	0		0	0	0	0
H_3	1	0	0	0	0		0	0	0
H_4	0	1	0	0	0	0		0	0
H_5	0	1	0	0	0	0	0		0
H_6	0	1	0	0	0	0	0	0	

$$\textbf{D}$$

(such matrices are particularly useful for computer manipulation). The two molecules clearly differ in connectivity and are constitutional isomers. These isomers may be distinguished by diffraction experiments, but simpler criteria, such as acidity, volatility (reflecting the presence or absence of hydrogen bonding), or spectroscopic criteria for the OH group in the alcohol, may be used more conveniently.

To draw a constitutional formula or to construct a connectivity matrix, one must decide whether or not a bond exists between two given atoms within a molecule of known structure. Since it is difficult to determine the electron density between the atoms in question experimentally (see also Section 2-5), bonds are usually drawn in such a way as to (a) satisfy the normal valencies of the atoms and (b) take into account what is known about bonded and nonbonded distances between atoms. Thus two carbon atoms may normally be considered bonded if their internuclear distance is less than 160 pm (1.6 Å) [the sum of the single-bond radii is 153 pm (1.53 Å); cf. Table 2.1] but not bonded if their distance exceeds 270 pm (2.7 Å) [the sum of the van der Waals radii is 340 pm (3.4 Å)]. Distances in the intermediate range are encountered rarely.

A more pragmatic definition has been given by Pauling:[7] "There is a chemical bond between two atoms or groups of atoms in case that the forces acting between them are such as to lead to the formation of an aggregate with sufficient stability to make it convenient for the chemist to consider it as an independent molecular species." This definition, which is in terms of energy rather than geometry, primarily applies to intermolecular bonds. Thus we may consider the hydrogen bridges in the acetic acid dimer (Fig. 2.4, **A**) as true bonds, since the dimer persists even in the vapor phase and affects the boiling point and the IR spectrum; in contrast, we may not wish to consider the hydrogen bridge between two molecules of methyl mercaptan, CH_3SH (Fig. 2.4, **B**) as a bond, since it scarcely manifests itself in the properties of the substance. In an intramolecular analogy, we may wish to pay no attention to the very weak hydrogen bond that links the two ends of an ethylenediamine molecule (Fig. 2.4, **C**) but we cannot disregard the hydrogen bonds between peptide units in a polypeptide (Fig. 2.4, **D**) since these bonds are responsible for the important secondary structure of proteins.

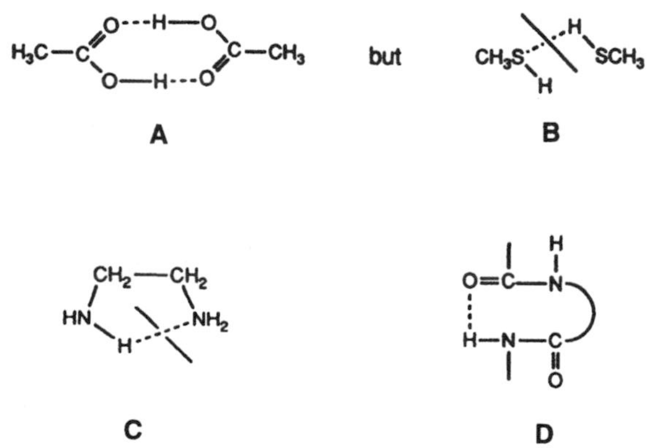

Figure 2.4. Hydrogen bonds.

In Section 2-1 we discussed the question of structural homogeneity. Constitutional isomerism provides examples involving a variety of time scales. For example, the keto **1** and enol **2** forms of acetoacetic ester, which coexist in equilibrium, display distinct NMR spectra and are separable by low-pressure distillation in quartz equipment. But when warmed in the presence of a trace of base they become "tautomeric," that is, rapidly equilibrating; under those circumstances the two species are no longer separable and even their NMR spectra may coalesce. Nevertheless, two molecular species still coexist, as can readily be verified by IR spectroscopy or by careful separation following neutralization of the base.

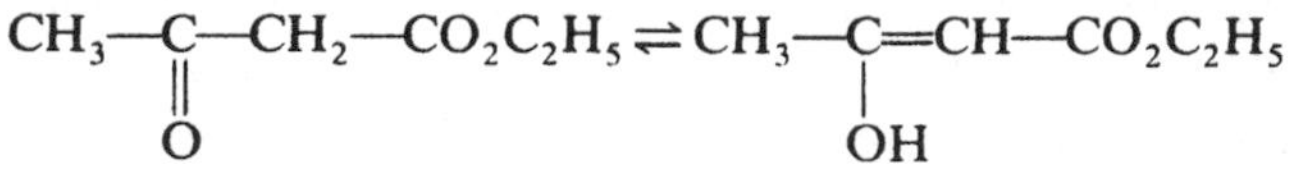

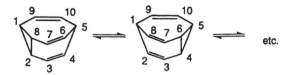

Figure 2.5. Bullvalene.

Another example of tautomerism called "valence tautomerism" is represented by bullvalene (Fig. 2.5). This molecule exists in 1,209,600 (10!/3) structures because of the possibility of bond migration, as shown in Figure 2.5. The structures are degenerate (i.e., superposable except for the numbering of the atoms) but, at room temperature, the carbon atoms may be distinguished in a ^{13}C NMR spectrum as being vinylic, allylic, cyclopropanoid, and so on. Upon heating to 100°C, however, bond migration or reorganization (fluxion) becomes so fast that the carbon atoms lose their distinctive character on the NMR time scale and only a single ^{13}C signal is seen. Nevertheless, the greater than 10^6 fluxional structures still coexist; the NMR spectrum corresponds to an average of these structures, not to an actual, unique molecule.[8]

2-3. CONFIGURATION

Molecules of identical constitution may yet differ in structure. Thus fluorochlorobromomethane (CHFClBr) has a unique connectivity but there are two enantiomers (nonsuperposable mirror-image structures; Fig. 2.6.[9–11] These molecules are said to differ in configuration. Again, when the substance 1,2-dichloroethane is viewed by IR spectroscopy,[12] it is found that at least two different isomers coexist, gauche and anti (Fig. 2.7); these isomeric structures are said to differ in conformation. Figure 2.7 shows the *three* staggered conformational isomers of 1,2-dichloroethane. Structures **A** and **C** are enantiomeric, and therefore indistinguishable by IR spectroscopy (or any other scalar observation; cf. p. 37).

In contrast to constitution, which can be represented by a 2D graph of atoms linked by bonds, configuration and conformation embody the three-dimensional (3D) or stereochemical part of structure. The most fundamental distinction one can make between configuration and conformation is to say that configurational differences

Figure 2.6. Enantiomers of CHFClBr.

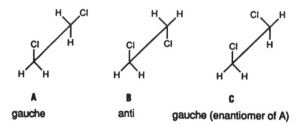

Figure 2.7. 1,2-Dichloroethane.

imply differences in bond angles, whereas conformational differences involve differences in torsion angles (including, in both cases, differences that are exclusively in sign). If one considers an assembly of four atoms, A, B, C, and D, one can envisage two different types of connectivity (constitution) shown in Figure 2.8. In array **a**, assuming constant bond distances A–B, B–C, and C–D and constant bond angles ABC and BCD, one can generate an infinite number of structures by rotating around the BC bond, that is, by changing the torsion angle ABC/BCD, ω(ABCD). These structures are said to differ in conformation. The conformational differences (cf. also Figs. 2.2 and 2.7) are independent of the nature of the atoms A, B, C, and D; conformation for the array can be defined even if all the atoms are the same (i.e., in A–A–A–A) as long as no three of them are collinear. Array **b**, in contrast, presents no conformational variability; but if B ≠ C ≠ D, that is, if atom A is chiral [or (see later discussion) stereogenic], there are two possible arrangements of the type shown in Figure 2.6. These two arrangements are said to differ in configuration.

If one views the molecule **b** and its mirror image (enantiomer) from the side of the B–C–D "tripod" (with A at the apex of the tripod), one sees the two different arrangements shown in Figure 2.9. In the former, the sequence B–C–D describes a clockwise array, in the latter a counterclockwise array. [The former array might be described as *R* in the Cahn–Ingold–Prelog system (cf. Chapter 5) and the latter as *S*.] This difference implies that the 3D angle subtended by the ligands B, C, and D at the pivot atom A is of opposite sign in the two (*R* or *S*) configurations.

Configuration and conformation as defined above are *not* delineated from each other by considerations of energy barriers. The energy barrier between molecules of

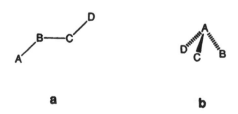

Figure 2.8. Four-atom assemblies.

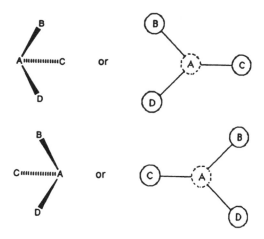

Figure 2.9. Tripodal array.

opposite configuration may be quite low, as in NMeEtPr [ethylmethylpropylamine, ca. 8 kcal mol^{-1} (33.5 kJ mol^{-1})], moderate, as in PMePrPh [methylphenylpropylphosphine, 32.1 kcal mol^{-1} (134 kJ mol^{-1})],[13,14] or large, as in a tetrasubstituted methane (where the structure shown in Fig. 2.8, **b** has been modified by addition of a fourth ligand). In all three cases the enantiomers are considered to differ in configuration.

Whereas molecules of type **b** are usually considered to fall under the definition of configurational isomerism, the picture with molecules of type **a** is less clear-cut. This type covers a wide range of species, as will be shown later.

2-4. CONFORMATION

Reference to Figure 2.1 discloses that definition of constitution and configuration does not suffice to locate the atoms of a molecule in space, since structural differences may be engendered by change of torsion angle ω, as exemplified in Figure 2.7. Thus to complete the description of structure one must specify the torsion angles. By "conformation" of a molecule of given constitution and configuration is meant the rotational arrangement about all bonds as defined by the magnitude and sign of all pertinent torsion angles. Conformation may be described exactly, by specifying the exact magnitude and sign of the torsion angles (cf. Fig. 2.2) or it may be described approximately, by a range of these angles. The latter approach is often preferable, since the exact values of the torsion angles, especially for molecules in the liquid and gaseous states, are frequently not known. An appropriate system of classification[15] is summarized in Table 2.2 and Figure 2.10.

Many conformations, such as the eclipsed conformations in ethane. do not correspond to energy minima. Those that do, such as the conformations of

TABLE 2.2. Specification of Torsion Angle (Klyne–Prelog)

Angle of Torsion (ω)	Designation	Symbol
−30° to +30°	synperiplanar	sp^a
+30° to +90°	+ synclinal	$+sc^b$
+90° to +150°	+ anticlinal	$+ac$
+150° to −150°	antiperiplanar	ap^c
−150° to −90°	− anticlinal	$-ac$
−90° to −30°	− synclinal	$-sc^b$

[a]The designation syn or eclipsed is often used for $\omega \approx 0°$.
[b]The designation gauche is frequently used for $\omega \approx 60°$.
[c]The designation anti (or, less property, trans) is often used for $\omega \approx 180°$.

Figure 2.10. Specification of torsion angle (Klyne–Prelog).

$ClCH_2CH_2Cl$ shown in Figure 2.7, may be called "conformational isomers" or "conformers." Conformation in acyclic molecules is considered in more detail in Chapter 10.

We shall end this discussion with two somewhat subtle cases. The ring reversal of chlorocyclohexane (Fig. 2.3) is a conformational change; the two isomers differ in the sense of the torsion angles but not in the sense of the bond angles (although some change of bond angles must occur during the flipping process). The trans and cis isomers of 1,4-dimethylcyclohexane (Fig. 2.11) differ in configuration but the ring-reversed forms (**A/B** or **C/D**, respectively) of each individual configurational isomer differ from each other in conformation only. The ring-reversed forms of *cis-*

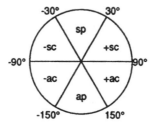

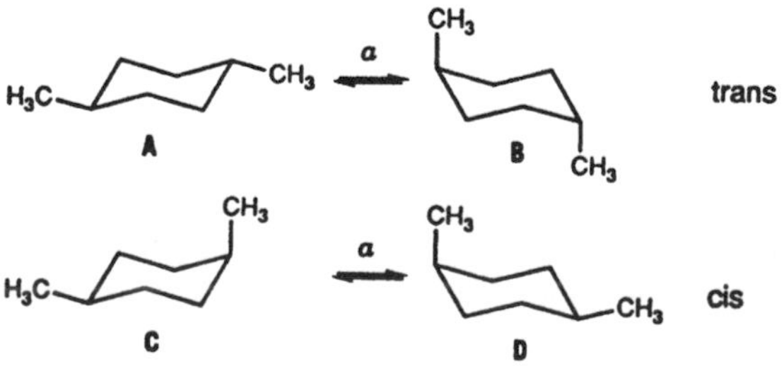

Figure 2.11. 1,4-Dimethylcyclohexanes. *a*: Conformational change. The cis and trans isomers differ in configuration.

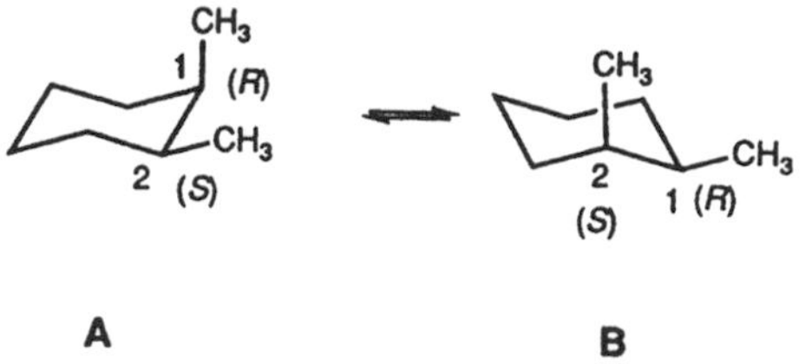

Figure 2.12. *cis*-1,2-Dimethylcyclohexane.

1,2-dimethylcyclohexane (Fig. 2.12, **A**, **B**) also differ in conformation, even though these two forms are enantiomers (cf. Section 11-4.c and Fig. 11.20).

2-5. DETERMINATION OF STRUCTURE

The structure of a molecule may be established by separate investigation of constitution, configuration, and conformation, or it may be determined all at once. We shall not deal in this book with the separate investigation of constitution; the determination of configuration will be treated in Chapter 5 and that of conformation in Chapters 10 and 11. Here we shall discuss briefly the most important methods for the integral determination of structure: X-ray and neutron diffraction analysis, electron diffraction, and microwave spectroscopy. The former two methods are applied in the solid state, the latter two in the gas phase. There are few methods for the direct determination of structure in the liquid or solution phase; IR spectroscopy combined with normal coordinate analysis is of very limited application and the method of lanthanide shift reagents in NMR, once thought to be quite powerful, is probably only of limited usefulness. Nuclear magnetic resonance in the nematic phase has also been used; the best method is probably 2D NMR with quantitative evaluation of nuclear Overhauser effects (NOESY, see pp. 19, 20).

   X-ray diffraction[16,17] is by far the most powerful technique for the structure determination of crystalline materials, from the smallest molecules that can conveniently be crystallized (the analysis may be carried out at low temperature if necessary) to species as large as proteins. In this method, X-rays of suitable wavelength are allowed to impinge on a single crystal, 0.1–1 mm in length, of the material to be investigated. The X-rays are scattered by the atoms (by their electrons rather than by the nuclei) and the interference of the scattered radiation is recorded as a diffraction pattern.

   If one knows the structure of a compound, one can unequivocally calculate the diffraction pattern, but the desired reverse process is not so straightforward. The "structure factor," which is a function (the Fourier transform) of the scattering density and which is crucial in the determination of the structure, has both magnitude and phase. The scattering pattern provides the magnitude of the structure factors but not (at least not in a straightforward way) their phases. It is necessary to obtain the missing phase information before the structure can be solved.

In a classical (if no longer much used) approach[18] the phase problem is solved by the introduction of a heavy atom (usually from the second complete row of the periodic system or heavier) in the molecule. The position of the heavy atom in the unit cell can usually be ascertained directly from the diffraction pattern and the phases of a number of the scattering amplitudes can then be determined. (The unit cell is the smallest geometric unit from which the complete 3D array of molecules is generated by translation along the axes. It usually contains more than one molecule.) This procedure leads to the establishment of a "trial structure" from which improved atomic coordinates and the phases of the remaining scattering amplitudes are obtained by a process of least squares "refinement." Refinement involves calculation of the diffraction pattern from the trial structure followed by adjustment of the latter until good agreement in terms of the R factor (see below) is attained. The introduction of the heavy atom usually involves the formation of a derivative, such as p-bromobenzoate, but it suffices even to have heavy atoms in the solvent of crystallization.[19]

As a result of the pioneering work of Karle and Hauptmann,[20] recognized by the 1985 Nobel Prize in Chemistry, it has become possible to solve the phase problem by "direct methods," which involve a mathematical manipulation of the intensity data (symbolic addition). The method[21,22] is now used routinely for molecules of moderate size, including quite large organic structures (concerning potential applications to macromolecules see ref. 23); it produces more accurate structural data than the heavy atom method because the potentially disturbing effect of the heavy atom is absent.

The availability of good crystals is a prerequisite for X-ray analysis. The crystals need not be large but must be single, not twinned. Since molecular vibrations occur in the crystal, diffraction analysis establishes an average position of the atoms. The thermal motion of the atoms can be assessed in the analysis; the "temperature factors" along the three axes measure the mean square amplitudes of the thermal vibration. In the general case, these vibrations are anisotropic (i.e., differ in magnitude along the three direction axes) and may be included in the drawings of the atomic positions in molecules as ellipsoids or "footballs." Sometimes molecular vibrations in certain parts of a molecule are so large that the positions of the atoms are not well defined; in that case one speaks of *dynamic disorder* in that part of the structure. More commonly, disorder (*static disorder*) is due to the coexistence of two different structures (e.g., two different conformations of the same molecule[24] in the crystal. It may lead to misinterpretations of structure.[25] Uncertainties in atomic position due to thermal motion can be minimized by performing the crystallography at low temperature.

The goodness of fit of an X-ray structure determination (important during refinement, see above) is expressed as the R factor (agreement factor), which measures the agreement between the observed structure amplitudes and those calculated from an intermediate or the final structure. The smaller R, the better the fit.

$$R = \left[\frac{\sum (|F_{obs}|^2 - |F_{calc}|^2)}{\sum |F_{obs}|^2} \right]^{1/2}$$

where the F values are the absolute values of the structure factors.

Neutron diffraction analysis[26,27] is sometimes used as a complement to X-ray diffraction. It requires the availability of a strong flux of neutrons from an atomic reactor and the crystals must be somewhat larger than for X-ray analysis. However, there are several advantages, of which the principal one is the much greater scattering power for neutrons (compared to X-rays) of light atoms, such as hydrogen, deuterium, or lithium. The locations of hydrogen atoms from X-ray diffraction patterns are inaccurate, being usually based on the difference of the observed electron density with that calculated disregarding the hydrogen atoms. In some instances the hydrogen atoms are located simply on the basis of model considerations. In contrast, neutron diffraction permits accurate location of hydrogen or deuterium atoms. Deuterated compounds are preferred because hydrogen absorbs neutrons strongly and thereby seriously reduces the intensity of the diffraction pattern.

Electron diffraction[28] is a means for determining structure in the vapor phase. The scattering of the impinging electrons is caused mainly by the atomic nuclei and the method yields internuclear distances between both bonded and nonbonded atoms. Bond distances are obtained with greater precision than by X-ray diffraction but the method is limited to molecules of relatively small size, not only because of the requirement for appreciable volatility but also because the difficulty of interpreting the diffraction pattern increases as the square of the number of atoms. Its scope may be enhanced by combination with other techniques, such as microwave or vibrational spectroscopy (see below) or a priori calculations (cf. Section 2-6).[28] The method, as we shall see in Chapter 11, is useful for assessing conformational isomerism.

Microwave spectroscopy[29] is an alternative method of structure determination in the vapor phase, applicable to small molecules that have dipole moments. This method primarily yields the dipole moment and the moments of inertia of the molecular species investigated. To obtain structural data, it is generally necessary to investigate several isotopically substituted species and combine the information obtained from them. Because of the need to use several different species, because of the anharmonicity of the potential function, and for other reasons, the bond distances obtained by microwave spectroscopy differ slightly but systematically from those obtained by diffraction methods; this problem has been discussed extensively.[30] Structure determinations by microwave spectroscopy, just as those by electron diffraction, may be enhanced through combination with theoretical calculations[31] (see below).

Normal coordinate analysis is a means to deduce structure from IR and Raman spectroscopic data by complete theoretical interpretation of all the vibrational frequencies. The technique is not often used to determine total structure and is limited to relatively small molecules for which extensive spectroscopic information is available.

The best currently available method for determining total structure in solution is undoubtedly 2D NMR. The principal (though not exclusive) tools are proton correlated spectroscopy (COSY) and determination of nuclear Overhauser effects by nuclear Overhauser and exchange spectroscopy (NOESY).[32] The most fruitful application of these tools has been in the structure analysis of peptides and small proteins and nucleic acids.[33] Only a brief discussion of the method can be given here.

A combination of COSY and NOESY is used to assign proton resonances to the protons belonging to individual amino acids in a peptide chain. Only geminal and vicinal protons will normally couple strongly enough to give cross-peaks in proton–proton COSY, though the method can be extended to certain more long-range couplings. Thus the low-field $-CH(NH_2)-$ proton of alanine can be uniquely recognized because it is the only such proton in a naturally occurring amino acid that is coupled to a methyl group: $CH_3-CH(NH_2)-CO_2H$, though coupling between an even lower field proton and a methyl group also occurs in threonine, $CH_3-CHOH-CH(NH_2)-CO_2H$. Nuclear Overhauser effects (NOEs) can be exploited to determine protons that are close to each other within one residue, such as the ortho and alpha protons in aromatic (Ar) residues $Ar-CH_2-$ and protons that are located 1,3 to each other ($H-C-C-C-H$). Its greater usefulness, however, lies in identifying protons that are close to each other in distinct residues. Thus NOEs are seen for protons in sequential residues, especially for CH/NH protons, as in $-CH-CO-NH-$, and will thus help in the establishment of sequence. The NOEs can also be found, however, in residues that are distant in sequence but close in space, as in α-helices and β-pleated sheets; thus they provide a tool for inferring conformation. Because the NOE falls off with the inverse sixth power of distance, protons must be within approximately 500 pm (5 Å) of each other for an NOE between them to be observed. Conclusions are beginning to be drawn as to the distance between NOE active protons on the basis of the intensity of the NOE between them.

Additional information can be obtained from NMR studies of peptides that are specifically ^{15}N or ^{13}C labeled, nitrogen labeling being particularly useful since it is the easier to effect, and since it is particularly helpful in heteronuclear COSY and NOESY. Sometimes information can be obtained from exchange rates of amide protons; while these are normally very fast, they can become slow on the NMR time scale for amide protons that are strongly hydrogen bonded in the secondary structure of peptides or are buried within the protein. Last, but not least, NOESY can also be used to demonstrate intermolecular association of peptides (as in the quaternary structure of proteins), of nucleic acids (as in double helices), of nucleic acids with proteins, and of nucleic acids with drugs.

Because it is often important for the chemist to ascertain the structure either of the molecules they are working with or, if that structure is not available, of chemically similar molecules, it is fortunate that a number of tabulations of molecules for which accurate structures have been determined are available[2,34,35] (see also ref. 36 concerning the Cambridge Crystallographic Database.)

2-6. A PRIORI CALCULATION OF STRUCTURE

When, hypothetically, one throws together an appropriate array of atoms to form a molecule, one should be able to calculate, by the principles of quantum mechanics, which arrangement of the atoms corresponds to the absolute minimum in the total energy of the resulting molecule. This arrangement should then correspond to the actual structure. (For an overview, see ref. 37.)

In practice, there are limitations to such an approach. In principle, structure, energy, and other properties are derivable from the Schrödinger equation. Exact solution of this equation is possible only for extremely small molecules, such as H_2, but for larger molecules, various kinds of approximation methods have been developed and can be very effective. Among the most powerful are what are termed *ab initio* methods,[38,39] as well as *density functional* methods.[40,41] Depending on details (size of basis set, extent of inclusion of electron correlation), the results may be more or less accurate; in general, the most accurate methods require the most computer time. Because of limitations of computer power (which, however, is constantly improving), the size of molecules that can be handled by any particular method is inversely related to the accuracy of the result. In the best cases, molecular energies can be calculated[42] to less than 1 kcal mol^{-1} (which is the accuracy needed for calculating differences between conformations; see Chapters 10 and 11) but the approaches needed for such accuracy are limited today to molecules of less than 100 atoms. Much larger molecules can be handled by so-called semiempirical methods[43] which—unlike the *ab initio* methods—incorporate parameters obtained from experimental results from a limited number of compounds; these methods are generally less reliable than *ab initio* methods.

It must be recognized that calculations refer to isolated molecules, that is, to the gas phase. Structure in the solid may be similar because the relative magnitude of packing forces is often small except when intermolecular hydrogen bonding or other strong electrostatic interaction occurs, but calculations from the vapor phase cannot usually be transferred to the liquid phase because of sizeable solvation as well as association energies. Efforts are being made to take the solvent into account either by considering specific interactions[44] or by summing up solvation energy in a large solvent shell.[45]

A different approach, which can be applied with much fewer restrictions as to the complexity of the molecules considered, was suggested by Westheimer and Mayer[46] and by Hill.[47] In this method[48–50] one computes, by a "mechanical" method, the excess energy ("steric energy") of a given array of atoms in a (as yet hypothetical) molecule over the minimum energy that the array would posses if certain kinds of interactions (see below) were "turned off."

One expresses this energy in terms of a number of contributions: bond stretching or compression strain, bond angle and torsion angle strain, nonbonded interaction, electrostatic interaction, and (in the liquid phase) solvation energy. One then changes these parameters in concert (using computer programs) so as to minimize the total energy. The geometry at the energy minimum is taken to be the actual predicted structure of the molecule in question, and the sum of the absolute minimum energy plus the residual steric energy at the calculated minimum is taken to be the heat of formation. Corrections must be made for zero point and thermal energies. Fortunately, often one is interested in calculating energy *differences* between isomeric configurations or conformations. In that case the absolute minimum energy term cancels out in the difference of energy between the two isomers.

The method permits the a priori calculation of both molecular structure and energy. It is sometimes called the "method of molecular mechanics," the "energy optimization method," or the "force field method." The first and preferred designation stems from the fact that the strain parameters are calculated in essentially a classically mechanical way. The force field designation comes from the idea that the array of atoms finds itself in a field of interatomic forces whose resultants must vanish for the equilibrium structure.

In detail, the total strain energy of the molecule V_{tot} may be set equal to

$$V_{tot} = V_r + V_\theta + V_\omega + V_{UB} + V_{nb} + V_E - V_S \tag{2.1}$$

where V_r is the energy due to bond stretching or compression (summed for all bonds), V_θ is the energy increment (summed for all angles) for bond angle deformations, V_ω is the sum total of the excess energy due to changes of torsion angles from their (energy) optima, V_{UB} stands for the (Urey–Bradley) nonbonded energy of atoms 1,3 to each other (this term is often subsumed in V_{nb}), V_{nb} is the sum total of the remaining nonbonded energy within the molecule (involving more distant atoms), V_E is the sum total of the intramolecular electrostatic energy, and V_S is the solvation energy of the molecule.

If one wants to find the structure of a molecule using Eq. 2.1, one begins by assuming a trial structure (e.g., one obtained from a molecular model; cf. Section 2-7) and calculates the energy from Eq. 2.1 using an appropriate computer program. One then allows the computer to change the coordinates and to recompute the energy. The computer thus explores the energies of all structures slightly deformed from the original trial structure; from these it selects the one that has the lowest energy and then repeats the exploration in the same way, always along the path of "steepest descent" in energy. The process continues until the structure found is at an energy minimum such that all further deformations lead to a higher energy. At this point the energy and the atomic coordinates (structure) are computed. A problem that arises not infrequently is that the structure exploration ends in a local minimum that may still lie well above the most stable structure.[50,51] Often this pitfall can be avoided by starting the search with different trial structures. Thus, if one started with a trial structure near the gauche form of butane (Section 10-1.a), one would probably find a slightly deformed version of the gauche form as minimum, but if one then restarted near the anti form, the latter would emerge as the true minimum in energy for butane.

For a more detailed discussion of this popular method appropriate reviews (see above) should be consulted.

Of the several terms of Eq. 2.1, two, the nonbonded interaction V_{nb} and the electrostatic term V_E, relevant to dipolar molecules (Eq. 2.1) deserve special mention. Experimental information for these terms is scarce; a variety of expressions have been used for V_{nb}.[52] One aspect all of them have in common is an attractive term that predominates at larger distances and a repulsive term that becomes dominant on close approach. The situation is depicted in Figure 2.13. At infinite distance, two nonbonded atoms do not interact ($V_{nb} = 0$). As the atoms approach each other, an attractive force,

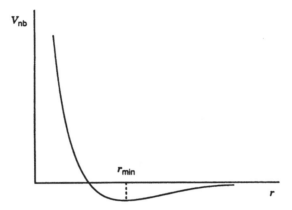

Figure 2.13. Nonbonded interaction.

the so-called London or dispersion force, comes into play leading to a lowering of the energy, The energy corresponding to this force is commonly taken as $-ar^{-6}$, where a is a constant and r is the internuclear distance. The coefficient a is proportional to the polarizability of the two atoms that approach each other; thus the attractive interaction will be greatest (other things being equal) for atoms in the lower right of the periodic table, such as S, Br, or I.

At a still closer distance of approach, a repulsive force (due to closed-shell repulsion) is encountered. The repulsive energy term may be expressed as an inverse twelfth power of the internuclear distance (br^{-12}, b being a constant). In that case the overall nonbonded potential V_{nb} is expressed by Eq. 2.2,

$$V_{nb} = -ar^{-6} + br^{-12} \tag{2.2}$$

called a 6-12 or Lennard–Jones potential (after its first proponent); it is plotted in Figure 2.13. Equation 2.2 implies an energy minimum at an internuclear distance that turns out to be an additive property of the two atoms A and B that approach each other. It is convenient to assign to these atoms radii (called van der Waals radii) r_A and r_B whose sum $r^* = r_A + r_B$ is approximately equal to the internuclear distance r_{min} at which the nonbonded energy is at a minimum. It should be realized that V_{nb} is negative (attractive) not zero, at that distance (in fact the net attraction is maximal at r_{min}). When $r < r_{min}$, V_{nb} increases, crosses the zero energy axis, and, at still shorter internuclear distances, becomes repulsive. Thus nonbonded interactions may be either attractive or repulsive, depending on the internuclear distance. Other potentials have been used.

The electrostatic term V_E (Eq. 2.3) is a Coulombic term

$$V_E = e_A e_B r^{-1} \varepsilon^{-1} \tag{2.3}$$

where e_A and e_B are the charges on the interacting atoms, r is the distance between them, as before, and ε is the dielectric constant. Assessing the magnitude of these charges is not straightforward; it has been customary to compute the charges from

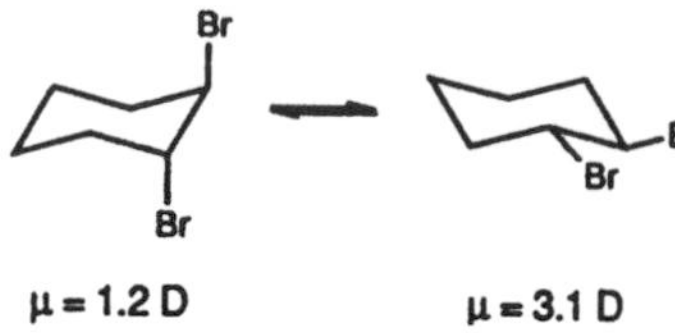

$$\mu = 1.2 \text{ D} \qquad\qquad \mu = 3.1 \text{ D}$$

Figure 2.14. *trans*-1,2-Dibromocyclohexane, where $\Delta G^0 = 0.65$ kcal mol^{-1} (2.7 kJ mol^{-1}) for CCl$_4$ and $\Delta G^0 = -0.30$ kcal mol^{-1} (−1.3 kJ mol^{-1}) for CH$_3$CN.

bond dipoles and locate them at the center of the pertinent atomic nuclei. This approach has been taken even for the C–H bond ($\mu = 0.3$ D) in some calculation.[53] A better procedure is to calculate the charges from quantum mechanics. Yet another alternative is to compute the energy from the electrostatic energy between two dipoles.[54]

The parameter ε in Eq. 2.3 is the effective dielectric constant whose value also presents problems. Since in a molecule, part of the space between the atomic point charges is occupied by the molecule itself and part by the solvent, ε corresponds neither to the dielectric constant of the solvent nor to that of the solute, but is somewhere in between.[55] It is customary to use an empirical value for ε somewhere between 2 and 5; this value may, in fact, be treated as a disposable parameter in model calculations on molecules where the experimental results are known.

That the V_E term is of considerable importance in dipolar molecules is seen, for example, in the conformational equilibrium of *trans*-1,2-dibromocyclohexane (Fig. 2.14). On purely steric grounds this molecule should prefer the diequatorial conformation but, in fact, 75% of the molecules are in the diaxial conformation in carbon tetrachloride ($\varepsilon = 2.2$), whereas the preference is for the diequatorial form (64%) in acetonitrile, CH$_3$CN ($\varepsilon = 36$), at room temperature.[56] The diaxial preference in CCl$_4$ clearly has its origin in the electrostatic repulsion of the dipoles when they are equatorially oriented (see also Chapter 11); it diminishes in higher-dielectric solvents (which also solvate the conformer of higher dipole moment better: V_S term in Eq. 21).

The molecular mechanics method as described can be applied only to systems that do not possess resonance energy (aromatics, conjugated alkenes, etc.) unless consideration is confined to a difference in energy between two structures for which delocalization energy remains constant, for example, the equatorial and axial forms of phenylcyclohexane.[57] However, it has been possible to calculate the energy of conjugated and aromatic systems by combining quantum mechanical and molecular mechanical calculations (the so-called MMP methods).[58]

One of the attractive aspects of molecular mechanics calculations is that they provide insight as to the source of experimentally observed steric interactions. Thus, for example, Allinger and Tribble[57] (see also ref. 59) showed that the calculated difference of 3 kcal mol^{-1} (12.6 kJ mol^{-1}) between the energy of equatorial and axial phenylcyclohexane (later found, experimentally, to amount[60] to 2.87 kcal mol^{-1}) is mainly due to the interaction of ortho hydrogen and carbon atoms of the benzene ring with the adjacent equatorial hydrogen atoms on the cyclohexane ring when the phenyl

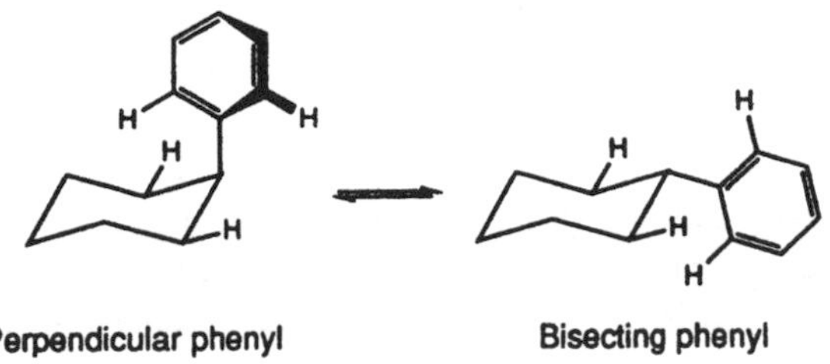

Perpendicular phenyl   Bisecting phenyl

Figure 2.15. Phenylcyclohexane.

is axial (Fig. 2.15), rather than to the compression of the synaxial hydrogen atoms, which is normally principally responsible for the instability of the axial conformers (cf. Chapter 11). Some caution must be exercised in the dissection of steric interactions as among angle bending, torsion, and nonbonded interactions; it must be remembered that in the Hendrickson–Wiberg–Allinger approach the parameters ascribed to these various interaction modes are picked empirically and somewhat arbitrarily. Therefore, two different types of calculation using one or the other of several available force fields can both give the right answers in terms of energy and structure, but both are not likely to be correct as to the exact physical origin of whatever interaction energies are found. By the same token, it is not surprising that force fields are constantly being modified and improved.

2-7. MOLECULAR MODELS

Only a very brief discussion of molecular models will be given here (for a detailed discussion, see ref. 61).

There are essentially three types of models:

1. Those that simply help to visualize the 3D architecture and stereochemistry but are not to scale.
2. Framework-type models that indicate correct bond distances and bond angles and can be used to measure distances between nonbonded atoms in molecules but do not show the atoms as such.
3. So-called space-filling models that provide a fairly realistic 3D representation of what the molecule actually looks like.

The first kind of models are the "ball-and-stick" kind, though the "balls" (wooden or plastic) are sometimes polyhedra; the "sticks" may be made of wood, of hard rubber, or of other polymeric material. These models are generally inexpensive and help beginning students visualize molecules in three dimensions, manipulate projection formulas, and count stereoisomers, but they do not show the dimensions of the molecule.

The second (framework-type) kind of model is exemplified by the well-known "Dreiding models," which are machined so that when the atomic models are put together, interatomic distances as well as bond angles are nearly correct. In these models, if the conformation is arranged properly (and this is a rather important condition, not easily fulfilled in flexible structures), distances between nonbonded atoms can be measured and nonbonded repulsions leading to strain can be sensed and even estimated as to their magnitude. The Kendrew Skeletal Molecular Models are also of this type.

The third kind, space-filling models, give a better picture of the true molecular shape and dimensions but are too congested to allow measurements of interatomic distances. The so-called Fisher–Hirschfelder–Taylor, Stuart–Briegleb, C-P-K (Cory–Pauling–Koltun), Catalin, Courtauld, and Godfrey models are examples of this type.

A serious shortcoming, common to virtually all molecular models, is that they have fixed bond angles and that rotation about single bonds is excessively facile, especially in the first two types of models. This is in contrast to the real situation (cf. the discussion of molecular mechanics, Section 2-6) of relatively easily deformable bond angles and substantial barriers to rotation about single bonds. Moreover, if one wants to measure an intramolecular bond distance between two atoms in a Dreiding model, one must fix the model in its actual conformation. For molecules having a number of single bonds, this may be quite inconvenient (even though mechanical devices to stop bond rotation are available) and it may be difficult to set the actual torsion angles with any kind of precision.

As a result of all these difficulties, molecular modeling in suitable computers in conjunction with appropriate displays has become a superior substitute for the use of mechanical models in situations where quantitative (as distinct from qualitative or semiquantitative) information about exact molecular shapes and intra- or intermolecular interactions is desired. The molecule can be input with standard coordinates (bond lengths, bond angles, and torsion angles) or may even be drawn on the screen by a "mouse" or other device and is then manipulated in the computer by stretching or compressing bonds, opening or closing bond angles, and changing torsion angles so as to minimize the energy (or in any other way desired); interatomic distances may then be read off directly on screen. The energy minimization is performed by a built-in molecular mechanics program, as described in Section 2-6.

REFERENCES

1. Moss, G. P. *Pure Appl. Chem.* **1996**, *68*, 2193.

2. Sutton, L. E., ed. *Tables of Interatomic Distances and Configuration of Molecules and Ions*, Special Publication No. 11, 1958 and Supplement No. 18, 1965, The Chemical Society, London.

3. Jensen, F. R. and Buschweller, C. H. *J. Am. Chem. Soc.* **1966**, *88*, 4279; *ibid.* **1969**, *91*, 3223.

4. Rouvray, D. H. *J. Chem. Educ.* **1975**, *52*, 768.

5. King, R. B., ed. *Chemical Applications of Topology and Graph Theory,* Elsevier, New York, 1983.

6. Wheland, G. W. *Advanced Organic Chemistry,* 3rd ed., Wiley, New York, 1960, p. 41.

7. Pauling, L. *The Nature of the Chemical Bond,* 3rd ed., Cornell University Press, Ithaca, NY, 1960, p. 6.

8. Saunders, M. *Tetrahedron Lett.* **1963**, 1699.

9. Hargreaves, M. K. and Modarai, B. *Chem. Commun.* **1969**, 16; *J. Chem. Soc. C* **1971**, 1013; Wilen, S. H., Bunding, K. A., Kascheres, C. M., and Wieder, M. J. *J. Am. Chem. Soc.* **1985**, *107*, 6997.

10. Doyle, T. R. and Vogl, O. *J. Am. Chem. Soc.* **1989**, *111*, 8510.

11. Costante, J., Hecht, L., Polavarapu, P. L., Collet, A., and Barron, L. D. *Angew. Chem. Int. Ed. Engl.* **1997**, *36*, 885.

12. Mizushima, S.-I. *Structure of Molecules and Internal Rotation,* Academic Press, New York, 1954, pp. 7ff.

13. Lehn, J.-M. *Top. Curr. Chem.* **1970**, *15*, 311.

14. Lambert, J. B. *Top. Stereochem.* **1971**, *6*, 19.

15. Klyne, W. and Prelog, V. *Experientia* 1960, *16*, 521.

16. Glusker, J. P. and Trublood, K. N. *Crystal Structure Analysis, A Primer,* 2nd ed. Oxford University Press, New York, 1985.

17. Dunitz, J. D. *X-Ray Analysis and the Structure of Organic Molecules,* Cornell University Press, Ithaca, NY, 1979.

18. Patterson, A. L. *Z. Kryst.* **1935**, *90*, 517.

19. Akimoto, H., Shioiri, T., Iitaka, Y., and Yamada, S.-i. *Tetrahedron Lett.* **1968**, 97.

20. Karle, J. and Hauptman, H. *Acta Crystallogr.* **1956**, *9*, 635; see also Karle, J. *Science* **1986**, *232*, 837.

21. Karle J. and Karle, I. "Application of Direct Methods in X-Ray Crystallography," in Robertson, J. M., ed., *MTP International Review of Science*, Physical Chemistry Series, 1, Vol. 11, Butterworths, London, 1972, p. 247.

22. Ladd, M. F. C. and Palmer, P. A. *Theory and Practice of Direct Methods in Crystallography*, Plenum Press, New York, 1980.

23. Hendrickson, W. A. *Science* **1991**, *254*, 51.

24. Dunitz, J. D., Eser, H., Bixon, M., and Lifson, S. *Helv. Chim. Acta* **1967**, *50*, 1572.

25. Ermer, O. *Angew. Chem. Int. Ed. Engl.* **1983**, *22*, 251.

26. Bacon G. E. *Neutron Diffraction,* 3rd ed., Oxford University Press, Oxford, UK, 1975.

27. Speakman, J. C. "Neutron Diffraction," in *Molecular Structure by Diffraction Methods*, A Specialist Periodical Report, The Chemical Society, London, Vol. 6, 1978, p. 117.

28. Hargittai, I. and Hargittai, M. *Stereochemical Applications of Gas-Phase Electron Diffraction*, VCH, New York, 1988. Part A, "The Electron Diffraction Technique"; Part B, "Structural Information for Selected Classes of Compounds."

29. Gordy, W. and Cook, R. L. *Microwave Molecular Spectra*, Wiley, New York, 1984.

30. Robiette, A. G. "The Interplay between Spectroscopy and Electron Diffraction," in *Molecular Structure by Diffraction Methods,* A Specialist Periodical Report, The Chemical Society, London, 1972, Vol. 1, p. 161.

31. Schäfer, L., Siam, K., Ewbank, J. D., Caminati, W., and Fantoni, A. "Ab initio Studies of Structural Features Not Easily Amenable to Experiment: Some Surprising Applications of Ab initio Geometries in Microwave Spectroscopic Conformational Analyses," in Maksić, Z. B., ed., *Modeling of Structure and Properties of Molecules*, Wiley, New York, 1987, p. 79.

32. Neuhaus, D. and Williamson, M. P. *The Nuclear Overhauser Effect in Structural and Conformational Analysis*, VCH, New York, 1989.

33. Wüthrich, K. *NMR of Proteins and Nucleic Acids*, Wiley-Interscience, New York, 1986.

34. Kennard O. and others, *Molecular Structure and Dimensions*, Crystallographic Data Center, Cambridge, UK, Vols. 1–15, 1970–1984.

35. Duax, W. L., Weeks, C. M., and Rohrer, D. C. *Top. Stereochem* **1976**, *9*, 271.

36. Allen, F. H. "The Cambridge Structural Data Base as a Research Tool in Chemistry," in Maksić Z. B., ed., *Modelling of Structure and Properties of Molecules*, Wiley, New York, 1987, 1987, p. 51. Allen, F. H., Kennard, O., and Taylor, R. *Acc. Chem. Res.* **1983**, *16*, 146.

37. Jensen, F. *Introduction to Computational Chemistry*, Wiley, New York, 1999.

38. Hehre, W. J., Radom, L., Schleyer, P. v. R., and Pople, J. A. *Ab Initio Molecular Orbital Theory*, Wiley, New York, 1986.

39. Simons, J. and Nichols, J. *Quantum Mechanics in Chemistry*, Oxford University Press, New York, 1997, Section 6.

40. Parr, R. G. and Yang, W. *Density-Functional Theory of Atoms and Molecules*, Oxford University Press, New York, 1989.

41. Koch, W. and Holthausen, M. C. *A Chemist's Guide to Density Functional Theory*, VCH-Wiley, New York, 2000.

42. Császár, A. G., Allen, W. D., and Schaefer, H. F. *J. Chem. Phys.* **1998**, *108*, 9751.

43. Sadlej, J., *Semi-Empirical Methods of Quantum Chemistry*, Wiley, New York, 1985.

44. Tomasi, G. J., Alagona, G., Bonaccorsi, R., and Ghio, C. "A Theoretical Method for Solvation—Some Applications to Biological Systems," in Maksić, Z. B., ed., *Modeling of Structure and Properties of Molecules*, Wiley, New York, 1987, p. 330.

45. Jorgenson, W. L. *Acc. Chem. Res.* 1 **1989**, *22*, 184.

46. Westheimer, F. H. and Meyer, J. E. *J. Chem. Phys.* **1946**, *14*, 733. Westheimer, F. H. "Calculation of the Magnitude of Steric Effects," in Newman, M. S., ed., *Steric Effects in Organic Chemistry*, Wiley, New York, 1956, p. 523.

47. Hill, T. L. *J. Chem. Phys.* **1946**, 14, 465.

48. Burkert, U. and Allinger, N. L. *Molecular Mechanics*, ACS Monograph 177, American Chemical Society, Washington, DC, 1982.

49. Allinger, N. L. "Molecular Mechanics," in Domenicano, A. and Hargittai, I., eds., *Accurate Molecular Structures*, Oxford University Press, New York, 1992, p. 336.

50. Rappé, A. K. and Casewit C. A. *Molecular Mechanics Across Chemistry*, University Science Books, Sausalito, CA, 1997.

51. Kollman, P. A. and Merz, K. M. *Acc. Chem. Res.* **1990**, *23*, 246.

52. Dunitz, J. D. and Bürgi, H. B. "Non-bonded Interaction in Organic Molecules," in *MTP International Reviews of Science, Physical Chemistry*, Series 2, Vol. 11, Butterworths, Boston, 1975, p. 81.

53. Warshel, A. and Lifson, S. *J. Chem. Phys.* **1970**, *53*, 582.

54. Eliel, E. L., Allinger, N. L., Angyal, S. J., and Morrison, G. A. *Conformational Analysis*, Interscience-Wiley, New York, 1965; reprinted by American Chemical Society, Washington, DC, 1981, p. 461.

55. Kirkwood, J. G. and Westheimer, F. H. *J. Chem. Phys*, **1938**, *6*, 506.

56. Abraham, R. J. and Brettschneider, E. "Medium Effects on Rotational and Conformational Equilibria," in Orville-Thomas, W., ed., *Internal Rotation in Molecules*, Wiley, New York, 1974, p. 481.

57. Allinger, N. L. and Tribble, M. T. *Tetrahedron Lett.* **1971**, 3259.

58. Allinger, N. L. and Sprague, J. T. *J. Am. Chem. Soc.* **1973**, *95*, 3893.

59. Hodgson, D. J., Rychlewska, U., Eliel, E. L., Manoharan, M., Knox, D. E., and Olefirowicz, E. M. *J. Org. Chem.* **1985**, *50*, 4838.

60. Eliel, E. L. and Manoharan, M. *J. Org. Chem.* **1981**, *46*, 1959.

61. Walton, A. *Molecular and Crystal Structure Models*, Wiley, New York, 1978.

3

STEREOISOMERS

3-1. NATURE OF STEREOISOMERS

a. General

In Chapter 2 isomers were defined as compounds having the same molecular formula but differing in structure. The isomers were subdivided according to whether they differ in constitution, in configuration, and/or in conformation. Isomers differing only in configuration and/or conformation were recognized as stereoisomers.

In this chapter we consider an alternative subdivision of stereoisomers: between enantiomers and diastereomers. Unlike the somewhat fuzzy (cf. Section 2-3) division between configuration and conformation, the dichotomy between enantiomers and diastereomers is unequivocal. Moreover, the two classifications are orthogonal: Enantiomers may differ in configuration or only in conformation; the same is true of diastereomers (diastereoisomers).

Enantiomers are pairs of isomers related as an object is to its mirror image. This relationship may stem from a configurational difference, as in CHFClBr (Fig. 2.6) or (depending on definition, see Chapter 2) in ethylmethylbenzylamine (Fig. 3.1), or from a conformational (sign of torsion angle) difference, as in a tetra-o-substituted biphenyl (Fig. 3.2), the gauche forms of 1,2-dichloroethane (Fig. 2.7, **A** and **C**), or the two chair forms of cis-1,2-dimethylcyclohexane (Fig. 2.12). The stability to interconversion of the enantiomers, which is high in the case of CHFClBr and appropriately tetra-o-substituted biphenyls (Chapter 13) but fleeting in the other three examples, is of no concern in the definition; conceptually, the structures that are compared are considered as being rigid.

Enantiomers must be isomers as well as mirror images; that is, they must not be superposable. One can draw many structures (some exemplified in Fig. 3.3) that bear

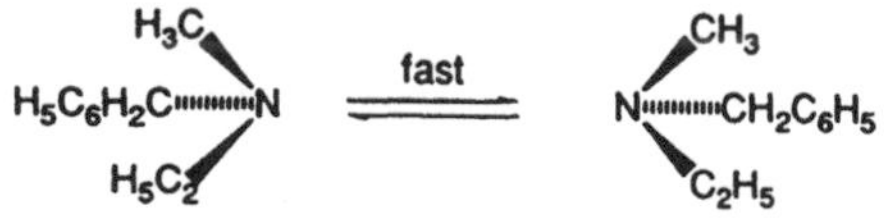

Figure 3.1. Enantiomers of ethylmethylbenzylamine.

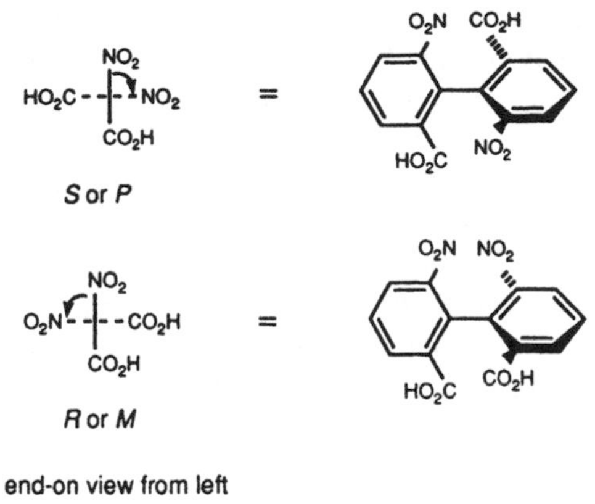

NO$_2$

HO$_2$C - - $\left[\begin{array}{c}\end{array}\right]$ NO$_2$

CO$_2$H

S or *P*

=

O$_2$N   CO$_2$H

HO$_2$C   NO$_2$

NO$_2$

O$_2$N - $\left[\begin{array}{c}\end{array}\right]$ - CO$_2$H

CO$_2$H

R or *M*

=

O$_2$N   NO$_2$

HO$_2$C   CO$_2$H

end-on view from left

Figure 3.2. Enantiomers of 6,6′-dinitro-2,2′-diphenic acid.

a mirror-image relationship but, upon rotation of the entire model (rigid rotation) around an appropriate axis, turn out to be superposable; such structures are identical (or homomeric, see below) not enantiomeric.

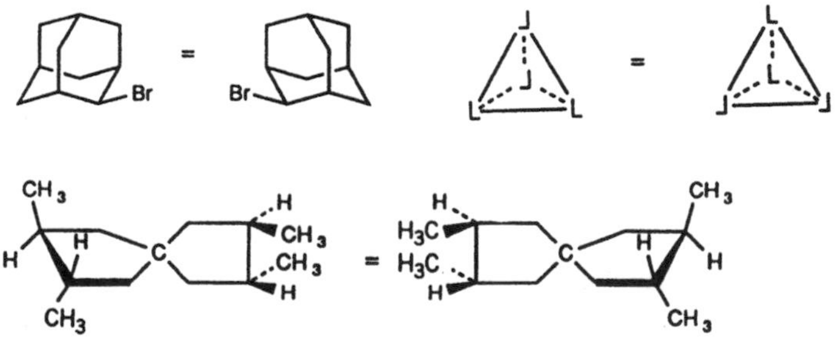

Figure 3.3. Identical mirror-image structures.

Diastereomers (or diastereoisomers) are stereoisomers (i.e., isomers of identical constitution but differing three-dimensional architecture) that do not bear a mirror-image relation to each other. Diastereoisomerism may be due to differences in configuration (or conformation) at several sites in the molecule, as in the tartaric acids (Fig. 3.4) or in appropriately substituted terphenyls (Fig. 3.5, **A**) or biphenyls (Fig. 3.5,

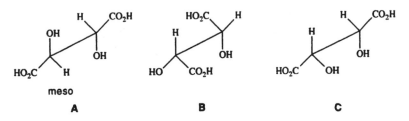

Figure 3.4. The tartaric acids. Enantiomers are **B**, **C**; diastereomers are **A**, **B**, and **A**, **C**.

B); we are dealing in the first case with configurational differences, in the second case with differences in conformation between the isomers, and in the third case with a combination of the two types. In these particular cases the barriers between the diastereomers are high and each isomer is stable, at least at room temperature. However, this is not essential: The rapidly interconverting isomers of tertiary amines shown in Figure 3.6 and the gauche and anti forms of 1,2-dichloroethane (Fig. 2.7, **A** and **B**) are also diastereoisomers.

The terms enantiomer and diastereomer relate to molecules as a whole. Thus, if two molecules have the same constitution (connectivity) but different spatial arrangements of the atoms (i.e., if they are stereoisomeric), they must either be related as mirror images or not: In the former case they are enantiomers, in the latter case they are diastereomers. The differentiation can be made without considering any particular part of the molecule. Nonetheless, in viewing and specifying (cf. Chapter 5) enantiomers and diastereoisomers one often focuses on particular sites in the molecule, such as the carbon atom in CHFClBr or the torsion axis containing the phenyl–phenyl bond in the biphenyls and terphenyls shown in Figures 3.2 and 3.5. In the tradition of van't Hoff, one "factorizes"[1] stereoisomerism by attributing it to a "chiral center" (cf. Chapter 1) or a torsion axis.

A chiral center (or, more properly, center of chirality) is a focus of chirality; in the case of carbon, at least, it corresponds to the asymmetric tetrahedral atom of van't Hoff as shown in Figure 3.7. The existence of enantiomers is usually, but not invariably, associated with at least one chiral center or chiral torsion axis (axis of chirality) as seen in the above examples (exceptions will be discussed in Chapter 13).

Diastereomers, as we have seen, often contain two or more chiral centers (Figs. 3.4 and 3.6, **A**), chiral (torsion) axes (Fig. 3.5, **A**), or a combination thereof (Fig. 3.5, **B**). However, this is not necessarily the case; Figures 3.6, **B** and 3.8 illustrate cases where diastereomers are neither chiral nor contain chiral centers; alkene diastereomers (Fig. 3.9) are also of this type. Thus diastereoisomerism is not necessarily associated with chiral centers or chiral torsion axes; a general scheme for factorizing diastereoisomerism must go beyond consideration of these chiral elements.

The starred carbon atoms (C*) in Figure 3.8 (which are the foci of the diastereoisomerism but are not chiral, since each bears two identical ligands) have been called "centers of stereoisomerism"[2] or "stereogenic centers."[3–5] Interchange of

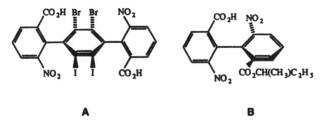

         A   **B**

Figure 3.5. Diastereoisomerism in terphenyls and biphenyls. **A**: Terphenyl capable of existing in diastereomeric forms. **B**: Diastereoisomers generated by combination of a chiral center and a chiral axis.

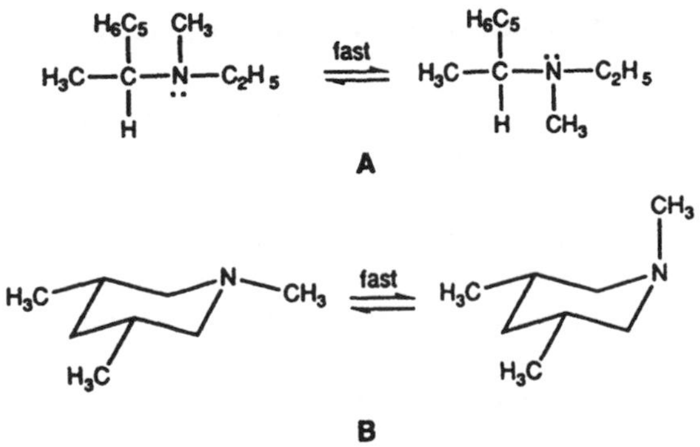

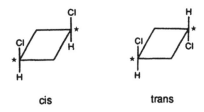

Figure 3.6. Rapidly interconverting diastereomers.

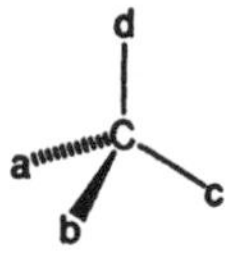

Figure 3.7. Center of chirality.

two ligands at a stereogenic center leads to a stereoisomer. By extension one might call the axis containing the olefinic double bond (dashed in Fig. 3.9) a "stereogenic axis." The axis containing the phenyl–phenyl bonds in the terphenyl shown in Figure 3.5, **A** is also a stereogenic axis. Stereogenic centers thus may be or may not be chiral (i.e., centers of chirality). Conversely, however, *all* chiral centers are stereogenic.

Figure 3.8. Diastereomeric 1,3-dichlorocyclobutanes.

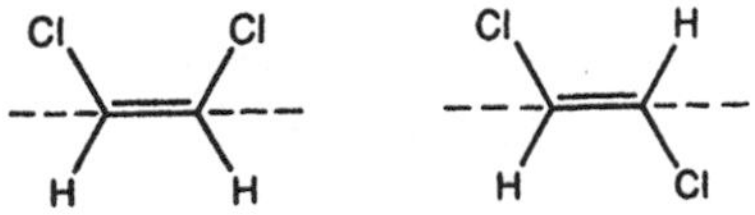

Figure 3.9. Diastereomeric alkenes.

b. Barriers Between Stereoisomers and Residual Stereoisomers

In the discussions so far, the question as to whether stereoisomers, (of whatever type) can or cannot be isolated or otherwise observed was considered immaterial. Defining stereoisomers in that fashion has advantages and disadvantages. The advantage is that the number of stereoisomers so counted is independent of the method of observation and, generally, of the temperature. Thus chlorocyclohexane has two dominant stereoisomers (Fig. 2.3), one with equatorial chlorine, the other with axial chlorine (the concentration of twist forms is small enough to be disregarded; cf. Chapter 11). With this insight one knows what one might expect upon physical examination of chlorocyclohexane by methods that give an "instantaneous" picture, such as IR spectroscopy or electron diffraction. The disadvantage is that the classification does not correspond to the experience of the preparative chemist who encounters, at or near room temperature, only a single substance called chlorocyclohexane in the laboratory. [This occurs because the rate of interconversion of the isomers (Fig. 2.3) is very high; the isomers are separated by energy barriers of much less than ca. 20 kcal mol^{-1} (84 kJ mol^{-1}) and thus are interconverted too rapidly for isolation at room temperature.]

A barrier ($\Delta G^{\ddagger}$) of 20 kcal mol^{-1} (84 kJ mol^{-1}) corresponds to an interconversion rate of 1.3×10^{-2} s^{-1} at 25°C, that is, a half-life ($t_{1/2}$) of about 1 min; a barrier of 25 kcal mol^{-1} (105 kJ mol^{-1}) indicates an interconversion rate of 2.9×10^{-6} s^{-1}; $t_{1/2}$ = 66 h at 25°C. (In general, $k = 2.084 \times 10^{10} \, Te^{-\Delta G^{\ddagger}/1.986T}$, where $\Delta G^{\ddagger}$ is in calories per mole; the exponent is $-\Delta G^{\ddagger}/8.315T$ when $\Delta G^{\ddagger}$ is in joules per mole.)

Even in NMR spectroscopy, the question of whether one sees one or two spectra of a substituted cyclohexane depends on temperature and instrumental frequency. At −100°C, the spectra of the axial and equatorial conformers would appear separately, but at room temperature a single (averaged) spectrum would be seen (cf. Chapter 11).

Thus in deciding whether structures are *effectively* enantiomeric (or diastereomeric) or not, one must make allowance for those stereochemical changes (rotations or inversions) that occur rapidly on the scale of the experiment under consideration (chemical transformation, spectroscopic observation, determination of a physical property). The resulting need for introducing a time scale presents an area of conflict between the definition of stereoisomers given above, which relates to rigid structures, and the more practical view relating to substances as they are encountered in the laboratory. The latter must necessarily include energy considerations, and this affects the isomer count.

A very useful concept introduced by the Mislow group,[6] namely, that of "residual stereoisomerism," has been generalized[7] to cover the above cases. We define as "residual stereoisomers" those subsets of the total set of stereoisomers that can be distinguished under specified conditions by a given technique. Thus 2-bromobutane (Fig. 3.10) has six stereoisomers but [in the limit of fast rotation around the C(2)–C(3) bond that applies to chemical manipulations in the laboratory at room temperature] only two residual stereoisomers with respect to isolation: the (+) and (−) enantiomers (Fig. 3.11). Similarly, N-methyl-1-phenylethylamine (4 stereoisomers, **A′**, **A″**, **B′**, **B″**, Fig. 3.12) has two residual stereoisomers (the two enantiomers **A**, **B** differing in

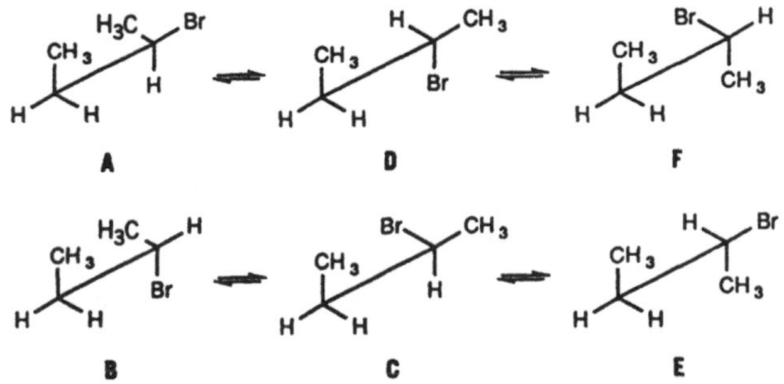

Figure 3.10. 2-Bromobutanes: a complete view.

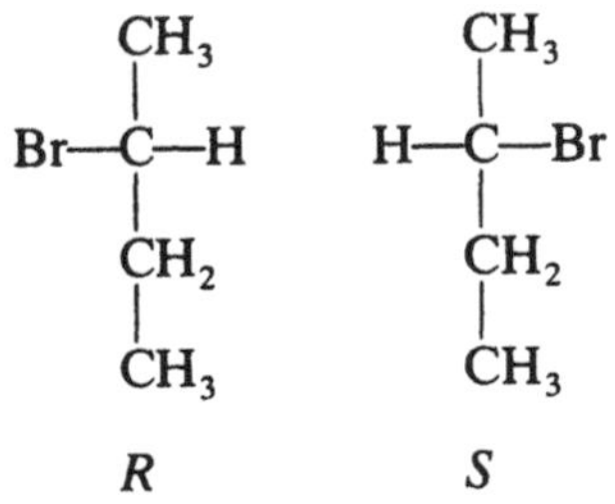

Figure 3.11. 2-Bromobutanes: a laboratory view.

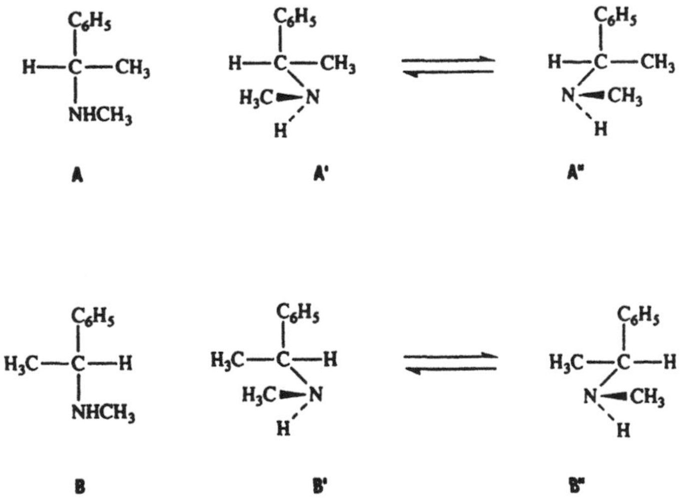

Figure 3.12. *N*-methyl-1-phenylethylamine.

Figure 3.13. 2-Bromobutane isomers.

configuration at the benzylic carbon). Chlorocyclohexane (Fig. 2.3) displays two stereoisomers (with equatorial and axial chlorine) when viewed by IR or Raman spectroscopy, when handled in the laboratory at −150°C, or when viewed by NMR spectroscopy at −100°C (see Chapter 11). At room temperature, the two stereoisomers can still be seen by vibrational spectroscopy, but as far as NMR or isolation is concerned, there is only a single substance, that is, a single residual species (since there is only one, it cannot be called an isomer). Thus, unlike stereoisomerism, which is a structural concept independent of the means of detection, residual stereoisomerism depends on the circumstances under which it is determined.

Returning to the 2-bromobutane case (Fig. 3.10), we may represent the complete set of six isomers as shown in Figure 3.13. Only under special circumstances (e.g., by IR spectroscopy in a chiral medium) might it be possible to distinguish all six members of the set. Under polarimetric investigation one may discern two subsets: A/D/F and C/B/E (separated by the horizontal line in Fig. 3.13) corresponding to the (R)-(+) and (S)-(−) isomers, respectively (cf. Fig. 3.11). These are the two residual stereoisomers, since the conformational isomers cannot be distinguished under the conditions of the experiment. On the other hand, spectroscopic investigation (e.g., by IR) in an achiral medium might disclose the presence of three residual stereoisomers separated, in Figure 3.13, by the vertical lines: conformers A/C, D/B, and F/E. Finally, chemical manipulation in an achiral medium would put into evidence only a single residual 2-bromobutane species (comprising the whole box in Fig. 3.13). This case shows not only that the number of residual isomers depends on the method of observation, but also that different submultiples (3 or 2 as well as 1) of the total number of isomers may appear as residual isomers or species under different circumstances.

3-2. ENANTIOMERS

Enantiomers are characterized by nonsuperposability and a mirror-image relationship. The absence of superposability may be tested directly in the model or, as further explained in Chapter 4, it may be recognized from symmetry considerations. Molecules having planes, centers, or alternating axes of symmetry are superposable with their mirror images; those lacking all such elements of symmetry are not.

As explained in Section 3-1, one is usually interested in residual enantiomerism and therefore disregards events that are rapid on the time scale of the experiment. Thus 1.2-dichloroethane (ClCH₂CH₂Cl) is generally considered a single achiral substance: Even though, as shown in Figure 2.7, there are, in principle, two enantiomers and a diastereomer, there is a single residual species when one considers chemical manipulation at room temperature, where rotation about the C–C bond is fast.

As indicated in Section 2-3, the simplest source of chirality in an organic molecule is a carbon atom with four different ligands. A historically much studied example is lactic acid (Fig. 3.14). Although a number of stereoisomers exist by virtue of rotation around the C–OH and C–CO_2H bonds, there are only two residual enantiomers, which are depicted in Fig. 3.14. Both occur in nature. The (+)-lactic acid is found in muscle fluid and results from the biochemical processes occurring when a muscle does work. It is sometimes called "sarcolactic acid" in the old literature (*sarcos* meaning muscle in Greek), melts at 25–26°C, and has a specific rotation $[\alpha]_D^{15}$ 3.8 in 10% aqueous solution. In contrast, anaerobic fermentation of glucose with *Leishmania brazilensis panamensis* gives enantiomerically pure (−)-lactic acid among other products[8]; the acid melts at 26–27°C and has the same magnitude of rotation, but in the opposite sense, as the (+) isomer.

Since most properties of matter are invariant to reflection ("scalar properties"), it is to be expected that (+)- and (−)-lactic acid will be identical in many respects. Not only do they have the same melting point, but they also have the same solubility, density, refractive index, IR, Raman, UV, and NMR spectra, and, at least very nearly, the same X-ray diffraction pattern. It is only with respect to properties or manipulations that change sign, but not magnitude, upon reflection ("pseudoscalar properties"), that (+)- and (−)-lactic acid differ. This is true for chiroptical properties (cf. Chapter 1), such as optical rotation, optical rotatory dispersion (ORD), and circular dichroism (CD), topics to be pursued further in Chapter 12.

Some fermentative processes generate a mixture of (+)- and (−)-lactic acid in equal amounts; that is, they yield the so-called (±)- or racemic lactic acid that does not rotate polarized light and melts at 18°C (see Chapter 6 for properties of racemates).

Various representations of the lactic acids are shown in Figure 3.14 [The question as to which model corresponds to the (+) and which to the (−) enantiomer will be taken

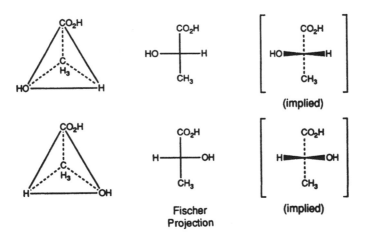

Figure 3.14. Lactic acids.

up in Chapter 5.] The representation of the tetrahedron (with the chiral carbon atom at the center not shown) is cumbersome. The model is depicted by way of an invitation to the readers to build their own. A common representation is the planar projection formula, first proposed by Fischer.[9] It is important to perform the Fischer projection correctly: The atoms pointing sideways must project forward in the model but those pointing up and down in the projection must extend toward the rear. There are 24 (= 4! this being the number of permutations of 4 ligands among 4 sites) ways of writing the projection formula; 12 correspond to one enantiomer and 12 to the other. Twelve of these, corresponding to the (+) enantiomer, are depicted in Figure 3.15; the 12 mirror images of those shown would represent the (-) isomer.

The 12 permutations shown in Figure 3.15 can be generated rather simply, either by permuting groups in threes (e.g., $1 \rightarrow 2 \rightarrow 3, 3 \rightarrow 4 \rightarrow 7$, or $4 \rightarrow 5 \rightarrow 6$) or by turning the formula by 180° ($1 \rightarrow 7, 6 \rightarrow 12$, etc.). The former permutation corresponds to holding the model at one of the ligands ($1 \rightarrow 2 \rightarrow 3$ and $4 \rightarrow 5 \rightarrow 6$, CO_2H; $3 \rightarrow 4 \rightarrow 7$, OH) and turning 120° around an axis linking that ligand to the chiral center. The latter change corresponds to turning the model upside down. (The reader is invited to carry out these operations with a model.)

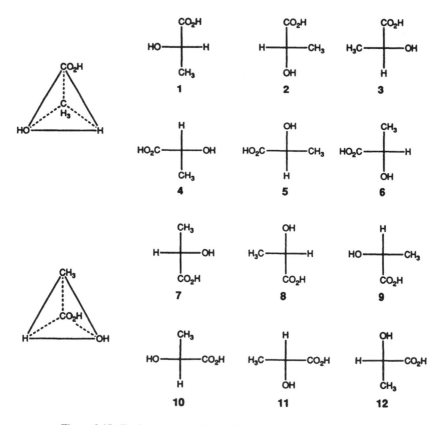

Figure 3.15. Twelve representations of (+)-lactic acid (Fischer projections).

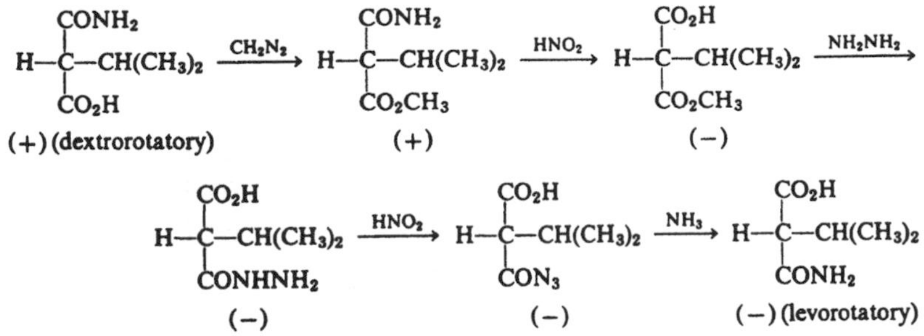

Figure 3.16. Systematic interconversion of enantiomers.

Not permitted are interchanges of two groups (such interchanges clearly lead from the molecule to its enantiomer) and 90° rotations of the formula. (Rotating the model 90° does not yield a projection formula rotated by 90°, since the sideways groups no longer point forward nor do those on top and bottom point backward.) However, two successive exchanges of two groups are permitted as converting the model to its enantiomer and then back to the original stereoisomer; such double exchanges (e.g., **1 → 5**) are, in all cases, equivalent to permutations already shown in Figure 3.15.

If the configuration (or sense of chirality) at the chiral (stereogenic) center is indeed responsible for the direction of optical rotation, then reversing the sense of chirality by chemical manipulation should reverse the rotation. This classical experiment was carried out by Fischer and Brauns[10] and is shown in Figure 3.16.

3-3. DIASTEREOMERS

a. General Cases

Diastereomers are stereoisomers that are not related as object and mirror image. Whereas a set of enantiomers can contain only two members, there is no such limitation for diastereomers. Since each stereogenic center can exist in one or the other of two configurations, there will be two stereoisomers for a single stereogenic center, $2 \cdot 2$ or 2^2 combinations for two such centers, $2 \cdot 2 \cdot 2$ or 2^3 for three centers, and, in general, 2^n for n centers. Since each stereoisomer has its enantiomer, the 2^n stereoisomers will exist as 2^{n-1} pairs of enantiomers, with each pair of enantiomers being diastereomeric with any other pair.

The aldose sugars (open-chain forms) illustrate this progression ($n = 1-4$). Glyceraldehyde, the simplest sugar, exists as a pair of enantiomers: For the tetroses, there are two enantiomer pairs; for the pentoses, four; for the hexoses, eight; and so on. Sugar chemists assign different names to the (diastereomeric) enantiomer pairs; within each pair they distinguish the enantiomers by giving the prefix D to the isomer that has the hydroxyl group on the right in the Fischer projection formula at the highest numbered chiral carbon atom. (The other enantiomer is given the prefix L.) In Figure

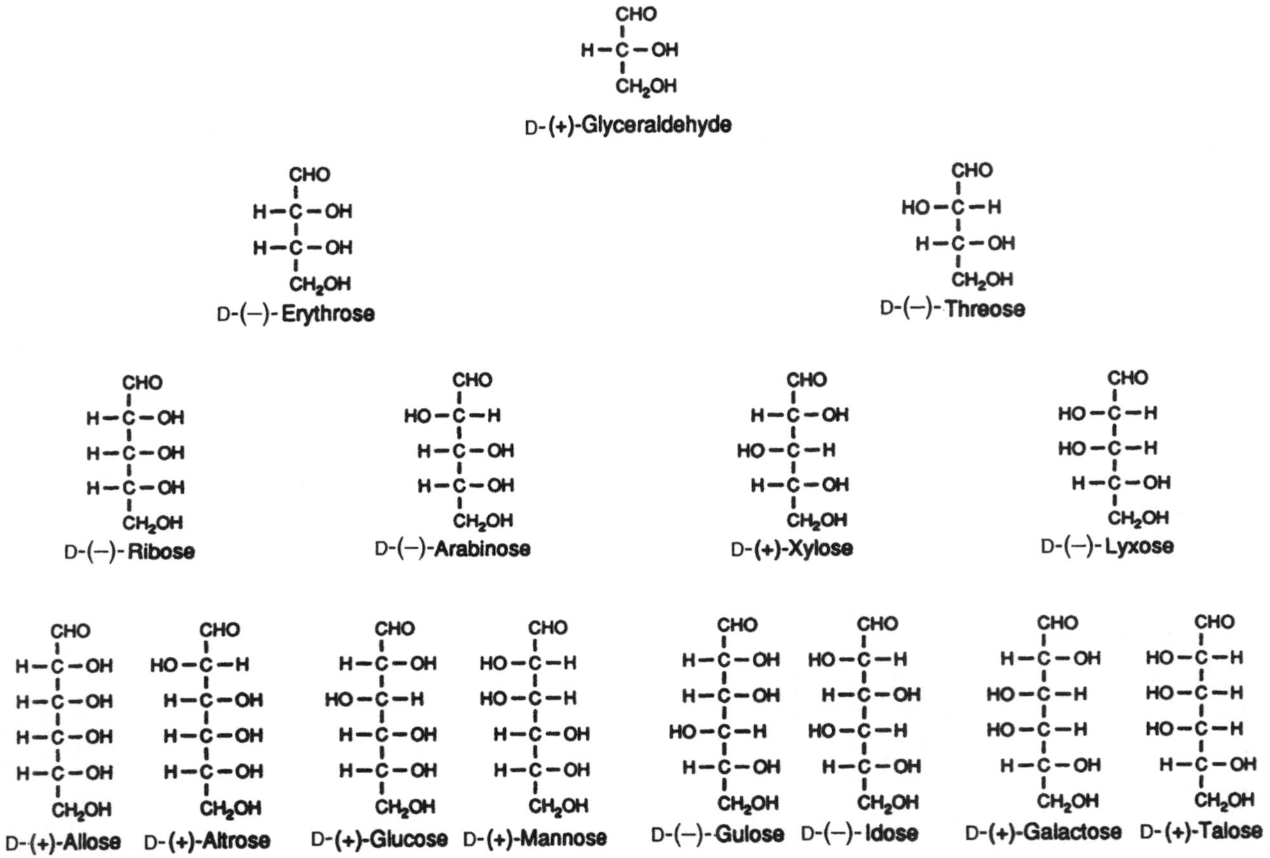

Figure 3.17. The aldose sugars.

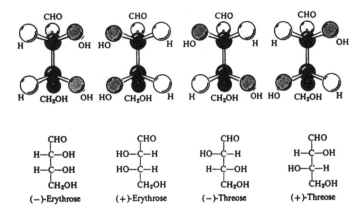

CHO	CHO	CHO	CHO
H—C—OH	HO—C—H	HO—C—H	H—C—OH
H—C—OH	HO—C—H	H—C—OH	HO—C—H
CH₂OH	CH₂OH	CH₂OH	CH₂OH
(−)-Erythrose	(+)-Erythrose	(−)-Threose	(+)-Threose

Figure 3.18. Projection formulas of the aldotetroses.

3.17 the D isomers of the aldoses are represented; the L enantiomers (not shown) would be generated by mirror imaging each structure (or, equivalently, exchanging H and OH at each chiral center). In interpreting the Fischer projection formulas the reader must keep in mind the points made in Section 3-2; in particular, the sideways groups all point forward, whereas the "backbone" of the sugars (the sequence of carbon atoms written vertically) curves backward into the plane of the paper (in horseshoe fashion) both at the top and at the bottom of the structure. The situation for the aldotetroses is shown in both three dimensions and Fischer projection formulas in Figure 3.18; it is customary to write the most highly oxidized (CHO) carbon atom (number 1 in IUPAC numbering) at the top.

It was pointed out in Section 3-2 that enantiomers are identical in all scalar properties (unless examined in a chiral environment; cf. Chapter 6). In contrast, diastereomers differ in most, if not all, physical and chemical properties; in fact, diastereomers tend to be as different from each other as many constitutional isomers. The basic reason for this difference is that enantiomers are "isometric"[11]; that is, for each distance between two given atoms (whether bonded or not) in one isomer there is a corresponding identical distance in the other. No such "isometry" exists in diastereomers or in constitutional isomers.

The practical consequences of the presence or absence of isometry are best seen in an example, the cyclopentane-1,2-diols shown in Figure 3.19. The two trans

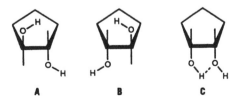

A   **B**   **C**

Figure 3.19. The cyclopentane-1,2-diols.

enantiomers (**A**, **B**) are isometric; as a result, the distances between the hydroxyl groups in the two are exactly the same. Neither isomer undergoes intramolecular hydrogen bonding,[12] since the hydroxyl groups are too far from each other, and both are oxidized by lead tetraacetate at the same slow rate.[13] In contrast, the cis diastereomer **C**, in which the hydroxyl groups are much closer to each other and which is thus not isometric with **A** and **B**, forms an intramolecular hydrogen bond and is oxidized by lead tetraacetate over 3000 times faster than **A** or **B**.

b. Degenerate Cases

In the instances discussed so far, the chiral elements were distinct. Degeneracies occur when this is not the case. The cyclopentanediols (Fig. 3.19) constitute an example: Diastereomer **C** is (on the laboratory time scale, when rotation about the C–O bonds is fast) achiral; hence, instead of four stereoisomers; (2^2) there are only three. A similar example involving two identical chiral centers is tartaric acid (Fig. 3.20). In this case also one of the diastereoisomeric sets contains two enantiomers, whereas the other contains a single, achiral, isomer. Instead of four isomers there are again only three. The achiral diastereomer is called a *meso* form, defined as an achiral member of a set of diastereomers that also contains chiral members. (Thus defined, the term "meso form" is not applied to the 1,3-dichlorocyclobutanes shown in Fig. 3.8, nor to the 1,2-dichloroethylenes in Fig. 3.9, since there are no chiral members in these sets.)

It might appear (Fig. 3.20) that *meso*-tartaric acid has a plane of symmetry. However, the (eclipsed) Fischer formulas do not ordinarily represent stable conformations; rather, *meso*-tartaric acid exists in the staggered conformations shown in Figure 3.4, **A**. Thus, in one of the conformers, there is a center of symmetry; this conformer is achiral. [The other staggered conformers (not shown) probably also contribute, but since they are enantiomeric and equally populated, they constitute a racemate: though each is chiral, there is no residual chirality just as there is no residual chirality in 1,2-dichloroethane (cf. Fig. 2.7 and earlier discussions).]

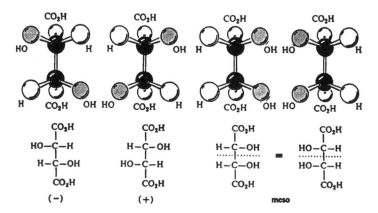

Figure 3.20. The tartaric acids.

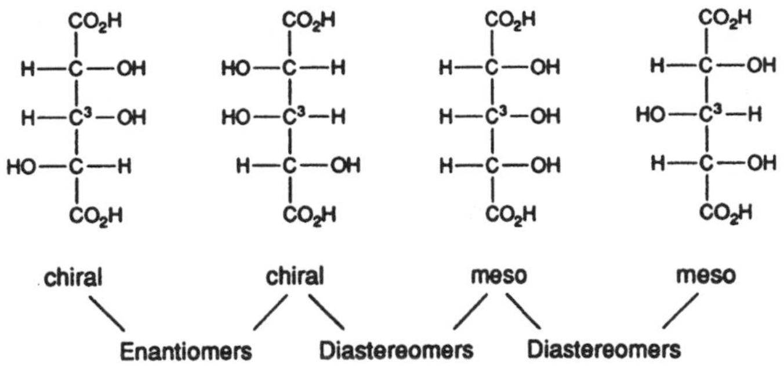

Figure 3.21. The trihydroxyglutaric acids.

As expected, *meso*-tartaric acid differs from its chiral diastereomers in physical properties; it melts at 140°C [the (+) and (−) acids melt at 170°C] and is less dense, less soluble in water, and is a weaker acid than the active forms. [The (+) and (−) isomers are, of course, identical to each other in all physical properties, other than chiroptical ones such as specific rotation (cf. Chapter 12).]

The acids obtained by oxidation of the aldopentose sugars (trihydroxyglutaric acids) constitute a somewhat more complex case (see Fig. 3.21). In this set of isomers there are two meso forms and one pair of enantiomers. Carbon (3) in these compounds is of particular interest. In the chiral members of the set, C(3) is not a stereogenic center since its two chiral ligands (CHOHCO$_2$H) are homomorphic (identical). [The reader will realize that transposing H and OH at C(3) in either enantiomer leads to an identical structure; this is best seen by turning the new structure by 180°, which is an allowed operation (see Fig. 3.15).] Carbon (3) in the meso isomers is a center of stereoisomerism (stereogenic center), but it is not a chiral center. It has been called a "pseudoasymmetric" center.

The case of the tetrahydroxyadipic or hexaric acids is depicted in Figure 3.22. Carbon (2) is constitutionally equivalent to C(5), and C(3) is constitutionally equivalent to C(4). This case is therefore sometimes said to be of the "ABBA" type (the trihydroxyglutaric or pentaric acid case, by the same notation, is "ABA," the tartaric acid case "AA"; the use of achiral letters here implies achiral ligands but not necessarily absence of chirality). The reader may work out that the hexaric acids exist as four racemic pairs and two meso forms; there are no pseudoasymmetric carbon atoms. The next higher homologue (heptaric acid) is of the "ABCBA" type and presents six racemic pairs and four meso isomers; here, in two of the six isomers that generate racemic pairs C(4) is not stereogenic, whereas in all the meso isomers it is pseudoasymmetric. In general,[14] in the degenerate cases with an even number n of

$$HO_2C—CHOH—CHOH—CHOH—CHOH—CO_2H$$

Carbon No.   2   3   4   5

Figure 3.22. Tetrahydroxyadipic or hexaric acids.

chiral centers, the number of chiral stereoisomers is 2^{n-1} (i.e., the number of enantiomer pairs is 2^{n-2}) and the number of meso forms is $2^{(n-2)/2}$. When the number of like chiral centers m is odd, the number of chiral stereoisomers is $2^{(m-1)} - 2^{(m-1)/2}$ (corresponding to $2^{m-2} - 2^{(m-3)/2}$ enantiomer pairs) and the number of meso forms is $2^{(m-1)/2}$. The above formulas apply only to cases where the chiral centers are arranged in a straight chain. The more general cases with chiral centers in branches have been treated by Senior[15]; see also Nourse.[16]

REFERENCES

1. Cahn, R. S., Ingold, C. K., and Prelog, V. *Angew. Chem. Int. Ed. Engl.* **1966**, *5*, 385.

2. Hirschmann, H. and Hanson. K. R. *J. Org. Chem.* **1971**, *36*, 3293.

3. McCasland, G. E. "A New General System for the Naming of Stereoisomers," Chemical Abstracts Service, Columbus, OH, 1953.

4. Mislow, K. and Siegel, J. *J. Am. Chem. Soc.* **1984**, *106*, 3319.

5. Eliel, E. L. *Top. Curr. Chem.* **1982**, *105*, 1.

6. Finocchiaro, P., Gust, D., and Mislow, K. *J. Am. Chem. Soc.* **1973**, *95*, 8172.

7. Eliel, E. L. *Israel J. Chem.* **1976/77**, *15*, 7.

8. Darling, T. N., Davis, D. G., London, R. E., and Blum, J. J. *Proc. Natl. Acad. Sci. USA* **1987**, *84*, 7129.

9. Fischer, E. *Ber. Dtsch. Chem. Ges.* **1891**, *24*, 2683.

10. Fischer, E. and Brauns, F. *Ber. Dtsch. Chem. Ges.* **1914**, *47*, 3181.

11. Mislow, K. *Bull. Soc. Chim. Belg.* **1977**, *86*, 595.

12. Kuhn, L. P. *J. Am. Chem. Soc.* **1952**, *74*, 2492.

13. Criegee, R., Büchner, E., and Walther, W. *Ber. Dtsch. Chem. Ges.* **1940**, *73B*, 571.

14. Landolt, H. *Optical Activity and Chemical Composition*, Whittaker & Co., New York, 1899.

15. Senior, J. K. *Ber. Dtsch. Chem. Ges.*, **1927**, *60B*, 73.

16. Nourse, J. G. *J. Am. Chem. Soc.* **1975**, *97*, 4594.

4

SYMMETRY

4-1. INTRODUCTION

Symmetry is an esthetically pleasing attribute of objects found in architecture and in various forms of art as well as in the realm of nature. It also plays an important part in science: in molecular spectroscopy, in quantum mechanics, as well as (in the present context) in the determination of structure and the understanding of stereochemistry. The essence of symmetry is the regular recurrence of certain patterns within an object or structure.[1-3]

As pointed out in Chapter 2, it is often convenient to think of molecules as idealized static entities that can be represented by rigid mechanical models. The symmetry relationships to be considered in this chapter will generally refer to such ideal molecules or molecular models (but see Section 4-4). One must also keep in mind that, since molecules are three-dimensional (3D), in general only 3D representations will be completely adequate as models. Thus the tetrahedron representing (+)-lactic acid or its perspective drawing (Fig. 3.14) is a proper model, but the Fischer projection (without the appropriate specifications as to which groups are in front and which are in back) is not; a novice contemplating the formulas of (+)- and (−)-lactic acids in their Fischer projections might come to the erroneous conclusion that these structures have a plane of symmetry and are superposable (as they would be if carbon were square planar rather than tetrahedral).

4-2. SYMMETRY ELEMENTS

Symmetry elements are the operators that generate the repeat pattern of symmetry. In finite objects they are the simple or proper axis of symmetry C_n, the plane of symmetry σ, the center of symmetry i, and the alternating or improper axis of symmetry S_n.

The axis (or simple axis) of symmetry, of multiplicity n, also called the "n-fold axis" and denoted by the symbol C_n, is an axis such that if one rotates the model (or molecule) around the axis by $360°/n$, the new position of the model is superposable with the original one. Examples are cis-(1R,3R)-sec-butylcyclobutane (Fig. 4.1, **A**), which has a twofold (C_2) axis and r-1,c-2,c-3,c-4-(1R,2R,3R,4R)-tetra-sec-butylcyclobutane (Fig. 4.1, **B**), which has a fourfold axis of symmetry (C_4).

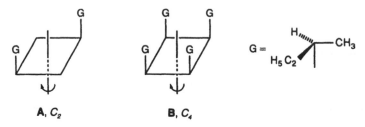

Figure 4.1. Examples of molecules with C_2 and C_4 symmetry axes. The (chiral) letter G stands for a chiral ligand, in this case a *sec*-butyl group. The inverse letter Ɔ will be used in Figures 4.2–4.5 as being enantiomorphous to G.

(Regarding nomenclature, cf. Chapters 5 and 11: *c* stands for *cis*, *t* for *trans*, and *r* for reference group.)

It is clear that any figure or model turned 360° around any axis will be superposable upon itself. The C_1 symmetry axis is thus a universal (and therefore trivial) symmetry element; it is equivalent to the "identity operation" (*E* or *I*) in group theory.

Rotation is a "real" operation in the sense that the points that are brought into superposition are actual, material points. Such an operation is called a "symmetry operation of the first kind" or "proper operation" (in contrast to the rotation–reflection operations discussed below). The presence of a symmetry axis does not preclude chirality; indeed, both of the molecules shown in Figure 4.1 are chiral. Hence chirality cannot be equated with asymmetry (i.e., the total absence of symmetry); in the older literature the word "dissymmetry" is often used as a synonym for what we now call chirality (cf. Chapter 1).

Planes, centers, and alternating axes of symmetry are elements corresponding to "symmetry operations of the second kind" or "improper operations" because, rather than bringing into coincidence material points, they bring into coincidence one material point with the reflection of another, a coincidence that might be termed virtual rather than real.

A plane of symmetry σ is a reflection plane by the operation of which each part of a model (each atom of a molecule) is brought into coincidence with a like part (or atom), located elsewhere in the model (or molecule). The compound *cis*-(1*R*,3*S*)-di-*sec*-butylcyclobutane (Fig. 4.2) provides an example; the plane shown is a plane of symmetry.

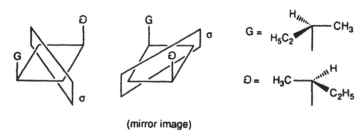

(mirror image)

Figure 4.2. Example of a molecule with a symmetry plane.

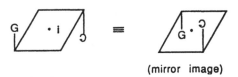

(mirror image)

Figure 4.3. Example of a molecule with a center of symmetry (*i*). See Figures 4.1 and 4.2 for explanation of G and Ɔ.

The center of symmetry *i* is a point such that if a line is drawn from any part of the model (or atom in a molecule) to that point and extended an equal distance on the other side, a like part (or atom) is encountered. An example of a molecule with a center of symmetry is *trans*-(1*R*,3*S*)-di-*sec*-butylcyclobutane (Fig. 4.3): The center of the four-membered ring is the center of symmetry, sometimes also called "point of inversion." If the point of inversion is taken as the origin of a Cartesian coordinate system, the effect of the inversion operation is to transpose a point of coordinates x, y, z into a point of coordinates $-x$, $-y$, $-z$.

Simple reflection (σ) changes one coordinate but not the other two (e.g., x, y, z to $-x$, y, z). Twofold rotation (C_2) changes two of the three coordinates (e.g., x, y, z to $-x$, $-y$, z).

The remaining symmetry element is the rotation–reflection or alternating axis of symmetry of order n, S_n. This is an axis such that if the model is turned $360°/n$ about the axis and then reflected across a plane perpendicular to that axis, each part of the original model encounters an equivalent part in the rotated–reflected model (and likewise for atoms in a molecule). An example with an S_4 axis [*r*-1, *t*-2, *c*-3, *t*-4-(1*R*,2*S*,3*R*,4*S*)-tetra-*sec*-butylcyclobutane] is displayed in Figure 4.4.

A $360°$ rotation–reflection operation about an S_n axis is necessarily equivalent to the identity (i.e., leads back to the original model) only if n is even. In that case, there is also a simple $C_{n/2}$ axis contained within the S_n axis.

An S_2 axis corresponds to a point of inversion ($S_2 = i$) located at the intersection of the axis and reflection plane and an S_1 axis is equivalent to a plane of symmetry ($S_1 = \sigma$). It can be proved in group theory that any structure that possesses a plane, center,

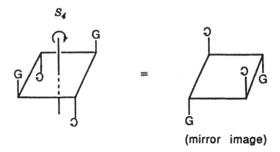

(mirror image)

Figure 4.4. Example of a molecule with a fourfold alternating axis of symmetry S_4. See Figures 4.1 and 4.2 for explanation of G and Ɔ.

or alternating axis of symmetry can be superposed with its mirror image; that is, such a structure is achiral. (By way of an exercise, the reader should note that structures shown in Figures 4.2–4.4 are superposable with their mirror images. Some of the structures may have to be rotated in space to make this evident.) It follows, as a corollary, that the presence of an S_n axis in a molecule is a necessary and sufficient condition for absence of chirality.

4-3.   SYMMETRY OPERATORS AND SYMMETRY POINT GROUPS

A symmetry operation is an operation that carries a molecule (or a model or, for that matter, any figure or set of points) into a position indistinguishable from (or equivalent to) the original position. It is brought about by the operation of one or several of the earlier mentioned symmetry elements that thus function as "symmetry operators." For example, successive operation of the C_4 axis in the molecule in Figure 4.1, **B** starting with the original molecule, gives rise to three additional, superposable species rotated around the C_4 axis by 90°, 180°, and 270°, respectively; rotation by 360° leads back to the original position. These symmetry operations so performed are called E, C_4^1, C_4^2, C_4^3; E being the identity operation.

The sum total of all possible symmetry operations defines a group. The number of different operations that can be performed in a group is called the *order* of the group. In the case of C_4, there are four operations: E, C_4^1, C_4^2, C_4^3, so the order of the group generated by the C_4 operator acting alone is four. (The orders of other point groups will be taken up as they are discussed.)

It quickly becomes clear that one cannot combine symmetry operators at random in a group. The existence of certain symmetry elements implies the simultaneous existence of certain others. For example, the existence of a C_2 axis and a σ plane perpendicular to it necessarily implies the presence of an S_2 axis (i.e., a center of symmetry, i). [The reverse is not true, however; that is, i (or S_2) can be present in the absence of C_2 and σ as exemplified in Fig. 4.3.] Thus $C_2 \times \sigma = S_2$ or, in words, C_2 followed by σ is equivalent to S_2. The point is exemplified by the molecule shown in Figure 4.5. It is also true that the existence or combination of certain symmetry elements excludes certain others. For example, a species that has a C_3 axis cannot have a collinear C_2 axis unless it also has a C_6 axis ($C_2 \times C_3 = C_6$). Nor can it have a single

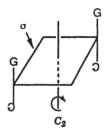

Figure 4.5. Molecule with C_2 and σ (implying $i = S_2$). See Figures 4.1 and 4.2 for explanation of G and Ɔ.

C_2 axis at right angles to the C_3; there must be either three such axes, at 60° angles to each other, or none. Thus we cannot combine symmetry elements indiscriminately but only according to the rules applicable to groups (see above). A proper combination of symmetry elements gives rise to what is known as a "symmetry point group." There is a finite number of point groups as will be shown in the sequel.

Chiral molecules necessarily belong to point groups $\mathbf{C_1}$, $\mathbf{C_n}$, or $\mathbf{D_n}$ (or rarely to $\mathbf{T}$, $\mathbf{O}$, or $\mathbf{I}$), that is, groups that have *only* proper axes. All other point groups, that is, those containing alternating (improper) axes, including reflection planes and inversion centers, are associated with achiral molecules. In these latter point groups, the number n of rotation operations for C_n (including $C_1 = E$) is always equal to the number of reflection operations (σ or S_n).

a. Point Groups Containing Chiral Molecules

Point Group C_1. This point group has the lowest degree of symmetry. It is represented by a molecule of the type Cabcd (e.g., CHFClBr, Fig. 2.6); such a molecule has no symmetry at all and is thus truly "asymmetric." The only symmetry element is the identity E (always present) or the equivalent onefold axis C_1 and this point group is therefore denoted as $\mathbf{C_1}$; its order is 1.

Point Groups C_n. In the $\mathbf{C_n}$ point groups, the only symmetry element is the C_n axis.

Point group $\mathbf{C_2}$ is of fairly common occurrence. Examples are (+)- or (−)-tartaric acid (Fig. 3.20), chiral biphenyls in which both rings bear the same substituents (e.g., Fig. 3.2), 1,3-dichloroallene (Fig. 4.6; the axis passes diagonally through the central carbon atom), and the cyclobutane derivative shown in Figure 4.1, **A**, as well as the gauche forms of 1,2-dichloroethane (Fig. 2.7).

Point group $\mathbf{C_3}$ is quite rare. Tri-*o*-thymotide (Fig. 4.7) constitutes an example in two of the four possible conformations attainable by tilting the rings. (The other two conformations are $\mathbf{C_1}$.) The optically active compound racemizes with an activation energy of about 22 kcal mol^{-1} (92 kJ mol^{-1})[4] by a flipping of the rings.

Another chiral $\mathbf{C_3}$ compound is *trans, trans, trans*- 3,7,11-trimethylcyclododeca-1,5,9-triene (Fig. 4.8), which is obtained along with its double-bond isomers and some head-to-head constitutional isomers upon trimerization of 1,3-pentadiene.[5] The compound may be obtained optically active either by asymmetric synthesis with a

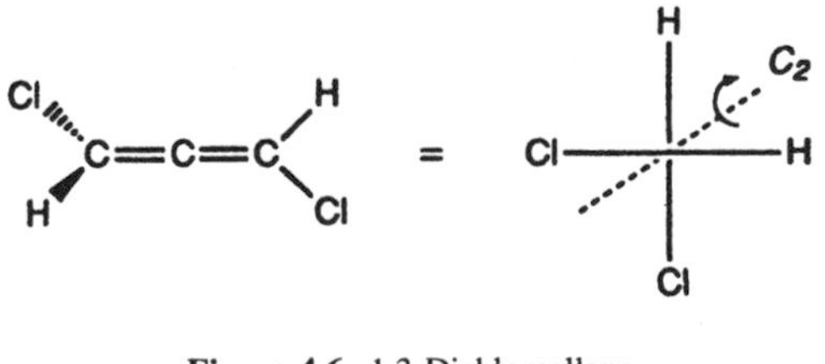

Figure 4.6. 1,3-Dichloroallene.

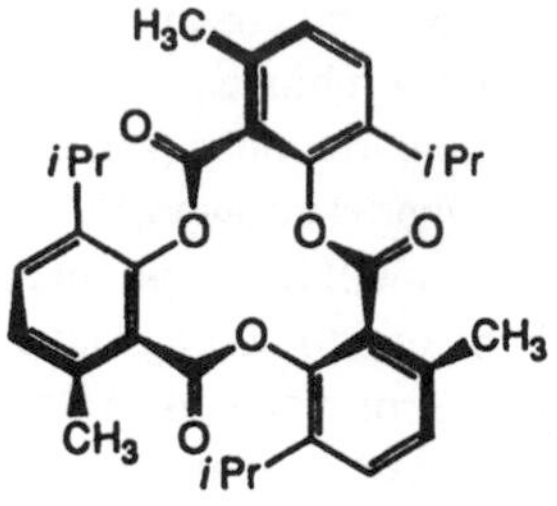

Figure 4.7. Tri-*o*-thymotide.

chiral titanium methoxide–Et$_2$AlCl catalyst or by partial asymmetric destruction with tetrapinanyldiborane (cf. Chapter 7).

Point group $\mathbf{C_6}$ is found[6] in cyclohexaamylose (Fig. 4.9), also called α-cyclodextrin.

Extensive lists of chiral compounds in the less common symmetry groups ($\mathbf{C_n}$, $n >$ 2; $\mathbf{D_n}$) have been compiled by Farina and Morandi[4] and by Nakazaki.[7] As we have already seen, the order of a $\mathbf{C_n}$ point group is n.

Point Groups D_n. These are the so-called "dihedral" point groups. These point groups are characterized by n C_2 axes perpendicular to the main C_n axis. The symmetry of these point groups is thus quite high; nevertheless they are chiral.

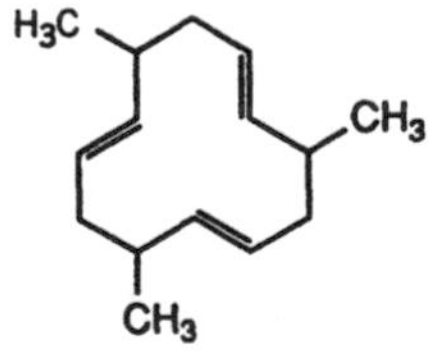

Figure 4.8. *trans,trans,trans*-3,7,11-Trimethylcyclododeca-1,5,9-triene.

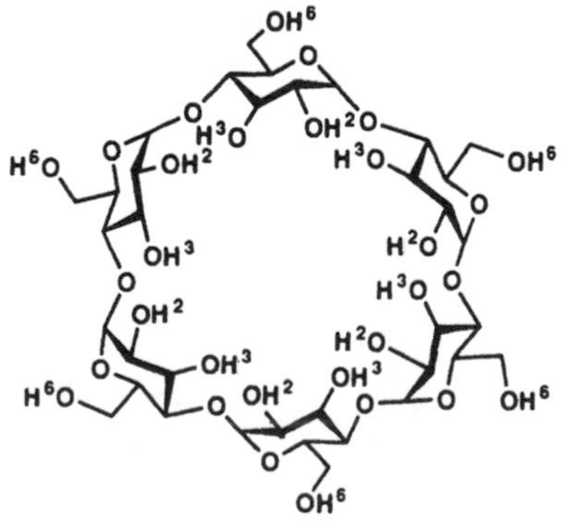

Figure 4.9. Cyclohexaamylose (α-cyclodextrin).

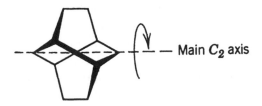

Figure 4.10. Twistane.

The $\mathbf{D_2}$ point group[4,7] displays three mutually perpendicular C_2 axes. Among examples is the intersting molecule twistane[8,9] shown in Figure 4.10. A number of additional examples of molecules in the $\mathbf{D_2}$ point group (allenes, spiranes, and biphenyls) will be found in Chapter 13.

The first organic $\mathbf{D_3}$ compound to be obtained in optically active form *trans-transoid-trans-transoid-trans*-perhydrotriphenylene (Fig. 4.11).[10] A common example of $\mathbf{D_3}$ symmetry is the (ephemeral) skew form of ethane (cf. Chapter 10), in a conformation that is neither staggered nor eclipsed.

The order of a $\mathbf{D_n}$ point group is $2n$. Thus for $\mathbf{D_2}$ the possible symmetry operations are E, C_2, C_2', and C_2'' (there are three C_2 axes, unprimed, primed, and double primed for distinction). For $\mathbf{D_4}$ the symmetry operations would be E, C_4^1, C_4^2 (or C_2), C_4^3, C_2', C_2'', C_2''', and C_2'''' (for the C_4 and four perpendicular C_2 axes, respectively), thus the order is 8.

b. Point Groups Containing Only Achiral Molecules

Point groups other than $\mathbf{C_n}$ and $\mathbf{D_n}$ generally have planes, a center, or an alternating axis of symmetry and are therefore achiral. These point groups will be discussed here in order of increasing number of symmetry elements.

Point Group C_s (or C_{1h}).   This point group has a symmetry plane σ *only* (no C_n). Examples are common among appropriately substituted alkenes, aromatics, and heterocyclic compounds. Chloroethylene, $CHCl{=}CH_2$, *m*-chlorobromobenzene, and furfural may be cited as examples. The compound *m*-chlorotoluene also has $\mathbf{C_s}$ symmetry provided either the conformation of the methyl group is such that one of the

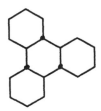

Figure 4.11. *trans-transoid-trans-transoid-trans*-Perhydrotriphenylene.

C–H bonds lies in the plane of the benzene ring, which then bisects the H–C–H angle of the other two, or the methyl group rotates fast enough on the time scale of the observation that it may be considered, *on the average,* to have a symmetry plane coincident with the plane of the benzene ring (cf. Section 4-4). With a corresponding proviso, methanol (CH_3OH) also belongs in this group. Molecules of the type CH_2XY or CR_2XY and aldehydes (RHC=O) are other common examples of C_s provided the R groups have a plane of symmetry either intrinsic (e.g., for R = Cl) or on the average (i.e., through fast rotation on the experimental time scale, see above). The order of the C_s point group is 2 (operations E, σ).

Point Group S_n.   Molecules in point group S_n have an n-fold alternating axis of symmetry. When n is even there is the additional condition that there must be no symmetry planes, but there will necessarily be a proper rotation axis $C_{n/2}$ coextensive with S_n. When $n = 4m + 2$ ($m = 0, 1, 2, \ldots$) there is also a center of symmetry, but when $n = 4m$ there is no center of symmetry. When n is odd, the S_n axis cannot exist by itself but must coexist with C_n and σ_h. Point groups in this category are customarily called C_{nh} (see below) rather than S_n (odd n).

An S_2 axis corresponds to a center of symmetry and so the point group S_2 may also be called C_i (cf. Fig. 4.3). Examples (provided one properly orients the methyl groups) are the anti conformation of *meso*-2,3-dichlorobutane (**A**), the *trans*-diketopiperazine derived from 1 mol of D-alanine and 1 mol of L-alanine (**B**), and the dibromo[2,2]paracyclophane (**C**) (cf. Chapter 13), all shown in Figure 4.12.

Dimerization of two alanine molecules gives either meso (C_i) dimer **B**, if the two alanine molecules are of opposite configuration (heterochiral) or a chiral (C_2) dimer (*cis*-diketopiperazine, diastereomeric to **B**) if they are of the same configuration (homochiral). One must not, however, conclude that dimerization of two identical homochiral molecules *necessarily* gives rise to a chiral dimer; this has been shown not to be the case.[11] One example where dimerization of homochiral molecules gives an achiral dimer is shown in Figure 4.13.[12]

The reverse process—"cutting" an achiral molecule into two homochiral fragments, the so-called "coupe du roi," (for the history of this intriguing operation, see ref. 11)— has also been realized.[13]

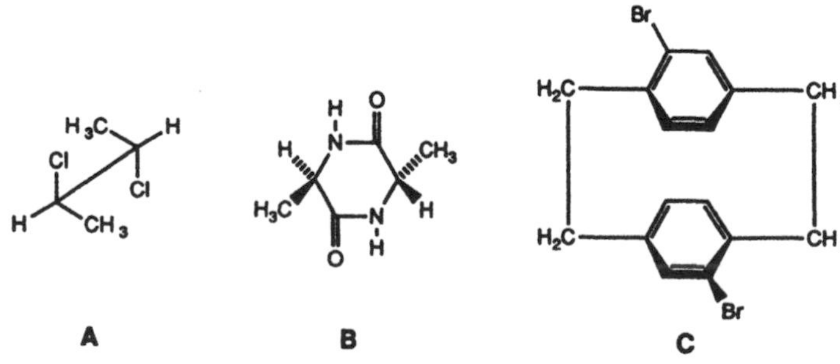

Figure 4.12. Molecules in S_2 point group (C_i).

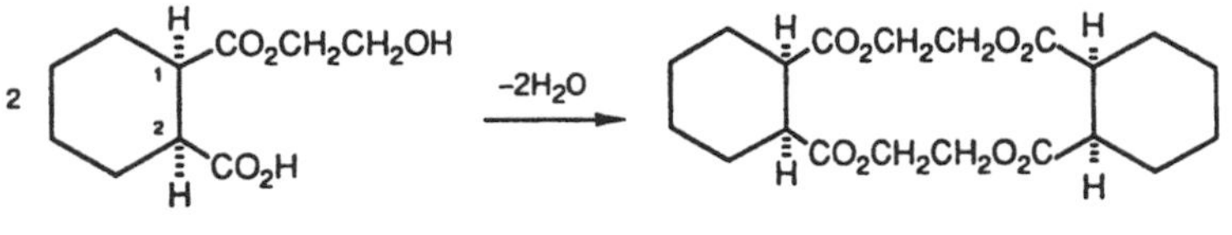

Figure 4.13. An example of dimerization of homochiral molecules yielding an achiral dimer.

It is worth noting that the stereochemical result of dimerizing pure enantiomers is substantially different from that of dimerizing the corresponding racemate. Thus, in the case of alanine, dimerization of the S enantiomer gives only the chiral (S,S)-diketopiperazine, but dimerization of (racemic) (RS)-alanine gives not only the racemic (RS,RS)-diketopiperazine but also the diastereomeric RS,SR-meso isomer (Fig. 4.12, **B**). Similarly, in Figure 4.13, dimerization of the pure $1R,2S$-enantiomer gives only the meso (C_{2v}) dimer shown (RS,SR), but dimerization of the corresponding racemic half-ester would give a mixture of the meso (C_{2v}) and meso (RS,RS, C_{2h}) dimers.

The order of the S_2 point group is 2, operators E and i.

A hypothetical example of the relatively rare S_4 point group is shown in Figure 4.4 (again one must properly orient the alkyl groups). Other examples (A^{14} and B^{15}) are shown in Figure 4.14. The order of S_4 is 4, operators E, S_4^1, C_2, and S_4^3.

Molecules belonging to the S_6, S_8, and S_{10} point groups are exemplified by the class of cyclopeptides.[16] The order of S_6 is 6, operators E, S_6^1, C_3^1, i, C_3^2, and S_6^5.

The remaining symmetry point groups have both axes and planes of symmetry. The planes are distinguished as either σ_v, σ_d symmetry planes that contain the main axis, or σ_h symmetry planes that are perpendicular to the main axis. Various combinations of these planes with axes of the C_n or D_n type generate most of the following groups, with the exception of some highly symmetric ones to be mentioned at the end.

Point Groups C_{nv}.   This point group has a single C_n axis and n symmetry planes σ_v, which all contain the axis and intersect at it. If the axis is assumed to be vertical, so are the planes, hence the symbol v.

A number of common planar molecules belonging to the C_{2v} point group are depicted in Figure 4.15. Any planar molecule also having a twofold axis in the plane

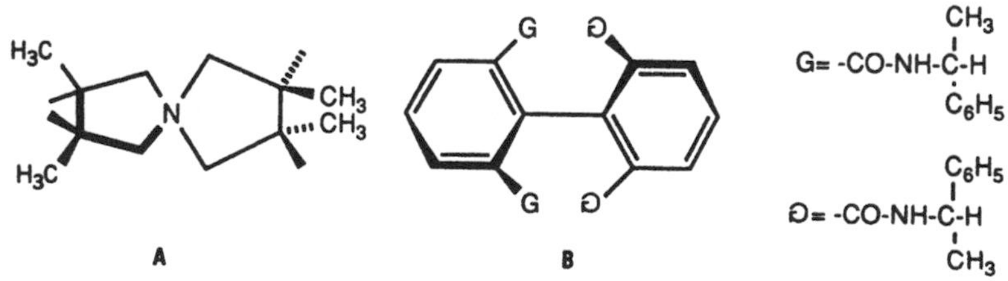

Figure 4.14. Molecules with S₄ symmetry.

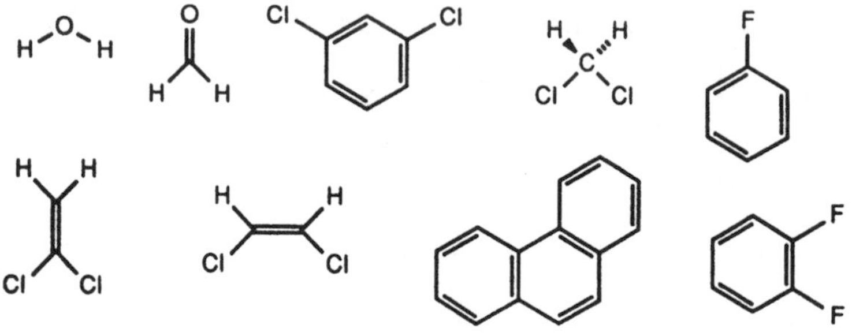

Figure 4.15. Molecules belonging to the **C₂ᵥ** point group.

will necessarily have a second plane of symmetry at right angles to the first plane and will belong to this group. The order of C_{2v} is 4, operators E, C_2, and $2\sigma_v$.

The point group C_{3v} is also common, with such representatives as CHCl₃, NH₃, eclipsed CH₃CCl₃, and C₆H₆Cr(CO)₃ (benzenechromium tricarbonyl). The order of C_{3v} is 6, operators E, C_3^1, C_3^2, and $3\sigma_v$. The octahedral molecule SF₅Cl is a representative of the C_{4v} group.

The order of C_{nv} is $2n$, as may readily be extrapolated from the specific cases above.

A C_∞ symmetry axis is an axis about which rotation by *any* angle (no matter how small) gives a molecule superposable with the original one. The point group $C_{\infty v}$ contains such an axis along with an infinite number of symmetry planes intersecting the axis. but no other symmetry elements. Hydrogen chloride (H–Cl), carbon monoxide (C=O), and chloroacetylene (H–C≡C–Cl) belong to the $C_{\infty v}$ point group. This type of symmetry is that displayed by a cone and is therefore often called "conical symmetry." The order of this group is infinite.

Point Groups Cₙₕ. The C_{nh} point groups have a C_n axis but only a single symmetry plane σ_h, which is perpendicular to the axis (i.e., horizontal, assuming the axis to be vertical).

Point group C_{2h} is the symmetry point group of *trans*-1,2-dichloroethylene and similarly substituted alkenes and of the *s*-trans forms of 1,3-butadiene (**A**) (cf. Chapter 9) and glyoxal (**B**) (Fig. 4.16). 1,4-Dichloro-2.5-dibromoblenzene (**C**) also belongs in this group. The order of this group is 4; operators E, C_2, σ, and i.

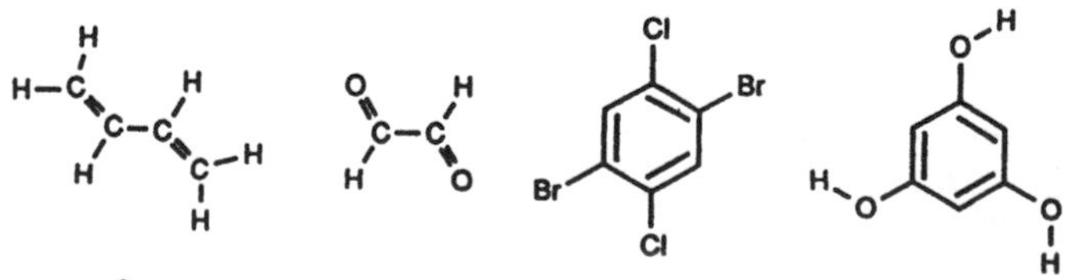

Figure 4.16. Molecules in point groups **C₂ₕ** and **C₃ₕ**.

Higher $\mathbf{C_{nh}}$ groups are rare and generally limited to specific conformations of the molecules in question. Phloroglucinol (Fig. 4.16 **D**) with all hydroxyl groups pointing in the same direction belongs to point group $\mathbf{C_{3h}}$ and an appropriate conformer of hexahydroxybenzene would similarly belong to $\mathbf{C_{6h}}$. An actual example of $\mathbf{C_{6h}}$ is found in hexaisopropylbenzene,[17] where "gearing" holds the isopropyl group in conformations in which the methine hydrogen atoms all point in the same direction. The order of $\mathbf{C_{3h}}$ is 6: operators E, C_3^1, C_3^2, σ, S_3^1, and S_3^2; in general, the order of $\mathbf{C_{nh}}$ is $2n$.

Point Groups D_{nd}. We have already seen that the dihedral point groups contain a C_n axis and n C_2 axes perpendicular to it. When, in addition, there are n symmetry planes intersecting in the principal axis, the symmetry point group is $\mathbf{D_{nd}}$. [The symmetry planes are considered to be diagonal (σ_d), since they do not contain the horizontal axes but rather the bisectors between two such axes.]

Point group $\mathbf{D_{2d}}$ is found in allene, certain spiranes, and the perpendicular conformation of biphenyl shown in Figure 4.17. As depicted these molecules have a vertical C_2 symmetry axis, two vertical symmetry planes, and two perpendicular horizontal (diagonal) symmetry axes not contained in the σ_d planes. There is also an S_4 axis. The order of $\mathbf{D_{2d}}$ is 8: operators E, C_2, $2C_2'$, $2\sigma_d$, S_4^1, and S_4^3.

Point group $\mathbf{D_{3d}}$ occurs in the staggered form of ethane and the chair form of cyclohexane as the most important representatives of this group. In addition to C_3, nC_2, and $3\sigma_v$ there is an S_6 axis in this group. The order is 12, operators E, C_3^1, C_3^2, $3C_2$, $3\sigma_v$, i, S_6^1, and S_6^5.

Molecules in higher $\mathbf{D_{nd}}$ groups are rare. Ferrocene in the staggered conformation is $\mathbf{D_{5d}}$ (Fig. 4.18), whereas in the eclipsed conformation it is $\mathbf{D_{5h}}$ (see below). Dibenzenechromium and uranocene (Fig. 4.18) may be considered as $\mathbf{D_{6h}}$ and $\mathbf{D_{8h}}$, or $\mathbf{D_{6d}}$ and $\mathbf{D_{8d}}$, respectively, according to whether they are eclipsed or staggered. The order of $\mathbf{D_{nd}}$ is $4n$, as exemplified above.

Point Group D_{nh}. This symmetry point group has symmetry elements similar to those of $\mathbf{D_{nd}}$ (except that σ_v planes, containing the horizontal axes, take the place of σ_d planes) *plus* a horizontal plane. It is more common than $\mathbf{D_{nd}}$.

Common representatives of the point group $\mathbf{D_{2h}}$ are ethylene, 1,4-dichlorobenzene, naphthalene, dibenzocyclobutadiene, and anthracene. Its order is 8, operators E, C_2, $2C_2'$, $2\sigma_v$, σ_h, and i.

Figure 4.17. Molecules in point groups $\mathbf{D_{2d}}$.

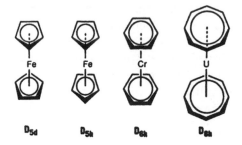

Figure 4.18. Ferrocene, dibenzenechromium, and uranocene.

Cyclopropane, 1,3,5-trichlorobenzene, boron trifluoride (BF_3), and triphenylene (Fig. 4.19) are found in the $\mathbf{D_{3h}}$ point group. Its order is 12, operators E, C_3^1, C_3^2, $3C_2$, $3\sigma_v$, σ_h, S_3^1, and S_3^2.

Point group $\mathbf{D_{4h}}$ represents the symmetry of the (average) planar form of cyclobutane. Square planar metal compounds with four equal ligands, such as $PtCl_4$, are also in this group. The order of this group is 16, operators E, C_4^1, C_2, C_4^3, $4C_2'$, $4\sigma_v$, σ_h, S_4^1, i, and S_4^3.

The symmetry point group $\mathbf{D_{5h}}$ represents planar (or average planar) cyclopentane. It is also seen in the eclipsed form of ferrocene (Fig. 4.18) and in the cyclopentadienyl anion. The symmetry point group $\mathbf{D_{6h}}$ is common; it is represented by benzene, hexachlorobenzene, coronene (Fig. 4.19). The symmetry point groups $\mathbf{D_{7h}}$, $\mathbf{D_{8h}}$, and $\mathbf{D_{9h}}$ are found in the tropylium cation ($\mathbf{D_{7h}}$); in uranocene (Fig. 4.18)[18,19]; in the cyclooctatetraenyl dianion ($\mathbf{D_{8h}}$)[20]; and in the cyclononatetraenyl anion ($\mathbf{D_{9h}}$).[21,22] The general class of corannulenes,[23] of which coronene (Fig. 4.19) is an example, belong to the $\mathbf{D_{nh}}$ point groups. The order of $\mathbf{D_{nh}}$ is $4n$, as exemplified for specific cases above.

The point group $\mathbf{D_{\infty h}}$, in addition to a C_∞ axis, contains an infinite number of symmetry planes intersecting in it and an infinite number of C_2 axes perpendicular to the main axis, and has a symmetry plane perpendicular to the C_∞ axis. This type of symmetry is called "cylindrical" since a cylinder displays it. Molecules in the $\mathbf{D_{\infty h}}$ group must be linear and end-over-end symmetrical, such as ethyne (H–C≡C–H), carbon dioxide (O=C=O), and diatomic molecules such as dihydrogen (H_2). The order of this group, as that of $\mathbf{C_{\infty v}}$, is infinite.

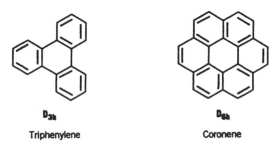

Figure 4.19. Structure of triphenylene and coronene.

Point Groups Corresponding to the Platonic Solids: T_d, O_h, I_h. We now come to the point groups of the most highly symmetric bodies, the tetrahedron, cube, octahedron, dodecahedron (12 faces), and icosahedron (20 faces), the so-called "Platonic solids" (mentioned in Plato's dialogue *Timaeus*). Examples of all of these bodies are now represented in the molecular realm.

The Tetrahedral Point Group (T_d). The point group of the regular tetrahedron is $\mathbf{T_d}$. It has four C_3 axes passing through each apex and the center of the opposite face, three C_2 axes passing through pairs of opposed edges (i.e., edges ending in noncommon vertices), and six σ_d planes, each containing one edge and bisecting the opposite one. Examples of molecules in group $\mathbf{T_d}$ are methane (**A**) and adamantane (**B**) (Fig. 4.20). The basic tetrahedral skeleton of tetrahedrane (Fig. 4.20, **C**; R = H) has not yet been obtained, but the corresponding tetra-*tert*-butyl derivative [**C**, R = C(CH$_3$)$_3$] has been synthesized by Maier's groups.[24] The order of $\mathbf{T_d}$ is 24 (identity, 2 operations for each of the 4 C_3 axes, one for each of the C_2 axes, plus 6 σ_v planes and 3 S_4 axes).

The Cubic Point Group O_h. The cube and the octahedron belong to the octahedral point group $\mathbf{O_h}$. This group has three C_4 axes (passing through the centers of opposite faces of the cube or opposite apices of the octahedron), four C_3 axes (the space diagonals of the cube or axes passing through the centers of opposite faces in the octahedron), and six C_2 axes (passing through the centers of opposite edges). In addition there are nine σ planes, three passing through the middle of opposite faces (bisecting the edges) and six passing diagonally through opposite faces. Octahedral symmetry is found in cubane (C_8H_8), first synthesized by Eaton and Cole[25] (see also Section 11-6.e), sulfur hexafluoride (SF_6), and octahedral coordination compounds with six equal ligands.

The order of $\mathbf{O_h}$ is 48 (E, 9 stemming from the 3 C_4 axes, 8 from the 4 C_3 axes, and 6 stemming from the C_2 axes; the resulting total of 24 is doubled by the existence of the symmetry planes).

The Icosahedral Point Group (I_h). The remaining regular polyhedra are the dodecahedron and the icosahedron, the former having 12 faces in the shape of regular pentagons, the latter 20 in the shape of equilateral triangles. (The icosahedron has 12 vertices, the dodecahedron has 20.) These two solids belong to the symmetry point

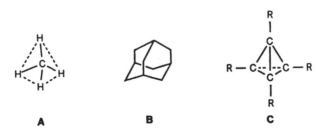

Figure 4.20. Molecules in point group $\mathbf{T_d}$.

Figure 4.21. Dodecahedrane.

group I_h, which has 6 C_5, 10 C_3, and 15 C_2 axes as well as 15 σ planes. The substance dodecahedrane ($C_{12}H_{12}$, Fig. 4.21), which has the shape of a regular dodecahedron, has been synthesized by Paquette and co-workers.[26] The icosahedron is found in the dodecaborane dianion $B_{12}H_{12}^{2-}$ and the corresponding halides $B_{12}X_{12}^{2-}$(X = Cl, Br, or I).[27] The order of the icosahedreal (I_h) point group is 120 and the group also has six S_{10} axes. The substance buckminsterfullerene (Section 11-6.e) corresponds to a truncated icosahedron and also has I_h symmetry.

The highest symmetry, spherical symmetry (K_h; K for German *Kugel*, meaning sphere), for obvious reasons is not represented by a finite point group.

The symmetry point group of a given molecule or model can be inferred systematically by an appropriate dichotomous tree.[1,3] However, the result can be arrived at faster by proceeding in nondichotomous fashion: Does the molecule have a finite symmetry axis, an infinite symmetry axis, or no symmetry axis at all? If the symmetry axis is infinite, the point group is $C_{\infty v}$ or $D_{\infty h}$, depending on the absence or presence of a perpendicular symmetry plane. If there is no symmetry axis, the possible point groups are C_1 (asymmetric), C_i (point of symmetry only), or C_s (plane of symmetry only). If there is at least one symmetry axis, what is its order, and are there other axes (C_2) perpendicular to it? If there are no other symmetry axes, we are dealing with the C_{nx} family; if there are other axes, we are dealing with D_{nx}, n being the order of the principal axis. If the family is C_{nx}, are there no planes of symmetry, n planes of symmetry containing the axis, or one plane of symmetry perpendicular to the axis? In the first case, the point group is C_n, unless there is also an alternating axis of order $2n$, in which case the point group is S_{2n}. In the second case the group is C_{nv}, in the third case C_{nh}. If the molecule belongs to the D_{nx} family, is it devoid of symmetry planes, does it have diagonal symmetry planes only, or does it also have a symmetry plane perpendicular to the principal axis? In the first case, the point group is D_n, in the second case it is D_{nd}, and in the third case D_{nh}. This leaves only the highly symmetric point groups T_d, O_h, and I_h, which are rare and readily recognized.

4-4. AVERAGED SYMMETRY

Averaged symmetry is that of a molecule which simulates higher symmetry than the symmetry present in any of its contributing structures because of a rapid interconversion of these structures on a time scale that is appreciably faster than that of the experiment designed to determine the symmetry. As discussed in Chapter 2, this situation is not uncommon in real molecules.

By way of an example that is intuitively easy to grasp, let us consider cyclohexane. Cyclohexane is known to exist in the chair form of symmetry $\mathbf{D_{3d}}$, and at $-100°C$ the NMR spectrum of the compound (cf. Chapter 11) is indeed appropriate for a molecule of that symmetry. At room temperature, however, due to rapid inversion of the chair, the ^1H NMR spectrum shows a single signal due to the equivalent averaged hydrogen atoms. This observation is what would be expected for planar cyclohexane ($\mathbf{D_{6h}}$) and it is therefore reasonable to assume that the average symmetry of cyclohexane is indeed $\mathbf{D_{6h}}$, even though the planar form is of very high energy and does not even correspond to the transition state for chair inversion (Chapter 11). A rigorous demonstration of this intuitive conclusion has been given.[28] Unfortunately, most cases of "nonrigid symmetry" cannot be treated in this intuitive fashion and are beyond the scope of this book.

4-5. SYMMETRY AND MOLECULAR PROPERTIES

In general, physical properties of matter are dependent on molecular structure and among these are properties that are predictable on the basis of symmetry arguments. We shall deal here with three such properties: the ability to rotate the plane of polarized light, the ability to display a permanent dipole moment, and the thermodynamically important entity called symmetry number. (We will take up the relation of symmetry properties to NMR spectra in Chapter 8, and their relation to chiroptical properties in Chapter 12.)

a. Rotation of Polarized Light

The relation of optical rotation (cf. Chapter 1) to symmetry is simple: Only substances whose molecules belong to chiral point groups can rotate polarized light. The presence of an S_n axis (including $S_1 = \sigma$ and $S_2 = i$) in a molecule precludes its displaying optical activity. Optical activity is a *pseudoscalar property*, which means the property remains invariant under a proper operation (rotation) but changes sign under an improper operation (reflection). In contrast, *scalar properties* are properties that are invariant to both proper and improper operations. Since an improper operation (reflection) transforms one enantiomorph into another, enantiomorphs (including enantiomeric substances) have identical scalar properties as already indicated in Chapter 3.

It follows immediately that only molecules in point groups $\mathbf{C_n}$ (including $\mathbf{C_1}$) and $\mathbf{D_n}$ (as well as the less common desymmetrized groups $\mathbf{T}$, $\mathbf{O}$, and $\mathbf{I}$), can give rise to optical rotation. The reverse does not necessarily follow. Molecules in a chiral point group will normally display optical rotation, but they may occasionally fail to do so because of what is called "accidental degeneracy."[29] We already saw (Chapter 1) that change of temperature, solvent, or pH may carry optical rotation from a positive to a negative value through a null, and the same is true of changes in wavelength (Chapter 12). Moreover, a mixture of two different chiral substances of opposite rotation (e.g., in an extract from a natural product) may accidentally display zero rotation by fortuitous compensation. Finally, the rotation of a chiral substance may, in some cases, be so small as to be unobservable, as a consequence of instrumental limitations. Examples are n-butylethyl-n-hexyl-n-propylmethane,[30] which has a specific rotation at 280 nm of less than 0.04, and neopentyl-1-d alcohol, $(CH_3)_3CCHDOH$, which displays no rotation above 300 nm in 80% acetone solution,[31] but does display a plain negative ORD curve for the S isomer in hexane solvent, albeit of low intensity.[32] Such molecules have been called *cryptochiral*.[29] A more trivial case of zero rotation in chiral molecules, concerned with racemates, is discussed in Chapter 12.

Mention of neopentyl-1-d alcohol brings up the general question of optical activity of chiral compounds, two of whose ligands differ only in isotopic composition. Clearly, compounds of the general type R_1R_2CXX', where X and X' are isotopically distinct, are chiral; the question is whether, if nonracemic, they display palpable optical activity. The earliest candidates for investigation were deuterated compounds of the type R_1R_2CHD and $R_1R_2CX^DX^H$ (X^D and X^H being otherwise identical ligands that differ in being substituted by one or more deuterium atoms taking the place of one or more hydrogen atoms). The reason for this was twofold: Both the early availability of deuterium and the fact that the relative difference between hydrogen and deuterium is greater than that of any other two stable isotopes made it relatively easy for optical activity, if any, to be detected in this case.

A number of optically active compounds of this type have, in fact, been found since 1949,[33] including $C_6H_5CHDCH_3$ (ref. 34) and $CH_3CHOHCD_3$ (ref. 35), which exemplify the two types of chiral deuterio compounds mentioned above. The configuration and maximum optical rotation of $(+)$-$C_6H_5CHDCH_3$ have been determined unequivocally[36]; the dextrorotating hydrocarbon is S (i.e., the earlier studied LiAlD$_4$ reduction[34] of $C_6H_5CHClCH_3$ proceeds with nearly complete inversion of configuration) and the specific rotation $[\alpha]_D^{20}$ of neat, enantiomerically pure material is 0.81 ± 0.01. The compound $(+)$-$CH_3CHOHCD_3$ is S and its $[\alpha]_D^{27}$ is 0.2 (c 4, $CHCl_3$). The compound $C_6H_5CHOHC_6D_5$ has also been obtained optically active[37]; $[\alpha]_D^{20}$ (c 16, $CHCl_3$) for enantiomerically pure material was calculated to be 1.00. For historical accounts, see refs. 37–39.

Cases of optical activity due to isotopes of other elements are few; compounds investigated with their respective rotations are shown in Figure 4.22, **A** (ref. 40), **B** (ref. 41), and **C** (ref. 42).

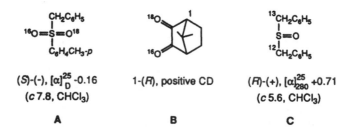

(S)-$(-)$, $[\alpha]_D^{25}$ -0.16
(c 7.8, $CHCl_3$)

1-(R), positive CD

(R)-$(+)$, $[\alpha]_{280}^{25}$ +0.71
(c 5.6, $CHCl_3$)

A   **B**   **C**

Figure 4.22. Optical activity due to isotopic differences other than H or D.

b. Dipole Moment

The (permanent) dipole moment of a chemical compound is due to an unbalanced charge distribution in its molecules. In principle, any two distinct atoms in a molecule (say, a grouping $-X-Y-$) will give rise to a local dipole, since X and Y will, to a greater or lesser extent, differ in electronegativity. However, the molecule as a whole will have a dipole moment only if the local dipole of an $-X-Y-$ grouping in one region is not compensated by an exactly opposite local dipole elsewhere in the molecule. Symmetry properties dictate whether this will or will not be the case. Specifically, if a molecule has a center of symmetry, to each local dipole there will correspond an equal and opposite one, and the overall dipole moment will be zero. Thus molecules in point groups S_n with even n (including C_i), among other point groups, cannot have a permanent dipole moment. If a molecule has a C_n axis, the component of the dipole moment vector at right angles to the axis averages to zero and the molecule can, at best, have a moment along the direction of the axis. However, this component of the moment will also vanish if (a) there is a symmetry plane perpendicular to the axis or (b) there is at least one other axis perpendicular to the first. Condition (a) eliminates C_{nh} and condition (b) eliminates all dihedral classes: D_n, D_{nd}, and D_{nh}. It follows that only molecules in symmetry point groups C_n (including C_1), C_s, and C_{nv} can have permanent dipole moments.

By way of an example we may consider the two chair conformations of *trans*-1,4- and *trans*-1,2-dibromocyclohexane, respectively (Fig. 4.23). Both conformational isomers of the *trans*-1,4 compound belong to symmetry point group C_{2h} and, therefore, necessarily have zero dipole moments. In contrast, *both* conformations of the *trans*-1,2 compound belong to symmetry point group C_2 and thus have dipole moments (the dipole of the diaxial conformer is erroneously claimed to be zero in some textbooks). From model studies [dipole measurements of 4-*tert*-butyl substituted analogues (cf. Chapter 11)] the moment of the diequatorial conformer is 3.3 D and that of the diaxial conformer 1.2 D. The latter moment is surprisingly large, the calculated value being only 0.37 D.[43,44]

Molecular symmetry must *not* be averaged in the decision whether a molecule does or does not have a dipole moment. Thus 1,2-dichloroethane (Fig. 2.7) has a dipole moment (1.12 D at 32°C in the gas phase) even though $\mu = 0$ for the centrosymmetric anti form and it might be thought (erroneously!) that the dipole moments of the equally

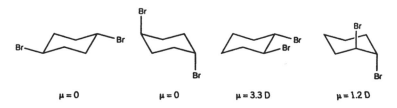

$\mu = 0$ $\qquad$ $\mu = 0$ $\qquad$ $\mu = 3.3\,D$ $\qquad$ $\mu = 1.2\,D$

Figure 4.23. Dipole moments of *trans*-1,4- and *trans*-1,2-dibromocyclohexane.

populated gauche forms cancel each other. The reason they do not is that the polarizations (i.e., squares of the dipole moments) average, not the moments themselves.

c. Symmetry Number

The symmetry number σ of a molecule is defined as the number of indistinguishable but nonidentical positions into which the molecule can be turned by rigid rotation. It is important in the computation of entropy. Symmetry reduces entropy, because it implies that arrangements of a molecule obtainable by virtue of rigid rotation (i.e., rotation of the molecule as a whole), and that would otherwise contribute to the rotational entropy, are superposable and are therefore thermodynamically indistin- guishable. Thus the number of distinct rotational arrays is diminished and the rotational entropy is thereby reduced.

To attain the appropriate indistinguishable positions (indistinguishable except for conceptual labeling that avails nothing in thermodynamics), it is easiest to operate the various symmetry axes in a molecule. Thus for water, operation of the C_2 axis (E, C_2) will interchange the position of the hydrogen atoms: The symmetry number is thus 2. A more complex case, that 1,3,5-trichlorobenzene (symmetry point group $\mathbf{D_{3h}}$, $\sigma = 6$) is shown in Figure 4.24. The top three arrangements are generated by rotation about the threefold axis and the bottom three from the first (E) by rotation about each of the twofold axes. If we remember the concept of the order of a symmetry point group (Section 4-3), we recognize that, for a chiral point group, the symmetry number is equal to the order of the groups, whereas for a finite achiral point group it is one-half the order (since only one-half of the symmetry operations are rotations, whereas the other one-half are operations of the second kind, which are not pertinent to the concept of symmetry number in that they do not produce a *real* coincidence of atoms).

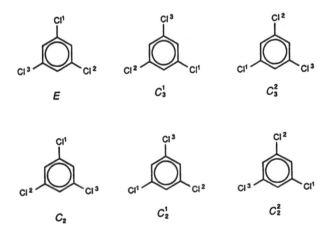

Figure 4.24. Rigid rotation of 1,3,5-trichlorobenzene.

TABLE 4.1. Point Groups, Their Order, and Their Symmetry Number

Group											
	C_1	C_n	D_n	C_s	$S_n{}^a$	C_{nv}; C_{nh}	$C_{\infty v}$	D_{nd}; D_{nh}	$D_{\infty h}$	T_d	O_h	I_h
Order	1	n	$2n$	2	n	$2n$	∞	$4n$	∞	24	48	120
σ	1	n	$2n$	1	$n/2$	n	1	$2n$	2	12	24	60

^{a}Includes $C_i = S_2$, order 2, $\sigma = 1$.

Exceptions are $C_{\infty v}$ and $D_{\infty h}$, whose orders are infinite, but whose symmetry numbers are 1 and 2, respectively.

Table 4.1 lists the order of various point groups and the corresponding symmetry numbers. We shall return to the use of symmetry numbers in computing entropy differences between stereoisomers in Chapters 10 and 11.

REFERENCES

1. Donaldson, J. D. and Ross, S. D. *Symmetry and Stereochemistry*, Halsted/Wiley, New York, 1972.

2. Heilbronner, E. and Dunitz, J. D. *Reflections on Symmetry,* VCH, New York, 1993.

3. Hargittai, I. and Hargittai, M. *Symmetry Through the Eyes of a Chemist*, 2nd ed., Plenum Press, New York, 1995.

4. Farina, M. and Morandi, C. *Tetrahedron* **1974**, *30*, 1819.

5. Furukawa, J., Kakuzen, T., Morikawa, H., Yamamoto, R., and Okuno, O. *Bull. Chem. Soc. Jpn.* **1968**, *41*, 155.

6. Hybl, A., Rundle, R. E., and Williams, D. E. *J. Am. Chem. Soc.* **1965**, *87*, 2779.

7. Nakazaki, M. *Top. Stereochem.* **1984**, *15*, 199.

8. Whitlock, H. W. *J. Am. Chem. Soc.* **1962**, *84*, 3412.

9. Adachi, K., Naemura, K., and Nakazaki, M. *Tetrahedron Lett.* **1968**, 5467.

10. Farina, M. and Audisio, G. *Tetrahedron* **1970**, *26*, 1827, 1839.

11. (a) Anet, F. A. L., Miura, S. S., Siegel, J., and Mislow, K. *J. Am. Chem. Soc.* **1983**, *105*, 1419. (b) Mislow, K. *Croatica Chem. Acta* **1985**, *58*, 353.

12. Nouaille, A. and Horeau, A. *C. R. Acad. Sc. Ser. II* **1985**, *300*, 335.

13. Cinquini, M., Cozzi, F., Sannicolo, F., and Sironi, A. *J. Am. Chem. Soc.* **1988**, *110*, 4363.

14. McCasland, G. E. and Proskow, S. *J. Am. Chem. Soc.* **1955**, 77, 4688; *ibid.* **1956**, *78*, 5646.

15. Helmchen, G., Haas, G., and Prelog, V. *Helv. Chim. Acta* **1973**, *56*, 2255.

16. Prelog, V. and Gerlach, H. *Helv. Chim. Acta* **1964**, *47*, 2288.

17. Siegel, J., Gutierrez, A., Schweizer, W. B., Ermer, O., and Mislow, K. *J. Am. Chem. Soc.* **1986**, *108*, 1569.

18. Streitwieser, A. and Müller-Westerhoff, U. *J. Am. Chem. Soc.* **1968**, *90*, 7364. Streitwieser, A., Müller-Westerhoff, U., Sonnichsen, G., Mares, F., Morrell, D. G., Hodgson, K. O., and Harmon, C. A. *J. Am. Chem. Soc.* **1973**, *95*, 8644.

19. Zalkin, A. and Raymond, K. N. *J. Am. Chem. Soc.* **1969**, *91*, 5667.

20. Katz, T. J. *J. Am. Chem. Soc.* **1960**, *82*, 3784.

21. Katz, T. J. and Garratt, P. J. *J. Am. Chem. Soc.* **1963**, *85*, 2852; *ibid.* **1964**, *86*, 5194.

22. LaLancette, E. A. and Benson, R. E. *J. Am. Chem. Soc.* **1963**, *85*, 2853; *ibid.* **1965**, *87*, 1941.

23. Agranat, I., Hess, B. A., and Schaad, L. J. *Pure Appl. Chem.* **1980**, *52*, 1399.

24. Maier, G., Pfriem, S., Schäfer, U., and Matusch, R. *Angew. Chem. Int. Ed. Engl.* **1978**, *17*, 520. Maier, G., Pfriem, S., Schäfer, U., Malsch, K. D., and Matusch, R. *Chem. Ber.* **1981**, *114*, 3965.

25. Eaton, P. E. and Cole, T. W. *J. Am. Chem. Soc.* **1964**, *86*, 963, 3157.

26. Ternansky, R. J., Balogh, D. W., and Paquette, L. A. *J. Am. Chem. Soc.* **1982**, *104*, 4503. See also Paquette, L. A. In Lindberg, T., ed., *Strategies and Tactics of Organic Synthesis*, Academic Press, New York, 1984, p. 175.

27. Muetterties, E. L. and Knoth, W. H. *Polyhedral Boranes,* Marcel Dekker, New York, 1968, pp. 21ff.

28. Leonard, J. E., Hammond, G. S., and Simmons, H. E. *J. Am. Chem. Soc.* **1975**, *97*, 5052.

29. Mislow, K. and Bickart, P. *Isr. J. Chem.* **1976/77**, *15*, 1.

30. Wynberg, H., Hekkert, G. L., Houbiers, J. P. M., and Bosch, H. W. *J. Am. Chem. Soc.* **1965**, *87*, 2635. Wynberg, H. and Hulshof, L. A. *Tetrahedron* **1974**, *30*, 1775.

31. Mosher, H. S. *Tetrahedron* **1974**, *30*, 1733.

32. Anderson, P. H., Stephenson, B., and Mosher, H. S. *J. Am. Chem. Soc.* **1974**, *96*, 3171.

33. Alexander, E. R. and Pinkus, A. G. *J. Am. Chem. Soc.* **1949**, *71*, 1786.

34. Eliel, E. L. *J. Am. Chem. Soc.* **1949**, *71*, 3970.

35. Mislow, K., O'Brien, R. E., and Schaefer, H. *J. Am. Chem. Soc.* **1960**, *82*, 5512.

36. Elsenbaumer, R. L. and Mosher, H. S. *J. Org. Chem.* **1979**, *44*, 600.

37. Makino, T., Orfanopoulos, M., You, T.-P., Wu, B., Mosher, C. W., and Mosher, H. S. *J. Org. Chem.* **1985**, *50*, 5357.

38. Arigoni, D. and Eliel, E. L. *Top. Stereochem.* **1969**, *4*, 127.

39. Verbit, L. *Progr. Phys. Org. Chem.* **1970**, *7*, 51.

40. Stirling, C. J. M. *J. Chem. Soc.* **1963**, 5741.

41. Kokke, W. C. M. C. and Osterhoff, L. J. *J. Am. Chem. Soc.* **1972**, *94*, 7583. Kokke, W. C. M. C. *J Org. Chem.* **1973**, *38*, 2989.

42. Andersen, K. K., Colonna, S., and Stirling, C. J. M. *J. Chem. Soc. Chem. Commun.* **1973**, 645.

43. Abraham, R. J. and Bretschneider, E. "Medium Effects on Rotational and Conformational Equilibria," in Orville-Thomas, W. J., ed., *Internal Rotation in Molecules*, Wiley, New York, 1974, pp. 481, 567.

44. Eliel, E. L., Allinger, N. L., Angyal, S. J., and Morrison, G. A. *Conformational Analysis*, Wiley-Interscience, New York, 1965; reprinted by American Chemical Society, Washington, DC. 1981.

5

CONFIGURATION

5-1. DEFINITIONS: RELATIVE AND ABSOLUTE CONFIGURATION

IUPAC[1] defines configuration as "the arrangement of atoms of a molecular entity in space that distinguishes stereoisomers, the isomerism between which is not due to conformation differences." This definition by exclusion is thus dependent on that of conformation[1]: "The spatial arrangement of the atoms affording distinction between stereoisomers which can be interconverted by rotations about formally single bonds. Some authorities extend the term to include inversion at trigonal pyramidal centers and other polytopal rearrangements." Implicitly, other authorities consider inversion at trigonal pyramidal centers as configurational changes, and there is also difficulty in defining a "formal single bond." Are the C–N bonds in the amides and thioamides depicted in Figure 5.1 to be considered single or partial double bonds, and are the cis and trans isomers of these compounds thus to be considered as differing in configuration, or are they simply conformational isomers of compounds of unique configuration? The same problem arises in biphenyls (Fig. 3.2; and Chapter 13) and in a number of other structures. As detailed in Chapter 3 in connection with the 2-bromobutanes (Fig. 3.10) the problem is related to the technique of observation. Physical chemists, whose methodology (e.g., vibrational spectroscopy) is capable of detecting species separated by quite low barriers, have, in fact, tended to define configuration *without* the exemption regarding rotation about single bonds. If this is done, configuration simply refers to the spatial arrangements of atoms in a molecule of given constitution, that is, to the stereochemical aspect of structure. The gauche and anti forms of butane then become configurational isomers. Few organic chemists use the term configuration in this sense. At the opposite extreme, it has been suggested that, in the definition of configuration (see above), no regard should be given to

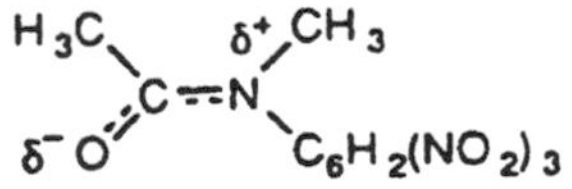

C–N barrier 21.0 kcal mol^{-1} 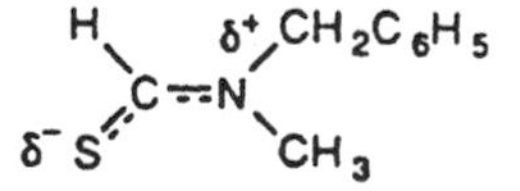 $\qquad$ C–N barrier 25.1 kcal mol^{-1}

Figure 5.1. Amide and thioamide barriers.

rotation about bonds of any order, including double bonds. Under such a point of view *cis-* and *trans*-2-butene would be conformers, not configurational isomers; it is understandable that such a view is excluded by the IUPAC definition.

In this book we shall use configuration in the sense of "the arrangement of the atoms in space of a molecule of defined constitution without regard to arrangements that differ only by rotation about one or more *single* bonds, provided that such rotation is so fast as not to allow isolation of the species so differing." Under this definition stereoisomers of appropriately substituted biphenyls (*atropisomers*, cf. Chapter 13) are considered configurational isomers as long as they can be isolated, and so, of course, are cis–trans isomers of alkenes (the question as to the delineation of single and double bonds is swept under the rug).

It is recognized that this definition is quite imperfect for several reasons, one being the just mentioned uncertainty as to what constitutes a single bond. Another is that it arbitrarily takes isolation (rather than some other factor, such as spectroscopic observability by one or other technique) as the criterion of existence of configurational isomers; that is, it deals with "residual isomers" employing the criterion of isolability (cf. Chapter 3). A more serious imperfection is that "isolability" is itself not a clearly defined term, in that it lacks reference to both the temperature of the experiment and the lifetime required to consider a species "isolable." Despite these shortcomings, chemists usually agree as to what constitutes a difference in configuration; in those few cases where there is doubt, the meaning of the term as used by a given author should be explicitly defined.

The determination and specification of configuration is an essential part of the structure determination of molecules displaying stereoisomerism, including chiral ones. Later in this chapter we shall explore how configuration is determined experimentally and how it is specified, either graphically or by descriptors attached to the chemical names (see also Chapters 9 and 11). Before this is done, however, it is desirable to distinguish between relative and absolute configuration. We shall illustrate this distinction with an analogy.

Suppose that a child who has not yet learned to distinguish right from left is given a right and a left shoe. The child will not be able to tell which of the shoes is right and which is left (i.e., their absolute configuration), but the child may well be able to tell that they are different in that one fits one of his or her feet (foot A) but not the other foot (B) and vice versa for the other shoe. Moreover, when provided with an assembly of shoes the child may (with some experimentation) be able to divide them into two kinds: one kind that fits foot A, the other that fits foot B. In other words, the child will be able to determine the configuration of the shoes relative to the feet and relative to each other (i.e., their relative configuration), even without knowing which shoe (or foot) is "right" and which is "left," that is, their absolute configuration.

In the terminology of Ruch,[2] the child can tell whether two shoes (or a foot and a shoe) are *homochiral* or *heterochiral* (referring to the same or opposite relative configuration) even though the child does not know the actual *sense of chirality* (absolute configuration) of either object.[3]

Prior to 1951, chemists were in the position of the child: The configurational correlation (i.e., relative configuration) of a large number of chiral substances with respect to others was known (cf. Section 5-4), but the absolute configuration or sense of chirality was not. [It was usual to correlate configuration with that of (+)- or (–)-glyceraldehyde ($HOCH_2CHOHCHO$)]. Also known was the relative configuration of the chiral centers within numerous molecules having more than one chiral center; this correlation may, for example, be established by X-ray structural determination. However (cf. Chapter 2), X-ray diffraction was thought to be unable to differentiate between enantiomers and thus unable to reveal which enantiomer is which. This situation was changed only in 1951, thanks to the ingenious application of anomalous X-ray scattering to the determination of absolute configuration, to be described in Section 5-3.[4]

5-2. ABSOLUTE CONFIGURATION AND NOTATION

Just as constitutional isomers are characterized by names as well as formulas, it is desirable to give differentiating symbols (so-called "descriptors") to enantiomers, such as those of the lactic acids shown, in three-dimensional (3D) or projection formulas, in Figure 3.14. Of course one can characterize the two acids experimentally by the observable signs of their rotation but one also wants to specify their 3D structure without resorting to a formula drawing. It is for the purpose of such specification that systems of configurational notation have been devised.

There is no 1:1 relation between the sign of optical rotation (+ or –) and configurational notation (e.g., *R* or *S*). In fact, the sign of rotation may change not only with wavelength λ (cf. Chapter 12) but also with solvent, or even with concentration (Chapter 1). The conditions of the polarimetric measurement (λ, temperature, solvent, and concentration) must thus always be specified.[5]

Perhaps by coincidence, the now universally used Cahn–Ingold–Prelog (CIP) system[6,7] for the specification of molecular configuration originated in the same year[8] in which absolute configuration was first experimentally determined by Bijvoet.[4] In this system, the (italicized) descriptors *R* and *S*, which are placed in parentheses and connected with a hyphen only when used as prefixes (*configurational descriptors*) of chemical names, are used to denote the configurations of enantiomers and diastereomers. To determine the proper descriptor for a chiral center, the *CIP chirality rule*[6b] is employed. In general, the four ligands attached to the chiral center can be arranged in a unique sequence; for the moment let us call these ligands A, B, D, and E with the proviso that the sequence (or priority) is A precedes B precedes D precedes E (A > B > D > E or A → B → D → E). The CIP chirality rule requires that the model be viewed from the side opposite to that occupied by the ligand E of lowest priority (Fig. 5.2). The remaining three ligands then present a tripodal array, with the legs extending toward the viewer. If the sequential arrangement or sense of direction of these three ligands (A → B → D) is then clockwise (as in Fig. 5.2), the configurational descriptor is *R* (for Latin *rectus* meaning right); if it is counterclockwise, it is *S* (for Latin *sinister* meaning left).

Figure 5.2. Chirality rule.

We next discuss the *sequence rules*,[6b,7] that is, the rules for arranging groups A, B, D, and E in order of priority. In the first instance, this is done on the basis of atomic number: Ligands of higher atomic number precede ligands of lower atomic number. Figure 5.3 shows the assignment of descriptors to the enantiomers of one of the simplest chiral molecules known, CHFClBr (Chapter 2).

The reader should refer to Figure 3.15 regarding the conversion of the 3D into a Fischer projection formula: If the projection formula is written with the atom of lowest priority at the bottom or at the top, clockwise array of the remaining three corresponds to the *R* configuration and counterclockwise array corresponds to the *S* configuration. This correspondence does not hold when the lowest sequence group is on the side of the Fischer projection (see Fig. 5.3(*b*), *R* configuration); in that case the opposite assignment is correct (see also p. 74).

To include pyramidal triligant (three-coordinate) atoms, such as N or P, in the scheme, the "absent" ligand (usually a lone pair) is considered to have atomic number zero and will thus automatically have the lowest precedence (Fig. 5.4).

When two or more atoms directly attached to the chiral center are the same, as in the case of lactic acid (Fig. 5.5), a principle of outward exploration is applied: If the proximal atom attached to the chiral center does not provide a decision, one proceeds outwardly to the second atom, then, if needed, to the third and so on. In the case of the lactic acids, since the carbon atoms of CO_2H and CH_3 provide no decision, one proceeds to the next attached atom and notes that O > H. The proper execution of this tree-graph exploration (Fig. 5.6) requires a well-defined hierarchy.[7] The most important principles are (a) *all* ligands in a given sphere (Fig. 5.6) must be explored before one proceeds to the next sphere. Thus in β-methoxylactic acid (Fig. 5.5), after one finds one oxygen atom in the second sphere attached to each carbon, one must not proceed to the third sphere, but one must explore the second sphere (Fig. 5.6) further

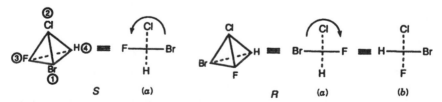

Figure 5.3. Configuration of CHFClBr. (*a*) Fischer projection as viewed from left face. (*b*) As seen from back face.

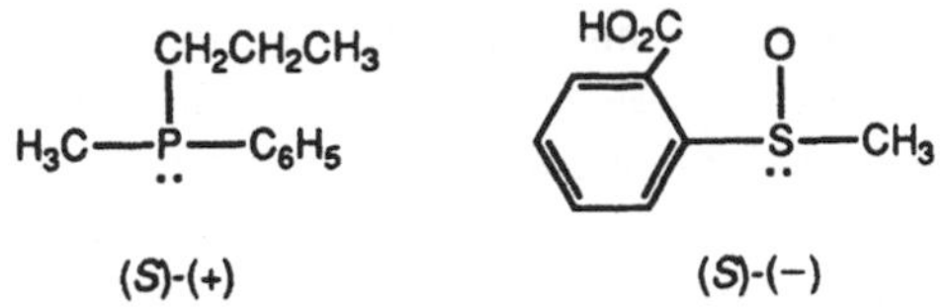

Figure 5.4. Configuration of triligant atoms.

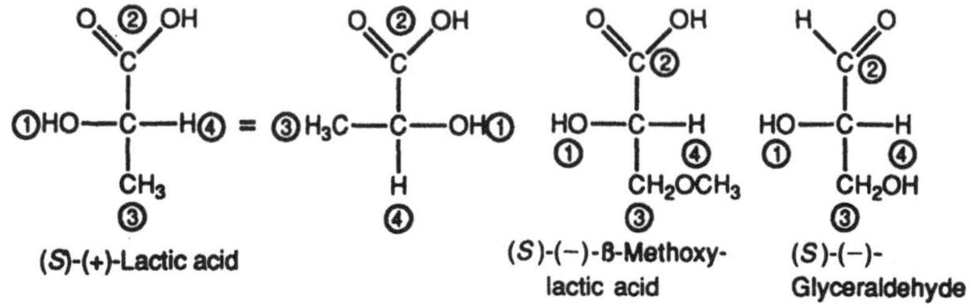

Figure 5.5. Applications of the sequence rule.

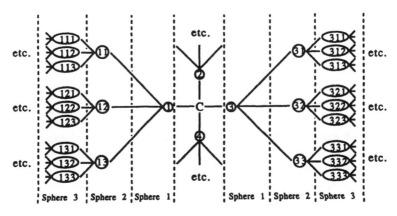

Figure 5.6. Tree graph.

and note that there is a second oxygen atom attached to the carbon in CO_2H but not in CH_2OCH_3 and since $O > H$, $CO_2H > CH_2OCH_3$. (b) In contrast, however, once a precedence of one path of exploration over another has been established in one sphere, that precedence carries to the next sphere. Suppose one wants to determine the sequence of the two ligands shown in Figure 5.7. Clearly, no decision is reached at the first (C), second (C, C, H), or third (C, H, F, C, H, Br) sphere. One therefore has to proceed to the fourth sphere, and one does so using the path of higher precedence at the third sphere ($Br > F$). The decision that ligand B precedes ligand A is therefore made on the basis that $Br > Cl$ in the fourth sphere of the preferred branch rather than on the basis of the observation that I is the highest atomic number atom in the fourth sphere (of the less preferred branch).

We next consider the case of glyceraldehyde (Fig. 5.5). Clearly, $OH > C > H$, but there is a question as to the priority of CH_2OH and $CH{=}O$. This question is resolved by the convention of regarding each atom at a multiple bond as being associated with a "phantom" (duplicate) atom or atoms of the species at the other end of the multiple bond. Thus

$$\underset{(O)}{-C}{=}O \quad \text{becomes} \quad -\underset{(O)}{\overset{|}{C}}-O-(C),$$

$$-C{=}C \quad \text{becomes} \quad -\underset{(C)}{\overset{|}{C}}-C-(C),$$

$$-C{\equiv}N \quad \text{becomes} \quad -\underset{(N)(C)}{\overset{(N)(C)}{\overset{|}{C}}}-\overset{|}{N}$$

$$\text{and} \quad -C{\equiv}C \quad \text{becomes} \quad -\underset{(C)(C)}{\overset{(C)(C)}{\overset{|}{C}}}-\overset{|}{C}$$

[The complemented (duplicate or phantom) atoms are enclosed in parentheses and considered to bear no further substituents.] It follows that in glyceraldehyde $C{=}O$ has precedence over $C-O$ (Fig. 5.5).

Ligand complementation is also required when a ligand is bidentate (or tri- or tetradentate) or when there is a cyclic or bicyclic component to a ligand.[7,9] In either case, each branch of the cyclic structure is severed at the branch point (where it doubles onto itself) and the atom at the branch point is then complemented at the end of the chain resulting from the disconnection. The case of a bidentate ligand is exemplified by the tetrahydrofuran (THF) derivative shown in Figure 5.8. The two branches of the five-membered ring are (separately) severed at the chiral center, which

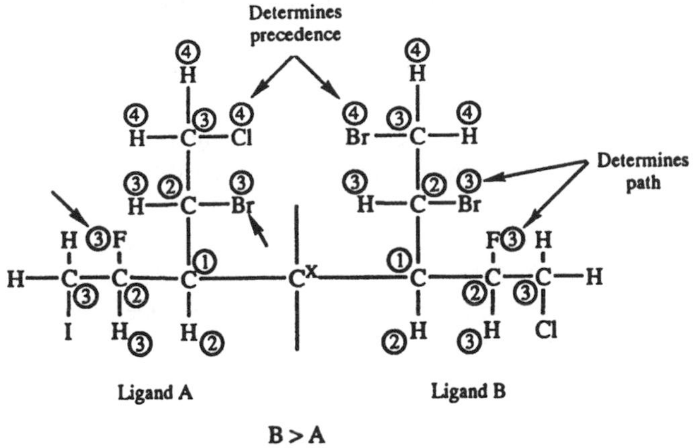

Figure 5.7. Hierarchy of paths.

is then complemented at the end of each of the two chains so created. It is seen that in disjuncture A the hypothetical ligand produced, $-CH_2OCH_2CH_2-(C)$, has precedence over the (real) acyclic ligand $CH_2OCH_2CH_3$ because of the presence of the phantom C at the end of the former ligand. In contrast, with disjuncture B the hypothetical ligand created, $-CH_2CH_2OCH_2-(C)$, falls behind the real one, $-CH_2CH_2OCH_2CH_3$, since the latter has three hydrogen atoms attached to the terminal carbon and the former has none. The configurational symbol is therefore S.

The related case of cutting a ring substituent is exemplified by the structure in Figure 5.9 in which **B** illustrates the treatment of the cyclohexyl ring (in **A**). The

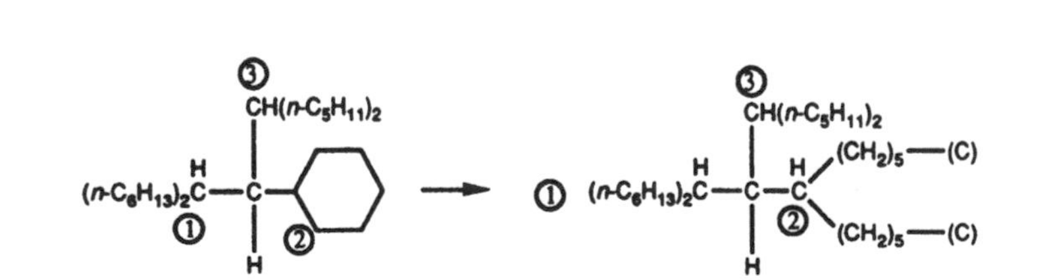

Figure 5.8. Stereochemical nomenclature for bidentate ligands showing precedence of disjuncture A.

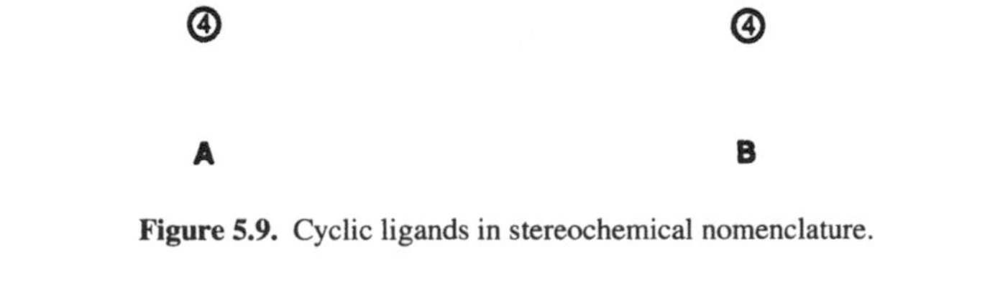

Figure 5.9. Cyclic ligands in stereochemical nomenclature.

proper sequence is therefore di-*n*-hexylcarbinyl > cyclohexyl > di-*n*-pentylcarbinyl > H and the configuration is *S*.

We are now ready to consider phenyl (Fig. 5.10, **A**). Since each of the six carbon atoms is doubly bonded (in either Kekulé structure) to another carbon, each ring carbon bears a duplicate carbon as a substituent. The ring so complemented (Fig. 5.10, **B**) is then opened according to the rules for cyclic systems to yield the representation in Figure 5.10, **C**.

The last rule relating to material or constitutional differences (as distinct from configurational ones) to be mentioned in this chapter applies to molecules in which ligands differ *only* as a result of isotopic substitution. The sequence rule here is "higher atomic mass precedes lower atomic mass isotope," for example, D > H and $^{13}C > ^{12}C$. Applications to $C_6H_5CHDCH_3$, $CD_3CHOHCH_3$, benzyl *p*-tolyl sulfone-^{16}O, ^{18}O, and L-valine-4-^{13}C are shown in Figure 5.11. On the other hand, 3-hexanol-2-*d* (Fig. 5.11) is an example where this rule does not apply because the propyl group has precedence over the ethyl group in the absence of isotopic substitution; the symbol is therefore the same as it would be for unlabeled 3-hexanol of corresponding configuration.

We consider now chiral compounds in which the differences between ligands are not material or constitutional but configurational. Most such compounds contain more than one chiral center and will be considered in Section 5-4. Here we are considering ligands that differ in (olefinic) cis–trans isomerism. According to Prelog and Helmchen[7] the olefinic ligand in which the substituent of higher sequence is on the *same* side of the olefinic double bond as the chiral center has precedence over the ligand in which the substituent of higher sequence is trans to the chiral center. This precedence is related to neither the classical cis–trans designation of double-bond configuration nor the *E–Z* designation. Examples are shown in Figure 5.12.

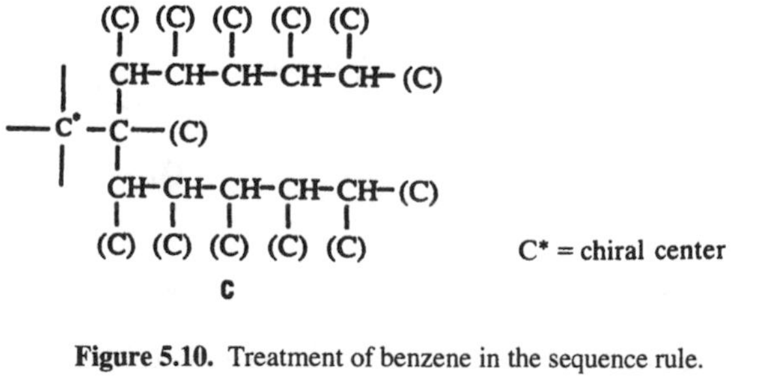

Figure 5.10. Treatment of benzene in the sequence rule.

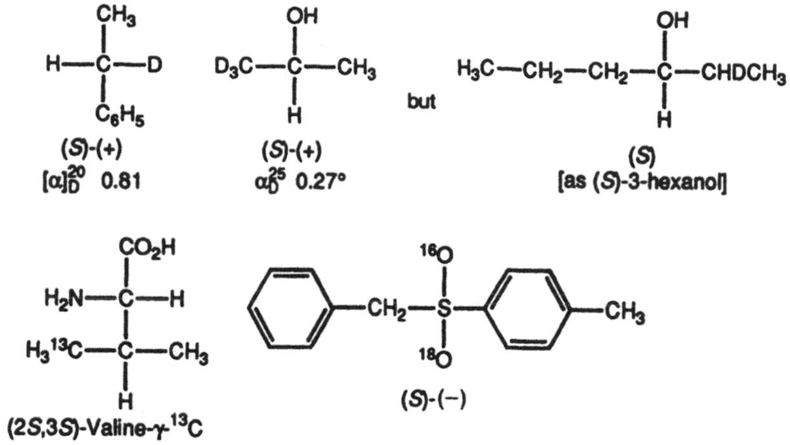

Figure 5.11. Configuration of chirally labeled compounds.

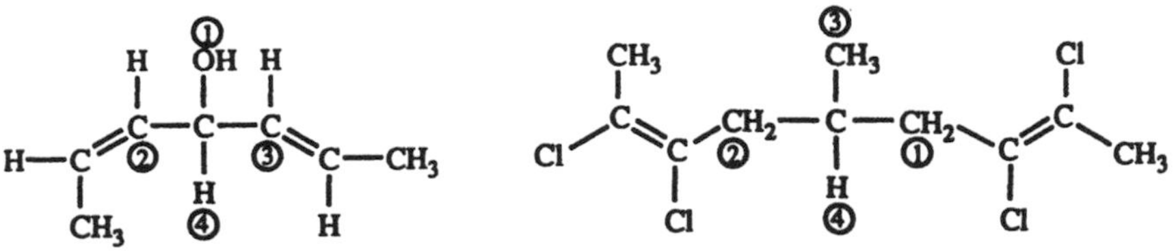

Figure 5.12. Configuration of chiral centers with cis–trans isomeric ligands.

One frequently encounters cases where a formula is depicted with the lowest sequence group in front rather than in back or with some group other than the lowest sequence group in back. The steroid shown in Figure 5.13 (methyl groups above the plane of the paper) will illustrate the former case; let us consider its configuration at C(10). The sequence of the attached atoms is C(9) > C(5) > C(1) > CH₃. It should be

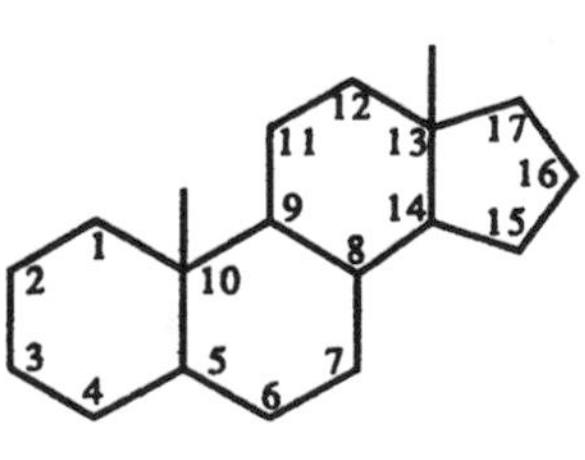

Figure 5.13. Steroid skeleton.

evident by now that CH > CH$_2$ > CH$_3$; the sequence C(9) > C(5) follows from the precedence of the attached atoms [C(8) vs. C(4) or C(6).] Since the lowest sequence group CH$_3$ is in front, the representation is opposite to that demanded in Figure 5.2. Therefore the descriptor R, which would be obtained by considering the (clockwise) array of the remaining three groups C(9), C(5), and C(1), must be reversed; the correct descriptor is S. One can, of course, arrive at the same conclusion by placing oneself *behind* the plane of the paper.

In general, there are eight possibilities that are summarized schematically in Figure 5.14.[9] The middle row indicates the CIP priority of the ligands that are in front (upper row) or in back (lower row) in the 3D formula. The signs in the upper and lower rows indicate whether the apparent array (clockwise, R; counterclockwise, S) of the remaining three ligands gives the correct (indicated by +) or reversed (−) descriptor. Clearly, when the No. 4 ligand is in the back (Fig. 5.2), the descriptor obtained by observing the array of the remaining three is correct, so the corresponding entry in the scheme is +. In contrast, in the steroid depicted in Fig. 5.13, the lowest-sequence ligand (methyl) is in front; the minus sign in Fig. 5.14 for ligand No. 4 in front implies that the clockwise array of C(9), C(5), C(4) must be reversed to give the descriptor S, as already mentioned above. For other ligands in front or in back the signs alternate, as shown in Figure 5.14, so the scheme can be reconstructed at a moment's notice when needed.

An additional application is shown in Figure 5.15. In the decalone derivative the sequence at C(4a) is OH > C(8a) > C(4) > C(5). Since the highest priority group OH is in back, the clockwise array of C(8a) → C(4) → C(5) does not give the correct descriptor (cf. Fig. 5.14, lower row, minus sign for ligand No. 1); the correct descriptor is therefore S. At C(8a), however, where the sequence is C(1) > CHO > C(4a) > C(8), the group of second priority (CHO) is in the back. This situation corresponds to a plus sign in Figure 5.14 (lower row), so the clockwise array C(1) → C(4a) → C(8) gives rise to the correct descriptor R. The compound is therefore 4aS, 8aR.

We conclude with a listing of some common substituents in order of priority in the CIP sequence (for a more extensive list, see refs. 10 and 11): −I, −Br, −Cl, −PR$_2$, −SO$_3$H, −SO$_2$R, −SOR, −SR, −SH, −F, −OTs, −OCOCH$_3$, −OC$_6$H$_5$, −OCH$_3$, −OH, −NO$_2$, −N$^+$(CH$_3$)$_3$, N(C$_2$H$_5$)$_2$, −N(CH$_3$)$_2$, −NHCOC$_6$H$_5$, −NHR, −NH$_2$, −CO$_2$R, −CO$_2$H, −COC$_6$H$_5$, −COCH$_3$, −CHO, −CH$_2$OR, −CH$_2$OH, −CN, −CH$_2$NH$_2$, −C$_6$H$_5$, −C≡CH, (CH$_3$)$_3$C−, cyclohexyl, *sec*-butyl, −CH=CH$_2$, isopropyl, −CH$_2$C$_6$H$_5$, −CH$_2$CH=CH$_2$, isobutyl, −C$_2$H$_5$, −CH$_3$, −D, −H.

The CIP system, though applicable to the vast majority of chiral compounds, is not perfect. The system has some inconsistencies, especially when the chiral element is a chiral axis or plane (see Chapter 13) or when chirality is due to the molecular

In front:	+	−	+	−
Ligand No:	1	2	3	4
In back:	−	+	−	+

Figure 5.14. Ligand permutation scheme.

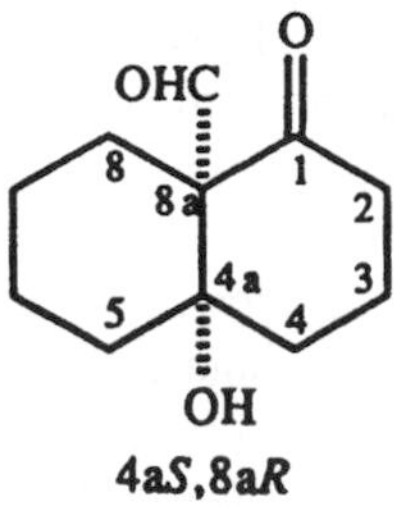

Figure 5.15. Assignment of configurational symbol.

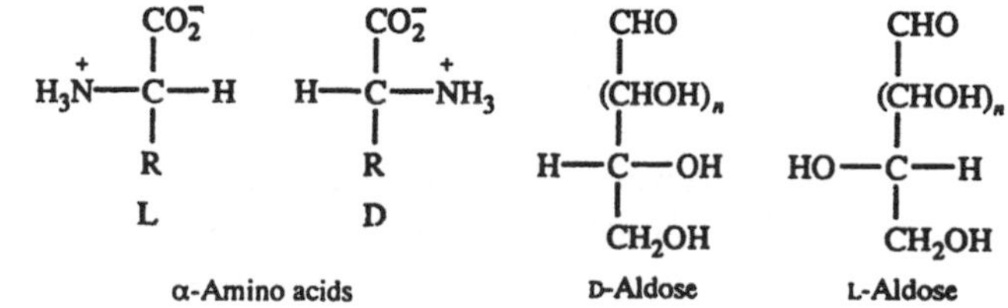

Figure 5.16. Configuration of α-amino acids and monosaccharides.

framework as a whole and cannot be readily factorized (e.g., twistane, Fig. 4.10). For a criticism of the system see ref. 12 and references cited therein.

Prior to the 1960s, it was common to denote the configuration at chiral centers with references to planar projection (Fischer) rather than 3D formulas; the descriptors used to this end are D and L.[13] At the present time, the DL system is used only for amino acids and carbohydrates, as shown in Figure 5.16 (see also Figs. 3.17 and 3.18). In α-amino acids, the configuration is L if the amino (or ammonio) group is on the left of the Fischer projection formula with the carboxylate group written on top; D for the enantiomer. For sugars, the configurational notation is based on the highest numbered (farthest from the carbonyl end) chiral CHOH group. If in this group (with the Fischer projection written so that the C=O is on top and the CH_2OH is on the bottom) OH is on the right, the configuration is D; if OH is on the left, the configuration is L.

A number of other configurational names and symbols are used in compounds having more than one chiral center, for example, in ring compounds and alkenes. These names and symbols will be introduced in later sections or chapters, where the appropriate compounds are treated.

5-3. DETERMINATION OF ABSOLUTE CONFIGURATION

a. Bijvoet Method

It was only in 1951 that the first experimental determination of the absolute configuration of any chiral molecule was achieved.[4] In ordinary X-ray crystallography the intensities of the diffracted beams depend on the distances between the atoms but not on the absolute spatial orientation of the structure. This comes about because the phase change due to scattering of the incident radiation is (nearly) the same for all atoms. Thus a chiral crystal and its enantiomorph produce the same X-ray patterns and cannot be distinguished from one another. As a simple example, the one-dimensional (1D) chiral array A–B cannot be distinguished from B–A because the interference pattern at any point I in space will be the same for the two 1D enantiomers since it depends only on the absolute value of the path length difference |AI – BI| = |BI – AI| (Fig. 5.17). Thus there will be no change in interference pattern when atoms A and B are interchanged. The same is true in the more complex case of a 3D chiral crystal: Since there is nothing to distinguish propagation of X-rays in a direction A–B from that in the opposite direction B–A (Friedel's law), X-ray diffraction patterns are centrosymmetric, whether the crystal under investigation has a center of symmetry or not (see ref. 14).

The solution to this dilemma was, in principle, found in 1930 when Coster et al.[15] used so-called anomalous X-ray scattering to determine the sequence of planes of zinc and sulfur atoms in a crystal of zinc blende (ZnS). The method involves using X-rays of a wavelength near the absorption edge of one of the atoms, in this case zinc. This results in a small *phase change* of the X-rays scattered by the zinc atoms relative to the sulfur atoms, which does *not* depend on their relative positions. If B in Figure 5.17 denotes the zinc atom, the effect of retarding, say, the phase of the diffracted X-ray is the same as if the path B–I were longer than A–I, by a hypothetical extra path length p, simulated by the phase change. Thus the scattered *intensities* for A–B and B–A are no longer the same, since the phase change is in opposite directions in the two cases. (In terms of the hypothetical quantity p, it appears as if the path length for A–B is |AI – BI| – p, but for B–A is |AI – BI| + p.) As a result, the diffraction pattern is no longer truly centrosymmetric; pairs of spots in the pattern that are related by the center of symmetry (cf. Chapter 4), which are now called *Bijvoet pairs* (see below), become unequal in intensity. Provided that the structure is known except for the absolute configuration, one can then calculate the relative intensities of the Bijvoet pairs for the R and S isomers, and by comparison with experiment one can tell which is which. Thus one can determine the sense of chirality by using X-rays on the absorption edge of one of the atoms in the structure. Usually, this is a relatively heavy atom (sulfur or larger) since the phase change generally increases with atomic mass.

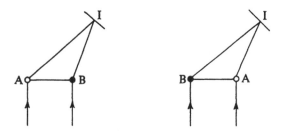

Figure 5.17. Determination of absolute configuration.

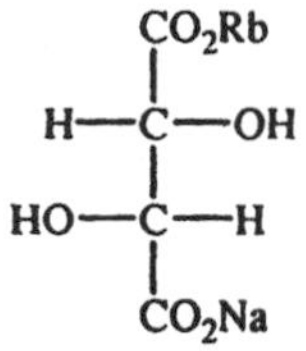

(+)-Tartaric acid
sodium rubidium salt
(*R*,*R*)

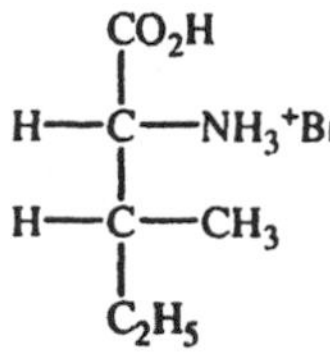

D-(−)-Isoleucine
Hydrobromide
(2*R*,3*R*)

Figure 5.18. Absolute configuration as determined by anomalous X-ray scattering.

This principle was applied by Bijvoet et al.[4] in 1951 by using zirconium K_α X-rays in the X-ray crystallographic study of sodium rubidium tartrate (Fig. 5.18). The (+)-tartrate anion was found to have the *R*, *R* configuration. Since (+)-tartaric acid had been correlated with many other chiral compounds, notably the sugars (Section 5-5), the determination of its configuration marked a major milestone in stereochemistry. Soon afterwards the absolute configuration of the hydrobromide of the amino acid D-(−)-isoleucine (Fig. 5.18) was determined utilizing the phase lag at bromine introduced by the use of uranium L_α radiation.[16]

b. Theoretical Approaches

A second (theoretical) approach to the determination of absolute configuration that actually antedated the experimental approach depends on the comparison of measured optical rotations with values computed by theory, an approach pioneered by Kuhn[17] and by Kirkwood[18] (see also ref. 19). More recent extensions include the comparison of experimental and theoretically computed ORD and CD curves as well as the exciton chirality method (cf. Chapter 12-4.d).[20]

c. Modification of Crystal Morphology in the Presence of Additives

The concept of polarity in crystals has not only contributed to the confirmation of the correctness of Bijvoet's method but also has given rise to another, independent method of determining absolute configuration experimentally.[21] Suppose that one examines a crystal of a given enantiomer. For the sake of simplicity we shall again assume 1D chirality, as in a molecule W–Y (enantiomer: Y–W). Let us further suppose that the disposition of the molecules in the crystal has been determined by ordinary X-ray diffraction and that they are oriented along the crystal axis as shown in Figure 5.19. As already pointed out (cf. Fig. 5.17) ordinary X-ray diffraction cannot distinguish between W–Y and Y–W arrangements. Thus, even though the two ends of the polar crystal (along the W–Y or Y–W axis) may be visually distinct (cf. the case of ZnS above), it is not known which direction in the crystal corresponds to which of the constituent molecules (W–Y or Y–W) and thus the configuration of the latter

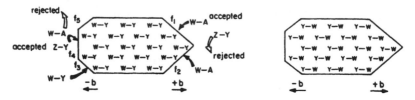

Figure 5.19. Polar crystal W–Y in the presence of impurities W–A or Z–Y. [Reprinted with permission from Berkovitch-Yellin et al., *Nature (London)* **1982**, *296*, 27.]

remains unknown. The breaking of this impasse rests on the observation that impurities or additives adsorbed on growing crystal faces tend to selectively stunt the growth of the crystal in the direction orthogonal to the faces in question. Thus impurity W–A in Figure 5.19, being adsorbed on the face exposing Y, would impede growth in the direction +b (i.e., along faces f_1 and f_2), whereas additive Z–Y will impede growth along –b by being adsorbed on exposed W at faces f_3, f_4, f_5. It remains to find additives that, in a defined way, attach themselves to the W or Y ends of the molecules (or rather, cf. Fig. 5.19, to the faces where W or Y groups, respectively, are exposed). Provided a consistent set of such additives can be found and if the stunting of growth in one or other direction along the crystal axis can be observed (this is relatively easy, since substantial changes in crystal habit are often seen), then the orientation of W–Y relative to the macroscopic crystal axis can be defined and the configuration of W–Y thus determined.

A 3D example is provided by L-lysine hydrochloride dihydrate (Fig. 5.20, **A**). In this crystal the carbon chain as written orients itself more or less along one of the crystal axes. Impurities tend to interfere with growth along directions where there are constitutional or configurational differences between the substrate and the impurity.

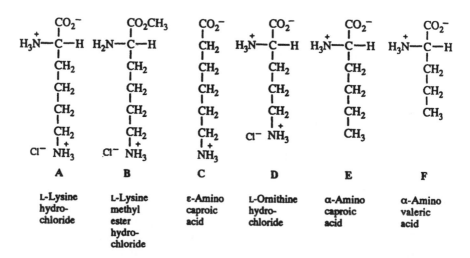

Figure 5.20. L-Lysine hydrochloride and addends used in the determination of its configuration.

Thus the first two extraneous substances (**B** and **C**) shown in Figure 5.20 are identical with L-lysine hydrochloride in the chain-end NH_3^+ but differ at the carboxylic amino acid end. These compounds will, in fact, inhibit growth of the L-lysine crystal in one and the same direction; this direction is assumed to be that along the CO_2^- end of the chain. The latter three impurities (**D–F**), in contrast, are all α-amino acid zwitterions but differ at the other end of the chain, either by chain length or by end group. These additives impede crystal growth in the opposite direction from the first two; this is taken to be the direction along the $-CH_2-NH_3^+$ moiety. The direction of the lysine molecule along the crystal axis and hence its absolute configuration as L are thus determined. It should be noted that one of the addends (ε-aminocaproic acid, **C**) is achiral; in this method the addends need not be chiral or, if they are chiral, their absolute configuration need not be known.

5-4. RELATIVE CONFIGURATION AND NOTATION

As was mentioned earlier, the term "relative configuration" may be used in two contexts: relative configuration of a chiral center of one compound with respect to that of another, or relative configuration of several chiral centers within one compound with respect to each other. We shall postpone the comparison of configuration of distinct compounds until Section 5-5. In this section we will deal with single compounds containing two or more chiral centers.

Fischer projection formulas for representation of compounds with more than one chiral center have already been discussed in Chapter 3 (e.g., Figs. 3.18 and 3.20). These formulas have the drawback, however, that they depict the molecule in the unrealistic eclipsed conformation (cf. Chapter 10). Two other systems, the so-called Newman projection formulas (Fig. 5.21),[22] and the so-called sawhorse formulas (Fig. 5.22) show the molecules in their staggered conformations. "Front-on view" formulas very similar to the Newman projections have been used much earlier[23,24] and are also shown in Figure 5.21. Figure 5.22 also indicates the interconversion of the various perspectives. Figure 5.23 shows the "flying wedge" and "zigzag" formulas that are useful to show the actual conformation (as well as configuration) of compounds having two or more chiral centers. (Newman and sawhorse formulas are most useful for compounds having no more than two chiral centers.) The flying wedge and zigzag formulas have the advantage of showing the molecular backbone in the (true) staggered conformation and with the backbone entirely in the plane of the paper. The

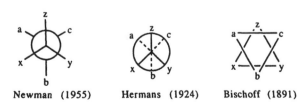

Figure 5.21. Front-on projection formulas.

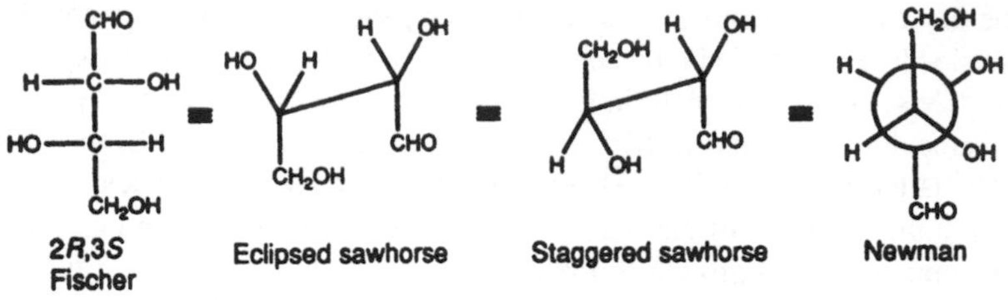

Figure 5.22. Fischer, sawhorse, and Newman projections of L-threose.

zigzag formula, although simpler, can be used only for chiral centers bearing a hydrogen atom, since this atom is assumed and not drawn. The flying wedge formula does not suffer from this limitation and is convenient for the representation of any number of chiral centers in a straight chain. In a molecule having two or more identical chiral centers, such as tartaric acid (Fig. 5.23), the fact that two corresponding groups (here OH) are on the same side of the molecular framework does not imply the presence of a symmetry plane; this is in contrast to the situation in the Fischer projection (Fig. 3.20).

The notation for compounds having two or more chiral centers should, in principle, be simple: Each center is given its proper CIP descriptor along with the usual positional number (locant). Thus L-threose (Fig. 5.22) is $(2R,3S)$-2,3,4-trihydroxybutanal and D-glucose (Fig. 5.23) is $(2R,3S,4R,5R)$-2.3,4,5,6-pentahydroxyhexanal. This system works well when the compounds are resolved (nonracemic) and their absolute configurations at each chiral center (i.e., their absolute as well as relative configurations) are known. But if, as frequently happens, one deals with racemates, or if only the relative but not the absolute configuration is known, the system requires modification.

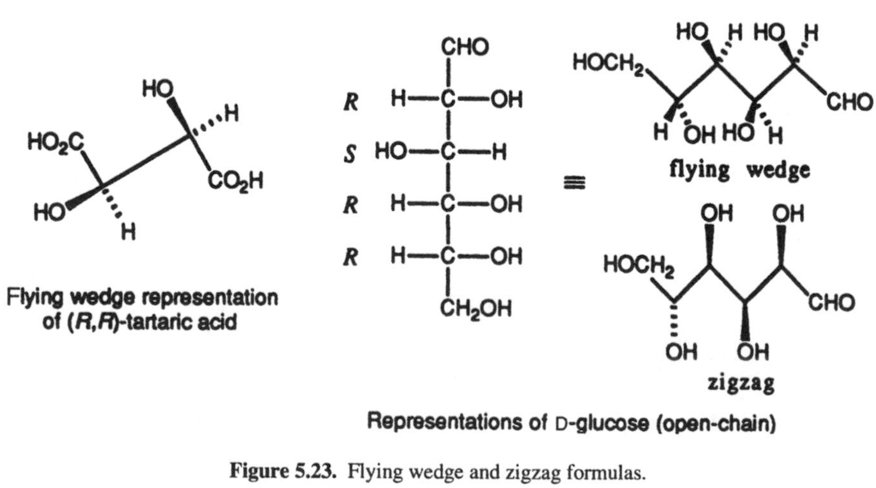

Figure 5.23. Flying wedge and zigzag formulas.

For racemates, the symbols *RS* or *SR* are used for the chiral centers. The lowest numbered (or sole) chiral center is automatically given the symbol *RS*, which thus serves as a reference; let us call this chiral center No. 1. The next chiral center No. 2 is then denoted by *RS* if, in the enantiomer where the No. 1 chiral center is *R*, No. 2 is *R* also (and, likewise, in the enantiomer where No. 1 is *S*, No. 2 is *S* also). Conversely, if the relative configuration is such that, when No. 1 is *R*, No. 2 is *S* (and, therefore, if No. 1 is *S*, No. 2 is *R*), the configuration of the second center is denoted by *SR*. The two racemic diastereomers are therefore 1*RS*,2*RS* and 1*RS*,2*SR*, the former denoting a racemic mixture of 1*R*,2*R* and 1*S*,2*S*, the latter of 1*R*,2*S* and 1*S*,2*R*. By way of examples, let us consider the racemates of the open-chain forms of threose depicted in Figure 5.22 and glucose shown in Figure 5.23. The former would be (2*RS*,3*SR*)-2,3,4-trihydroxybutanal and the latter (2*RS*,3*SR*,4*RS*,5*RS*)-2,3,4,5,6-pentahydroxyhexanal. Had we started from the descriptor of the enantiomeric D-threose, 2*S*,3*R*, we should still have arrived at the symbol 2*RS*,3*SR* (and not 2*SR*,3*RS*) for the racemate, since, by convention, the lowest numbered chiral center is always *RS*.

On occasion, only the relative configuration of several chiral centers, not the absolute configuration of any one of them, is known. Since the *RS/SR* system only implies relative configuration, it may be used in this case also. Alternatively, a system used by *Chemical Abstracts* (CA) may be employed: The lowest numbered chiral center is arbitrarily assumed to be *R* and the configuration of the others is denoted as R^* or S^*, relative to it. In order to signal that the configurations are relative, not absolute (i.e., the configuration of the first chiral center is assumed arbitrarily), all symbols are starred. Thus, prior to 1951, when the absolute configuration of glucose was not known (though its relative configuration had been determined by Fischer; cf. Section 5-6) its configurational descriptor would have been $2R^*$, $3S^*$, $4R^*$, $5R^*$. Beilstein uses *rel*-(2*R*,3*S*,4*R*,5*R*) for the same case of unknown absolute configuration. It might be noted here that CA uses separate symbols for absolute and relative configuration; thus the configurational notations of CA for the tartaric acids (Fig. 3.20) are [S-(R^*,R^*)]-(−), [R-(R^*,R^*)]-(+), and (R^*,S^*) (meso). The descriptor for racemic tartaric acid is (R^*,R^*)-(±).

Because of the very frequent occurrence of compounds with two chiral centers, chemists have wanted to have words as well as symbols for the relative chirality of the two centers involved without individually specifying the configuration of either one. It is, of course, possible to do this on the basis of the CIP system; thus Seebach and Prelog[25] proposed the symbol *l* (for like) and *u* (for unlike) when the CIP descriptors for the two chiral centers are like (*R,R* or *S,S* or *RS,RS*) or unlike (*R,S* or *S,R* or *RS,SR*), respectively. While this notation is as unambiguous as the CIP system itself, it is not particularly convenient, because it requires ascertaining the CIP descriptors for both chiral centers before the *l* or *u* descriptor can be assigned. What most chemists want is an "at a glance" designation for a system of two chiral centers.

Such an "at a glance" nomenclature was coined many years ago; it is based on the names of the four-carbon sugars erythrose and threose (Fig. 3.18). As shown in Figure

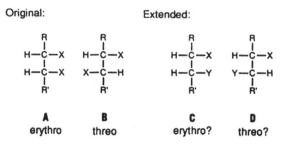

Original: Extended:

<pre>
        R   R   R   R
        | | | |
   H—C—X   H—C—X   H—C—X   H—C—X
        | | | |
   H—C—X   X—C—H   H—C—Y   Y—C—H
        | | | |
        R'   R'   R'   R'

        A   B   C   D
     erythro   threo   erythro?   threo?
</pre>

Figure 5.24. Erythro–threo nomenclature.

5.24, **A** and **B** (R ≠ R′), isomers with the two identical ligands attached to a carbon chain on the same side of the Fischer projection formula (as in erythrose) were called erythro and those with the identical ligands on opposite sides (as in threose) were called threo. (Although individual enantiomers are depicted in Fig. 5.24, the same nomenclature applies to diastereomeric racemates.)

An alternative way of defining threo and erythro is in terms of the configuration of the fragments R–CHX–Cβ and R′–CHX–Cβ: If the configurations of these two fragments in isolation correspond (R′ taking the place of R), the compound is threo; if they are opposite, the configuration is erytho.[26] This nomenclature would appear simple enough and is quite unequivocal. However, complications set in when attempts were made to extend it to more complex cases (Fig. 5.24, **C** and **D**). Here, identical (or opposite) configurations of R–CHX–Cβ and R′–CHY–Cβ can no longer be defined unequivocally, because now *two* substitutions or comparisons, R/R′ and X/Y, must be made and it is not always clear which group is to be substituted for which. The nomenclature still works when X and Y are heteroatoms, such as OR, NRR′, or halogen, and when R and R′ are alkyl or aryl. However, when X or Y, also, are alkyl or aryl, there may be ambiguity. The R and R′ groups have been defined as being part of the *main chain*.[13] Unfortunately, this may lead to a conflict with the definition given earlier, as shown in Figure 5.25, **A**. There may also be a problem with defining the main chain, for example, in case **B**, Figure 5.25.[27] The root of the problem is that erythro/threo, like D/L, is based

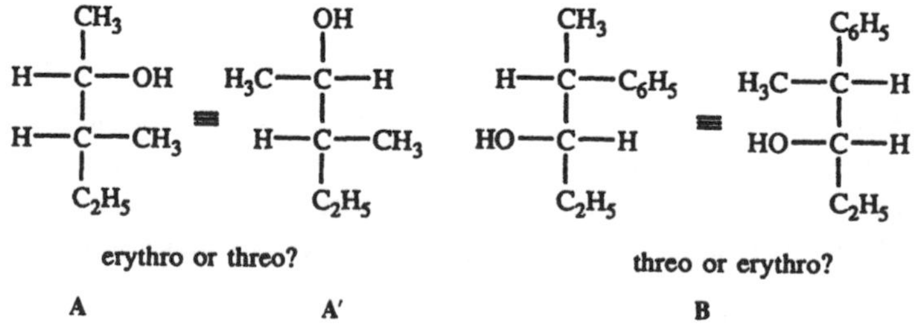

<pre>
        CH3   OH   CH3   C6H5
        | | | |
   H—C—OH   H3C—C—H   H—C—C6H5   H3C—C—H
        | ≡ | | ≡ |
   H—C—CH3   H—C—CH3   HO—C—H   HO—C—H
        | | | |
        C2H5   C2H5   C2H5   C2H5

     erythro or threo?   threo or erythro?

        A   A'   B
</pre>

Figure 5.25. Conflicts in erythro–threo nomenclature.

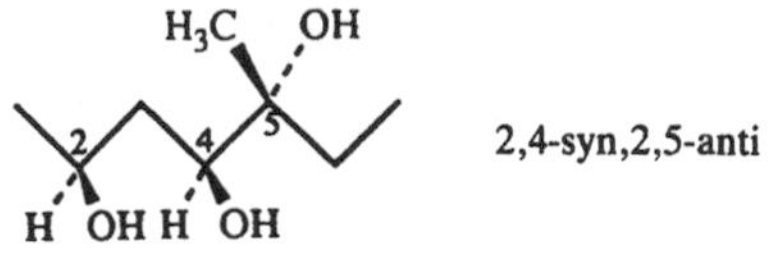

2,4-syn,2,5-anti

Figure 5.26. The syn–anti nomenclature for diastereomers.

on Fischer projection formulas and that, moreover, there is no clear definition of ligand precedence.

A better system is based on zigzag formulas. If one can define a main chain in a molecule (and this may sometimes be troublesome), it then requires ordering of only two substituents at each of the chiral centers to define relative configuration; such ordering can usually be performed at a glance. An example is shown in Figure 5.26 for a compound with three chiral centers; the designation syn–anti proposed by Masamune et al.[28] is now preferred for such compounds. The ordering of the two substituents at each chiral center of the chain is best done by the CIP system (thus OH > H and OH > CH$_3$ in Fig. 5.26), though other suggestions have been made.[29]

Nomenclature for Pseudoasymmetric Centers.   Figure 5.27 represents the four stereoisomers of 2,3,4-trihydroxyglutaric acid, which were discussed earlier in Section 3-3.b (Fig. 3.21). Structures **A** and **B** are mirror images and thus represent a pair of enantiomers 2*R*,4*R* and 2*S*,4*S*. Carbon (3) is not a stereogenic center in **A** and **B** because it bears two identical ligands; the structure generated by interchange of H and OH at C(3) can be converted into the original structure by a 180° rotation. In contrast, C(3) in the meso isomers **C** and **D** *is* stereogenic.[30] In fact, **C** and **D** differ only in configuration at that center. Since reflection of **C** or of **D** converts it into a superposable structure, neither **C** nor **D** is chiral and so the stereogenic C(3) is not appropriately called a chiral center and so cannot be described as being *R* or *S*. [Chiral centers change configuration and configurational descriptors upon reflection of the molecule as a whole. The C(3)

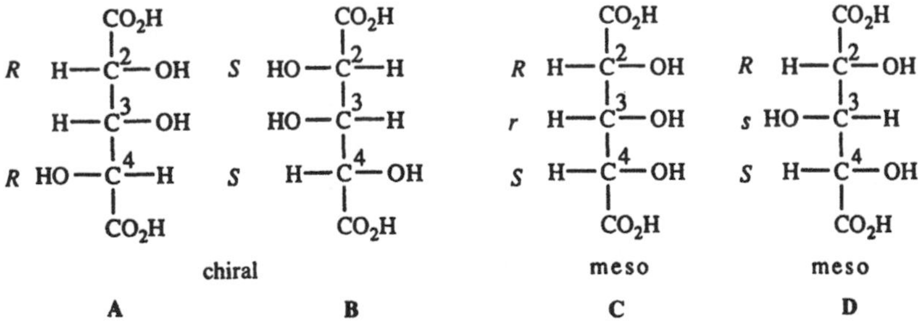

Figure 5.27. 2,3,4-Trihydroxyglutaric acids.

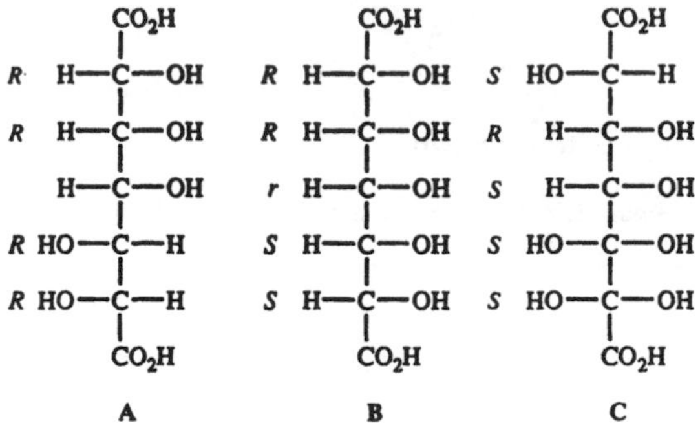

$$
\begin{array}{ccc}
& \text{CO}_2\text{H} & \\
R & \text{H}-\text{C}-\text{OH} & \\
R & \text{H}-\text{C}-\text{OH} & \\
& \text{H}-\text{C}-\text{OH} & \\
R & \text{HO}-\text{C}-\text{H} & \\
R & \text{HO}-\text{C}-\text{H} & \\
& \text{CO}_2\text{H} &
\end{array}
$$

Figure 5.28. Representative pentahydroxypimelic acids.

atom in **C** and **D** is reflection invariant.] The name "pseudoasymmetric center" has historically been used for such an achiral stereogenic center (cf. Chapter 3); the notation for such centers is lower-case r or s[6b] to distinguish them from chiral centers. Figure 5.27 (isomers **C** and **D**) also implies that in the ordering of the enantiomorphous CHOHCO$_2$H ligands, the sequence $R > S$ is followed.

No particular nomenclature problem is presented by compounds with four, pairwise corresponding chiral centers, such as the tetrahydroxyadipic acids (Fig. 3.22). However, the next higher homologue, of which three diastereomers are shown in Figure 5.28, presents six racemic pairs and four meso forms. In two of the racemic pairs C(4) is achiral (type **A**), and in all of the meso forms C(4) is pseudoasymmetric (type **B**). Structure **C** is representative of the remaining four pairs of enantiomers in which C(4) is chiral. To assign it a descriptor (here S), the rule *like precedes unlike*[7] is used; in the case of **C**, $S,S > S,R$. (This rule, where applicable, has precedence over $R > S$.)

5-5. DETERMINATION OF RELATIVE CONFIGURATION OF SATURATED ALIPHATIC COMPOUNDS

There are many methods for correlating configurations, that is, for determining the configuration of one chiral center relative to another either in the same molecule or another.[31,32] This is fortunate, because the determination of absolute configuration by the Bijvoet method is limited to substances that are crystalline and possess an appropriate "heavy" atom; only a small percentage of all chiral organic compounds has had its configuration determined either in this way or by the examination of crystal morphology in the presence of tailor-made additives (Section 5-3). On the other hand, a very large number of correlations have been effected and catalogued,[33] and the configurations of compounds with a single chiral center at carbon have been tabulated.[34]

Methods used to determine relative configuration are (a) X-ray structure analysis, (b) chemical interconversion not affecting bonds to the chiral atom, (c) methods based on symmetry properties, (d) correlation via diastereomers ("confrontation correlation"), (e) correlation via quasi-racemates, (f) chemical correlations affecting bonds to a chiral atom in a "known" way, (g) correlations by asymmetric synthesis of "known" course, and (h) various spectroscopic and other physical methods, among which ORD–CD and NMR techniques should be singled out. We shall deal with the first seven methods in this chapter, though methods (b), (e), and (f) are now largely of historic interest. The application of asymmetric synthesis (g), a somewhat uncertain method for configurational assignment at best, will be discussed briefly. Spectroscopic methods (h) will be dealt with elsewhere in this book, principally in Chapters 10 and 11 (NMR) and 12 (ORD–CD), as will be the configuration of alkenes in Chapter 9, and of compounds with axial and planar chirality or helicity in Chapter 13.

a. X-Ray Structure Analysis

X-ray structure determination yields the relative configuration of all chiral centers in a molecule, though, in the absence of Bijvoet anomalous scattering, it does not yield the absolute configuration of the molecule as a whole. Nevertheless, if the absolute configuration of one chiral center in the molecule is known, that of all the others follows. Thus the structure determination of the natural product (+)-S-methylcysteine sulfoxide (Fig. 5.29, **A**) revealed[35] that the configuration at sulfur is S, given that the product can be related to natural L- or (R)-cysteine by methylation and oxidation of the latter (so that the configuration at carbon must be R) and that the relative configuration of the two chiral centers is evident from the X-ray analysis.

Mathieson[36] subsequently pointed out that the X-ray correlation method can be used to determine the absolute configuration of a chiral compound of unknown configuration by linking it chemically (covalently or otherwise) to another of known absolute configuration, and then subjecting the product to X-ray analysis. An example is the determination of the absolute configuration of (R)-(−)-1,1′-dimethylferrocene-3- carboxylic acid (Fig. 5.29, **B**) by X-ray structure analysis of its quinidine salt.[37] The absolute configuration of quinidine is known and that of the ferrocene moiety was thus deduced to be R as shown in Figure 5.29. In this case the result was confirmed by an independent Bijvoet analysis of the salt, which yielded the absolute configuration directly. Another example is the determination of

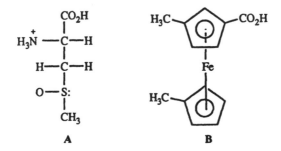

Figure 5.29. Compounds studied by the X-ray correlation method.

configuration of Tröger's base (Fig. 7.27) through its (–)-1,1'-binaphthyl-2,2'-diyl hydrogen phosphate salt.[38] The dextrorotatory base has the S,S configuration, contrary to what had been deduced from an exciton chirality calculation.

b. Chemical Interconversion Not Affecting Bonds to the Stereogenic Atom

If a compound Cabde can be converted to another, Cabdf, without severing the C–e (or C–f) bonds, the configurations of the two compounds are unequivocally correlated: Figure 5.30, **A**. If we join the three common ligands a, b, and d by a triangle (Fig. 5.30, **B**), we note that the new ligand, f, is located on the same (back) face of the triangle as the original one, e: We may say that Cabde and Cabdf are "homofacial" molecules (see also p. 91). Stepwise correlation of molecules of this type is exemplified[33a] in Figure 5.31 for the correlation of (R,R)-(+)-tartaric acid [whose absolute configuration is known by Bijvoet's original experiment (see above)] with (+)-glyceraldehyde and (–)-lactic acid whose configurations are thus determined to be R in both cases.

A correlation involving chiral deuterated compounds is shown in Figure 5.32: The configuration of (–)-ethanol-1-d was shown to be S by chemical correlation (not affecting the chiral center) with (S)-(–)-glycolic-d acid[39] (see also ref. 40).

c. Methods Based on Symmetry Considerations

Emil Fischer's famous proof of the configuration of glucose, summarized in many undergraduate textbooks of organic chemistry, relies heavily on symmetry properties.

For example, the diastereomeric tartaric acids [(+), (–), and meso] shown in Figure 3.20 may be distinguished in that the (+) or (–) compounds are optically active (or can be resolved if encountered as a racemate), whereas the meso form is inactive. The relative configuration of the CHOH groups follows immediately as being R^*,R^* in the active or resolvable species and R^*,S^* in the meso or nonresolvable species. The relative configuration of the tetroses shown in Figure 3.18 can then be established by oxidation (HNO_3) to the corresponding tartaric acids: meso from erythrose, (+) or (–) from threose.

The absolute configuration of (–)-threose ($2S,3R$, or D) can thus also be deduced since its oxidation leads to (–)-tartaric acid, the enantiomer of the acid found to be R,R by the Bijvoet method (cf. Figs. 5.18 and 3.20). Obviously, the absolute configuration of (–)-erythrose cannot be determined similarly, since its oxidation gives the achiral

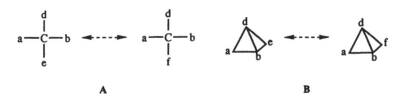

Figure 5.30. Principle of chemical interconversion not affecting bonds to the chiral atom.

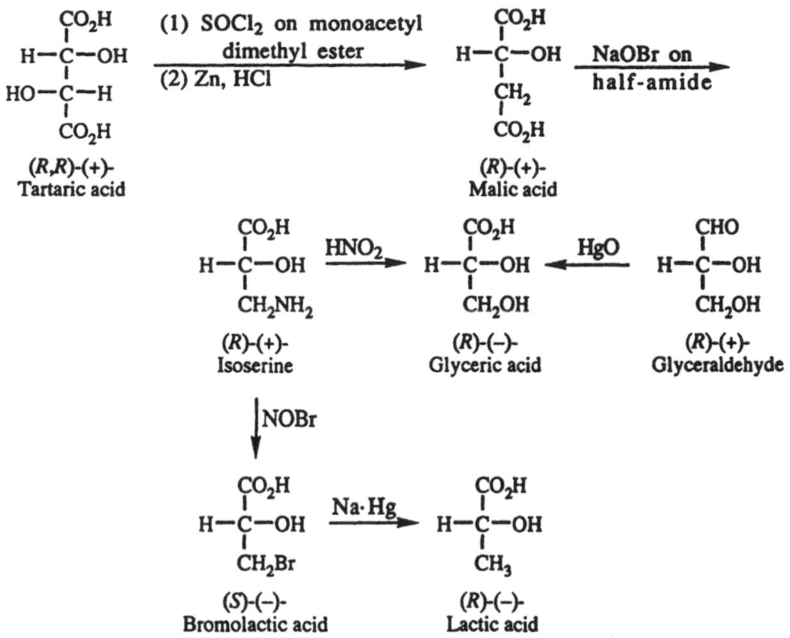

CO₂H
|
H—C—OH
|
HO—C—H
|
CO₂H

(R,R)-(+)-
Tartaric acid

(1) SOCl₂ on monoacetyl
    dimethyl ester
⟶
(2) Zn, HCl

CO₂H
|
H—C—OH
|
CH₂
|
CO₂H

(R)-(+)-
Malic acid

NaOBr on
half-amide ⟶

CO₂H
|
H—C—OH
|
CH₂NH₂

(R)-(+)-
Isoserine

HNO₂ ⟶

CO₂H
|
H—C—OH
|
CH₂OH

(R)-(-)-
Glyceric acid

⟵ HgO

CHO
|
H—C—OH
|
CH₂OH

(R)-(+)-
Glyceraldehyde

↓ NOBr

CO₂H
|
H—C—OH
|
CH₂Br

(S)-(-)-
Bromolactic acid

Na·Hg ⟶

CO₂H
|
H—C—OH
|
CH₃

(R)-(-)-
Lactic acid

Figure 5.31. Application of chemical interconversion not affecting the bond to the chiral atom.

meso acid, but it can be correlated with (−)-threose through their common origin in the homologation (addition of a carbon to the chain by well-established synthetic procedures) of (R)-(+)-glyceraldehyde (cf. Fig. 3.17).

Similar principles can be applied to the trihydroxyglutaric acids (Fig. 3.21): The chiral acid must be $2R^*,4R^*$. Referring to Figure 3.17, one notes that arabinose and lyxose will be oxidized to the chiral acid, whereas ribose and xylose give rise to one or the other of the meso acids, which cannot be distinguished by symmetry arguments alone.

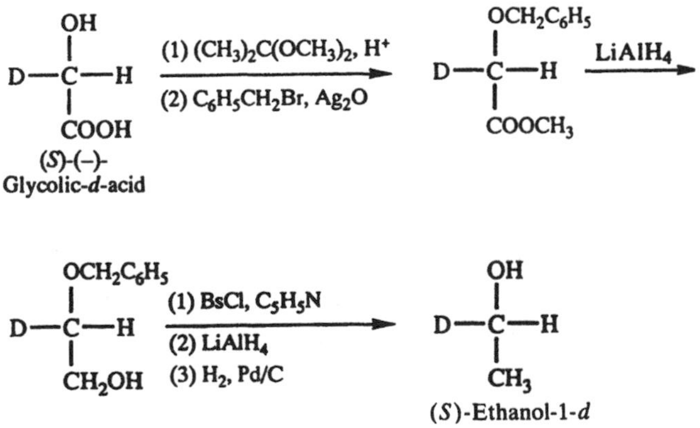

OH
|
D—C—H
|
COOH

(S)-(-)-
Glycolic-d-acid

(1) (CH₃)₂C(OCH₃)₂, H⁺
⟶
(2) C₆H₅CH₂Br, Ag₂O

OCH₂C₆H₅
|
D—C—H
|
COOCH₃

LiAlH₄ ⟶

OCH₂C₆H₅
|
D—C—H
|
CH₂OH

(1) BsCl, C₅H₅N
⟶
(2) LiAlH₄
(3) H₂, Pd/C

OH
|
D—C—H
|
CH₃

(S)-Ethanol-1-d

Figure 5.32. Correlation of configuration of deuterated molecules. BsCl represents p-bromo-benzenesulfonyl chloride.

Arabinose and lyxose can be distinguished in that degradation (i.e., removal of the aldehyde carbon and conversion of the adjacent carbon to an aldehyde group) of arabinose gives erythrose, whereas lyxose similarly gives rise to threose (Fig. 3.17); these two-carbon sugars are distinguished as explained above. To complete the assignment of relative configuration of glucose [the absolute configuration of (+)-glucose was, of course, not known in Fischer's time but was arbitrarily assumed to be D], it was synthesized by homologation of arabinose; this gave rise to a mixture of glucose and mannose (Fig. 3.17). These two hexoses, in turn, were distinguished by symmetry arguments applied to the corresponding dicarboxylic acids (Fig. 5.33): The diacid **A** obtained from mannose has C_2 symmetry but **B**, which is derived from glucose, belongs to the C_1 point group. Both acids are chiral but can be distinguished because **A** can *only* be obtained by oxidation of D-mannose, whereas **B** can be obtained *not only* by oxidation of D-glucose *but also* by oxidation of L-gulose (Fig. 5.33). Today **A** and **B** can be more simply distinguished by ^{13}C NMR: **A**, having a C_2 axis, displays only three signals whereas **B** is expected to display six (see also Chapter 8).

In summary, symmetry properties may be utilized to distinguish diastereomers that have C_2, C_s, or higher symmetry from those that lack axes or planes of symmetry (for a more detailed discussion, see Chapter 8). The presence of symmetry planes leads to lack of chirality and resolvability; axes lead to a reduction in the number of possible derivatives (such as esters, see above); both axes and planes lead to degeneracies in ^{1}H and ^{13}C NMR spectra. Compounds possessing C_2 axes of

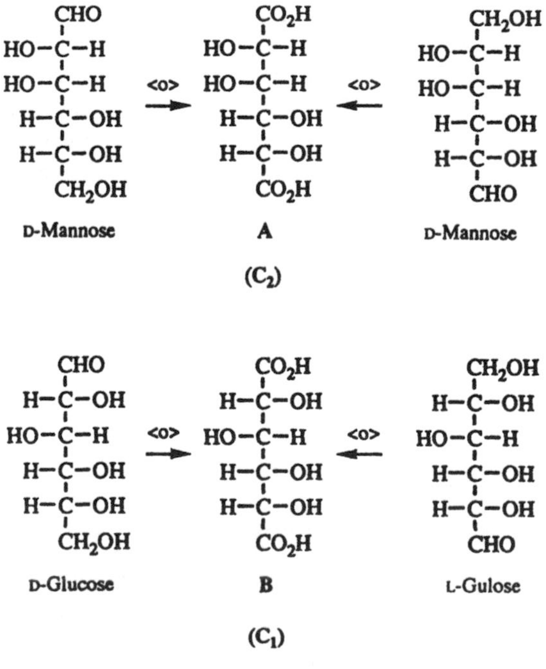

Figure 5.33. Hexaric acids from mannose and glucose.

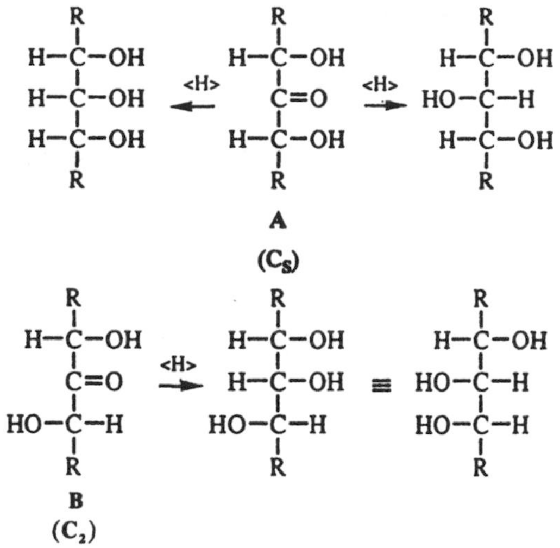

Figure 5.34. Differentiation between isomers of C_s and C_2 symmetry.

symmetry cannot usually be distinguished by NMR spectroscopy from those possessing symmetry planes (see also Chapter 8) but sometimes other means of distinction are available. Thus (Fig. 5.34) ketone **A** of C_s symmetry gives rise to two triols on reduction, whereas **B** (C_2) can give rise to only one because the two faces of the carbonyl group are equivalent (homotopic) (Chapter 8).

d. Correlation Via Compounds with Chiral Centers of Two Types

The term "confrontation analysis" has been coined by Brewster[32] for this type of analysis. It involves comparing one type of chiral center (say, RCabR') with another (say, RCxR'R", where the R's represent alkyl groups and a,b,x represent hydrogen or heteroatoms or groups) and determining their relative configuration after building the two chiral centers into one and the same compound (confronting them). Then, if both centers can be carved out without loss of chirality, and the configuration of one of them (say, RCabR') is known, that of the other (say, RCxR'R") can be deduced. We have already seen this method in operation in cases where X-ray structure analysis was used to correlate the relative configuration of two chiral centers (cf. Fig. 5.29). However, other correlation methods may, in principle, be used. A number of examples have been presented.[31,32]

The four methods discussed so far are compelling: If they are properly carried out, and if a meaningful result is obtained, the answer is not in doubt. In contrast, the four remaining methods do not have the same certitude: Since they depend on chemically less well established principles, they yield probable (sometimes highly probable) but not compelling answers (e.g., see ref. 41 for the disproof of an earlier correlation). It must be pointed out that, since a random guess at configuration has a 50% chance of being correct, even an unreliable method can repeatedly give the correct result.

e. The Method of Quasi-Racemates

This method, introduced by Timmermans,[42] based on earlier observations by Centnerszwer[43], has been used most extensively by Fredga,[44] who has reviewed it in detail (see also ref. 32). It rests on the phase behavior of conglomerates and racemic compounds (Chapters 6 and 7).

When two *chemically very similar* chiral substances, such as (+)-chlorosuccinic and (−)-bromosuccinic acid, which are heterofacial (see above), are mixed, a phase behavior similar to that seen in racemic compounds may be observed; that is, a molecular compound in the solid state *may* be formed.[43] Such a compound is called a *quasi-racemate* and shows typical compound behavior in the phase diagram. In contrast, the homofacial (+)-chlorosuccinic acid and (+)-bromosuccinic acid will show the eutectic behavior similar to that of a conglomerate. Figure 5.35 shows melting point phase diagrams for a pair of this type.

The quasi-racemate method has been little used in recent years (but see refs. 45 and 46); it is, of course, limited to crystalline solids and many other procedures applicable to both liquids and solids are now available.

f. Chemical Correlations Affecting Bonds to a Chiral Atom in a "Known" Way (For an overview, see ref. 32.)

Nucleophilic Displacement Reactions. In principle, a transformation Cabde + f → Cabdf + e (where e and f are ligands *directly* attached to the chiral carbon atom or other chiral center) could be used to establish the absolute configuration of Cabdf given that of Cabde, provided the stereochemical course of the replacement of e by f is *reliably* known. Thus, if the replacement of e by f proceeds homofacially (i.e., with *retention* of configuration, see also p. 91), Cabde and Cabdf will be *homofacial* (i.e., have the *same* configuration), whereas they will be *heterofacial* (i.e. have *opposite* configuration) if the reaction proceeds heterofacially (with *inversion* of configuration). Indeed these may be considered to be the definitions of "retention" and

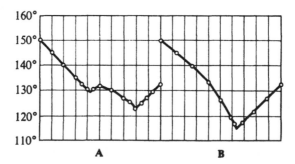

Figure 5.35. Phase diagrams for examples of mixtures of related heterofacial (**A**) and homofacial (**B**) compounds. Mixture **A** is shown to form a quasi-racemic compound. (From "The Svedberg" Memorial Volum, Almquist & Wiksells, Uppsala, Sweden, 1944. By permission of the publishers.)

"inversion." The problem, which will be addressed in this section, is to *know* whether a given reaction proceeds with retention or inversion and also to assess the reliability of that knowledge.

The terms *homofacial* and *heterofacial* were suggested by Ruch[47]; these terms require that three of the four ligands at the chiral centers of the molecules compared be identical (cf. Fig. 5.30). These terms must not be confused with the terms *homochiral* and *heterochiral* (p. 66), which imply that chiral molecules have either the same (homo) or the opposite (hetero) sense of chirality and strictly require comparison of isometric molecules (see p. 41), although Ruch[2b] used the terms in a slightly broader sense (e.g., all right-shoes are homochiral). It is common to say that homofacial molecules *have the same configuration* whereas heterofacial molecules *have opposite configurations*.

The primitive notion that all displacement reactions proceed homofacially was dispelled by Walden,[48] who, at the end of the nineteenth century, performed the sequence of reactions shown in Figure 5.36. The enantiomeric (–)-chlorosuccinic and (+)-chlorosuccinic acids obviously have opposite configurations; thus if the AgOH reaction proceeds homofacially with retention, the KOH and PCl$_5$ reactions must proceed heterofacially with inversion. Conversely, if the AgOH reaction involves inversion, the other two reactions must proceed with retention. Unfortunately, it was not known in Walden's time which of the two possibilities was the correct one and it was only a quarter of a century later[49] that the stereochemical course of *any* chemical reaction proceeding at a stereogenic center was elucidated, as shown in Figure 5.37.

Of the four reactions involved in this correlation, three (acetylation, *p*-toluenesulfinylation, and oxidation of the sulfinate to the sulfonate) do not involve breaking of bonds to the chiral center and, therefore, must proceed homofacially (with retention of configuration). It follows that (–)-2-octyl acetate and (–)-2-octyl *p*-toluenesulfonate (tosylate) have the same configuration (i.e., are homofacial) and that the reaction of the latter with acetate to give (+)-2-octyl acetate therefore proceeds with inversion of configuration (heterofacially; indicated by a looped arrow in Fig. 5.37). The conclusion that nucleophilic displacement reactions of the type shown in Figure 5.37 proceed with inversion of configuration was later extended, by analogy,

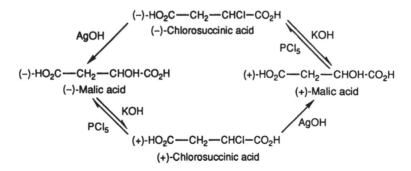

Figure 5.36. Walden inversion.

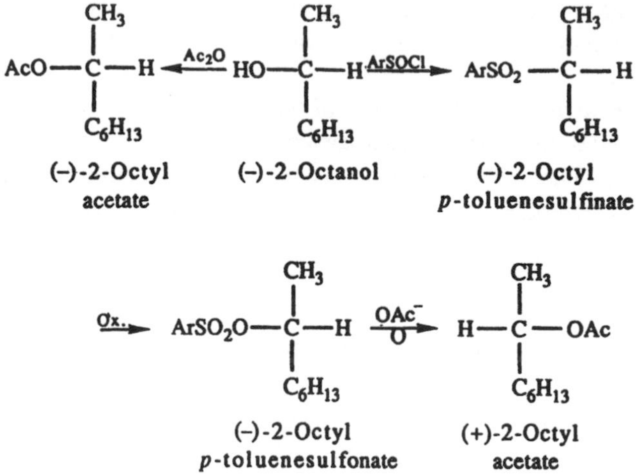

$$\underset{\substack{\text{(−)-2-Octyl}\\\text{acetate}}}{\text{AcO}-\overset{\overset{\displaystyle CH_3}{|}}{\underset{\underset{\displaystyle C_6H_{13}}{|}}{C}}-H} \xleftarrow{\text{Ac}_2\text{O}} \underset{\substack{\text{(−)-2-Octanol}}}{\text{HO}-\overset{\overset{\displaystyle CH_3}{|}}{\underset{\underset{\displaystyle C_6H_{13}}{|}}{C}}-H} \xrightarrow{\text{ArSOCl}} \underset{\substack{\text{(−)-2-Octyl}\\p\text{-toluenesulfinate}}}{\text{ArSO}_2-\overset{\overset{\displaystyle CH_3}{|}}{\underset{\underset{\displaystyle C_6H_{13}}{|}}{C}}-H}$$

$$\xrightarrow{\text{Ox.}} \underset{\substack{\text{(−)-2-Octyl}\\p\text{-toluenesulfonate}}}{\text{ArSO}_2\text{O}-\overset{\overset{\displaystyle CH_3}{|}}{\underset{\underset{\displaystyle C_6H_{13}}{|}}{C}}-H} \xrightarrow{\underset{\displaystyle O}{\text{OAc}^-}} \underset{\substack{\text{(+)-2-Octyl}\\\text{acetate}}}{\text{H}-\overset{\overset{\displaystyle CH_3}{|}}{\underset{\underset{\displaystyle C_6H_{13}}{|}}{C}}-\text{OAc}}$$

Figure 5.37. Inversion in the reaction of (−)-2-octyl p-toluenesulfonate with acetate.

to the reaction of 2-octyl tosylate with halide ions[50] and the configurations of 2-halooctanes were thus assigned.

However, this finding is clearly not the whole story. In Walden's scheme (Fig. 5.37) some reactions are evidently heterofacial (i.e., they involve inversion), whereas others are homofacial (i.e., they proceed with retention). If the nucleophilic displacement reaction is to be used as a tool for assigning configuration, one must know which reaction course is followed in a specific instance. This problem was addressed by the group of Hughes and Ingold in further studies of the mechanism of nucleophilic displacement reactions. It was concluded that inversion occurs reliably only in bimolecular nucleophilic displacement (S_N2) reactions[51]; this postulate is sometimes called the S_N2 rule. Thus to assure that a reaction of the type R–X + Y⁻ (or R–X⁺ + Y⁻, or R–X + Y, or R–X⁺ Y) → R–Y + X⁻ (or R–Y + X, or R–Y⁺ + X⁻, or R–Y⁺ + X) proceeds heterofacially, one must also study its kinetics; inversion of configuration is assured only if the reaction is bimolecular (generally second order in substrate and nucleophile). Figure 5.38 shows the overall picture of the S_N2 reaction,

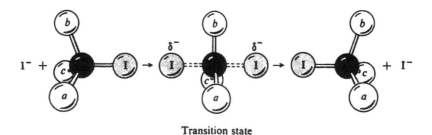

Transition state

Figure 5.38. Mechanism of the S_N2 reaction.

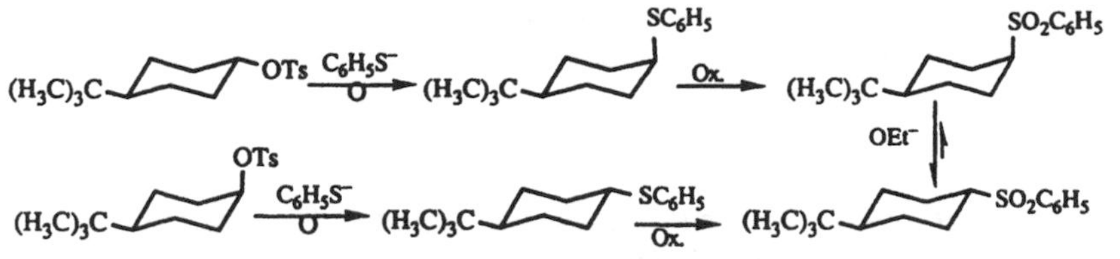

Figure 5.39. The S_N2 reaction of *trans-* and *cis-4-tert-*butylcyclohexyl *p*-toluenesufonates.

in which the incoming nucleophile engages the σ^* orbital of the atom that is being inverted.[52,53] (For an extensive discussion of the detailed course of the S_N2 reaction, both in solution and in the gas phase, see ref. 54.)

It might appear from the above that the S_N2 rule rests on a very limited experimental basis. In fact, however, quite apart from the theoretical underpinnings,[53] there are a number of other instances where the configuration of the starting material and that of the product are independently known to be opposite, that is, where starting material and product are heterofacially related. In some of these cases the configurations have been determined by independent methods discussed elsewhere in this chapter. In others, the substrate had two (or more) stereogenic centers of which only one was inverted in the nucleophilic displacement. In such cases correlation of the relative configuration of the pertinent stereogenic center with that of other stereogenic centers in the same molecule (whose configuration is not affected) will document the occurrence of inversion. An example is the reaction (shown to be second-order kinetically) of *trans-4-tert-*butylcyclohexyl *p*-toluenesulfonate with thiophenolate to give, as the major product, *cis-4-tert-*butylcyclohexyl phenyl sulfide[55] (Fig. 5.39). The cis configuration of the product was inferred from the S_N2 rule but was independently confirmed by oxidizing the sulfide to the corresponding sulfone and demonstrating that the presumed axial sulfone could be epimerized, by base, to its more stable equatorial isomer (Fig. 5.39); it was later corroborated by ^{1}H NMR spectroscopy.[56]

An application of the S_N2 rule to the configurational correlation of (+)-α-methylbenzyl chloride and (−)-α-methylbenzylamine via the azide is shown in Figure 5.40.[57] The kinetics of the azide displacement shows it to be bimolecular, so inversion occurs and the chloride and amine shown have opposite configurations.

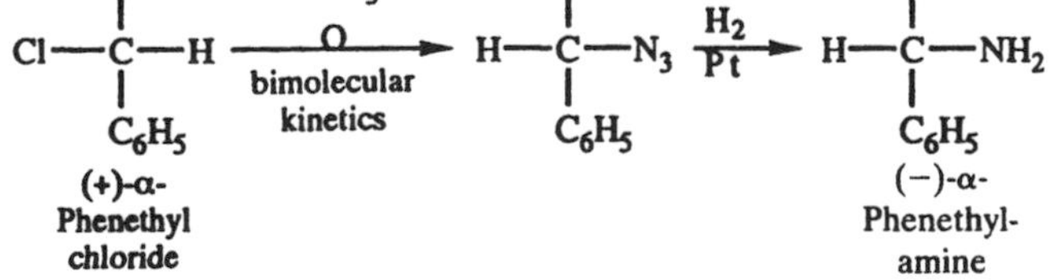

Figure 5.40. Correlation of configuration by application of the S_N2 rule.

Displacement by azide followed by reduction rather than displacement by amide or ammonia was chosen, because the azide reaction is cleaner and more easily studied kinetically.

Rearrangement Reactions at Saturated Carbon Atoms.   The   reactions   to   be discussed here are of the general type shown in Figure 5.41. A chiral carbon atom, zyxC–, migrates from carbon to a heteroatom X [in the case of the Stevens rearrangement (see below) the migration is from nitrogen to carbon]; thus if the configuration of the starting chiral center, zyxC*–C– is known, that of the product zyxC*–X, where X may be nitrogen or oxygen, can be inferred, provided the steric course of the migration (homofacial or heterofacial with respect to the migrating group –Cxyz) is known. Both on theoretical grounds and on the basis of many experiments it has been concluded that most reactions of this type proceed homofacially, that is, with retention of configuration at the migrating group. A number of examples, involving migration from carbon to nitrogen [Hofmann bromamide, Curtius, Schmidt, Lossen, and Beckmann (cf. p. 352) rearrangements], from carbon to oxygen (Baeyer–Villiger rearrangement), and the similar Stevens rearrangement (nitrogen to carbon) have been summarized elsewhere.[13,32] The steric course of these rearrangements was first elucidated in cases where the configuration of both the starting materials and products was known. The reactions were subsequently used to correlate configurations of products and starting materials where only one of these configurations was known (cf. ref. 58). A less than routine example[59] is shown in Figure 5.42; it provides one of three configurational correlations of the (–)-ethanol-1-*d* obtained by enzymatic reduction of CH_3CDO (see discussion on p. 86). The overall correlation involves three different principles: the confrontation method (see above), correlation of chiral centers by nucleophilic displacement, and the Baeyer–Villiger reaction. Deuteride reduction of (+)-2.3-epoxybutane gives levorotatory 2-butanol-3-*d*.[60] Since C(3), chiral by virtue of the presence of deuterium, contributes little to the rotation of the alcohol, it is safe to assume that the latter's configuration at C(2) is *R*, the same as that of (–)-2-butanol, which has been correlated with tartaric acid (cf. ref. 33a). The configuration of the starting epoxide is therefore unequivocally shown to be 2*R*,3*R*. (It cannot be 2*R*,3*S*, because that would constitute an inactive meso form.) Since opening of aliphatic secondary epoxides with nucleophiles involves inversion of configuration,[61] the configuration of **A** at C(3) must thus be *S*. The carbinol C(2) and CHD C(3) chiral centers in **A** are thus "confronted" as shown in Figure 5.42. The carbinol center is subsequently destroyed and the CHD center incorporated in ethanol-1-*d* by Baeyer–Villiger reaction and saponification as shown. An additional feature of the correlation is that the final product, (*S*)-ethanol-1-*d*, was characterized not by optical rotation (which is small and

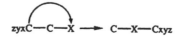

Figure 5.41. Intramolecular 1,2-rearrangements.

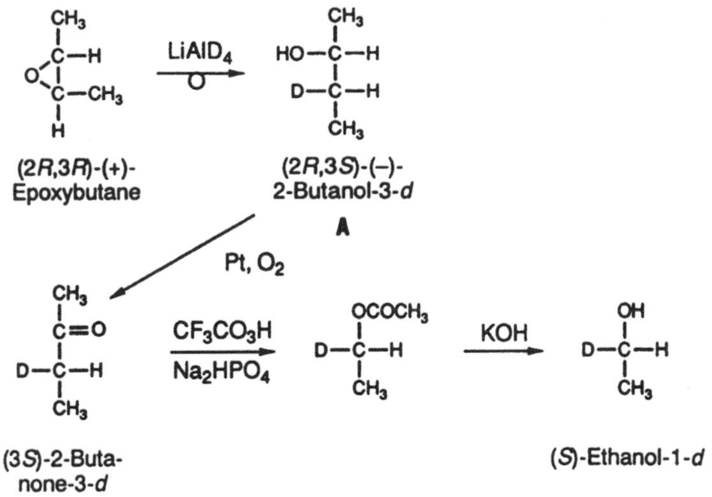

Figure 5.42. Configuration of (S)-$(-)$-ethanol-1-d by Baeyer–Villiger reaction.

therefore subject to falsification by small amounts of chiral impurities), but by enzymatic oxidation that resulted largely in retention of the deuterium atom in the acetaldehyde (CH_3CDO) so formed.[59] This behavior is characteristic of levoratory ethanol-1-d (Fig. 8.35), which is thus shown to be S.

g. Correlation by Stereoselective Synthesis of "Known" Stereochemical Course

A stereoselective synthesis involves introducing a stereogenic center in a molecule in such a way that one stereoisomer of the product is formed in preference to the other or others. The synthesis may be either enantioselective or diastereoselective; in the former case one of two enantiomers is formed preferentially, in the latter case one of two or more diastereomers is formed. Both types will be discussed here, though in the case of diastereoselective syntheses we shall confine ourselves to cases where the newly introduced stereogenic center is chiral. (Cases involving achiral stereogenic centers, such as the reduction of 4-*tert*-butylcyclohexanone to *cis*- or *trans*-4-*tert*-butylcyclohexanol, will be taken up in Chapter 11.) In a typical case, reaction of an achiral precursor with a chiral reagent (or in the presence of a chiral auxiliary substance, sometimes called a *chiral adjuvant*) will give a chiral product. If the configuration of the reagent (or chiral auxiliary) is known, and if there is a unique configurational correlation between the configuration of the reagent or auxiliary and that of the newly created chiral center, then the configuration of the latter is correlated with that of the former. The main problem with this type of correlation is that the steric course of stereoselective "asymmetric" syntheses is even more uncertain than that of reactions at chiral centers discussed in Section 5-5.f. Although an asymmetric synthesis may have taken the same steric course in n known cases (where n is usually a relatively small number), it cannot be guaranteed to take the same course in the (n +

1)st case, perhaps because some steric factor in the substrate alters the course of the reaction. Therefore correlations of this type can never be considered entirely secure, especially in the (frequent) absence of detailed mechanistic information. Nevertheless, we shall present here four examples of correlations of this kind, the first two of which have been frequently used in assigning configuration: Horeau's, Prelog's, Cram's, and Sharpless' rule.

Horeau's Rule.[62a]   This topic has been reviewed by Brewster[32] and by Horeau.[62b] If an optically active secondary alcohol (RR'CHOH) is esterified with a racemic acid (R*–CO$_2$H) the transition states [(–)-RR'CHOH.(–)-R*-CO$_2$H]‡ and [(–)-RR'CHOH. (+)-R*-CO$_2$H]‡ are diastereomeric and thus unequal in energy. Therefore the activation energies and hence reaction rates of esterification with the (+) and (–) acids will be unequal (the energies of the initial states are the same, since the alcohol is one and the same in the two combinations and the acids are enantiomeric, and therefore equienergetic). Thus, if an excess of acid is used, one enantiomer of the acid will be preferentially incorporated in the ester and the other will be predominantly left behind (kinetic resolution, see Chapter 7).

Horeau[62a] introduced inactive 2-phenylbutyric anhydride (or sometimes the corresponding chloride) as the reagent for an optically active alcohol of unknown configuration. It matters not that the anhydride is a mixture of *RR/SS* (racemic) and *RS* (meso) species: One of the chiral C$_6$H$_5$CH(C$_2$H$_5$)CO– moieties will react faster with the alcohol than the enantiomeric one. The residual acid (after hydrolysis of the left-over anhydride) is submitted to polarimetry: If it is levorotatory, the configuration of the optically active secondary alcohol RR'CHOH is such that, in its Fischer projection, the hydroxy group is down, the hydrogen atom up, and the larger of the two remaining substituents is on the right (Fig. 5.43, **A**). The reverse obviously applies if the recovered acid is dextrorotatory (Fig. 5.43, **B**). Evidently the (–) acid reacts more slowly with alcohol **A** and vice versa for the (+) acid.

The method and its limitations have been discussed in detail in a review by Horeau.[62b] It is possible to start with racemic alcohol and use optically active 2-phenylbutyric anhydride to lead, at the same time, to kinetic resolution (partial) of the alcohol and to determination of the configuration of the alcohol so resolved.[63a] Thus if (S)-(+)-2-phenylbutyric anhydride is used with an excess of racemic RR'CHOH, the left-over alcohol (whose sign of rotation needs to be determined) will have configuration **B** in Figure 5.43.

Modifications of Horeau's procedure[63b,64] allow determination of the configuration of optically active alcohols on a microscale, where measurement of the optical

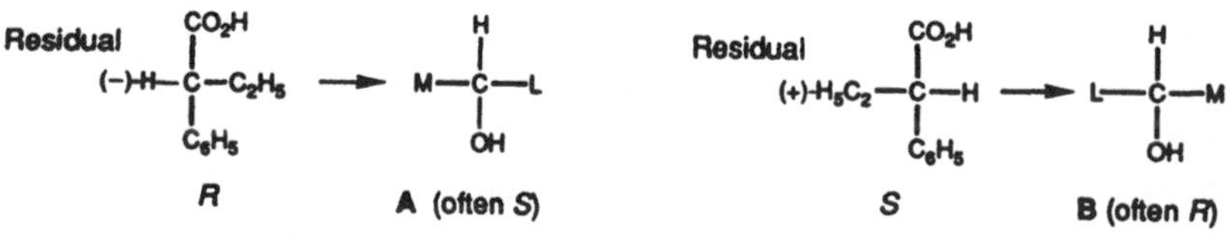

Figure 5.43. Horeau's rule.

rotation of the recovered acid is not feasible. Thus the Horeau method can be used for microscale determinations provided a highly sensitive method (GC, HPLC, MS, or CD) is available to determine the composition of the diastereomeric esters formed; the detection method must be chosen so as to be independent of the chemical nature of the alcohol moiety of the esters.

Prelog's Rule.[65]   This rule[66] relates to the course of asymmetric synthesis when a Grignard reagent is added to an α-ketoester (frequently a pyruvate) of a chiral secondary or tertiary alcohol, SMLC–OH (S, M, and L stand for the small, medium-sized, and large substituents, respectively; S may be H). Schematically, the reaction generally follows the stereochemical course outlined in Figure 5.44.[65] If the ketoester is in the conformation in which the two carbonyl groups are antiperiplanar and in which the L group occupies the same plane as the two carbonyl groups and the alkyl oxygen, the Grignard reagent (RMgX) will approach the keto carbonyl function from the side of the S group of the two remaining alcohol substituents. This rule correlates the configuration of the α-hydroxyester moiety (and therefore of the α-hydroxyacid obtained by hydrolysis thereof, Fig. 5.44) with that of the starting alcohol SMLCOH incorporated in the α-ketoester.

It will be noted that if the alcohol is *S* (on the above premise of CIP sequence), the Grignard reagent will approach the keto carbonyl from the *Re* face (the face above the paper in Fig. 5.44, cf. Chapter 8). Since *S* and *Re* are "opposite," the reaction is of the "*ul*" (unlike) rather than of the "*lk*" (like) type, according to the notation introduced by Seebach and Prelog.[25] That the reaction is *ul* holds, of course, even if the starting alcohol is of the *R* configuration (as long as the CIP sequence is as indicated above); in that case the picture in Figure 5.44 must be mirrored and the reagent will attack from the rear or *Si* face of the carbonyl.

Prelog's rule has been used to assign the configuration of numerous alcohols, generally secondary ones (S = H).[66] In most cases, the addition studied is that of phenylmagnesium bromide to the pyruvate of the alcohol under study; this produces an atrolactate. The atrolactic acid obtained after saponification is identified polarimetrically as (S)-(+)- or (R)-(−)-C$_6$H$_5$(CH$_3$)COHCO$_2$H and the configuration of the alcohol is deduced therefrom (*S* acid → *R* alcohol, and vice versa). Some precautions need to be taken: It is important that the saponification of the atrolactate be quantitative, otherwise the results may be falsified by kinetic resolution (Chapter 7). Thus, if the predominant ester product is that of the (+) acid, but the (diastereomeric) ester of the (−) acid is saponified more rapidly, it might appear, if saponification is incomplete, that the (−) acid predominated in the product; a false

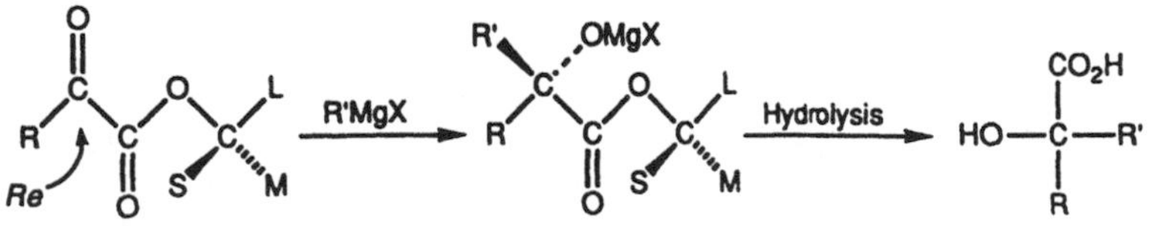

Figure 5.44. Prelog's rule.

conclusion would thus be drawn. This problem is avoided by saponifying the ester completely.

Cram's Rule and Sharpless' Rule.   Cram's rule[67-69] refers to the diastereomer ratio formed in the addition of organometallic reagents to chiral ketones of the type R′COCHXR, whereas Sharpless' rule[70] deals with the ratio of enantiomers formed in epoxidation of RR′C=CR″CH$_2$OH with *tert*-butyl hydroperoxide and titanium tetraalkoxide in the presence of an *R,R* or *S,S* dialkyl tartrate.

h. Chiroptical, Spectroscopic, and Other Physical Methods

These methods will be discussed in other chapters (see especially Chapter 13) as appropriate.

5-6. CONCLUSION: NETWORK ARGUMENTS

In concluding this chapter, it must be stressed that many chiral compounds have been correlated with others by multiple pathways (cf. ref. 33). Changing the configurational assignment of one compound in such a network means changing the configuration assignment of all the others correlated with it (unless it can also be shown that *all* the individual correlations with the compound under scrutiny are wrong). Thus it is incumbent on anyone who claims a change in configurational assignment of a given compound to investigate all the other changes that would entail and rationalize or explain all the concomitant changes. Of course, the configurations of some compounds are more firmly anchored than those of others because they are more tightly tied into the network by multiple correlations, whereas others may have been correlated by only a single method and possibly not a very reliable one. This point must be kept in mind when configurational assignments are discussed.

REFERENCES

1. Moss, G. P. (for IUPAC). *Pure Appl. Chem.* **1996**, *68*, 2193.

2. (a) Ruch, E. *Theor. Chim. Acta* **1968**, *11*, 183, 462; (b) *Acc. Chem. Res.* **1972**, *5*, 49.

3. Kelvin, Lord (W. Thomson). *Baltimore Lectures on Molecular Dynamics and Wave Theory of Light*, C. J. Clay, London, 1904.

4. Bijvoet, J. M., Peerdeman, A. F., and van Bommel, A. J. *Nature* **1951**, *168*, 271; Bijvoet, J. M. *Endeavour* **1955**, *14*, 71.

5. Lowry, T. M. *Optical Rotatory Power*, Longmans Green & Co., New York, 1935, Chap. 7.

6. (a) Cahn, R. S. Ingold, C. K., and Prelog, V. *Experientia* **1956**, *12*, 81; (b) Cahn, R. S., Ingold, Sir C., and Prelog, V. *Angew. Chem Int. Ed. Engl.* **1966**, *5*, 385.

7. Prelog, V. and Helmchen, G. *Angew. Chem. Int. Ed. Engl.* **1982**, *21*, 567.

8. Cahn, R. S. and Ingold, C. K. *J. Chem. Soc.* **1951**, 612.

9. Eliel, E. L. *J. Chem. Educ.* **1985**, *62*, 223.

10. Cross, L. C. and Klyne, W. *Pure Appl. Chem.* **1976**, *45,* 13; see also *J. Org. Chem.* **1970**, *35*, 2849.

11. IUPAC. *Nomenclature of Organic Chemistry*, Pergamon Press, New York, 1973, p. 473.

12. Dodziuk, H. and Mirowicz, M. *Tetrahedron Asym.* **1990**, *1*, 171.

13. Eliel, E. L. *Stereochemistry of Carbon Compounds*, McGraw-Hill, New York, 1962, pp. 22, 88–92, 119–121.

14. Dunitz, J. D. *X-Ray Analysis and Structure of Organic Molecules,* Cornell University Press, Ithaca, NY, 1972.

15. Coster, D., Knol, K. S., and Prins, J. A. *Z. Phys.* **1930**, *63*, 345.

16. Trommel, J. and Bijvoet, J. M. *Acta Crystallogr.* **1954**, *7*, 703.

17. Kuhn, W. *Z. Phys. Chem.* **1935**, *B31*, 23; *Z. Elektrochem.* **1952**, *56*, 506.

18. Kirkwood, J. G. *J. Chem. Phys.* **1937**, *5*, 479.

19. Wood, W. W., Fickett, W., and Kirkwood, J. G. *J. Chem. Phys.* **1952**, *20*, 561.

20. Berova, N., Nakanishi, K., and Woody, R., eds. *Circular Dichorism: Principles and Applications*, Second Ed., Wiley-VCH, New York, 2000.

21. Addadi, L., Berkovitch-Yellin, Z., Weissbuch, I., Lahav, M., and Leiserowitz, L. *Top. Stereochem.* **1986**, *16*, 1. Addadi, L., Berkovitch-Yellin, Z., Weissbuch, I., van Mil, J., Shimon, L. J. W., Lahav, M., and Leiserowitz, L. *Angew. Chem. Int. Ed. Engl.* **1985**, *24*, 466.

22. Newman, M. S. *J. Chem. Educ.* **1955**, *32*, 344.

23. Bischoff, C. A. *Ber. Dtsch. Chem. Ges.* **1891**, *24*, 1085.

24. Hermans, P. H. *Z. Phys. Chem.* **1924**, *113*, 337.

25. Seebach, D. and Prelog, V. *Angew. Chem. Int. Ed. Engl.* **1982**, *21*, 654.

26. Gielen, M. *J. Chem. Ed.* **1977**, *54*, 673.

27. Cram, D. J. *J. Am. Chem. Soc.* **1952**, *74*, 2152.

28. Masamune, S., Ali, Sk. A., Snitman, D. L., and Garvey, D. S. *Angew. Chem. Int. Ed. Engl.* **1980**, *19*, 557.

29. Carey, F. A. and Kuehne, M. E. *J. Org. Chem.* **1982**, *47*, 3811.

30. Mislow, K. and Siegel, J. *J. Am. Chem. Soc.* **1984**, *106*, 3319.

31. Klyne, W. and Scopes, P. M. "Stereochemical Correlations," in Aylett, B. J. and Harris, M. M., eds., *Progress in Stereochemistry*, Vol. 4, Butterworths, London, 1969.

32. Brewster, J. H. "Assignment of Stereochemical Configuration by Chemical Methods," in Bentley, K. W. and Kirby, G. W., eds. *Elucidation of Organic Structures by Physical and Chemical Methods*, Vol. IV, Part III, Wiley-Interscience, New York, 1972, pp. 1–249.

33. (a) Klyne, W. and Buckingham, J. *Atlas of Stereochemistry*, Vols. I and II, 2nd ed., Chapman & Hall, London, 1978. (b) Buckingham, J. and Hill, R. A. *Atlas of Stereochemistry—Supplement*, Chapman & Hill, New York, 1986.

34. Jacques, J., Gros, C., and Bourcier, S. "Absolute Configuration of 6000 Selected Compounds with One Asymmetric Carbon Atom," in Kagan, H. B., ed., *Stereochemistry, Fundamentals and Methods*, Vol. 4, Georg Thieme Publishers, Stuttgart, Germany, 1977.

35. Hine, R. and Rogers, D. *Chem. Ind.* **1956**, 1428.

36. Mathieson, A. McL. *Acta Crystallogr.* **1956**, *9*, 317.

37. Carter, O. L., McPhail, A. T., and Sim, G. A. *J. Chem. Soc. A* **1967**, 365.

38. Wilen S. H., Qi, J. Z., and Willard, P. G. *J. Org. Chem.* **1991**, *56*, 485.

39. Weber, H. Ph. D. Dissertation No. 3591, Eidgenössische Technische Hochschule, Zurich, Switzerland, 1965.

40. Arigoni, D. and Eliel, E. L. *Top. Stereochem.* **1968**, *4*, 160.

41. Anderson, R. C. and Fraser-Reid, B. *J. Org. Chem.* **1985**, *50*, 4781.

42. Timmermans, J. *Recl. Trav. Chim. Pays-Bas* **1928**, *48*, 890.

43. Centnerszwer, M. *Z. Phys. Chem.* **1899**, *29*, 715.

44. Fredga, A. *Tetrahedron* **1960**, *8*, 126; *Bull. Soc. Chim. Fr.* **1973**, 173.

45. Patterson, K. *Ark. Kemi* **1954**, *7*, 347.

46. Gronowitz, S. and Larson, S. *Ark Kemi* **1955**, *8*, 567. Gronowitz, S. *ibid.*, **1957**, *11*, 361.

47. Ruch, E. personal communication to Prelog, V. and Helmchen, G., *Helv. Chim. Acta* **1972**, *55*, 2581.

48. Walden, P. *Ber. Dtsch. Chem. Ges.* **1896**, *29*, 133; *ibid.* **1897**, *30*, 3146.

49. Phillips, H. *J. Chem. Soc.* **1923**, *123*, 44; *ibid.* **1925**, *127*, 2552.

50. Houssa, A. J. H., Kenyon, J., and Phillips, H. *J. Chem. Soc.* **1929**, 1700.

51. Ingold, C. K. *Structure and Mechanism in Organic Chemistry*, 2nd ed., Cornell University Press, Ithaca, NY, Chap. VII.

52. Meer, N. and Polanyi, M. *Z. Phys. Chem.* **1932**, *B19*, 164.

53. Olson, A. R. *J. Chem. Phys.* **1933**, *1*, 418.

54. Shaik, S. S., Schlegel, H. B., and Wolfe, S. *Theoretical Aspects of Physical Organic Chemistry. The S_N2 Mechanism*, Wiley, New York, 1992.

55. Eliel, E. L. and Ro, R. S. *J. Am. Chem. Soc.* **1957**, *79*, 5995.

56. Eliel, E. L. and Gianni M. H. *Tetrahedron Lett.* **1962**, 97.

57. Brewster, P., Hiron, F., Hughes, E. D., Ingold, C. K., and Rao, P. A. D. S. *Nature* **1950**, *166*, 179.

58. Fiaud, J. C. and Kagan, H. B. "Determination of Stereochemistry by Chemical Correlation Methods," in H. B. Kagan, ed., *Stereochemistry, Fundamentals and Methods,* Vol. 3, Georg Thieme Publishers, Stuttgart, Germany, 1977, p. 1.

59. Weber, H., Seibl, J., and Arigoni D. *Helv. Chim. Acta* **1966**, *49*, 741.

60. Skell, P. S., Allen, R. G., and Helmkamp, G. K. *J. Am. Chem. Soc.* **1960**, *82*, 410.

61. Wohl, R. A. *Chimia* **1974**, *28*, 1.

62. (a) Horeau, A. *Tetrahedron Lett.* **1961**, 506. (b) Horeau, A. "Determination of Configuration of Secondary Alcohols by Partial Resolution," in Kagan, H. B., ed., *Stereochemistry, Fundamentals and Methods*, Vol. 3, Georg Thieme Publishers, Stuttgart, Germany, 1977, Chap. 3.

63. Weidman, R. and Horeau, A. (a) *Bull. Soc. Chim. Fr.* **1967**, 117; (b) *Tetrahedron Lett.* **1973**, 2979.

64. Horeau, A. and Nouaille, A. *Tetrahedron Lett.* **1990**, *31*, 2707.

65. Prelog, V. *Helv. Chim. Acta* **1953**, *36*, 308.

66. Fiaud, J. C. "Prelog's Methods" in Kagan, H. B., ed. *Stereochemistry, Fundamentals and Methods*, Vol. 3, Georg Thieme, Publishers, Stuttgart, Germany, p. 19.

67. Cram, D. J. and Abd Elhafez, F. A. *J. Am. Chem. Soc.* **1952**, *74*, 5828. Cram, D. J. and Kopecky, K. R. *J. Am. Chem. Soc.* **1959**, *81*, 2748.

68. See also Morrison, J. D. and Mosher, H. S. *Asymmetric Organic Reactions*, Prentice-Hall, Englewood Cliffs, NJ, 1971; reprinted by American Chemical Society, Washington, DC, 1976, Chap. 3; Eliel, E. L. "Application of Cram's Rule: Addition of Achiral Nucleophiles to Chiral Substrates," in J. D. Morrison, ed., *Asymmetric Synthesis*, Vol. 2, Academic Press, New York, 1983, pp. 125–155.

69. See also Mengel, A. and Reiser, O. *Chem. Rev.* **1999**, *99*, 1191.

70. Katsuki, T. and Sharpless, K. B. *J. Am. Chem. Soc.* **1980**, *102*, 5974; Finn, M. G. and Sharpless, K. B. "On the Mechanism of Asymmetric Epoxidation with Titanium-Tartrate Catalysts," in Morrison, J. D., ed., *Asymmetric Synthesis*, Academic Press, New York, 1985, pp. 247–308.

6

PROPERTIES OF STEREOISOMERS AND STEREOISOMER DISCRIMINATION

6-1. INTRODUCTION

This chapter will focus on the physical properties of enantiomer pairs and on methods for the determination of enantiomer composition. We begin by examining some of the ways in which the enantiomers of a chiral substance interact with one another. The properties of enantiomerically pure compounds are contrasted with those of the corresponding racemates, and also with those of unequal mixtures of the enantiomers. We shall see that the properties of chiral substances are affected by the proportion of the enantiomers in the mixture and by the state of the system. Knowledge of such properties and of their differences is essential to the design of efficient separation methods (Chapter 7). In addition, the study of these properties (and those of diastereomers) is useful in such diverse applications as the determination of enantiomeric and diastereomeric purity and the generation of enantiomeric bias in Nature.

6-2. STEREOISOMER DISCRIMINATION

The word racemate describes an equimolar mixture of a pair of enantiomers. An important question that will concern us through the first half of this chapter is whether such a mixture behaves ideally, that is, does this mixture behave in the same manner as the individual constituent enantiomers. Therefore, the question is whether homochiral and heterochiral interactions do or do not differ significantly. Once it was believed that, in the liquid and gaseous phases, these interactions do not differ, but we shall see that this is not true in general.

The terms homochiral and heterochiral, which describe the relatedness of chiral classes a compounds, were first used by William Thomson (Lord Kelvin) a century ago.[1–3] "Homochiral interactions," interactions between homochiral molecules (cf. Chapter 5), are defined as intermolecular nonbonded attractions or repulsions present in assemblies of molecules having like chirality sense. For the R enantiomers of a given substance and *in the absence of chemical reaction*, homochiral interactions might be represented as in Eq. 6.1:

$$R + R \rightleftharpoons R \cdots R \tag{6.1}$$

Homochiral interactions must be identical for either enantiomer. The phrase "heterochiral interactions" is a shorthand term for interactions between molecules of unlike chirality sense (heterochiral molecules), again in the absence of a chemical reaction; for example,

$$R + S \rightleftharpoons R \cdots S \tag{6.2}$$

The nonbonded interactions in question (Eqs. 6.1 and 6.2) comprise van der Waals and electrostatic interactions, hydrogen bonding, π-complex formation, and other forms of electron donation and acceptance that are readily reversible. The forces involve long-range dispersion (London forces) and have inductive and permanent multipolar components.[4] The nature and magnitude of these forces is partially dependent on the symmetry of the molecules (see Section 4-5). Although Eqs. 6.1 and 6.2 describe only dimeric interactions, the possible formation of higher aggregates is not excluded.

The surrounding medium, not unexpectedly, affects the interactions implicit in Eqs. 6.1 and 6.2; for example, an alcohol solvent often overwhelms, and thereby destroys, solute–solute hydrogen-bonding interactions and a high dielectric solvent diminishes dipolar interactions.[5]

Homochiral (Eq. 6.1) and heterochiral (Eq. 6.2) interactions among molecules of like constitution are unlikely ever to be exactly equal in magnitude ($\Delta G_{homo} \neq \Delta G_{hetero}$) because the two types of aggregates are *anisometric* (diastereomeric). The difference $\Delta\Delta G$ between homochiral and heterochiral interactions ($R \cdots R$ vs. $R \cdots S$) is responsible for measurable differences in physical properties exhibited by racemates on one hand, and by the corresponding enantiomerically pure compounds on the other. This difference is a manifestation of *enantiomer discrimination.*

Interactions between a given enantiomer of one compound and the two enantiomers of another species give rise to diastereomeric pairs, such as $R_I \cdots R_{II}$ and $R_I \cdots S$: the difference between these interactions is called *diastereomer discrimination.*[6,7] A comprehensive term for the two types of discrimination is *stereoisomer discrimination.*

Figure 6.1 is a scheme describing the relationships between enantiomeric interactions and the diastereomeric interactions that exist among the enantiomers of two different compounds, **I** (R_I and S_I) and **II** (R_{II} and S_{II}). Only the horizontally related enantiomeric interactions are necessarily equal in magnitude (by symmetry) whether the systems behave ideally or not [if $M(X \cdots Y)$ represents the numerical magnitude of any physical property measured in a bulk sample, then $M(R_I \cdots R_{II}) = M(S_{II} \cdots S_I)$ and $M(R_{I'} \cdots R_{II}) - M(R_I \cdots S_{II}) = M(S_I \cdots S_I) - M(R_{II} \cdots S_I)$ and, of course, $M(R_I \cdots R_I) = M(S_I \cdots S_I)$].[7]

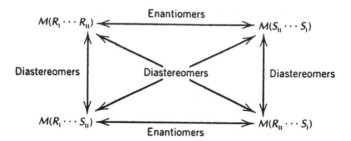

Figure 6.1. Schematic relationship of enantiomeric and diasteromeric interactions for two pairs of enantiomers. [Reprinted with permission from Stewart, M. V. and Arnett, E. M. *Top. Stereochem.* **1982,** *13,* 195. Copyright © 1982 John Wiley & Sons, Inc.]

Stereoisomer discrimination is strongly phase dependent. At one extreme, enantiomer discrimination in the solid state is responsible for the occurrence of well-defined racemic compounds having properties that are significantly different from those of enantiomerically pure samples of the same substance. Moreover, the physical properties of a given chiral substance in bulk evidently depend on the proportion of the enantiomers, that is, on the *enantiomer composition,* of the sample. At the other extreme, stereoisomer discrimination in the gaseous state (Section 6-4.n) is rarely observed. The situation in liquids and solutions is intermediate.

The magnitude of enantiomer discrimination in aqueous solution may be estimated from calorimetric data for the mixing of solutions of the enantiomers of a given chiral compound. For the tartaric acid enantiomers, the heat of mixing, $\Delta H^m = 0.48$ cal mol^{-1} (2.0 J mol^{-1}), that is, the homochiral combination is enthalpically preferred; for the threonine enantiomers: $\Delta H^m = -1.3$ cal mol^{-1} (−5.5 J mol^{-1}) at 25.6°C. The two substances exhibit opposite senses of homochiral versus heterochiral enthalpic preference.[8] For fenchone, dilution of one enantiomer with the other gives $\Delta H^m = -1.1$ cal mol^{-1} (−4.5 J mol^{-1}); the corresponding values for α-methylbenzylamine are $\Delta H^m = +1.7$ cal mol^{-1} (+7.3 J mol^{-1}) at 30°C. It appears that, for the amine, ΔH^m changes sign in the vicinity of 67°C.[9] Thus the nonbonded attractive interactions in the homochiral versus the heterochiral aggregates responsible for enantiomer discrimination in solution may be either greater or smaller, according to conditions, and these interactions are quite small.

Calorimetric measurements of the heats of solution, ΔH^{soln}, of crystalline racemates and of the corresponding crystalline enantiomer components (in water) illustrate the much greater magnitude of enantiomer discrimination in the solid state relative to that in solution. For tartaric acid, $\Delta H^{soln}_{rac} - \Delta H^{soln}_{enant} = \Delta\Delta H^{soln} = 2.2 \pm 0.07$ kcal mol^{-1} (9.4 ± 0.3 kJ mol^{-1}) and for threonine $\Delta\Delta H^{soln} = 0.04 \pm 0.03$ kcal mol^{-1} (0.17 ± 0.13 kJ mol^{-1}) at 25°C.[10]

The importance of these enthalpies is best appreciated from the thermodynamic cycle illustrated in Figure 6.2. In this cycle, ΔH^{form} is the enthalpy of formation of the racemic compound from the crystalline enantiomers.

From the cycle, we see that $\Delta H^{soln}_{rac} - \Delta H^{soln}_{enant} = \Delta H^m_{enant} - \Delta H^{form}_{rac}$. For a conglomerate, $\Delta H^{form}_{rac} = 0$; hence the equation reduces to

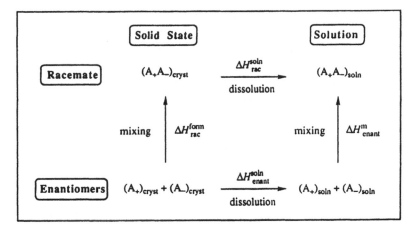

Figure 6.2. Thermodynamic cycle for the mixing of enantiomers.

$$\Delta\Delta H^{\text{soln}} = \Delta H^{\text{soln}}_{\text{rac}} - \Delta H^{\text{soln}}_{\text{enant}} = \Delta H^{\text{m}}_{\text{enant}} \tag{6.3}$$

While enantiomer discrimination may not be measurable in the gas phase or in dilute solution (ideal behavior obtains), differences in physical properties reflecting enantiomer discrimination will be increasingly manifest as the concentration of solutions is increased (see above), as measurements are compared in the neat liquid state, and as they are carried out in liquid crystals and in the solid state.[11,12]

Ideal behavior is understood to mean that the enthalpy of mixing of enantiomer pairs in the liquid (or gaseous) state is equal to zero for any mixture of enantiomers, and the corresponding entropy of mixing is equal to $R \ln 2 = 1.38$ cal mol^{-1} K^{-1} (5.77 J mol^{-1} K^{-1}). In fact, the enthalpy of mixing ΔH^{m} is often nonzero in the liquid state and in solution (for a summary, see ref. 13, p. 47ff). The values are typically quite small [ca. 0.5–50 cal mol^{-1} (ca. 2–200 J mol^{-1})] and they vary over quite a range depending on the substances that are mixed.

Enantiomer discrimination in the solid state can also be assessed by comparing heats of sublimation ΔH^{s} of enantiomerically pure and racemic samples of a given substance. The magnitude of this energy difference is about 1000 times greater than that found for the liquid state.[14] The origin, sense, and theoretical basis of these energy discriminations have been discussed.[6,15]

A major consequence of the nearly ideal behavior of neat liquid mixtures of enantiomers is that enantiomeric enrichment by distillation is, in general, not practical. An unequal mixture of enantiomers ($x_R \neq x_S$, where x is the mole fraction) could, in principle, be enriched with respect to one of the enantiomers. Such separations of enantiomers from racemates have, in fact, been achieved by liquid chromatography on *achiral* columns (see Section 6-4.m).

Enantiomer enrichment based on manipulation of solid samples of chiral substances, in contrast, is entirely practical. The substrate at hand may be racemic ($x_R = x_S$) or it may already be partially enriched ($x_R \neq x_S$). In the latter case, further enrichment, which may ultimately lead to enantiomeric purity, is possible (in fact, may be unavoidable!) by conventional (i.e., excluding the use of optically active solvents or reagents) crystallization, sublimation, extraction (even washing), or by combinations of these processes.[16] In a racemic sample, separation of the enantiomers (resolution) is possible without undertaking chemical reactions or conventional chromatography provided only that the solid racemate is a conglomerate (see below).

To summarize, stereoisomer separations are possible without the use of chiral reagents or chiral solvents. Diastereomers are more or less easy to separate from one another by conventional crystallization, distillation, or sublimation. To some extent, this is also true for enantiomer pairs. An enantiomerically enriched substance can, for example, be further enriched by these procedures, and especially by crystallization. Such separations depend on the difference between homochiral and heterochiral interactions. In general, separation procedures that depend on the crossing of phase boundaries rather than on chemical transformations are likely to succeed if the entities involved (enantiomers, racemates, and diastereomers) differ by energies of the order of kilocalories per mole (kilojoules per mole) and to fall if these energies are only of the order of small calories per mole (or joules per mole).

6-3. THE NATURE OF RACEMATES

Three types of crystalline racemates are known.[17]

1. The racemate is simply a 1:1 mechanical mixture or *conglomerate* of crystals of the two enantiomers, each crystal being made up of homochiral molecules. [The term racemic mixture has often been used to describe this type of racemate in the literature. Since "racemic mixture" has also been used to describe 1:1 mixtures of enantiomers of unspecified type (in this book called racemate), we will not use the term here.]

2. The racemate consists of crystals in each of which the (+) and (−) enantiomers are present in a 1:1 ratio down to the unit cell level. This corresponds to the formation of a solid compound, called a *racemic compound*. [The term true racemate, or simply racemate, has occasionally been used to describe a racemic compound. We prefer the latter term for reasons of clarity; we use "racemate" for any 1:1 mixture of enantiomers.)

3. The racemate consists of a solid solution of the two enantiomers, that is, a single homogeneous phase in which a 1:1 stoichiometric mixture of the two enantiomers is present unordered in the solid phase. This type is called *pseudoracemate*.

Samples of chiral compounds having no observable optical activity are generally assumed to comprise 1:1 mixtures of the enantiomers, that is, to be racemates.

However, one must not ignore the possibility that such a measurement may be accidental. At a different temperature, wavelength, or in a different solvent, optical activity could be revealed (cf. Chapter 12), and consequently the sample in question is not racemic; it is said to be *cryptochiral.*

One of the simplest ways of recognizing a racemate type is by inspection of binary phase diagrams. Such characterization was first undertaken by Roozeboom[17] as an application of the phase rule. Typical phase diagrams that relate the composition to the melting point for the three racemate classes are shown in Figure 6.3.

The most common of these racemate types is seen in Figure 6.3*b*; that is, the majority of racemates of chiral organic compounds (ca. 90%) exist in the form of racemic compounds, only about 10% as conglomerates (Section 7-2.b). The third type (Fig. 6.3*c*) is relatively rare. While conglomerate behavior is less common than racemic compound formation, the simplicity of enantiomer separation in the case of conglomerates gives this category special importance. The utility of these phase diagrams will be taken up in the following section.

The above classification is not meant to imply that the racemate of a given chiral substance may exist in only one crystalline form. Many compounds are known whose racemate may crystallize in either of two (racemic compound or conglomerate) of the three principal forms depending on conditions. When this occurs, one of the forms is metastable (within a given range of temperature and pressure). In the solid state, the metastable form may often be maintained for a long time without change.

Whenever two crystalline forms of a given substance can be isolated, there is also a fair chance that they can be interconverted. So too with racemate types. For example, 1,1'-binaphthyl (Fig. 7.5, **9**) crystallizes as a racemic compound (mp 154°C) that is converted to the thermodynamically more stable conglomerate (mp 159°C) on heating.[18,19] Numerous other examples of such transformations are known.[13]

The existence of a given substance in more than one crystalline form is called polymorphism. Polymorphs of the same compound differ in melting point and in crystal structure; polymorphs can sometimes be transformed into one another by changing the temperature or pressure. The incidence of two crystalline forms of racemates is but a special case of polymorphism. Differently solvated forms of a given compound have sometimes loosely been called polymorphs as well, even though this is incorrect.

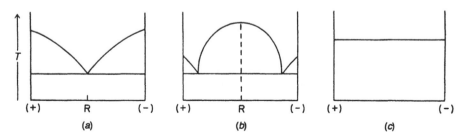

Figure 6.3. Binary phase diagrams describing the melting behavior of the common types of racemates. (*a*) Conglomerate, (*b*) racemic compound, and (*c*) ideal solid solution (pseudoracemate).

The two pure enantiomers of a chiral substance always form enantiomorphous (oppositely handed) crystals in the solid state, though this may not be visually discernible. For a given enantiomer, the sample may well be enantiopure, that is, consisting of molecules of only one chirality sense (within experimental limits). In this connection we use the following terms: (a) the older adjective *enantiomorphous* is used to describe appropriate chiral crystals; for molecules, we use *enantiomeric*; (b) a molecule, or an object, is either chiral or achiral. In contrast, a macroscopic chiral sample (loosely, a substance) is either racemic or *nonracemic*. The term nonracemic conveys the concept of a sample being made up mainly of homochiral molecules (rather than being racemic); nonracemic is not synonymous with "enantiomerically pure" (abbreviated to enantiopure), however. The latter expression relates to an experimentally determined fact [implying 100% enantiomeric excess (ee); see also Section 6-5.a].

> In the description of the enantiomer composition of macroscopic samples, the terms nonracemic and/or enantiopure are to be used in preference to "homochiral," a term meaning "of the same chirality sense," as in (all) right-handed shoes; use of the term homochiral (see p. 66) should be reserved for the description of a chiral class.[3]

A useful finding in this connection is that the frequency of conglomerate formation among salts appears to be two to three times as great as for covalent chiral compounds.[20] The consequences of this fact are taken up in Chapter 7. The oppositely charged counterions in ion pairs, for example, α-methylbenzylammonium cinnamate, form helical columns of hydrogen-bonded ion pairs that appear to be incompatible with centrosymmetric space groups; such salts exhibit conglomerate behavior (see also Section 7-2.c).[21]

6-4. PROPERTIES OF RACEMATES AND OF THEIR ENANTIOMER COMPONENTS

a. Introduction

Elementary textbooks in organic chemistry uniformly stress the fact that pairs of enantiomers have identical physical properties, save the sign of their optical rotation. This emphasis is misleading: It tends to imply that optical activity is the only significant measurement available for characterizing chirality-related properties of substances; and it ignores the fact that *all* properties of chiral substances differ according to their enantiomeric composition. The textbook emphasis is on comparison of pure (+) with pure (−) enantiomers. However, many and perhaps most samples of optically active materials encountered in the laboratory are not enantiomerically pure. Naturally occurring optically active substances isolated from their original source and purified conventionally are generally regarded as being enantiomerically pure.

Since the optically active materials that one works with in the laboratory (or in the factory) are often not enantiopure, their properties should be routinely contrasted not just with those of the pure enantiomers, but also with those of the corresponding racemate. Indeed, it is often helpful to think of enantiomerically impure samples as mixtures of single enantiomers and of their racemate.

A comparison of the properties of racemates and of their enantiomer components in the solid state is equivalent to the comparison of properties of enantiomorphous crystals (of pure enantiomers as well as of conglomerates) with those of crystalline racemic compounds. This section examines such properties as densities, melting points, and various types of spectra of pure enantiomers and of enantiomer mixtures.

b. Optical Activity

Given a sample of a chiral substance, one often wants to know whether the sample is optically active. The answer to this question is a function of the sampling as well as of the nature of the racemate and the state of the system. Provided that the sensitivity of measurement is adequate and that the sample is not fortuitously cryptochiral, a clearcut answer is possible for substances in the liquid or gaseous states. Not so for crystalline solids, however. Thus measurement of the optical rotation of a single well-formed crystal of a conglomerate (either in solution or on the crystal itself) would reveal optical activity even in a sample that was globally racemic as established by a rotation measurement carried out on a larger, homogeneous, portion.

It has been demonstrated that crystal pieces taken from opposite ends of a single crystal isolated from a globally racemic sample exhibit optical activity and circular dichroism (CD) with one end of the crystal being enriched in one enantiomer and the other end enriched in the second enantiomer.[22,23]

c. Crystal Shape

Crystallography played a prominent role in the early phases of stereochemical experimentation. Ever since the description of the first experiments of Pasteur, it has been recognized that individual enantiomers can sometimes be recognized from the outward appearance (morphology or habit) of their crystals.

All enantiomerically pure solid samples inhabit enantiomorphous crystals; that is, they inhabit one of the 11 noncentrosymmetric (chiral) crystal classes. Should it not be possible to differentiate individual crystals of the two enantiomers of *any* chiral substance by inspection of their outward shape? To put it another way, can one tell whether two crystals of the same chiral substance have the same or opposite handedness? In fact, such recognition is possible only when the crystals are hemihedral and possess hemihedral faces (hemihedry, see Chapter 1).

The presence of hemihedral faces is related to the crystal class and space group of the crystals being investigated. The 65 space groups comprising the 11 enantiomorphous crystal classes are unequally represented in the set of chiral substances. Two of these space groups account for about 80% of chiral crystals among organic compounds, namely, space groups and $P2_1$ and $P2_12_12_1$.[13,24] The probability of finding hemihedral faces is good in $P2_1$ and poor in $P2_12_12_1$.[19] Unfortunately, hemihedral faces are present less often in the crystal classes typically found among chiral organic compounds than one might like.

The utility of all of this lies in the possibility of carrying out a resolution by mechanical separation (triage) of crystals from a conglomerate à la Pasteur (see Section 7-2.a). It becomes evident that the rarity of this type of resolution relates not just to the probability of finding chiral crystals (conglomerate formation) but on the more severe requirement of observing hemihedry. The existence of hemihedry is the significant and fortuitous circumstance that made possible the famous experiment wherein Pasteur separated crystals of (+)- from (−)-sodium ammonium tartrate.[25,26]

There are other, no less fortuitous, circumstances that made this first resolution possible: The racemic double salt that Pasteur prepared was a conglomerate, and it crystallized as such rather than as a racemic compound because the crystallization happened to take place at a temperature lower than 28°C. Above this temperature, the conglomerate (a tetrahydrate) is converted to a monohydrate, which is a racemic compound. Crystals of the latter, called Scacchi's salt, are neither enantiomorphous nor hemihedral.

d. Density and Racemate Type

Solid racemic compounds differ significantly in density from the corresponding pure enantiomers. The difference can be as large as 5%; for example, the densities of racemic and enantiomerically pure *trans*-1,2-cyclohexanecarboxylic acid are 1.43 and 1.38, respectively ($\Delta d = 3.6\%$).[13] This difference stems from the enthalpy change associated with the following reaction, that is, the formation of a racemic compound from the enantiomers in the solid state:

$$A_+ + A_- \rightarrow A_+A_-$$

The difference in density is reflected in the magnitude of the melting point difference between that of the racemate (A_+A_-) and that of the pure enantiomer A_+ (or A_-) as well as in the shape of the melting point phase diagram (Sections 6-3 and 6-4.e).

There remains a widespread notion that enantiomers can be packed into crystals more tightly in a heterochiral manner than homochirally, in other words, that racemic compounds are more stable than the corresponding enantiomers. There may indeed be a "genuine tendency" for racemic compounds to be more stable (and therefore be slightly denser) than the corresponding enantiomers. This would account, in part, for the greater incidence of racemic compounds over conglomerates among racemates.[13] However, it has also been suggested that this tendency is not thermodynamic in origin but rather that it reflects either kinetic factors dealing with nucleation and growth of crystals from racemic solution and/or packing arrangements in crystallographic space groups that favor racemic molecular crystals over those of pure enantiomers.[15,27]

In principle, a racemic solution can crystallize either as a racemic compound or as a conglomerate. Conglomerates necessarily crystallize in one of the 65 chiral space groups (Sections 6-4.c and 6-4.k) whereas racemic compounds crystallize (with rare exception) in one of the 165 racemic (centrosymmetric) space groups. The increase in packing possibilities afforded by the larger number of centrosymmetric space groups may account, in part, for the greater number of racemates crystallizing as racemic compounds than as conglomerates.[28]

e. Melting Point

The melting point of a solid chiral substance is a highly revealing physical property. For the purpose of extracting stereochemically relevant information, the racemate type must be taken into account, and both the beginning and the termination of melting must be known. Conventional melting point determination usually does not suffice when both of these temperatures must be determined. The preferred techniques that provide the necessary data are either differential thermal analysis (DTA) or differential scanning calorimetry (DSC).

The necessary information is conveniently summarized by means of symmetrical binary phase diagrams that relate the composition to the melting point. The first of the three basic types (identified in Section 6-3), that of conglomerate behavior, is illustrated in Figure 6.4. This system implies no mutual solubility of the enantiomers in the solid state. Composition E (eutectic) consists of a mixture of two kinds of crystals: those of the (+) enantiomer and those of the (−) enantiomer. The melting point of the pure enantiomer is given by T_A^f, while that of the racemate, corresponding to the eutectic in the melting point diagram, is T_E (or T_R). It is characteristic of conglomerate systems that the racemates *always* melt below the corresponding pure enantiomers.

Mixtures of intermediate composition, for example, M [rich in, say, the (+) enantiomer], begin melting at T_E. Crystals of both enantiomers disappear during the melting process so as to produce a liquid that is racemic (the eutectic disappears first)

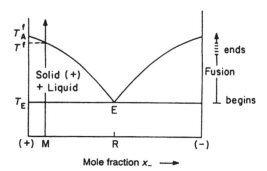

Figure 6.4. Melting of a mixture M in a conglomerate system (E = eutectic; R racemate). [Adapted with permission from Jacques, J., Collet, A., and Wilen, S. H. *Enantiomers, Racemates and Resolutions*, p. 44. Copyright © 1981 John Wiley & Sons, Inc.]

and the temperature stays constant at T_E. After all the racemate has melted [i.e., no (–) enantiomer is left in the solid phase], the remaining (+) enantiomer gradually melts in a liquid of varying composition with the last bit of solid disappearing at T^f. The same diagram also describes the reverse process, namely, crystallization from the melt; the key point is that it is (+) solid alone [the pure (+) enantiomer] that crystallizes beginning at temperature T^f when a liquid mixture of composition M is cooled. This is true also when solvent is present.

The composition of mixture M (Fig. 6.4) is given by the ratio MR/(+)R and the ee $= 2x_+ - 1$, where $x_+ =$ mole fraction of the dominant enantiomer (see Section 6-6). An alternative way of describing mixture M is as a mixture consisting of $2x - 1$ mol of the (+) enantiomer and $2(1 - x)$ mol of racemate (= eutectic). That such a description is not a figment of the imagination is revealed by examination of a mixture of enantiomers by DSC. In DSC, one measures the enthalpy absorbed by a sample as a function of temperature (Fig. 6.5). The DSC trace for a mixture of composition M actually shows separately the absorption of energy during the melting of the eutectic E and of the (+) enantiomer with increasing temperature. The area of the first event (peak) is directly proportional to the heat necessary to melt the racemate present in the mixture. In the case of a conglomerate, if one knows the molar heat of fusion of the racemate (ΔH_R) and the total weight of the sample, the proportion of racemate and of pure enantiomer, hence also the enantiomer composition, are easily determined from the DSC trace provided that the peaks are reasonably well separated.

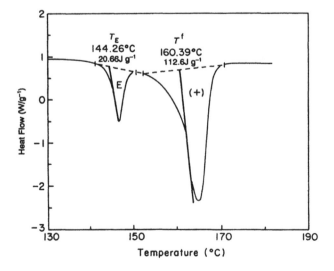

Figure 6.5. Differential scanning calorimetry trace of the melting of an enantiomer mixture (approximating composition M in Fig. 6.4) of α-methylbenzylammonium cinnamate (3.1-mg sample) exhibiting conglomerate behavior. The "peak" at left represents the enthalpy of melting of the eutectic E (racemate) while the larger "peak" to the right is the enthalpy of melting of the (+) enantiomer in the mixture. The temperatures shown represent those symbolized in Figure 6.4.[29]

The molar heats of fusion of the racemate ΔH_R and of the pure enantiomers ΔH_A are unequal. That this is so even in a conglomerate system follows from the fact that the melting points of the two forms differ while the specific heats do not. Consequently, the relative amounts of the racemate and the enantiomers are not directly given by the areas of the DSC peaks.[13]

Some examples of compounds that behave in the way just described are ($\pm$)-hydrobenzoin **1** ($T_A^f = 147.5°C$ and $T_R^f = 121°C$; $\Delta T = 27.5°C$), 1-phenyl-1-butanol **2** ($T_A^f = 50°C$ and $T_R^f = 16°C$; $\Delta T = 34°C$), and hexahelicene **3** ($T_A^f = 265–267°C$ and $T_R^f = 231–233°C$; $\Delta T = 34°C$; Fig. 6.6). Phenylglycine **4** (dec $> 200°C$) also behaves in this way but precise temperatures cannot be obtained due to decomposition during melting. This result is a significant limitation of the DSC technique. The examples reveal an important characteristic of conglomerate systems, namely, that the racemate melts at a much lower temperature than the pure enantiomer. The difference ranges from 25 to 35°C (cf. Table 7.2).[13]

A second type of racemate (Fig. 6.3*b*) corresponds to the case in which the two enantiomers of a given compound coexist in the same unit cell; that is, the enantiomer pair forms a well-defined racemic compound. Figure 6.7 describes such cases, which correspond to the vast majority of organic racemates. The eutectics in racemic compounds (Fig. 6.7*a*) may be treated as mixtures of crystalline enantiomer and of a racemic compound. Thus E represents a mixture of $2x_E – 1$ mol of crystalline (+) enantiomer and $2(1 – x_E)$ mol of racemic compound R.

Racemic compound systems are characterized by the fact that addition of a small amount of pure enantiomer to the racemate invariably *lowers* the melting point of the racemate (contrary to what obtains in conglomerate systems). On heating a mixture of composition N [poorer in the (+) enantiomer than the eutectic E^+; Fig. 6.7*a*], melting

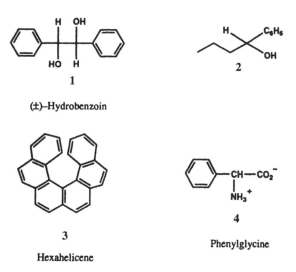

Figure 6.6. Compounds exhibiting conglomerate behavior.

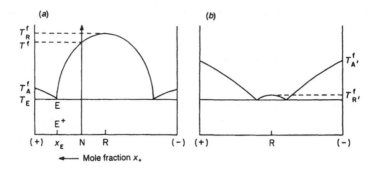

Figure 6.7. Racemic compound formation. The melting point of the racemic compound is T_R^f (or $T_{R'}^f$). [Adapted with permission from Jacques, J., Collet, A., and Wilen. S. H. *Enantiomers, Racemates and Resolutions*. p. 90. Copyright © 1981 John Wiley & Sons, Inc.]

begins at T_E and it continues until all the eutectic in the sample becomes liquid. The solid phase remaining above T_E is the racemic compound. To ascertain whether a given mixture lies to the right or the left of the eutectic, it suffices to add a small amount of racemate to the mixture. Redetermination of the melting point of the mixture shows where the composition lies (on the racemate side of the eutectic if the melting point rises and on the pure enantiomer side of the eutectic if it drops).

Racemic compound formation may be illustrated by dimethyl tartrate, **5** (Fig. 6.8) ($T_A^f = 43.3°C$ and $T_R^f = 86.4°C; \Delta T = -46.1°C$) (see Fig. 6.7a), and by mandelic acid, **6** (Fig. 6.8) ($T_A^f = 132.8°C$ and $T_R^f = 118.0°C; \Delta T = 14.8°C$) (see Fig. 6.7b).

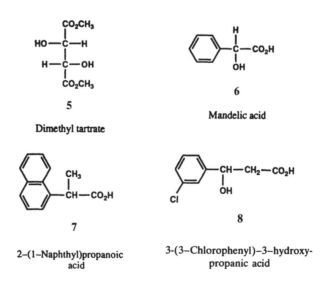

Figure 6.8. Chiral substances forming racemic compounds.

Occasionally, extreme cases are observed, such as 2-(1-naphthyl)propanoic acid **7** (Fig. 6.8) (see Fig. 6.7*a*), where the eutectic is so close to the pure enantiomer that it is hardly detectable[30] and 3-(*m*-chlorophenyl)-3-hydroxypropanoic acid **8** (Fig. 6.8) (see Fig. 6.7*b*), where the racemic compound is difficult to detect by inspection of the phase diagram.[31]

Only a few racemates exhibit solid solution behavior (Fig. 6.3*c*) in which the two enantiomers are miscible in the solid state (p. 107).

Ideal solid solution behavior is exemplified by camphor whose racemate and pure enantiomers each melt at T = ca. 178°C. The phase diagram describing this type of behavior is shown in Figure 6.9*a*. Solid solution behavior is particularly prevalent among molecules forming plastic crystals and among those substances forming rotationally disordered crystals made up of molecules of spheroidal shape.[15] Whenever the racemate and one of the pure enantiomers are reported to melt at the same temperature, existence of a solid solution may be suspected.

Figure 6.9*b*, which describes positive deviations from ideality, and Figure 6.9*c*, which describes negative deviations from ideality, illustrate even rarer cases than solid solutions behaving ideally; only 19 examples of these rare cases are listed in the inventory of Jacques et al.[13] Note that there is no danger of confusing a pseudoracemate system of the type illustrated by Figure 6.9*c* with a conglomerate system. Mixtures of composition P (Fig. 6.9*c*) exhibit only one peak in the DCS trace (no eutectic peak is present).[32]

f. Solubility

Stereoisomer discrimination is most obviously manifested in the fusion and solubility properties of chiral compounds. The former properties were examined in Section 6-4.e immediately preceding. The latter are revealed in the measurement of heats of solution and in solubility properties of the enantiomers relative to those of the racemates. Differences in the heats of solution. corresponding to the enthalphy of mixing of the enantiomers of a given compound in the solid state, are summarized for amino and hydroxy acids (in water at 25°C) in Table 6.1.[10] The most revealing aspect of these data are their large magnitude relative to that found for comparable mixing in solution (Section 6-2), where averaging of interactions obtains.[8]

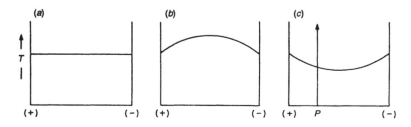

Figure 6.9. Melting point phase diagrams of solid solutions of enantiomers.

TABLE 6.1   Differences in Heats of Solution Between Enantiopure Solid Chiral Compounds and the Corresponding Racemates[a,b]

Compound	(kcal mol^{-1})	(kJ mol^{-1})
Alanine	0.268	2.00 ± 0.21
Glutamic acid	0.98	4.1 ± 0.02
Histidine	0.36	1.5 ± 0.2
Threonine	0.041	0.17 ± 0.13
Valine	0.547	2.29 ± 0.01
Tartaric acid	2.2	9.4 ± 0.3

[a]At 298.15 K.
[b]Reference 10.

Most chiral compounds exhibit significantly different solubilities for the racemate and the corresponding pure enantiomer. This fact forms the basis of a relatively simple enantiomeric enrichment process that may be applied when a nonracemic but enantiomerically impure sample is available. However, rational application of such a process requires a knowledge of the solubility behavior of the racemate–enantiomer system; that is, the ternary phase diagram (at least its essential features) must be known.

Knowledge of the solubility of enantiomers in solution furnishes information about the crystallization of the enantiomers from solution. This fact is entirely analogous to the use of binary phase diagrams in understanding both the melting of a sample and the crystallization of a chiral nonracemic sample from the melt. Evaluation of solution properties requires consideration of an additional variable, namely, the solvent, leading to the use of ternary phase diagrams. Solvent presents a constraint in that changes in concentration as well as in temperatures can no longer be represented all at once in two dimensions. Representation of temperature as well as three concentrations would require a 3D diagram (a triangular prism), which is awkward to say the least. For most purposes, it suffices to examine an isothermal slice of the triangular prism corresponding to a ternary diagram representing the solubility of (+) and (−) enantiomers in solvent S at a given temperature T_0.

Conglomerate systems are illustrated by Figure 6.10. In such a diagram, the two sides of the equilateral triangle represent the concentrations of (+) and (−) enantiomers in solvent, while the bottom describes the mole fraction of the enantiomer mixture, just as in the case of the binary phase diagram. In this example, composition E corresponds to the solubility of the racemate (eutectic) at T_0 [the concentrations of the two enantiomers in solvent may be expressed as mole fractions, or conveniently as weight/weight (w/w) percentages; these two concentration modes do not give rise to identical phase diagrams].

The proportion of solvent in E is given by segment (−)a, that of (+) by Sb, and that of (−) by (+)c; with respect to E, each of these segments is measured in directions parallel to the sides of the triangle. Figure 6.10 represents the behavior of a conglomerate in solution under equilibrium conditions. Note that the solubility

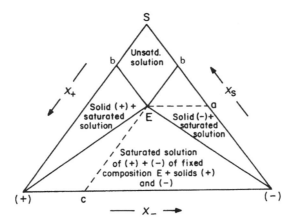

Figure 6.10. Solubility diagram for a conglomerate system at fixed temperature T_o. The saturated solutions in equilibrium with pure (+)- and (−)-enantiomers, respectively, have variable composition. The dashed lines show that point E has a composition given by concentrations a, b, and c measured along the sides.

"curve" bEb looks remarkably like the melting curve in the corresponding binary phase diagram (Fig. 6.4). The solubility of each enantiomer in solvent S is given by segment Sb for the (+) enantiomer and by segment (−)b for the mirror-image isomer (note the direction of the arrows that specify increasing concentrations). These solubilities are, of course, equal; the values are for temperature T_o, the temperature of an isothermal slice of the triangular prism alluded to above. The solubility of the racemate is given by E (segment Sa, measured on either side of the diagram).

From simple thermodynamic considerations, it is evident that the solubility of the conglomerate is greater than that of the individual enantiomers [this is equally true of any mixture of (+) and (−)]. This situation brings to mind the old empirical double-solubility rule rationalized by Meyerhoffer,[33] that is, that the racemate is twice as soluble as the enantiomers. This rule is applicable to neutral (nondissociable) organic compounds having the same thermodynamic constants and behaving in an ideal way (implying that solvent properties do not enter into the calculation of solubilities). The same double-solubility conclusion may be reached in a more precise way by application of the Schröder–van Laar equation.[34] Effectively, this relationship requires that, for an ideal solution, liquidus curve T_A^f E (Fig. 6.4) is unchanged when (−) is replaced by (−) plus solvent.

The solubility–concentration relations are given by straight lines that are parallel to the solvent–enantiomer sides of the diagram provided that the composition is expressed as mole fraction. Experimental results are in accord with the foregoing statements and the solubility ratio α, defined as α = solubility of racemate/solubility of enantiomer, is indeed approximately equal to 2 for those covalent compounds whose solubilities have been measured quantitatively.[35] The precise value 2 is found if (and only if) solubilities are expressed as mole fractions x: $\alpha = 2$, where $\alpha_x = x_d/x_l$. By way of example, the solubilities of the racemate and one of the pure enantiomers

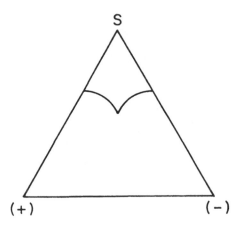

Figure 6.11. Solubility of a fully dissociated conglomerate ($\alpha = \sqrt{2}$).

of α-methylbenzyl 3,5-dinitrobenzoate are 27.4 g% (g 100 g^{-1} solution) and 13.2g%, respectively.[36]

The solubility ratio for fully dissociated solutes (+1/–1 salts) is $\alpha = \sqrt{2}$ and the lines describing the solubility (and the supersaturation) in the ternary phase diagram are no longer straight and parallel to the sides due to the common ion effect of the counterion (Fig. 6.11).

For racemic compound systems, the following general statements may be made: (a) The solubility of the racemates is less clearly related to that of the corresponding enantiomers (unlike the case of conglomerates). The racemic compound solubility may be either greater or lower than that of the enantiomer. (b) Eutectic compositions apparently are close to those found in binary phase diagrams for the corresponding (+) plus (–) mixture. (c) Even in unfavorable cases, enrichment is possible. Consider that when crystallization of a mixture of low enantiomeric purity yields a precipitate of racemic compound, enrichment takes place in the mother liquor.

Figure 6.12 illustrates the three types of pseudoracemate solubility behavior. The similarity to the melting point diagrams (Fig. 6.9) is again evident. It suffices to point out for this relatively rare type that solubility is little affected by the enantiomer

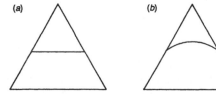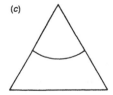

Figure 6.12. Solubility diagrams of pseudoracemate systems.

composition. For the type illustrated by Figure 6.12*a* (the common type), the solubility is completely independent of enantiomer composition. Consequently, purification by recrystallization of a mixture with $x_+ \neq x_-$ exhibiting this type of behavior is effectively precluded.

g. Vapor Pressure

Just as the melting points and the solubilities of racemates differ from those of their enantiomer constituents, so too do the vapor pressures. This statement refers specifically to the vapor pressures of solids; for differences in the vapor pressure of liquid enantiomers and the corresponding racemates, see Section 6-2. Such a difference was first suggested in the case of dimethyl tartrate by Adriani.[37]

Differences between the heats of sublimation (ΔH_{sublim}) of enantiopure samples and of the corresponding racemic compounds (Eq. 6.4) have been measured in the case

$$\text{versus} \left. \begin{array}{l} (+)A_{solid} \rightarrow (+)A_{vapor} \\ (\pm)A_{solid} \rightarrow (\pm)A_{vapor} \end{array} \right\} \Delta\Delta H^{25}_{sublim} \qquad (6.4)$$

of just a few compounds (Table 6.2).[14] The fact that both negative and positive values of $\Delta\Delta H_{sublim}$ have been found implies that heterochiral interactions between the solid

TABLE 6.2 Differences Between Heats of Sublimation of Enantiopure Compounds and the Corresponding Racemates[a,b]

	$\Delta\Delta H_{sublim}$		
	(kcal mol^{-1})	(kJ mol^{-1})	mp (°C)[c]
Carvone oxime	2.6	10.9	71(91)
$CH_3OOC-C-C-COOCH_3$ (with OH H / H OH)	8.4	35.1	49(87)
Menthol	−4.1	−17.2	43(28)

[a]At 25°C.
[b]Reference 14.
[c]Melting poing of (+)- or (−)-enantiomer (mp of racemate).

enantiomers (Section 6-2) can be either stronger or weaker than the corresponding homochiral interactions.

A systematic study of the sublimation of mandelic acid [$C_6H_5CH(OH)CO_2H$] was carried out by Garin et al.[38] The latter group of investigators as well as Kwart and Hoster[39] called attention to the high efficiency of separation by sublimation. Indeed, it appears that sublimation may be a significantly more efficient process than recrystallization in effecting enantiomer enrichment in enantiomer mixtures.

The principal limitations are that the samples need to have reasonable volatility at the operating temperatures, which need to be below the melting points of the lowest melting crystalline form, and that these temperatures not be so high as to lead to thermal decomposition. Intuitively, there must exist a fairly close relationship between the phase diagrams describing melting and the sublimation of enantiomer mixtures. After all, the heat necessary to pass from the solid state to the gas state, ΔH_{sublim}, corresponds to the heat of fusion plus the heat of vaporization (the latter being the heat required to pass from the liquid to the vapor state) and we have already seen (Section 6-2) that the latter is practically the same for enantiomers and for their mixtures. This analogy must, however, take into account the fact that, in the case of sublimation, a new variable can be introduced, namely, pressure.

Calculated pressure–temperature) phase diagrams have since been reported[40]; vapor pressure of a conglomerate is exactly twice that of the corresponding pure enantiomers. On the other hand, the vapor pressure of a racemic compound can be either higher or lower than that of the corresponding enantiomers; it depends chiefly on the enthalpy of decomposition of the racemic compound and on the temperature at which this decomposition takes place.[40]

h. Infrared Spectra

While differences in solid state IR spectra of enantiomers and the corresponding racemates were observed long ago,[41,42] the origin of these differences was sometimes ignored.[43,44] It is now clear that such differences are not to be expected between spectra (measured either as KBr disks or as mulls) of either enantiomer and of the corresponding racemate when the latter is a conglomerate. On the other hand, IR spectra of racemic compounds usually differ significantly from those of the corresponding enantiomers.[13] Infrared spectroscopy is often the diagnostic procedure of choice in ascertaining whether or not a racemate is a conglomerate. However, in case of doubt (see below), it would seem best to utilize more than one criterion to determine the racemate type (for a discussion of the several methods available, see Section 7-2.b).

The observed differences in IR spectra reveal that certain molecular interactions are modified. In carboxylic acids, for example, strong hydrogen bonds are usually present. Eliel and Kofron[42] observed that such hydrogen bonding was much more important in the racemate than in the individual enantiomers of the hydrogen phthalate ester **9** that they studied (Fig. 6.13). The hydrogen bonds are responsible for dimer formation and the IR spectrum reveals that the dimer is a racemic compound.

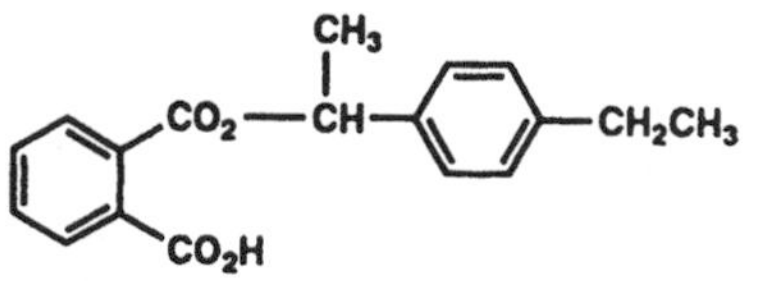

9

Figure 6.13. α-Methyl-(*p*-ethylbenzyl) hydrogen phthalate.

Conversely, in the enantiomer, intermolecular hydrogen bonding, as revealed by the position of the OH and C=O stretching frequencies, is less important. This difference is then responsible for the contrast in IR spectra between the two stereoisomer forms.

Differentiation between racemates and the corresponding enantiomers in the case of hydrogen-bonded carboxylic acids (mandelic, lactic, and tartaric acids), carbohydrates, for example, arabinose and salts (sodium ammonium tartrate), is readily effected by IR spectroscopy.[45] Brockmann and Musso[46] observed differences in IR spectra of racemic amino acid derivatives and the corresponding enantiomers and have correlated them with differences in solubility.

When no intermolecular hydrogen bonding is present, the IR spectral differences between enantiomers and racemates may be quite small. For example, in the case of a chiral sulfide, the two spectra were superposable even though the racemate is clearly a racemic compound.[39]

i. Electronic Spectra

(1*S*)-Dicarbonylrhodium(I) 3-trifluoroacetylcamphorate (Fig. 6.14, **10**), a chiral planar d^8 metal complex, is a yellow solid, mp 134°C. Admixture of this substance with the independently prepared (1*R*) enantiomer in solution and evaporation of the solvent leaves a red-green (dichroic) solid, mp 130.5°C, which can easily be identified as a racemic compound (a mixture melting point with a small amount of the enantiomer leads to melting point lowering).[47] This striking color change constitutes unequivocal visual proof that electronic absorption spectra of racemic compounds and of the corresponding enantiomers in the solid state can differ significantly. Diffuse reflectance spectra of the two forms also differ markedly. In contrast, the colors of the corresponding melts were alike (brown).

Analogous differential behavior in solid state emission spectra (fluorescence), as well as differences in color, has been observed in several typical organic compounds, for example, **11** and **12** (Fig. 6.14), whose racemates had been shown to be racemic compounds by X-ray analysis.[48,49] The racemic compounds exhibited characteristic α-type structureless excimer emission while the enantiomers have γ-type monomer emission. The utility of this differential behavior in the determination of enantiomeric purity has been demonstrated.[49] Enantiomer discrimination is observed (by fluorescence) in the dissociation of intermolecular homochiral and heterochiral excimers of methyl *N*-acetyl-1-pyrenylalaninate (Fig. 6.14, **12**), but not in the

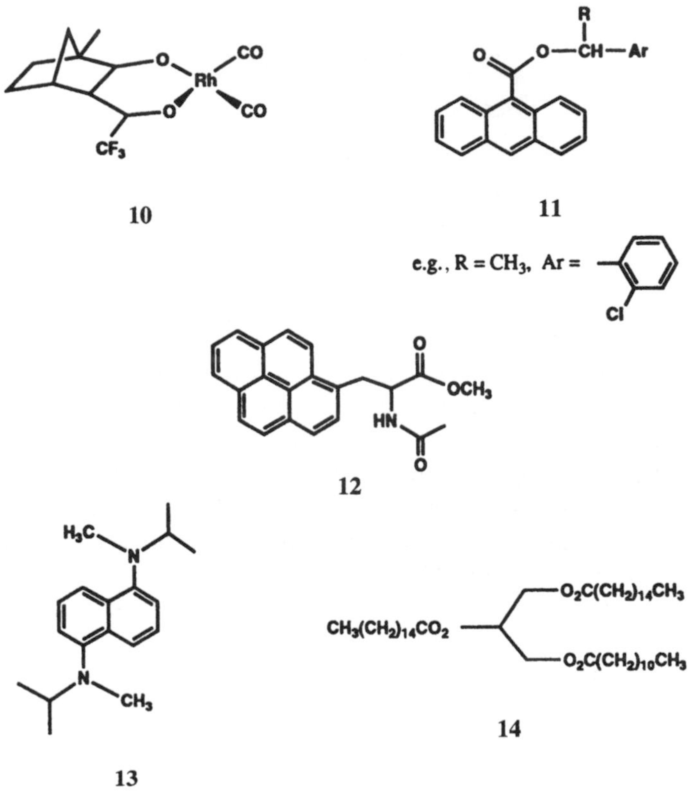

Figure 6.14. Structures 10–14.

formation of the excimers. Differences in hydrogen bonding favor the D–L (heterochiral) excimer more than the L–L (homochiral) excimer in CH_3CN but not in N,N'-dimethylformamide (DMF), the latter solvent being a hydrogen-bond acceptor able to interfere with excimer formation.[50]

j. Nuclear Magnetic Resonance Spectra

Differences in solid state ^{13}C NMR spectra of the enantiomers and the racemic compound have been demonstrated for tartaric acid.[51] Instrumental facilities for measuring high-resolution solid state NMR spectra by a combination of cross-polarization (CP) and magic angle spinning (MAS) are widely available; hence such measurements can be important in identifying racemate types. Since enantiopure and racemic samples of the same substance in the solid state can exhibit different chemical shifts (as in tartaric acid), when conglomerates are not formed, solid state NMR may serve to determine the enantiomer purity of a solid sample (see Section 6-6). Enantiomer discrimination in solution is taken up in Section 6-4.1.

Solid state ^{13}C NMR spectra of inclusion compounds obtained when solutions of tri-*o*-thymotide (TOT; Fig. 7.4, **7**) in *rac*-2-halobutanes ($CH_3CHXCH_2CH_3$) or 2-butanol are allowed to evaporate demonstrate stereoisomer discrimination. Virtually all the carbon signals observed, of host as well as guest molecules, exhibit anisochrony. All signals due to host carbon atoms become triplets (TOT in the solid state loses its threefold symmetry) and signals due to guest molecules (when not obscured by peaks arising from the host) are doublets that reveal the diastereomeric environments of the two guest enantiomers. Solid state NMR provides a way of measuring the enantiomer composition of the guest; however, allowance must be made for the CP time-dependent line intensities.

Application of the CP–MAS technique sometimes permits the differentiation of meso from racemic *chiral* diastereomers. In the solid state, the ^{13}C NMR spectrum of 2,3-dimethylsuccinic acid yields two signals (for the methyl and the methine carbon atoms) in the case of the meso diastereomer, whereas in the (±) diastereomer each of these signals is split into a doublet since the pairs of CH and CH_3 carbon atoms are not related by a plane or center of symmetry. This finding is equivalent to stating that, in contrast to what obtains in solution (cf. Chapter 8), the solid chiral diastereomer exists as two noninterconverting conformers whose nuclei are externally diastereotopic. It must be noted that this result is possible because the meso acid adopts a single symmetric (anti) conformation in the solid state, a situation that does not universally apply (e.g., it does not in *meso*-tartaric acid). Solid state CP–MAS NMR spectra may also be useful in differentiating meso and chiral conformers that cannot be separated physically, for example, **13** (Fig. 6.14), in which the methyl and isopropyl groups are twisted with the methyl groups on the same or opposite sides of the naphthalene plane yielding two diastereomeric conformations.[52]

k. X-Ray Spectra

We have already seen that X-ray diffraction furnishes two types of stereochemical information: Relative configuration of chiral centers, both intra- and intermolecular, can be determined (cf. Section 5-5), and absolute configuration can be established when the wavelength of the X-rays is chosen to be close to the absorption wavelength of an inner-shell electron of one of the atoms in the crystal (Section 5-3). There is yet more stereochemical information extractable from X-ray spectra.

Diffraction experiments begin with the assignment of the space group of the crystal (which reflects the symmetry of its constituent molecules) by examination of the diffraction pattern.[53] This assignment is important because enantiopure samples of chiral molecules necessarily crystallize in noncentrosymmetric space groups. The space group [together with information about the number of molecules (Z) per unit cell] reveals the presence or absence of a center of symmetry; that is, the space group (together with the value of Z) tells us whether the crystal is centrosymmetric or noncentrosymmetric. The latter category (65 out of 230 space groups, cf. Section 6-4.c) comprises all enantiomorphous crystals. In other words, determination of the space group (and of Z) reveals the presence or absence of enantiopure crystals; for racemic samples, this is equivalent to determining whether or not the compound exists as a conglomerate.[54] Racemic compounds crystallize mostly in centrosymmetric space groups. A few examples of racemic compounds crystallizing in chiral space groups are known,[13,27] whereas conglomerates necessarily crystallize in noncentrosymmetric space groups.

The space group also reveals the presence or absence of a polar axis, a crystal attribute that is, in turn, responsible for interesting and useful chemical and physical properties of numerous common chemicals.[55] The utility of the polar axis in the determination of absolute configuration through crystal habit modification was described in Chapter 5.

Finally, in a strictly empirical manner, comparison of X-ray powder diffraction spectra (Debye–Scherrer diagrams) measured on powders rather than on single crystal[56] shows distinct bands for enantiopure samples and for the corresponding racemates but only when the latter are racemic compounds. Thus X-ray powder spectra can reveal the nonracemic character cryptochiral samples, such as the triglyceride 1-laurodipalmitin (Fig. 6.14, **14**).[57]

l. Liquid State and Interfacial Properties

It should now be quite clear that the properties, both physical and chemical (though the latter have not been described), of enantiomeric systems will often be unequal whenever one compares two samples differing in enantiomeric purity [ranging from total (= pure enantiomer) to zero (= racemate)]. Such differences, which manifest themselves most strongly in the solid state, will be evident to some extent in other states of matter as well.

Among the most striking illustrations of enantiomer discrimination in the liquid state are those found in solution NMR spectra. In solution, enantiopure and racemic samples of chiral compounds usually exhibit identical NMR spectra, but small differences in the chemical shifts are sometimes observed.[58] Moreover, nonracemic but not enantiopure samples of compounds that strongly self-associate, for example, hydroxyamines, amides, and carboxylic acids (either as neat liquids or in solutions of *achiral* nonpolar solvents), sometimes exhibit anisochrony (split signals) reflecting the enantiomer composition even in the absence of chiral inducing agents, such as nonracemic chiral solvents or shift reagents.[59]

Williams, Uskokovic, and co-workers discovered (in 1969) that a racemic solution of dihydroquinine (Fig. 6.15, **15**; ca. 0.3 M in $CDCl_3$) exhibits an ^1H NMR spectrum different from that of the optically active compound (at 100 MHz). Chemical shifts were affected throughout the spectrum, and peak intensities of partially resolved material depended on the enantiomeric purity of the sample. (The spectra of the racemate and of the optically active sample tended to become identical at high dilution.) Accordingly, the enantiomer composition of the mixture may be deduced without resorting to a chiral solvent or to a shift reagent. That the observed effect is a manifestation of enantiomer discrimination is made evident by the following

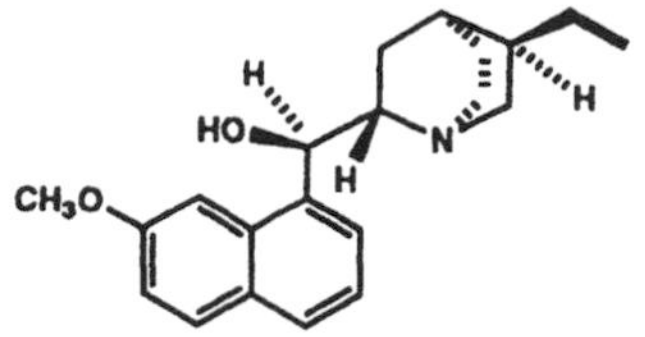

15   **16**

Dihydroquinine

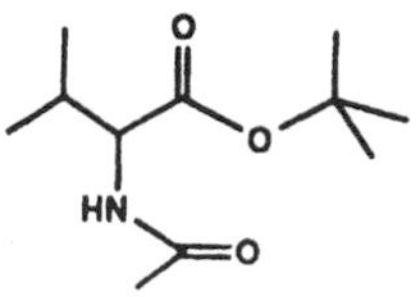

Figure 6.15. Examples of compounds exhibiting enantiomer discrimination in their NMR spectra.

additional facts: (a) The additional peaks disappear on acetylation of the hydroxyl group; and (b) replacement of the aprotic solvent by CH_3OH leads to identical spectra for enantiopure and racemic **15** as well as for all mixtures of (+)- and (−)-**15**, presumably by destroying the hydrogen-bonded dimeric or oligomeric "associates" responsible for the effect.

The self-induced anisochrony requires formation of diastereomeric associates and consequently one would expect to observe this otherwise unexpected phenomenon only when hydrogen bonding or other strong association is present. The time-average environments of [(−)-**15**·(−)-**15**, (+)-**15**·(+)-**15**], and [(−)-**15**·(+)-**15**] dimeric associates (as well as oligomeric associates) are different in the absence of extraneous influences such as high temperature and polar solvents able to interfere with the self-association.[60] The anisochrony occurs under conditions of fast exchange and is favored by low temperature and high concentration. A linear relationship exists between $\Delta\delta$ and the proportion of the two enantiomers present indicating that the proportion of homochiral and heterochiral aggregates formed is under statistical control in this case.[58,59] Other compounds have been reported to display this type of enantiomer discrimination in NMR spectra.

The search for solutes exhibiting enantiomer discrimination in NMR spectra has been influenced by the observation of chelated diastereomeric "solvates" that are formed between chiral solvating agents (CSAs) and solutes.[61,62] Such solvates are responsible for anisochrony in NMR spectra (Section 6-5.c) and for separation in high-performance liquid chromatography (HPLC) when the CSAs are added to chromatographic mobile phases (Section 6-5.d).

In the case of **16**, evidence for $NH \cdots O{=}C$ hydrogen bonding (involving both amide and ester groups) between solute molecules was obtained from NMR chemical shifts and from IR spectra in solution. Consequently, enantiomer discrimination was ascribed to the formation of diastereomeric dimers as illustrated in Figure 6.16. Dimer **A** results from homochiral interaction (Eq. 6.1), whereas dimer **B** results from heterochiral interaction (Eq. 6.2).[58] In contrast, in the case of carboxamides lacking additional hydrogen-bonding acceptor sites, the observed enantiomer discrimination cannot be ascribed to cyclic dimers in view of the preferred Z conformation (CO and NH anti) of the typical carboxamide functionality (Section 10-2.a). Instead, the observed discrimination is ascribed to linear hydrogen-bonded associates.[63]

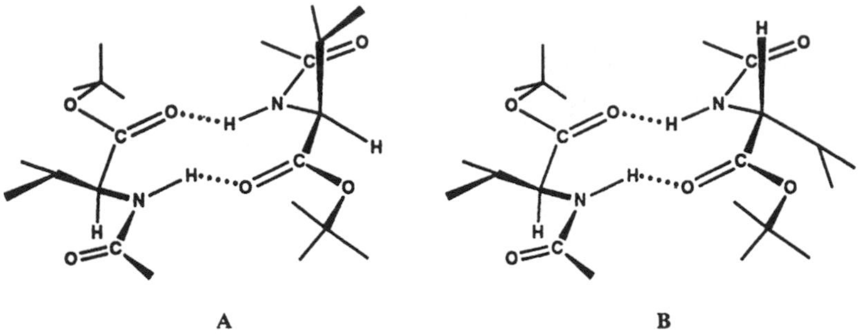

A   B

Figure 6.16. Proposed diastereomeric dimers (**A** = homochiral: **B** = heterochiral) formed from **16** [Adapted with permission from Dobashi, A., Saito, N., Motoyama, Y., and Hara, S. *J. Am. Chem. Soc.* **1986**, *108*, 307. Copyright © 1986 American Chemical Society.]

If the requirement for observable enantiomer discrimination is aggregation giving rise to homochiral and heterochiral associates, then reaction with achiral reagents promoting such aggregation may increase the tendency for the development of anisochrony in NMR spectra. This possibility has been demonstrated with nonracemic mixtures of chiral 1,2-diols whose enantiomer composition can be measured by ^{13}C NMR spectroscopy following their quantitative conversion to dioxastannolanes[64] by reaction with (achiral) dibutyltin(IV) oxide in CDCl$_3$ or in other nonpolar solvents (Fig. 6.17). In the case of 1,2-propanediol, anisochrony ($\Delta\delta$) of the methine carbon was observable on mixtures ranging from 20% to 89% ee with the intensities of the pairs of signals closely reflecting the proportion of *R* and *S* diol enantiomers present in the analyte. Similar analysis is also possible with dioxastannolanes formed in situ (with Bu$_2$SnCl$_2$ in CDCl$_3$).

Critical analysis of "self-discrimination" by Luchinat and Roelens[64] reveals that (a) systems unable to form heterochiral dimers (heterodimers; Eq. 6.2) do not give rise to enantiomer discrimination, (b) maximal enantiomer discrimination is exhibited by systems favoring formation of heterochiral over homochiral dimers, and (c) calculation of $\Delta\delta$ values must take into account the K_{equil} of the *SS* + *RR* $\rightleftharpoons$ 2*RS*

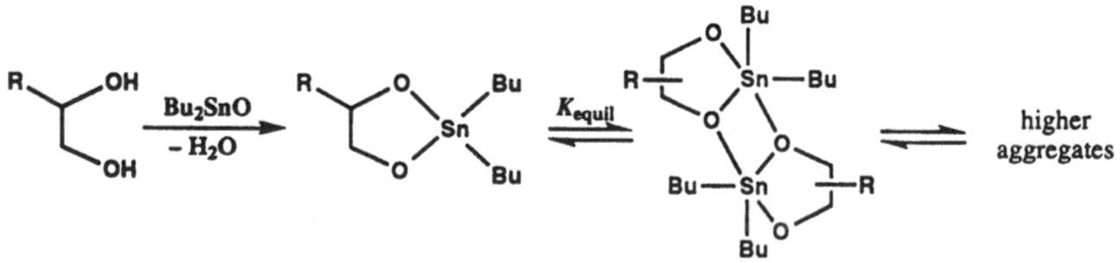

Figure 6.17. Formation of dioxastannolanes (R = CH$_3$, C$_6$H$_5$) for the enantiomeric purity determination of chiral 1,2-diols without chiral auxiliaries. [Adapted with permission from Luchinat, C. and Roelens, S. *J. Am. Chem. Soc.* **1986**, *108*, 4873. Copyright © 1986 American Chemical Society.]

process; $\Delta\delta$ is a linear function of the enantiomeric excess only for the case of $K = 4$ (corresponding to the measured value in the case of dioxastannolane formation with 1,2-propanediol and 1-phenyl-1,2-ethanediol). The value 4 is, in fact; statistically expected:

$$SS + RR \rightleftharpoons 2RS$$

$$K = \frac{[RS]^2}{[RR][SS]} = \frac{[RS]^2}{[SS]^2} = \frac{2^2}{1^2} = 4$$

Pasquier and Marty[65] translated the enantiomer discrimination potentially demonstrable in a nonracemic sample into diastereomer discrimination in order to increase the possibility of observing the former. The diastereomer discrimination has been achieved by use of the coupling (duplication) method of Horeau,[66] see also Section 6-5.c and Section 7-4). Reaction of (±)-1-diphenylphosphino-2-propanethiol, $(C_6H_5)_2PCH_2CH(CH_3)SH$, with nickel nitrate in solution leads to formation of a mixture of meso and chiral diastereomeric *trans*-Ni(thiol)$_2$ complexes whose composition is easily analyzable by ^{31}P NMR spectroscopy. Provided that the diastereoselectivity of the reaction is thermodynamic in origin (reversible reaction) and its extent {$s = ([m]/[e])_{racemate}$, where [m] is the concentration of the achiral meso species and [e] is the concentration of the chiral (±) complex} is known from reaction of the bifunctional Ni(II) with *racemic* substrate then, if no free (uncomplexed) analyte remains, the enantiomer composition of an enantiomerically enriched sample, the phosphinopropanethiol, can be calculated directly by means of Eq. 6.5:

$$\%ee \approx \frac{\sqrt{K^2 - s^2}}{K + 1} \times 100 \qquad (6.5)$$

where K is the measured [m]/[e] ratio of the product.

The involvement of dimeric and oligomeric species in enantiomer discrimination is demonstrated even in aqueous solution by "mere addition" of lanthanide ions. Thus the methyl resonance of a 80(*S*):20(*R*) mixture of sodium lactate was resolved in the presence of EuCl$_3$ (3:1 or 2:1 ligand to EuCl$_3$ ratio).[67] Anisochrony in the NMR spectrum of *racemic* lactate was observed in the presence of another chiral (nonracemic) ligand, for example, (*R*)-(+)-malate or (*S*)-(+)-citramalate; here the discrimination is more obviously diastereomeric in nature.

In any event, anisochrony resulting from "self-discrimination" can no longer be considered exceptional; the statement often made that enantiomers and racemates (or mixtures of intermediate enantiomer composition) have identical NMR spectra in solution is not universally correct.

Stereoisomer discrimination may also be responsible for differences in reaction rates and product distribution according to whether chiral reactants are pure enantiomers or racemates.[12]

Horeau and Guetté[61] summarized attempts to demonstrate enantiomer discrimination in solution through measurements of classical physical properties:

surface tension, index of refraction, viscosity, and so on. The effects are uniformly very small, barely, if at all, detectable. However, careful experimentation coupled with improved instrumentation is beginning to reveal such effects; for example, IR spectra of alcohols and of amino acid derivatives in solution give clear evidence of intermolecular association (hydrogen-bonded dimers and oligomers) that differentiates enantiomers from racemates (see above).

m. Chromatography

It has generally been assumed that, in order for enantiomers to be separated (or for the enantiomer composition of a nonracemic sample to be modified) by chromatography, either the stationary phase or the mobile phase must be nonracemic.[13] The following results make it clear that the preceding assumption is not generally justified.

High-performance liquid chromatography of ^{14}C radiolabeled racemic nicotine on an ordinary reversed phase column gives rise, as expected, to but a single peak in the chromatogram. However, in the presence of varying amounts of unlabeled (S)-(−)-nicotine (Fig. 6.18, **17**), the isotopically labeled nicotine exhibits two radioactive peaks in the chromatogram as monitored by a radioactivity sensing detector. A similar study has demonstrated the chromatographic separation (on ordinary silica gel) of ^{14}C radiolabeled racemic N-acetylvaline tert-butyl ester (Fig. 6.16, **16**) diluted with (−)-**16** into two peaks.[68] The order of emergence of the two radiolabeled enatiomers was determined by coinjection experiments with nonradioactive (−)-**16** and (+)-**16**.

Chromatography of nonracemic (but entiomerically impure) samples of N-lauroylvaline tert-butylamide (Fig. 6.18, **18**) on silica gel eluted with the usual achiral solvents (hexane + ethyl acetate mixtures) furnished fractions differing in their melting points. Careful chromatography on Kieselgur 60 of a sample having an initial

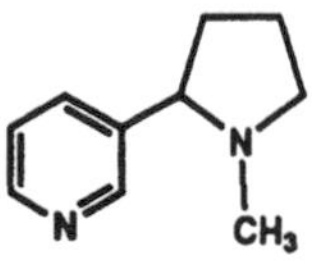

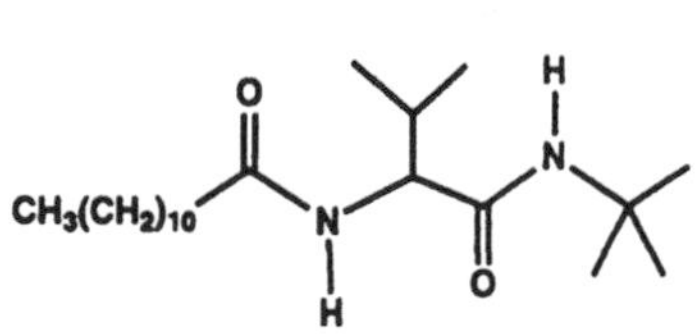

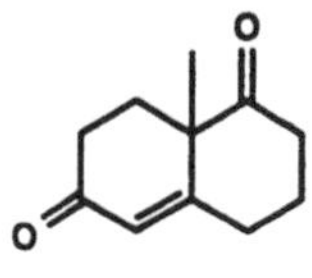

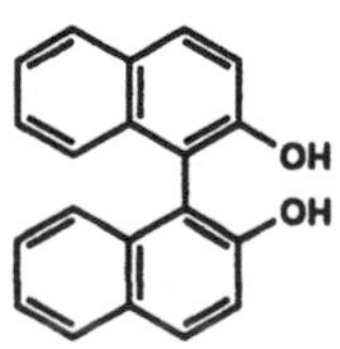

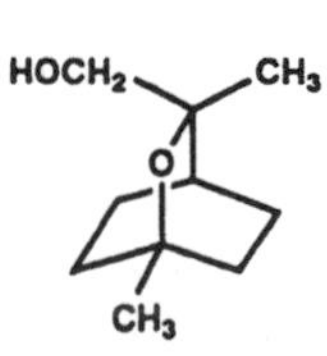

Figure 6.18. Compounds exhibiting enantiomer discrimination in chromatography on achiral stationary phases.

enantiomer composition L/D = 87:13 (74% ee) gave a fraction (ca. 30% of the sample) of lower enantiomeric purity (46% ee) than that of the analyte. Later fractions exhibited enrichment (as high as 97% ee) in the L enantiomer; this is consistent with the easier elution of the racemate relative to the predominant enantiomer.

Similarly, fractionation of a sample of Wieland–Miescher ketone (Fig. 6.18, **19**) of 65% ee by chromatography on silica gel led to 10 fractions differing significantly in enantiomer composition: The initial fraction had ee = 84% and the final fraction ee = 51%. Control experiments confirmed the fact that the observed effect was not due to decomposition of the sample, to racemization, or to impurities in the sample or on the column. The observation of enantiomer discrimination in ketone **19** supports the contention that this effect is not limited to molecules that are able to form hydrogen bonds.[69]

Enantiomeric enrichment has been observed also on chromatography of nonracemic samples of binaphthol (Fig. 6.18, **20**), specifically on aminopropyl silica gel (and not on silica gel itself), and most recently in cineole metabolites (Fig. 6.18, **21**) isolated from the urine of female Australian brushtail possum.[70]

It is evident from the foregoing results that chromatography of enantiomerically impure (but nonracemic) samples, including natural products in their native state, on achiral stationary phases and with achiral mobile phases can lead to fractions in which enantiomeric enrichment is observed, sometimes in the early chromatographic fractions and sometimes in the later ones.[71]

The attribution of separation to the formation of diastereomeric associates on the surface of the stationary phase is consistent with the observation that, when well-resolved peaks are obtained, one peak contains optically active material whereas the other represent inactive material as measured polarimetrically.[72] Thus, when chromatography of nonracemic samples on strongly bonding achiral stationary phases (aminopropyl silica gel) leads to separation, the fractions are the dominant enantiomer and the racemate and not the enantiomers themselves.

n. Mass Spectrometry

The preferential vaporization of a single solid enantiomer relative to its solid racemate discovered by mass spectrometry (MS) was one of the first bits of evidence of potential enantiomer enrichment based on sublimation (Section 6-4.g). The separation was revealed in differences in the relative abundance of the two molecular ions produced in the electron impact (EI) mass spectrum of an enantiomer mixture as a function of time as compared with that of the isotopically labeled, and thus distinct, pure enantiomer.[73] The same effect has been observed in the chemical ionization (CI) mass spectrum (presumably with CH_5^+ ions) of mixtures of unequal amounts of dimethyl-d_6(2S,3S)-tartrate (M_{d_6}) and dimethyl-d_0(2R,3R)-tartrate (M_{d_0}) in the MH$^+$ peaks, where m/z = 185 and 179, respectively.[74]

An additional and very interesting finding in the work of Fales and Wright,[74] made on inspection of the protonated dimer (2M + H)$^+$ peaks, is that the combined abundance of $(2M_{d_6} + H)^+$ and $(2M_{d_0} + H)^+$ homodimers exceeds that of the heterodimer $(M_{d_6} + M_{d_0} + H)^+$. Instead of finding equal intensities for the sum of two homodimer $(2M+H)^+$ peaks and for the heterodimer peak [as expected for a 50:50 mixture of the quasi-enantiomeric ions, and as observed in the case of a 1:1 d_0-S + d_6-S mixture (1:2:1 triplet), thus ruling out isotope effects in the ionization], the central peak of the triplet (m/z = 363 due to the heterodimer) has an intensity only 78% of that calculated. The heterochiral $(2M + H)^+$ ion is seen to be destabilized relative to the homochiral ions.

The relative stability of the homochiral versus the heterochiral protonated dimer ions of diisopropyl tartrate has been measured by CI mass spectrometry (with $C_4H_9^+$ ions) on a mixture of d_0-S + d_{14}-R isotopically labeled quasi-enantiomers. The ratio of virtual equilibrium constants K_{SS}/K_{SR} (= K_{RR}/K_{SR}) ≈ 1.6 corresponding to $-\Delta\Delta G$ ≈ 0.29 kcal mol^{-1} (1.2 kJ mol^{-1}), a value similar to those found for enantiomer discrimination of strongly hydrogen-bonded compounds in solution.[75,76] A similar result has been obtained by Fourier transform ion cyclotron resonance mass spectrometry.[77]

o. Interaction with Other Chiral Substances

Sections 6-4.a–n dealt mostly with properties resulting from enantiomer discrimination. Much better known, however, are properties that reflect diastereomer discrimination, that is, those involving reversible interaction of a chiral substance. whether racemic or enantiomerically enriched, with a second chiral substance in which one enantiomer predominates (Section 6-2). Such interactions are responsible for the NMR anisochronies and chromatographic separations that permit the determination of enantiomer composition without conversion of enantiomer mixtures into diastereomer mixtures. These applications of diastereomer discrimination are described in detail in Section 6-5. Diastereomer discrimination in biological systems is treated in Section 6-4.p.

Diastereomer discrimination in the solid state is manifested by significant differences in heats of fusion, in melting points, in heats of solution, and in solubilities of diastereomers; for example, $\Delta\Delta H_{fus}$ of the diastereomeric α-methylbenzyl-ammonium and ephedrinium mandelates R,R or S,S versus R,S exceed 5 kcal mol^{-1} (≥20 kJ mol^{-1}).[78–80] These differences between crystalline diastereomers form the basis for their separation in the classical Pasteurian resolutions, as well as for resolution by inclusion compound formation (Section 7-3.c).

Diastereomer discrimination in the liquid state is several orders of magnitude larger than liquid state enantiomer discrimination.[15] Titration (neutralization) of either α-methylbenzylamine (b) enantiomer with each mandelic acid (a) enantiomer in water produced an equal amount of heat (ΔH_{neut}). However, in dioxane or in dimethyl sulfoxide (DMSO), thermometric titrations reveal differences of the order of 0.25 kcal mol^{-1} (1 kJ mol^{-1}) between the two nonenantiomeric pairs of reactants (R)-a, (R)-b and (R)-a, (S)-b. Similar results have been obtained in the reactions of the mandelic acid enantiomers with the enantiomers of ephedrine and of pseudoephedrine (Fig. 6.19), respectively. Diastereomer discrimination is observed also in the measurement

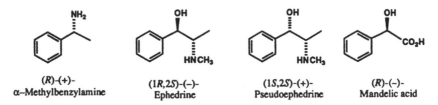

<table>
<tr><td>(R)-(+)-
α–Methylbenzylamine</td><td>(1R,2S)-(-)-
Ephedrine</td><td>(1S,2S)-(+)-
Pseudoephedrine</td><td>(R)-(-)-
Mandelic acid</td></tr>
</table>

Figure 6.19. Compounds exhibiting diastereomer discrimination in acid–base reactions (in the liquid state).

of enthalpies of solution and of dissociation [average values of the several processes measured in solution for the three above-mentioned sets of salts lie between 0.20 and 0.70 kcal mol^{-1} (0.84–2.9 kJ mol^{-1})] and in NMR spectra (in chemical shifts as well as in vicinal spin–spin coupling constants) on the salts.[78,81]

Differences in solubility between the enantiomers of chiral organic compounds exist, as is readily demonstrated by the chromatographic separation of enantiomers on chiral stationary phases. Such differences in retention behavior of the enantiomers may be ascribed to differences in their solubilities in chiral stationary phases. Use of moderately high temperatures (as in GC) is no impediment to such separation. Even modest diastereomer discrimination is subject to amplification in efficient chromatographic columns as the partition process is repeated numerous times (see Section 6-5.d). Diastereomer discrimination also operates in monolayers.[7]

Diastereomer discrimination in solution also is revealed by the differential shielding or deshielding of magnetic nuclei from large external magnetic fields when such nuclei (e.g., ^{1}H, ^{13}C, and ^{31}P) are contained in chiral molecules that are dissolved in optically active solvents; these solvents are specifically chosen for their ability to strongly interact intermolecularly. This type of NMR experiment is described in Section 6-5.c in connection with the determination of enantiomer purity. Information about the configuration of chiral solute molecules may also be obtained from such experiments (see Chapter 5).

The application of diastereomer discrimination has made possible the visual distinction between the enantiomers of chiral compounds.[82] The idea was to combine a chiral crown ether (host) capable of exhibiting enantioselective complexation (Section 7-6) with a chromophore so as to elicit a change in color that would differ according to the absolute configuration of the chiral guest. Success was achieved by linking an achiral monobenzo-18-crown-6 to cholesterol (Fig. 6.20), thereby endowing the crown ether simultaneously with chirality and with liquid crystal properties (cf. Section 12-4.e).

Addition of alkali metal mandelates to the cholesteric mesophase shown in Figure 6.20, leads to enantioselective complexation with an attendant change in the helical pitch of the liquid crystals. In the case of potassium mandelate, the difference in wavelength of maximum reflection for incident light, $\Delta\lambda_{Refl}$ [$\lambda_{Refl} = nP$, where n is the mean index of reflection and P is the helical pitch of the cholesteric mesophase (Section 12-4.e)], amounts to 61 nm (for a given ratio of potassium mandelate to

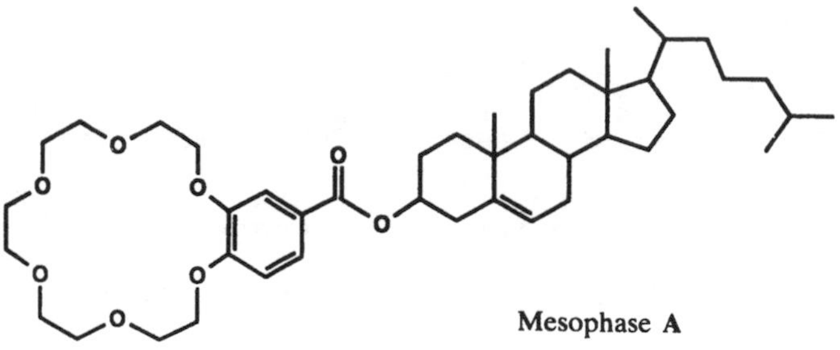

Figure 6.20. Mesophase A.

mesophase **A**, in CHCl$_3$), which is detectable as a change in color, blue in the case of the S enantiomer and green in the case of the R enantiomer.[83] The principle underlying this remarkable result is that the organization of the liquid crystal system (Section 12-4.e) serves to amplify the small diastereomer discrimination that attends the enantioselective complexation of the guest by the chiral host and translates the discrimination into one detectable with the naked eye.[84]

Diastereomer discrimination may also be responsible for stereoselectivity in chemical reactions. Consider, for example, the chlorination (at −60°C) of 2,2-diphenylaziridine with *tert*-butyl hypochlorite. When the reaction is carried out in a chiral solvent [CH$_2$Cl$_2$ containing (S)-(+)-2,2-trifluoro-1-(6-anthryl)ethanol], optically active 1-chloro-2,2-diphenylaziridine (optical purity > 29%) is produced (Eq. 6.6)[85]:

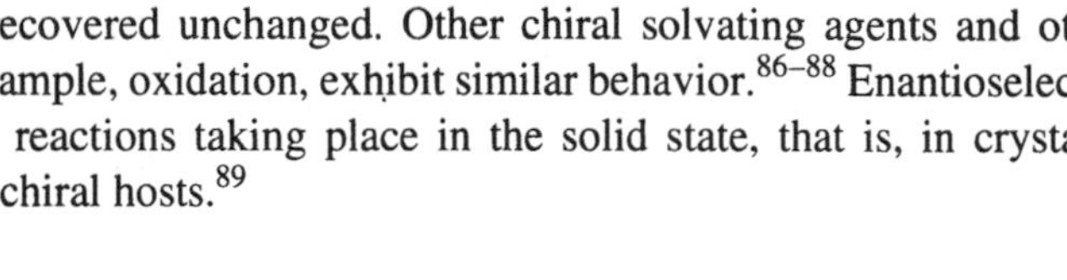

$$(6.6)$$

The chiral solvating agent responsible for the asymmetric induction observed is apparently recovered unchanged. Other chiral solvating agents and other reaction types, for example, oxidation, exhibit similar behavior.[86–88] Enantioselectivity is also exhibited in reactions taking place in the solid state, that is, in crystal lattices of enantiopure chiral hosts.[89]

p. Biological Properties

Diastereomer discrimination is especially striking in biological systems (biological recognition or biodiscrimination), where it is responsible for the differences in taste, in odor, and in other physiological responses to the individual enantiomers of a given substrate, and to racemates as compared to their corresponding pure enantiomers.[6,90]

To the extent that these stereoisomers interact with chiral receptors. biodiscrimination is diastereomer discrimination. One of the first reports of biodiscrimination is that of Piutti (1886) who reported the isolation of dextrorotatory asparagine, $HO_2CCH(NH_2)CH_2CONH_2$, as having a sweet taste, whereas the naturally occurring levorotatory asparagine is tasteless. In his presentation of Piutti's work before the Académie des Sciences de Paris, Pasteur (1886) interpreted this difference in taste as arising from differential interaction of the two enantiomers with dissymmetric nerve tissue (matière nerveuse).[91]

Stereochemical differences affecting the human senses are quite common, as in the case of many amino acids,[92,93] but are by no means universal. For example, it has been reported that for some of the monosaccharides, both enantiomers have virtually the same sweetness.[94] In contrast, of the four stereoisomers of *N*-aspartylphenylalanine methyl ester, it is the L,L-isomer (Fig. 6.21, **22**) that is marketed as a synthetic sweetening agent (under the name aspartame; it is more than 100 times as sweet as sucrose); the L,D-diastereomer, for example, is bitter.[95]

Stereoisomer discrimination in odor perception is also well recognized.[90,96,97] It has become evident that chirality plays a role in the olfactory properties of perfumes and fragrances; cases are known in which the two enantiomers of a pair posses significantly different olfactory properties.[98,99] The experimental results on the carvone and limonene enantiomers (Fig. 6.21, **23** and **24**, respectively) are particularly

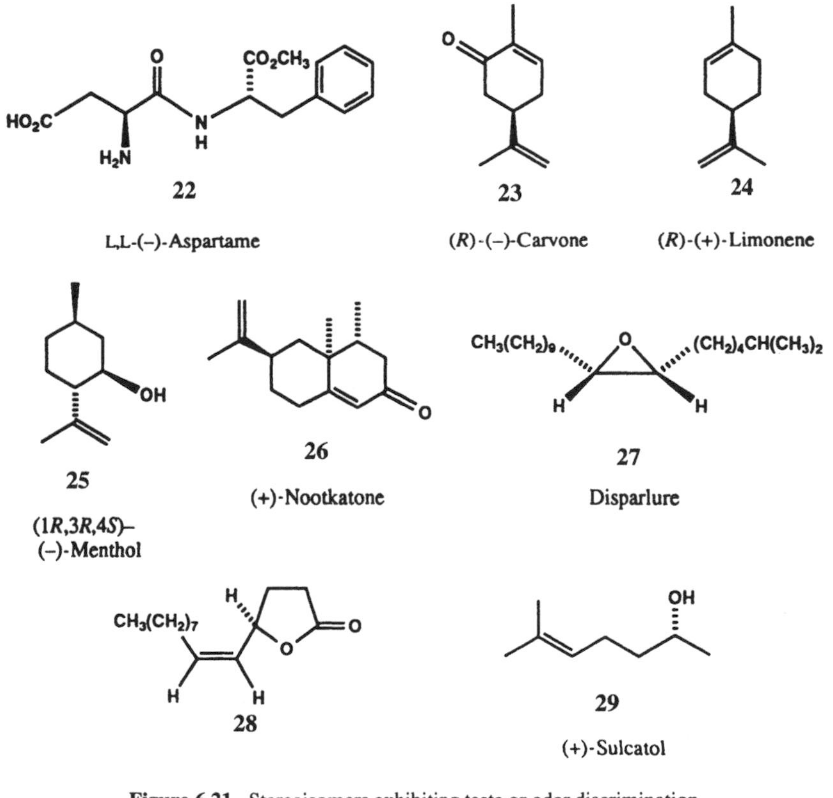
O
CO2CH3
HO2C
N
H
H2N
22
L,L-(–)-Aspartame
O
23
(R)-(–)-Carvone
24
(R)-(+)-Limonene
OH
25
(1R,3R,4S)-
(–)-Menthol
O
26
(+)-Nootkatone
CH3(CH2)9
O
(CH2)4CH(CH3)2
H
H
27
Disparlure
H
CH3(CH2)7
O
O
H
H
28
OH
29
(+)-Sulcatol

Figure 6.21. Stereoisomers exhibiting taste or odor discrimination.

striking[100–102]: (S)-(+)-carvone possesses the odor perception of caraway while (R)-(−)-carvone has a spearmint odor; (R)-(+)-limonene has an orange odor while its enantiomer has that of lemons. Clearly, differences of this kind (stereoisomer discrimination) have commercial consequences. A specific example is that only the (−)-menthol enantiomer (Fig. 6.21, **25**; cf. Section 7-5.b) exhibits the desirable cooling effect in tobacco smoke, as well as the lower threshold (concentration at which the effect is perceived).[103]

Not only does odor quality show considerable differences between many, but not all, enantiomeric compounds but so does their potency. Thus the odor threshold of (+)-nootkatone (Fig. 6.21, **26**; 0.8 ppm), which is responsible for the aroma of grapefruit, is some 750 times lower than that of its enantiomer (600 ppm).[98] There is also evidence that, with respect to some substances, anosmia (loss of the sense of smell) may be stereochemistry dependent.[104]

Numerous studies on insect pheromones revealed that "olfactory" communication among insects is subject to stereoisomer discrimination, for example, in disparlure (Fig. 6.21, **27**), the sex attractant of the gypsy moth.[105] As little as 1% of the "wrong" enantiomer of lactone **28** (Fig. 6.21), the Japanese beetle pheromone, can significantly reduce the biological activity,[106] and, in the case of sulcatol (Fig. 6.21, **29**), the aggregation pheromone of the ambrosia beetle (a timber pest), the racemate is more active than either enantiomer; that is, the response to the enantiomers is synergistic.[107] In some cases, the "wrong" enantiomer can even be repellent or counterproductive.[108]

Naturally occurring chiral foodstuffs all have the "correct" stereochemistry relative to those of the enzymes that catalyze the conversion of polymeric nutrients to monomeric constituents of living cells and that burn up nutrients as fuel for energy production. The issue of stereoisomer discrimination does not arise in the use of naturally occurring foodstuffs. On the other hand, it has been found that while L-glucose is comparable in sweetness to the natural D-enantiomer[94] microorganisms do not metabolize L-glucose.[109] Taking advantage of the inability of enzymes to metabolize L-sugars, the production of nonnutritive sweeteners (L-hexoses having sweetness comparable to D-hexoses or to sucrose) has been patented.[110]

L-Amino acids, such as lysine, are food supplements (in cereals) both for human beings and for poultry and cattle. The quantities of amino acids required for this purpose are large enough that some must be produced synthetically. The matter of stereoisomer discrimination becomes relevant since only the L-enantiomer is usable; some of the strategies used in the economical synthesis of enantiomerically pure amino acids on a large scale are dealt with in Chapter 7. A large and specialized industry has evolved in Japan to provide for the production of amino acids for this purpose, for example, L-lysine ($> 10^4$ tons per year).[111]

D-Amino acids do occur in free and peptide-bound form in Nature.[112] Cell walls and capsules of bacteria and fungi contain peptides constructed partially with "unnatural" D-amino acids, for example, the pathogenic anthrax bacillus capsule consists entirely of poly-D-glutamate.[113,114] The deleterious effect (virulence) of these microorganisms on human beings and on farm animals may stem from the inability of phagocytes to digest bacteria containing D-amino acids in their cell walls and capsules.[115]

On the other hand, D-amino acid oxidase (found in human neutrophilic leucocytes), one of the few enzymes able to process D-amino acids, catalyzes the oxidation of D-amino acids derived from ingested bacteria in the presence of myeloperoxidase. The by-product of the oxidation, H_2O_2, is what actually kills the invading bacteria.[116] This by-product is an instance of a specific defense mechanism that relies on diastereomer discrimination for its effect.

D-Amino acids also are found in oligopeptide antibiotics, such as Gramicidin S.[117] D-Amino acids resulting from the racemization of the naturally occurring L-enantiomers have been found, for example, in aged wine[117] and in processed foodstuffs.[119] The effect of the incorporation of D-amino acids in peptide hormones has been studied.[120]

More recently, an enzyme consisting of a 99 amino acid polypeptide chain, HIV-1 protease, synthesized from D-amino acids only, has been shown to cleave only D-amino acid peptides, whereas the analogous enzyme synthesized from L-amino acids cleaves only L-amino acid peptides.[121,122] The two enantiomers of a chiral inhibitor exhibit similar specificity toward the corresponding enzyme forms. The D-form of the enzyme exhibited equal and opposite CD to that of the L-form consistent with the mirror-image folding of the two enantiomeric enzyme forms (cf. Section 12-4.f).

In recent years, numerous meetings, articles, and reviews have addressed the pharmacological implications of stereoisomerism, especially those of chirality.[123–128] The increased interest has been fueled by several factors among which are the present availability of sensitive analytical methods permitting the monitoring of the enantiomer composition of chiral medicinal agents and their metabolites at therapeutic concentrations in physiological fluids, and the increased facility in synthesizing enantiomerically pure organic compounds. Beginning in 1988 the Food and Drug Administration (FDA) explicitly required the submission of information about the enantiomer composition of chiral substances in new drug applications.[129]

It is by now quite clear that diastereomer discrimination abounds in the domain of medicinal chemistry and pharmacology.[90,124,130–135] A few additional examples must suffice here. Only (*S*)-(–)-3-(3,4-dihydroxyphenyl)-alanine (L-DOPA) (Fig. 6.22, **30**) is active chemotherapeutically in Parkinson's disease. The toxicity of naturally occurring (–)-nicotine (Fig. 6.22, **31**) is much greater than that of the dextrorotatory enantiomer. (–)-Morphine (Fig. 6.22, **32**), and not the synthetic (+)-enantiomer, is the one with the analgesic activity.

The biologically more active isomer of a stereoisomeric pair has been named *eutomer*; the corresponding less potent or inactive isomer is then called the *distomer*.[133] The ratio of the activities (eutomer:distomer), the so-called eudismic ratio, is a measure of the degree of stereoselectivity of the biological activity. Eudismic ratios greater than 100 are not uncommon among chiral medicinal agents.[136]

It must not be thought that enantiomers behave only in the manner of one having biological activity and the other serving only as "stereochemical ballast" when a drug

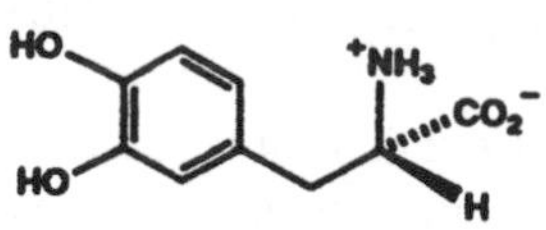

L-DOPA, 30

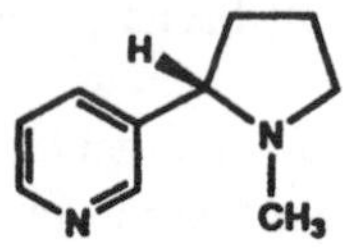

(S)-(−)-Nicotine, 31

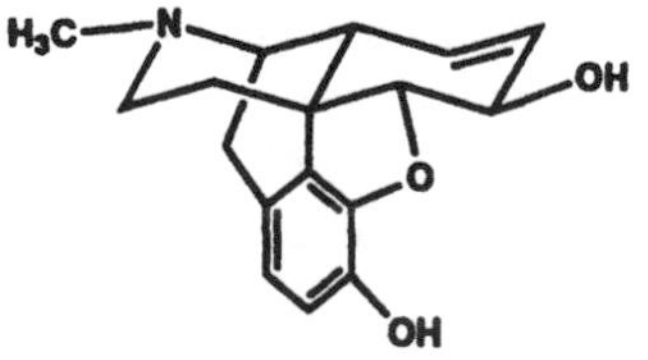

Morphine, 32

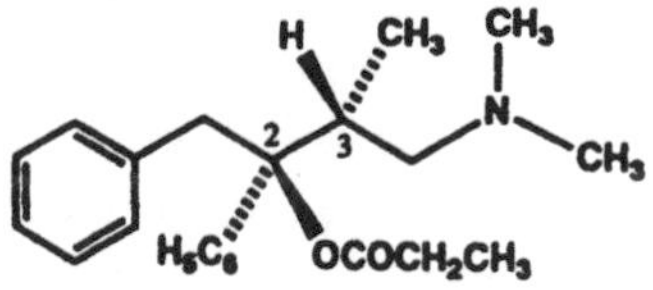

(2S,3R)-(+)-Propoxyphene, 33

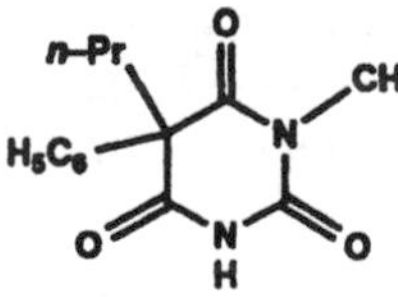

34

1-Methyl-5-phenyl-5-propylbarbituric acid

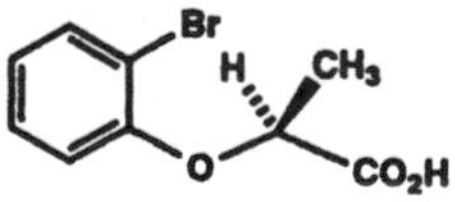

(S)-(+)-α-(2-Bromophenoxy)-
propionic acid, 35

Figure 6.22. Biodiscriminating stereoisomers.

is marketed and consumed as a racemate.[136] In the case of drugs, the distomer can be a eutomer with respect to a different type of activity. Propoxyphene (Fig. 6.22, **33**), a synthetic compound structurally related to morphine and to methadone, exemplifies this phenomenon. The (+)-enantiomer is an analgesic while the (−)-enantiomer is an antitussive. Both enantiomers are marketed (individually). Barbiturate enatiomers, such as those of 1-methyl-5-phenyl-5-propylbarbituric acid (Fig. 6.22, **34**), exhibit opposite effects on the central nervous system (CNS). One enantiomer is a (useful) sedative (hypnotic or narcotic, according to the dosage); the other is an (undesirable) convulsant.[137]

An interesting feature of some stereoisomeric pairs of biologically active compounds is the possibility that their activities will cancel one another.[130] An example is α-(2-bromophenoxy)propionic acid, whose (S)-(+)-enantiomer (Fig. 6.22, **35**) is a plant growth stimulant (auxin) while the (−)-enantiomer is an antagonist thereof (anti-auxin).[138] On the other hand, in some cases, both enantiomers must be present for physiological activity to be manifest.[139]

With respect to chiral compounds in general, four different types of behavior may obtain: (a) The desired biological activity resides entirely in one of the enantiomers, whereas the other is essentially without effect; (b) the enantiomers have identical (or nearly identical) qualitative and quantitative pharmacological activity; (c) the activity is qualitatively identical but quantitatively different between the enantiomers; and (d) the activities of the enantiomers are qualitatively different.[140] For additional examples and summaries of the types of relationship possible between biological activity and the two enantiomers of a given medicinal agent see refs. 132 and 136. In spite of the high degree of biodiscrimination exhibited by stereoisomers, some chiral synthetic medicinal agents are still marketed as racemates.[135] On the other hand, the possibility of patenting an enantiopure drug even when the corresponding racemate is already patented is rapidly changing this situation.[127] One commercial advantage is that patent protection on the drug is indirectly extended.

Biodiscrimination occurs when a chiral compound having a "messenger" function binds to a specific site (or sites) in a receptor molecule, thus being activated so as to elicit a response. In the case chiral compounds, discrimination has tended to be explained in terms of three-point interaction models. These models (pioneered by Easson and Stedman in 1933;[141] see also Chapter 8) stipulate that binding eliciting optimal recognition of the stereochemistry of the reagent (the eutomer) requires interaction at three complementary sites in the messenger and in the receptor molecules (Fig. 8.7).[142-144] A stereoisomer of the messenger (the distomer) might well bind with the eutomer receptor at two or at only one site with the result that the response relative to that of the eutomer is decreased or annihilated. On the other hand, the distomer may bind very well (better than the cited eutomer) with a different receptor with the consequence that the eutomer–distomer roles are reversed at the latter receptor.[135] The greater the affinity of a biological agent toward its receptor the greater its stereoselectivity. This common relationship is known as Pfeiffer's rule.[145]

An interesting aspect of biodiscrimination is the possibility that a given bioactive enantiomer (eutomer) may racemize in solution. An example is (S)-(–)-hyoscyamine (an anticholinergic agent; eudismic ratio (–)/(+) = 200). In view of its rapid racemization, it is administered as the racemate.[135] The antiinflammatory agent ibuprofen (Fig. 6.23, **36**) is also administered as the racemate because the distomer [the (R)-(–)-enantiomer, as measured in vitro] undergoes inversion to the eutomer [(S)-(+)-enantiomer] in vivo.[131,146]

Chiral herbicides, pesticides, and plant growth regulators, chemicals widely used in agriculture, are also subject to biodiscrimination. Examples are the insecticide malathion (Fig. 6.23, **37**), which is applied as a racemate, and paclobutrazol (Fig. 6.23, **38**). In the case of paclobutrazol, both enantiomers are biologically active, as fungicides active against cereal mildew and rust and as plant growth regulators (e.g., in apple seedlings). However, the (2S,3S)-(–)-enantiomer has the greater activity as a growth regulator, whereas the (2R,3R)-(+)-enantiomer is the eutomer with respect to the fungicidal activity.[135] In contrast, of the eight possible stereoisomers of the insecticide deltamethrin, the only two having insecticidal activity are the (1R,3R,αS)- and (1R,3S,αS)-diastereomers (see Section 7-5.a).

Biodiscrimination is, of course, not limited to compounds having carbon stereocenters. The four major nerve gases, including sarin, tabun, and soman, which are potential chemical warfare agents, all are endowed with a stereogenic phosphorus atom having the general structure $R(R'O)P(=O)X$. Although the four have been produced as racemates, it has been demonstrated that the enantiomers (diastereomers in the case of soman) differ in their anticholinesterase activity and lethality.[147]

Stereoisomer discrimination even has significant legal implications. By way of example, forensic chemists employed by law enforcement agencies have had to defend in court analytical work leading to the identification of controlled substances, such as cocaine (Fig. 6.23, **39**). Lawyers for defendants have argued that the identification of the CNS active substance was not carried out "beyond a reasonable doubt" when the analysis, for example, by TLC or by nonchiroptical spectroscopic techniques, could not be expected to differentiate between (–)-cocaine (the naturally occurring and psychoactive isomer) and (+)-cocaine (presumed to be inactive and apparently not a controlled substance) (*People v. Aston*).[148]

q. Origins of Enantiomeric Homogeneity in Nature

While nonracemic and enantiomerically pure compounds found in Nature are understood to arise mainly from chemical reactions catalyzed by enantioselective catalysts (i.e., enzymes), the original source of the latter and of their components, enantiomerically pure amino acids, eludes us. This mystery is a fascinating aspect of science that lends itself to much speculation and indirect experimentation; yet by its very nature the experimentation has led to few definitive conclusions. Part of the fascination stems from the question of the origin of life on earth that is invariably linked to that of the origin of enantiomerically pure compounds.

A very thorough review of this subject is that by Bonner with an update by Avalos and coworkers.[149–151] Bonner classifies all theories bearing on the origin of chiral homogeneity as being *biotic* or *abiotic*. Biotic theories presuppose that life originated

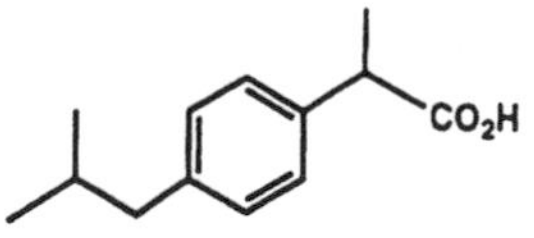

Ibuprofen, **36**

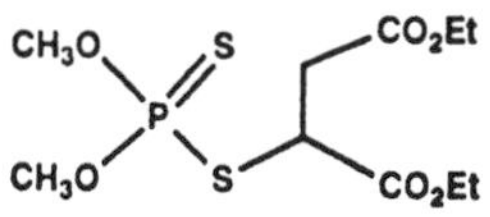

Malathion, **37**

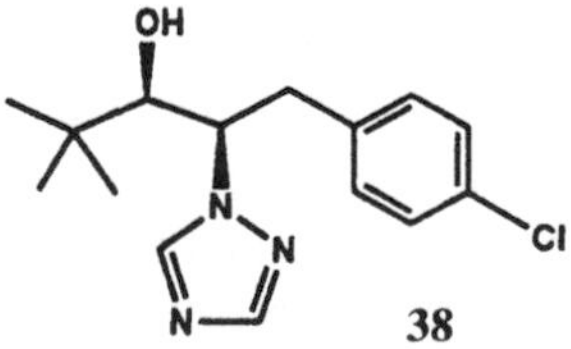

(2*R*, 3*R*)-(+)-Paclobutrazol **38**

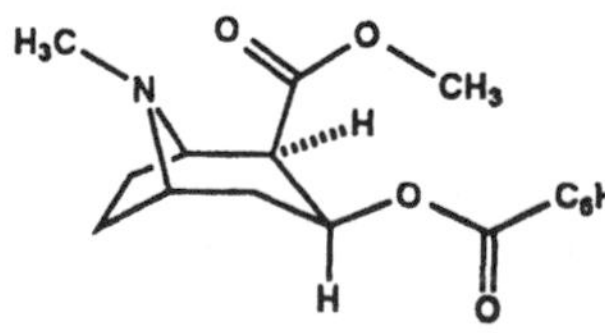

Cocaine, **39**

Figure 6.23. Structures **36–39**.

at an advanced stage of chemical evolution in the presence of numerous racemic building blocks. These theories are consistent with the notion that competing life forms gradually *selected* one enantiomer (the L-amino acids and D-sugars) as being more efficient to survival than their enantiomers. A corollary of such theories is that prebiotic enantiomeric homogeneity is not a *prerequisite* for the origin of life. Bonner classifies biotic theories as speculative and probably impossible to verify experimentally.

Abiotic theories, on the contrary, presuppose that life originated *after* the initial establishment of enantiomeric excess, that is, that the molecules that characterize life processes (e.g., RNA, proteins, and/of DNA) could not have originated or evolved in the absence of the predominance (albeit small) of one of the enantiomeric forms of the requisite precursor molecules. The "abiotic" establishment of an enantiomerically enriched chemical environment could have taken place by chance or in a determinate way. Mechanisms conforming to the chance hypothesis (in any process, the probability of forming either enantiomer of a pair is equal) include the spontaneous crystallization of conglomerates (Section 7-2) and the occurrence of asymmetric transformations, that is, the spontaneous crystallization of conglomerates of easily racemized compounds (Section 7-2.d). Other chance mechanisms include the occurrence of chemical reactions in chiral crystals under lattice control, in cholesteric (liquid crystal) phases, and adsorption or catalysis on chiral solid supports, such as quartz crystals of a given configuration. Although the operation of chance mechanisms is quite plausible, Bonner concluded on the basis of statistical considerations that the random establishment of chiral homogeneity at several sites on earth would not likely have led to the dominance of one of these over time.[149]

Determinate mechanisms assume that some chiral physical force acting on (or during formation of) racemates is responsible for establishment of an initial enantiomer excess. The latter is subsequently amplified (e.g., by polymerization mechanisms; see below) until enantiomeric homogeneity obtains and the resulting enantiopure compounds are converted to those molecules that we recognize as being essential to life processes. Experimental efforts aimed at the verification of determinate mechanisms have been the subject of numerous and fruitful studies over the past three decades.[151-153] The principal mechanisms studied involve the violation of parity and the interaction of matter with chiral radiation.

> The notion of universal dissymmetric forces that pervade the world and are responsible for optically active natural products stems from the 1850s and was pioneered by Pasteur. Many wrong (i.e., theoretically impossible) forces have been suggested based on dubious experiments.[154]

In 1956 Lee and Yang proposed that parity is not conserved in the weak interactions (e.g., in β decay).[155] Experimental verification of their proposal was provided shortly thereafter; Wu and co-workers demonstrated that electrons emitted during the decay of ^{60}Co nuclei were longitudinally polarized in a left-handed way to a greater extent than in a right-handed way.[156] These results constituted the first experimental evidence that the parity principle (that the laws of Nature are invariant under spatial reflection) is not conserved for weak interactions. Since the parity principle also requires that elementary particles be present in mirror-image forms,[157] the violation of parity bespeaks an imbalance between the quantities of matter and antimatter.[15,158]

A consequence of the nonconservation of parity is the prediction that the two enantiomers of a chiral compound are of slightly unequal energy.[159] This prediction arises from the parity-violating neutral current interaction between the two enantiomers of a compound. The evaluation of parity-violating energy differences as a possible source of chiral homogeneity in Nature by different authors has been reviewed.[149,160] Attempts have been made to produce enantiomeric excesses by the intervention of electric, magnetic, and gravitational fields on stereoselective syntheses; the results have been highly controversial.[149]

In contrast to the above very small and highly controversial effects, absolute asymmetric synthesis under the influence of circularly polarized light (cpl, a "truly chiral" physical force) has been observed numerous times. The most successful experiments are of two types: photochemical asymmetric synthesis and asymmetric photolysis. The former type is exemplified by the cyclization of diarylethylenes to helicenes (e.g., Fig. 6.24). Optically active hexahelicene (Fig. 6.6, **3**), $[\alpha]_{436}^{23} - 30.0$ (CHCl$_3$) was obtained by irradiation of compound **40** with right circularly polarized light (rcpl), while **3** having $[\alpha]_{436}^{23} + 30.5$ (CHCl$_3$) resulted on irradiation with left cpl (ee < ca. 0.2%).[161] Similar results were obtained by Bernstein, Calvin, and Buchardt.[162,163] In such syntheses, the enantiomeric purity of the product is independent of the conversion and is related to the differential absorbance of the cpl: $\Delta\varepsilon/2\varepsilon = g/2$ (for a definition of the γ number,[164] see Section 12-4.a; see also Chapter 13.

The second of the mentioned processes, asymmetric photodestruction, is the more important one from the standpoint of the abiotic generation of enantiomerically enriched compounds. Three reasons for the importance attached to asymmetric photodestruction are that (a) the process is widely applicable to virtually any type of racemic organic compound possessing a chromophore; (b) the enantiomeric excess attainable can be significantly higher and can approach 100% if the γ number (Section 12-4.a) is relatively high and the conversion is appreciable (cf. Section 7-5); and (c) a source of cpl is available on earth since cpl is known to be generated in the sky by reflection and scattering from aerosols.[149] The asymmetric photolysis of a biologically

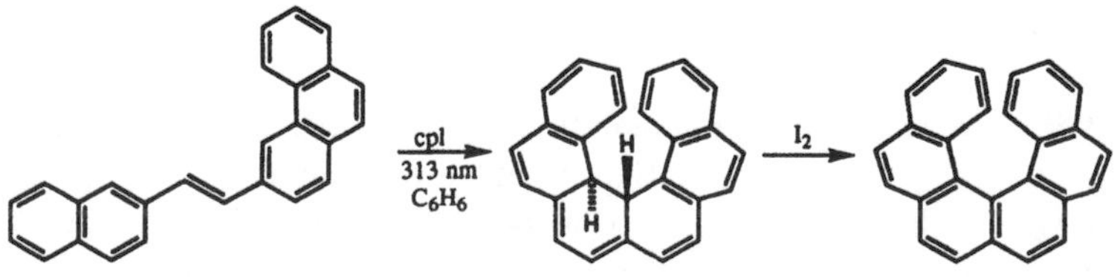

Figure 6.24. Photochemical asymmetric synthesis of hexahelicene.

significant compound, DL-leucine, with rcpl (at 212.8 nm; the light was obtained from a laser source) generated samples of 2% ee (L-leucine > D-leucine) at a conversion of 59% and of 2.5% ee (D > L) on 75% conversion with lcpl without concurrent photoracemization.[165]

The operation of such determinate mechanisms provides for formation of small enantiomer excesses of chiral compounds. Amplification processes must be operative if the latter are to yield, ultimately, enantiomerically pure compounds. A number of such processes are described in Chapter 7, namely, total spontaneous resolution (Section 7-2.d) and amplification during incomplete reaction (Section 7-4). However, since these processes are not general ones, one could hardly claim that they are the likely ones to have led to the enantiomeric homogeneity that characterizes the important biomolecules found nowadays in living systems.

One set of amplification experiments that is quite pertinent to the origin of biomolecular chiral homogeneity is that which occurs during polymerization. Incomplete polymerization of enantioenriched alanine N-carboxyanhydride (Fig. 6.25, **41**, R = CH$_3$; L > D) leads to preferential incorporation of the predominant enantiomer at the beginning of the polymerization.[166] The selectivity has been attributed to the configuration of the amino acid at the growing end of the polymer chain as well as by a cooperative amplification resulting from formation of an α-helix conformation in accord with an earlier proposal by Wald.[167] The latter rationale is reminiscent of the formation of partially isotactic polymers by *stereoelective* polymerization of nonracemic, enantiomerically pure monomers.[168]

Analogous results were found for leucine N-carboxyanhydride (Fig. 6.25, **41**, R = i-Bu) but not for valine (Fig. 6.25, **41**, R = i-Pr), the latter being unable to form an α-helix due to steric hindrance.[169,170] Subsequently, Blair, Dirbas, and Bonner demonstrated that enantiomeric enrichment also took place during partial hydrolysis of polypeptides, for example, with poly(DL-leucine) hydrolyzing faster than poly(L-Leu) or poly(D-Leu).[171] Consequently, enantiomeric enrichment takes place during partial hydrolysis of poly(Leu) prepared from nonracemic leucine, for example, 45.4% ee, with the recovered peptide containing leucine having as much as 54.9% ee.

Also pertinent is the synthesis of nonracemic poly(triphenylmethyl methacrylate), an isotactic linear polymer whose chiroptical properties are due solely to the generation of a preferred helicity sense (cf. Section 6-5.d). The latter arises during the polymerization of (achiral) triphenylmethyl methacrylate monomer by butyllithium in the presence of (−)-sparteine (monomer/initiator ratio ≥40:1). It is evident that one

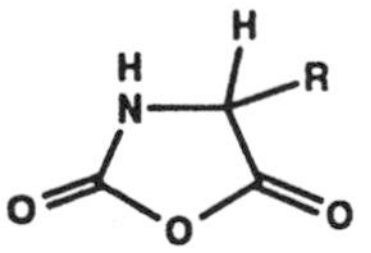

41

Figure 6.25. Alanine N-carboxyanhydride (R = CH$_3$).

chiral molecule in the anionic polymerization initiator (a molecule that is not even incorporated in the polymer) is responsible for the generation of a preferred helicity sense in a polymer molecule incorporating as many as 200 monomers (at least during the initial phases of the polymerization).[172]

Cooperative effects responsible for the amplification of enantiomeric bias is also seen during the copolymerization of achiral hexyl isocyanate with as little as 0.12% of nonracemic chiral isocyanate. The resulting polyisocyanate copolymer consists of a 56:44 mixture of mirror-image helices (at −20°C).[173] The even more dramatic influence of minor chiral perturbations is seen in the polymerization of (R)-1-deuterio-1-hexyl isocyanate (n-$C_5H_{11}CHD$—NCO) giving rise to a helical polymer having a large optical rotation, $[\alpha]_D^{10}$ − 450 ($CHCl_3$). This rotation has been interpreted as arising from an excess of one of the helical senses and not from a structural or conformational perturbation. The amplification mechanism giving rise to the helical sense excess is due to a conformational equilibrium isotope effect in which the energy difference per deuterium, hence per polymer residue, is very small [ca. 1 cal mol^{-1} (ca. 4 J mol^{-1})].[174,175]

These model experiments suggest that the secondary structure of a polymer is implicated in the amplification of enantiomeric bias during polymer synthesis and that it can protect the stereochemical integrity of the polymer once formed during subsequent partial degradation. The relevance of model experiments demonstrating amplification of low enantiomeric enrichment in polymers to the broader question of the origin of enantiomeric bias in Nature is evident because of the important role that chiral biopolymers (polysaccharides, polynucleotides, and proteins) play in the maintenance of the enantiomeric homogeneity in Nature.[149,154]

6-5. DETERMINATION OF ENANTIOMER AND DIASTEREOMER COMPOSITION

a. Introduction

Let us review the terms that are used to describe the enantiomer composition (i.e., the proportion of enantiomers) of chiral samples. Although the adjective *chiral* has often been used descriptively to imply that a sample is nonracemic (commonly optically active),[176] it should be reserved to characterize, in a conceptual way, molecules (and by extension, compounds), crystals, or objects that are not superposable with their mirror images (cf. Section 1-2).

The enantiomer composition of macroscopic samples of chiral compounds, requires its own descriptive adjectives. The term racemic (1:1 proportion of enantiomers) is time-honored, unequivocal, and universally accepted. Note that racemic and chiral are not incompatible adjectives: only a chiral substance can give rise to a racemate. A term is clearly needed to describe the composition of samples that until now have been called optically active. The latter expression suffers from being too closely linked to the measurement of chiroptical properties that are gradually being deemphasized for the purpose of enantiomer purity determination. Moreover, it is now much better understood than heretofore that enantiomerically enriched (even pure) samples need not be optically active (at a given concentration, temperature, wavelength, or in a given solvent; see Section 12-2). In this book, we have used the adjective nonracemic to describe samples of chiral compounds whose enantiomer composition is somewhere between 50:50 and 100:0.[177] It is thus correct, albeit imprecise, to speak of a nonracemic sample even if it is enantiomerically pure.

The enantiomer composition of a sample may be described as a dimensionless mole ratio (or as mole percent of the major enantiomer) and this is, in fact, the most generally useful way to describe the composition of all types of stereoisomer mixtures. A second and very common term used in this connection is *enantiomer excess,* usually expressed as a percentage. The latter expression describes the excess of one enantiomer over the other. The percentage enantiomer excess, ee = $100(x_R - x_S)/(x_R + x_S)$, where $x_R > x_S$. Alternatively, ee = $100(2x - 1)$, where x is the mole fraction of the dominant enantiomer in a mixture. For a mixture that is rich in one enantiomer, for example, 80:20, $x = 0.8$ but the enantiomer excess is only 60%. The converse relation is $x_R = (ee_R + 100)/200$, where the dominant enantiomer is R.

Enantiomer excess is defined so as to correspond to the older expression, *optical purity*: op = $([\alpha]_{obs}/[\alpha]_{max}) \times 100\%$. The maximum or absolute rotation $[\alpha]_{max}$ is that of an enantiomerically pure sample.[178] For the use of an analogous expression, diastereomer excess, see Thaisrivongs and Seebach.[179]

Although in the late 1980s the use of rotation measurements to determine stereoisomer composition has decreased relative to other methods (see above), the expression optical purity continues to be used even when rotation is not the basis of the determination. This usage is not only anachronistic but it is occasionally inappropriate, since it stems from the generally held belief that enantiomer excess and optical purity are numerically equal,[180] whereas experimental studies (to be described) have shown that this is not necessarily the case.[181] In consequence, we do not use the two expressions synonymously in this book.

The principal methods for the determination of enantiomer composition are summarized in Table 6.3. The table gives the basis of the measurement and its experimental nature. It indicates whether the measurement is undertaken on the intact mixture or whether pretreatment of the original mixture has taken place, that is, quantitative conversion of an enantiomer mixture to a mixture of diastereomers, or quantitative conversion to a different enantiomer mixture to allow a particular analysis to be carried out (e.g., conversion of a liquid to a solid mixture). Note that complete separation of the stereoisomers is required only in the case of the chromatographic methods.

Few functional group types are not now amenable to analysis by one of the methods shown in Table 6.3. Even chiral hydrocarbons (at least some of them) respond to one or more of these methods. The principal limitation in methodology is that a very low enantiomer excess (below ca. 2%) or a very high one (above ca. 98%) remains difficult to measure with precision, that is, reproducibly in different laboratories.

TABLE 6.3   Methods for the Determination of Enantiomer Composition

Basis of Measurement	Nature of Measurement	Treatment[a]	Species Examined[b]
1. Chiroptical	A. α, φ, or Δε	I[c]	E or D
	B. Circular polarization of emission	I	E
2. Diastereotopicity (external comparison) (see Chap. 8)	A. NMR of diastereomers in achiral solvents[d]	Der	D
	B. NMR in chiral solvents (chiral solvating agents)	I	E
	C. NMR with chiral shift reagents	I	E
3. Diastereomeric interactions (separation)	A. Chromatography on diastereoselective stationary phases		
	i. GC	Der	D
	ii. HPLC	Der	D
	iii. HPLC with chiral solvent	I	E
	iv. TLC	Der	D
	B. Chromatography on enantioselective stationary phases	I	E
	i. GC		
	ii. HPLC		
	iii. TLC		
	C. Electrophoresis with enantioselective supporting electrolyte	I	E
4. Kinetics	Product composition	I or Der	E or D
5. Enzyme specificity	Quantitative enzyme-catalyzed reaction	I	E
6. Fusion properties	Differential scanning calorimetry	I or Der	E or D
7. Isotope dilution	Isotope analysis	I[e]	E
8. Potentiometry	Potential of an electrochemical cell	I	E

[a]I = intact mixture; Der = analysis on a diastereomeric derivative.
[b]E = original enantiomer mixture: D = diastereomer mixture prepared from enantiomer mixture to be examined.
[c]A derivative prepared with an achiral, chromophoric reagent may be used.
[d]Also, NMR in the solid state.
[e]This method requires reisolation of a sample following the dilution procedure.

The choice of method to be used depends on a variety of factors, not the least of which is convenience and the availability of the necessary instrumentation. Moreover, it is necessary to take into account the purpose of the measurement in choosing the method to be used. Very precise analyses would not normally be required while monitoring the progress of a resolution.

Great care must be taken not to modify the enantiomer composition of a mixture prior to the analysis lest the results be invalidated. Chemical purification, as in the workup of a reaction mixture, involving washing, crystallization, or sublimation of solids will change the enantiomer ratio of the mixture. Even chromatography of a partially resolved chiral sample on an achiral stationary phase may occasionally modify the enantiomer composition (Section 6-4.m) especially if carried out carefully, that is, when small fractions are individually analyzed. This possibility follows from the principles outlined in Section 6-4.

It is fair to say that no one method for the determination of enantiomer composition, whether spectroscopic, chromatographic, or other technique is universally applicable. Each new case requiring such analysis must be examined individually and a choice of method made based on the structure of the chiral analyte, the state of the art at the moment, local resources and experience, and the required precision of measurement. A comprehensive survey of the better known and most useful methods comprises an entire volume.[88] Reviews dealing with individual methods are cited in the appropriate sections that follow.

b. Chiroptical Methods

These methods, involving the measurement of optical rotation typically at a single wavelength, provide results (optical purity) rather fast but the information is often not very precise, nor is it necessarily very accurate. The measurement of optical rotation is the traditional method for assessing enantiomer composition; it requires a knowledge of the specific rotation, or molar rotation, of the pure compound under investigation, op $= 100[\alpha]/[\alpha]_{max}\%$. Numerous examples are known wherein the enantiomer composition determined in this way in earlier years has been found to be incorrect. Such errors arose principally because the assumption of complete resolution of the sample used for determining $[\alpha]_{max}$ was invalid[182] or because the sample examined contained solvent residues or other impurities.

We must also remember that the optical rotation, and hence the optical purity determination, may be affected by numerous variables: wavelength of light used, presence or absence of solvent, the nature of the solvent used in the rotation measurement, concentration of the solution (even though this variable is factored into the calculation of the specific rotation, Eq. 1.1 and Section 12-5.a), temperature, and presence of impurities; the effect of these variables will be analyzed in detail in Chapter 12. This is a long list of variables that potentially modify the specific rotation to be compared to $[\alpha]_{max}$. The latter, whether obtained directly or indirectly by calculation, may have been determined in another laboratory and under circumstances that are not sufficiently well specified; for optimal results, the rotations to be compared should, of course, be measured under precisely identical conditions (see also Section 1-3). Since this is difficult to do, if not impractical in most circumstances, and since the precision of polarimetric measurements is typically no better than 1–2%, the optical purity will, in many cases, give only a rough estimate of the enantiomer composition of a mixture. New developments in polarimetry are taken up in Section 12-5.a.

Asymmetric hydroformylation of styrene leads to optically active 2-phenyl-propanal with an optical purity of about 95% (max) as determined by comparison of the rotation of purified product $[\alpha]_D^{21} + 224.8$ measured in benzene (c 1.5–20) with the maximum specific rotation reported earlier $[\alpha]_D^{25} + 238$ (neat).[183] Subsequent reevaluation of the optical purity revealed a significant difference in the specific rotation of the product as measured in benzene and in the absence of solvent (neat) as well as a concentration effect, $[\alpha]_D^{21} + 214.7$ (c 1.5, benzene) versus $[\alpha]_D^{21} + 182.2$ (c 46.4 benzene) with all measurements being made using a sample of fixed enantiomer composition (op 68%) (see page 5 for units of c). Note the substantial increase in specific rotation as the concentration is decreased. Consequently, the actual optical purity of the product obtained as described above needed to be recalculated: op ca. 73% instead of 95%.[184]

The concentration effect arises as a consequence of association phenomena that vary according to the functional groups that are present and the solvent that is used. It is pronounced when polar molecules (especially alcohols and carboxylic acids) are measured in nonpolar solvents. Polyfunctional molecules, for example, diols and hydroxy acids, are also prone to exhibit concentration effects as intramolecular aggregates give way to intermolecular ones when fewer solvent molecules are present.

These data suggest that the measured optical purity may not always be numerically equal to the enantiomer excess. The discrepancies discussed so far occur because the two specific rotations being compared were not measured under strictly identical conditions. However, there is another possible source of nonequivalence of enantiomeric excess with optical purity, namely, enantiomer discrimination in mixtures of enantiomers of composition intermediate between 50–50 and 100%. Such discrimination would undermine the assumption of strict additivity of specific rotations that is the basis of the optical purity determination.

The nonequivalence of enantiomer and optical purities was first observed by Horeau in α-ethyl-α-methylsuccinic acid (Fig. 6.26, **42A**).[181] For the pure acid $[\alpha]_D^{22} + 4.4$ (c 15, $CHCl_3$), hence Horeau calculated $[\alpha]_D^{22} + 2.2$ for a 75:25 enantiomer mixture (ee = 50%). He actually observed $[\alpha]_D^{22} + 1.6$ (c 15, $CHCl_3$), corresponding to op ca. 36%. The effect was only observable in weakly polar solvents (CH_2Cl_2, $CHCl_3$, and C_6H_6); the discrepancy between enantiomeric excess and optical purity was found to disappear in polar solvents (ethanol, pyridine, diglyme, acetonitrile). The structurally related α-isopropyl-α-methylsuccinic acid (Fig. 6.26, **42B**) has also been shown to exhibit the *Horeau effect* as this lack of linearity between specific rotation and enantiomer purity has come to be known. It shows an even larger divergence of measured $[\alpha]_D^{22} + 7.3$ (c 0.8, $CHCl_3$) for an enantiomerically pure sample from that calculated from the rotation of sample of op ca. 25%: $[\alpha]_D^{22} + 17$. The discrepancy amounts to 118%.[16,61]

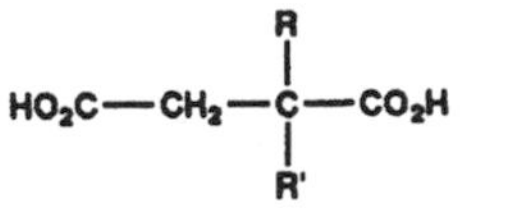

$$HO_2C—CH_2—\overset{\overset{\displaystyle R}{|}}{\underset{\underset{\displaystyle R'}{|}}{C}}—CO_2H$$

42

A : R = Et ; R' = CH_3

B : R = *i*-Pr ; R' = CH_3

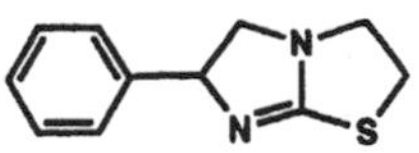

43

Figure 6.26. Compounds exhibiting the Horeau effect.

The departure from linearity is influenced by factors such as the wavelength at which the specific rotation is measured and by the concentration (note that for enantiopure acid **42A**, whose rotation is strongly concentration dependent in chloroform, $[\alpha]_D^{22}$ is 0 at c 6.3, CHCl$_3$). The effect is the more important the smaller the enantiomer excess of the sample; it vanishes in the vicinity of 100% ee and as the ee tends to 0%. A striking demonstration of the Horeau effect is given by the finding that for acid **42A** the pure R enantiomer is dextrorotatory, $[\alpha]_D^{22} + 0.6$ for c 7.5, while a 75:25 enantiomer mixture is levorotatory, $[\alpha]_D^{22} - 1.5$ at the same concentration.[61] In the case of **43** (Fig. 6.26), the observed Horeau effect has also been shown to depend on temperature.[185]

The Horeau effect appears not to have been observed in other covalent organic compounds. It has been observed in ionic complexes where it occurs even in aqueous solution (mandelic acid in the presence of ammonium molybdate,[186] and mixtures of cupric and potassium tartrates[187]).

c. NMR Methods Based on Diastereotopicity

NMR of Diastereomers and Chiral Derivatizing Agents.   The first NMR technique to be applied in the determination of enantiomer composition was the analysis of covalent diastereomer mixtures.[188–192] This approach is exemplified by the derivatization of chiral alcohols and amines with optically active acids.

In the pioneering work of Raban and Mislow,[193,194] alcohols and amines were converted to esters and amides, respectively, by reaction with the *chiral derivatizing agent* (CDA) (R)-(−)-O-methylmandeloyl chloride (acid chloride of **44**, Fig. 6.27) (Eq. 17)[195,196]:

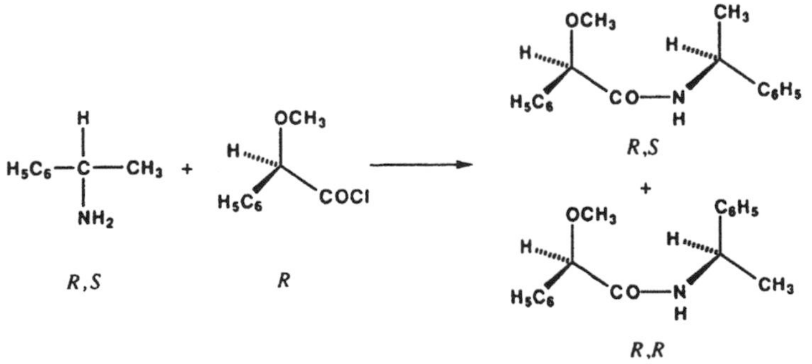

(6.7)

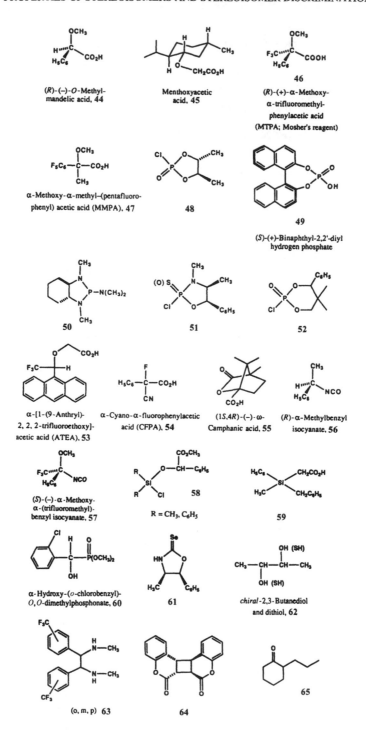

Figure 6.27. Chiral derivatizing agents for NMR analysis of diastereomers.

[For the application of (−)-menthoxyacetic acid (Fig. 6.27, **45** to such determinations, see Galpin and Huitric[197] and Cochran and Huitric.[98]] Raban and Mislow pointed out the advantage of having easily identifiable diastereotopic nuclei in the CDA, which, in some instances, might serve as a resolving agent (see Chapter 7 for the latter usage) as well as serving to monitor the resolution by NMR spectroscopy.[180] If the NMR resonances contributed by the diastereomer moiety originating in the analyte are not sharp and well resolved, so as to permit accurate integration, then diastereotopic nuclei contributed by the CDA, for example, the isopropyl CH_3 groups or the side chain acetate CH_2 groups in the menthoxyacetates or menthoxyacetamides might serve this purpose.

Subsequent developments were prompted, in part, by the observation that some *O*-methylmandelate esters and *O*-methylmandelamide derivatives are subject to epimerization at the hydrogen alpha to the carbonyl. This factor leads to incorrectly low values of the enantiomer composition.[199] α-Methoxy-α-trifluoromethylphenylacetic acid, *Mosher's reagent* (MTPA reagent; Fig. 6.27, **46**), was designed to avoid this problem by eliminating an α-hydrogen atom in the CDA.[200] At the same time, incorporation of fluorine by way of a CF_3 group made possible analysis of the ester and amide derivatives by means of ^{19}F NMR spectroscopy. The latter often simplifies the analysis since there is much less likelihood of finding overlapping peaks; the number of peaks is smaller and they are better separated than are comparable peaks in ^1H NMR (the chemical shift dispersion of ^{19}F NMR is greater than that of ^1H NMR).[201]

Parallel application of such diastereomer derivatives to enantiomer composition determination by GC (see Section 6-5.d) led to the development of other CDAs including MMPA (Fig. 6.27, **47**).[202] In a similar way, application of the *Anderson–Shapiro reagent* (Fig. 6.27, **48**) to the analysis of nonracemic samples of primary and secondary alcohols by ^{31}P NMR spectroscopy is simple in view of the absence of interfering bands in the spectrum of the phosphorus ester mixture.[203] Since the phosphorus atom is nonstereogenic as a consequence of the C_2 symmetry axis of the parent glycol moiety, derivatization of a given analyte enantiomer with **48** gives rise to but one diastereomer irrespective of the mechanism of bond formation to phosphorus, retention, or inversion. The binaphthylphosphoric acid **49** (Fig. 6.27), a C_2 symmetric reagent, may be a useful CDA for chiral alcohols.[204,205] More recently, a C_2 symmetric diazaphospholidine (Fig. 6.27, **50**) has been shown to be a powerful and easy to use CDA for the determination of enantiomeric purity of a wide range of alcohols and thiols, even in the presence of other functionalities.[206]

Several 1,3,2-oxazaphospholidine sulfides and oxides derived from (−)-ephedrine (Fig. 6.27, **51**) have been applied as CDAs in the determination of the enantiomeric purity of alcohols and amines.[205] Here the phosphorus atom is chiral and, in contrast to C_2-symmetric CDAs (above), an enantiopure alcohol can give rise to more than one diastereomeric phosphorus ester according to the mechanism of reaction. In fact, mixed mechanisms have been observed in reactions of **51** (especially in the presence of pyridine) leading to loss of stereochemical integrity, and this reduces the utility of this CDA.[207] A 1,3,2-dioxaphosphorinane oxide (Fig. 6.27, **52**)[208] is an excellent CDA for the determination of the enantiomer composition of amines, even though it suffers from the same problem.

Dale and Mosher[199,209] proposed that the spectral anisochrony (chemical shift difference, $\Delta\delta$) observed between the two diastereomers produced from enantiomeric substrate molecules is dependent on a combination of steric and electronic nonbonded interactions enhanced by the anisotropy of the aromatic ring in the CDA. These interactions tend to populate conformations that are quite distinct for the two diastereomers. In consequence, selected nuclei in these conformations exhibit significant chemical shift differences for the two diastereomers.

Another development is the application of the ATEA reagent (Fig. 6.27, **53**) to the determination of enantiomer composition of amines, alcohols, and thiols.[210] The design of this CDA is based on the hypothesis that conformationally more rigid diastereomers would engender greater anisochrony. More specifically, the observed anisochronies are weighted averages of effects caused by all conformations present in the diastereomer mixtures, both those that exhibit nonequivalence of chemical shifts and those that do not. Reduction of conformational mobility might increase the population of the former (see below for another way of achieving this end).

In order that analysis of the amide (or other diastereomer) mixture accurately reflect the substrate enantiomer composition, it is essential that reaction be complete. The CFPA reagent, α-cyano-α-fluorophenylacetic acid (Fig. 6.27, **54**), has recently been proposed as an alternative CDA to the widely used MTPA reagent (see above) particularly for hindered substrates, such as 3,3-dimethyl-2-butanol, whose reaction with MTPA chloride is relatively slow and may be incomplete under a given set of conditions.[211] In cases tested thus far, diastereomers prepared with the CFPA reagent also have been found to have greater $\Delta\delta$ values than for the corresponding MTPA diastereomers in both ^{19}F and ^{1}H NMR spectra.[212] In addition, neither reaction partner must be prone to racemization during reaction and neither equilibration nor alteration of the diastereomer ratio must be allowed to take place following reaction.

A significant limitation of the method is that the nonracemic CDA *must* be enantiomerically pure (see Section 6-5.d for a justification); it is wise to check this before using the reagent, for example, by combining it with a known enantiomerically pure substrate and ascertaining that only one product diastereomer results. In spite of this disadvantage relative to some of the other analytical methods, diastereomer analysis by NMR is very popular. Also confirmation of the validity of the method (by observation of anisochronous signals) must always be undertaken by reaction of the CDA with a racemic sample of the analyte. If the racemate is unavailable (as in the case of natural products), then the validity of the method could be ascertained by analysis of a derivative obtained on reaction of the analyte with the enantiomeric CDA. Such has been the case with $(1S,4R)$-$(-)$-ω-camphanic acid.[213-215]

One may take advantage of two other means of enhancing the anisochrony between diastereomers: solvent effects and addition of achiral shift reagents. Determination of NMR spectra in benzene (also pyridine, trifluoromethylbenzene, and halogen-substituted aromatics[216]), instead of in CCl_4 or $CDCl_3$, often leads to increased peak separations (aromatic solvent-induced shifts, ASISs) that have facilitated configurational assignments. The improved resolution may also permit or simplify the quantitative analysis of diastereomer mixtures.[217,218]

The second mentioned enhancement process is an application of *achiral* shift reagent techniques.[190] For example, addition of Eu(dpm)$_3$ (dpm = dipivaloylmethane; see Fig. 6.31) to the camphanates of chiral α-deuterated primary alcohols leads to a chemical shift difference of as much as 0.5 ppm between the diastereotopic α-methylene protons.[214,219] An example involving enhanced anisochrony in ^{19}F spectra is given by Merckx.[220] It has been postulated that achiral lanthanide shift reagents [e.g., Eu(fod)$_3$; Fig. 6.31] enhance the separation (Δδ) of specific peaks, such as the OCH$_3$ in the MPTA moiety (see **46** in Fig. 6.27) of MTPA esters by reducing conformational mobility within the diastereomeric derivatives.[221]

The structures of the most useful CDA reagents are shown in Figure 6.27. In addition to the CDA already mentioned, α-methylbenzyl isocyanate (**56** has been applied to the quantitative analysis of chiral amines by formation of dieastereomeric ureas[222,223]; an analogous reagent is **57**.[224] The chiral silyl reagent **58** has been applied to the analysis of chiral alcohols.[225] A second silyl reagent (**59**) incorporating a stereogenic silicon atom has been applied as CDA for alcohols and amines.[226,227] Diastereomeric esters or amides prepared with **59** generate signals in the Si-CH$_3$ region of the ^{1}H NMR spectrum that are generally free of interference by other signals, thus simplifying spectral interpretation.

Chiral carboxylic acids can often be analyzed as amides with optically active α-methylbenzylamine (Fig. 6.19).[228–230] Compound **60** has been proposed as a reagent for the determination of the enantiomer composition of carboxylic acids.[231] 2-Chloropropionic acid has been applied to the enantiomeric purity analysis of amino acids.[232] More recently, a selenium reagent, oxazoildine-2-selone **61** (Fig. 6.27), having a very high [77]Se chemical shift sensitivity, has been shown to be a useful reagent for the measurement of enantiomeric purity of carboxylic acids even when the latter bear stereocenters remote from the carboxyl group.[233]

Ketones have been analyzed by ^{13}C NMR following conversion to diastereomeric ketals with *threo*-2,3-butanediol, or with *threo*-2,3-butanedithiol (Fig. 6.27, **62**) to thioketals[234–236] and aldehydes by conversion to imidazolidines with reagent **63**.[237] The latter three CDA possess a C_2 axis (see also Fig. 6.27, **64**, a CDA useful in the analysis of amines).[238] In the absence of this symmetry element, an additional stereogenic center (hence more than two diastereomers) would be created during formation of the derivative, unless the ketone itself possesses a C_2 axis. The reciprocal process has been applied to the analysis of glycols as ketals formed with optically active 2-propylcyclohexanone (Fig. 6.27, **65**).[239]

While the CDAs thus far described have all involved formation of covalent derivatives, it was observed by Horeau and Guetté[240] and Guetté et al.[186] that ionic derivatives could also serve as probes for the determination of enantiomer composition. The induction of nonequivalence in NMR spectra of diastereomeric salts, that is, chiral ions in the presence of chiral counterions, is illustrated in Figure 6.28.

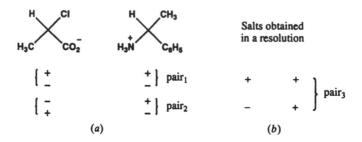

Figure 6.28. Nuclear magnetic resonance nonequivalence in diastereomeric salts.

When racemic acids and bases are combined to form diastereomeric salt mixtures, each ion in solution is surrounded by both of the enantiomeric counterions. Since the ions change partners at a rate rapid on the NMR time scale, chemical shift differences of groups are averaged out and the mixture of the four salts (two sets of enantiomeric pairs; Fig. 6.28a) behaves as if it were a single substance. However, when one of the counterions is present in only one enantiomeric form, as in the resolution of *rac*-2-chloropropanoic acid with (+)-α- methylbenzylamine, diastereomeric ion pairs are obtained (Fig. 6.28b) and persist regardless of the rate of counterion exchange. One might consider that the amine present in only one enantiomeric form [that generated from the (+) amine in the example, Fig. 6.28b] serves as a chiral solvating agent (CSA). The resulting anisochrony of the diastereomeric ion pairs provides a very simple means for determining the enantiomer composition during resolutions by diastereomeric salt formation. The same considerations and limitations listed for CSA apply (Section 6-5.c); for the effect of temperature, concentration, and stoichiometry, see Fulwood and Parker.[241]

Finally, Feringa, Smaardijk, and Wynberg[231,242] emphasize the utility of the "duplication" method (Section 7-4) as an analytical technique for determining the enantiomer composition of alcohols. This method eschews the need for a nonracemic chiral reagent. By way of an example, alcohols, such as 2-octanol, are quantitatively converted to mixtures of diastereomeric O,O-dialkylphosphonates by reaction with PCl_3 (Eqs. 6.8 and 6.9) whose ratio must be directly related to the proportion of

$$3\ ROH\ +\ PCl_3\ \xrightarrow[-\ HCl]{20^{\circ}C}\ P(OR)_3\ \xrightarrow{HCl}\ HP(OR)_2\ +\ RCl \qquad (6.8)$$

$$H_{13}C_6CH(OH)CH_3 \xrightarrow[\text{pyridine}]{PCl_3} \qquad (6.9)$$

C_6H_{13}	C_6H_{13}	C_6H_{13}
H——CH$_3$	H——CH$_3$	H——CH$_3$
O	O	O
H—P=O	H—P=O	O=P—H
O	O	O
H$_3$C——H	H——CH$_3$	H——CH$_3$
C_6H_{13}	C_6H_{13}	C_6H_{13}
RR.SS	RS^1	RS^2
(± pair)	meso	meso

enantiomers in the substrate. Analysis of these diastereomers by ^{31}P NMR gives rise to but three peaks in a ratio of 2:1:1 in the case of racemic 2-octanol, whereas enantiomerically pure (S)-(−)-2-octanol exhibits only one peak (that for the S,S diastereomer).

NMR in Chiral Solvents and Chiral Solvating Agents. Following the prediction of anisochrony in NMR spectroscopy with the use of nonracemic chiral solvents,[193] Pirkle promptly demonstrated the possibility of distinguishing enantiomers by recording the ^{19}F NMR spectrum of racemic 2,2,2-trifluoro-1-phenylethanol, TFPE (Fig. 6.29, **66**) in (−)-α-methylbenzylamine (PEA, Fig. 6.29).[243] The trifluoromethyl groups of the two enantiomers were nonequivalent with a chemical shift difference of 2 Hz (at 56 MHz). In the same year, Burlingame and Pirkle found that such nonequivalence was exhibited also in ^{1}H spectra.[244] The observation that the two enantiomers of a compound may exhibit different NMR spectra when the measurement is carried out in nonracemic solvents is one of the most direct manifestations of diastereomer discrimination (Section 6-2). It is not necessary to use a chiral solvent; a nonracemic CSA (i.e., a chiral reagent that complexes, at least to some extent, with the substrate) in an achiral solvent is an often used alternative (see below). The application of chiral solvating agents in the determination of enantiomer composition has been reviewed by Pirkle and Hoover,[245] by Weisman,[60] and by Parker.[191]

The structures of the most frequently used CSAs are shown in Figure 6.29. A wide variety of solutes (alcohols, amines, amino acids, ketones, carboxylic acids, lactones, ethers, oxaziridines, sulfoxides, and amine and phosphine oxides) have been analyzed by NMR spectroscopy in the presence of CSAs for the determination of enantiomer purity. Enantiopure 2,2,2-trifluoro-1-(anthryl)-ethanol, TFAE (**67**) is one of the most widely used CSAs in the determination of enantiomer composition,[246] for example, of lactones.[247] Compound **68** has been applied to the analysis of chiral methyl sulfoxides[248,249] and phospine oxides.[250] Based on an earlier finding that carboxylic acids complex strongly with sulfoxides,[251] the determination of enantiomeric composition of chiral sulfoxides (including quasi-symmetrical dialkyl sulfoxides) has been effected with (S)-(+)-O-methylmandelic acid (Fig. 6.27, **44**).[252]

Compounds **69**, **70**,[253,254] **71** (quinine),[255] and Tröger's base (**72**)[256] are more recently introduced CSAs, the latter two for the analysis of alcohols. Compound **73**, a dicarboxamide CSA, has been applied to the determination of the enantiomer composition of carboxamides and hence of their precursors, amines and acids.[257] The application of quinine as a CSA [also for the analysis of binaphthyl derivatives and for N-(3,5-dinitrobenzoyl)-β-hydroxy-β-phenethylamines] is noteworthy because of its ready availability and low cost.[258] The utility of β- and γ-cyclodextrins (Section 7-3.c) as CSAs in the determination of the enantiomeric composition of amine salts (e.g., propranolol hydrochloride) *in water* has been described.[259] Spectral nonequivalence generated by cyclodextrins is due to formation of diastereomeric inclusion compounds *in solution* (Section 7-3.c).

It was first recognized by Mislow and Raban in 1965[193] that interaction of chiral solute molecules with a chiral nonracemic environment transforms the previous enantiotopic relationships of specific nuclei into diastereotopic ones (both by internal and by external comparison; cf. Chapter 8). The differences in magnetic environment of nuclei are generated by diastereomeric solvates that form and dissociate rapidly on the NMR time scale [Eqs. 6.10 and 6.11, where (R)-A and (S)-A stand for the enantiomers of the substrate (analyte) and (R)-X stands for a chiral solvating or shift reagent]. The NMR shifts observed are thus average shifts of the solvated (right-hand

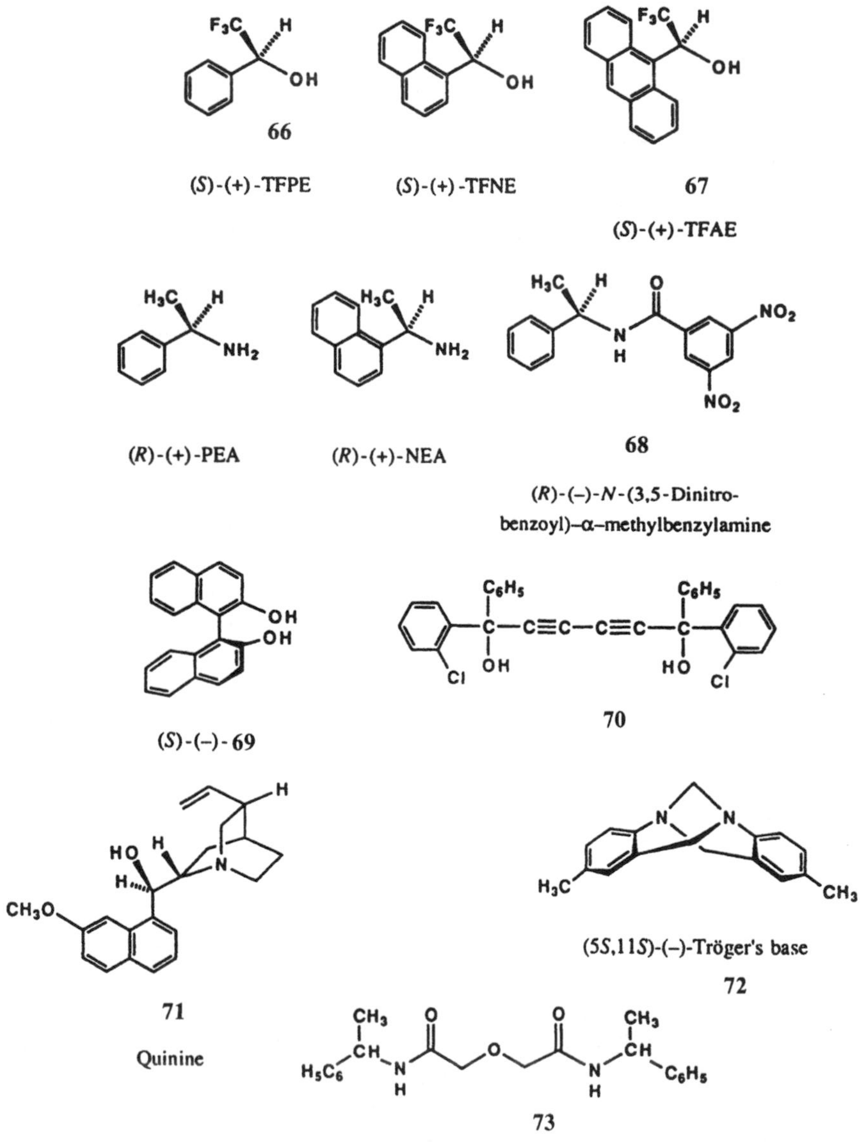

Figure 6.29. Chiral solvating agents (CSAs).

side of Eqs. 6.10 and 6.11) and unsolvated species (left side of the equations), and since the shifts for the former, though not the latter, differ for the enantiomers, so does the average.

$$(R)\text{-A} + (R)\text{-X} \underset{K_{RR}}{\rightleftharpoons} (R)\text{-A} \cdot (R)\text{-X} \tag{6.10}$$

$$(S)\text{-A} + (R)\text{-X} \underset{K_{SR}}{\rightleftharpoons} (S)\text{-A} \cdot (R)\text{-X} \tag{6.11}$$

There are two reasons why the equilibria in Eqs. 6.10 and 6.11 taking place in the presence of CSAs lead to differently shifted signals: (a) Intrinsic: (R)-A·(R)-X and (S)-A·(R)-X are diastereomers and therefore have, at least in principle, different chemical shifts. Of course, (R)-A and (S)-A, being enantiomers, have the same shift, so the existence of rapid equilibrium diminishes the observed shift difference (by averaging) but does not obliterate it. (b) Differential stability of diastereomeric solvates: $K_{RR} \neq K_{SR}$, since the species on the right-hand side of the two equations are diastereomers and hence may be unequally stable (the species on the left-hand side of Eqs. 6.10 and 6.11, in contrast, have equal stability). Thus complexing may not be the same for the two enantiomers; other things being equal, the more complexed enantiomer is likely to present the larger chemical shift change by the CSA or lanthanide shift reagent (LSR). However, complexing to unequal extents cannot be the only reason for the shift differences between the two enantiomers. Explanation (a) is also significant; this follows from the observation of shift differences of enantiotopic nuclei *within one molecule* (by use of CASs or LSRs; see also p. 319). Here differential solvation or complexation cannot come into play.

The observed chemical shift differences ($\Delta\Delta\delta$) vary with spectrometer frequency but are typically quite small (0–10 Hz for ^{1}H NMR at 100 MHz). Sharp, well-resolved signals are required, particularly for quantitative work. The limitation of inconveniently small shift differences can be overcome in several ways. One way is by increasing the spectrometer frequency. Another way is to lower the temperature, thereby increasing solvate formation; however, this possibility is subject to solubility limitations. The magnitude of the NMR spectral nonequivalence can also be increased by addition of achiral shift reagents, for example, Eu(fod)$_3$ (Fig. 6.31, p. 158) (mixed CSA–shift reagent system).[260] The effects observed upon such addition have been interpreted as arising from the differential stability of the solute–CSA solvates[261]; shift reagent molecules can, in some cases, preferentially displace solute molecules from the least stable solvate.[262] The enhancement has been observed especially in the case of analytes that associate with lanthanides independently of the presence of CSAs, such as **68** (Fig. 6.29), that themselves associate weakly, if at all, with lanthanide ions.[263] The enhancement is also seen with cyclodextrins acting as CSAs (see above).

In general, analytical methods that involve formation of transient diastereomers or dynamic diastereomer systems (ion pairs, charge-transfer complexes, solvates, and complexes with shift reagents) in which the diastereomeric partners and the enantiomer substrates undergo rapid exchange do not require that the chiral detecting agent (solvent, shift reagent, or stationary phase; Section 6-5.d) be enantiomerically pure. In the absence of complete enantiomer purity of the reagent, the observed anisochrony or separation is simply attenuated. With racemic CSA, the chemical shift differences disappear altogether, as a consequence of the rapid statistical averaging of solvate formation.[13,264]

Analysis in the presence of CSA has also been carried out for the assignment of absolute configuration. The latter use follows from the possibility of relating the sense of chemical shift nonequivalence (upfield vs. downfield shift of a signal) for one enantiomer of a compound dissolved in a given CSA, to the configuration of that enantiomer. This configurational correlation depends on the application of models that specify the number and kind of interactions possible between solute and solvent (e.g., see Fig. 6.30); the correlation is subject to the limitation of mechanistic understanding of all such models (cf. Chapter 5).

The dibasic solute model[245] requires that CSAs and solute molecules have complementary functionalities permitting formation of chelate-like solvates (Fig. 6.30). In appropriate solutes (e.g., amines), there is simultaneous interaction of two basic sites (B_1 and B_2) in the solute with acidic sites in the CSA. The interaction accounts for the widespread use of TFPE and of TFAE (Fig. 6.29, **66** and **67**, respectively) with their acidic hydroxyl and carbinyl hydrogen atoms both of which may participate in hydrogen bonding. The aromatic rings serve as anisotropic shift perturbers (enhancers), thus providing yet a third site for interaction between CSA and solute. A size difference between sites B_1 and B_2 is not essential. At 100 MHz, the CH_3 protons of the *RS* solvate (Fig. 6.30) are shifted to lower field (2.8 Hz) while the carbinyl hydrogen and the CH_3' (ester) group of this solvate are both shifted upfield (by 2.3 and 1.3 Hz), respectively.[246]

Compounds that act as inclusion hosts, for example, **69** and **70** (Fig. 6.29) and **94** (Fig. 7.25), also serve as CSAs. The observation of ^1H NMR differential shifts, for example, of α-methylbenzylamine (Fig. 6.29, PEA) with these CSAs, has been attributed to formation of 2:1 molecular complexes (clathrates; Section 7-3.c) in solution.

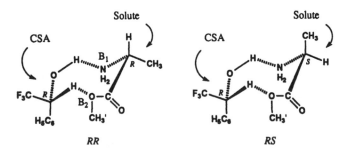

Figure 6.30. Dibasic solute model for diastereomeric solvates. Effect of (*R*)-(−)-TFPE on ^1H NMR of methyl alanate enantiomers (100 MHz, 29°C, $CFCl_3$ diluent): B_1 = hydrogen bond receptor and B_2 = carbinyl hydrogen bond receptor.[246]

NMR with Chiral Shift Reagents

Shift Reagents. Lanthanide shift reagents (LSRs) are compounds prepared by reaction of certain transition metal (Eu, Pr, or Yb) salts with β-diketones. The LSRs themselves are tris-chelates of coordination number 6 that behave as weak Lewis acids. In nonpolar solvents (e.g., $CDCl_3$, CCl_4, or CS_2), these paramagnetic salts are able to bind Lewis bases, especially amides, amines, esters, ketones, and sulfoxides, by expansion of the coordination sphere of the metal. The organic solutes rapidly exchange between bound and unbound states.

As a result, protons, carbon atoms, and other nuclei have averaged magnetic environments and are chemically shifted (usually deshielded relative to their positions in the uncomplexed substrates) to an extent that depends on the strength of the complex and how far the nuclei are from the paramagnetic metal atom. Thus different nuclei are shifted to different extents, thereby enhancing spectral dispersion (much as a higher magnetic field would) and leading to spectral simplification.

Achiral LSRs (Fig. 6.31) are occasionally used to enhance anisochrony in the spectra of *diastereomer mixtures*, facilitating quantitative analysis of such mixtures.[265]

When the permanent coordination sphere of the metal in the LSR is chiral (through complexation with nonracemic β-diketones, such as 3-acylcamphor or dicampholylmethane), chiral shift reagents (CSRs) are created. Addition of such reagents to racemic organic solutes in solution gives rise to diastereomeric complexes. Individual groups in the organic solute are differentially shielded from the external magnetic field just as they are when achiral LSRs are used; the peaks are spread over a wide range of chemical shifts leading to spectral simplification. Superimposed upon this is the anisochrony of externally enantiotopic groups that have become diastereotopic under the influence of the chiral LSR (cf. Chapter 8). If peaks due to the two enantiomers of a racemic solute are well separated, quantitative analysis of enantiomer composition becomes feasible.

Lanthanide-induced shifts (LISs) due to chiral LSRs were first observed by Whitesides and Lewis in 1970 [Fig. 6.31, Eu(t-cam)$_3$].[266] Shortly thereafter, Fraser and Goering described CSRs that had a much greater range of applications.[267–269] The application of chiral LSRs to the determination of enantiomer composition has been reviewed.[191,261,264,270–272] The structures of the most frequently used CSRs are given in Figure 6.31.

The following particulars apply to the use of CSRs. Substrate concentrations range from 0.1 to 0.25 M while the CSR/substrate ratio generally is in the range of 0.5–1. Observed ^1H chemical shift differences $\Delta\Delta\delta$ between corresponding groups in the two enantiomers are in the 0.1–0.5-ppm range (occasionally as large as 4 ppm), that is, 10–50 times as large as the nonequivalences observed with CSAs. CSRs are usable with ^1H, ^{13}C, and ^{19}F probe nuclei, and the accuracy of enantiomer compositions are unaffected if the CSRs are not enantiomerically pure, though the resolution of the enantiotopic signals decreases with decreasing enantiomeric purity of the CSR. The choice of metal for the CSR determines the magnitude of the LIS, the direction (upfield or downfield) of the LIS, and the extent of peak broadening resulting from the paramagnetic character of the metal. The specific CSR selected for a particular use is often a compromise of several factors including accessibility and cost of the reagent.[261,271]

The CSRs have many advantages over CSAs or the use of CDAs in the determination of enantiomer composition by spectroscopy. Among these are the convenience of the method and the ease of interpretation and manipulation of the spectroscopic data; and, in comparison with CDAs, the absence of the need for enantiomerically pure reagents and of the need for quantitative derivatization. The CSRs are applicable to the analysis of most "hard" organic bases. One of the most effective CSRs in differentiating nuclei that are enantiotopic by external comparison is Eu(dcm)$_3$ (Fig 6.31).[273]

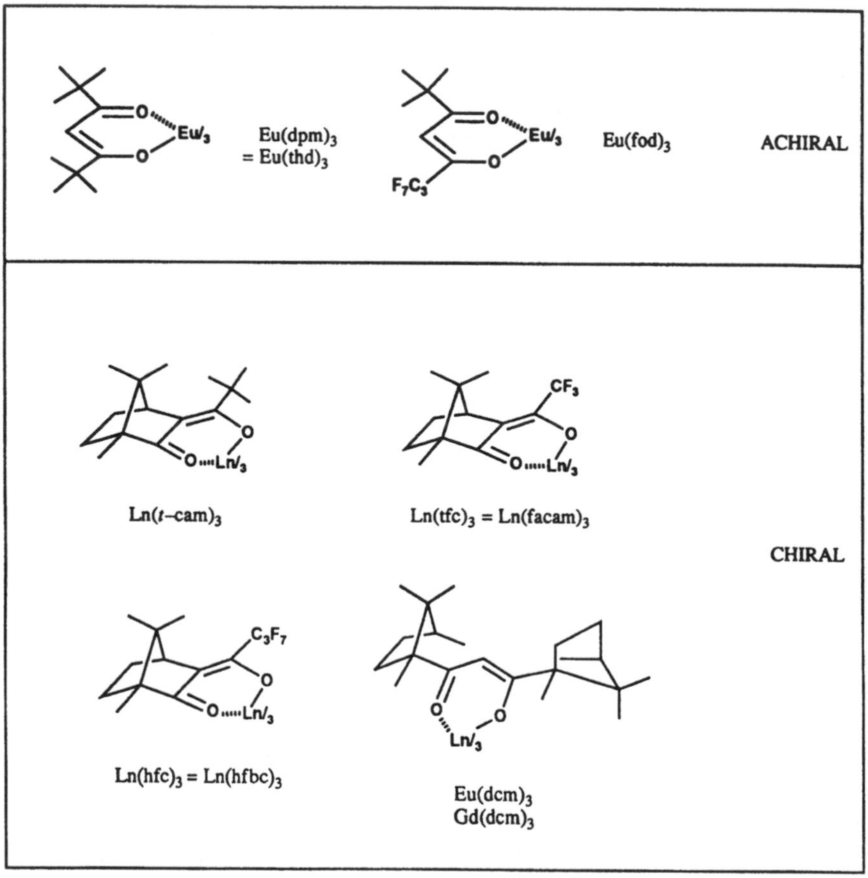

Figure 6.31. Lanthanide shift and relaxation reagents (Ln = a lanthanide metal). Eu(dpm)$_3$ = Eu(thd)$_3$: tris(dipivaloylmethanato)europium(III); Eu(fod)$_3$: tris(6,6,7,7,8,8,8-heptafluoro-2,2-dimethyl-3,5-octane-dionato-O,O′)europium(III); Eu(t-cam)$_3$ (Ln = Eu): tris(3-tert-butylhydroxymethylene-d-camphorato)-europium(III); Eu(tfc)$_3$ = Eu(facam)$_3$ (Ln = Eu): tris(3-trifluoroacetyl-d-camphorato)europium(III); Eu(hfc)$_3$ = Eu(hfbc)$_3$ (Ln = Eu): tris(3-heptafluorobutyryl-d-camphorato)europium(III); Eu(dcm)$_3$: tris(d,d-dicampholylmethanato)europium(III).

However, the application of CSRs is not always successful. In most instances only a few nuclei in the analyte exhibit stereoisomer discrimination in their shifts. Insufficient peak resolution, signal broadening, chemical degradation of the CSR, and inattention to instrumental conditions may thwart success in the use of this technique. An example is the determination of the enantiomer composition of mevalonolactone.[274] Computer processing of NMR spectra (rephasing, baseline correction, and Gaussian line narrowing) has been shown to increase the precision of the analyses.[275] The use of high-field (even 300 MHz) NMR spectrometers is detrimental to the application of LSRs since, under the typically prevailing fast exchange conditions, signal broadening is proportional to the square of the magnetic field strength.[192] This problem is particularly acute with compounds exhibiting large LISs (e.g., alcohols). The broadening can be reduced and useful anisochrony restored on warming the sample[263] or on application of NMR spin-echo techniques.[276]

Chemical exchange is the origin of the signal broadening effect. The two principal variables affecting the signals due to species undergoing exchange are field strength, which affects line separation, and temperature, which affects the exchange rate. It is the interplay between these variables that is responsible for both the line broadening at high field and its reduction on warming. The fast exchange approximation $\delta\nu = \pi\nu_{AB}^2/2k$, where k is the rate constant for exchange, shows that linewidth ($\delta\nu$) is proportional to the square of the separation (chemical shift difference ν_{AB}) between two lines; this separation is, in turn, proportional to the square of the field.[277] The problem becomes more pronounced at higher frequencies.

Combination of CSRs with soft acids (e.g., $AgNO_3$ or $AgOCOCF_3$)[278] or with an achiral LSR [e.g., Ag(fod)] gives rise to so-called binuclear lanthanide(III)−silver(I) shift reagents. The latter sometimes permits the determination of enantiomer composition of weak or "softer" bases, such as alkenes, arenes, and halogen compounds,[279−282] and allenes[283,284] with which they complex.[261,285]

Enantiomer mixtures may need to be converted to derivatives to obtain reasonable peak dispersion with CSRs. Optically active glycols, for example, have been converted to mixtures of epimeric 1,3-dioxolanes on condensation with benzaldehyde. The analysis of the mixture by NMR [by addition of Eu(hfc)$_3$; Fig. 6.31] has been described by Eliel and Ko.[28] Subsequently, determination of the enantiomer composition of polar substrates (diols, triols, or glycidol) was shown to be possible in polar solvents, such as CD_3CN, acetone-d_6, and $CDCl_3$, with Eu(facam)$_3$ and with Eu(hfbc)$_3$ (Fig. 6.31).[287]

The reagent Eu(hfc)$_3$ has been applied to the determination of the enantiomer composition of compounds whose chirality depends only on the presence of H and D or CH_3 and CD_3 at the stereogenic atom.[288] Chiral shift reagents permitting enantiomer purity analysis of amino and hydroxy acids and of carboxylic acids in aqueous solutions have also been described.[67,289−291] Chiral shift reagents can also help to differentiate *meso*- from *rac*-diastereomers (cf. Chapter 8)[292] and to induce chemical shift differences between internally enantiotopic nuclei such as the two benzylic protons of benzyl alcohol.[293]

An achiral dinuclear lanthanide shift reagent, $Pr(tpip)_3$, has been applied to the enantiomeric purity determination of carboxylic acids.[294] The complexes formed in the presence of carboxylic acid salts contain two carboxylate ligands; hence, when racemic salts are analyzed, two diastereomeric complexes are formed giving widely separated signals. This finding is yet another example of the duplication method described in Section 6-5.c (see also Section 7-4).

Two possible and probably not independent explanations for the spectral nonequivalence observed in the presence of CSRs have been advanced. These explanations have much in common with the anisochrony observed with CSAs (Section 6-5.c): (a) In the presence of CSRs, transient time-averaged diastereomeric complexes of different stabilities are formed that equilibrate with the unbound enantiomers of the substrate (alcohol, amine, etc.); see Eqs. 6.10 and 6.11.[271] (b) The geometries of the CSR–chiral substrate diastereomeric complexes will differ, and hence the magnetic environments of selected nuclei in the complexes may differ sufficiently to lead to anisochrony.[268,273,275] The CSRs may also generate transient diastereomeric species of more than one type of stoichiometry,[267,269] which further affects the position of equilibrium of complexed and uncomplexed species as well as the intrinsic shifts in the complexed species.

d. Chromatographic and Related Separation Methods Based on Diastereomeric Interactions

Chromatographic methods, methods that depend on the total separation (direct or indirect) of the enantiomers of a chiral substance, are among the most powerful available for the determination of enantiomer composition. The earliest determinations of this type required prior conversion of enantiomer mixtures to diastereomeric derivatives. The latter were then analyzed by GC on achiral stationary phases. Subsequently, HPLC has been applied to such analyses.

A second type of chromatographic process applied to enantiomer purity determinations is the separation of enantiomers on chiral stationary phases (CSPs). Both GC and HPLC techniques have been used in this way. Most recently, enantiomeric composition analyses have been carried out by HPLC on achiral stationary phases with the aid of chiral mobile phases.

All of the separation processes mentioned rely on the intervention of stable or transient diastereomeric species whose different solubilities, stabilities, or adsorption characteristics are responsible for the separation of the stereoisomers.

Although the focus of this section is on separations (resolutions) effected for analytical purposes, it will be evident that small scale preparative chromatographic resolutions use identical methodology and even conditions. Although preparative chromatographic resolutions are described in Chapter 7, much of what follows here is pertinent to such macroscopic separations.

Chromatography on Diastereoselective Stationary Phases

Gas Chromatography.   Since the report by Casanova and Corey[295] that the resolution of camphor could be effected by means of GC of its dioxolane derivatives with (−)-2,3-butanediol, it has been apparent that GC of diastereomer mixtures offers a simple route to the assessment of enantiomer composition of chiral substances. An even earlier report describes the analysis of *rac*-phenylalanine by GC of its L-alanyl derivatives [as the *N*-trifluoroacetyl (TFA) methyl esters.[296]

Gas chromatographic analysis of diastereomer mixtures has the same limitations as NMR analysis, including the requirement that the CDA must be enantiomerically pure. With an enantiomerically impure CDA, the minor enantiomer generates a diastereomer mixture that is enantiomeric with the diastereomer mixture produced by reaction of the. dominant CDA enantiomer with the analyte. Since the two (enantiomeric) diastereomer mixtures are not separable in an achiral stationary phase (just as they do not give rise to separate signals on NMR analysis in achiral solvents), the analysis is falsified.

By way of example, consider an analyte **A** consisting of 99.5% (+) and 0.5% (−) enantiomers (99% ee) that is derivatized with enantiomerically impure CDA (+)-**B** [the latter consisting of 99% (+)-**B** and 1% (−)-B]. Following quantitative derivatization, four diastereomeric products are present: I (+)-A·(+)-**B** (98.5%), II (−)-A·(+)-**B** (0.5%), III (+)-A·(−)-**B** (1.0%), IV (−)-A·(−)-**B** (0.0%). Chromatographic analysis of the mixture on an achiral stationary phase yields but two peaks: I + IV (these being enantiomers), and II + III (also enantiomers) with the mixture having an apparent enantiomer composition of 98.5% (+) and 1.5% (−), hence 97% ee.[297]

Care must be taken during the preparation and isolation of the derivative to avoid accidental fractionation lest the diastereomer ratio not be equal to that of the enantiomers prior to derivatization.[298] In addition, the derivatives must be thermally stable and reasonably volatile if the analysis is not to be excessively time consuming due to long retention times. Here, as in NMR, a control experiment on derivatives of the racemate or of a synthetic mixture of the two enantiomers of the analyte must be carried out to ensure that absence of a second GC peak reflects enantiomeric homogeneity and not the inability of the column to resolve the diastereomeric derivatives.

As pointed out by Karger,[299] the degree of separation of the diastereomeric derivatives, and hence the efficiency of the separation, depends principally on the CDAs rather than on the (achiral) stationary phase.[300] In consequence, many CDAs have been examined for their potential in the determination of enantiomer composition. The number and variety of CDAs used with GC (and analogous HPLC techniques, to be described below) is so large that only a limited number of examples can be cited here (see Fig. 6.32).[297]

Chiral derivatizing agents derived from amino acids have been applied to the GC analysis of chiral secondary alcohols,[301] of amines,[302] and of amino acids.[303] Natural

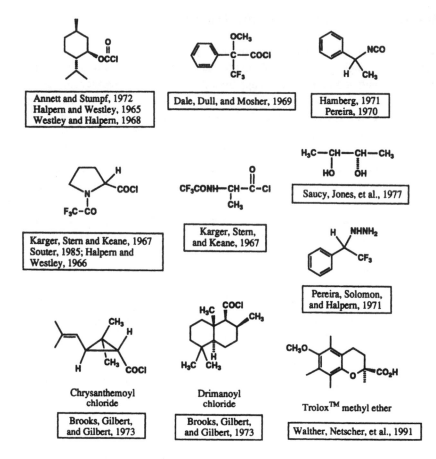

Figure 6.32. Common CDAs useful in the determination of enantiomer composition by GC.

product derived CDAs (other than amino acids) have been used for the GC determination of enantiomer purity of alcohols, such as borneol and menthol (e.g., as their tetra-O-acetyl glucosides). In turn, menthol, as the chloroformate (Fig. 6.32), has been applied to the enantiomer purity determination of amino acids (e.g., as the N-TFA derivatives) as well as of alcohols and α-hydroxy acids.[304–307] A chromanecarboxylic acid CDA derived (following resolution) from the racemic commercial antioxidant Trolox™ (Fig. 6.32) has been applied to the enantiomeric purity determination of chiral 1° and 2° alcohols.[308] As an alternative, 1° alcohols have been analyzed following chromic acid oxidation to carboxylic acids; the oxidation is known to proceed without racemization. The latter are derivatized to amides with α-methylbenzylamine.[309,310]

Lactones have been analyzed as ortho esters following their derivatization with chiral nonracemic 2,3-butanediol.[311] Mono- and sesquiterpenoids (chrysanthemic and drimanoic acids) have been applied to the analytical resolution of alcohols and amines by GC.[312] A chiral hydrazine reagent (Fig. 6.32) provides a means for the enantiomer purity determination of ketones via conversion of the latter to hydrazones followed by

GC resolution.[313] Diels–Alder adducts of dienes with α,β-unsaturated aldehydes have been analyzed for their enantiomer content as acetals following derivatization with chiral nonracemic 2,4-pentanediol.[314]

High-Performance Liquid Chromatography. While the analysis of diastereomer mixtures by GC will continue to be a useful method for a long time to come, it is limited to compounds that are volatile and reasonably thermally stable. In recent years, it has been complemented, and often superseded, by the analogous HPLC methods and by methods that employ chiral stationary phases. The versatility of HPLC, its wider range of application, and the wider choice of process variables (adsorption vs. partition mode, choice of mobile as well as stationary phase, etc.) is responsible for the increase in the use of this modern technique.

Among the earliest applications of HPLC to the determination of enantiomer composition is the report of the chromatographic separation of lactates and mandelates of alcohols.[315] Application of the HPLC technique became more common in the mid-1970s when HPLC instruments and columns filled with microparticulate stationary phases became routinely available.

The principal types of chiral derivatizing agents used in the preparation of diastereomeric derivatives used in HPLC are acylating agents, amines, isocyanates and isothiocyanates, and alkylating agents (Fig. 6.33).[297,316,317] Some of these reagents have been developed specifically to resist racemizatiom prior to or during the derivatization reaction.[318] It is fair to say that *all* CDAs should be checked for enantiomer purity before use.

The formation and chromatographic resolution of stable diastereomeric platinum complexes have been described in connection with the determination of enantiomer composition of chiral alkenes and sulfoxides by HPLC.[319–321] Selection of CDAs for liquid chromatographic resolution of acids has been influenced by the observation that diastereoisomeric amides are more strongly hydrogen bonded to silica gel and to alumina than are diastereomeric esters. Diastereoselectivity is enhanced by the presence of suitably located aromatic and polar functional group, such as hydroxyl.[322,323] The determination of the enantiomer composition of both carboxylic acids and of amines can readily be effected by prior conversion to diastereomeric amides. Hydroxyamides, such as those derived from phenylglycinol or from β-phenylbutyrolactone (Fig. 6.33) have been shown by Helmchen to be especially good CDAs for both analytical and preparative purposes.[322,324,325] Separation factors, α, which are a measure of the ease of separation of the diastereomers (Section 7-3.d), are quite large ($\alpha > 2.5$) for diastereomeric amide pairs derived from phenylglycinol (Fig. 6.33), whereas α values greater than 2 are rarely found in other typical diastereomer systems.

Diastereomeric carbamates, ureas, and allophanates (derived, e.g., from nonracemic isocyanate, 2-oxazolidone, and ureide CDAs) have been applied to the chromatographic resolution and assessment of enantiomer purity of alcohols, amines, and lactams. Reviews of the extensive literature dealing with resolution and determination of ee by liquid chromatography (LC), especially on achiral stationary phases, are available.[323,326–332]

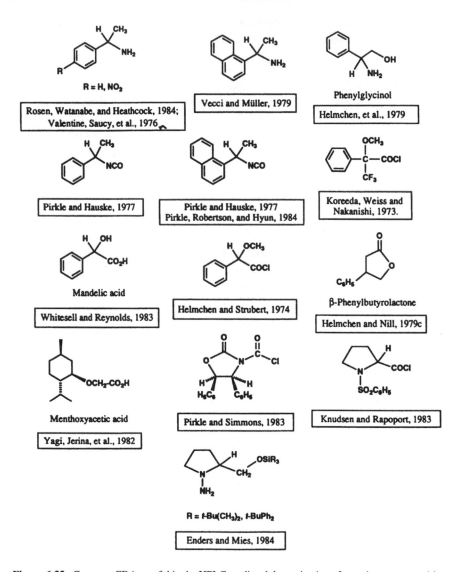

Figure 6.33. Common CDAs useful in the HPLC mediated determination of enantiomer composition.

High-Performance Liquid Chromatography with Chiral Solvents.   One of the more recent developments in the liquid chromatographic determination of enantiomer purity is the application of chiral mobile phases (eluents) together with achiral stationary phases.[333,334]

The possibility of effecting such analyses was first demonstrated by Pirkle and Sikkenga in 1976 when they partially resolved a racemic sulfoxide by LC on silica gel.[335]   The mobile phase consisted of CCl_4 to which nonracemic 2,2,2-trifluoro-1-(9anthryl)ethanol (Fig. 6.29, **67**) had been added. Other simple chiral mobile phase additives, such as (+)-*N*,*N*-diisopropyltartramide, have been applied to analytical chromatographic resolutions of amino alcohols, glycols, hydroxy ketones, and hydroxy carboxylic and amino acids.[62,336,337] In these instances, separation may depend on the differential stability of diastereomeric intermolecularly hydrogen-bonded dimers.

Analyses of chiral amines (β-blocking drugs), such as propranolol (Fig. 6.34), have been effected by addition of (+)-10-camphorsulfonic acid to a mobile phase of low polarity. Formation of diastereomeric ion pairs that are bound both by electrostatic forces and by hydrogen bonds, and their differential migration accounts for the resolution.[338] Chiral acids (e.g., naproxen, Fig. 6.34) have similarly been analyzed with solvents containing alkaloids.[339]

The particular simplicity and sensitivity of diastereomeric complexes is appealing. Since switching the chirality sense of chiral eluents is far easier than switching the sense of chirality of enantioselective columns, this technique easily lends itself to the distinction of achiral artifacts from true enantiomers.[340] Nevertheless, the number of applications of chiral eluents has been limited possibly due to the required quantities and costs of the reagents together with the impracticality of recovering them.[343]

Chromatography on Enantioselective Stationary Phases   The direct resolution of enantiomers by chromatography on nonracemic chiral stationary phases is a goal that has tantalized chemists virtually since chromatography was recognized as a potentially valuable separation technique. Early trials with such columns were oriented to preparative resolution (see Section 7-3.d) and were not very successful. No resolution, or only partial resolution, was achieved. Some of the early trials are described in Eliel[344] and in Feibush and Grinberg.[345] The advent of modern chromatography equipment (GC and HPLC) with its high-resolution columns and sensitive detectors made analytical resolutions with enantioselective columns and their application to enantiomer purity determinations possible.

Numerous enantioselective stationary phases have been developed over a 35-year period (1966–2000); the end of this development is nowhere in sight. A summary of the various types is given in Table 6.4 along with the preferred modes of use (GC, HPLC, and TLC).[346] While the principles of separation are also given for most cases, there is no absolute certainty about these. It is likely that in many instances more than one mechanism applies.[297,326,329,342,347,348]

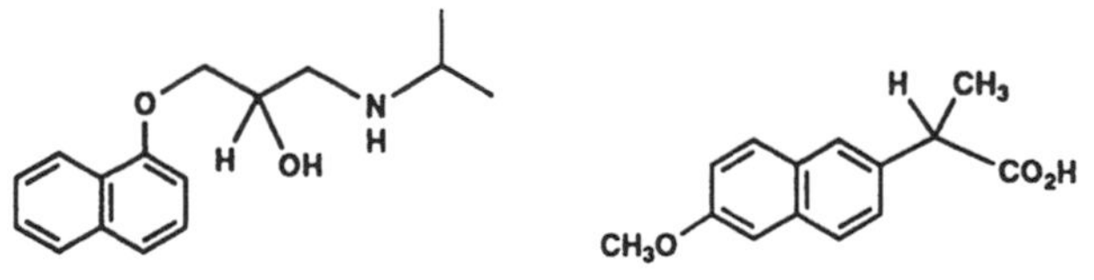

Propranolol   Naproxen

Figure 6.34. Propranolol and naproxen.

TABLE 6.4   Principal Types of Enantioselective Chromatographic Stationary Phases

	Type	Principle of Separation	Mode
I[a]	Amide	Attractive interactions, hydrogen bonding, π-interaction	GC, HPLC
I	Fluoro alcohol	Dipole attraction, charge transfer	HPLC
I	π Acid		
II	Carbohydrate	Attractive interactions + inclusion	HPLC
III	Cyclic hexose oligomers	Inclusion compound formation	GC, HPLC
III	Crown ether	Inclusion	HPLC
IV	Metal chelates	Ligand exchange	GC, HPLC, TLC
V	Proteins	Hydrophobic and polar interactions in a protein	HPLC
	Ureide	Interaction with mesophases	GC[b]

[a]Categories I–V as defined by Wainer.[349,350]
[b]Below selector melting point.

Gas Chromatography.   The earliest nonracemic chiral stationary phases were patterned after structural elements present in enzymes; they contained –CO– and –NH– moieties present in the vicinity of stereocenters. It was anticipated that hydrogen-bond formation between the stationary phase and chiral amino acid derivatives would provide a small degree of enantioselectivity that might be sufficient for quantitative analysis of the enantiomer composition provided that the effect could be amplified. The use of long capillary columns provided for the amplification.[342] The first successful CSP incorporating these elements was N-trifluoroacetyl-L-isoleucine dodecyl ester.[341] This phase was able to resolve simultaneously the enantiomers of several TFA-α-amino acid isopropyl esters by GC.

The following terms have been suggested for describing the two chromatographic partners in the stereoisomer discrimination phenomenon: *selector* (CSP = resolving agent) and *selectand* (chiral solute = analyte).[351]

The structural similarity and complementary functionalities of the analyte and of the chiral component of the stationary phase reveals the basis of the column design, formation of transient diastereomeric complexes.[352] Subsequent developments in analytical GC included the application of dipeptide, diamide, and similar phases permitting multiple contact points between selectand and selector molecules (Fig. 6.35).[326] Enantioselective capillary GC columns are capable of separating enantiomers with separation factors [α = $(t_2 - t_1)/(t_1 - t_0)$, where t_2 and t_1 are the retention times of the second and first enantiomers eluted, respectively, and t_0 is that of an unretained substance; see also Section 7-3.d] of 1.05 (or even less) corresponding to a difference of only $\Delta\Delta G^0 = 29$ cal mol^{-1} (121 J mol^{-1}) between the transient diastereomeric solvates.[297,298,353–356]

Higher thermal stability and lower volatility than are available with columns containing amino acid or dipeptide derivatives were achieved by linking such compounds covalently to polymer backbones. An example of such enantioselective stationary phases usable to at least 220°C in the form of an open tubular column is N-proplonyl-L-valine t-butylamide bound to a copolymer of dimethylsiloxane and

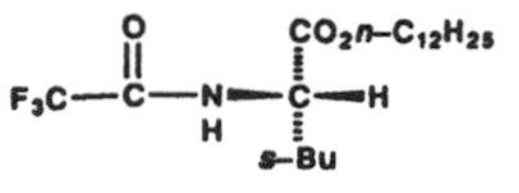

N–TFA–L–isoleucine lauryl ester [341]

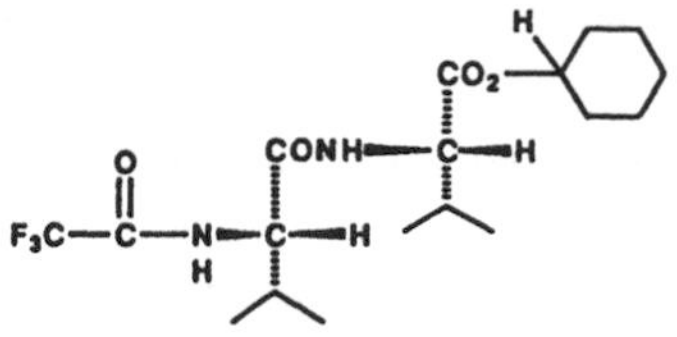

A typical dipeptide phase [342]

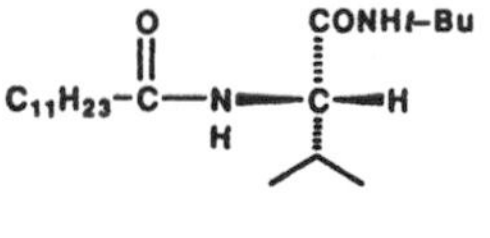

A typical diamide phase

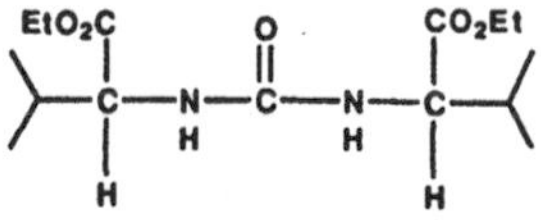

Carbonyl-bis(L-valine ethyl ester). A typical chiral mesophase

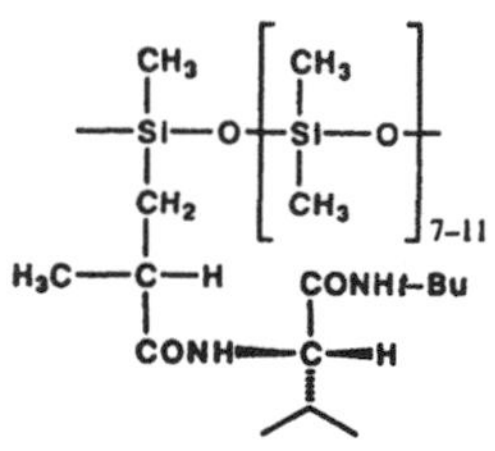

Chirasil-Val™
(a chiral polysiloxane)

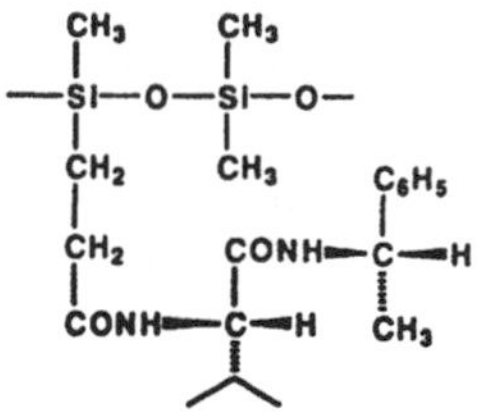

Chrompack™
(a chiral phase based on polysiloxane XE-60)

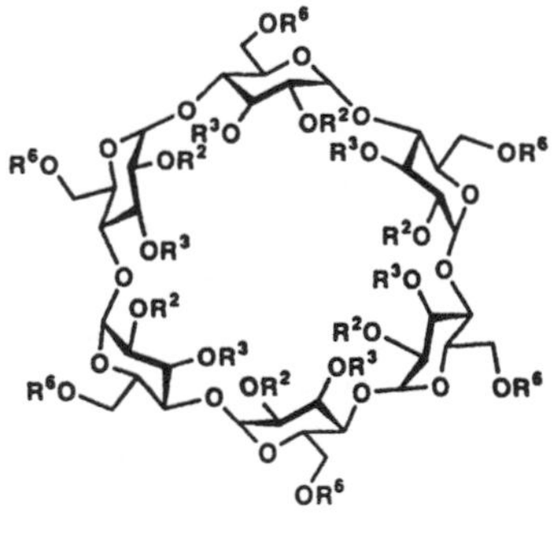

α-Cyclodextrin = cyclohexaamylose
OR^2, OR^3, and OR^6 = OH

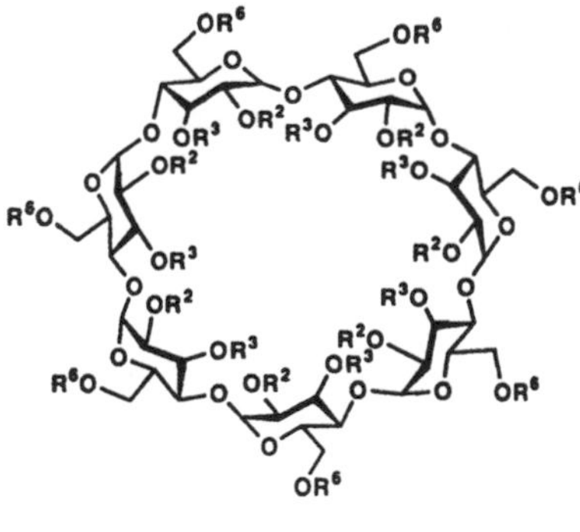

β-Cyclodextrin = cycloheptaamylose
OR^2, OR^3, and OR^6 = OH

Lipodex™
OR^2 and OR^6 = O-pentyl
OR^3 = O-pentyl or O-acyl

Peroctylated α-CD
OR^2, OR^3, and OR^6 = O-octyl

Figure 6.35. Basic structural types of gas chromatographic enantioselective stationary phases.

carboxyalkylmethylsiloxane units (Chirasil-Val[TM], Fig. 6.35).[357] Dipeptides linked to polysiloxanes (e.g., Chrompack[TM]) exhibit a wider range of utility. Analytical GC resolution of sugar derivatives and of ketones (as their oximes) is possible on such a column.[354] It is now recognized that the precise nature of the functional group and other polar sites present in the derivatized analytes strongly affects the degree of separation of the enantiomers. In the case of chiral ketones, neither the free ketones nor the less polar O-methyloximes or O-trimethylsilyloximes could be separated.[354] While hydrogen bonding between analyte and chiral selector molecules is very likely to play a role in the enantioselectivity of the columns described,[355] other types of intermolecular forces (e.g., dipole–dipole interactions or van der Waals forces) must be involved since dipeptide chiral selectors lacking NH can fully separate racemic N-TFA-proline isopropyl ester. There are no sites for hydrogen bonding in such systems.[358]

Methylated cyclodextrins (*CD*s) dissolved in achiral liquid supports, liquid nonpolar alkyl derivatives [e.g., hexakis-(2,3,6-tri-O-methyl)-β-cyclodextrin (Cyclodex-B[TM])] and mixed regioselectively alkylated and acylated α- and β-cyclodextrins [e.g., Lipodex[TM], Fig. 6.35; Lipodex A[TM] is hexakis-(2,3,6-tri-O-pentyl)-α-cyclodextrin] have been applied, since about 1987, as stationary phases in the capillary GC determination of enantiomer purity.[346,359–363] These commercially available phases are remarkably thermally stable (to 220°C). A wide range of structural types are baseline separated on these columns consistent with the presumed operation of an inclusion mechanism: alcohols, alkenes, alkyl halides (e.g., the chiral inhalation anesthetic halothane, $CF_3CHBrCl$, is fully resolved on a Lipodex A[TM] column at 30°C)[364] amines, carbohydrates (e.g., as O-trifluoro-acetates), ethers, ketones, lactones, and even saturated hydrocarbons. However, other mechanisms involving interaction of the analytes with the chirotopic periphery of the derivatized cyclodextrins cannot be excluded.

Polar hydrophilic *CD* derivatives [e.g., permethyl-(R)-2-hydroxypropyl-β-cyclodextrin] are also useful in the resolution of a wide range of compounds including many that are devoid of aromatic groups. The configuration of the chiral 2-methoxypropyl side chains does not affect the resolution order of analyte enantiomers.[365] On the other hand, it has been demonstrated in the case of 2-amino-1-propanol as the TFA derivative that the elution order of its enantiomers can be reversed on shifting between derivatives in the same β-*CD* series, for example, from the less polar trifluoroacetylated 2,6-di-O-pentyl-β-cyclodextrin to the above mentioned polar *CD* stationary phase.[366]

An interesting modification of enantiomer purity determinations by GC involves the use of the enantiomer of the dominant chiral species to spike the analyte in the role of an "internal standard".[355] Thus, for the enantiomeric purity determination of L-amino acids the D-amino acids are used in lieu of a true internal standard. The procedure, known as "enantiomer labeling," is recommended quite generally for checking the reliability and accuracy of chromatographic (and NMR) enantiomer purity determinations, especially in those cases in which the proportion of the dominant enantiomer is greater than about 95%.[367,368] It serves to indicate qualitatively that the minor peak is, in fact, due to the enantiomer and not to something else. Quantitatively, if one adds 2.5% (by weight) of the minor enantiomer to a weighed sample of a compound having 95% ee (enantiomer ratio = 97.5:2.5), the resulting mixture should show an enantiomeric excess of about 90% (ca. 95:5). If this is not found, then something is wrong.

High-Performance Liquid Chromatography.   High-performance liquid chromatography is a second instrumental chromatographic technique permitting enantiomer purity determinations to be carried out relatively simply. Here a much wider range of enantioselective columns is available (in part because thermal stability is not a limitation) and, moreover, the liquid (mobile) phase can play an important role in facilitating resolution.[297,323,327–329,347,369–377] As of 1992, more than 50 enantioselective LC stationary phases were available.[346]

Enantiomer purity determinations by either GC or HPLC with chiral stationary phases have the advantage of being absolute; that is, both techniques give correct enantiomer purity information even with stationary phases that are not enantiomerically pure as recognized by Davankov.[378] With reduced enantiomer purity, the separation factor α is reduced:

$$\alpha^* = \frac{\alpha(P + 100) + (P - 100)}{(P + 100) + \alpha(P - 100)} \tag{6.12}$$

Equation 6.12 permits the calculation of α^*, the separation factor that would be obtained if P [the enantiomeric purity (%ee)] of the stationary phase were 100%.[379,380]

In the late 1970s, Pirkle and co-workers began to adapt NMR chiral solvating agents that exhibit stereoisomer discrimination in solution (Section 6-5.c) to chromatographic resolutions.[323,381–385] A CSA analogous to TFAE (Fig. 6.29; 10-methyl-TFAE) was anchored to γ-mercaptopropyl silanized silica to produce CSP **A** (Fig. 6.36), useful in the enantiomer purity determination of sulfoxides, amines, alcohols, thiols, amino and hydroxy acids, and lactones.[386,387] A later development also based on the general structure **E** and incorporating *N*-(2-naphthyl)-α-amino acids (alanine and valine) as the chiral elements in the CSP is exemplified by **D**. Use of the latter requires that the selectands (amines, alcohols, etc.) be converted to their 3,5-dinitrobenzoyl derivatives prior to analysis.[388,389]

This and other CSPs were designed on the basis of the three-point recognition model (cf. Fig. 8.7). Since, the above cited functional groups in the analyte provide only two interaction sites, for example, a basic site (NH, OH, etc.) and a carbinyl hydrogen atom (the latter being postulated as being sufficiently acidic to interact with a basic site in the selectand), the compounds are derivatized to provide the required third site. This process typically involves incorporation of a π-acid, for example, a 3,5-dinitrobenzoyl group that is able to form a charge-transfer complex with the aryl group of the CSP. The three interactions, postulated as giving rise to the more stable of the two possible diastereomeric solvates, are illustrated in Figure 6.37.

Additional CSPs have been designed on the basis that reciprocity obtains: If racemate **A** can be resolved by nonracemic **B**, then racemate **B** should be resolvable

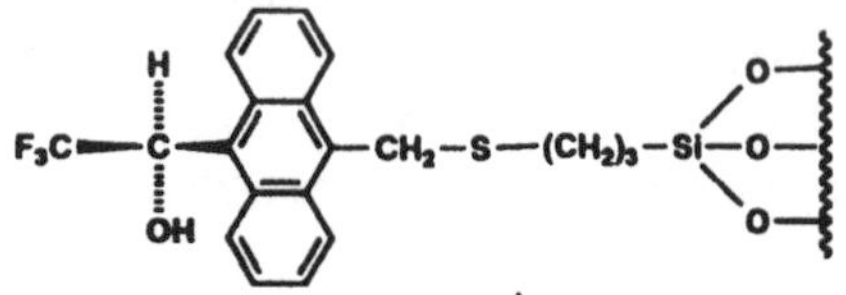

A

(R)-2,2,2-Trifluoro-1-[9-(10-α-thio-methyl) anthryl]ethanol-derived CSP.
Covalently bonded (1979)

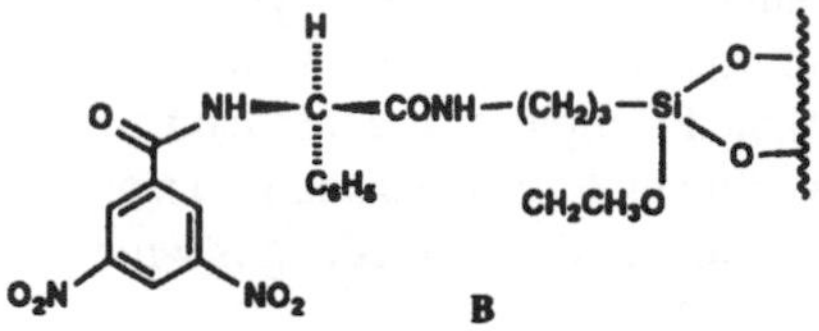

B

(R)-N-(3,5-Dinitrobenzoyl)phenyl-glycine-derived CSP.
Covalently bonded (1980)

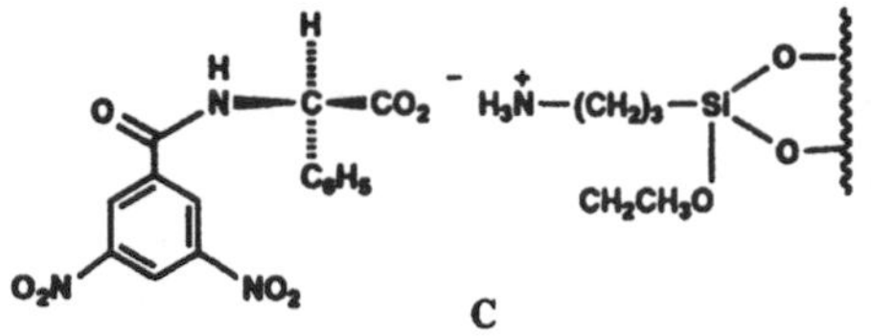

C

(R)-N-(3,5-Dinitrobenzoyl) phenyl-glycine-derived CSP.
Ionically bonded (1981)

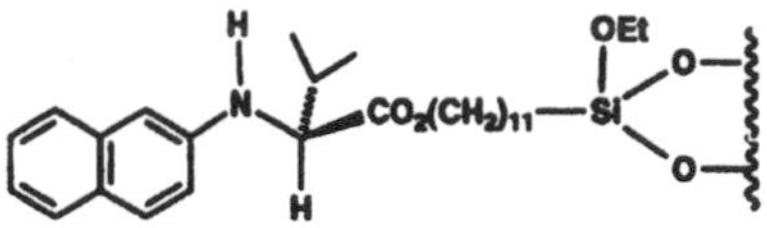

D

Covalently bonded CSP derived
from N-(2-naphthyl)-α-amino
acids (1986)

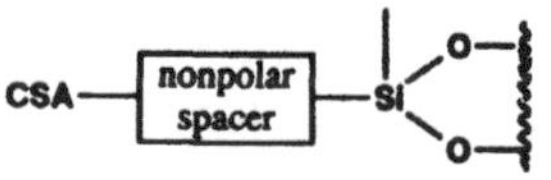

CSA—[nonpolar spacer]—Si

E

Generalized structure of
covalently bonded CSP
based on silica gel

Figure 6.36. Chiral stationary phases for enantiomeric purity determination by HPLC: Pirkle columns.

by nonracemic **A** (see Section 7-3.a). Linking N-(3,5-dinitrobenzoyl) phenylglycine to γ-aminopropyl-derivatized silica gel covalently gives rise to CSP **B** (Fig. 6.36)[387] that can resolve alcohols. A similar but ionically bonded CSP (**C**) has overlapping but not identical applications.[328,390] Numerous chiral heterocyclic systems are amenable to resolution with the ionic Pirkle CSP[391] as are underivatized alcohols possessing at

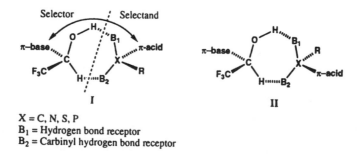

X = C, N, S, P
B₁ = Hydrogen bond receptor
B₂ = Carbinyl hydrogen bond receptor

Figure 6.37. Three-point interaction model responsible for the preferential relative retention of solutes on chiral stationary phases. Fluoroalcohol model. Diastereomeric solvates **I** and **II**.

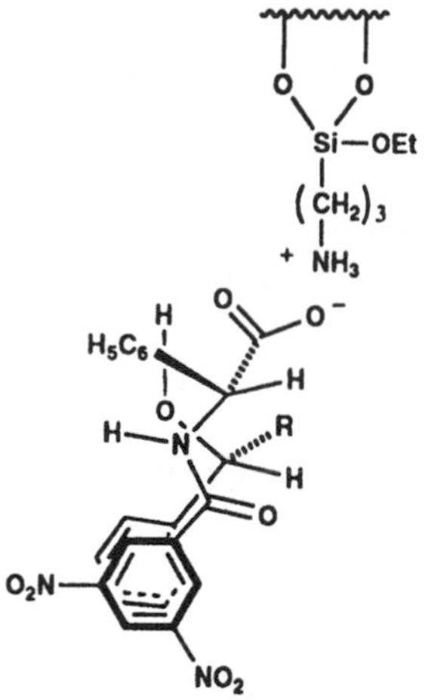

Figure 6.38. Application of the general "recognition" model to HPLC of a chiral alkyl aryl carbinol on CSP C (Fig. 6.36). The model shows the interaction of (R)-CSP with the more strongly retained (R) enantiomer of the alcohol (corresponding to the stabler diastereomeric solvate). [Adapted with permission from Pirkle, W.H., Finn, J.M., Hamper, B.C., Schreiner, J., and Pribish, J.R. *Asymmetric Reactions and Processes in Chemistry*, ACS Symposium Series, No. 185. Copyright © 1982 American Chemical Society.]

least one aryl group able to serve as a π-base[392] and atropisomeric (cf. Chapter 13) binaphthols and analogues.[393] The diastereomeric solvate responsible for preferential retention of the enantiomer having the longest retention time is illustrated in Figure 6.38.

Both the covalent and the ionic Pirkle CSPs are commercially available. Amines may be analyzed on the covalent Pirkle CSP (Fig. 6.36, **B**) following acylation, for example, with α-naphthoyl chloride.[394–396] The CSPs of this type have been applied to the assessment of enantiomer composition of chiral diastereomer mixtures whose enantiomer purity determination on achiral stationary phases was unsuccessful.[397] The Pirkle CSPs (**B** and **C**) depend on a combination of hydrogen bonding, π–π (charge-transfer) interactions, dipole–dipole interactions (dipole stacking)[398] and steric interactions to achieve both absolute and enantioselective retention of analyte molecules. The likely interaction sites in a typical selector are identified in Fig. 6.39.

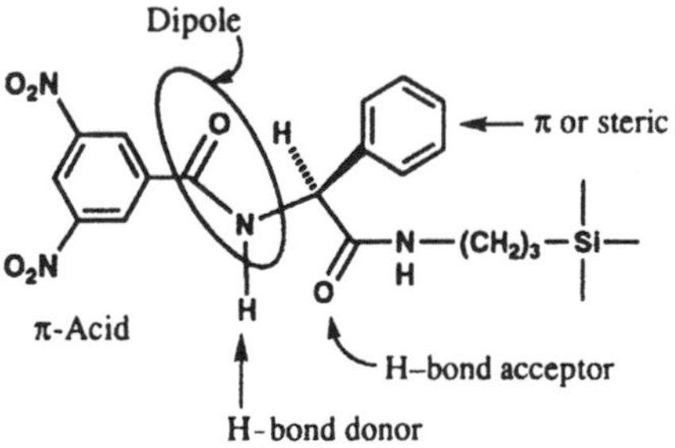

Figure 6.39. Enantioselective Pirkle Type **B** CSP showing interaction sites and types [Adapted with permission from Cook, C.E. *Pharm. Int.* **1985**, *6*, 302.]

Yet other CSPs have been developed based on the foregoing considerations. These selectors incorporate chiral ureas[399]; chiral amino acid derivatives, such as N-formyl-L-valine[400]; tartramide[401]; α-(1-naphthyl)alkylamine analogues[402]; and hydantoins[403] (see the generalized CSP structure **E** in Fig. 6.36).[404]

In some instances, chemoselective derivatization of the analyte may be necessary to promote adequate enantioselectivity. In amino acid analysis on (N-formyl-L-valylamino)propyl (FVA) silica gel, O-alkylation by bulky groups (as in tert-butyl esters) is more helpful for increasing the selectivity than are changes in the N-acyl substituent.[400]

Cinchona alkaloids (e.g., quinine, Fig. 6.29, **71**), incorporated in CSPs by reaction with a mercaptopropyl silanized silica gel (by free radical addition of the thiol group to the quinine vinyl group), are effective in the resolution of alcohols and binaphthol derivatives and acylated amines.[258,405] It is noteworthy that, with the analogous selector based on quinidine, inversion of the elution order of selectands is observed [C(8) and C(9) in quinidine are "quasi-enantiomeric" with the corresponding chiral centers in quinine.[258]

Chiral selectors based on proteins have been developed based on accumulated knowledge of protein binding (multiple bonding sites and enantioselective binding).[297] One of these stationary phases consists of bovine serum albumin (BSA) bonded to silica gel.[406,407] More recently, an enantioselective column based on α_1-acid glycoprotein (AGP), another serum protein, has been investigated.[297,408–411] Both protein columns are commercially available (Resolvosil[TM] and Enantiopac[TM], respectively).

In consequence of the mixed interactions (hydrophobic, electrostatic, or steric) that are likely to obtain between proteins and typical analytes and the high sensitivity of the binding sites to small changes in the mobile phase, especially pH and chemical modifiers (acids, bases, or alcohols), these columns can be used to analyze a very wide range of compounds, for example, amino acid derivatives, amines, alcohols, sulfoxides, and carboxylic acids including numerous medicinal agents.[412]

Microcrystalline cellulose triacetate (triacetylcellulose, TAC) is a very useful type of optically active polymer used in the analysis of mixtures of racemates including some that are devoid of functional groups.[297,413–418] When prepared by heterogeneous acetylation of cellulose in benzene suspension, TAC retains regions of crystallinity. Cavities within these regions allow enantioselective inclusion of solutes, especially those incorporating substituent-free phenyl groups. The resulting "inclusion chromatography" is effective with a wide variety of chiral compounds: hydrocarbons including atropisomers, heterocycles, and medicinal agents.[374,419,420] A proposed model for this inclusion is shown in Figure 6.40. A comparative study of TAC and poly[ethyl (S)-N-acryloyl phenylalaninate] shows that polyamide and TAC stationary phases have complementary utility.[421]

It has been observed that when TAC is solubilized and reprecipitated, its resolving power is largely lost due to a breakdown of the original crystalline structure.[413] However, TAC, cellulose tribenzoate, and the phenylcarbamates of cellulose,

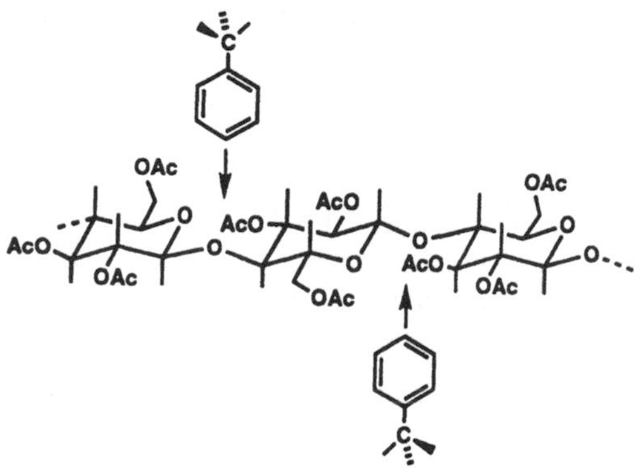

Figure 6.40. Model for inclusion chromatography of chiral analytes in microcrystalline cellulose triacetate. [Adapted with permission from Blaschke, G. *Angew. Chem. Int. Ed. Engl.* **1980**, *19*, 113.]

amylose, and other polysaccharides exhibit good resolving power when adsorbed from solution onto silica gel. Here too, it was found that none of the cited CSPs was effective for all solute types tried.[422,423] Cellulose ester, carbamate, and ether columns are commercially available [Daicel: Chiralcel™ OA, OB (esters); OC (carbamate); OE, OK (ether); see Fig, 6.41].

Liquid chromatography chiral stationary phases have also been prepared by a process called *molecular imprinting*.[297,424,425] The process consists of three steps: (1) An achiral monomer is covalently bonded to a nonracemic compound serving as a template (also called a print molecule); (2) copolymerization of the modified monomer in the presence of a cross-linking agent gives rise to a rigid polymeric network; and (3) the rigid polymer is freed of the chiral template moiety by hydrolysis. There results a polymer that retains the ability to recognize the template molecule if the latter is passed in solution through a bed containing the rigid polymer. The stationary phase evidently contains chiral cavities that are retained over long periods of time in spite of repeated washing with solvent. These

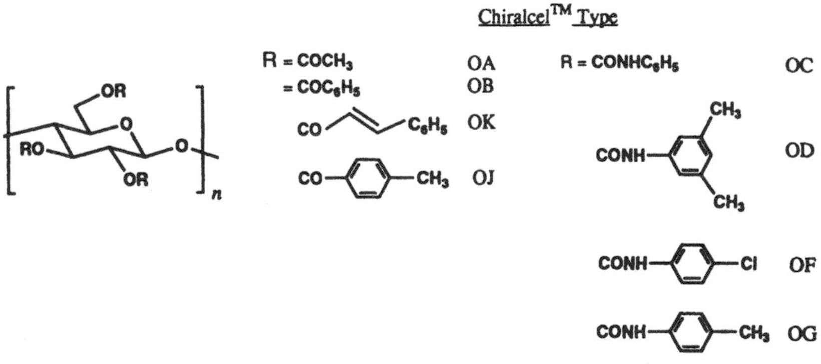

Figure 6.41. Cellulose-based CSPs (Daicel Chemical Industries, Ltd.).

cavities are responsible for the stereoisomer discrimination by the column in which either the template molecule enantiomer, an analogue thereof, or the enantiomer of the template molecule is preferentially retained during chromatography.[426–428] The process is illustrated in Fig. 6.42 with L-*N*-propionyl-2-amino-3-(4-hydroxyphenyl)-1-propanol **A** serving as the template molecule.

Inclusion complex formation with β-cyclodextrin (Fig. 6.35) bonded to silica gel by means of a covalently bonded spacer[429–432] is widely applicable to the determination of the enantiomer composition of, for example, chiral amines, amino alcohols (medicinal agents), and binaphthyl derived crown ether by HPLC.[431,433,434]

While inclusion of analyte molecules in the selector is required for diastereomer discrimination to take place, this mechanism is not the only factor contributing to differential chromatographic retention; the order of capacity factors ($k' = A_s/A_m$, where A_s and A_m represent the amounts of analyte in the stationary and mobile phases, respectively) does not always mimic that of binding constants (K_b) in a series of measurements.[431]

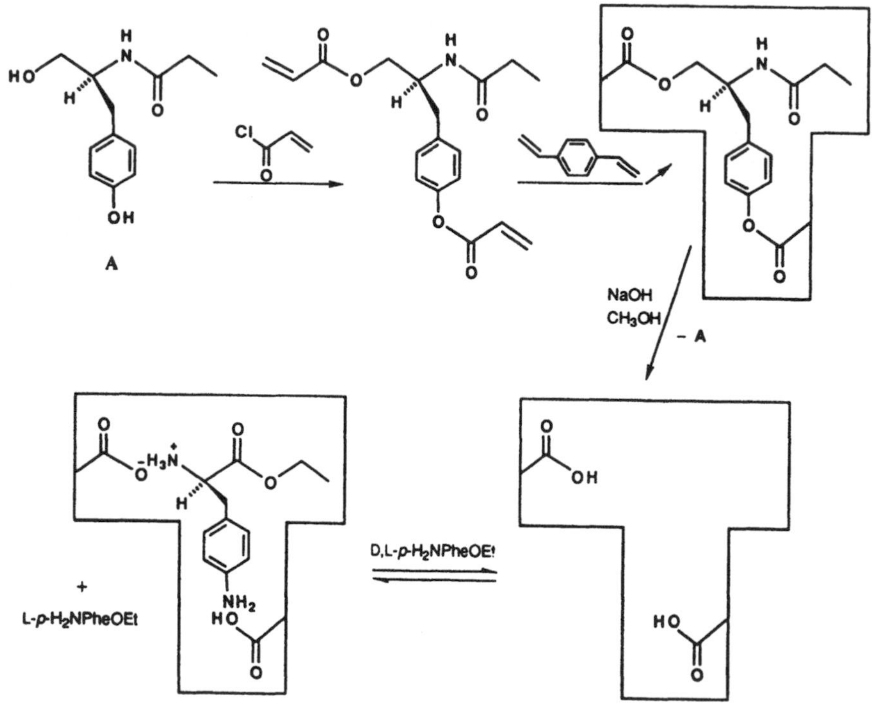

Figure 6.42. Model for chromatographic resolution of enantiomer mixtures by molecular imprinting. In the example, L-propionyl-2-amino-3-(4-hydroxyphenyl)-1-propanol (**A**) serves as the template (or print) molecule. The imprinted polymer discriminates between the enantiomers of *p*-aminophenylalanine ethyl ester [*p*-H₂NPheOEt] [Adapted with permission from Sellergren, B. and Andersson, L., *J. Org. Chem.* **1990**, *55*, 3381. Copyright © 1990 American Chemical Society.]

Chiral isotactic poly(triphenylmethyl methacrylate) (PTrMA) is the preeminent example of a cooperative CSP, one in which the stereoisomer discrimination does not arise from interaction between analyte residues and specific functional moieties in the chiral selector acting independently.[323,374] In the latter, the polymer molecules as a whole are chiral without possessing side chains bearing stereocenters; such polymers have a chiral helical backbone. Cooperative phenomena (those involving numerous chirotopic sites in the selector acting simultaneously) may account for the stereoisomer recognition and resolution that is observed in this type of enantioselective stationary phase. Poly(triphenylmethyl methacrylate) is prepared by enantioselective polymerization of triphenylmethyl methacrylate monomer by a nonracemic anionic catalyst [(–)-sparteine/*n*-butyl lithium]. The chirality of the resulting isotactic polymer is due solely to its helicity.[172,435–438]

Many hydrophobic substances including hexahelicene (Fig. 6.6, **3**), tetramesitylethylene,[439] and Tröger's base (Fig. 6.29) are resolved on PTrMA columns with separation factors α ranging from 1.1 to > 2 (α > 13 for hexahelicene). Virtually all have structures bearing one or more aromatic or heteroaromatic rings. The utility of PTrMA is comparable (in some cases superior) to TAC.

Capillary Electrophoresis with Enantioselective Supporting Electrolytes.   Formation of diastereomeric complexes by ligand exchange is the basis of a highly sensitive analytical method for the determination of enantiomeric purity by high-voltage electrophoresis in capillary columns.[440,441] The analyte, for example, a chiral amino acid [derivatized with 5-(dimethylamino)naphthalene-1-sulfonyl (dansyl, DNS) chloride] is dissolved in a support electrolyte consisting of copper(II)-L-histidine (or copper(II)–aspartame) complex. The sample migrates by a combination of electroosmosis and electrophoretic action under the influence of a strong (300 V cm^{-1}) electric field. Analysis of the separated DNS–amino acids is carried out with a laser–fluorescence detector.

The principle of the method is analogous to that which obtains in ligand-exchange chromatographic resolutions with chiral mobile phases where diastereomeric complexes are formed that migrate at different rates (Section 6-5.d). In the present method, diastereomeric complexes are also formed. Differences in complexation constants cause these transient charged species to acquire different mobilities under the influence of the applied electric field (electrokinetic separation). Note that there is no mobile phase per se in electrophoresis. The method permits rapid (ca. 10 min) analysis of the enantiomer composition of femtomolar amounts of analyte.[442,443] Diastereomer discrimination is also observed in differences in the intensity of the fluorescence signals of the diastereomeric DNS–amino acid complexes.

Dansylated amino acids have also been electrophoretically separated in the presence of ionic cyclodextrin derivatives.[444] Selective inclusion and inclusion complex migration is responsible for the separation. While typically electrophoresis is limited to ionic analytes (in this context mixtures of enantiomeric ions and of diastereomeric ions), neutral analytes can also be electophoretically analyzed by this method in the presence of surfactants that convert conventional electrolytes to micellar ones with ionic micelles acting as carrier.[444,445] The enantiomers of 1,1'-binaphthalene-2,2'-diol (Fig. 6.18, **20**) and of the related binaphthylphosphoric acid (Fig. 6.27, **49**) are resolved by capillary electrophoresis in the presence of bile salts, for example, sodium deoxycholate.

e. Kinetic Methods

Several analytical methods are based on differences in rates of reaction of enantiomers as their mixtures react with nonracemic chiral reagents. Preparative methods based on this principle are called kinetic resolutions (Section 7-5). The rate differences stem from the fact that diastereomeric transition states are formed at rates that reflect the differing free energies of activation.[446]

Methods Based on Enzyme Specificity.   Reactions of chiral substrates brought about by enzymes are often subject to enormous rate differences for the two enantiomers so that, to all intents and purposes, only one of them reacts. The enzyme, for example, hog kidney acylase,[447] may be said to be "specific" to one of the enantiomeric substrates. The rate difference may be made the basis of a powerful analytical method for the determination of enantiomer composition.[448] Moreover, in many cases, pairs of enzymes of opposite stereospecificity are known, so enantiomerically impure samples may be analyzed with either enzyme with one analysis serving as a check on the other since the two results must sum to 100%. For example, the proportion of enantiomers in α-hydroxy acids may be determined by oxidation of the acid with β-nicotinamide adenine dinucleotide (NAD$^+$) in the presence of either D- or L-lactic acid dehydrogenase (LDH) depending on the information and precision desired.[449–451]

A particularly precise way of determining enantiomer purities employs enzymes that catalyze the reactions of the *minor* enantiomer present in an enantiomerically impure sample. In the following example, the enantiomeric purity of (R)-lactate is determined enzymatically by oxidation of the minor (S)-lactate enantiomer with (S)-lactate dehydrogenase in the presence of the stoichiometric cofactor NAD$^+$, which is concomitantly converted to its reduced form NADH as shown in Eq. 6.13:

$$\underset{\substack{| \\ \text{H}_3\text{C}-\text{CH}-\text{CO}_2\text{H}}}{\overset{\text{OH}}{}} + \text{NAD}^+ \longrightarrow \underset{\substack{\| \\ \text{H}_3\text{C}-\text{C}-\text{CO}_2\text{H}}}{\overset{\text{O}}{}} + \text{NADH} + \text{H}^+ \qquad (6.13)$$

The formation of NADH is monitored by UV spectroscopy at 340 nm, the wavelength at which it absorbs.[452] In such analyses, care must be taken to insure that the reaction goes to completion, or else a correction must be applied that requires knowledge of the equilibrium constant measured under identical conditions.

While enzymatic methods have been applied mostly to the analysis of amino acids,[92,453] other functional groups lend themselves to this approach: alcohols,[454] halogenated carboxylic acids,[455] and carbohydrates (e.g., galactose).[456]

Enzymatic methods are the principal ones available for the determination of enantiomeric purity of compounds containing chiral methyl CHDT groups.

Nonenzymatic Methods.   Kinetic resolutions are governed by the relative rates of two competing reactions having rate constants k_R and k_S (i.e., those of the two enantiomers of the substrate; Section 7-5). The enantiomeric purity attainable in such a reaction is dependent on the conversion C. The relation between C, the relative rate k_R/k_S (the stereoselectivity factor **s**) and the ee of the unreacted substrate is given by Eq. 6.14. It is

$$s = \frac{\ln(1 - C)(1 - ee)}{\ln[(1 - C)(1 + ee)]} \tag{6.14}$$

evident that the enantiomeric purity of the resolved sample (unreacted kinetic resolution substrate) may be determined if **s** and C ($C < 1$) are known. An alternative equation (Eq. 6.15),

$$[S] - [R] = 0.5(e^{-k_S t} - e^{-k_R t}) \tag{6.15}$$

where $[S]$ and $[R]$ refer to the concentrations of the enantiomers shows that the enantiomer composition of the substrate may be determined at a given *time t* after the start of the reaction from a knowledge of the relative rate constants.[457] In either of these approaches, the enantiomeric purity is not measured directly, rather it is calculated from the conversion or time. Applications of these equations for the purpose of determining enantiomer compositions are few in number (e.g., to hydrocarbons, Fig. 6.43, **74**), which were kinetically resolved by enantioselective rearrangement to achiral indenes **75** (Fig. 6.43) in the presence of a chiral catalyst.[458] Equations 6.14 and 6.15 serve mostly to guide preparative kinetic resolutions (see Section 7-6).

In this connection, Kagan, Bergson, and others explicitly pointed out that two separate kinetic resolutions on the same substrate, when carried out to known conversions, suffice to determine the degree of stereoselectivity (**s**) of the reaction and the absolute rotation of the substrate, $[\alpha]_{max}$, and hence, the enantiomer composition of the substrate (assuming that the ratio $ee_1/ee_2 = [\alpha]_1/[\alpha]_2$ for the two reactions with $[\alpha]$ being that of the residual substrates).[458–461]

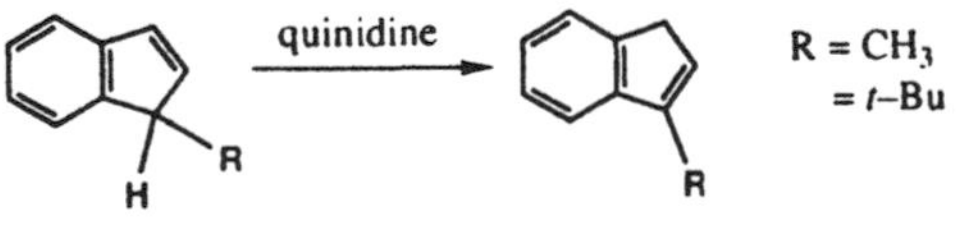

Figure 6.43. Kinetic resolution of indenes.

In 1964, Horeau described an elegant but elaborate kinetic method for determining enantiomer compositions that requires two partial consecutive resolutions to be carried out.[462] This method, which also permits the determination of configuration in some cases, is relative in that it correlates an unknown enantiomer composition with that of another substance whose enantiomer composition is known.[180,460] Subsequently, Horeau described a second kinetic method permitting both the enantiomeric enrichment of a chiral sample and the assessment of its composition just as in an enzymatic (kinetic) resolution (see above).[463]

When two racemic chiral substances R,S and D,L, respectively, react with one another in a stereoselective process, the diastereomeric products $R_D + S_L$ and $R_L + S_D$ (**A** and **B**, respectively) are formed at different rates, $k_A/k_B = K$ (when $k_A > k_B$). This ratio may easily be measured by carrying out the reaction with racemic substrates and reagents, and measuring the ratio of the racemic diastereomers formed. At the same time, when the reactant samples are nonracemic, knowledge of K and of the conversion allows one to calculate the maximum rotatory power of the remaining (slower reacting) substrate enantiomer. With this information, the optical purity of the original substrate mixture may be calculated.

f. Miscellaneous Methods

Calorimetric methods[13,464,465] and isotope dilution[466,467] have been used to determine enantiomer purity, especially at high levels of purity. The enantiomer composition of chiral ions (e.g., in ephedrinium salts) may be measured potentiometrically in an electrochemical cell fitted with two liquid polyvinyl chloride (PVC) membranes each containing one enantiomer of an electrically neutral chiral ionophore, for example, enantiopure (R,R and S,S)-5-nonyl tartrate (Fig. 6.44).[468]

Each of the membranes selectively extracts one enantiomer of the analyte forming diastereomeric complexes that formally permeate across the membrane. An electric potential difference is established between the analyte solution and the reference solution; strictly speaking, this potential difference is the sum of two phase boundary potentials and a potential within the membrane. The potential difference, which is also affected by the analyte concentration and by the necessarily unequal analyte enantiomer ratio, differs for the left-hand side and right-hand side of the electrochemical cell (Fig. 6.44). Calibration is effected separately with solutions of fixed concentration of each analyte enantiomer.[468] Along similar lines, potentiometric ion-selective electrodes incorporating peroctylated α-cyclodextrin (Fig. 6.35) [e.g., in a membrane containing PVC, and bis(butylpentyl)adipate] have been shown to be useful in the determination of the enantiomer composition of ephedrine (Fig. 6.19) in the presence of serum cations.[469,470]

A promising analytical method for the determination of enantiomer composition, especially in complex biological fluids and at very low concentrations, is radioimmunoassay (RIA). The possibility of applying RIA methodology derives from the finding (ca. 1929) that serum reactions are enantioselective,[471] that is, specific

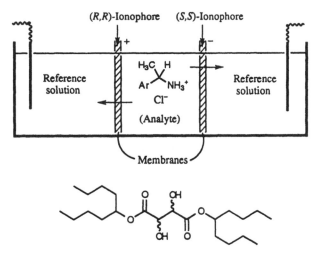

Ionophore: $(R,R / S,S)$-5-nonyl tartrate

Figure 6.44. Electrochemical cell for the potentiometric determination of the enantiomer composition of chiral ionic compounds.

antibodies (antisera) may be produced in living organisms "against practically any type of organic compound, including enantiomers of any chiral molecule".[472–474]

The process requires that typically small molecules, *haptens* (from the Greek απτειν, to fasten), radioisotopically labeled for easy detection, be conjugated, that is, covalently bound to a macromolecule (e.g., a protein), to form immunogens capable of stimulating the formation of antibodies.

Immunization of rabbits with (–)-propranolol (Fig. 6.34) conjugated with BSA generates antisera that have very low affinity[473] for the enantiomer of the hapten, that is (+)-propranolol. Analyses consist of the addition of biological fluids containing an unknown amount of (–)-propranolol to the labeled (–)-propranolol–antibody complex from which labeled (–)-propranolol is quantitatively displaced. Following separation of free analyte from the complex (by electrophoresis, precipitation, or chromatography), the concentration of the radiolabeled material is measured in a scintillation counter. As little as 10 pg of (–)-propranolol was recognized by the antiserum.[475]

Yet another analytical method, one involving isotope labeling of one of the enantiomers of a pair, permits the quantitative assessment of in vivo enantioselective effects. Administration to a living system of a 1:1 mixture of (+) and (–) enantiomers, one of whose components is labeled with a stable isotope [the mixture is regrettably misnamed pseudoracemate in the original paper (see Section 6-3)], permits the determination of the enantiomer composition of the recovered substrate by mass spectrometric analysis of isotope ratios. An essential requirement of the method is that isotope effects be known not to bias the results of the analysis.[318,476–478]

REFERENCES

1. Kelvin, Lord. *The Second Robert Boyle Lecture* in *J. Oxford Univ. Scientific Club* **1894**, No. 18, 25. Kelvin, Lord (W. Thomson). Baltimore *Lectures on Molecular Dynamics and the Wave Theory of Light,* C. J. Clay & Sons, London, 1904.

2. Ruch, E. *Acc. Chem. Res.* **1972**, *5*, 49.

3. Ruch, E. *Theor. Chem. Acta* **1968**, *11*, 183.

4. Craig., D. P. and Schipper, P. E. *Proc. R. Soc. London Ser. A* **1975**, *342*, 19.

5. Craig, D. P. and Elsum, I. R. *Chem. Phys.* **1982**, *73*, 349.

6. Craig, D. P. and Mellor, D. P. *Top. Curr. Chem.* **1976**, *63*, 1.

7. Stewart, M. V. and Arnett, E. M. *Top. Stereochem.* **1982**, *13*, 195.

8. Takagi, S., Fujishiro, R., and Amaya, K. *Chem. Commun.* **1968**, 480.

9. Atik, Z., Ewing, M. B., and McGlashan, M. L. *J. Chem. Thermodyn.* **1983**, *15*, 159.

10. Matsumoto, M. and Amaya, K. *Bull. Chem. Soc. Jpn.* **1980**, *53*, 3510.

11. Wynberg, H. *Chimia* **1976**, *30*, 445.

12. Wynberg, H. and Feringa, B. *Tetrahedron* **1976**, *32*, 2831.

13. Jacques, J., Collet, A., and Wilen, S. H. *Enantiomers, Racemates and Resolutions,* Wiley-Interscience, New York, 1981.

14. Chickos, J. S., Garin, D. L., Hitt, M., and Schilling, G. *Tetrahedron* **1981**, *37*, 2255.

15. Mason, S. F. *Molecular Optical Activity and the Chiral Discriminations,* Cambridge University Press, Cambridge, UK, 1982.

16. Horeau, A. "Safety on the Routes to Asymmetric Syntheses," Lecture presented at La Baule, France, 1972.

17. Roozeboom, H. W. B. *Z. Phys. Chem.* **1899**, *28*, 494.

18. Wilson, K. R. and Pincock, R. E. *J. Am. Chem. Soc.* **1975**, *97*, 1474.

19. Kress, R. B., Duesler, E. N., Etter, M. C., Paul, I. C., and Curtin, D. Y. *J. Am. Chem. Soc.* **1980**, *102*, 7709.

20. Jacques, J., Leclercq, M., and Brienne, M.-J. *Tetrahedron* **1981**, *37*, 1727.

21. Saigo, K., Kimoto, H., Nohira, H., Yanagi, K., and Hasegawa, M. *Bull. Chem. Soc. Jpn.* **1987**, *60*, 3655.

22. Vaida, M., Shimon, L. J. W., Weisinger-Lewin, Y., Frolow, F., Lahav, M., Leiserowitz, L., and McMullan, R. K. *Science* **1988**, *241*, 1475.

23. Weissbuch, I., Addadi, L., Lahav, M., and Leiserowitz, L. *Science* **1991**, *253*, 637.

24. Mighell, A. D., Ondik, H. M., and Molino, B. B. *J. Phys. Chem. Ref. Data* **1977**, *6*, 675.

25. Pasteur, L. C. *R. Hebd Seances Acad. Sci.* **1848**, *26*, 535.

26. Kauffman, G. B. and Myers, R. D. *J. Chem. Educ.* **1975**, *52*, 777.

27. Brock, C. P., Schweizer, W. B., and Dunitz, J. D. *J. Am. Chem. Soc.* **1991**, *113*, 9811.

28. Heilbronner, E. and Dunitz, J. D. *Reflections on Symmetry in Chemistry . . . and Elsewhere,* VHCA, Verlag Helvetica Chimica Acta, Basel, Switzerland, 1993.

29. Wilen, S. H. and Toporovsky, I. Unpublished data, 1992.

30. Sjöberg, B. *Ark Kemi* **1957**, *11*, 439.

31. Collet, A. and Jacques, J. *Bull. Soc. Chem. Fr.* **1972**, 3857.

32. Pella, E. and Restelli, R. *Microchim. Acta* **1983**, *1*, 65.

33. Meyerhoffer, W. *Ber. Dtsch. Chem. Ges.* **1904, 37**, 2604.

34. Eliel, E. L. and Wilen, S. H. *Stereochemistry of Organic Compounds*, Wiley, New York, 1994, p. 170.

35. Collet, A., Brienne, M.-J., and Jacques, J. *Chem. Rev.* **1980**, *80*, 215.

36. Brienne, M.-J., Collet, A., and Jacques, J. *Synthesis* **1983**, 704.

37. Adriani, J. H. *Z. Phys. Chem.* **1900**, *33*, 453.

38. Garin, D. L., Greco, D. J. C., and Kelley, L. *J. Org. Chem.* **1977**, *42*, 1249.

39. Kwart, H. and Hoster, D. P. *J. Org. Chem.* **1967**, *32*, 1867.

40. Farina, M. *J. Chem. Soc. Chem. Commun.* **1987**, 1121.

41. Wright, N. *J. Biol. Chem.* **1937**, *120*, 641; *ibid.* **1939**, *127*, 137.

42. Eliel, E. L. and Kofron, J. T. *J. Am. Chem. Soc.* **1953**, *75*, 4585.

43. Rao, C. N. R. *Chemical Application of Infrared Spectroscopy*, Academic Press, New York, 1963.

44. Avram, M. and Mateescu, Gh. D. *Infrared Spectroscopy*, Wiley, New York, 1972.

45. Wirzing, G. *Z. Anal. Chem.* **1973**, *267*, 1.

46. Brockmann, H., Jr. and Musso, H. *Chem. Ber.* **1956**, *89*, 241.

47. Schurig, V. *Angew. Chem. Int. Ed. Engl.* **1981**, *20*, 807.

48. Lahav, M., Laub, F., Gati, E., Lelserowitz, L., and Ludmer, Z. *J. Am. Chem. Soc.* **1976**, *98*, 1620.

49. Ludmer, Z., Lahav, M., Leiserowitz, L., and Roitman, L. *J. Chem. Soc. Chem. Commun.* **1982**, 326.

50. López-Arbeloa, F., Goedeweeck, R., Ruttens, F., De Schryver, F. C., and Sisido, M. *J. Am. Chem. Soc.* **1987**, *109*, 3068.

51. Hill, H. D. W., Zens, A. P., and Jacobus, J. *J. Am. Chem. Soc.* **1979**, *101*, 7090.

52. Casarini, D., Lunazzi, L., and Macciantelli, D. *J. Org. Chem.* **1988**, *53*, 177.

53. Dunitz, J. D. *X-Ray Analysis and the Structure of Organic Molecules*, Cornell University Press, Ithaca, NY, 1979.

54. Shoemaker, D. P., Donohue, J., Schomaker, V., and Corey, R. B. *J. Am. Chem. Soc.* **1950**, *72*, 2328.

55. (a) Paul, I. C. and Curtin, D. Y. *Science* **1975**, *187*, 19. (b) Curtin, D. Y. and Paul, I. C. *Chem. Rev.* **1981**, *81*, 525.

56. Glusker, J. P. and Trublood, K. N. *Crystal Structure Analysis, A Primer*, 2nd ed., Oxford University Press, New York, 1985.

57. Schlenk, W., Jr. *Angew. Chem. Int. Ed. Engl.* **1965**, *4*, 139.

58. Dobashi, A., Saito, N., Motoyama, Y., and Hara, S. *J. Am. Chem. Soc.* **1986**, *108*, 307.

59. Kabachnik, M. I., Mastryukova T. A., Fedin, E. I., Vaisberg, M. S., Morozov, L. L., Petrovskii, P. V., and Shipov, A. E. *Usp. Khim.* **1978**, *47*, 1541; *Russ. Chem. Rev.* **1978**, *47*, 821.

60. Weisman, G. R. "Nuclear Magnetic Resonance Anal sis Using Chiral Solvating Agents," in Morrison, J. D., ed., *Asymmetric Synthesis*, Academic Press, New York, 1983, Chap. 8.

61. Horeau, A. and Guetté, J.-P. *Tetrahedron* **1974**, *30*, 1923.

62. Dobashi, A. and Hara, S. *Anal. Chem.* **1983**, *55*, 1805.

63. Jursic, B. S. and Goldberg, S. I. *J. Org. Chem.* **1992**, *57*, 7172.

64. Luchinat, C. and Roelens, S. *J. Am. Chem. Soc.* **1986**, *108*, 4873.

65. Pasquier, M. L. and Marty, W. *Angew. Chem. Int. Ed. Engl.* **1985**, *24*, 315.

66. Vigneron, J. P., Dhaenens, M., and Horeau, A. *Tetrahedron* **1973**, *29*, 1055.

67. Reuben, J. *J. Am. Chem. Soc.* **1980**, *102*, 2232.

68. Dobashi, A., Motoyama, Y., Kinoshita, K., Hara, S., and Fukasaku, N. *Anal. Chem.* **1987**, *59*, 2209.

69. Tsai, W.-L., Hermann, K., Hug, E., Rohde, B., and Dreiding, A. S. *Helv. Chim. Acta* **1985**, *68*, 2238.

70. Carman, R. M. and Klika, K. D. *Aust. J. Chem.* **1991**, *44*, 895.

71. Martens, J. and Bhushan, R. *J. Liq. Chromatogr.* **1992**, *15*, 1.

72. Matusch, R. and Coors, C. *Angew. Chem. Int. Ed. Engl.* **1989**, *28*, 626.

73. Záhorsky, U.-I. and Musso, H. *Chem. Ber.* **1973**, *106*, 3608.

74. Fales, H. M. and Wright, G. J. *J. Am. Chem. Soc.* **1977**, *99*, 2339.

75. Winkler, F. J., Stahl, D., and Maquin, F. *Tetrahedron Lett.* **1986**, *27*, 335.

76. Baldwin, M. A., Howell, S. A., Welham, K. J., and Winkler, F. J. *Biomed Environ. Mass Spectrom.* **1988**, *16*, 357.

77. Nikolaev, E. N., Goginashvili, G. T., Tal'rose, V. L., and Kostyanovsky, R. G. *Int. J. Mass Spectrom. Ion Processes* **1988**, *86*, 249.

78. Zingg, S. P., Arnett, E. M., McPhail, A. T., Bothner-By, A. A., and Gilkerson, W. R. *J. Am. Chem. Soc.* **1988**, *110*, 1565.

79. van der Haest, A. D., Wynberg, H., Leusen, F. J. J., and Bruggink, A. *Recl. Trav. Chem. Pays-Bás* **1990**, *109*, 523.

80. Valente, E. J., Zubrowski, J., and Eggleston, D. S. *Chirality* **1992**, *4*, 494.

81. Arnett, E. M. and Zingg, S. P. *J. Am. Chem. Soc.* **1981**, *103*, 1221.

82. Vögtle, F. and Knops, P. *Angew. Chem. Int. Ed. Engl.* **1991**, *30*, 958.

83. Shinkai, S., Nishi, T., and Matsuda, T. *Chem. Lett.* **1991**, 437.

84. Nishi, T., Ikeda, A., Matsuda, T., and Shinkai, S. *J. Chem. Soc. Chem. Commun.* **1991**, 339.

85. Bruckner, S., Forni, A., Moretti, I., and Torre, G. *J. Chem. Soc. Chem. Commun.* **1982**, 1218.

86. Forni, A., Moretti, I., Prosyanik, A. V., and Torre, G. *J. Chem. Soc. Chem. Commun.* **1981**, 588.

87. Bucciarelli, M., Forni, A., Moretti, I., and Torre, G. *J. Chem. Soc. Perkin Trans 1* **1980**, 2152.

88. Morrison, J. D. and Mosher, H. S. *Asymmetric Organic Reactions,* Prentice-Hall, Englewood Cliffs, NJ, 1971, and American Chemical Society (reprint), Washington, DC.

89. Toda, F. *Bioorg. Chem.* **1991**, *19*, 157.

90. Holmstedt, B., Frank, H., and Testa, B., eds. *Chirality and Biological Activity,* Liss, New York, 1990.

91. Pasteur, L. C. R. *Hebd. Seances Acad. Sci.* **1886**, *103*, 138.

92. Greenstein, J. P. and Winitz, M. *Chemistry of the Amino Acids,* Vols. 1 and 2, Wiley, New York, 1961.

93. Solms, J., Vuataz, L., and Egli, R. H. *Experientia* **1965**, *21*, 692.

94. Shallenberger, R. S., Acree, T. E., and Lee, C. Y. *Nature (London)* **1969**, *221*, 555.

95. Mazur, R. H., Schlatter, J. M., and Goldkamp, A. H. *J. Am. Chem. Soc.* **1969**, *91*, 2684.

96. Bentley, R. *Molecular Asymmetry in Biology,* Vol. 1, Academic Press, New York, 1969.

97. Ohloff, G. *Experientia* **1986**, *42*, 271.

98. Ohloff, G., Vial, C., Wolf, H. R., Job, K., Jégou, E., Polonsky, J., and Lederer, E. *Helv. Chim. Acta* **1980**, *63*, 1932.

99. Ohloff, G. *Riechstoffe und Geruchssinn. Die molekulare Welt der Düfte*, Springer, Berlin, 1990, Chap. 2.3.7.

100. Russell, G. F. and Hills, J. 1. *Science* **1971**, *172*, 1043.

101. Friedman, L. and Miller, J. G. *Science* **1971**, *172*, 1044.

102. Leitereg, T. J., Guadagni, D. G., Harris, J., Mon, T. R., and Teranishi, R. *Nature (London)* **1971**, *230*, 455.

103. Emberger, R. and Hopp, R. *Spec. Chem.* **1987**, *7*, 193.

104. Theimer, E. T. and McDaniel, M. R. *J. Soc. Cosmet. Chem.* **1971**, *22*, 15.

105. Beroza, M., ed. *Chemicals Controlling Insect Behavior*, Academic Press, New York, 1970.

106. Tumlinson, J. H., Klein, M. G., Doolittle, R. E., Ladd, T. L., and Proveaux, A. T. *Science* **1977**, *197*, 789.

107. Bordon, J. H., Chong, L., McLean, J. A., Slessor, K. N., and Mori, K. *Science* **1976**, *192*, 894.

108. Silverstein, R. M. "Enantiomer Composition and Bioactivity of Chiral Semiochemicals in Insects," in Ritter, F. J., ed., *Chemical Ecology: Odour Communication in Animals,* Elsevier, Amsterdam, The Netherlands, 1979.

109. Bruton, J., Horner, W. H., and Russ, G. A. *J. Biol. Chem.* **1967**, *242*, 813.

110. Levin, G. V. U.S. Patent 4 262 032, Apr. 14, 1981; *Chem. Abstr.* **1981**, *95*, 78771h.

111. Kaneko, T., Izumi, Y., Chibata, I., and Itoh, T., eds. *Synthetic Production and Utilization of Amino Acids,* Koansha, Tokyo and Wiley, New York, 1974.

112. Corrigan, J. J. *Science* **1969**, *164*, 142.

113. Zwartouw, H. T. and Smith, H. *Biochem. J.* **1956**, *63*, 437.

114. Glwysen, J. M., Strominger, J. L., and Tipper, D. J. "Bacterial Cell Wall," in Florkin, M. and Stotz, E. H., eds., *Comprehensive Biochemistry,* Vol. 26A, Elsevier, Amsterdam, The Netherlands, 1968.

115. Brubaker, R. R. *Ann. Rev. Microbiol.* **1985**, *39*, 21.

116. Cline, M. J. and Lehrer, R. I. *Proc. Natl. Acad Sci. USA* **1969**, *62*, 756.

117. Abraham, E. P. "The Antibiotics," in Florkin, M. and Stotz, E. H., eds. *Comprehensive Biochemistry,* Vol. 11, Elsevier, Amsterdam, The Netherlands, 1963, p. 181.

118. Chaves das Neves, H. J., Vasconcelos, A. M. P., and Costa, M. L. "Racemization of Wine Free Amino Acids as a Function of Bottling Age," in Holmstedt, B., Frank, H., and Testa, B., eds. *Chirality and Biological Activity,* Liss, New York, 1990, Chap. 12.

119. Brückner, H. and Hausch, M. "D-Amino Acids in Food: Detection and Nutritional Aspects," in Holmstedt, B., Frank, H., and Testa, B., eds., *Chirality and Biological Activity,* Liss, New York, 1990, Chap. 11.

120. Geiger, R. and Konig, W. "Configurational Modification of Peptide Hormones," in Holmstedt, B., Frank, H., and Testa, B., eds., *Chiraliry and Biological Activity,* Liss, New York, 1990, Chap. 21.

121. Milton, R. C. deL., Milton, S. C. F., and Kent, S. B. H. *Science* **1992,** *256,* 1445.

122. Petsko, G. A. *Science* **1992,** *256,* 1403.

123. Ariëns, E. J. "Stereoselectivity of Bioactive Agents: General Aspects," in Ariëns, E. J., Soudijn, W., and Timmermans, P. B. M. W. M., eds., *Stereochemistry and Biological Activity of Drugs,* Blackwell, Oxford, UK, 1983.

124. Wainer, I. W. and Drayer, D. E., eds., *Drug Stereochemistry. Analytical Methods and Pharmacology,* Marcel Dekker, New York, 1988.

125. Borman, S. *Chem. Eng. News* **1990,** *68,* 9.

126. Stinson, S. C. *Chem. Eng. News* **1992,** *70* (Sept. 28, 1992), 46.

127. Stinson, S. C. *Chem. Eng. News* **1993,** *71* (Sept. 27, 1993), 38.

128. Stinson, S. C. *Chem. Eng. News* **1998,** *76,* (Sept. 21, 1998) 83.

129. De Camp, W. H. *Chirality* **1989,** *1,* 2.

130. Ariëns, E. J., Simonis, A. M., and van Rossum, J. M. "Drug-Receptor Interaction: Interaction of One or More Drugs with One Receptor System," in Ariëns, E. J., ed., *Molecular Pharmacology. The Mode of Action of Biologically Active Compounds,* Vol. 1, Academic Press, New York, 1964, p. 119.

131. Sastry, B. V. R. *Annu. Rev. Pharmacol.* **1973,** *13,* 253.

132. Patil, P. N., Miller, D. D., and Trendelenburg, U. *Pharmacol. Rev.* **1975,** *26,* 323.

133. Lehmann, F. P. A., Rodrigues de Miranda, J. F., and Ariëns, E. J. *Prog. Drug Res.* **1976,** *20,* 101.

134. Witiak, D. T. and Inbasekaran, M. N. "Optically Active Pharmaceuticals," in Grayson, M., ed. *Kirk–Othmer Encyclopedia of Chemical Technology,* 3rd ed., Vol. 17, Wiley, New York, 1982.

135. Ariëns, E. J. "Stereospecificity in Bioactive Agents: General Aspects," in Ariëns, E. J., van Rensen, J. J. S., and Welling, W., eds., *Stereoselectivity of Pesticides: Biological and Chemical Problems,* Elsevier, Amsterdam, The Netherlands, 1988, Chap. 3.

136. Ariëns, E. J. *Trends Pharmacol. Sci.* **1986,** *7,* 2000.

137. Knabe, J., Rummel, W., Buech, H. P., and Franz, N. *Arzneim.-Forsch./Drug Res.* **1978,** *28*(II), 1048.

138. Draber, W. and Stetter, J. "Plant Growth Regulators," in *Chemistry and Agriculture,* Spec. Publ. No. 36, The Chemical Society, London, 1979.

139. Barfnecht, C. F. and Nichols, D. E. *J. Med Chem.* **1972,** *15,* 109.

140. Powell, J. R., Ambre, J. J., and Ruo, T. I. "The Efficacy and Toxicity of Drug Stereoisomers," in Wainer, I. W. and Drayer, D. E., eds., *Drug Stereochemistry. Analytical Methods and Pharmacology,* Marcel Dekker, New York, 1988.

141. Easson, L. H. and Stedman, E. *Biochem. J.* **1933,** *27,* 1257.

142. Bergman, M. *Science* **1934,** *79,* 439.

143. Ogston, A. G. *Nature (London)* **1948,** *162,* 963.

144. Dalgliesh, C. E. *J. Chem. Soc.* **1952,** 3940.

145. Pfeiffer, C. C. *Science* **1956,** *124,* 29.

146. Kumkumian, C. S. "The Use of Stereochemically Pure Pharmaceuticals: A Regulatory Point of View," in Wainer, I. W. and Drayer, D. S., eds. *Drug Stereochemistry. Analytical Methods and Pharmacology,* Marcel Dekker, New York, 1988, Chap. 12.

147. Benschop, H. P. and De Jong, L. P. A. *Acc. Chem. Res.* **1988**, *21*, 368.

148. *People* v. *Aston*, J. G. *California Appellate Reports, Third Dist.* **1984**, *154*, 818.

149. Bonner, W. A. *Top Stereochem.* **1988**, *18*, 1. See also Bonner, W. A. "Origins of Molecular Chirality," in Ponnamperuma, C., ed., *Exobiology,* North-Holland, Amsterdam, The Netherlands, 1972.

150. Avalos, M., Babiano, R., Cintas, P., Jimenez, J. L., and Palacios, J. C. *J. Chem. Soc. Chem. Commun.* **2000**, 887.

151. Avalos, M., Babiano, R., Cintas, P., Jiménez, J. L., Palacios, J. C., and Barron, L. D. *Chem. Rev.* **1998**, *98*, 2391.

152. Podlech, J. *Angew. Chem. Int. Ed. Engl.* **1999**, *38*, 477.

153. Feringa, B. L. and van Delden, R. A. *Angew. Chem. Int. Ed. Engl.* **1999**, *38*, 3418.

154. Mason, S. F. *Nouv. J. Chem.* **1986**, *10*, 739. See also Mason, S. F. "The Origin of Biomolecular Chirality in Nature," in Krstulovic, A. M., ed., *Chiral Separations by HPLC. Applications to Pharmaceutical Compounds,* Horwood Chichester, UK, 1989, Chap. 1.

155. Lee, T. D. and Yang, C. N. *Phys. Rev.* **1956**, *104*, 254.

156. Wu, C. S., Ambler, E., Hayward, R. W., Hoppes, D. D., and Hudson, R. P. *Phys. Rev.* **1957**, *105*, 1413.

157. Ulbricht, T. L. V. *Q. Rev. Chem. Soc.* **1959**, *13*, 48.

158. Mislow, K. and Bickart, P. *Isr. J. Chem.* **1976**, *15*, 1.

159. Rain, D. W. *J. Mol. Evol.* **1974**, *4*, 15.

160. MacDermott, A. J. and Tranter, G. E. *Croat. Chem. Acta* **1989**, *62*, 165.

161. Moradpour, A., Nicoud, J. F., Balavoine, G., Kagan, H., and Tsoucaris, G. *J. Am. Chem. Soc.* **1971**, *93*, 2353.

162. Bernstein, W. J., Calvin, M., and Cubhardt, O. *J. Am. Chem. Soc.* **1972**, *94*, 494.

163. Kagan, H. B. and Fiaud, J. C. *Top. Stereochem.* **1978**, *10*, 175.

164. Buchardt, O. *Angew. Chem. Int. Ed. Engl.* **1974**, *13*, 179.

165. Flores, J. J., Bonner, W. A., and Massey, G. A. *J. Am. Chem. Soc.* **1991**, *99*, 3622.

166. Matsuura, K., Inoue, S., and Tsurata, T. *Makromol. Chem.* **1965**, *85*, 284.

167. Wald, G. *Ann. N.Y. Acad. Sci.* **1957**, *69*, 152.

168. Farina, M. *Top. Stereochem.* **1987**, *17*, 1.

169. Akaike, T., Aogaki, Y., and Inoue, S. *Biopolymers* **1975**, *14*, 2577.

170. Blair, N. E. and Bonner, W. A. *Origins Life* **1980**, *10*, 255.

171. Blair, N. E., Dirbas, F. M., and Bonner, W. A. *Tetrahedron* **1981**, *37*, 27.

172. Yuki, H., Okamoto, Y., and Okamoto, I. *J. Am. Chem. Soc.* **1980**, *102*, 6356.

173. Green, M. M., Reidy, M. P., Johnson, R. J., Darling, G., O'Leary, D. J., and Wilson, G. *J. Am. Chem. Soc.* **1989**, *111*, 6452.

174. Green, M. M., Andreola, C., Muñoz, B., Reidy, M. P., and Zero, K. *J. Am. Chem. Soc.* **1988**, *110*, 4063.

175. Green, M. M., Lifson, S., and Teramoto, A. *Chirality* **1991**, *3*, 285.

176. Nógrádi, M. *Stereochemistry,* Pergamon, Oxford, UK, 1981.

177. Halevi, E. A. *Chem. & Eng. News* **1992**, *70* (Oct. 26), 2.

178. Farmer, R. F. and Hamer, J. *J. Org. Chem.* **1966**, *31*, 2418.

179. Thaisrivongs, S. and Seebach, D. *J. Am. Chem. Soc.* **1983**, *105*, 7407.

180. Raban, M. and Mislow, K. *Top. Stereochem.* **1967**, *2*, 199.

181. Horeau, A. *Tetrahedron. Lett.* **1969**, 3121.

182. Guetté, J.-P., Perlat, M., Capillon, J., and Boucherot, D. *Tetrahedron Lett.* **1974**, 2411.

183. Pittman, C. U., Jr., Kawabata, Y., and Flowers, L. I. *J. Chem. Soc. Chem. Commun.* **1982**, 473.

184. Consiglio, G., Pino, P., Flowers, L. I., and Pittmann, C. U., Jr. *J. Chem. Soc. Chem. Commun.* **1983**, 612.

185. Ács, M. "Chiral Recognition in the Light of Molecular Associations," in Simonyi, M., ed., *Problems and Wonders of Chiral Molecules,* Akadémiai Kiadó, Budapest, Hungary, 1990, pp. 111–123.

186. Guetté, J. P., Boucherot, D., and Horeau, A. *C. R. Seances Acad. Sci. Ser. C* **1974**, *278*, 1243.

187. Morozov, L. L., Vetrov, A. A., Vaisberg, M. S., and Kuz'min, V. V. *Dokl. Akad Nauk. SSSR* **1979**, *247*, 875 [Enal. Trans. 1980, 655].

188. Gaudemer, A. "Determination of Configurations by NMR Spectroscopy Methods," in Kagan, H. B., ed., *Stereochemistry. Fundamentals and Methods*, Vol. 1, Thieme, Stuttgart, Germany, 1977.

189. Rinaldi, P. L. *Prog. Nucl. Magn. Res. Spectrosc.* **1982**, *15*, 291.

190. Yamaguchi, S. "Nuclear Magnetic Resonance Analysis Using Chiral Derivatives," in Morrison, J. D., ed., *Asymmetric Synthesis,* Vol. 1, Academic Press, New York, 1983, Chap. 7.

191. Parker, D. *Chem. Rev.* **1991**, *91*, 1441.

192. Parker, D. and Taylor, R. J. "Analytical Methods: Determination of Enantiomeric Purity," in Aitken, R. A. and Kilényi, S. N., eds., *Asymmetric Synthesis,* Blackie, London, 1992, Chap. 3.

193. Raban, M. and Mislow, K. *Tetrahedron Lett.* **1965**, 4249.

194. Jacobus, J., Raban, M., and Mislow, K. *J. Org. Chem.* **1968**, *33*, 1142.

195. Jacobus, J. and Raban, M. *J. Chem. Educ.* **1969**, *46*, 351.

196. Jacobus, J. and Jones, T. B. *J. Am. Chem. Soc.* **1970**, *92*, 4583.

197. Galpin, D. R. and Huitric, A. C. *J. Org. Chem.* **1968**, *33*, 921.

198. Cochran, T. G. and Huitric, A. C. *J. Org. Chem.* **1971**, *36*, 3046.

199. Dale, J. A. and Moser, H. S. *J. Am. Chem. Soc.* **1973**, *95*, 512.

200. Dale, J. A., Dull, D. L., and Mosher, H. S. *J. Org. Chem.* **1969**, *34*, 2543.

201. Guerrier, L., Royer, J., Grierson, D. S., and Husson, H.-P. *J. Am. Chem. Soc.* **1983**, *105*, 7754.

202. Pohl, L. R. and Trager, W. F. *J. Med. Chem.* **1973**, *16*, 475.

203. Anderson, R. C. and Shapiro, M. J. *J. Org. Chem.* **1984**, *49*, 1304.

204. Kato, N. *J. Am. Chem. Soc.* **1990**, *112*, 254.

205. Johnson, C. R., Elliott, R. C., and Penning, T. D. *J. Am. Chem. Soc.* **1984**, *106*, 5019.

206. Alexakis, A., Mutti, S., and Mangeney, P. *J. Org. Chem.* **1992**, *57*, 1224.

207. Cullis, P. M., Iagrossi, A., Rous, A. J., and Schilling, M. B. *J. Chem. Soc. Chem. Commun.* **1987**, 996.

208. ten Joeve, W. and Wynberg, H. *J. Org. Chem.* **1985**, *50*, 4508.

209. Dale, J. A. and Mosher, H. S. *J. Am. Chem. Soc.,* **1969**, *90*, 3732.

210. Pirkle, W. H. and Simmons, K. A. *J. Org. Chem.* **1981**, *46*, 3239.

211. Takeuchi, Y., Itoh, N., and Amaya, K. *J. Chem. Soc. Chem. Commun.* **1992**, 1514.

212. Takeuchi, Y., Itoh, N., Satoh, T., Koizumi, T., and Yamaguchi, K. *J. Org. Chem.* **1993**, *58*, 1812.

213. Raban, M. and Mislow, K. *Tetrahedron Lett.* **1966**, 3961.

214. Gerlach, H. and Zagalak, B. *J. Chem. Soc. Chem. Commun.* **1973**, 274.

215. Gerlach, H. *Helv. Chem. Acta* **1966**, *49*, 2481.

216. Morrill, T. C. "An Introduction to Lanthanide Shift Reagents," in Morrill, T. C., ed., *Lanthanide Shift Reagents in Stereochemical Analysis,* VCH, Deerfield Beach, FL, 1986, Chap. 1.

217. Laszlo, P. *Prog. NMR Spectrosc.* **1967**, *3*, 231.

218. Kalyanam, N. *J. Chem. Educ.* **1983**, *60*, 635.

219. Armarego, W. L. F., Millow, B. A., and Pendergast, W. *J. Chem. Soc. Perkin Trans. 1* **1976**, 2229.

220. Merckx, E. M,. Lepoivre, J. A., Lemiére, G. L., and Alderweireldt, F-C. *Org. Magn. Reson.* **1983**, *21*, 380.

221. Yamaguchi, S., Yasuhara, F., and Kabuto, K. *Tetrahedron* **1976**, *32*, 1363.

222. Rice, K. and Brossi, A. *J. Org. Chem.* **1980**, *45*, 592.

223. Hauser, F. M., Rhee, R. P., and Ellenberger, S. R. *J. Org. Chem.* **1984**, *49*, 2236.

224. Nabeya, A. and Endo, T. *J. Org. Chem.* **1988**, *53*, 3358.

225. Chan, T. H., Peng, Q.-J., Wang, D., and Guo, J. A. *J. Chem. Soc. Chem. Commun.* **1987**, 325.

226. Terunuma, D., Kato, M., Kamei, M., Uchida, H., and Nohira, H. *Chem. Lett.* **1985**, 13.

227. Saigo, K., Sugiura, I., Shida, I., Tachibana, K., and Hasegawa, M. *Bull. Chem. Soc. Jpn.* **1986**, *59*, 2915.

228. Mamlok, L., Marquet, A., and Lacombe, L. *Tetrahedron Lett.* **1971**, 1039.

229. Paquette, L. A., Ley, S. V., and Farnham, W. B. *J. Am. Chem. Soc.* **1974**, *96*, 312.

230. Rosen, T., Watanabe, M., and Heathcock, C. H. *J. Org. Chem.* **1984**, *49*, 3657.

231. Smaardijk, A. D. Ph.D. Dissertation, University of Groningen, Groningen, The Netherlands, 1986, Chap. 4.

232. Kruizinga, W. H., Bolster, J., Kellogg, R. M., Kamphuis, J., Boesten, W. H. J., Meijer, E. M., and Schoemaker, H. E. *J. Org. Chem.* **1988**, *53*, 1826.

233. Silks, L. A., III, Peng, J., Odom, J. D., and Dunlap, R. B. *J. Org. Chem.* **1991**, *56*, 6733.

234. Hiemstra, H. and Wynberg, H. *Tetrahedron Lett.* **1977**, 2183; see also Hiemstra, H. Ph.D. Dissertation, University of Groningen, Groningen, The Netherlands, 1980, p. 62.

235. ten Hoeve, W. and Wynberg, H. *J. Org. Chem.* **1979**, *44*, 1508.

236. Meyers, A. I., Williams, D. R., Erickson, G. W., White, S., and Druelinger, M. *J. Am. Chem. Soc.* **1981**, *103*, 3081.

237. Cuvinot, D., Mangeney, P., Alexakis, A., Normant, J.-F., and Lellouche, J.-P. *J. Org. Chem.* **1989**, *54*, 2420.

238. Saigo, K., Sekimoto, K., Yonezawa, N., Ishii, F., and Hasegawa, M. *Bull. Chem. Soc. Jpn.* **1985**, *58*, 1006.

239. Meyers, A. I., White, S. K., and Fuentes, L. M. *Tetrahedron Lett.* **1983**, *24*, 3551.

240. Horeau, A. and Guetté, J.-P. *C. R. Seances Acad. Sci. Ser. C* **1968**, *267*, 257.

241. Fulwood, R. and Parker, D. *Tetrahedron: Asymmetry* **1992**, *3*, 25.

242. Feringa, B. L., Smaardijk, A., and Wynberg, H. *J. Am. Chem. Soc.* **1985**, *107*, 4798.

243. Pirkle, W. H. *J. Am. Chem. Soc.* **1966**, *88*, 1837.

244. Burlingame, T. G. and Pirkle, W. H. *J. Am. Chem. Soc.* **1966**, *88*, 4294.

245. Pirkle, W. H. and Hoover, D. J. *Top. Stereochem.* **1982**, *13*, 263.

246. Pirkle, W. H. and Beare, S. D. *J. Am. Chem. Soc.* **1969**, *91*, 5150.

247. Strekowski, L., Visnick, M., and Battiste, M. A. *J. Org. Chem.* **1986**, *51*, 4836.

248. Deshmukh, M., Duñach, E., Juge, S., and Kagan, H. B. *Tetrahedron Lett.* **1984**, *25*, 3467.

249. Toda, F., Toyotaka, R., and Fukuda, H. *Tetrahedron: Asymmetry* **1990**, *1*, 303.

250. Duñach, E. and Kagan, H. B. *Tetrahedron Lett.* **1985**, *26*, 2649.

251. Nishio, M. *Chem. Pharm. Bull.* (Tokyo) **1969**, *17*, 262.

252. Buist, P. H. and Marecak, D. M. *J. Am. Chem. Soc.* **1992**, *114*, 5073.

253. Toda, F., Mori, K., and Sato, A. *Bull. Chem. Soc. Jpn.* **1988**, *61*, 4167.

254. Toda, F., Mori, K., Okada, J., Node, M., Itoh, A., Oomine, K., and Fuji, K. *Chem. Lett.* **1988**, 131.

255. Rosini, C., Uccello-Barretta, G., Pini, D., Abete, C., and Salvadori, P. *J. Org. Chem.* **1988**, *53*, 4579.

256. Wilen, S. H., Qi, J. Z., and Williard, P. G. *J. Org. Chem.* **1991**, *56*, 485.

257. Jursic, B. S. and Goldberg, S. I. *J. Org. Chem.* **1992**, *57*, 7370.

258. Salvadori, P., Pini, D., Rosini, C., Bertucci, C., and Uccello-Barretta, G. *Chirality* **1992**, *4*, 43.

259. Greatbanks, D. and Pickford, R. *Magn. Reson. Chem.* **1987**, *25*, 208.

260. Jennison, C. P. R. and Mackay, D. *Can. J. Chem.* **1973**, *51*, 3726.

261. Wenzel, T. J. *NMR Shift Reagents*, CRC Press, Boca Raton, FL, 1987.

262. Pirkle, W. H. and Sikkenga, D. L. *J. Org. Chem.* **1975**, *40*, 3430.

263. Wenzel, T. J., Morin, C. A., and Brechting, A. A. *J. Org. Chem.* **1992**, *57*, 3594.

264. Sullivan, G. R. *Top. Stereochem.* **1978**, *10*, 287.

265. Corfield, J. R. and Trippett, S. *J. Chem. Soc. D.* **1971**, 721.

266. Whitesides, G. M. and Lewis, D. W. *J. Am. Chem. Soc.* **1970**, *92*, 6979; *ibid.* **1971**, *93*, 5914.

267. Fraser, R. R., Petit, M. A., and Saunders, J. K. *J. Chem. Soc. Chem. Commun.* **1971**, 1450.

268. Goering, H. L., Eikenberry, J. N., and Koermer, G. S. *J. Am. Chem. Soc.* **1971**, *93*, 5913.

269. Goering, H. L., Eikenberry, J. N., Koermer, G. S., and Lattimer, C. J. *J. Am. Chem. Soc.* **1974**, *96*, 1493.

270. Reuben, J. and Elgavish, G. A. "Shift Reagents and NMR of Paramagnetic Lanthanide Complexes," in Gschneidner, K. A., Jr. and Eyring, L., eds., *Handbook on the Physics and Chemistry of Rare Earths*, Vol. 4, Elsevier, Amsterdam, The Netherlands, 1979.

271. Fraser, R. R. "Nuclear Magnetic Resonance Analysis Using Chiral Shift Reagents," in Morrison, J. D., ed. *Asymmetric Synthesis*, Vol. 1, Academic Press, New York, 1983, Chap. 9.

272. (a) Schurig, V. *Kontakte (Darmstadt)* **1985**, [1], 54; (b) *ibid.* **1985**, [2], 22.

273. McCreary, M. D., Lewis, D. W., Wernick, D. L., and Whitesides, G. M. *J. Am. Chem. Soc.* **1974**, *96*, 1038.

274. Wilson, W. K., Scallen, T. J., and Morrow, C. J. *J. Lipid Res.* **1982**, *23*, 645.

275. Peterson, P. E. and Stepanian, M. *J. Org. Chem.* **1988**, *53*, 1907.

276. Bulsing, J. M., Sanders, J. K. M., and Hall, L. D. *J. Chem. Soc. Chem. Commun.* **1981**, 1201.

277. Anet, F. A. L. and Bourri, A. J. R. *J. Am. Chem. Soc.* **1967**, *89*, 760.

278. Ho, T.-L. *Hard and Soft Acids and Bases Principle in Organic Chemistry*, Academic Press, New York, 1977.

279. Offermann, W. and Mannschreck, A. *Tetrahedron Lett.* **1981**, *22*, 3227.

280. Wenzel, T. J. and Sievers, R. E. *Anal. Chem.* **1981**, *53*, 393.

281. Wenzel, T. J. and Sievers, R. E. *J. Am. Chem. Soc.* **1982**, *104*, 382.

282. Wenzel, T. J. and Lalonde, D. R., Jr. *J. Org. Chem.* **1983**, *48*, 1951.

283. Mannschreck, A., Munninger, W., Burgemeister, T., Gore, J., and Cazes, B. *Tetrahedron* **1986**, *42*, 399.

284. Peterson, P. E. and Jensen, B. L. *Tetrahedron Lett.* **1984**, *25*, 5711.

285. Wenzel, T. J. "Binuclear Lanthanide (III)–Silver (I) NTMR Shift Reagents," in Morrill, T. C., ed., *Lanthanide Shift Reagents in Stereochemical Analysis,* VCH, Deerfield Beach, FL, 1986, Chap. 5.

286. Eliel, E. L. and Ko, K.-Y. *Tetrahedron Lett.* **1983**, *24*, 3547.

287. Sweeting, L. M., Crans, D. C., and Whitesides, G. M. *J. Org. Chem.* **1987**, *52*, 2273.

288. Eliel, E. L., Alvarez, M. T., and Lynch, J. E. *Nouv. J. Chem.* **1986**, *10*, 749.

289. Peters, J. A., Vijverberg, C. A. M., Kieboom, A. P. G., and van Bekkum, H. *Tetrahedron Lett.* **1983**, *24*, 3141.

290. Kabuto, K. and Sasaki, Y. *J. Chem. Soc. Chem. Commun.* **1984**, 316.

291. Kabuto, K. and Sasaki, Y. *J. Chem. Soc. Chem. Commun.* **1987**, 670.

292. Goe, G. L. *J. Org. Chem.* **1973**, *38*, 4285.

293. Fraser, R. R., Petit, M. A., and Miskow, M. *J. Am. Chem. Soc.* **1972**, *94*, 3253.

294. Alvarez, C., Barkaoui, L., Goasdoue, N., Daran, J. C., Platzer, N., Rudler, H., and Vaissermann, J. *J. Chem. Soc. Chem. Commun.* **1989**, 1507.

295. Casanova, J., Jr. and Corey, E. J. *Chem. Ind. (London)* **1961**, 1664.

296. Weygard, F., Kolb, B., Prox, A., Tilak, M., and Tomida, I. *Z. Physiol. Chem.* **1960**, *322*, 38.

297. Allenmark, S. G. *Chromatographic Enantioseparation: Methods and Applications*, 2nd ed., Ellis Horwood, New York, 1991.

298. Schurig, V. "Gas Chromatographic Methods," in Morrison, J. D., ed. *Asymmetric Synthesis*, Vol. 1, Academic Press, New York, Chap. 5.

299. Karger, B. L., Stern, R. L., and Keane, W. *Anal. Chem.* **1967**, *39*, 228.

300. Rose, H. C., Stern, R. L., and Karger, B. L. *Anal. Chem.* **1966**, *38*, 469.

301. Ayers, G. S., Mossholder, J. H., and Monroe, R. E. *J. Chromatogr.* **1970**, *51*, 407.

302. Halpern, B. and Westley, J. W. *Chem Commun.* **1966**, 34.

303. Halpern, B. and Westley, J. W. *Chem. Commun.* **1965**, 246.

304. Halpern, B. and Westley, J. W. *Chem. Commun.* **1965**, 421.

305. Westley, J. W. and Halpern, B. *J. Org. Chem.* **1968**, *33*, 3978.

306. Hauser, F. M., Coleman, M. L., Huffman, R. C., and Carroll, F. I. *J. Org. Chem.* **1974**, *39*, 3426.

307. Hirota, K., Koizumi, H., Hironaka, Y., and Isumi, Y. *Bull Chem. Soc. Jpn.* **1976**, *49*, 289.

308. Walther, W., Vetter, W., Vecchi, M., Schneider, H., Müller, R. K., and Netscher, T. *Chimia* **1991**, *45*, 121.

309. Sonnet, P. E. *J. Org. Chem.* **1987**, *52*, 3477.

310. Högberg, H.-E., Hedenstrom, E., Fägerhag, J., and Servi, S. *J. Org. Chem.* **1992**, *57*, 2052.

311. Saucy, G., Borer, R., Trullinger, D. P., Jones, J. B., and Lok, K. P. *J. Org. Chem.* **1977**, *42*, 3206.

312. Brooks, C. J. W., Gilbert, M. T., and Gilbert, J. D. *Anal. Chem.* **1973**, *45*, 896.

313. Pereira, W. E., Jr., Solomon, M., and Halpern, B. *Aust. J. Chem.* **1971**, 24, 1103.

314. Furuta, K., Shimizu, S., Miwa, Y., and Yamamoto, H. *J. Org. Chem.* **1989**, *54*, 1481.

315. Leitch, R. E., Rothbart, H. L., and Rieman, W. M., III. *Talanta* **1968**, *15*, 213.

316. Lindner, W. "Indirect Separation of Enantiomers by Liquid Chromatography," in Zief, M. and Crane, L. J., eds., *Chromatographic Chiral Separations,* Marcel Dekker, New York, 1988, Chap. 4.

317. Ahnoff, M. and Einarsson, S. "Chiral Derivatization," in Lough, W. J., ed., *Chiral Liquid Chromatography,* Blackie, Glasgow, UK, 1989, Chap. 4.

318. Cook, C. E. *Pharm. Int.* **1985**, *6*, 302.

319. Goldman, M., Kustanovich, Z., Weinstein, S., Tishbee, A., and Gil-Av, E. *J. Am. Chem. Soc.* **1982**, *104*, 1093.

320. Köhler, J. and Schomburg, G. *Chromatographia* **1981**, *14*, 559.

321. Köhler, J. and Schomburg, G. *J. Chromatogr.* **1983**, *255*, 311.

322. Helmchen, G., Nill, G., Flockerzi, D., Schuhle, W., and Youssef, M. S. K. *Angew. Chem. Int. Ed. Engl.* **1979**, *18*, 62. Helmchen, G. Nill, G., Flockerzi, D., and Youssef, M. S. K. *ibid.* **1979**, *18*, 63. Helmchen, G. and Nill, G. *ibid.* **1979**, *18*, 65.

323. Pirkle, W. H. and Finn, J. "Separations of Enantiomers by Liquid Chromatographic Methods," in Morrison, J. D., ed., *Asymmetric Synthesis*, Vol. 1, Academic Press, New York, 1983, Chap. 6.

324. (a) Helmchen, G., Ott, R., and Sauber, K. *Tetrahedron Lett.* **1972**, 3873. (b) Helmchen, G., Völter, H., and Schuhle, W. *Tetrahedron Lett.* **1977**, 1417.

325. Helmchen, G. and Strubert, W. *Chromatographia* **1974**, *7*, 713.

326. Lochmüller, C. H. and Souter, R. W. *J. Chromatogr.* **1975**, *113*, 283.

327. Lindner, W. *Chimia* **1981**, *35*, 294.

328. Wainer, I. W. and Doyle, T. D. *LC, Liq. Chromatogr. HPLC Mag.* **1984**, *2*, 88.

329. Souter, R. *Chromatographic Separations of Stereoisomers,* CRC Press, Boca Raton, FL, 1985.

330. Testa, B. and Jenner, P. "Stereochemical Methodology," in Garrett, E. R. and Hirtz, J. L., eds., *Drug Fate and Metabolism: Methods and Techniques,* Vol. 2, Marcel Dekker, New York, 1978.

331. Gal, J. *LC-GC* **1987**, *5*, 106.

332. Görög, S. "Enantiomeric Derivatization," in Lingeman, H. and Underberg, W. J. M., eds., *Detection-Oriented Derviatization Techniques in Liquid Chromatography,* Marcel Dekker, New York, 1990, Chap. 5.

333. Szepesi, G. "Ion-Pairing," in Lough, W. J., ed., *Chiral Liquid Chromatography,* Blackie, Glasgow, UK, 1989, Chap. 11.

334. Pettersson, C. "Formation of Diastereomeric Ion-Pairs," in Krstulovic, A. M., ed., *Chiral Separations by HPLC Applications to Pharmaceutical Compounds,* Horwood, Chichester, UK, 1989, Chap. 6.

335. Pirkle, W. H. and Sikkenga, D. L. *J. Chromatogr.* **1976**, *123*, 400.

336. Dobashi, Y., Dobashi, A., and Hara, S. *Tetrahedron Lett.* **1984**, *25*, 329.

337. Dobashi, Y. and Hara, S. *J. Am. Chem. Soc.* **1985**, *107*, 3406.

338. Pettersson, C. and Schill, G. *J. Chromatogr.* **1981**, *204*, 179.

339. Pettersson, C. *J. Chromatogr.* **1984**, *316*, 553.

340. Hare, P. E. and Gil-Av, E. *Science* **1979**, *204*, 1226.

341. Gil-Av, E., Feibush, B., and Charles-Sigler, R. *Tetrahedron Lett.* **1966**, 1009.

342. Gil-Av, E. *J. Chromatogr. Libr.* **1985**, *32*, 111.

343. Testa, B. *Zenobiotica* **1986**, *16*, 265.

344. Eliel, E. L. *Stereochemistry of Carbon Compounds,* McGraw-Hill, New York, 1962.

345. Feibush, B. and Grinberg, N. "The History of Enantiomeric Resolution," in Zief, M. and Crane, I. J., eds., *Chromatographic Chiral Separations,* Marcel Dekker, New York, 1988, Chap. 1.

346. Armstrong, D. W. *LC-GC* **1992**, *10*, 249.

347. Krstulovic, A. M., ed. *Chiral Separations by HPLC. Applications to Pharmaceutical Compounds,* Horwood, Chichester, UK, 1989.

348. Allenmark, S. *J. Biochem. Biophys. Methods* **1984**, *9*, 1.

349. Wainer, I. W. *Trends Anal. Chem.* **1987**, *6*, 125.

350. Wainer, I. W. *LC-GC* **1989**, *7*, 378.

351. Mikes, F., Boshart, G., and Gil-Av, E. *J. Chem. Soc. Chem. Commun.* **1976**, *99*; *J. Chromatogr.* **1976**, *122*, 205.

352. Feibush, B. and Gil-Av, E. *Tetrahedron* **1970**, *26*, 1361.

353. König, W. A. *The Practice of Enantiomer Separation by Capillary Gas Chromatography,* Hüthig, Heidelberg, Germany, 1987.

354. König, W. A. *HRC CC, J. High Resolut. Chromatogr. Chromatogr. Commun.* **1982**, *5*, 588.

355. Schurig, V. *Angew. Chem. Int. Ed. Engl.* **1984**, *23*, 747.

356. Schurig, V. *Kontakte (Darmstadt)* **1986**, [1], 3.

357. Frank, H., Nicholson, G. J., and Bayer, E. *J. Chromatogr. Sci.* **1977**, *15*, 174.

358. Stölting, K. and König, W. A. *Chromatographia* **1976**, *9*, 331.

359. König, W. A., Lutz, S., and Wenz, G. *Angew. Chem. Int. Ed. Engl.* **1988**, *27*, 979.

360. König, W. A., Lutz, S., Wenz, G., Görgen, G., Neumann, C., Gäbler, A., and Boland, W. *Angew. Chem. Int. Ed. Engl.* **1989**, *28*, 178.

361. König, W. A. *Nach. Chem. Tech. Lab.* **1989**, *37*, 471.

362. König, W. A. and Lutz, S. "Gas Chromatographic Enantiomer Separation with Modified Cyclodextrins," in Holmstedt, B., Frank, H., and Testa, B., eds., *Chirality and Biological Activity*, Liss, New York, 1990, Chap. 4.

363. Schurig, V. and Nowotny, H.-P. *Angew. Chem. Int. Ed. Engl.* **1990**, *29*, 939.

364. Meinwald, J., Thompson, W. R., Pearson, D. L., König, W. A., Runge, T., and Francke, W. *Science* **1991**, *251*, 560.

365. Armstrong, D. W., Li, W., Chang, C.-D., and Pitha, J. *Anal. Chem.* **1990**, *62*, 914.

366. Armstrong, D. W., Li, W., and Pitha, J. *Anal. Chem.* **1990**, *62*, 214.

367. Bonner, W. A. *J. Chromatogr. Sci.* **1973**, *11*, 101.

368. Frank, H., Nicholson, G. J., and Bayer, E. *J. Chromatogr.* **1978**, *167*, 187.

369. Zief, M. and Crane, L. J., eds. *Chromatographic Chiral Separations,* Marcel Dekker, New York, 1988.

370. Lough, W. J., ed., *Chiral Liquid Chromatography,* Blackie, Glasgow, UK, 1989.

371. Ahuja, S., ed., *Chiral Separations by Liquid Chromatography,* ACS Symposium Series 471, American Chemical Society, Washington, DC, 1991.

372. Krull, I. S. *Adv. Chromatogr.* **1978**, *16*, 175.

373. Audebert, R. *J. Liq. Chromatogr.* **1979**, *2*, 1063.

374. Blaschke, G. *Angew. Chem. Int. Ed Engl.* **1980**, *19*, 13.

375. Davankov, V. A., Kurganov, A. A., and Bochkov, A. S. *Adv. Chromatogr.* **1983**, *22*, 71.

376. Pirkle, W. H. and Pochapsky, T. C. *Chem. Rev.* **1989**, *89*, 347.

377. Wainer, I. W. *Chromatogr. Forum* **1986**, *1*, 55.

378. Davankov, V. A. *Adv. Chromatogr.* **1980**, *18*, 139.

379. Beitler, U. and Feibush, B. *J. Chromatogr.* **1976**, *123*, 149.

380. Davankov, V. A. *Chromatographia* **1989**, *27*, 475.

381. Finn, J. M. "Rational Design of Pirkle-Type Chiral Stationary Phases," in Zief, M. and Crane, L. J., eds., *Chromatographic Chiral Separations,* Marcel Dekker, New York, 1988, Chap. 3.

382. Doyle, T. D. "Synthetic Multiple-Interaction Chiral Bonded Phases," in Lough, W. J., ed., *Chiral Liquid Chromatography,* Blackie, Glasgow, UK, 1989, Chap. 6.

383. Macaudière, P., Lienne, M., Tambuté, A., and Caude, M. "Pirkle-Type and Related Chiral Stationary Phases for Enantiomeric Resolutions," in Krstulovic, A. M., ed,. *Chiral Separations by HPLC. Applications to Pharmaceutical Compounds,* Horwood, Chichester, UK, 1989, Chap. 14.

384. Pirkle, W. H. and Pochapsky, T. C. *Chromatogr. Sci.* **1990**, *47*, 783.

385. Perrin, S. R. and Pirkle, W. H. "Commercially Available Brush-Type Chiral Selectors for the Direct Resolution of Enantiomers," in Ahuja, S., ed., *Chiral Separations by*

Liquid Chromatography, ACS Symposium Series 471, American Chemical Society, Washington, DC, 1991, Chap. 3.

386. Pirkle, W. H. and House, D. W. *J. Org. Chem.* **1979**, *44*, 1957.

387. Pirkle, W. H., House, D. W., and Finn, J. M. *J. Chromatogr.* **1980**, *192*, 143.

388. Pirkle, W. H. and Pochapsky, T. C. *J. Am. Chem. Soc.* **1986**, *108*, 352.

389. Pirkle, W. H., Pochapsky, T. C., Mahler, G. S., Corey, D. E., Reno, D. S., and Alessi, D. M. *J. Org. Chem.* **1986**, *51*, 4991.

390. Pirkle, W. H. and Finn, J. M. *J. Org. Chem.* **1981**, *46*, 2935.

391. Pirkle, W. H., Finn, J. M., Hamper, B. C., Schreiner, J., and Pribish, J. R. "A Useful and Conveniently Accessible Chiral Stationary Phase for the Liquid Chromatographic Separation of Enantiomers," in Eliel, E. L. and Otsuka, S., eds., *Asymmetric Reactions and Processes in Chemistry*, ACS Symposium Series, No. 185, American Chemical Society, Washington, DC, 1982, Chap. 18.

392. Pirkle, W. H., Finn, J. M., Schreiner, J. L., and Hamper, B. C. *J. Am. Chem. Soc.* **1981**, *103*, 3964.

393. Pirkle, W. H. and Schreiner, J. L. *J. Org. Chem.* **1981**, *46*, 4988.

394. Pirkle, W. H. and Welch, C. J. *J. Org. Chem.* **1984**, *49*, 138.

395. Pirkle, W. H., Welch, C. J., and Hyun, M. H. *J. Org. Chem.* **1983**, *48*, 5022.

396. Pirkle, W. H., Welch, C. J., Mahler, G. S., Meyers, A. I., Fuentes, L. M., and Boes, M. *J. Org. Chem.* **1984**, *49*, 2504.

397. Evans, D. A., Mathre, D. J., and Scott, W. L. *J. Org. Chem.* **1985**, *50*, 1830.

398. Pirkle, W. H. *Tetrahedron Lett.* **1983**, *24*, 5707.

399. *Supelco Reporter* IV (2) 1, Supelco, Inc., Bellefonte, PA, 1985.

400. Dobashi, A., Oka, K., and Hara, S. *J. Am. Chem. Soc.* **1980**, *102*, 7122.

401. Dobashi, A., Dobashi, Y., and Hara, S. "Liquid Chromatographic Separation of Enantiomers by Hydrogen-Bonding Association," in Ahuja, S., ed., *Chiral Separations by Liquid Chromatography,* ACS Symposium Series 471, American Chemical Society, Washington, DC, 1991, Chap. 10.

402. Pirkle, W. H. and Hyun, M. H. *J. Org. Chem.* **1984**, *49*, 3043.

403. Pirkle, W. H. and Hyun, M. H. *J. Chromatogr.* **1985**, *322*, 309.

404. Pirkle, W. H., Hyun, M. H., and Bank, B. *J. Chromatogr.* **1984**, *316*, 585.

405. Rosini, C., Bertucci, C., Pini, D., Altemura, P., and Salvadori, I. *Tetrahedron Lett.* **1985**, *26*, 3361.

406. Allenmark, S. *LC, Liq. Chromatogr. HPLC Mag.* **1985**, *3*, 348, 352.

407. Allenmark, S., Bomgren, B., and Boren, H. *J. Chromatogr.* **1983**, *264*, 63.

408. Hermansson, J. *J. Chromatogr.* **1983**, *269*, 71.

409. Hermansson, J. *J. Chromatogr.* **1984**, *298*, 67.

410. Hermansson, J. *J. Chromatogr.* **1985**, *325*, 379.

411. Hermansson, J. and Schill, G. "Resolution of Enantiomeric Compounds by Silica Bonded α_1-Acid Glycoprotein," in Zief, M. and Crane, L. J., eds., *Chromatographic Chiral Separations,* Marcel Dekker, New York, Chap. 10.

412. Wainer, I. W., Barkan, S. A., and Schill, G. *LC, Liq. Chromatogr. HPLC Mag.* **1986**, *4*, 422.

413. Hesse, G. and Hagel, R. *Chromatographia* **1976**, *6*, 277.

414. Hesse, G. and Hagel, R. *Chromatographia* **1976**, *9*, 62.

415. Hesse, G. and Hagel, R. *Justus Liebigs Ann. Chem.* **1976**, 996.

416. Shibata, T., Okamoto, I., and Ishii, K. *J. Liq. Chromatogr.* **1986**, *9*, 313.

417. Ichida, A. and Shibata, T. "Cellulose Derivatives as Stationary Chiral Phases," in Zief, M. and Crane, L. J., eds., *Chromatographic Chiral Separations,* Marcel Dekker, New York, 1988, Chap. 9.

418. Johns, D. M. "Binding to Cellulose Derivatives," in Lough, W. J., ed., *Chiral Liquid Chromatography,* Blackie, Glasgow, UK, 1989, Chap. 9.

419. Frejd, T. and Klingsted, J. *Chem. Soc. Chem. Commun.* **1983**, 1021.

420. Pettersson, I. and Berg, U. *J. Chem. Res. Synop.* **1984**, 208.

421. Blaschke, G., Kraft, H. P., and Markgraf, H. *Chem. Ber.* **1983**, *116*, 3611.

422. Okamoto, Y., Kawashima, M., and Hatada, K. *J. Am Chem. Soc.* **1984**, *106*, 5357.

423. Okamoto, Y., Kawashima, M., Yamamoto, K., and Hatada, K. *Chem. Lett.* **1984**, 739.

424. Wulff, G., Sczepan, R., and Steigel, A. *Tetrahedron Lett.* **1986**, *27*, 1991.

425. Wulff, G. and Minarik, M. "Tailor-Made Sorbents: A Modular Approach to Chiral Separations," in Zief, M. and Crane, L. J., eds., *Chromatographic Chiral Separations,* Marcel Dekker, New York, 1988, Chap. 2.

426. Sellergren, B. and Andersson, L. *J. Org. Chem.* **1990**, *55*, 3381.

427. Wulff, G. and Schauhoff, S. *J. Org. Chem.* **1991**, *56*, 395.

428. Fischer, L., Müller, R., Ekberg, B., and Mosbach, K. *J. Am. Chem. Soc.* **1991**, *113*, 9358.

429. Armstrong, D. W. and DeMond, W. *J. Chromatogr. Sci.* **1984**, *22*, 411.

430. Armstrong, D. W. U.S. Patent 4 539 399, Sept. 3, 1985; *Chem. Abstr.* **1985**, *103*, 226754f.

431. Armstrong, D. W., Ward, T. J., Czech, A., Czech, B. P., and Bartsch, R. A. *J. Org. Chem.* **1985**, *50*, 5556.

432. Ward, T. J. and Armstrong, D. W. "Cyclodextrin-Stationary Phases," in Zief, M. and Crane, L. J., eds., *Chromatographic Chiral Separations,* Marcel Dekker, New York, 1988, Chap. 5.

433. Armstrong, D. W., DeMond, W., and Czech, B. P. *Anal. Chem.* **1985**, *57*, 481.

434. Hinze, W. L., Riehl, T. E., Armstrong, D. W., DeMond, W., Alak, A., and Ward, T. *Anal. Chem.* **1985**, *57*, 237.

435. Okamoto, Y., Suzuki, K., Ohta, K., Hatada, K., and Yuki, H. *J. Am. Chem. Soc.* **1979**, *101*, 4763.

436. Okamoto, Y. *CHEMTECH* **1987**, *17*, 176.

437. Okamoto, Y. and Hatada, K. "Optically Active Poly(Triphenylmethyl Methacrylate) as a Chiral Stationary Phase," in Zief, M. and Crane, L. J., eds., *Chromatographic Chiral Separations,* Marcel Dekker, New York, 1988, Chap. 8.

438. Johns, D. M. "Binding to Synthetic Polymers," in Lough, W. J., ed., *Chiral Liquid Chromatography,* Blackie, Glasgow, UK, 1989, Chap. 10.

439. Gur, E., Kaida, Y., Okamoto, Y., Biali, S. E., and Rappoport, Z. *J. Org. Chem.* **1992**, *57*, 3689.

440. Wallingford, R. A. and Ewing, A. G. "Capillary Electrophoresis," in Giddings, J. C., Grushka, E., and Brown, P. R., eds., *Advances in Chromatography*, Vol. 29, Marcel Dekker, New York, 1989, Chap. 1.

441. (a) Gübitz, G. and Schmid, M. G. *Enantiomer* **2000**, *5*, 5. (b) Camilleri, P., Brown, R., and Okafo, G. *Chem. Brit.* **1992**, *28*, 800.

442. Gassmann, E., Kuo, J. E., and Zare, R. N. *Science* **1985**, *230*, 813.

443. Gozel, P., Gassmann, E., Michelsen, H., and Zare, R. N. *Anal. Chem.* **1987**, *59*, 44.

444. Terabe, S. *Trends Anal. Chem.* **1989**, *8*, 129.

445. Nishi, H. and Terabe, S. *Electrphoresis (Weinheim)* **1990**, *11*, 691.

446. Mislow, K. *Introduction to Stereochemistry,* Benjamin, New York, 1965.

447. Price, V. E. and Greenstein, J. P. *J. Biol. Chem.* **1948**, *175*, 969.

448. Bergmeyer, H. U., ed., *Methods of Enzymatic Analysis,* 3rd ed., Vols. VI–VIII, VCH, Weinheim, Germany, 1984–1985.

449. Gawehn, K. "D-(−)-Lactate," in Bergmeyer, H. U., ed., *Methods of Enzymatic Analysis,* 3rd ed., Vol. 6, VCH, Weinheim, Germany, 1984.

450. Matos, J. R., Smith, M. B., and Wong, C.-H. *Bioorg. Chem.* **1985**, *13*, 121.

451. Wong, C.-H. and Matos, J. R. *J. Org. Chem.* **1985**, *50*, 1992.

452. Wong, C.-H. and Whitesides, G. M. *J. Am. Chem. Soc.* **1981**, *103*, 4890.

453. Hinkkanen, A. and Decker, A. "α-Amino Acids," in Bergmeyer, H. U., ed., *Methods of Enzymatic Analysis,* 3rd ed., Vol. 8, VCH, Weinheim, Germany, 1985, p. 329.

454. Caspi, E. and Eck, C. R. *J. Org. Chem.* **1977**, *42*, 767.

455. Motosugi, K., Esake, N., and Soda, K. *Anal. Lett.* **1983**, *16*, 509.

456. Whyte, N. N. C. and Englar, J. R. *Carbohydr. Res.* **1977**, *57*, 273.

457. Kagan, H. B. and Fiaud, J. C. *Top. Stereochem.* **1978**, *18*, 249.

458. (a) Meurling, L. and Bergson, G. *Chem. Scr.* **1974**, *6*, 104. (b) Meurling, L., Bergson, G., and Obenius, U. *Chem. Scr.* **1976**, *9*, 9.

459. Balavoine, G., Moradpour, A., and Kagan, H. B. *J. Am. Chem. Soc.* **1974**, *96*, 5152.

460. Schoofs, A. R. and Guetté, J.-P. "Competitive Reaction Methods for the Determination of Maximum Specific Rotations," Morrison, J. D., ed., *Asymmetric Synthesis*, Vol. 1, Academic Press, New York, 1983, Chap. 3.

461. Hawkins, J. M. and Meyer, A. *Science* **1993**, *260*, 1918.

462. Horeau, A. *J. Am. Chem. Soc.* **1964**, *86*, 3171.

463. Horeau, A. *Tetrahedron* **1975**, *31*, 1307.

464. McNaughton, J. L. and Mortimer, C. T. "Differential Scanning Calorimetry," in Buckingham, A. D., ed., *IRS (International Review of Science)*, Physical Chemistry Series 2, 1975, *Vol. 10, Thermochemistry and Thermodynamics*, Skinner, H. A., ed., Butterworths, London, 1975.

465. Wildmann, G. and Sommerauer, H. *Am. Lab.* **1986**, *20*, 107.

466. Berson, J. A. and Ben-Efraim, D. A. *J. Am. Chem. Soc.* **1959**, *81*, 4083.

467. Andersen, K. K., Gash, D. M., and Robertson, J. D. "Isotope-Dilution Techniques," in Morrison, J. D., ed., *Asymmetric Synthesis*, Vol. 1, Academic Press, New York, 1983, Chap. 4.

468. Bussmann, W. and Simon, W. *Helv. Chem. Acta* **1981**, *64*, 2101.

469. Bates, P. S., Kataky, R., and Parker, D. *J. Chem. Soc. Chem. Commun.* **1992**, 153.

470. Kataky, R., Bates, P. S., and Parker, D. *Analyst* **1992**, *117*, 1313.

471. Landsteiner, K. *The Specificity of Serological Reaction,* rev. ed., Dover, New York, 1962, p. 172.

472. Huhtikangas, A., Lehtola, T., Virtanen, R., and Peura, P. *Finn. Chem. Lett.* **1982**, 63.

473. Cook, C. E. "Enantiomer Analysis by Competitive Binding Methods," in Wainer, I. W. and Drayer, D. E., eds., *Drug Stereochemistry. Analytical Methods and Pharmacology,* Marcel Dekker, New. York, 1988, pp. 45–76.

474. Porter, W. H. *Pure Appl. Chem.* **1991**, *63*, 1119.

475. Kawashima, K., Levy, A., and Spector, S. *J. Pharmacol. Exp. Ther.* **1976**, *196*, 517.

476. McMahon, R. E. and Sullivan, H. R. *Res. Commun. Chem. Pathol. Pharmacol.* **1976**, *14*, 631.

477. Weinkam, R. J., Gal, J., Callery, P., and Castagnoli, N., Jr. *Anal. Chem.* **1976**, *48*, 203.

478. Howald, W. N., Bush, E. D., Trager, W. F., O'Reilly, R. A., and Motley, C. H. *Biomed Mass Spectrom.* **1980**, *7*, 35.

7

SEPARATION OF STEREOISOMERS, RESOLUTION, AND RACEMIZATION

7-1. INTRODUCTION

The production of nonracemic samples of chiral organic compounds from achiral or from racemic precursors by whatever means encompasses two broad categories. The focus of this chapter is on resolution methods. A resolution is the separation of a racemate into its two enantiomer constituents. In resolution methods, the point of departure is a racemate; therefore the maximum yield of each enantiomer is 50%. Resolution methods may involve (a) physical processes only (see Section 7-2) or (b) chemical reactions. In resolutions mediated by chemical reactions, diastereomeric transition states or diastereomeric products usually intervene. In using these methods, one may take advantage of the operation of either thermodynamic or kinetic control.

It is implied that whatever be the nature of the chemical transformation that takes place in connection with a resolution, the reaction is either reversible or, in some other way, the process ultimately leads back to the starting material separated into its enantiomer components. Hybrid processes in which the resolved products are derivatives of the starting racemates are also known and, occasionally, only one of the enantiomers is recovered from the separation. These processes do not neatly fit the above classification but they are nonetheless called resolutions.

TABLE 7.1 Resolutions

Crystallization of enantiomer mixtures. Mechanical separation of enantiomers
Processes under thermodynamic control
Processes under kinetic control
Chemical separation
Conversion to diastereomers
Thermodynamic control
Kinematic control
Intervention of diastereomeric transition states or excited states
Kinetic (including enzymatic) resolutions
Asymmetric destruction or transformation with circularly polarized light

We begin our survey (cf. Table 7.1) by examining one of the oldest and most fascinating resolution methods: the crystallization of enantiomers from solutions of racemates in the absence of resolving agents.

7-2. SEPARATION OF ENANTIOMERS BY CRYSTALLIZATION

a. Crystal Picking and Triage

When crystallization of a racemate leads to the formation of a conglomerate then, by definition, the substance is said to be spontaneously resolved. This means that during the crystallization of the racemate under equilibrium conditions, whether this takes place spontaneously at a slow rate or more rapidly when induced with seed crystals, both enantiomers of the substance deposit in equal quantities as enantiomorphous crystals.

The manual sorting of conglomerate crystals into fractions whose solutions are dextrorotatory and levorotatory (as measured by polarimetry or circular dichroism, CD) is called triage. Palpation of the crystal handedness can speed up the process and this is possible when the crystals are well formed and endowed with hemihedral faces. Two examples of such resolutions are that described by Pasteur[1,2] and that of asparagine.[3] Various stratagems for distinguishing crystals giving rise to oppositely signed solutions without relying on the presence of hemihedral faces (Section 6-4.c) have been discussed by Jacques, Collet, and Wilen[4] and Kaneko et al.[5]; see also Sections 7-2.b and 7-2.d.

b. Conglomerates

In order for resolutions of enantiomer mixtures to be possible without the use of chiral reagents or of enantioselective chromatographic columns, the racemate must be a conglomerate under the conditions of the crystallization. While only several hundred conglomerates have been explicitly described,[4] many more must exist in the domain of chiral organic compounds. A sampling of 1308 neutral compounds drawn from the Beilstein Handbook reveals that conglomerates constitute between 5% and 10% of the totality of chiral organic solids.[6]

Even if a given racemate does not crystallize as a conglomerate, it may be possible to reversibly convert it to a derivative that does. For example, while few of the common amino acids are spontaneously resolved, many derivatives are known that are conglomerates (cf. Section 7-2.c). Alanine, leucine, and tryptophan crystallize as racemic compounds, but their benzenesulfonate salts are conglomerates; histidine hydrochloride is a conglomerate, and there are many other cases.[7] In general, it has been shown that the frequency with which such derivatization leads to a conglomerate is significantly increased when the derivative is a salt rather than a covalent compound.[6] Thus while ($\pm$)-tartaric acid crystallizes as a racemic compound, the first substance ever resolved, Pasteur's salt (racemic $NaNH_4$ tartrate·$4H_2O$), is a conglomerate below 28°C, as is "Seignette's salt" (racemic NaK tartrate·4H$_2$O) below −6°C; these double salts are spontaneously resolved, These two cases also illustrate the possibility of transforming some conglomerates into less highly solvated racemic compounds as the temperature is raised[4] and vice versa.

Seeding with homochiral crystals of one of the enantiomers moves the process into the domain of nonequilibrium crystallizations. Here, a preponderance of one of the enantiomers of the substance may be deposited in a mother liquor whose optical rotation changes during the crystallization. *Preferential crystallization* is a practical variant of such nonequilibrium crystallizations carried out under carefully specified conditions designed to prevent the spontaneous crystallization of the undesired enantiomer during the controlled crystallization of its chiral partner (Section 7-2.c).

If one has a sample (Fig. 7.1) of enantiomerically enriched but impure **1** or **2**, one of the simplest ways of transforming it into an enantiomerically pure one is to convert the substance to one of the derivatives (**3**, **4**, or **5**). One crystallization should suffice to remove the minor enantiomer (in the mother liquor), leaving an enantiomerically pure crystalline sample of the major constituent. This result is possible *no matter what* the initial enantiomeric composition is. It is for this reason, in particular, that conglomerate-forming derivatives are so useful.

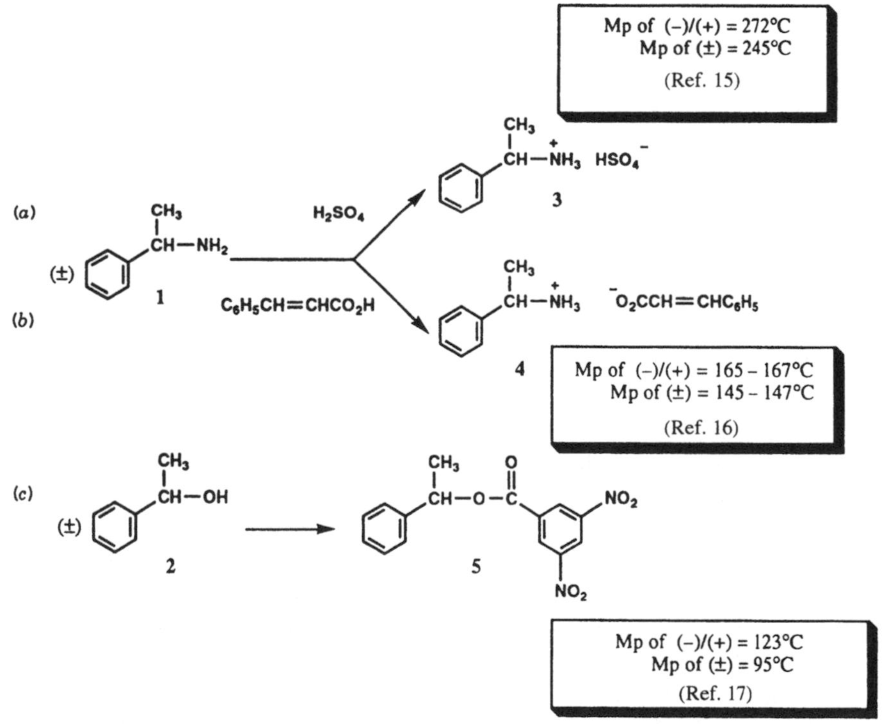

Figure 7.1. Derivatives exhibiting conglomerate behavior.

Since conglomerate systems are so much easier to resolve and to enrich than racemic compound systems, it is important to be able to identify conglomerates with confidence. In general, when a racemate that must ultimately be resolved is produced in the laboratory, knowledge of the racemate type should simplify subsequent manipulation. The more common methods for identifying conglomerates are summarized in Table 7.2.[4]

Method 1 serves for a first screening only; it must be followed by one of the other tests. However, in the absence of a significant melting temperature difference, or if the racemate melting point is *higher* than that of one of the pure enantiomers, conglomerate behavior is precluded.

Method 2 requires the measurement of optical activity of a solution prepared from a single crystal of the racemate. This method will be successful except when twinning occurs or when the crystals are too small to make it practical. Thus absence of optical activity as measured on a single crystal is not a sufficient criterion for excluding conglomerate behavior. Of course, with a small crystal, the magnitude of rotation may be too small to be observed.

Method 3 is effectively a variant of Method 2. The induced cholesteric behavior is an alternative and effective way of revealing optical activity in an extremely small sample (e.g., a single crystal).[4] In fact, such behavior would be revealed even on a cryptochiral sample (i.e., in the absence of optical activity).

Analysis of binary phase diagrams (Method 4) has been taken up in Section 6-4. The contact method of Kofler is a simpler way of obtaining the essential information conveyed by the phase diagram, namely, whether the diagram contains one or two eutectics, (see Fig. 6.3). The method of Kofler can be carried out on the hot stage of a polarizing microscope.[4]

TABLE 7.2 Methods for the Identification of Conglomerates

1.	First screening: If the melting point of the enantiomer exceeds that of the racemate by 25°C or more, the probability of conglomerate formation is high (the difference in melting point is generally in the range 25–35°C though differences of 20°C or even less are not uncommon).
2.	Demonstration of spontaneous resolution via measurement of a finite optical rotation on a solution prepared from a single crystal of the racemate.
3.	Solution in a nematic phase of a single crystal taken from a racemate; conversion of the nematic phase to a cholesteric liquid crystal.
4.	a. Analysis of the binary phase diagram of the enantiomer mixture. b. Application of the contact method of Kofler.
5.	Solubility behavior of one of the enantiomers in the saturated solution of the racemate. Insolubility is indicative of conglomerate behavior.
6.	Determination of the crystal space group and of Z (number of asymmetricm units per unit cell) by X-ray diffraction.
7.	Comparison of the *solid state* IR spectra (also of solid state NMR spectra or of the X-ray powder diffraction patterns) of the racemate with that of one of the enantiomers; identity of the spectra is characteristic of conglomerate behavior.

Method 5 deals with the solubility behavior of one of the enantiomers in a saturated solution of the racemate. An excess of one of the enantiomers is insoluble in such a solution if the racemate is a conglomerate. This phenomenon follows as a consequence of the shape of the solubility diagram (see Fig. 6.10).

Method 6 is a direct method, applicable if a well-formed crystal is available. If the crystal *space group* found is one of the noncentrosymmetric kinds, then (with rare exception; see Section 6-4.k) the sample is a conglomerate. *This analysis does not require a full structure determination by X-ray diffraction.*

Perhaps the simplest of the methods is the comparison of the IR spectra in the solid state of one of the enantiomers and of the corresponding racemate (Method 7). If the IR spectra of the two forms, as measured by the KBr pellet method or in a mull, are completely superposable, then the racemate is a conglomerate. If there is a significant difference between the spectra, then the racemic sample is most probably a racemic compound.

Three examples illustrating conglomerate behavior are shown in Figure 7.1 (compounds **3–5** with their melting points). Conglomerate behavior is confirmed (for **4** and **5**) by complete superposability of the IR spectra of the nonracemic and racemic forms.

Methods 1, 4, 5, and 7 require samples of both racemate and enantiomer (for Method 1, literature values of melting points would suffice) while Methods 2, 3, and 6 can be carried out with samples of racemate alone.

In order for a spontaneous resolution to have practical significance, the crystals must be separated. We say "mechanically separated" though what we really mean in most cases is manually separated, for no mechanical device is yet available that will unerringly pick all dextrorotatory crystals out of the crystalline matrix of the racemate and segregate them from the remaining levorotatory ones (however, see below for separation by sifting). Even for manual separation, there remains the question of how to identify the sense of chirality of any one crystal.

Two variants can transform the simultaneous crystallization of the two enantiomers of a compound into practical resolution methods. The first consists in the deposition of the two enantiomers at different places in the racemic solution under the influence of localized seeds, the (+)-seeds forming large homochiral (+)-crystals while the (–)-seeds induce formation of homochiral (–)-crystals elsewhere in the solution whose liquid-phase homogeneity may be maintained by stirring or by use of a circulating system. Several devices have been described to facilitate the localization of crystallization.[7]

In the second variant, one encourages the formation of large crystals of one of the enantiomers while the other enantiomer spontaneously deposits in the form of small crystals. Relatively large seeds of the first enantiomer are required to initiate what has been called "differentiated" crystallization. The two sizes of enantiomeric crystals may then be separated by sifting.[7]

c. Preferential Crystallization

When the resolution of a racemate is initiated by inoculation of a saturated or supersatured solution of the racemate with seeds (crystals) of one of the enantiomers, crystallization ensues in a nonequilibrium process called *preferential crystallization,* since it is only the seeded enantiomer that deposits.

Preferential crystallization works only for substances that are conglomerates.[8] More precisely, it may work also with compounds existing either as conglomerates or as racemic compounds within an accessible temperature range, provided that the racemic compound does not crystallize during the operation. The latter qualification is to allow for the possible existence of metastable racemic compounds that may impede a preferential crystallization.[7]

Examples of compounds resolvable by preferential crystallization above or below a known transition temperature, including the classic Pasteur case, are shown in Figure 7.2. The common case is that of a racemic compound that gives rise to a conglomerate on heating; the converse case (Pasteur's salt; phenylglycine sulfate; 2,2'-diamino-1,1'-biphenyl), though rarer, is also known.[4]

Conglomerates are sometimes obtainable in metastable form outside the temperature range in which they are normally stable, for example, 1,1'-binaphthyl by crystallization at room temperature.[9] The preferential crystallization of a conglomerate as a metastable polymorphic form in the presence of additives, for example, that of histidine·HCl at 25°C, is described in Section 7-2.d.

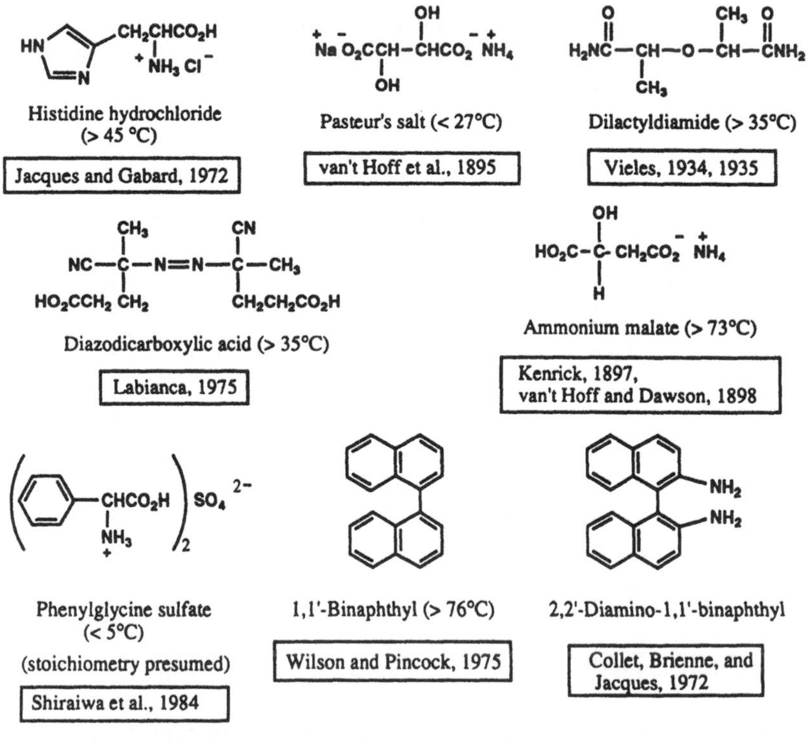

Figure 7.2. Compounds resolvable by preferential crystallization as a function of temperature.

In a preferred procedure, the excess enantiomer [e.g., (+)] crystallizes (either spontaneously or with seeds) from a solution of a conglomerate in which some of the (+) enantiomer has been dissolved by heating. When the solution is cooled, it becomes supersaturated with respect to the (+) enantiomer and the latter crystallizes. Moreover, the crystalline (+) enantiomer *entrains* the crystallization of that enantiomer so that more of this enantiomer crystallizes than was originally added to the racemate at the beginning of the experiment.[5,7]

This concept gives rise to the name *entrainment* that has been used as an alternative term for the process. Preferential crystallization was developed by Amiard, Velluz, and co-workers at Roussel-Uclaf, Paris[10–12] and applied with verve by Japanese investigators. It is based on a discovery by Gernez (1866)[13] who was a student of Pasteur's. The efficacy of the process was recognized by Duschinsky[14] to whom we owe the first practical resolution by preferential crystallization, that of histidine·HCl.

To emphasize the distinction between preferential crystallization and the spontaneous resolutions described in Sections 7-2.a and 7-2.b, the reader is reminded that in the latter (equilibrium processes), the rotation of the mother liquor remains zero all the while that crystallization takes place. Consider the following example involving preferential crystallization of α-methylbenzylamine cinnamate (Fig. 7.1*b*, **4**.[11] Racemic cinnamate (**4**, 15.5 g) is dissolved in aqueous CH_3OH by heating it along with 12.1 g (±)-α-methylbenzylamine and aq. HCl (12 *M*). The cooled racemic solution is inoculated with 20 mg of pure (+) salt. After 1 h the white crystals obtained are washed and dried. Yield: 3.4 g of (+)-**4** salt (3.4/7.75 × 100 = 44% of theory); ee = 75%. The mother liquor is levorotatory. Racemic salt is now added in a quantity roughly equal to the solid collected (3.5 g), dissolved by heating, cooled, and seeded with 20 mg of pure (–) salt. After 1 h, 5.4 g (57% of theory) of precipitated (–)-**4** salt is collected; ee = 85%.

If enantiomerically pure products are required, one recrystallization of each of the enantiomers (in the above example, from CH_3OH) suffices to produce pure (+)- and (–)-salt, respectively. The second part of the process leading to the (–)-salt suggests that, if the beginning solution is enriched in one enantiomer, the yield of crystals depositing within a fixed period of time will be higher than when the starting solution is racemic [after the second batch of racemic salt is added, the solution from which (–)-salt deposits is more highly supersaturated with respect to this salt]. Note that the yield of (–)-salt (5.4 g) is larger than the excess present at the onset of its preferential crystallization (3.4 g). This enhancement in yield is the essence of the entrainment. In any event, many resolutions by preferential crystallization are initiated with nonracemic solutions.

Preferential crystallization may be carried out in a cyclic, repetitive, manner as exemplified for hydrobenzoin. Initial crystallization takes place from a stirred solution containing a slight excess of one of the enantiomers [0.37 g of (–) and 11.0 g of racemate in 85 g of 95% ethanol]. In a typical case, the solution was cooled to 15°C, seeded with 10 mg of pure (–)-hydrobenzoin, and crystallization was allowed to proceed for 20 min. The crystals collected (0.87 g) were about double in weight with respect to the excess levorotatory isomer taken. *rac*-Hydrobenzoin was added to the mother liquor in an amount (0.9 g) equal to the (–)-crystals collected and the solution heated to dissolve the solid. Cooling the solution and seeding it with 10 mg of pure (+)-hydrobenzoin gave 0.9 g of (+)-crystals after 20 min. After removal of the latter, addition of 0.9 g of racemate to the mother liquor and warming reconstituted the solution whose preferential crystallization was repeated (8 cycles) as just described. The products [6.5 g of (–)- and 5.7 g of (+)-hydrobenzoin] had an enantiomeric composition approaching 99:1. A similar cyclic resolution process has been described for 1-phenylethyl 3,5-dinitrobenzoate (Fig. 7.1, **5**) on a larger scale.[17] Examples of even larger scale preferential crystallizations (kilogram lots) carried out for 30–40 cycles may be found in the literature.[18] The process has also been used in the industrial scale resolution of *threo*-1-(*p*-nitrophenyl)-2-amino-1,3-propanediol, whose (–) enantiomer is an intermediate in the synthesis of the antibiotic chloramphenicol.

The efficiency of preferential crystallization is expressed by a resolution index (RI) defined as the ratio of resolution product weight (calculated as pure enantiomer) to that of the initial excess of this enantiomer taken, that is, $[W_{recovered} \times ep - W_{seeds}]/E_{excess\ taken}$, where ep (enantiomeric purity) is equivalent to ee. Since RI = 1 corresponds just to recovery of the excess enantiomer taken, a value of RI $\geq$ 2 should obtain if the alternate crystallization of each enantiomer by preferential crystallization is to be considered efficient.[19] The value of RI calculated for the first cycle of the hydrobenzoin resolution described above (assuming a product ee of 97%) is [(0.87)(0.97) – 0.010]/0.37 = 2.2.

The factors favoring preferential crystallization[7] are as follows: (a) Racemic salts (hydrochlorides, sulfonates, etc.) are more prone to resolution by preferential crystallization than are covalent racemates.[20,21] (b) A solubility ratio, $\alpha_x = S_R/S_A$ (where S_R and S_A are the solubilities of the racemate and of one of the enantiomers, respectively, expressed as mole fractions) less than 2 is preferable to > 2. Contrary to intuition, the reduced solubility of the racemate relative to that of the enantiomers gives rise to a more favorable situation by enlarging the part of the solubility diagram usable for preferential crystallization. (c) The stirring rate can be optimized to increase the rate of crystal growth as well as to promote the growth of homochiral crystals.[22,23] However, increasing the stirring rate indiscriminately is counterproductive since spontaneous nucleation of the wrong enantiomer may ensue. (d) It is desirable to use seeds of uniform small size and composition. (e) Sterilization of the solution by boiling prior to the cooling and inoculation step may eliminate unwanted small seeds (nuclei) invisible to the naked eye. (f) It has been suggested that generation of microcrystalline seeds effected by ultrasonic irradiation (10–100 kHz) may reproducibly lead to a very rapid crystallization.[24]

Numerous papers and patents attest to the potential and actual economic importance of preferential crystallization as a technique for resolution on a large scale.

d. Asymmetric Transformation of Racemates and Total Spontaneous Resolution

The yield of a resolution is necessarily limited to 50%, that is, 100% of one enantiomer of the pair being resolved. If the compound being resolved is to be used in a synthesis,

50% of the resolution substrate may be wasted. There are, however, two ways in which this limitation may be circumvented: (a) Both enantiomers may conceivably be used in an enantioconvergent synthesis,[25] and (b) the undesired enantiomer may conceivably be racemized and subjected again to resolution. If the resolution could be coupled to the racemization step (Section 7-7), then a yield exceeding 50% of one enantiomer could be achieved efficiently essentially in one step. Such a process, corresponding to an asymmetric transformation (see below), actually has been described in a number of cases.[26]

Many chiral substances are known that are configurationally labile (i.e., their enantiomers interconvert) in solution. Displacement of the equilibrium in solution as a result of (a reversible) interaction with some external chiral entity corresponds to an *asymmetric transformation of the first kind*[27] a process that is more commonly observed in diastereomeric systems (Section 7-3.e). In the more recent literature, the term *enantiomerization*, applied to the interconversion of enantiomeric conformations by torsional motion or by inversion (e.g., of nitrogen compounds) in nonplanar environment[28] has often come to replace the older expression asymmetric transformation of the first kind (see below).[29] This type of stereoisomerization is manifested principally in NMR spectra where diastereotopic nuclei become enantiotopic (and consequently become isochronous; cf. Section 8-4.d) as the temperature is raised.[30,31] A few examples of photochemical enantiomerizations, either under the influence of circularly polarized light (cpl) or of chiral photosensitizers have been described.[32]

A beautiful example of an enantiomerization, that of a chiral polymer, has been described.[33] The polyisocyanates such as poly(n-hexyl isocyanate)–(N(n-C$_6$H$_{13}$)–C(=O)–)$_n$, are polymers devoid of chiral centers whose chirality is due to formation of helical conformations of the polyisocyanate backbone.[34] The absence of optical activity in such polymers stems from the equal probability of forming right (P) and left (M) handed helical forms during typical syntheses. The dynamic equilibrium existing in solution between the two enantiomeric conformations of the polymer has been both revealed and perturbed by dissolving the racemic polymer in a chiral (nonracemic) solvent, (R)-2-chlorobutane. Induction of positive CD (see Section 12-4.e) at 250 nm, a spectral region in which the solvent is transparent, indicates that (R)-2-chlorobutane (91% ee) displaces the (1:1) $P \rightleftharpoons M$ equilibrium in the direction of excess P helix (i.e., the polymer undergoes enantiomerization).

When configurationally labile racemates are crystallized rapidly, the enantiomeric composition of the liquid-phase equilibrium is reproduced in the solid state: the crystalline mass is racemic. This solid racemate may exist either as a racemic compound or it may be a conglomerate just as is found with configurationally stable chiral compounds. With conglomerate systems, however, *slow* crystallization occurring spontaneously, or induced by means of enantiopure seeds, may result in exclusive deposition of only one of the enantiomers.

For systems consisting of racemates, the descriptive term "crystallization-induced asymmetric disequilibration" has been suggested to describe this process. An even better term, applicable to both enantiomers and to diastereomers (see Section 7-3.d) is

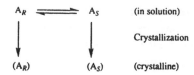

Figure 7.3. Crystallization-induced asymmetric transformation of a racemate.

crystallization-induced asymmetric transformation.[4] In a system containing enantiomeric molecules, a crystallization-induced asymmetric transformation is a process for converting a racemate into a single (pure) enantiomer of the substrate (Fig 7.3). In that sense, the process is the converse of racemization. The hallmark of such a transformation is the separation of one enantiomer from a racemate in a yield greater than 50%.

Two early discovered examples of compounds resolved by crystallization-induced asymmetric transformations are shown in Figure 7.4. When *rac*-*N,N,N*-allylethyl-methylanilinium iodide **6** is allowed to crystallize slowly from chloroform it forms enantiopure crystals. A characteristic of total spontaneous resolutions is that the mother liquor remains racemic because the enantiomer that crystallizes is constantly being replenished by a (+) ⇌ (−) equilibration (50:50) in solution.[35]

Crystallization of tri-*o*-thymotide (TOT. Fig. 7.4, **7**) leads to formation of clathrate inclusion complexes by inclusion of solvent molecules. These inclusion complexes are conglomerates.[36,37] Individual crystals of the conglomerate form containing either achiral or chiral guest molecules surrounded by several TOT molecules are homochiral. However, dissolution of these crystals regenerates racemic TOT as a result of interconversion of the TOT enantiomers by bond rotations.

The dithia-hexahelicene **8** (Fig. 7.5) undergoes spontaneous resolution on crystallization in benzene. When optically active **8** is dissolved in chloroform it

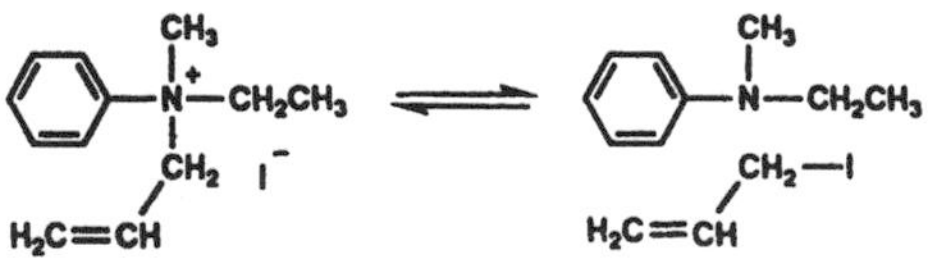

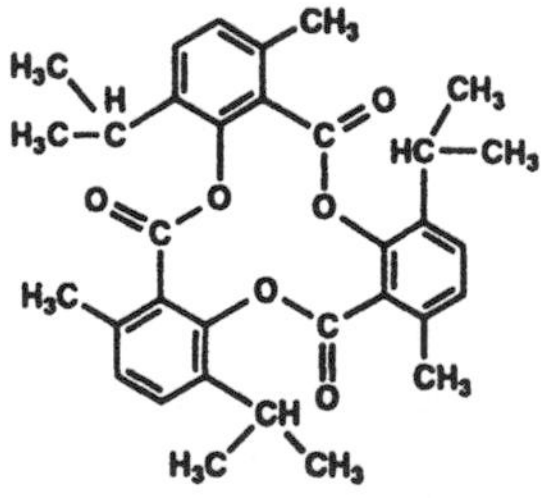

Figure 7.4. Compounds that undergo crystallization-induced asymmetric transformations. The enantiomers of **6** interconvert by dissociation into the achiral amine and alkyl halide precursors of the salt; those of TOT (**7**) are interconverted by bond rotations.

racemizes at room temperature with $t = 230$ min ($t = t_{1/2}$; see Section 7-7); this implies that crystallization of **8** is attended by asymmetric disequilibration.[38]

Spontaneous resolution has been observed in rac-1,1′-binaphthyl (Fig. 7.5, **9**) on crystallization either from solution or from the melt. Optical activity as high as $[\alpha]_D$ ± 233 (presumably in benzene) corresponding to 95% ee has been measured in individual samples.[9,39] The extent of resolution is very sensitive to the length and temperature of storage of the sample, to the extent of grinding of the crystals, and to inadvertent seeding. These facts have been rationalized by the finding that the racemic compound (mp 145°C) and the conglomerate (mp 158°C) polymorphic forms of **9** can coexist over a wide range of temperatures. The conglomerate is more stable than the racemic compound at temperatures greater than 76°C but a metastable conglomerate can easily be maintained even at room temperature. In solution, optical activity is gradually lost; $t_{1/2}$ for racemization is about 10 h (25°C) in several solvents and in the melt $t_{1/2}$ = about 0.5 s at 150°C with interconversion of the enantiomers taking place by rotation about the 1,1′ bond. Spontaneous resolution is thus understood to follow from the ease of racemization and from the presence of the conglomerate form that allows one of the enantiomers to grow at the expense of the racemate.

Chiral ketones with a stereogenic atom alpha to the carbonyl group undergo crystallization-induced asymmetric transformations in basic medium provided that the stereocenter bears a hydrogen atom. Examples are p-anisyl α-methylbenzyl ketone (Fig. 7.5 **10a**),[40] and the ketone precursor **12** (Fig. 7.5) of the plant growth regulator paclobutrazol.[41]

A sample of (±)-methyl α-(6-methoxy-2-naphtyl)propionate (naproxen methyl ester: Section 7-3.a; Fig. 7.5, **11**), was fused at 70°C, and sodium methoxide was added. The mixture was supercooled (to 67°C), seeded with (+)-**11**, then allowed to crystallize; (+)-**11** was obtained in 87% yield.[42] The interpretation of this experiment

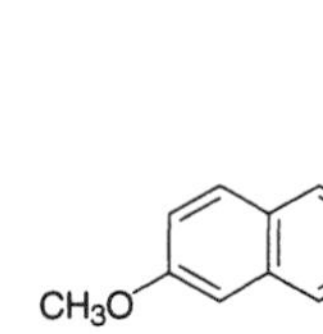

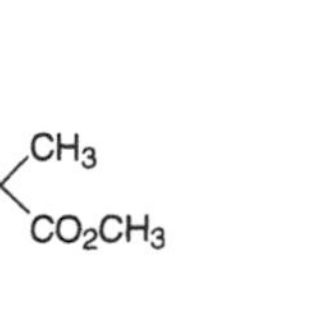

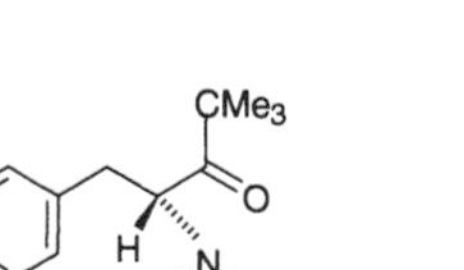

10a   R = OCH₃
10b   R = SCH₃

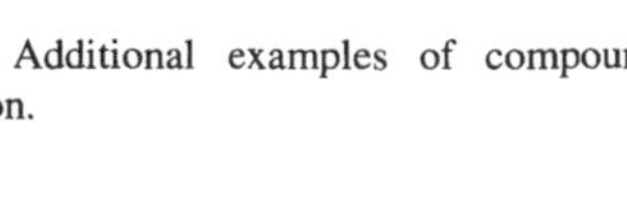

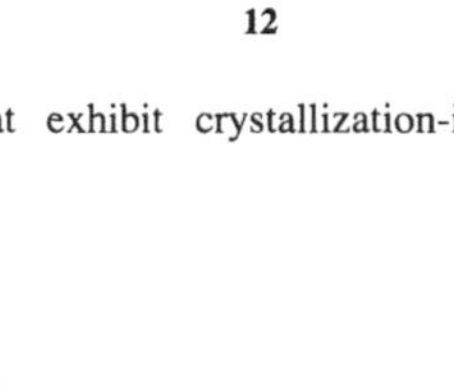

Figure 7.5. Additional examples of compounds that exhibit crystallization-induced asymmetric transformation.

is that ester **11** is a conglomerate; crystallization of one enantiomer takes place simultaneously with racemization of the other. The result is a crystallization-induced asymmetric transformation from the melt. A similar transformation, with a yield in excess of 90%, takes place in solution on the ethylamine salt of naproxen.[43]

The enzymatic hydrolysis of racemic amino acid derivatives is a fairly common kinetic resolution method wherein one enantiomer of the racemate is recovered as the unhydrolyzed derivative while the other may be recovered from the hydrolysate as a pure amino acid or as a new amino acid derivative (Section 7-5). In a number of instances, the yield of the latter exceeds 50% signaling concomitant racemization and the occurrence of an asymmetric transformation.[44] Since no crystallization is involved and a chiral catalyst is required to effect the process, we are not dealing here with a crystallization-induced asymmetric transformation. The process has much in common with asymmetric transformations of diastereomers in that diastereomer discrimination is operative during the hydrolysis. Nevertheless, since the substrate (both starting and recovered) is a racemate and the product is a single enantiomer, it is appropriate to describe examples of such processes in this section.

Hydantoins substituted in the 5 position are well-known precursors of amino acids. For example, 5-phenylhydantoin **13** is readily hydrolyzed in the presence of hydropyrimidine hydrase (a hydantoinase) isolated from calf liver. If the reaction medium is slightly alkaline (pH ca. 8), the hydantoins undergo racemization. Under the conditions of the reaction, racemic **13** is completely converted to (R)-carbamoylphenylglycine **14**, a phenylglycine precursor. No (S)-**14** is found in the reaction mixture since the unreacted (S)-**13** undergoes spontaneous racemization during the hydrolysis of the R enantiomer for which the enzyme is specific (Fig. 7.6).[45] In the presence of N-carbamoyl-D-amino acid amidohydrolase (or of the organism *Agrobacterium radiobacter* from which the enzyme can be isolated) *rac*-**14** yields 100% D-phenylglycine directly.[46] Similar enzymatic hydrolysis of 5-(4-hydroxyphenyl)hydantoin has been reported.[47,48]

Asymmetric transformations may also take place in other contexts. The inversion of configuration of the enantiomers of 2,6-dioxaspiro[4.4]nonane (Fig. 7.7a, **15** has been demonstrated during the chromatographic resolution of the racemate on an enantioselective stationary phase, where it is presumed that the process takes place via

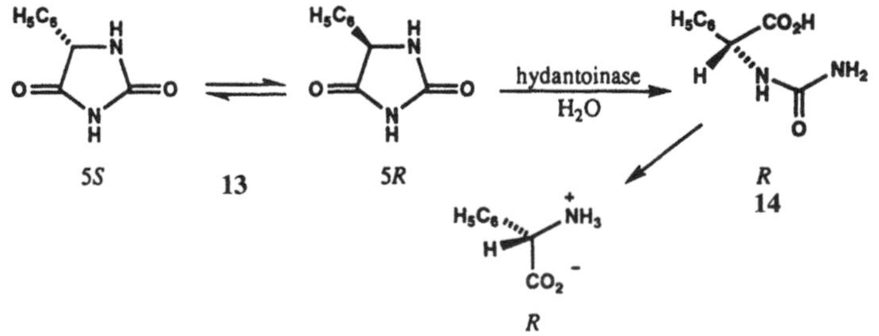

Figure 7.6. Enzyme catalyzed kinetic resolution of *rac*-5-phenylhydantoin **13** with concomitant racemization of (5S)-**13**.

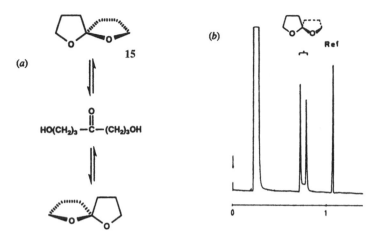

Figure 7.7. (*a*) Enantiomerization of spiroketal **15**. (*b*) Peak coalescence of **15** during its GC analysis on nickel(II) bis[(1*R*)-3-(heptafluorobutyryl)camphorate] (0.1*M* in squalane) at 70°C [Reprinted with permission from Schurig, V. and Bürkle, W. *J. Am. Chem. Soc.* **1982**, *104*, 7573. Copyright © 1982 American Chemical Society.]

a didydroxyketone intermediate (Fig. 7.7a).[49] The enantiomerization is revealed by a distinctive signature in the chromatogram (Fig. 7.7b). The "plateau" bridging the peaks due to the two enantiomers is caused by molecules that have inverted their configuration *in the column* and travel with velocities approaching those of their enantiomers. The process results in tailing of the leading peak and fronting of the peak due to the second appearing enantiomer.[50] Since the horizontal equilibration of Figure 7.3 now takes place in a nonracemic chiral environment, this type of enantiomerization has much in common with asymmetric transformations of diastereomers. The latter are described in Section 7-3.e.

7-3. CHEMICAL SEPARATION OF ENANTIOMERS VIA DIASTEREOMERS

a. Formation and Separation of Diastereomers; Resolving Agents

The largest number of recorded resolutions has been effected by conversion of a racemate to a mixture of diastereomers. In this type of reaction, the substrate to be resolved is treated with one enantiomer of a chiral substance (the resolving agent). The first such resolution, described by Pasteur in 1853,[51] is outlined in Figure 7.8.[4] Diastereomer pairs prepared in connection with resolutions may be ionic (diastereomeric salts), covalent, charge-transfer complexes, or inclusion compounds. The latter two types of diastereomers are discussed in Section 7-3.c.

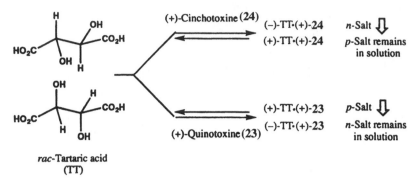

Figure 7.8. First resolution via diastereomers. Tartaric acid resolution with cinchotoxine and quinotoxine (Fig. 7.9) as resolving agents.[51] The *n* and *p* symbols are defined on p. 215.

The vast majority of resolutions mediated by diastereomers (diastereomeric salt mixtures, in particular) have been based on solubility differences of solids; however, in the contemporary literature, covalent diastereomer separations based on chromatography in all of its variants are used with great frequency. Chromatography has freed resolutions from the constraint of dependency on crystallization as the technique on which diastereomer separation has traditionally depended. As a result, resolutions in general are much more successful at present than they were in the past.

Not infrequently oily covalent diastereomer mixtures eventually crystallize and their resolution may then be performed in the more traditional way by taking advantage of solubility differences. Separation of diastereomeric salt mixtures by chromatography is also possible (Section 7-3.d). Because of this interplay between ionic and covalent structure and the several ways of separating diastereomer mixtures, we have chosen in this section not to treat resolving agents separately according to whether they form covalent or ionic diastereomer mixtures.

In this section, our analysis focuses on resolving agents with emphasis on the recent literature and with examples of their use. The desirable characteristics of a good resolving agent are[52]:

(a) Ready availability
(b) Stability of supply
(c) Stability in use and in storage
(d) Low price or ease of preparation
(e) Ease of recovery and reuse
(f) Low molecular weight
(g) Availability in high enantiomeric purity
(h) Availability of both enantiomers
(i) Low toxicity
(j) Reasonable solubility

α-Methyl-β-phenylethylamine, $C_6H_5CH_2CH(NH_2)CH_3$, illustrates the application of feature (a). This amine is a potentially useful resolving agent.[53] However, the amine (amphetamine) is a central nervous system (CNS) active compound, and accordingly it is a controlled substance. Like all such substances (e.g., deoxyephedrine and morphine) it is difficult to obtain. The acquisition of controlled substances for use as resolving agents is so complicated and time consuming (at least in the United States) that their use for this purpose is essentially precluded.

The supply of resolving agents that are derived from natural sources, such as brucine and 10-camphorsulfonic acid, may be shut off by economic or political problems that impede access to the sources [feature (b)].

Some resolving agents are awkward to use and to store without precaution. Liquid primary amines, such as α-methylbenzylamine and dehydroabietylamine (Fig. 7.9), readily form solid carbamates on exposure to air.[54] It may consequently be desirable to store these amines as salts [feature (c)]; if so, one may profitably choose salts that are conglomerates, since enantiomer purification of such salts would be concomitant with chemical purification (e.g., during recovery). α-Methylbenzylamine hydrogen sulfate (Fig. 7.1) and α-(1-naphthyl)ethylamine phenylacetate are examples of salts that are conglomerates.[4] All other things being equal, high expense is a negative feature in the choice of a resolving agent, although this feature (d) may be mitigated by the possibility of recovery and reuse [feature (e)]. When preparation of a resolving agent is required, the yield and complexity of the synthesis is likely to be a consideration.

Since resolving agents are purchased by weight but are used on a molar basis, low molecular weight [feature (f)] is an advantage. This is a significant consideration especially in resolutions carried out on an industrial scale. Unfortunately, many naturally occurring resolving agents, notably alkaloids, have high molecular weights (e.g., brucine, MW 394.4); this is less likely to be the case for synthetic ones (for lists of resolving agents giving molecular weights, see Jacques et al.[4] Moreover, synthetic resolving agents are usually obtainable in both enantiomeric forms and this feature (h) is advantageous, since it permits the preparation of both enantiomers of a compound by means of mirror-image resolutions (*Marckwald principle*[55]; type a in Table 7.3). A fair number of such pairs of enantiomers are available commercially (e.g., α-methylbenzylamine, ephedrine, tartaric acid, or 10-camphorsulfonic acid). Some synthetic resolving agents have been designed that explicitly incorporate many of the features listed above.[56]

Use of synthetic resolving agents requires their prior resolution. This requirement leads us to discuss the possibility of effecting *reciprocal resolutions*: If *rac-N*-benzyloxycarbonylalanine [(±)-Z-Ala] is resolvable with (−)-ephedrine [(−)-Ephl, then, as is often (but not invariably) the case, the resolving agent (±)-Eph will be resolvable with either (+)- or (−)-Z-Ala (type b in Table 7.3).[4,57]

Although it is not implicit in Table 7.3 (type a), only *one* enantiomer of the substrate is readily obtained in conventional (hence, also in reciprocal) diastereomer-mediated resolutions, for example, (−)-Z-Ala in the resolution of (±)-Z-Ala with (−)-Eph, and (+)-Eph in the reciprocal resolution of (±)-Eph with

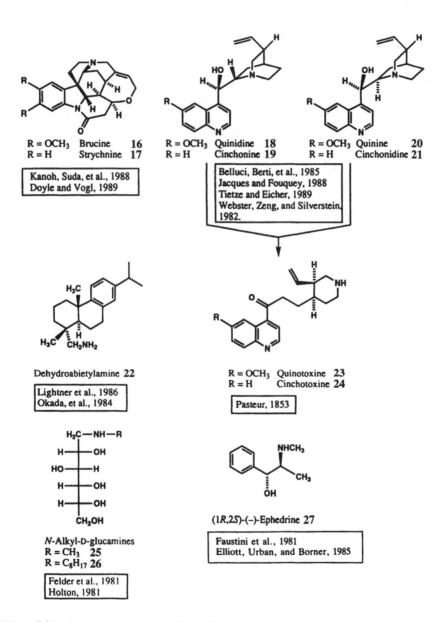

Figure 7.9(a). Resolving agents for acids. In the case of the naturally occurring agents, the absolute configurations shown are in agreement with those given in the *Atlas of Stereochemistry* (Klyne and Buckingham, 2nd ed., Vols. I and II, Oxford University Press, New York, 1978). In the case of synthetic resolving agents, though both enantiomers may have been used in resolutions, only one is shown.

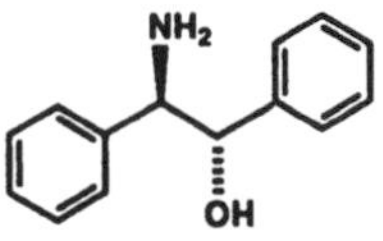

(1S,2R)-2-Amino–
1,2-diphenylethanol **28**

Saigo et al., 1982

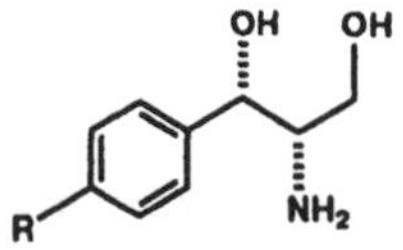

R = H (1S,2S)-(+)-2–Amino-1-phenyl-
1,3-propanediol **29**

ten Hoeve and Wynberg, 1985

R = NO$_2$ **30**

Kawanami, Katsuki, et al., 1984
Krause and Meinicke, 1985

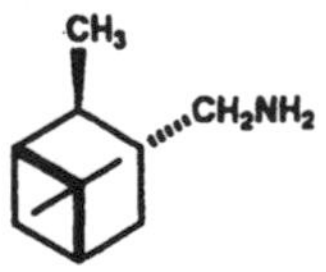

31

(+)-3-Aminomethylpinane

Himmele and Siegel, 1976

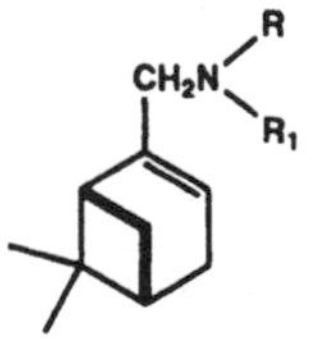

32

Pinenylamines, e.g., R = H, R' = n-Bu

Markowicz, 1979

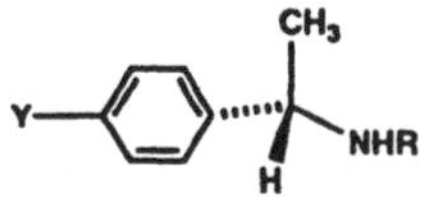

(S)-(–)-α-Methylbenzylamine

R = H, Y = H   **33** | Whitesell, et al., 1983
                          | Ács et al., 1990

R = CH$_2$C$_6$H$_5$, Y = H   **34**   de Heij, 1981

R = H, Y = (CH$_3$)$_2$CH   **35**   Saigo et al., 1985

R = H, Y = NO$_2$ **36** | Perry, Brossi, et al., 1977
                               | Schönenberger and Brossi, 1986

Analogues: Nicholson and Tatum, 1981
Matsumoto et al., 1986

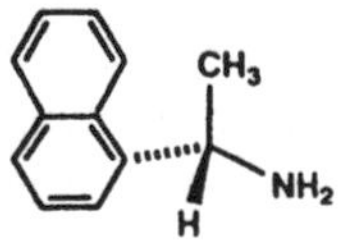

37

(S)-(–)-α-(1-Naphthyl)ethylamine

Corey, et al., 1978
Mori, et al., 1981
Sato, et al., 1980

Figure 7.9(b).

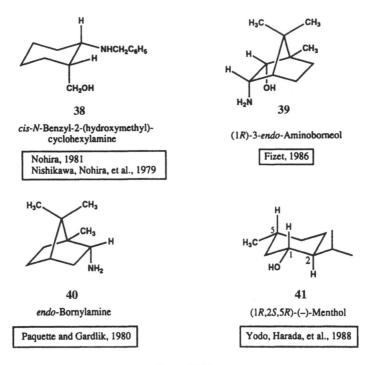

38

cis-N-Benzyl-2-(hydroxymethyl)-
cyclohexylamine

Nohira, 1981
Nishikawa, Nohira, et al., 1979

39

(1*R*)-3-*endo*-Aminoborneol

Fizet, 1986

40

endo-Bornylamine

Paquette and Gardlik, 1980

41

(1*R*,2*S*,5*R*)-(−)-Menthol

Yodo, Harada, et al., 1988

Figure 7.9(*c*).

(+)-Z-Ala (a and b in Table 7.3). In either case, it is only the enantiomer incorporated in the less soluble product that is readily obtained. A change in resolving agent or use of the enantiomeric resolving agent [(+)-Eph as in Table 7.3 (type a), second line (Marckwald principle)] is usually required to obtain the other enantiomer of the substrate to be resolved.[60] On the other hand, crystallization of solutions containing equivalent amounts of *racemic* substrate and *racemic* "resolving agent" may permit the isolation of either enantiomer of the material to be resolved and simultaneously either enantiomer of the "resolving agent" provided that the racemic salt is a conglomerate. The possibility of effecting such mutual resolutions by alternately seeding racemic solutions containing the four possible salts with crystals of one of the less soluble salts and then with crystals of the enantiomeric salt has been demonstrated.[59] In each case, the salt that precipitated had the same composition as that of the seeds; the enantiomers of only one of two possible diastereomeric salts crystallized (type d in Table 7.3). As expected, on admixture, the two precipitated enantiomers formed a conglomerate. Hence mutual resolution, though performed on diastereomeric salts, has the attributes of preferential crystallization.

The thermodynamic properties of pairs of diastereomeric salts formed from racemic acid and racemic base, corresponding to "*rac*-resolution substrate" plus "rac-resolving agent" (formation of such pairs is a special case of "reciprocal salt-pairs" studied a century ago by Meyerhoffer)[61] have been studied. Two kinds

TABLE 7.3 Types of Diastereomer-Mediated Resolutions^a

Type of Resolution^a	Resolution Substrate	Resolving Agent	Diastereomeric Products	
			Less Soluble	More Soluble
a. Normal	(±)-**Z-Ala**	+ (−)-Eph →	(−)-**Z-Ala**·(−)-Eph	+ (+)-Z-Ala·(−)-Eph
Marckwald	(±)-**Z-Ala**	+ (+)-Eph →	(+)-**Z-Ala**·(+)-Eph	+ (−)-Z-Ala·(+)-Eph
b. Normal	(±)-**Z-Ala**	+ (−)-Eph →	(−)-**Z-Ala**·(−)-Eph	+ (+)-Z-Ala·(−)-Eph
Reciprocal	(±)-**Eph**	+ (+)-Z-Ala →	(+)-Z-Ala·(+)-**Eph**	+ (+)-Z-Ala·(−)-**Eph**
c. Mutual	(+)-**Z-Ala**^{b}	+ (±)-Eph	(+)-**Z-Ala**·(+)-**Eph**	+ (+)-Z-Ala·(−)-Eph
d. Mutual	(±)-**Z-Ala**	+ (±)-**Eph**	$\begin{vmatrix}(+)\text{-}\textbf{Z-Ala}\cdot(+)\text{-}\textbf{Eph}^c + (-)\text{-Z-Ala}\cdot(-)\text{-Eph}^e\\(-)\text{-}\textbf{Z-Ala}\cdot(-)\text{-}\textbf{Eph}^d + (+)\text{-Z-Ala}\cdot(+)\text{-Eph}^e\end{vmatrix}$	

^{a}Types of resolutions: (a) normal and Marckwald resolutions; (b) normal and reciprocal resolutions; (c) mutual resolution; see text; (d) mutual resolutions. Resolution substrates and products are in boldface.
^{b}This resolution was carried out on partially resolved Ala enriched in (+)-Z-Ala.
^{c}On seeding with (+,+) salt.
^{d}On seeding with (−,−) salt.
^{e}The (+)-Z-Ala·(−)-Eph and (−)-Z-Ala·(+)-Eph diastereomic salts did not crystallize.

Z-Ala = N-Benzyloxycarbonylalanine
$CH_2OCNHCHCOOH$ with O, CH₃

Eph = (1R,2S)-(−)-Ephedrine

of behavior are possible: (a) formation of a simple eutectic as exemplified by the cases cited above and by atrolactic acid, $CH_3C(C_6H_5)(OH)CO_2H$, with α-methylbenzylamine, permitting mutual resolution to take place under conditions of kinetic control (seeding); and (b) formation of a double salt (addition compound; phase diagram as in Fig. 6.3b) in which mutual resolution is precluded. More specifically, in the former case, one salt pair (the p salt-pair in the case of atrolactic acid plus α-methylbenzylamine) is more stable and is the one precipitated on seeding; the less stable n salt-pair (that can be independently prepared) is not manifested. Mandelic acid plus deoxyephedrine is an example of the second type of behavior; here, both p and n salt-pairs are addition compounds (with eutectics having differing composition).

Diastereomeric salts are designated p if the acid and base reaction partners giving rise to the salts have the same sign of rotation (p_+ if both are + and p_- if both are −) and n if the rotation signs of the diastereomer constituents are unlike.[4]

In the case of resolutions, does it matter whether one uses enantiomerically pure resolving agents? Provided that the resolving agent is nearly, even if not completely, enantiopure, the answer is no. Contrary to what was earlier believed,[27] the enantiomer purity of a substance resolved via crystallization of diasteromeric derivatives may *exceed* that of the resolving agent. This result is evident from inspection of the relevant phase diagram provided only that the enantiomer purity of the relevant less soluble crystalline diastereomer is greater than that of the eutectic. Nevertheless, use of resolving agents of low enantiomeric purity is not desirable since the number of recrystallization steps necessary to attain stereochemical homogeneity, hence also the yield, is dependent on the original enantiomer composition of the resolving agent.[4]

What structural features are desirable for resolving agents? Polyfunctional compounds are generally preferred over monofunctional ones and aromatic compounds over aliphatic ones. Thus 2-amino-1-butanol and α-methylbenzylamine are more frequently used as resolving agents than is *sec*-butylamine. It would seem that multiple sites of (nonbonding) interaction between the counterions of diastereomeric salts or between different groups in covalent diastereomers enhance solubility differences between diastereomer pairs and thus favor their separation.[52] Multiple interaction sites in covalent diastereomers are also responsible for "anchoring" the latter selectively to chromatographic adsorbents or for "dissolving" them in chromatographic liquid stationary phases in such a manner as to permit differential elution. The constitutional and conformational rigidity of diastereomer molecules may be responsible for selectively "fending off" one of the diastereomers from the adsorbent thereby contributing to separation.[63] Rigidity may also play a role in the separation of diastereomeric salts on the basis of solubility differences though the reasons for this are less evident. The alkaloid brucine has a rigid heptacyclic skeleton (**16**, Fig. 7.9) that is remarkably effective in "conferring" low solubility on carboxylate salts; brucine is one of the most widely used basic resolving agents,[4] however, see below.

Resolutions mediated by salts are more likely to succeed when the acidic and basic functional groups whose interaction result in salt formation are closer "to those factors that render each asymmetric," for example, the chiral centers.[64] Strongly acidic and basic resolving agents are normally preferred over weak ones. The latter may yield salts of low stability relative to dissociation, or salt formation may be altogether precluded. High acid or base strength is one of the advantages of synthetic resolving agents, of amines in particular, which are often stronger than the typical alkaloidal resolving agents; for example, α-methylbenzylamine, pK_b 4.5 versus brucine, pK_b 5.9 in 95% ethanol (10°C).[62] A number of relatively new synthetic acidic resolving agents designed with this structural feature in mind are described below (e.g., in Fig. 7.11).

The structures of the principal resolving agents of modern usage are illustrated in Figures 7.9–7.15 and 7.17, arranged according to resolution substrate type, together with references (mostly after 1979) that illustrate their use. No attempt has been made to present an exhaustive list of resolving agents from the large literature on resolutions but the compounds whose structures are given are representative. More extensive lists of resolving agents, as well as their properties and methods of preparation, may be found in the compilations of Newman[65] and of Wilen.[66] These compilations also contain descriptions of numerous resolutions. For applications, also see reviews.[4,52,67,68]

Resolving Agents for Acids and Lactones.   These agents are listed in Figures 7.9 and 7.10. Figure 7.9 brings together a wide variety of naturally occurring bases and compounds derived from them, alkaloids (**16–24**), and terpene derivatives in particular (**31, 32, 39, 40**), as well as synthetic amines (**27–30, 33–38**). In this figure, no distinction is made between resolutions whose diastereomeric products are separated by crystallization or by chromatography. Few carbohydrate derivatives have seen use as resolving agents. A significant exception is the use of inexpensive *N*-alkyl-D-glucamines, for example, R = CH$_3$ (**25**), C$_8$H$_{17}$ (**26**) (Fig. 7.9) in the resolution of naproxen **72** (Fig. 7.13). Amino acids and derived bases figure prominently as resolving agents for carboxylic acids. These acids are exemplified by compounds **42–48** (Fig. 7.10).

Basic resolving agents have generally been used to transform racemic covalent substrates (carboxylic, sulfonic, and a variety of phosphorus acids) into

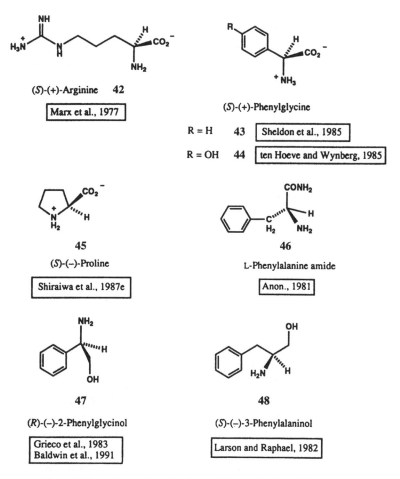

Figure 7.10. Amino acids and amino acid derived basic resolving agents.

diastereomeric salts that are separated by crystallization. Increasingly, resolutions of acids and of lactones have involved formation of covalent diastereomers, notably amides, for example, by compounds **47** and **48** (Fig. 7.10), and esters (for the latter, see menthol **41** in Fig. 7.9c) that are separated by either crystallization or by chromatography.[4] Resolving agents that have been applied to the resolution of lactones via diastereomeric amide formation are exemplified by **31**, **33**, and **39**. Amides are generally more easily separated chromatographically than are esters. The major limitation in amide-mediated resolutions has been the harsh conditions often required to hydrolyze the separated amide diastereomers. Numerous examples of resolutions in which such hydrolysis may be facilitated (e.g., by neighboring group participation), say, by a suitably placed hydroxyl group (e.g., Fig. 7.9c, **39** and Fig. 7.26), when permitted by the structure of either the resolution substrate or the resolving agent, or with the latter expressly modified (labilized) for the purpose following the resolution are available.[4,69–74]

Resolving Agents for Bases.   Tartaric acid **49** and its acyl derivatives (e.g., dibenzoyl- and di-*p*-toluyltartaric acids **50**; R = H and CH$_3$, respectively) continue to be widely used in resolutions (Fig. 7.11a). The latter are more acidic than the parent acid and the presence of aroyl groups may enhance diastereomer discrimination by providing additional anchoring points for the resolution substrate. One of the first applications of dibenzoyltartaric acid as a resolving agent (in the 1930s) was the resolution of *rac*-lobeline, which has been used as a deterrent to smoking.[75] Resolutions effected with polybasic acids, such as **49** and **50**, are generally carried out with a 1:1 stoichiometry with respect to the resolution substrate; the diastereomeric salts isolated are acid salts.[4]

Mandelic acid **53**, *O*-acetylmandelic acid **54**, and *O*-methylmandelic acid **55** (Fig. 7.11a), figure prominently as resolving agents among resolutions of amines (1° and 2°) reported in the last decade. To a lesser extent, this generalization applies also to the analogous Mosher's acid (MTPA acid, Fig. 7.11a, **56**). The 1,1′-binaphthylphosphoric acid **59** (Fig. 7.11b) (pK$_a$ = 2.50) has proved to be a very powerful resolving agent, even for the difficult resolution of 3° amines.[76] A second synthetic and strongly acidic resolving agent (introduced by ten Hoeve and Wynberg)[56] that has a wide range of utility is the cyclic phosphoric acid **60** (Fig. 7.11b; R = *o*-Cl).

For the most part, the cited acidic resolving agents (and others whose structures are given in Fig. 7.11) are involved in diastereomeric salt-mediated resolutions. In contrast, compounds **55**, **62**, and **63** (Fig. 7.11) have been applied to amide-mediated resolutions; isocyanate **64** (Fig. 7.11c) has been applied to the resolution of amines by urea formation (diastereomeric ureas are formed even by chiral hydroxyamines,[77] and menthol-derived **65** (Fig. 7.11c) to resolution of amines via carbamate formation. Acid **57** (Fig. 7.11a) has the interesting advantage of not bearing a hydrogen atom at the stereocenter, hence avoiding the danger of racemization during its recovery and recycling particularly if these processes are carried out under basic conditions.

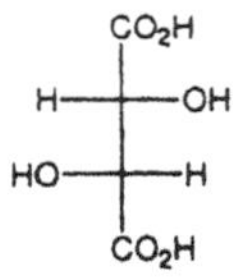

49

(2*R*,3*R*)-(+)-Tartaric acid

Corey et al., 1989
Geue, McCarthy, and Sargeson, 1984
Suda et al., 1979

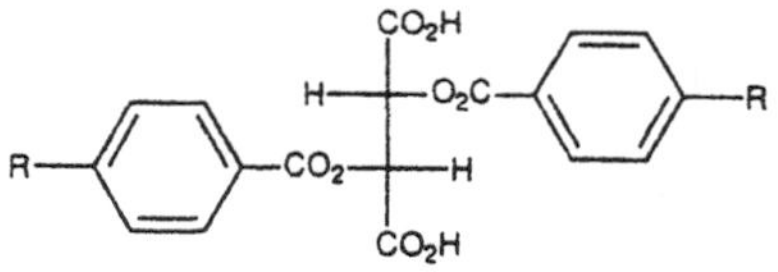

50

R = H   *O,O'*-Dibenzoyltartaric acid
R = CH₃ *O,O'*-di-*p*-Toluoyltartaric acid

Blaschke and Walther, 1987
Dumont, Brossi, and Silverton, 1986
Abu Zarga and Shamma, 1980; Smith et al., 1983

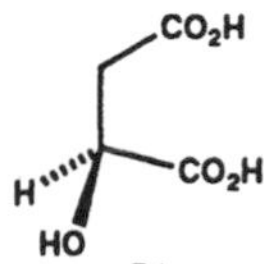

51

(*S*)-(−)-Malic acid

Anon., 1983

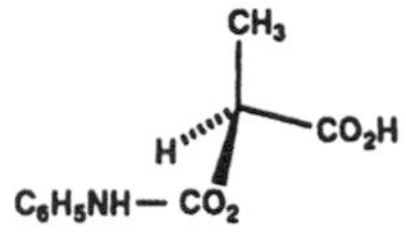

52

(*S*)-(−)-Carbamalactic acid

Brown, Viot, and Le Floc'h, 1985a

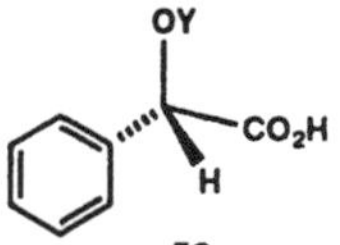

53

Y = H (*S*)-(+)-Mandelic acid

Fitzi and Seebach, 1988
Ohgi, Kondo, and Goto, 1979
Whitesell et al., 1988

54 Y = COCH₃ *O*-Acetylmandelic acid
Corey et al., 1987

55 Y = CH₃ *O*-Methylmandelic acid

Hecker and Heathcock, 1986
Nilsson and Hacksell, 1988

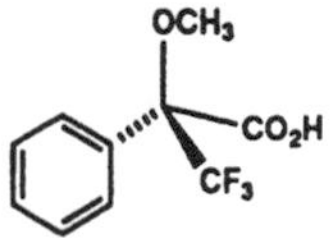

56

α-Methoxy-α-(trifluoromethyl)-
phenylacetic acid [Mosher's acid; MTPA]

Jacob, III, 1982

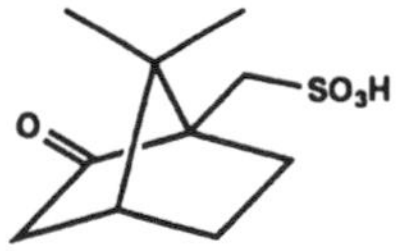

57

(+)-10-Camphorsulfonic acid

Brown, Berry, and Murdoch, 1985
Dumont, Brossi, and Silverton, 1986
Smith et al., 1983

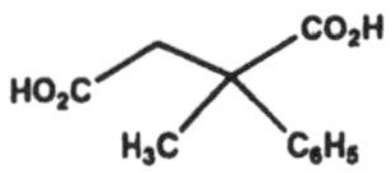

58

2-Methyl-2-phenyl-
butanedioic acid

Gharpure and Rao, 1988

Figure 7.11(*a*). Resolving agents for bases. In the case of synthetic resolving agents, though both enantiomers may have been used in resolutions, only one is shown.

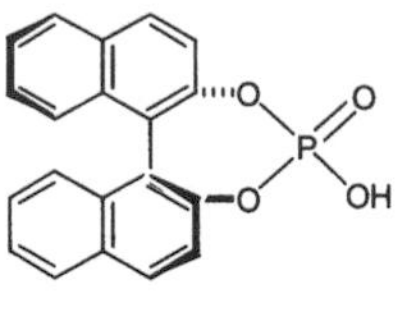

59	**60**	**61**
(S)-(+)- 1,1'-Binaphthyl-2,2'-diyl hydrogen phosphate	4-Aryl-5,5-dimethyl -2-hydroxy-1,3,2-dioxaphosphorinane 2-oxide	(−)-Diisopropylidene-2-keto-L-gulonic acid
Bey et al., 1979 Imhof, Kyburz and Daly, 1984 Jacques and Fouquey, 1988 Wilen et al., 1991	ten Hoeve and Wynberg, 1985 Vriesema, Wynberg, et al., 1986	Brossi and Teitel, 1970 Fitzi and Seebach, 1988

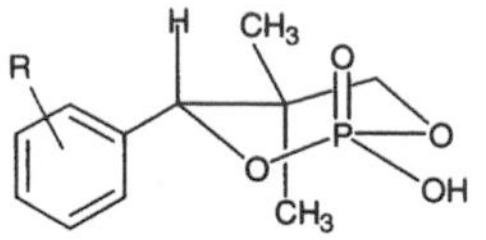

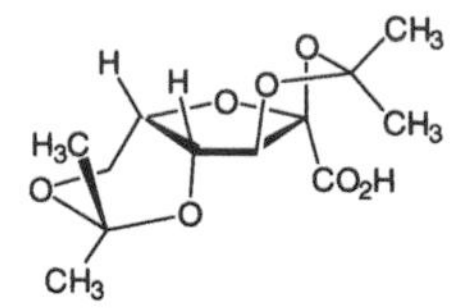

Figure 7.11(b).

Resolving Agents for Amino Acids. Amino acids are most often resolved in "protected" form, mainly via derivatization of the amino group.[4,52] The groups that figure prominently in the methodology are N-acetyl, N-formyl, N-benzoyl, N-tosyl, N-phthalyl, N-carbobenzyloxy, and N-(p-nitrophenyl)sulfenyl. These amino acids may be resolved by diastereomeric salt formation with basic resolving agents, such as brucine **16**, quinine **20**, ephedrine **27** (Fig. 7.9a) and pseudoephedrine, and chloramphenicol base **29** (Fig. 7.9b) among others and with tyrosine hydrazide (Fig. 7.12, **66**).[78] Lactone **68** (a protected sugar alcohol) has been applied to the resolution of the N-phthalimido derivatives of γ-aminobutyric acid analogues via ester formation.[79]

A widely applicable and attractive method that is apparently specific to amino acids is the simultaneous derivation and salt formation exemplified by Eq. 7.1:[80]

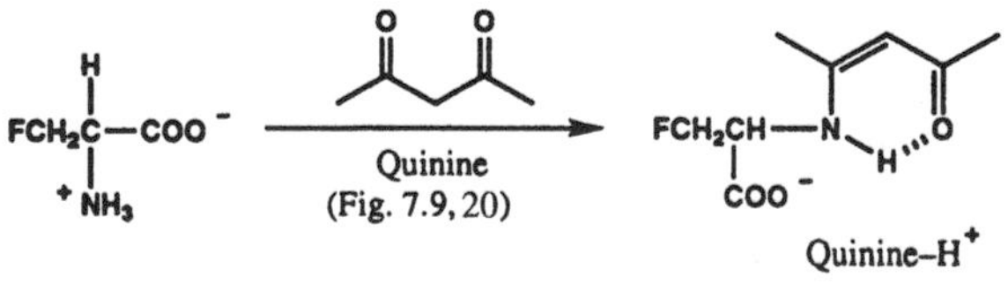

$$\tag{7.1}$$

Resolution of amino acids with acidic resolving agents is fairly common. Amino acids in unprotected form may be resolved: with mandelic acid **53** via amide formation[81] or via salt formation.[82,83] Other representative examples are terleucine $(CH_3)_3CCH(NH_3^+)COO^-$ with 10-camphorsulfonic acid **57** (Fig. 7.11a)[84]; o-tyrosine[85] and homomethionine[86] with 1,1'-binaphthylphosphoric acid **59** (Fig. 7.11b); and arginine[87] and p-hydroxyphenylglycine[88] with acid **67** (Fig. 7.12).

The resolution of amino acid esters as Schiff's bases, with ketone **69** (Fig. 7.12), an α-pinene derivative, has been reported.[89]

Resolving Agents for Alcohols, Diols, Thiols, Dithiols, and Phenols. Twenty-five years ago most alcohols requiring resolution were transformed to hydrogen phthalate derivatives, following which these acidic derivatives were converted to diastereomeric salts by reaction with alkaloidal bases.[4,66,90–92]

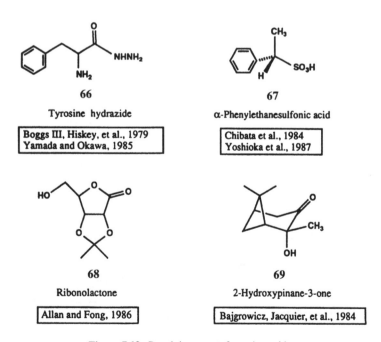

62

(−)-*N*-Butyloxycarbonyl-L-phenylglycine

Rittle et al., 1987

63

(S)-(−)-α-Methyl-α-
phenylsuccinic anhydride

Gharpure and Rao, 1988

64

α-Methylbenzyl isocyanate

Schönenberger and Brossi, 1986

65

Menthyl chlorocarbonate

Nohira and Hauske, 1977

Figure 7.11(*c*).

66

Tyrosine hydrazide

Boggs III, Hiskey, et al., 1979
Yamada and Okawa, 1985

67

α-Phenylethanesulfonic acid

Chibata et al., 1984
Yoshioka et al., 1987

68

Ribonolactone

Allan and Fong, 1986

69

2-Hydroxypinane-3-one

Bajgrowicz, Jacquier, et al., 1984

Figure 7.12. Resolving agents for amino acids.

Today, the phthalate method is rarely used. Rather, alcohols are most often resolved by conversion to diastereomeric esters; the latter are separated by crystallization or chromatography. What has changed most significantly is the widespread application of synthetic and semisynthetic acidic resolving agents to this process and the simplicity with which diastereomeric ester mixtures may be separated chromatographically. Resolving agents applied to the resolution of alcohols and diols as diastereomeric esters (in the case of diols often as the mono ester) are mandelic and O-acetylmandelic acids **53** and **54**, respectively (Fig. 7.11a,[93] O-methylmandelic acid **55**,[94–96] Mosher's acid **56** (Fig. 7.11a),[97] ω-camphanic acid **70** (Fig. 7.13) (also for the resolution of chiral phenols[98] and *myo*-inositol derivatives[99,100]; ref. 100 makes clear the possibility of recovering **70** following resolution), *cis*-caronaldehyde **71**, naproxen **72**, *trans*-1,2-cyclohexanedicarboxylic anhydride **73**, and tetrahydro-5-

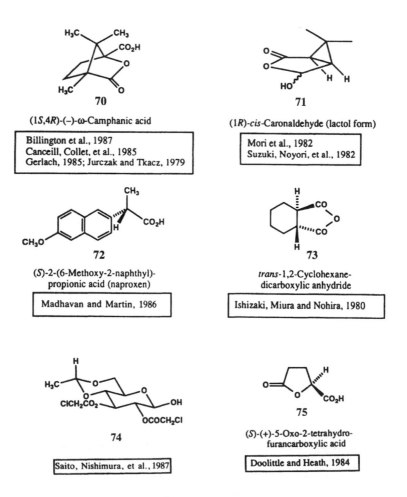

70

(1S,4R)-(–)-ω-Camphanic acid

Billington et al., 1987
Canceill, Collet, et al., 1985
Gerlach, 1985; Jurczak and Tkacz, 1979

71

(1R)-*cis*-Caronaldehyde (lactol form)

Mori et al., 1982
Suzuki, Noyori, et al., 1982

72

(S)-2-(6-Methoxy-2-naphthyl)-
propionic acid (naproxen)

Madhavan and Martin, 1986

73

trans-1,2-Cyclohexane-
dicarboxylic anhydride

Ishizaki, Miura and Nohira, 1980

74

Saito, Nishimura, et al., 1987

75

(S)-(+)-5-Oxo-2-tetrahydro-
furancarboxylic acid

Doolittle and Heath, 1984

Figure 7.13. Resolving agents and resolution adjuvants for alcohols, diols, thiols, dithiols, and phenols.

oxofurancarboxylic acid **75**. Alcohols may be resolved also as 10-camphorsulfonate esters.[101]

Urethane (carbamate) esters have seen repeated application to the resolution of alcohols (even 3° alcohols).[102] Specific resolving agents for this purpose are α-methylbenzyl isocyanate (Fig 7.11c, **64**)[103,104] and 1-(1-naphthyl)ethyl isocyanate.

D-Glucose (and even L-glucose) in a protected form (Fig. 7.13, **74**) has been applied as a resolving agent in the resolution of podophyllotoxin, the aglycone alcohol of an antineoplastic glucoside. The diastereomeric glycosides were separated chromatographically. The analogous 2-amino-2-deoxyglucopyranoses have also served as resolving agents.[105]

Resolving Agents for Aldehydes and Ketones.   Formulas of resolving agents for the direct resolution of aldehydes and ketones are shown in Figure 7.14. Among the more obvious types are those leading to diastereomeric acetals and ketals, namely, tartaric acid derivatives **76** and **77** (Fig. 7.14a) and (R,R)-2,3-butanediol.[106] α-Methylbenzylsemioxamazide (Fig. 7.14a, **78**) is a carbonyl derivatizing agent that has been applied to the resolution of ketones. Schiff's bases of carbonyl compounds prepared with a variety of amines or amino acid esters (e.g., **79**) are separable by crystallization[107] or thin-layer or high-performance liquid chromatography (HPLC). The potential antifertility agent gossypol (Fig. 7.15), a dialdehyde, has been resolved in this way.[108,109]

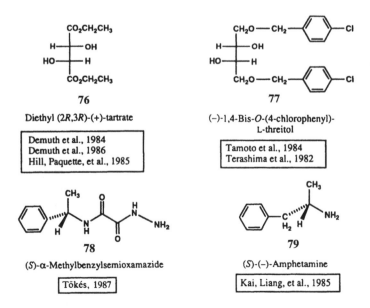

Figure 7.14(a). Resolving agents for aldehydes and ketones. Resolution auxiliaries for carbonyl compounds.

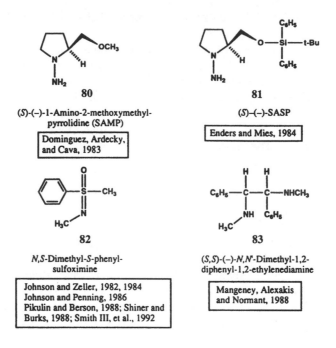

Figure 7.14(*b*).

(*R*)- and (*S*)-1-Amino-2-methoxymethylpyrrolidine (RAMP and SAMP reagents, Fig. 7.14*b*, **80**), derived from (*S*)-proline and from (*R*)-glutamic acid, respectively[110] and applied principally as chiral adjuvants in diastereoselective alkylations, aldol condensations, and Michael additions,[111] have also seen use as resolving agents via formation of diastereomeric hydrazones. The analogous SASP reagent **81** (Fig. 7.14*b*) has been advanced as a specific reagent for the resolution of aldehydes.

N,S-Dimethyl-*S*-phenylsulfoximine **82** is a versatile reagent for the resolution of ketones,[10] the example shown in Eq. 7.2 reveals the fact that application of reagent **82**

$$\text{Resolved ketone } + \textbf{82} \quad \xleftarrow{\quad 80 - 120^\circ C \quad} \qquad\qquad\qquad (7.2)$$

represents an exception to the usual rule that formation of additional stereocenters during a resolution is to be avoided (a new stereocenter is formed during the lithium salt addition, hence four diastereomeric products are possible). Since only two

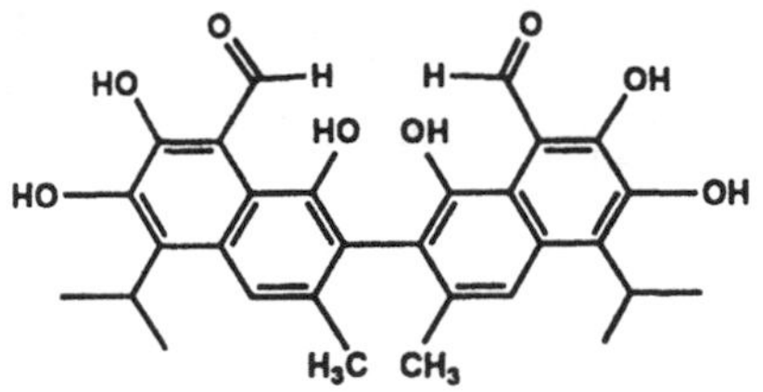

Figure 7.15. Gossypol.

diastereomers have been observed in some cases, the addition often takes place with high diastereoselectivity. Reciprocal resolution of **82**, for example, with (–)-menthone has been reported.[112]

Additional stereocenters are also generated when aldehydes and ketones are resolved through the separation of diastereomeric oxazolidines, which are prepared by reaction with ephedrine **27** (Fig. 7.9a),[113] and through bisulfite addition compounds formed by reaction of chiral ketones with SO_2. An example is given in Eq. 7.3. The diastereomeric salt mixtures may be separated by crystallization as well as, remarkably, by chromatography on silica gel.

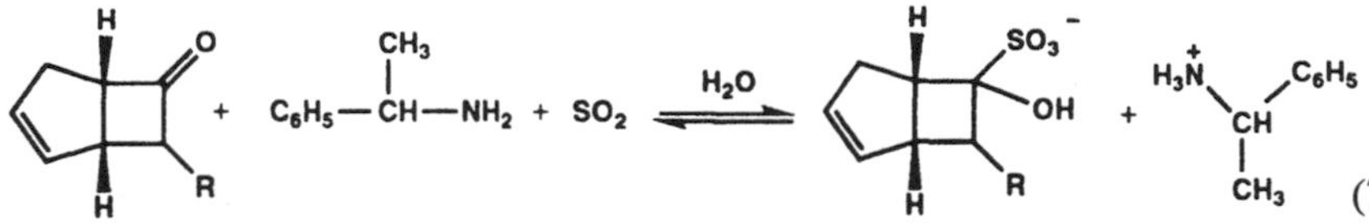

$$(7.3)$$

Compound **83** (Fig. 7.14b) has been applied specifically to the resolution of aldehydes via formation and chromatographic separation of diastereomeric imidazolidines (Eq. 7.4). The absence of additional stereocenters in the imidazolidines illustrates the advantage of building C_2 symmetry into the structure of a resolving agent.[114]

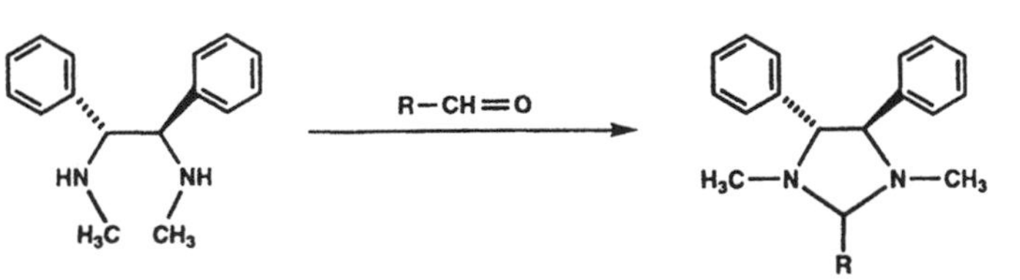

$$(7.4)$$

Miscellaneous Resolving Agents. Chiral triazolidinedoines, such as **84** (Fig. 7.16), derived from *endo*-bornylamine (Fig. 7.9c, **40**)[115] undergo facile Diels–Alder reaction with a variety of dienes. Separation of the diastereomeric urazole adducts followed by Diels–Alder reversal by hydrolysis–oxidation has been made the basis of the resolution of chiral dienes, such as substituted cyclooctatetraenes.

The resolution of an α,β-unsaturated ketone (4-*tert*-butoxycyclopent-2-one) by a novel reversible Michael addition–elimination is noteworthy (Eq. 7.5). The resolving agent, derived from 10-camphorsulfonic acid, is (–)-10-mercaptoisoborneol (Fig. 7.16, **85**). The success of this resolution hinges on several factors all of which tend to

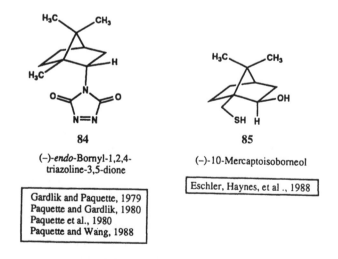

$$(7.5)$$

Two isomers (1:1)
(one crystallizes)

keep the increase in the number of stereocenters generated both in the sulfide and the sulfoxide adducts from producing intractable mixtures of diastereomers. The Michael addition occurs exclusively in trans fashion and the oxidation of the sulfide adducts is highly stereoselective.[116]

The optical activation of *rac*-menthone **86** illustrates a type of combined resolution and stereoselective synthesis that is not unique. Conversion of the ketone into a 1,3-dithiolane is followed by enantioselective oxidation under Sharpless conditions with diethyl (*R,R*)-tartrate (DET) (cf. Section 7-5). The mixture of diastereomeric 1,3-dithiolane-*S*-oxides is separable by low-pressure chromatography. Deoxygenation of the separated oxides and cleavage of the nonracemic dithiolanes affords menthone, ee = 93%[114] (Fig. 7.17). The analogy to the resolution shown in Eq. 7.5 is

84

(–)-*endo*-Bornyl-1,2,4-
triazoline-3,5-dione

Gardlik and Paquette, 1979
Paquette and Gardlik, 1980
Paquette et al., 1980
Paquette and Wang, 1988

85

(–)-10-Mercaptoisoborneol

Eschler, Haynes, et al ., 1988

Figure 7.16. Miscellaneous resolving agents.

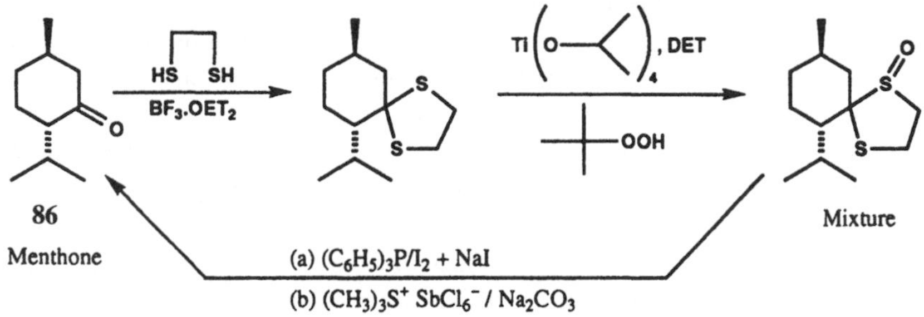

Figure 7.17. Optical activation of menthone.

obvious; however, note that no resolving agent is used in the process illustrated by Figure 7.17.

b. Resolution Principles and Practice

It is fair to say that few chiral compounds are totally resistant to resolution. Yet guidelines for successfully choosing resolving agents and for the selection of the most appropriate technique are mostly still qualitative in nature. In this section, we identify efforts that have been made to explain why one resolving agent may be better than another; features of resolutions that are known to impede success; and some studies aimed at understanding why a specific resolving agent forms a more stable (or less soluble) diastereomeric salt with one enantiomer of the resolution substrate than with the other. We also describe some recent studies designed to show how diastereomeric-salt-mediated resolutions may be optimized with respect to yield and the enantiomer purity of the desired enantiomer.

It has long been recognized, in a qualitative way, that multiple interactions between the resolution substrate and the resolving agent are beneficial, if not essential, to efficient resolutions.[52] This application of the three-point interaction model (Section 6-4.p and Fig. 8.7) may be responsible, in part, for the effectiveness of resolving agents incorporating aromatic rings and additional functional groups beyond the acid or basic functionality, as in O,O'-dibenzoyltartaric acid **50** and in mandelic acid **53** (Fig. 7.11a). The model also has served implicitly as a guide in the design of new resolving agents.

A statistical study of the resolution of phenylglycine (Fig. 7.10, **43**) derivatives with tartaric acid has shown that the primary influence on resolution results (yield and enantiomer composition of the resolution product) is the structure of the racemate (resolution substrate), notably the nature of the substituent on the benzene ring.[118] This type of study is suggestive and, if refined, may be of real use in the future.

Resolutions that depend on the formation and chromatographic separation of covalent diastereomers depend to a large extent on the nature and degree of interaction between stationary phases and mobile phase constituents. Some of these factors are described in Section 7-3.d. It must be emphasized that in the formation of these covalent diastereomers either an excess of resolving agent or a very long reaction time is required, or else reaction will be incomplete and a kinetic resolution may ensue as evidenced by the formation of diastereomer mixtures in proportions other than 1:1.[95]

Resolutions mediated by diastereomeric salts depend principaliy on solubility differences. This fact very clearly illustrated by a study of the reaction between mandelic acid (Fig. 7.11a, 53) and α-methylbenzylamine (Fig. 7.9b, 33). When carried out in aqueous solution, the first precipitate is practically pure n salt (Section 7-3.a)[119] This fact is consistent with the later finding that the p salt formed between 53 and 33 is much more soluble in water than the n salt, at 10 and 30°C.[120]

The resolving ability of a resolving agent may be measured in terms of a parameter S (which ranges from 0 for no resolution to 1 when 100% of one enantiomer having 100% ee is obtained):

$$S = k \times p = \frac{k_p - k_n}{0.5\,C_0} \qquad (7.6)$$

where k = the chemical (diastereomer) yield (with $k = 2$ for a 100% yield; however, in principle, S cannot exceed 1 since the maximum yield in a resolution, absent an asymmetric transformation, does not exceed 50% of enantiopure material) and p = the enantiomer purity (p = 1 for 100% ee) of the recovered resolution substrate. As shown in Eq. 7.6, S is related to the solubilities k_p and k_n of the p and n salts, respectively, as well as to the initial concentration of the resolution substrate C_0. Equation 7.6 permits one to calculate the maximum yield in a resolution from the measured solubilities. Figure 7.18a demonstrates that the maximum value of S is attained when $0.5\,C_0 = k_p$ (saturation of the p salt) in the case where the p salt is the more soluble diastereomer.

The resolving abilities S of a series of chiral phosphoric acids (Fig. 7.11b, 60, and analogues) on rac-ephedrine were correlated with differences in enthalpies of fusion between the diastereomer pairs. The greater the difference, the better the resolving

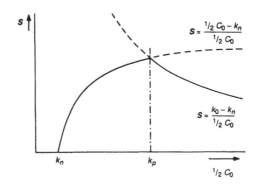

Figure 7.18(a). Idealized plot of resolving ability S as a function of the initial concentration C_0 of product when the p salt is more soluble than the n salt (for definitions of p and n, see page 215) [Adapted from the van der Haest, A. D., Wynberg, H., Lausen, F. J. J., and Bruggink, A. *Recl. Trav. Chim. Pays-Bas* **1990**, *109*, 523 with permission of the Royal Netherlands Chemical Society.]

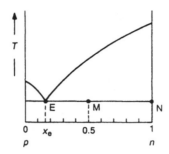

Figure 7.18(b). Typical phase diagram for the melting of a mixture of two diastereomers (E = eutectic).

ability. These authors observed that, in the resolution of ephedrine with acid **60** and five of its analogues, **S** is almost independent of the alcohol solvent used; the resolving ability is, however, highly dependent on the difference in lattice energy of the crystalline diastereomeric salts.

Binary mixtures of diastereomers obtained in a resolution often form simple eutectics; that is, their melting point behavior is summarized by unsymmetrical phase diagrams, such as that shown in Figure 7.18b.[4] Since the maximum yield of a pure diastereomer (e.g., *n*) is given by EM/EN, the "best" systems are those in which the eutectic lies far from composition M (the 50:50 mixture obtained in a diastereomer-mediated resolution.[121] The efficiency of a diastereomer-mediated resolution, as reflected in parameter **S**, can be calculated from the eutectic composition x_e of the binary phase diagram (Fig. 7.18b). If crystallization (under equilibrium conditions) is stopped when the mother liquor reaches the eutectic composition, then pure diastereomer is obtained (p = 1, Eq. 7.6) and

$$S = \frac{1 - 2x_e}{1 - x_e} \qquad (7.6a)$$

Preferential "complexation" (diastereomer salt formation) of (*S*)- versus (*R*)-**87** (Fig. 7.19), based on conformational factors, towards Lasalocid A were of the order of less than 2.5 kcal mol^{-1}.[121]

For resolutions that depend on solubility differences between diastereomers, there are two essential requirements for success[120]: (a) At least one of the pair of

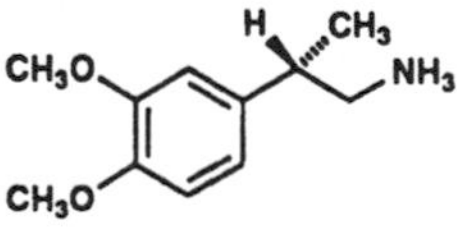

87

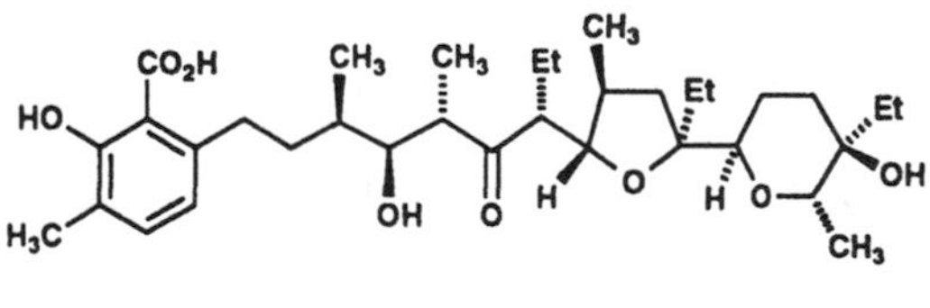

Lasalocid A

Figure 7.19. Structure **87** and Lasalocid A.

diastereomers formed must be crystalline and (b) the solubility difference between the two must be significant. The solvent plays a significant role in these requirements though often it does so in unpredictable ways.[4]

> The resolution of 2,2'-dimethyl-6,6'-biphenylcarboxylic acid with brucine **16** (Fig. 7.9*a*) as resolving agent (in an unspecified solvent) was previously reported to fail. Repetition of the resolution in methanol–acetone (7:3v/v) provided the (*S*)-(+) isomer with about 100% ee after one crystallization; crystallization in acetone of the salt recovered from the mother liquor afforded the (*R*)-(–)-isomer (ca. 99% ee).[122]

Two potential problems may forestall success: (c) The two diastereomers may interact to form a double salt or addition compound. This result leads both diastereorners to precipitate together. Recovery of the resolution substrate yields the racemate. (d) The two diastereomers may be partially miscible in one another in the solid state forming solid solutions. The inability of one of the diastereomeric counterions to fit in the crystal lattice of the other diastereomer tends to minimize this possibility.[4]

On a more practical level, in addition to choice of resolving agent (see above), the outcome of a diastereomeric salt-mediated resolution, that is, the yield and the enantiomeric enrichment of the initial precipitate, are significantly affected by variables, such as the nature and volume of solvent, stoichiometry, temperature,[123] and p*H*. One of the reasons for the sometimes dramatic effect of the nature of the solvent on a resolution is that the solvent can solvate the two *crystalline* diastereomeric species differently. The consequence is that the relative solubilities of the two species is changed. Examples of the effect of solvent, and of the effect of water in particular, on the relative solubilities of *p* and *n* salts are given in Jacques et al.[4] The crystal structure of the solvated salt can sometimes provide insight into the otherwise mysterious role that water plays in facilitating resolutions.[124]

The increasingly common use of nonstoichiometric amounts of resolving agent (e.g., 1:0.5; resolution substrate/resolving agent) is principally a way of manipulating the p*H* of a system of dissociable diastereomeric salts.[64,125–128] The main advantages of such manipulations are the economy in amount of resolving agent and solvent required and in the yield of resolved product obtained.[4] When the acid–base ratio in a "nonstoichiometric" resolution is restored to 1:1 by addition of an achiral acid or base,[129] additional salts are evidently present in the reaction mixture. This situation has been analyzed in detail in the case of the camphorate salts of the anorectic medicinal agent fenfluramine and a methodology described that permits the yield of desired diastereomer to be optimized as a function of the substrate/resolving agent ratio.[130]

The p*H* dependence on the yield and on the enantiomer purity of the diastereomeric salts formed in the resolution of *cis*-permethrinic acid (Fig. 7.20), an intermediate in the synthesis of the insecticide permethrin, with (*S*)-(+)-2-(*N*-benzylamino)-1-butanol, has been investigated.[131] Observed yields, enantiomeric purities, and optical yields could be correlated with thermodynamic parameters (solubility and

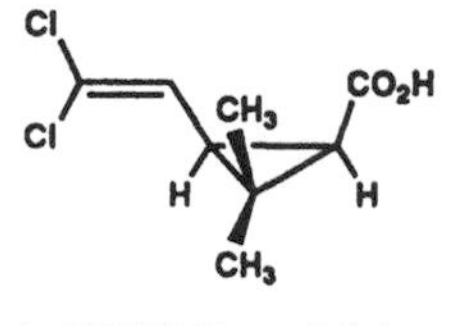

cis-(1R,3R)-Permethrinic acid

Figure 7.20. cis-Permethrinic acid.

dissociation constants, heats of fusion) by means of a previously proposed thermodynamic equilibrium model[125] that is valid if the diastereomeric salts do not form aggregates. This condition is fulfilled if these salts exhibit simple eutectic behavior [i.e., if they do not form an addition compound (see above)].

In view of the nearly equal solubilities of the diastereomeric salts (solubility constants K_{sR} and K_{sS} = 1.5 and 1.2 mol dm^{-3} in H$_2$O at 25°C for the p and n diastereomers, respectively), it could be postulated that resolution of cis-per-methrinic acid (Fig. 7.20) with 2-benzylaminobutanol is practically impossible. However, adjustment of the pH by addition of NaOH (optimally 0.45 equivalent) affects the yield [49% maximum; $S = k \times p = 0.49 \times 0.98 = 0.48$ (see Eq. 7.6)] *provided* that the dissociation constants of the two salts differ significantly, as was observed in this case.[131] A similar study permitted the optimization of the resolution of trans-permethrinic acid with the same resolving agent.[132]

While the resolution of racemates by distillation of diastereomeric esters has been previously reported and patented,[133,134] this approach has, to our knowledge, rarely been applied. However, the resolution of carboxylic acids, for example, tetrahydrofuran-2-carboxylic acid, by fractional distillation of the diastereomeric amides formed with, for examples, methyl esters of (S)-valine or (S)-leucine on a kilogram scale has been described.[135] The resolved acids as well as the amino acid resolving agents were recovered following acid hydrolysis of the separated amides.

c. Separation Via Complexes and Inclusion Compounds

Compounds devoid of conventional functionality (e.g., chiral alkenes and arenes, sulfoxides, and phosphines), for which alternative resolution routes are limited (see also Section 7-5 on kinetic resolution) may be resolved by incorporation into diastereomeric metal complexes or by reaction with chiral π-acids.[4]

Cope and co-workers first described the resolution of trans-cyclooctene by incorporation of the alkene enantiomers into diastereomeric trans platinum(0) square planar complexes[136,137] that are separable by low-temperature crystallization (Fig. 7.21). Other alkenes dienes, for example, spiro[3.3]hepta-1,5-dienes,[138] allenes, arsines, phosphines, and sulfoxides have analogously been resolved. 2-Vinyltetrahydropyran **88** has been resolved by formation of 4 diastereomeric platinum complexes (e.g., Fig. 7.21). Displacement of the pyran from one crystalline complex with excess ethylene gave (S)-(−)-**88** (92% ee).[139] An example of the

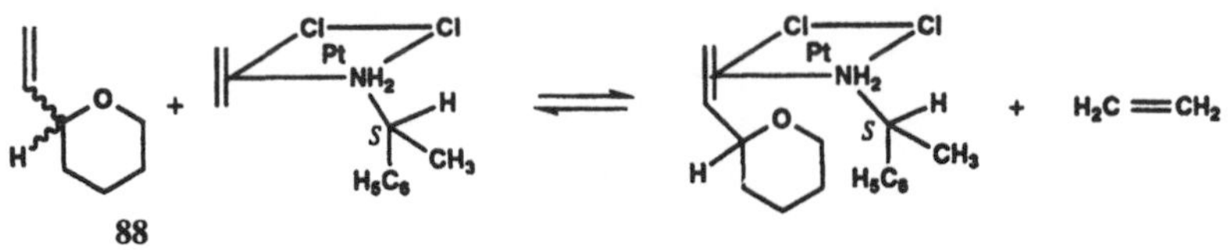

Figure 7.21. Resolution of chiral alkenes via neutral platinum coordination compounds (only one diastereomeric complex is shown).

resolution of an atropisomeric phosphinamine by separation of diastereomeric palladium complexes has been reported.[140] A chromatographic variant, called ligand-exchange chromatography, wherein one of the chiral ligands of the complex is incorporated into a polymer, has been applied mainly to the resolution of amino acids.

Diastereomeric π- (or charge-transfer) complexes, especially between the π-acid α-(2,4,5,7-tetranitro-9-fluorenylideneaminooxy)propionic acid **89** (TAPA; Fig. 7.22)[141] and chiral Lewis bases, may be reasonably strong. Such complexes may be separated by crystallization.[142,143] Chiral aromatic hydrocarbons, heteroaromatic compounds, and aromatic amines whose basicity is too low to permit resolution with conventional acidic resolving agents have been resolved with TAPA.

In cases of borderline stability, one of the diastereomeric π-complexes crystallizes while the other remains in solution largely dissociated. The converse situation is illustrated by the classic resolution of hexahelicene (Fig. 6.6, **3**) with (–)-TAPA; (–)-hexahelicene crystallizes whereas the diastereomeric π-complex containing the (+)-hexahelicene remains in solution.[4,142,143] Chromatographic resolutions based on TAPA are described in Section 7-3.d. An analogous resolution method that depends on coordination of chiral Lewis bases to nonracemic organometallic compounds (complexation GC) is described in Section 6-5.d.

Inclusion of molecules within others is a widespread phenomenon and the application of the "inclusion method" to resolutions, pioneered by Schlenk,[144] can no longer be considered to be a mere curiosity. We distinguish two broad classes of inclusion compounds: (a) the *cavitates* in which the chiral substrate to be resolved (the guest molecule) is partially or entirely enclosed within a second chiral substance (the

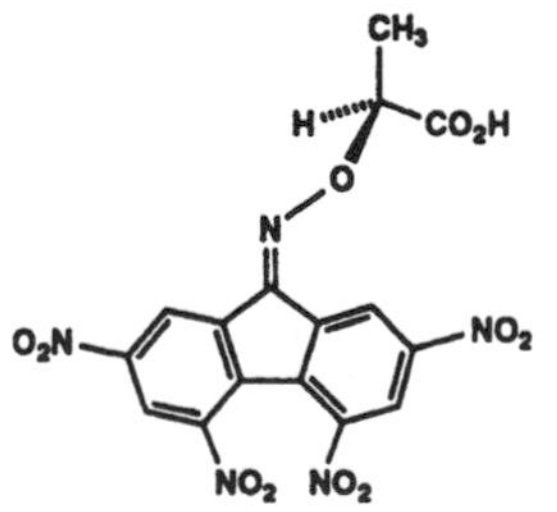

89

(+)-TAPA

Figure 7.22. TAPA.

resolving agent and host molecule) that is endowed with a chiral cavity, and (b) those in which the guest molecules are surrounded by several molecules of the resolving agent forming a cage or channel; these are often called *clathrates*.[146–150]

Preferential inclusion of one substrate enantiomer is the result of hydrogen bonding and van der Waals attraction between guest and host molecules. However, in addition, there must be a match between the size and shape of the guest molecule and a corresponding size and shape of the host cavity. While the size and shape of guest molecules may vary if they are conformationally flexible, there may be more play in the space available in the host molecules particularly in the case of lattice inclusion compounds. Separation of the diastereomeric inclusion compounds may depend on differential stability or solubility. Recovery of the resolution substrate may involve melting, dissolution, or extraction of the enantiomerically enriched guest.

In the first category (cavitates), inclusion of one enantiomer of amino acid derivatives in the cavity of racemic cyclic polyethers (chiral crown ethers)[151] leads to crystallization of only the more stable diastereomeric complex in enantiomerically pure form.[152] A resolution of a chiral crown ether comprising only one stereogenic element has been described.[153] The reciprocal resolution of racemic phenylglycine (as the perchlorate) by complex formation with one of the crown ether enantiomers is also possible (see Section 7-6).

The cyclodextrins, water-soluble macrocyclic glucose oligomers having six or more glucopyranose units linked α-1,4, form crystalline 1:1 cavitate inclusion complexes in solution. Alternatively, the complexes can be formed by grinding together (or sonicating) solid guest and host reaction partners. Partial resolution of a variety of functionalized compounds (carboxylic acids, esters, and alcohols) by inclusion in β-cyclodextrin (cycloheptaamylose; Fig. 7.23, **90**) was first described by Cramer and Dietsche.[154,155] Though the enantiomeric enrichment was low (<11% ee), this is not inherent in the process. Enrichment of the order of 60–70% ee is feasible in

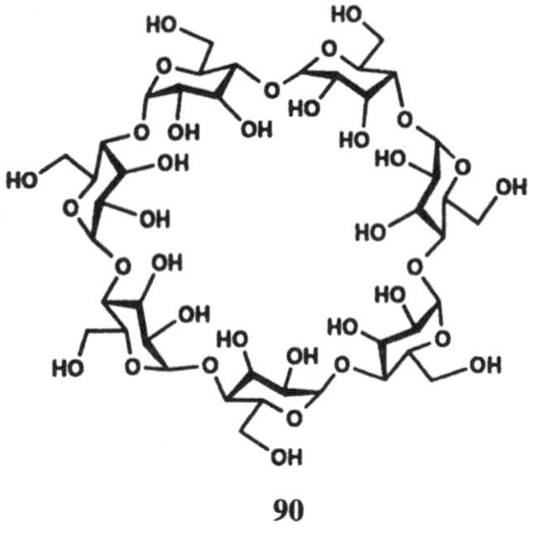

90

β-Cyclodextrin (β-CD; cycloheptaamylose)

Figure 7.23. β-Cyclodextrin.

specific cases, for example, for phosphinates.[156] Moreover, the inclusion process may be repeated to enhance the enantiomeric enrichment. Derivatized cyclodextrins form the basis of enantioselective GC stationary phases that are able to chromatographically resolve numerous racemates (see Section 6-5.d).

In the second category of inclusion complexes (clathrates, in which guest molecules are surrounded by many host molecules) it is sometimes the very presence of guest molecules that induces the formation of channels within the host crystal lattice that perforce contains the included guest molecules. Remarkably, lattice inclusion compounds useful in resolutions are formed with urea, an achiral compound, and with TOT (Fig. 7.4, **7**), a chiral compound that easily racemizes in solution (Section 7-2.d).[148,157] While free urea crystallizes in a tetragonal lattice, in the, presence of various guest substances, it is induced to crystallize in an hexagonal lattice in which the guest molecules are included in tubular channels having either right-handed or left-handed helical configurations.[158,159]

The most useful of these types of hosts thus far employed for resolution by inclusion compound formation is TOT (Fig. 7.4, **7**).[160,161] Though the TOT crystallizes in guest-free form as a racemic compound, in the presence of any of numerous guest molecules, it crystallizes as a conglomerate. The latter crystals may contain TOT in either the (P)-$(+)$ (right-handed propeller) or (M)-$(-)$ (left-handed propeller) configuration and the cavities wherein reside the guest molecules are correspondingly chiral. On dissolution of the separate guest-containing crystal batches at room temperature, the optical rotation falls nearly to zero as the TOT molecules undergo enantiomerization [interconversion $P \rightleftharpoons M$, $\Delta H^{\ddagger}$ ca. 21 kcal mol^{-1} (ca. 88 kJ mol^{-1})]. Cage clathrates (space group $P3_121$; host–guest ratio 2:1) are formed with small molecules (up to six nonhydrogen atoms; guest molecule length < 9 Å) and the enantiomeric enrichment observed ranges from less than 1% ee to at least 83% ee.[162,163] Although 3.5 kcal mol^{-1} (15 kJ mol^{-1}) separates the two diastereomeric complexes in the case of the 2-halobutanes [X = Br and Cl; the (P)-$(+)$-TOT/(S)-2-halobutanes are the preferred stereoisomers], twofold disorder of the guest molecules in the lattice may be responsible for the limited stereoselectivity observed; there are two R and one S guest molecules in each of three cages comprising one unit cell (32–45% ee observed for 2-chloro- and 2-bromobutanes). The very low enantiomeric enrichment observed with 2-butanol (<5% ee) is associated with an additional (torsional) degree of freedom of the guest molecule.

Lattice inclusion compounds are also formed with other chiral hosts. Among the better known examples are the alkaloids brucine and sparteine.[148,149] In 1981, it was observed that chiral tertiary acetylenic alcohols form stable 1:1 molecular complexes with brucine (Fig. 7.9a, **16**).[164] The alcohols recovered from the crystalline complexes formed in acetone exhibited high enantiomeric purity (Eq. 7.7).

$$\text{91} + \text{Brucine} \xrightarrow{\quad} \text{1:1 complex} \Downarrow \xrightarrow{\text{HCl}} \begin{array}{c} (R)\text{-}(+)\text{-} \mathbf{92} \\ 100\% \text{ ee} \end{array} \qquad (7.7)$$

(S)-$(-)$-**92** (93% ee)
(isolated from the
mother liquor)

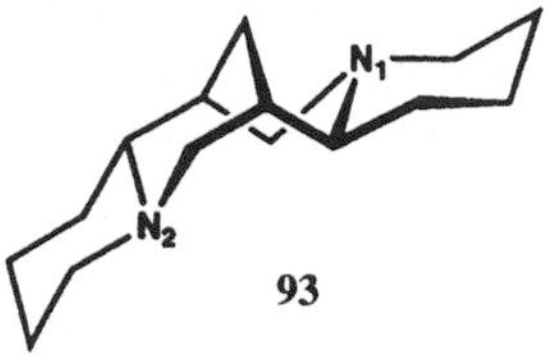

93

(–)-Sparteine

Figure 7.24. Sparteine.

Strong hydrogen bonds between the alcohol and the brucine were evidenced by IR spectroscopy and by X-ray crystallography and are responsible in part for the efficient diastereomer discrimination attending the resolution of the alcohol. Similar resolutions of diols, allenic alcohols, and cyanohydrins with brucine have been achieved.[149] It was subsequently discovered that resolutions of the same type of tertiary acetylenic alcohols takes place with (–)-sparteine serving as inclusion host (Fig. 7.24, **93**).[165] It has been observed that the nonracemic methosulfate salt of Tröger's base (Fig. 7.27, **98**) can serve as a resolving agent for alcohols (e.g., 2-phenyl-1-propanol) by inclusion compound formation.[166]

With the exception of brucine and of the cyclodextrins, both of which are natural products, other host compounds require resolution prior other to their use. This limitation has now been circumvented by application of compound **94** (Fig. 7.25), derived from diethyl (2*R*,3*R*)-tartrate (made from natural tartaric acid), to the

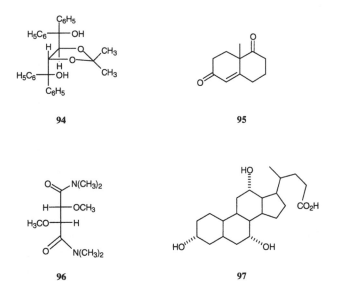

94

95

96

97

Figure 7.25. Structures 94–97.

resolution of Wieland–Miescher ketone **95**[167]; compound **96** (and analogues), also derived from diethyl (2R,3R)-tartrate, have been shown to be useful hosts for the resolution of chiral compounds with C_2 symmetry. Another host not requiring prior resolution is cholic acid **97** (Fig. 7.25); the latter has been shown to resolve lactones by inclusion.[168]

The high efficiency and simplicity of these resolutions is as remarkable as is the fact that they are effective with neutral compounds that are otherwise difficult to resolve. It has become apparent that the diastereomer discrimination responsible for the resolution is due to favorable packing of guest and host molecules in channel inclusion complexes. In some cases, the discrimination can be attributed to specific hydrogen bonding between guest and host.[169]

The heterogeneous brucine-mediated resolution of alkyl halides (1,2-dibromo-propane and 3-chloro-2-butanol) was discovered by Lucas and Gould[170] and applied also to (±)-2,3-dibromobutane.[171] Since the recognition that this type of resolution does not involve a kinetic resolution, that is, dehydrohalogenation,[171,172] but instead is due to complexation (the stoichiometry appears to be 1:1), the alcohol resolutions of Toda and co-workers, by inclusion in brucine (see above),[169] and those of the haloalkanes have been rationalized by the finding that the brucine crystal lattice possesses large channels of variable height that are capable of including solvent molecules, guest ions as well as neutral guest molecules.[124] That is, all of these resolutions with brucine (even resolutions of chiral acids by salt formation) are due to inclusion compound formation (cf. Section 7-3.b). The incursion of stereoselective inclusion of the halogen compound in the brucine lattice in the resolution of α-hexachlorocyclohexane with brucine (cf. Section 7.4.c)[173] must be considered a possibility.

d. Chromatographic Resolution

The vast majority of the numerous chromatographic resolutions reported in the literature have been analytical in nature; that is, their purpose has been the determination of stereoisomer and, in particular, enantiomer composition. These have been surveyed in Section 6-5.d. The present section deals with preparative chromatographic resolutions.

Preparative chromatographic resolutions are often performed on covalent diastereomer mixtures (Section 7-3.a). In numerous cases, chromatographic resolution of diastereomers on silica gel succeeds even with the rapid and simple flash chromatography technique[174] with the guidance of preliminary thin-layer chromatographic (TLC) experiments in which solvent composition is the only variable.[175] The resolution of gossypol (Fig. 7.15) as Schiff's bases with various amino acid esters (e.g., methyl L-phenylalaninate) by TLC on silica gel could be scaled up to separate 5 g (and even larger) samples on silica gel columns.[109] The chromatographic resolution of alcohols as diastereomeric carbamates in an automated liquid chromatograph (on 1-g samples) has been described.[110]

The ease with which covalent diastereomers are separated chromatographically serves to emphasize the anisometric character of such isomers. Failures of such chromatographic resolutions are probably the result of not building into the structures a sufficient number of intermolecular binding sites (or sites permitting sufficiently strong intermolecular binding) to permit selectivity to manifest itself as the diastereomeric molecules travel along the stationary phase. Chromatographic resolutions involving covalent diastereomers have been reviewed.[4,50,176–178] The reader is reminded of an important limitation of such resolutions: Separation via covalent diastereomers will necessarily be incomplete if the resolving agent is not enantiomerically pure.

When carried out on chromatographic columns of high efficiency (>10,000 theoretical plates), baseline separations are easily achieved when separation factors attain or exceed $\alpha = 1.2$ [$\alpha = (t_2 - t_1)/(t_1 - t_0)$, where t_2 and t_1 are the retention times of the second and first diastereomers eluted, respectively, and t_0 is that of an unretained substance, e.g., pentane]. In contrast, when $\alpha = 2.0$ [corresponding to $-\Delta(\Delta G)$ of 0.4 kcal mol^{-1} (1.7 kJ mol^{-1}) between the diastereomers], as is possible with some diastereomeric amides, the separations can be carried out even on relatively "primitive" columns, for example, the resolution of 2-phenylpropionic acid with phenylglycinol (Fig. 7.10, **47**) by chromatography of the diastereomeric amides (Fig. 7.26).[69]

The most versatile and powerful chromatographic resolutions are those that take place on enantioselective stationary phases; with such columns, it has been suggested that preparative resolutions are feasible when $\alpha \geq 1.4$.[179] The various types of enantioselective columns have been described in Section 6-5.d. Numerous examples of such resolutions on a scale of $10–10^2$ mg, but fewer examples of resolutions of 1 g or more of a racemate by one of these methods, have been described in the literature.[50,180–183]

The very earliest chromatographic resolutions (e.g., of Tröger's base, Fig. 7.27, **98**) were carried out on readily available natural products (lactose in the case of **98**)[184] and potato starch is still being used for the purpose (for applications of cellulose derivatives, see below).[185,186] Later, synthetic resolving agents impregnated on achiral supports, such as silica gel, were applied.

The first of the chiral stationary phases (CSPs) incorporating a π-base allowing for charge-transfer interaction as well as having hydrogen bonding capabilities is illustrated in Figure 7.28. This CSP was based on Pirkle's TFAE chiral solvating agent (Fig. 6.29, **67**) covalently bonded to mercaptopropyl silica. The structure of a generalized resolution substrate derivatized for optimal interaction with this stationary

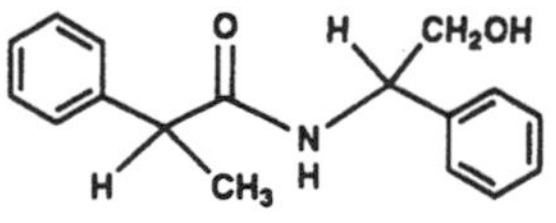

Figure 7.26. Diastereomeric amides separated by chromotography on "primitive" columns; $\alpha = 2.56$ (column = silica gel 60; mobile phase = hexane/EtOAc, 1:1).

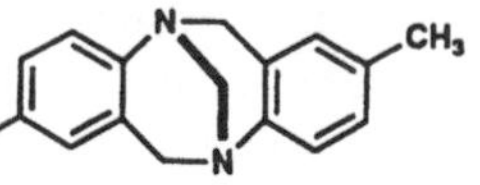

98

Figure 7.27. Tröger's base.

phase is given in Figure 7.28 (**99**). Chiral amines (Q = NH), amino acid esters (Q = NH, B = CO$_2$CH$_3$), alcohols (Q = O), sulfoxides and lactones, among others, could be resolved with this stationary phase following derivatization with 3,5-dinitrobenzoyl chloride, the latter to provide a π-acid bonding site complementary to the π-base incorporated in the stationary phase molecules.[187]

On the assumption that reciprocal resolution would be successful, a structure analogous to **99** (Fig. 7.28) was incorporated in a new stationary phase that is ionically bonded to 3-aminopropyl silica (as in Fig. 6.36, **C**). This CSP is effective in resolving chiral arylalkylcarbinols[188] and many other types of organic compounds as well (sulfoxides, amides, and heterocycles), all possessing at least one aryl group. Subsequently, CSPs derived from N-3,5-dinitrobenzoylphenylglycine and N-3,5-dinitrobenzoylleucine covalently bonded to 3-aminopropyl silica were applied to chromatographic resolutions (Fig. 6.36, **B**).[189] For a discussion of applications of Pirkle enantioselective stationary phases and stereoisomer discrimination models rationalizing their use, see Section 6-5.d.

Resolutions on a gram scale (up to 8 g in some cases) were reported with large columns (30 in. × 2 in.) filled with ionic phase **C** (Fig. 6.36)[190] and with the covalently bonded phenylglycine phase.[191] Preparative resolution by flash chromatography on the two covalently bonded enantioselective phases has also been reported.[192] Even larger scale resolutions (up to 125 g of racemate in a single run) with Pirkle enantioselective phases have been carried out.[193] More recently, an enantioselective stationary phase derived from N-(1-naphthyl)leucine having separation factors as large as α = 60 toward some analytes has been described.[194]

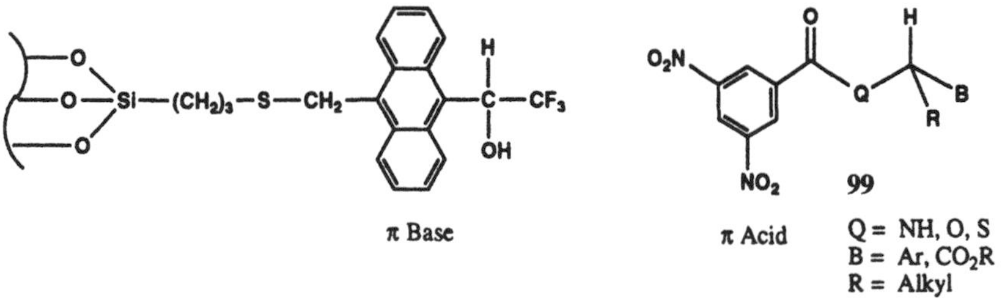

π Base

π Acid   **99**   Q = NH, O, S
              B = Ar, CO$_2$R
              R = Alkyl

CSP   Resolution substrate

Figure 7.28. Complementary enantioselective stationary phase (CSP) and resolution substrates incorporating, respectively, π-base and π-acid moieties.

Among the least expensive and most widely used chiral polymeric stationary phases are those based on cellulose and amylose (Section 6-5.d). Numerous preparative resolutions on the 100-mg scale have been reported with microcrystalline cellulose triacetate (TAC; Fig. 6.40). An advantage of such phases is that they lend themselves to resolution of nonfunctionalized (or nonderivatizable) racemates (e.g., perchlorotriphenylamine, **100**), whose partial resolution on TAC has been reported.[195] Stable atropisomers (e.g., Fig. 7.29, **101**) have been resolved on TAC[196] as has the strained hydrocarbon **102** (Fig. 7.29) that adopts a chiral *syn*-conformation.[197]

Underivatized carboxylic acids (e.g., mandelic acid) have been chromatographically resolved on both cellulose and amylose tris(3,5-dimethylphenyl-carbamates (**106** and **107**, respectively; Fig. 7.30) analogous to TAC.[198] Compound **103** (Fig. 7.29, in which the $CHCl_2$ groups are geared) that slowly undergoes

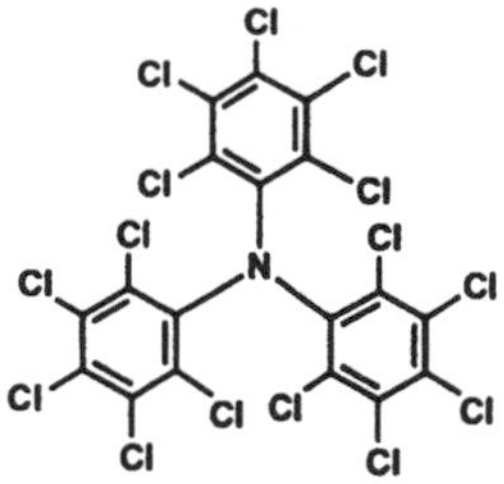

100

Perchlorotriphenylamine

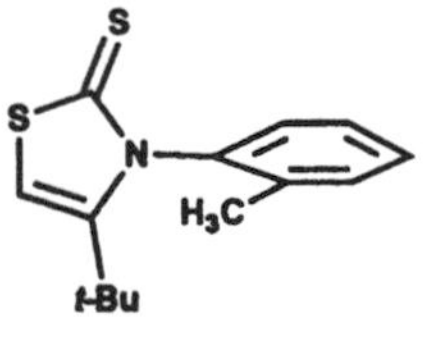

101

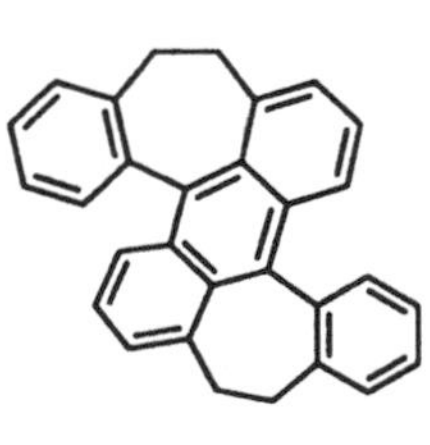

102

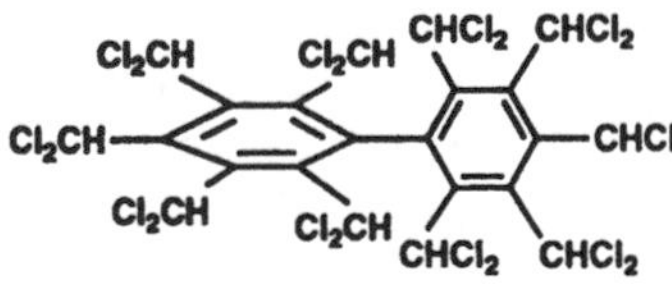

103

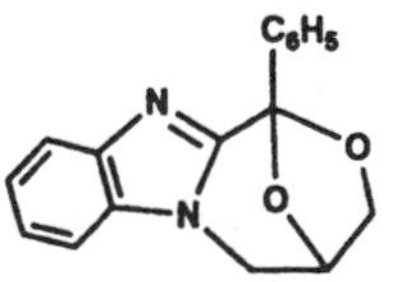

104

Oxapadol

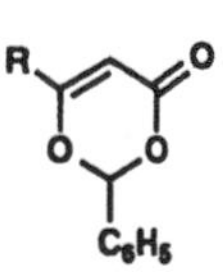

105

R = CH₃, H

Figure 7.29. Structures **100–105**.

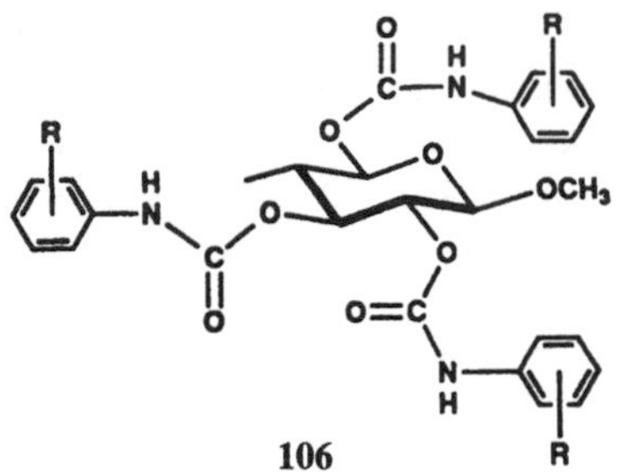

106   **107**

R = 3, 5-(CH₃)₂   R = 3, 5-(CH₃)₂

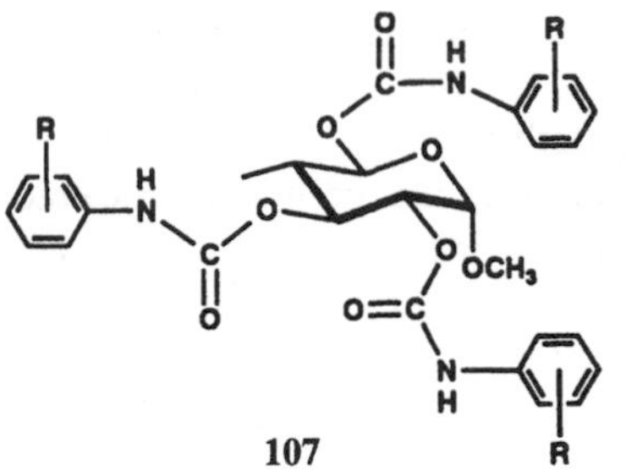

Figure 7.30. Cellulose (**106**) and amylose (**107**) based enantioselective stationary phases.

enantiomerization (by rotation of the CHCl₂ groups) has been preparatively resolved on stationary phase **106**.[199] Other resolutions have been carried out on polysaccharide stationary phases.[50,181,200,201] In spite of the low efficiency of TAC, resolution of gram and multigram samples (one report mentions a separation of up to 200 g per pass on TAC)[202] with this CSP have been reported, for example, oxapadol, **104**,[203] and 2-phenyl-1,3-dioxin-4-ones **105** (Fig. 7.29).[204]

Another polymeric CSP reported to have been used in preparative chromatographic resolutions is poly(triphenylmethyl methacrylate) (PTrMA, Section 6-5.d). Complete (baseline) resolution of propeller-shaped amine **100** (Fig. 7.29), with separation factor α = 2.9 (albeit only on ca. 1-mg scale) has been reported on PTrMA[205] as well as the resolution of Tröger's base (Fig. 7.27).[206] α-Methylbenzyl alcohol was partially resolved on PTrMA on a 219-mg scale.[207]

e. Asymmetric Transformations of Diastereomers

We have already seen that it is sometimes possible to obtain more than 50% of one enantiomer of a substance by a process whose departure point is a racemate (Section 7-2.d). Asymmetric transformation, the most general name applied to such processes, is also possible with diastereomers. Indeed, asymmetric transformations of diastereomers, either diastereomeric salts or covalent diastereomers, are more common than are transformations of enantiomeric species.

Figure 7.31 summarizes the process that obtains when components of diastereomeric mixtures, such as those formed in resolutions, are equilibrated. When equilibration, in solution or in the molten state, is rapid, mixtures are obtained that reflect the relative thermodynamic stability of the diastereomers. This spontaneous equilibration of stereoisomers in solution is often called an *asymmetric transformation of the first kind* (for a discussion of the history of the term, see Eliel).[27]

Asymmetric transformations of the first kind are the result of epimerization (a change in configuration at just one of several chiral centers present in a molecule) of both diastereomers present in a mixture. When a single diastereomer

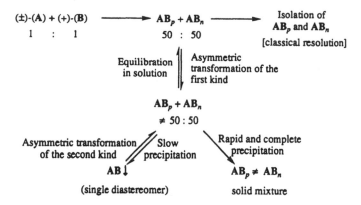

Figure 7.31. Asymmetric transformations of chiral diastereomers; **A** represents a racemate and **B** a resolving agent. The p and n subscripts are defined on page 215.

crystallizes from a solution containing several equilibrating diastereomers in a yield exceeding that given by its solution concentration, the process is termed *asymmetric transformation of the second kind*.[27,208] Here again, the term "crystallization-induced asymmetric transformation" has been suggested as a more descriptive alternative.[4]

Asymmetric transformations are usually observed during resolutions when more than 50% of one diastereomer is isolated. This finding is illustrated by the optical activation of the bridged biphenylcarboxylic acid (Fig. 7.32, **108**) with quinidine (Fig. 7.19a, **18**). Both the yield of quinidine salt (theoretical yield of one diastereomeric salt 50%; found 79%) and the fact that the acid recovered from the salt is optically labile ($t_{1/2} = 53$ min in o-xylene at 50°C) are indicative of the operation of an asymmetric transformation.[209]

In contrast to epimerization, the term asymmetric transformation is limited to processes in which the ultimate product is necessarily enantiomerically enriched (Eq. 7.8 in which the chiral acid H–R is nonracemic). However, the epimerization of *rac-cis*-1-decalone **111** to *trans*-1-decalone under the influence of an achiral catalyst or reagent is not considered an asymmetric transformation.

111

H–R

(7.8)

ee > 0   ee > 0

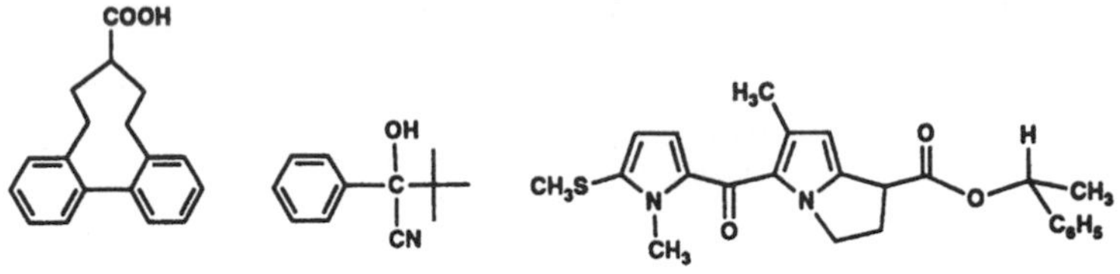

Figure 7.32. Compounds that undergo crystallization-induced asymmetric transformations.

The occurrence of asymmetric transformations among mixtures of diastereomers is much more common than is supposed. These occurrences may arise in diastereomeric salts whenever one of the ions, or the molecule that the ion can dissociate into, is labile (i.e., the ion can racemize). Such racemization can be spontaneous or induced, for example, by heating or on changing the pH. Asymmetric transformations are also observed in covalent diastereomeric systems. In both cases, the change in configuration at the labile stereocenter is induced by a chiral agent (by the counterion in the case of ionic diastereomers, and by the resolving agent in the case of covalent diastereomers). Asymmetric transformations less commonly take place when racemates are crystallized in optically active solvents. Epimerization of α- and β-anomers, representing diastereomer equilibration in aldohexoses, is dealt with in Chapters 10 and 11. In this section we principally survey those asymmetric transformations that have appeared in the literature since about 1979.[26]

A beautiful example of a crystallization-induced asymmetric transformation is that of the 3-amino-1,4-benzodiazepin-2-one (Fig. 7.33, **112**), and intermediate in the synthesis of a selective antagonist to the gastrointestinal hormone cholecystokinin (CCK). Resolution of **112** was effected with 0.5 equiv of (1S)-(+)-10-camphorsulfonic acid (CSA; Fig. 7.11, **57**) in isopropyl acetate to afford about 40% of the (3S)-amine-CSA salt (>99.5% ee). From the observation that the undesired (3R)-**112** racemizes at 90°C in the presence of the resolving agent (this process may, alternatively, be viewed as an epimerization), a system was devised to enhance the acidity of the α-hydrogen atom of **112** so that the epimerization could take place under mild conditions.

The epimerization was achieved by including a catalytic amount of an aromatic aldehyde (e.g., 3,5-dichlorosalicylaldehyde, 3 mol%) in the resolution reaction mixture, with the result that epimerization of the **112**·CSA diastereomers became possible at room temperature. The enhancement in the speed of the epimerization is attributed to formation of imine **113** (in low concentration). The ensuing crystallization-induced asymmetric transformation is driven by the insolubility of the (3S)-**112**·(1S)-CSA diastereomer. With all conditions optimized, there results a one-pot process (equivalent to a combined resolution–racemization): (±)-**112** (23 mol) treated with (1S)-(+)-CSA (21 mol) and (3S)-**112**·CSA seed (10 g) in isopropyl acetate/CH$_3$CN followed by addition of 3,5-dichlorosalicylaldehyde (0.69 mol). In

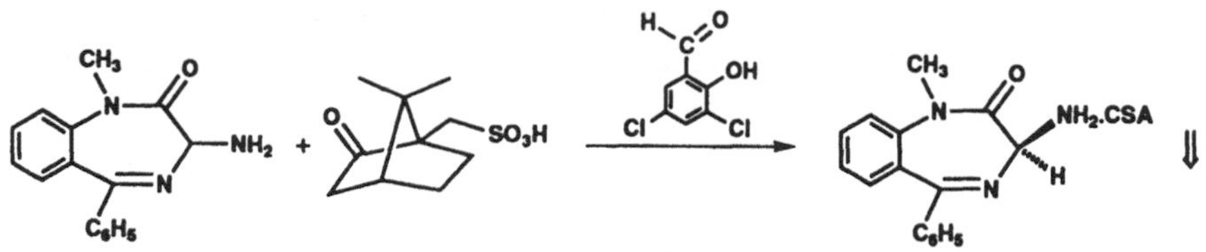

112   57 (CSA)

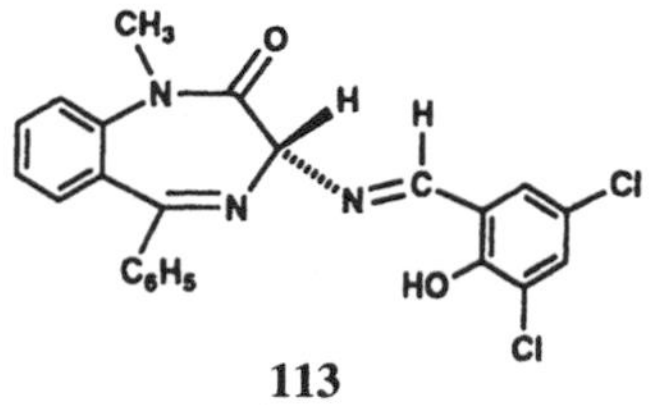

113

Figure 7.33. Asymmetric transformation of a benzodiazepinone.

this way, greater than 90% of the racemic substrate is converted to enantiomerically pure (3*S*)-**112** (1*S*)-CSA. In the reaction, epimerization is actually effected by a small amount of unprotonated base **112** (present as a result of an 8 mol% deficiency of resolving acid) generating an achiral enolate from **113** (Fig. 7.33).[210]

During the resolution of cyanohydrin **109** (Fig. 7.32) by inclusion in brucine, it was observed that the yield of the precipitated 1:1 complex exceeded 50%. Moreover, the cyanohydrin recovered from the filtrate was found to be optically inactive. (*Note*: This is another hallmark of an asymmetric transformation of the second kind.) On slow crystallization, the yield of cyanohydrin could be raised to 100%. In this crystallization-induced asymmetric transformation, the stereocenter is labilized by reversible dissociation of the cyanohydrin into a ketone and HCN, two achiral precursors.[211,212]

The resolution of Tröger's base (**98**, Fig. 7.27) with binaphthylphosphoric acid (Fig. 7.11*b*, **59**) is attended by a crystallization-induced asymmetric transformation; the yield of enantiopure TB obtained is 93%.[166] The attending acid-catalyzed racemization of TB is described in Section 7-7.a.

Resolution of (±)-phenylglycinate esters [e.g., $C_6H_5CH(NH_2)CO_2C_2H_5$] with (+)-(2*R*,3*R*)-tartaric acid in ethanol containing 10% dimethyl sulfoxide (DMSO) is attended by an asymmetric transformation (the yield of the less soluble ethyl (2*R*)-phenylglycinate diastereomer is 74%; >90% ee).[213] Resolution of (±)-methyl phenylglycinate and substituted phenylglycine esters $ArCH(NH_2)CO_2CH_3$ with (+)-(2*R*,3*R*)-tartaric acid (1 equiv) in the presence of 1 equiv of benzaldehyde speeds up the asymmetric transformation process; (2*R*)-$ArCH(NH_3^+)CO_2CH_3$ hydrogen (2*R*,3*R*)-tartrate (Ar = C_6H_5; 99% ee) is formed in 85% yield (reuse of the filtrate raises the yield of D salt to 95%.[214] Epimerization has also been carried out on

L-**43**·(+)-**67** by heating in acetic acid in the presence of salicylaldehyde (see Section 7-7b for a discussion of the function of aldehydes in the asymmetric transformations of amino acid derivatives).[88]

More commonly, asymmetric transformations of amino acids require prior derivatization of the resolution substrate. Derivatization permits a wider range of resolving agents or resolution methods to be used. That of phenylglycine (Fig. 7.10, **43**) may be carried out on the N-benzoyl derivative concomitant with the resolution [resolving agent: excess (−)-α-methylbenzylamine (Fig. 7.9b, **33**)]. The yield of D-N-benzoylphenylglycine is 95–98%.[215] A similar asymmetric transformation has been reported for N-acetylphenylglycine.[216]

Asymmetric transformations of α-amino acids may further be "promoted" by formation of N-acyl derivatives and complexation with transition metal ions. In a system designed to mimic the mechanism of action of a racemase enzyme (i.e., labilization of the α-hydrogen atom), the Schiff's bases of amino acids [e.g., (±)-alanine with (−)-menthyl 3-(2-hydroxybenzoyl)propionate] was treated with cobalt(II) acetate. A mixture of diastereomeric salts was formed from which one precipitated. D-Alanine·HCl (22% ee; L-enantiomers in the case of leucine and valine) was isolated from the precipitate (Fig. 7.34).[217]

An unusual example involving epimerization at two dissimilar stereocenters is that shown in Figure 7.35. On heating with (−)-10-camphorsulfonic acid (Fig. 7.11a, **57**) in EtOAc for 22.5 h, a racemic diastereomer mixture (**114**) yields a single dextrorotatory isomer in 80% yield.[218] The resolving agent catalyzes the equilibration of **115** at C(11b) (by reversible Mannich reaction) and at C(3) (by reversible enolization) and selectively causes precipitation of the least soluble salt.[219]

In the second approach, a racemic substrate is converted to an *achiral* intermediate or derivative (e.g., an enolate) and the latter is enantioselectively protonated by an optically active acid as shown for benzoin in Eq. 7.9.

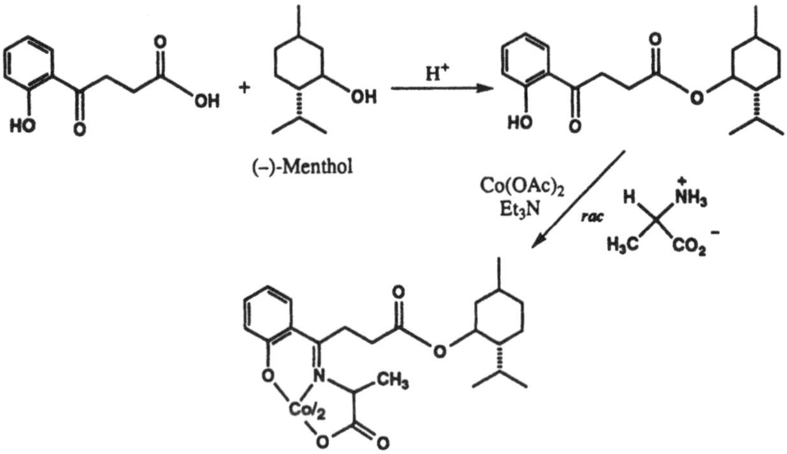

Figure 7.34. Synthesis of a Schiff's base derivative of alanine complexed with Co(II) to facilitate the asymmetric transformation of the amino acid.

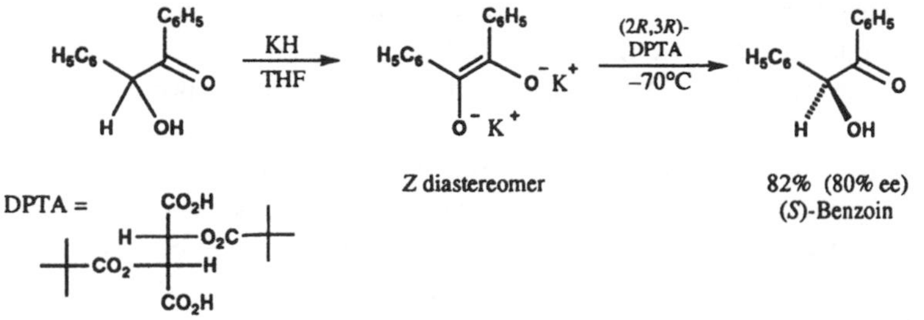

$$\text{(7.9)}$$

DPTA =

On isolation, the benzoin crystallizes in enantiomerically pure (100% ee) form.[220] While the "deracemization" of (±)-benzoin, a kinetically controlled enantioselective protonation, has features in common with crystallization-induced asymmetric transformations, the process is more usefully viewed as an asymmetric (i.e., enantioselective) synthesis.

When the enolate of ketone **116**, generated with lithium diisopropylamide (LDA) is protonated with (−)-mandelic acid, with ephedrine, or with (−)-menthol, the recovered **116** is not enantiomerically enriched. However, if the enolate is generated with lithium (S,S)-α,α'-dimethyldibenzylamide, hydrolysis regenerates **116** with 48% ee (Eq. 7.10).[221]

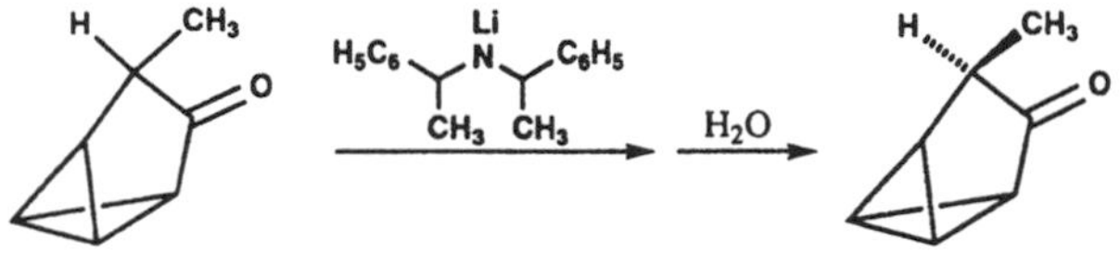

$$\text{(7.10)}$$

116   48% ee

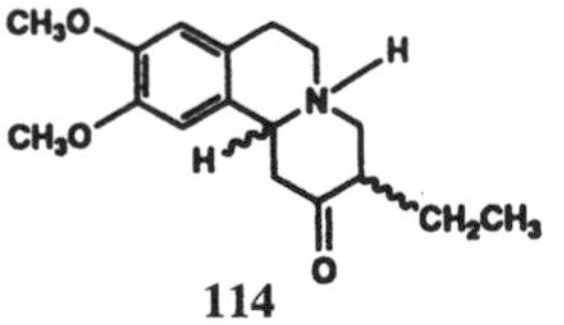

114

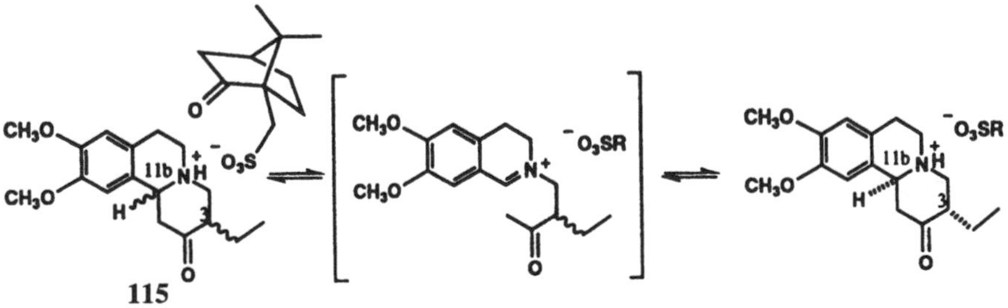

115

Figure 7.35. Asymmetric transformation involving epimerization at two stereocenters [Reprinted from Oppolzer, W. *Tetrahedron* **1987**, *43*, 1969. Copyright © 1987, with permission from Pergamon Press, Oxford, UK.]

The enantioselective protonations are complicated by the presence of more than one enamine or enolate stereoisomer, and by equilibration between stereoisomeric iminium salts (Fig. 7.36). Available evidence suggests that the principal driving force in these reactions is the relative rate of protonation (i.e., enantioselective protonation), whether R in the iminium salts or A in HA are stereogenic.[222–224]

An asymmetric transformation of the first kind has been described that is mediated by the formation of diastereomeric complexes. N-(±)-(3,5-Dinitrobenzoyl)leucine butyl thioester **117** (0.045 M) was allowed to stand in mixed cyclohexane/CH$_2$Cl$_2$ solvent containing triethylamine (0.18 M) and 10-undecenyl (R)-N-(1-naphthyl)alaninate (Fig. 7.37, **118**; 0.20 M). Periodic chromatographic analysis on an enantioselective stationary phase analogous to **118** (cf. also Fig. 6.36, **D**) revealed that thioester **117** was slowly undergoing an asymmetric transformation, with the enantiomeric enrichment reaching 78% ee of (R)-**117** after 28 days.[225] The choice of a thioester rather than an ordinary ester serves to increase the acidity of the α-hydrogen atom so as to enhance the rate of the Et$_3$N-catalyzed epimerization.

The diastereomer discrimination that provides the driving force for the asymmetric transformation of the thioester is similar to that operative in chromatographic resolutions with enantioselective stationary phases **B** and **C** (Fig. 6.36). Preferential formation of the R,R' transient complex as the result of one additional hydrogen bond (Fig. 7.37) is supported by nuclear Overhauser effect (NOE) measurements.[226]

f. General Methods for the Separation of Diastereomers

Diastereoisomer mixtures other than those obtained in resolutions frequently need to be separated. These are typically covalent compound mixtures of epimers, anomers, meso and chiral diastereomers and stereoisomers differentiated by permutations at

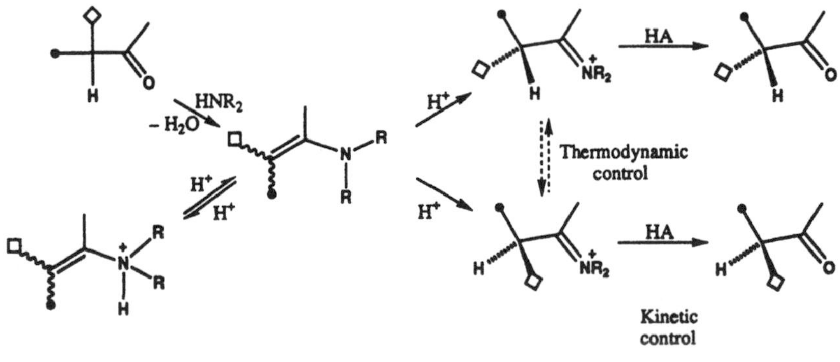

Figure 7.36. Mechanism of deracemization of chiral carbonyl compounds. [Adapted from Duhamel, L., Duhamel, P., Launay, J.-C., and Plaquevent, J.-C. *Bull. Soc. Chim. Fr.* **1984**, II-421. Reprinted with permission of the Société Francaise de Chime.]

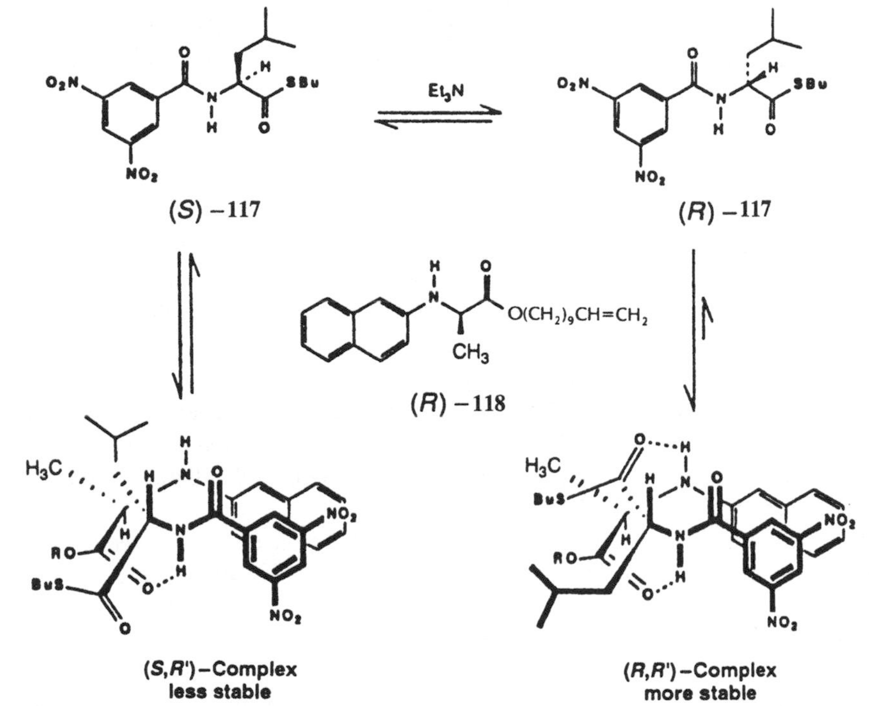

Figure 7.37. Asymmetric transformation on an amino acid derivative. [Adapted with permission from Roush, W. R. *Chemtracts: Org. Chem.* **1988**, *1*, 136. Copyright © 1988 by Data Trace Chemistry Publishers, Inc.]

several stereocenters. Mixtures of compounds differing in geometry at double bonds require the same treatment for their separation as do those above-mentioned in which chirality (though often present) is not at issue; separation of chiral (and racemic) or achiral diastereomers from their mixtures is probably more common in organic laboratory practice than is the separation of optically active diastereomers during a resolution.

The components of diastereomer mixtures, like mixtures of constitutional isomers, bear anisometric relationships to one another. Such mixtures may be separated by whatever method seems most expedient. Yet some of the most common and obvious separation methods are of little use in the separation of stereoisomers, to wit, extractive separation based upon acid–base properties. This fact follows from the very character of diastereomers: All members of the set possess the same kind and number of functional groups. While the acid–base properties do differ somewhat, the differences are rarely as large as those, say, between fumaric and maleic acids ($K_{a_1} = 1.0 \times 10^{-3}$ and 1.5×10^{-2}, respectively) and even this difference is difficult to exploit for separation.

Distillation on the other hand, though unexciting and relatively slow, is a perfectly useful and satisfying technique for separating stable stereoisomers differing in boiling

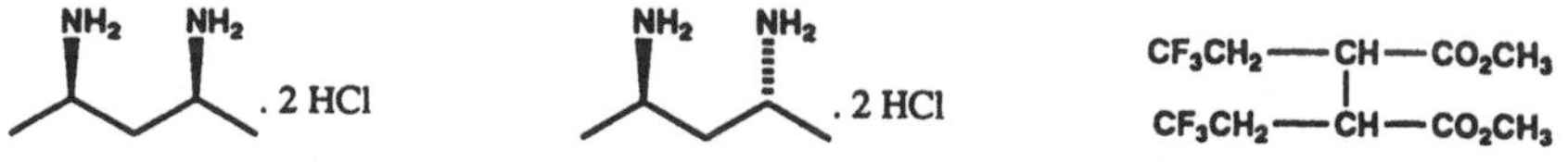

Figure 7.38. Structures **119–121**.

point by about 5°C or more (even less if a good spinning band column is at hand) or at least for purifying them. The pivotal demonstration of the existence of two decalin stereoisomers that helped to convince organic chemists that cyclohexane is not planar[27] depended on the ability of Hückel and co-workers to separate *cis-* from *trans*-decalin by distillation (the boiling point difference is 8°C).[227] Resolution has been effected by distillation with inclusion compounds.[228] Some additional examples of such separations are that of *cis-* and *trans*-3-isopropylcyclohexanols,[229] and the separation of *meso-* from *rac*-2,4-pentanediol by fractional distillation of the cyclic sulfite esters, bp 72 and 82°C/12 torr, respectively.[230]

Diastereomers may differ significantly in solubility and this difference may be exploited in the separation of their mixtures. For example, the *meso*-2,4-pentanediamine·2HCl **119** (Fig. 7.38) has a solubility of 3.3 g/100 mL in boiling EtOH. In contrast, the chiral diastereomer **120**, dissolves in the same medium only to the extent of 0.1 g/100 mL. This large solubility difference has been exploited in a separation described by Bosnich and Harrowfield.[231] Electrooxidative trifluoromethylation of methyl acrylate furnishes a 1:1 mixture of *meso-* and *rac*-dimethyl 2,3-bis(2,2,2-trifluoroethyl)succiniate **121**. The meso isomer was isolated in pure form on repeated recrystallization of the mixture from hexane. Analogous recrystallization of the mixture from pentane afforded the chiral diastereomer.[232]

Even trituration may suffice if the solubility difference is large enough, for example, in the case of the epimeric pair **122** (Fig. 7.39) with boiling toluene.[233] In

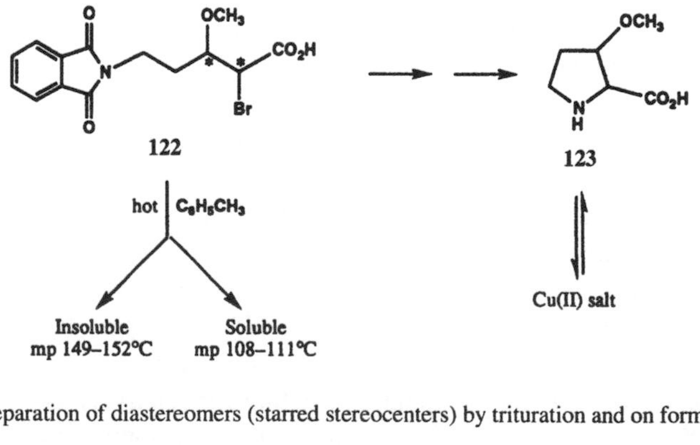

Figure 7.39. Separation of diastereomers (starred stereocenters) by trituration and on formation of Cu(II) salts.

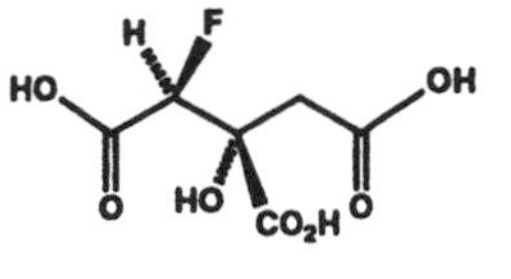

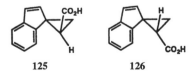

Figure 7.40. Separation of diastereomeric acids by fractional crystallization of their cyclohexyl-ammonium salts: *threo*-**124** (2*S*,3*R* enantiomer shown) and *erythro*-**124** (2*R*,3*R* enantiomer shown).

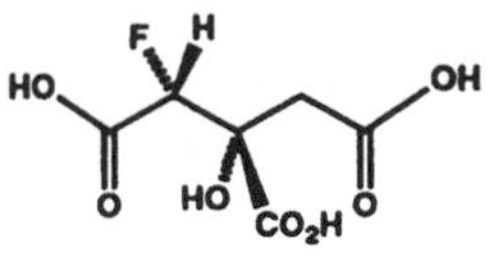

the same report there is a description of the separation of the *rac-trans-* and *cis*-3-methoxyprolines **123** by conversion to Cu(II) salts. Following the crystallization of their salts from ethanol, the *cis*-**123** compound is regenerated from the less soluble copper salt by passage through an ion exchange column.

Separation of relatively soluble substances, such as polycarboxylic acids, is facilitated by their conversion to derivatives, such as amine salts (cf. Section 7-4). The *threo-* and *erythro*-2-fluorocitric acids (Fig. 7.40, **124**) are separated by fractional crystallization of the cyclohexylammonium salts of the mixed diethyl 2-fluorocitrates. The erythro isomer is the less soluble one.[234]

Very efficient and rapid separation of stereoisomeric alcohols from mixtures may be achieved by selective complexation with inorganic cations. For example, either $CaCl_2$ or $MnCl_2$ form alcoholates with *trans-* but not *cis*-4-*tert*-butylcyclohexanol, with *erythro-* but not *threo*-3-phenyl-2-butanol, and with only one of four possible decahydro-1-naphthols. Ethanol serves as a catalyst in the complexation.[235] In the partial separation of *meso-* and *rac*-2,5-hexanediols, much of the achiral isomer is extracted from a chloroform solution of the two isomers by contacting it with anhydrous powdered $CaCl_2$.[93] Several Lasalocid A stereoisomers (Fig. 7.19) are complexed and isomerized concomitantly in the presence of barium hydroxide.[236]

Cooperative separation by serial application of two techniques is not infrequent. A 2:1 mixture of epimeric 2-spiro[cyclopropane-1,1-indene]carboxylic acids (**125**, trans, and **126**, cis, Fig. 7.41) was separated by crystallization from hexanes containing ethyl acetate (10:1). The major isomer (trans) was isolated in reasonably pure form (97% trans + 3% cis). Resolution of the trans acid (with quinine) freed it of the cis isomer. The latter accumulated in the resolution mother liquor.[237]

Figure 7.41. Epimeric 2-spiro[cyclopropane-1,1-indene]carboxylic acids.

It is not uncommon to find that one stereoisomer in a mixture crystallizes spontaneously from solution or on standing of the solvent-free mixture. Many classic isolation experiments depend on such fortuitous occurrences. The selective crystallization of one sugar anomer from a mixture consisting of several anomers and alternative ring types (furanoses and pyranoses) exemplifies such separations. Ordinary D-glucose crystallized from aqueous solution is a single species, α-D-glucopyranose (see Section 11-4).

However, the modern literature is replete with examples of diastereomer mixtures that do not crystallize at all, even following separation. Such mixtures are nowadays separated by chromatography. Just a few examples of such separations follow.

There are numerous reports of the application of flash chromatography to stereoisomer separation including the example given in the original report (Fig. 7.42, **127**).[174] Two fairly representative additional examples are the adamantanone isomers **128** that could not be readily separated and in any event epimerized upon chromatography on silica gel. These isomers were separated as ketals **129** (Fig. 7.42) by flash chromatography on basic alumina.[238] The *meso-* and *rac-*2,4-pentanediol diastereomers (see above) can also be separated by chromatography of their benzaldehyde acetals.[239]

Specific separation methods for stereoisomers take advantage of molecular symmetry or proximity of functional groups. The two diastereomeric 1-methoxybicyclo[2.2.2]oct-5-ene-2-carboxylic acids **130** are separated by taking advantage of the ease with which the endo isomer forms a lactone (by iodolactonization). Unreacted exo acid **130** is separated from lactone **131** by extraction with base (Fig. 7.43).[240]

Differences in reaction rates may also permit diastereomers to be separated. As an example, 1-methyl-*t*-4-*tert*-butylcyclohexane-*r*-carboxylic acid (Fig. 7.44, **132**) is esterified completely with 10% BF$_3$ in CH$_3$OH in 5 min (steam bath); the cis stereoisomer was similarly esterified only on refluxing for 1.5 h[241]; for an example involving amide hydrolysis (of **133**) at different rates, see Whitesell et al.[242]

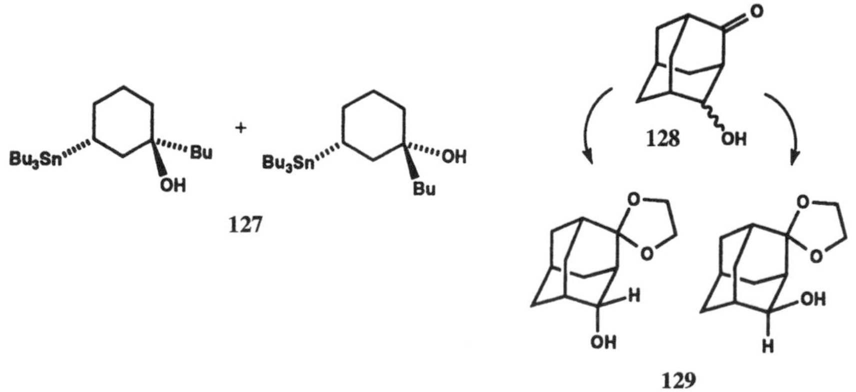

Figure 7.42. Structures **127–129**.

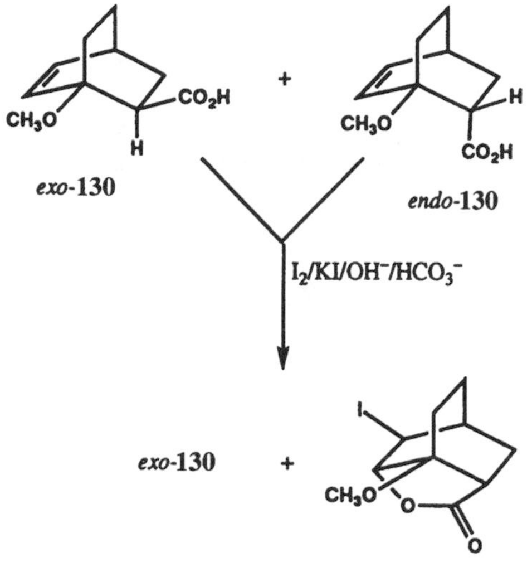

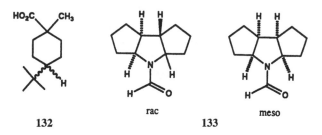

exo-**130**   *endo*-**130**

I₂/KI/OH⁻/HCO₃⁻

exo-**130**   +

131

Figure 7.43. Separation of the diastereomeric 1-methoxybicyclo[2.2.2]oct-5-ene-2-carboxylic acids.

Derivatization of diols or diol ethers with formation of volatile products (Fig. 7.45, **134**)[243] or of insoluble ones (Fig. 7.46, **135**)[244] similarly permits facile isolation of relatively homogeneous stereoisomers and recovery of the less volatile or less reactive epimers, respectively.

Separation is built into a system through the operation of a crystallization-induced asymmetric transformation (Section 7-3.e). As a consequence of rapid equilibrium among the several possible diastereomers in solution, crystallization of one of these as triggered adventitiously or purposefully by seeding can lead, in principle, to total precipitation of one of the components. One can imagine that other diastereomers could be crystallized instead by changing conditions (e.g., changing the solvent and/or using seeds of the other diastereomers). Changing the solvent need not result in a change in the equilibrium composition. Only the relative solubilities of the diastereomers need be affected.

132   *rac*   **133**   *meso*

Figure 7.44. Diastereomer separation based on rate differences.

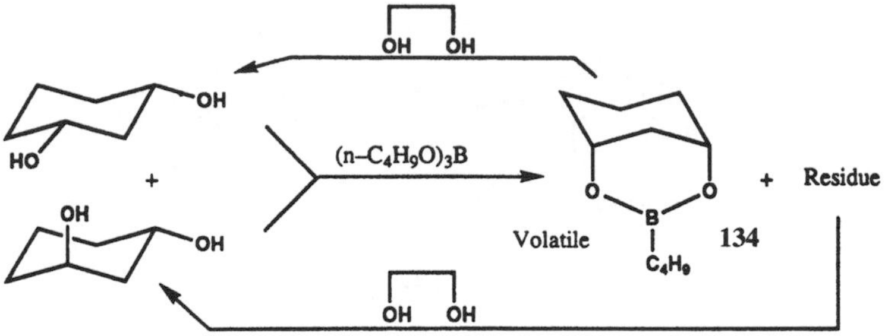

Figure 7.45. Separation of the diastereomeric 1,3-cyclohexanediols.

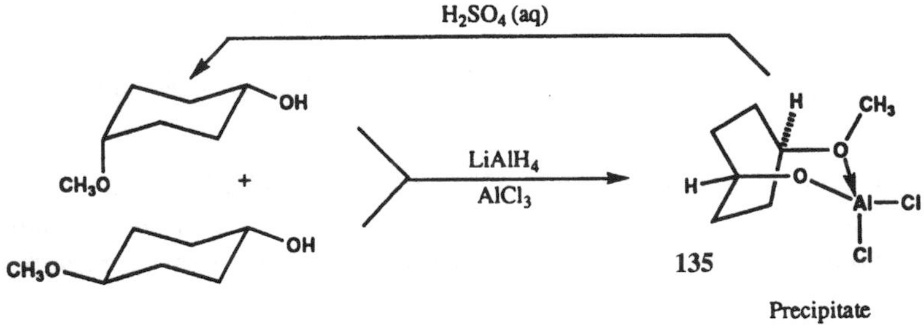

Figure 7.46. Separation of the diastereomeric 4-methoxycyclohexanols. Decomposition of the crystalline chelate with H_2SO_4 affords the cis isomer (> 95% cis).

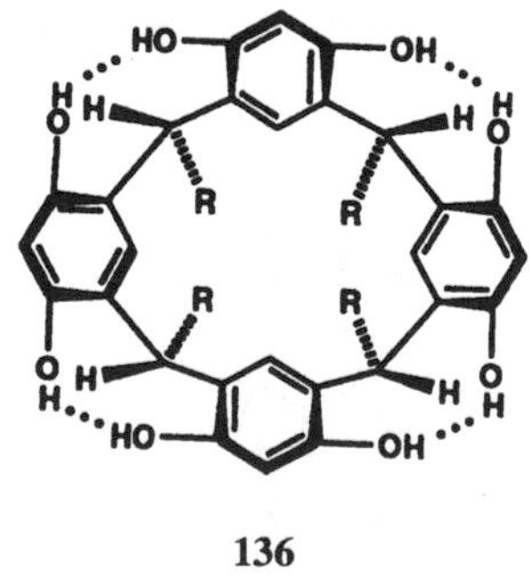

136

Figure 7.47. Diastereomer separation by inclusion compound formation and by clathration. [Structure **136** is reproduced with permission from Aoyama, Y., Tanaka, Y., and Sugahara, S. *J. Am. Chem. Soc.* **1989**, *111*, 5397. Copyright © 1989 American Chemical Society.]

The two equilibria involved, one homogeneous and the other heterogeneous, are not equally affected by changes in conditions. The sugar series has numerous examples of such separations (e.g., glucose), which is present in solution as an equilibrium mixture of α and β forms. The relative concentrations are little affected by the solvent. However, crystallization from ethanol affords the α anomer while crystallization from warm pyridine leads to the β-anomer.

Water-soluble diastereomers may be separated by clathrate formation with polyhydroxy macrocycles. Extraction of a 1,4-cyclohexanediol mixture (cis/trans = 53:47) with **136** (Fig. 7.47, R = $(CH_2)_{10}CH_3$) in CCl_4 yields a solution containing diol in a cis/trans= 83:17 ratio; the complexation of diol with the cyclotetramer is stereoselective. Aldohexose diastereomers are also selectively extracted from H_2O into CCl_4 with **136**.[245]

7-4. ENANTIOMERIC ENRICHMENT AND RESOLUTION STRATEGY

Numerous statements may be found in the literature that describe the enantiomeric enrichment of nonracemic and enantiomerically impure samples. Enantiomerically impure naturally occurring compounds as well as nonracemic samples prepared by resolution or stereoselective synthesis may be subjected to enantiomeric enrichment by crystallization, by sublimation (Section 6-4.g), by chromatography on enantioselective stationary phases (Section 7-3.d), and in some cases even by chromatography on achiral stationary phases (Section 6-4.m). Alternatively the nonracemic samples may be converted to diastereomer mixtures and the latter may be separated by the methods described in Sections 7-3 and 7-6. In this section we address the basis on which enantiomeric enrichment by crystallization takes place.[119]

We begin with the finding that vitamin K_3 epoxide **137** prepared by epoxidation of the unsaturated quinone precursor in the presence of a catalytic amount of the chiral phase-transfer agent, benzylquininium chloride **138**, is optically active. The product of this reaction (Eq. 7.11) had $[\alpha]_{587}^{21}$ 0; $[\alpha]_{436}^{21}$ − 6.8 ($CHCl_3$ or acetone). Recrystallization of the product (4–10 times) consistently produced enantiopure material $[\alpha]_{436}^{21}$ − 124 ± 5. It was discovered only after the purification efforts that the enantiomer excess of the product originally obtained from the epoxidation reaction was only 5–10%! Prior estimation of the enantiomer composition "might well have discouraged any but the most optimistic chemists from attempting purification by crystallization".[246]

(7.11)

137

The enantiomeric enrichment (to ee ca. 100%) of other products with initial enantiomeric purity as low as 50% (or less) by means of a few recrystallizations is perhaps rare.[247-249] It is significant that the enantiomeric enrichment by means of a few recrystallizations remains sufficiently remarkable that authors explicitly describe such findings in wonderment.

Enantiomeric enrichment by crystallization from the melt has been demonstrated in the case of the so-called Corey-lactone II (Fig. 7.48) whose pure enantiomer melts at 46 °C. The solid product obtained on cooling a molten sample with ee = 29.2% to 5°C was enriched to 72.5% ee. Cooling the melt derived from the same sample to 10°C led to recovery of a sample with ee = 90.7% ee.[250]

The enantiomeric enrichment of liquids or oily products is easily effected by their conversion to solid derivatives, for example, 1,2-propanediol as the ditosylate[251]; arylalkylcarbinols as 3,5-dinitrobenzoates or phenylcarbamates; see also Figure 7.1. A noteworthy instance of enantiomeric enrichment of a liquid is that of α-pinene. This enrichment is effected most simply by crystallization from pentane at −120°C.[253]

During the recrystallization of the Wieland–Miescher ketone (Fig. 7.25, **95**) prepared by enantioselective Robinson annulation, it was observed that enantiomeric enrichment took place if the initial enantiomer composition exceeded 75:25 (50% ee), whereas if the initial ee < 50%, the recrystallized product had a lower ee. In the latter case, product isolated from the mother liquor was found to be enantiomerically enriched.[254]

All of the above observations may be rationalized by reference to phase diagrams for the racemate–enantiomer system under study.[120] Ideally, a ternary diagram including the solvent should be used.[120] Ternary diagrams are rarely available, however. A binary phase diagram (e.g., Fig. 6.7) is often sufficient since the composition of the eutectics found in both the binary and the ternary diagrams are close to one another.[4] Even melting point information often suffices to establish whether the racemate is a conglomerate or a racemic compound and, if the latter, whether the racemic compound is very stable (high melting point relative to that of the pure enantiomer; cf. Fig. 6.7a). These facts are indicative of the general form of the binary (and by implication also the ternary) phase diagram and roughly of the location of the eutectics; for example, the fact that (+)-ketone **95** (Fig. 7.25) has mp 50–51°C while its racemate has an almost identical melting point (mp 49–50°C) suggests that the phase diagram may have eutectics in the vicinity of enantiomeric compositions 75:25 and 25:75, that is, 50% ee, with respect to either enantiomer (Fig. 7.49), which is precisely the break point where the outcome of its recrystallization undergoes reversal (see above).[254]

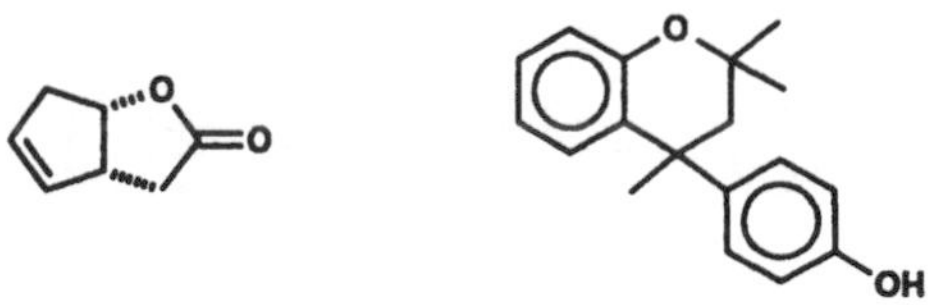

Figure 7.48. "Corey-lactone II" (left) and Dianin's compound (right).

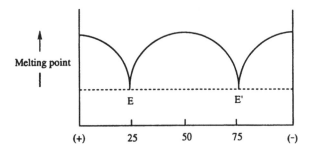

Figure 7.49. Proposed phase diagram of compound **95** (Fig. 7.25).

The following generalizations account for the results observed in enrichment by recrystallization: (a) If the racemate is a conglomerate, then enrichment is observed for *any* initial enantiomer composition (Section 6-4.f). (b) More frequently, the racemate is a racemic compound. In such cases, if the initial purity is low, with an enantiomer composition less enriched than that of the eutectic, that is, lying on the *racemate* branch of the phase diagram (EE′ in Fig. 7.49), then recrystallization *reduces* the enantiomeric purity of the initial product. A mixture lying on the enantiomer branch [i.e., (+)E or E′(−)] *can* be enriched during recrystallization. If sufficient solvent is taken, then *one* recrystallization suffices to reach 100% ee.[4] The latter statement suggests that there is necessarily a trade-off between yield and enantiomer purity; moreover, during the enrichment process it is essential to monitor the enantiomer composition (preferably by a method independent of $[\alpha]_D$).

Purification of carboxylic acids by recrystallization of their ammonium salts has the advantage of furnishing highly crystalline derivatives. Acids that are enantiomerically impure are often incidentally enriched during the recrystallization process[255]; for the use of dicyclohexyl- and dibenzylammonium salts in the enantiomeric enrichment of carboxylic acids see Kikukawa et al.[256] Whereas nonracemic 2-phenoxypropionic acids (ep 75–88%) cannot be purified by recrystallization, their *n*-propyl-, cyclohexyl-, or dicyclohexylammonium salts can. Note that all of these are salts of chiral carboxylic acids with *achiral* amines, that is, the salts are those of the enantiomers. Comparison of the binary phase diagrams of the acids with those of the ammonium salts reveals the reason for the difference in recrystallization outcome. The composition of the eutectics shifts from a range encompassing 10:90 and 1:99 (79–97% ee) in the case of the acids to one encompassing 20:80 and 50:50 (61–0% ee; i.e., one of the salts is a conglomerate) for the *n*-propyl- or cyclohexylammonium salts studied.[257]

If an enantiomer mixture has a composition corresponding to that of the eutectic (E in Fig. 7.49), then enrichment by crystallization becomes quite difficult. Such a situation is encountered in the asymmetric transformation of 2,2′-dihydroxy-1,1′-binaphthyl. Recrystallization of enantioenriched samples leads to a product that does not exceed about 82% ee (evidently, the composition of the eutectic). During the crystallization trials it was observed that two types of crystals were formed, one being the enantiopure diol (faster crystallizing), the other the racemic diol (the latter, crystallizing more slowly, is a racemic compound). By taking advantage of the different rates of crystallization of the two types of crystals on recrystallization of the eutectic mixture (kinetic control, with or without seeding with enantiopure diol), it was possible to surmount the "thermodynamic obstacle" and to obtain the enantiopure material.[258]

Purification of mixtures of diastereomers (by chromatography or by recrystallization) is the alternative to the above process, that is, diastereomer enrichment leading ultimately to enantiomer enrichment when the diastereomers are cleaved. When chromatography is employed, the result is straightforward: this process has been examined in Section 7-3.d. Diastereomer enrichment by recrystallization is usually effective because the shape of the unsymmetrical phase diagrams (either binary or ternary) is most often analogous to those of conglomerate systems, that is, there is but a single eutectic present (as in Fig. 7.18b).[4]

Two special procedures involving chemical reactions exist for the attainment of enantiomeric purity from enantiomerically impure samples. The first, called "duplication," requires that two molecules of the impure sample be chemically combined into one new molecule by whatever process seems best for the case at hand; for example, two amine molecules having opposite chirality sense may be incorporated in a neutral *achiral* nickel complex. Following the removal of the complex, enrichment is observed in the uncomplexed amine.[259] This type of process has been generalized by Horeau especially for the enantiomeric enrichment of enantiomerically impure alcohols (through formation of carbonate, malonate, or phthalate esters, e.g., Eq. 7.12,[260] and of carboxylic

$$RS \quad \text{meso (liquid)} +$$
$$RR/SS \quad \text{chiral (solid)}$$

acids, through formation of the anhydride.[261] The enantiomeric purity p' (ee) of the alcohol recovered from the chiral RR/SS diastereomer following separation (by chromatography or by crystallization) of the latter from the concomitantly formed meso RS diastereomer is given by Eq. 7.13, where p is the enantiomeric purity (ee) of the starting nonracemic alcohol[4,262]:

$$p' = \frac{2p}{1 + p^2} \qquad (7.13)$$

One of the more noteworthy applications of this enrichment process is to the purification of α-pinene. The liquid terpene (bp 155–156°C; 91% ee) was hydroborated to crystalline tetra-3-pinanyldiborane [commonly called diisopinocampheylborane (Ipc$_2$BH) ignoring the dimeric character of the borane] in the presence of 25% excess of α-pinene (Fig. 7.50). Liberation of the α-pinene by reaction of the crystalline [Ipc$_2$BH]$_2$ with benzaldehyde is catalyzed by BF$_3$·OEt$_2$. The isolated α-pinene, $[\alpha]_D^{23}$ + 51.4 (neat) has ee = 99.6%.[263–266]

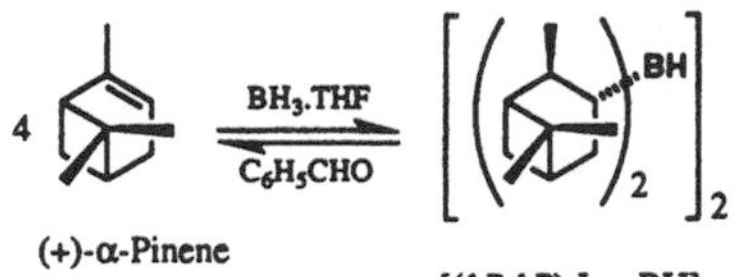

Figure 7.50. Purification of α-pinene by duplication to tetra-3-pinanyldiborane.

This improvement in enantiomeric purity of α-pinene is actually due to the equilibration and selective crystallization of the chiral tetra-3-pinanyldiborane diastereomer from a mixture containing the latter in admixture with the meso diastereomer.[4] In other words, the observed enantiomeric enrichment is an instance of the duplication process of Horeau (calculation of p' by means of Eq. 7.13 gives a value of 99.6%, precisely the value found by Brown and Joshi).[263]

The amplification of small enantiomer excesses by the foregoing mechanism is one possible route for the prebiotic generation of nonracemic compounds in nature. Briaucourt and Horeau demonstrated experimentally that starting from reactants (an alcohol and an acyl chloride) having an enantiomer purity not exceeding 0.1% each, successive incomplete esterifications generated residual nonracemic products whose enantiomeric purity reached 98%.[267]

7-5. KINETIC RESOLUTION

A kinetic resolution is a chemical reaction of a racemate in which one of the enantiomers forms a product more rapidly than the other. The rate difference arises from a difference in E_a required to reach the transition states (for the respective enantiomers of the substrate) of the reaction (Eq. 7.14). Recovery of the unreacted

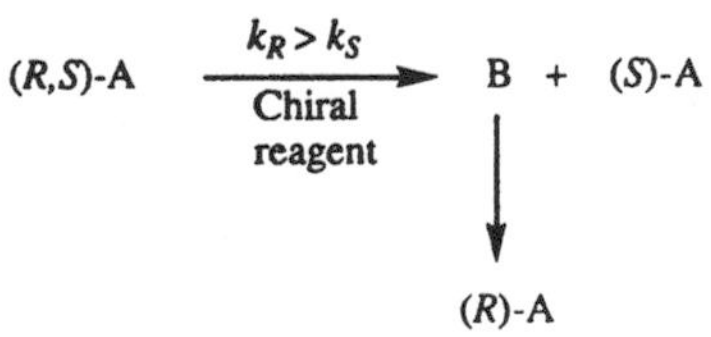

$$(R,S)\text{-}A \xrightarrow[\substack{\text{Chiral} \\ \text{reagent}}]{k_R > k_S} B + (S)\text{-}A$$
$$\downarrow$$
$$(R)\text{-}A \qquad\qquad (7.14)$$

enantiomer, (S)-A, in nonracemic form constitutes a resolution. The other enantiomer (R)-A may often be recovered as well (from product B) in nonracemic form by reversing the original enantioselective reaction in a nonselective manner (Eq. 7.14; k_R and k_S are rate constants).

The following conditions apply to kinetic resolutions:

1. If A and the chiral reagent are in stoichiometric ratio and sufficient time is allowed, both enantiomers of A are converted to product B and no resolution is effected. In order to be of practical use, it is essential that the reaction be stopped at some point short of 100% conversion. Either adjustment of the

reaction stoichiometry or of the reaction time (together with monitoring of the enantiomeric composition of the unreacted resolution substrate) may be used to control the extent of conversion.

2. Product B may be either chiral or achiral. If B is achiral, the reaction still constitutes a kinetic resolution but then only one enantiomer of the original racemate can be recovered in the resolution.

3. A chiral and nonracemic reagent, a chiral solvent, or a chiral physical force must be present in order for a kinetic resolution to take place, but the reagent need not be present in stoichiometric amount; that is, it can be a chiral catalyst.

Extensive and general reviews of kinetic resolutions are available.[27,262,268,269]

a. Theory and Stoichiometric and Abiotic Catalytic Kinetic Resolution

In this section, we describe the mathematical relationships that govern kinetic resolutions and then examine examples of stoichiometric and catalytic kinetic resolutions. Enzymatic resolutions, a subset of the latter type, are described in Section 7-5.b.

The efficiency of kinetic resolutions depends on the conversion (C) $(0 < C < 1)$ and the rate constants of the two competing reactions, k_R and k_S. More precisely, it is the relative rate of reaction of the two enantiomers $(k_R/k_S = s$, the stereoselectivity factor) that is governing. The efficiency is measured by the enantiomer excess of the reaction substrate [unreacted starting material A (Eq. 7.14)] following the resolution. The fundamental relationship between these three variables is given by Eq. 7.15 (for a derivation, see Kagan and Fiaud).[262]

$$s = \frac{\ln[(1 - C)(1 - ee)]}{\ln[(1 - C)(1 + ee)]} \tag{7.15}$$

This equation is valid when the competing reactions giving rise to B and unreacted A (Eq. 7.14) are first order with respect to A_R and A_S and regardless of the order in the reagent (chiral or achiral but catalyzed by a chiral catalyst). A related equation (Eq. 7.16) gives the enantiomer composition as a function of the rate constants and of time (t):

$$[S] - [R] = 0.5(e^{-k_s t} - e^{-k_R t}) \tag{7.16}$$

Such an equation had been derived and experimentally verified as early as 1958 by Newman, Rutkin, and Mislow.[270]

The evolution of enantiomerically enriched product resulting from a kinetic resolution is most clearly appreciated from plots of data calculated by means of Eq. 7.15 (Fig. 7.51). Since Eq. 7.15 is applicable to all types of reactions (see, however, Section 7-5.b), kinetic resolution appears to be a practical and general route to the optical activation of chiral compounds; this would seem to be true especially for

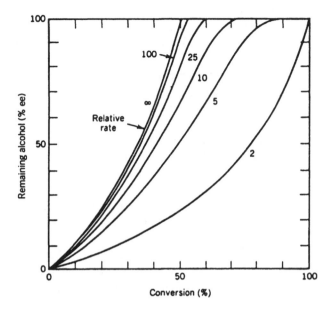

Figure 7.51. Enantiomeric purity of unreacted resolution substrate in kinetic resolutions. Effect of stereoselectivity factor, s, and of conversion (C) on the efficiency of resolution. [Adapted with permission from Martin, V. S., Woodward, S. S., Katsuki, T., Yamada, Y., Ikeda, M., and Sharpless, K. B. *J. Am. Chem. Soc.* **1981**, *103*, 6237. Copyright © 1981 American Chemical Society.]

reactions having a stereoselectivity factor **s** > 10. Stated in words, in order to obtain unreacted resolution substrate having 99% ee, a kinetic resolution having a relative rate ratio of 10 would have to be taken to 72.1% conversion; that is, the yield of the unreacted substrate would be 27.9% (the latter value must, of course, be compared to the maximum yield of one enantiomer obtainable in any resolution, i.e., 50%).

Equation 7.17 relates C and **s** to the enantiomeric purity, ee′, of the product B (Eq. 7.14) of a kinetic resolution when B is chiral and the fast reacting enantiomer (R)-A gives rise to product (R')-B:

$$\mathbf{s} = \frac{\ln[1 - C(1 + ee')]}{\ln[1 - C(1 - ee')]} \tag{7.17}$$

Combination of Eqs. 7.15 and 7.17 gives Eq. 7.21.

$$\frac{ee}{ee'} = \frac{C}{1 - C} \tag{7.18}$$

The latter equation illustrates the fact that the enantiomeric purities of the unreacted substrate and chiral product of a kinetic resolution are necessarily related and independent of the stereoselectivity factor. As the enantiomeric purity of the starting material goes up, so must that of the product go down. From Eq. 7.18 it also follows that it is impossible to maximize both the enantiomeric purity of the unreacted substrate *and* its yield.[262] These facts are also appreciated best in a graphical presentation (Fig. 7.52).

In a kinetic resolution, the enantiomer composition of chiral products is governed by s even when the enantiomers of the resolution substrate are in equilibrium with one another (in accord with the Curtin–Hammett principle; cf. Chapter 9).

In kinetic resolutions that are second order with respect to reactant (*S*)- or (*R*)-A, Eq. 7.14 no longer applies (as is true in stereo*elective* polymerizations).[271] In the second-order cases, the unreacted starting material does not generally attain as high an enantiomeric excess for comparable conversions as is true of the more common first-order (or pseudo-first-order) cases.[262]

Kinetic resolutions have been known for nearly a century[272] yet, except in the area of enzymatic resolutions (cf. Section 7-5.b), they had not been considered really useful optical activation methods.[67] The reasons for this are twofold: The reactivity ratios of most of the cases described were small (s < 10) and the underlying theory of the method was not widely known. Kinetic resolution also suffered from the same factors limiting the early development of efficient stereoselective synthesis, for example, the

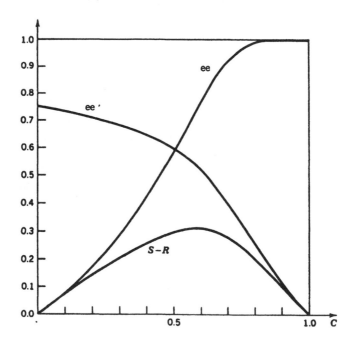

Figure 7.52. Relationship between the extent of reaction (conversion, *C*) and the enantiomeric enrichment of unreacted starting material (ee) and of chiral product (ee′) in a kinetic resolution. The *S*-*R* curve is the overall enrichment of the *S* configuration as a function of conversion. Computed curve (s = 7) for a kinetic resolution in which the reactants exhibit pseudo-first-order kinetics. [Reprinted with permission from Kagan, H. B. and Fiaud, J. C. *Top. Stereochem.* **1988**, *18*, 249. Copyright © 1988 John Wiley & Sons, Inc.]

inability until the 1960s of accurately measuring enantiomer compositions by nonchiroptical methods. A significant discovery by Sharpless and co-workers in 1981, that of the highly efficient enantioselective epoxidation of allylic alcohols coupled to the efficient kinetic resolution of the unreacted starting material, also helped to change the perception that kinetic resolution was not a very practical optical activation route (Fig. 7.53).[273] The unreacted (R)-**139** isolated ($C_{estimated} = 0.55$; $s_{experim.} = 104$) has ee > 96%. The epoxide products of the reaction (erythro/threo ratio 97:3) are also optically active (>96% ee for **140**, when $C = 0.52$). The epoxide products have been shown to be convertible to additional (R)-**139** by reaction of the methanesulfonate ester with telluride ion (Te^{2-}).[274]

An analogous process begins with an enantioselective synthesis whose substrate is *achiral* but endowed with prostereoisomeric groups or faces. In the presence of a nonracemic catalyst, one or more stereoisomers; of the product begin to accumulate (rate constants = $k_1, k_2, k_3, \ldots$). Since this product is partially racemic, reaction of the *product* (rate constants = $\beta_1, \beta_2, \beta_3, \ldots$) under the influence of the same catalyst constitutes a kinetic resolution. Such a system of consecutive reactions, one enantioselective synthesis and the other a kinetic resolution, has been analyzed in detail by Schreiber et al.[275]; see also Kagan and Fiaud.[262]

The achiral divinylcarbinol (Fig. 7.54), **141**, for example, is epoxidized with Sharpless reagent (Fig. 7.53) to give two of four possible monoepoxidation products, **142** and **143**; the major product **142** is formed as a result of a combination of selectivity for one enantiotopic group (*pro-S* vinyl group) and one diastereotopic (Re) face ($k_1 > k_3$).

Products **142** and **143** are further oxidized under the influence of the already present Sharpless catalyst. If the epoxy groups have little influence on the rates of addition to the remaining alkenyl groups, then it follows that **143** is oxidized faster to a diepoxide than **142** ($\beta_3 > \beta_1$). The ratio **142:143**, hence the enantiomeric purity of the monoepoxide, increases as the reaction proceeds and it can become as large as desired (with a correspondingly large conversion). For the case in which an excess of

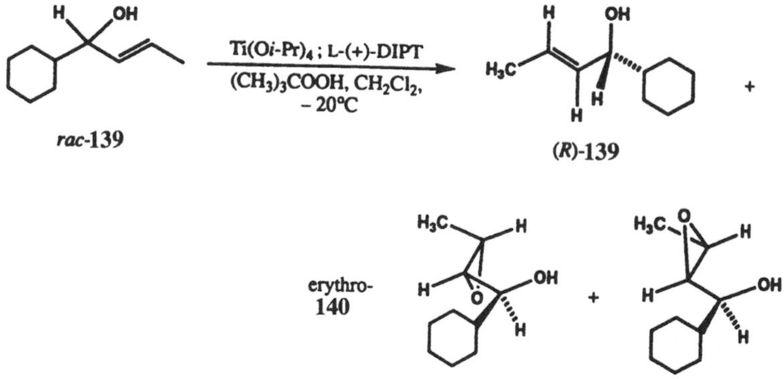

Figure 7.53. Kinetic resolution of a racemic allylic alcohol. Epoxidation of racemic E-**139** with titanium tartrate catalyst prepared from diisopropyl L-tartrate (DIPT).

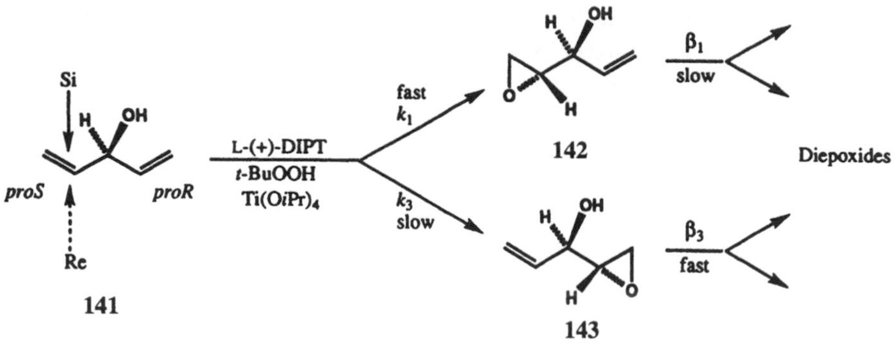

Figure 7.54. Epoxidation of an achiral divinylcarbinol with Sharpless L-(+)-DIPT reagent. Consecutive enantioselective synthesis and kinetic resolution. The minor diastereomeric monoepoxide products formed and the diepoxides subsequently formed from them are not shown.[275]

reagent (*tert*-butyl hydroperoxide) is present, an expression relating the enantiomer ratio to the rate constants at a given conversion was developed

$$\frac{X_1}{X_3} = \left[\frac{\delta_1(\delta_3 + \delta_4)}{\delta_3(\delta_1 + \delta_2)}\right] \left[\frac{s^{-(\delta_1+\delta_2)} - 1}{s^{-(\delta_3+\delta_4)} - 1}\right] \tag{7.19}$$

where $\delta_i = k_i/\Sigma\, k_i$ (fractional rate constant) and $s = [S]/[S]_{init}$ (fractional substrate concentration). Reaction of **141** with 4.8 equivalents of *tert*-butyl hydroperoxide in the presence of Sharpless catalyst derived from (2*R*,3*R*)-(+)-DIPT (Fig. 7.53) for 24 h (−25°C) afforded **142** with 93% ee over **143** and a diastereomer ratio of 99.9:0.1.[275]

Kinetic resolution may evidently be applied to the further enantiomeric enrichment of nonracemic samples. Two stratagems for doing this have been described in Section 7-4. Kinetic resolution has also been applied to the determination of configuration (Section 5-5.g, Horeau's method) and to the determination of enantiomeric purity (ee) (Section 6-5.e); for an analysis of kinetic resolutions occurring during polymerizations, see Kagan and Fiaud.[262]

Kinetic resolutions of preparative interest are of two types: stoichiometric and catalytic (including enzymatic). Examples of both types, excluding cases leading to low enantiomeric enrichments, follow.[262]

Stoichiometric Reactions. The simplicity of kinetic resolutions as a general approach to optical activation is illustrated by application to alcohols, such as α-methylbenzyl alcohol, by way of esterification with a deficiency of a nonracemic acid. Dicyclohexylcarbodiimide-mediated formation of the ester with one-half equivalent of (*R*)-2,4-dichlorophenoxypropanoic acid yields the corresponding ester

(75:25 diastereomer ratio) and unreacted (*S*) alcohol having 43% ee.[276] For this reaction, one calculates a stereoselectivity factor **s** = 3.6 (Eq. 7.15).

A useful kinetic resolution of secondary alcohols consists of reaction with a deficiency of L-valine, $(CH_3)_2CHCH_2CH(NH_3^+)CO_2^-$, in the presence of *p*-toluenesulfonic acid. The products, tosylate salts of (+) alkyl esters of L-valine (obtained in 60–80% yields with respect to one alcohol enantiomer), were recrystallized and hydrolyzed to afford enantiomerically homogeneous alcohols.[277,278]

In a method patterned after Horeau's method (Section 5-5.g), kinetic resolution of alcohols by incomplete reaction with, for example, *O,O'*-dibenzoyltartaric acid (Fig. 7.11, **50**), affords alcohols of ee up to 48%.[279] The reciprocal process, the kinetic resolution of acids (e.g., Fig. 7.13*a*, **72**), by reaction of its racemic anhydrides (without prior separation of the meso and threo diastereomers) with alcohols, for example, with nonracemic 1-(4-pyridyl)ethanol has also been observed.[280,281]

Glycols, such as 3-chloro-1,2-propanediol **144** (Fig. 7.55) may be resolved by reaction with D-camphorquinone **145**. Four monoketals are formed under kinetic control (excess of diol) in the proportion 27:45:17:12. Crystallization of the dominant isomer and recovery of the diol from the latter affords enantiomerically pure (*R*)-(–)-**144**. The latter is a precursor of nonracemic epichlorohydrin.[282]

The kinetic resolution of amines is occasionally practiced, for example, by partial reaction with D-10-camphorsulfonyl chloride[283] and by reaction of 1° amines with (*S*)-2-phenylbutyric anhydride.[284]

Kinetic resolution of ketones (e.g. **146**, Fig. 7.55) by reaction with an insufficient amount of nonracemic chiral primary amine, such as dehydroabietylamine (Fig. 7.9*a*, **22**), has been described; the product is a Schiff's base. Hydrolysis of the Schiff's base product following distillation of the optically active unreacted ketone gave the enantiomeric ketone.[285] Ketones, such as 2-methylcyclohexanone, (±)-**154** (Eq. 7.20), can be resolved via selective cleavage of their diastereomeric acetals [formed by acid-catalyzed reaction with (2*R*,4*R*)-(–)-2,4-pentanediol] by triisobutylaluminum (Eq. 7.20); the kinetically controlled step (a diastereoselective elimination) is the acetal cleavage to an enol ether. The (2*R*)-(–)-ketone **154** is obtained at –20°C (5 h; >95% ee); at 0°C (3 h), (+)-**154** is recovered from unreacted acetal.[286]

(7.20)

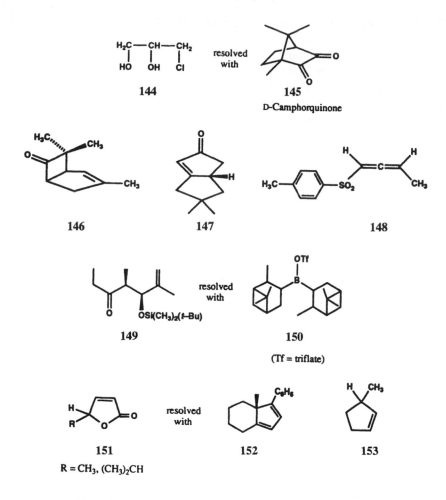

Figure 7.55. Compounds resolved by stoichiometric kinetic resolution.

Kinetic resolution of α,β-unsaturated ketone **147** (Fig. 7.55) takes place on addition of a nonracemic sulfoxide enolate.[287] Allenic sulfones (e.g., Fig. 7.55, **148**), having a double bond highly activated toward nucleophilic addition, may be efficiently resolved by partial reaction with chiral amines, such as **33** (Fig. 7.9b).[288]

The aldol condensation has been adapted to kinetic resolutions with foreknowledge of the degree of diastereofacial selectivity of a given system. *rac*-Ketone **149** is converted to its enolate by reaction with nonracemic di-(3-pinanyl)borane triflate (Fig. 7.55, **150**). Ketone (+)-**149** is isolated (ee > 95% with C ca. 75%) following partial diastereoselective reaction of the enolate mixture with methacrolein. Kinetic resolution also attends the enolization but with a lower efficiency.[289]

(±)-α,β-Unsaturated lactones, for example, **151** (Fig. 7.55) can be kinetically resolved when they react as dienophiles in enantioselective Diels–Alder reactions with chiral nonracemic diene **152**. Both the unreacted lactone and its enantiomer isolated by thermal retro-Diels–Alder reaction were found to be enantiomerically enriched.[290]

Among the most useful applications of stoichiometric kinetic resolutions are the preparation of nonracemic alkenes and allenes whose resolution by other means is often problematic. Hydroboration of racemic alkenes with a deficiency of tetra-3-pinanyldiborane (derived from α-pinene) leaves unreacted alkenes, such as **153** (Fig. 7.55), in an enantiomerically enriched state (up to 65% ee).[291] Allenes may be similarly resolved.[264] Racemic boranes may themselves be kinetically resolved by reaction with a deficiency of a nonracemic Lewis base.[292]

Kinetic resolutions have not always been monitored by examination of unreacted resolution substrate (see above), nor have nonracemic samples of the latter always been the objectives of such resolutions. Mixtures of covalent diastereomers obtained from racemates under kinetically controlled conditions have in many cases been directly converted to products of interest rather than to enantiomerically enriched starting materials. Although not strictly meeting the definition of resolution, the processes nevertheless are often considered to be examples of kinetic resolutions, for example, application of the sulfoximine reagent to kinetic resolution[293] and application of the ene reaction of N-sulfinylcarbamates.[294]

Catalytic Reactions.   Organic and organometallic catalysts have increasingly been studied as reagents for effecting kinetic resolutions;[295,296] preparative procedures have lagged somewhat behind the mechanistic studies. A more stringent limitation of some preparative catalytic kinetic resolutions relative to the stoichiometric ones is that separation of unreacted starting materials from products may be more difficult to carry out, for example, unsaturated alcohol starting materials from saturated alcohol products. Nevertheless, observations of selectivities s of the order of 10^1–10^2 in some kinetic resolutions catalyzed by "abiotic" catalysts (with recovered starting material >99% ee) make these types of kinetic resolutions very attractive.

The principal application of nonracemic catalysts to kinetic resolution has been that of the Sharpless reagent (see above) applied to the resolution of allylic alcohols.[297,298] The process, as originally described in 1981, required the use of a stoichiometric amount of catalyst. However, in the presence of molecular sieves, the process becomes truly catalytic, for example with 5–10 mol% of the $Ti^{(IV)}$ reagent in the presence of (+)-dicyclododecyl tartrate (recovered alcohols: >98% ee, C = 0.52–0.66).[299] With bulky β substituents on the double bond, selectivities as high as **s** = 700 have been found.[300] β-Amino alcohols are susceptible to kinetic resolution with the Sharpless reagent, as the result of their enantioselective conversion to N-amine oxides[301] and so are 2-furylcarbinols.[302] Kinetic resolution of a chiral sulfide, the antioxidant **155** (Fig. 7.56), by enantioselective oxidation to the sulfoxide (catalyzed by a water-modified Sharpless reagent) has been described.[303] Other applications of organometallic catalysts to kinetic resolutions are summarized and illustrated in Table 7.4 and Figure 7.57.

TABLE 7.4 Examples of Abiotic Catalytic Kinetic Resolutions

Reaction Type	Examples	Catalyst[a]	s	ee(%)[b]	References
Homogeneous hydrogenation		[(Dipamp)Rh]$^+$	4.5	>90% (c = 0.70)	304
		Ru(BINAP)(OAc)$_2$	74	>99% (c = 0.55)	305
Isomerization		Rh(BINAP)(OCH$_3$)$_2$$^+$	5	?	306
Cyclization		Rh(CHIRAPHOS)$^+$Cl$^-$	?	?	307
Allylic alkylation		"Ferrocenylbiphosphine" + "π-allyl Pd" **156**	14	>99% (c=0.80)	308

[a] Catalyst structures are given in Figure 7.57.
[b] Recovered starting material.

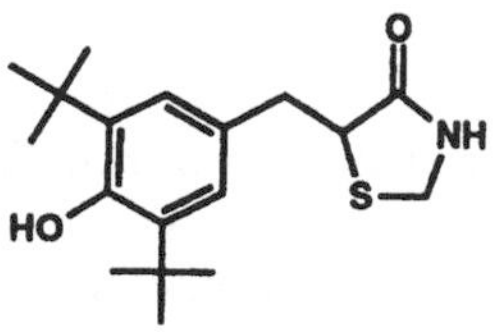

155

Figure 7.56. Structure 155.

A number of kinetic resolutions are catalyzed by alkaloids. Quinidine **18** (Fig. 7.9*a*), catalyzes the isomerization of *rac*-1-methylindene **157**, to 3-methylindene (Fig. 7.58); the former is recovered (by GC) with up to 73% ee ($C = 0.65$; $s \cong 5$).[309] Kinetic resolution also attends the Wittig rearrangement of the atropisomeric ether **158** triggered by butyllithium. (–)-Sparteine (Fig. 7.24, **93**), is the chiral inductor. Unreacted ether (Fig. 7.58, **158**), both product alcohols, and by-product pentahelicene all exhibit high rotations.[310]

Polymerization reactions may also be used to effect kinetically controlled resolutions.[262] Diethylzinc catalyzed polymerization of (±)-propylene oxide and of (±)-propylene sulfide (Fig. 7.58, **159**, Y = S), in the presence of (–)-3,3-dimethyl-1,2-butanediol, gives rise to optically active polymers (stereo*elective* polymerization).[271] The unreacted monomers are enantiomerically enriched.[311,312] Kinetic resolution of chiral lactone (Fig. 7.58, **160**) attends its polymerization to a nonracemic polyester in the presence of a chiral initiator.[313] *rac*-α-Methylbenzyl methacrylate (Fig. 7.58, **161**) undergoes anionic polymerization (to a mostly isotactic polymethacrylate) in the presence of achiral Grignard reagents modified by (–)-sparteine (Fig. 7.24, **93**). Unreacted **161** monomer exhibited up to 83% ee with cyclohexylmagnesium bromide-**93** initiator.[314]

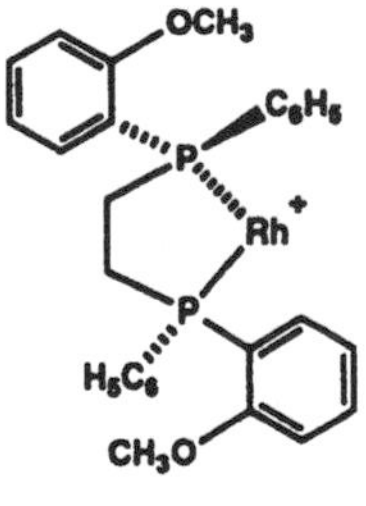

[(Dipamp)Rh]⁺ 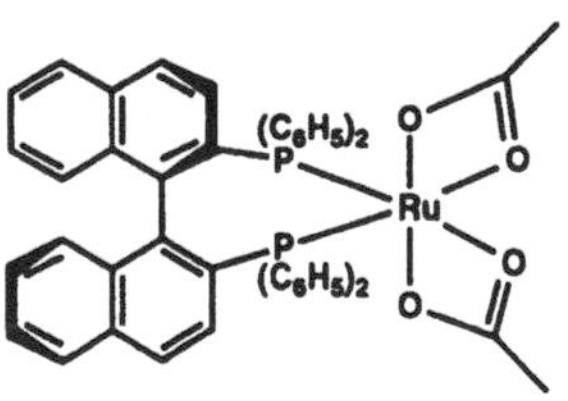   [Ru(BINAP)](OAc)₂

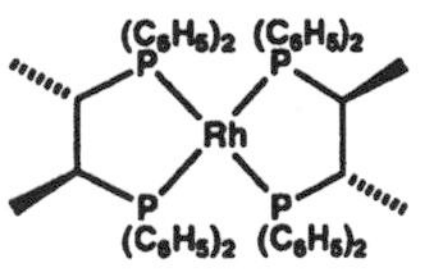

Rh[CHIRAPHOS]₂⁺Cl⁻

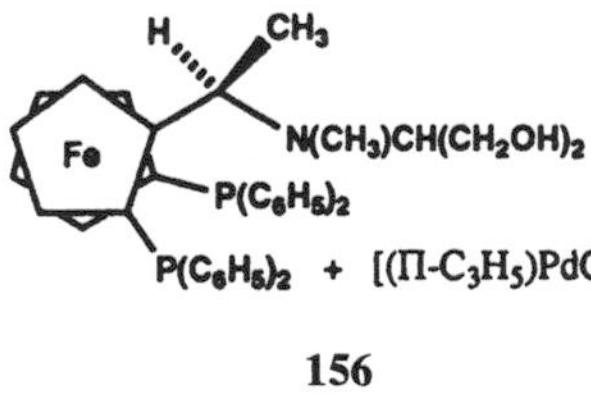

156

Figure 7.57. Organometallic catalysts and cocatalysts used in kinetic resolutions.

b. Enzymatic Resolution

Resolutions catalyzed by enzymes have been exploited to a larger degree than any other kind of kinetic resolution. The very first kinetic resolution discovered was of this type (Pasteur's third method in terms of chronology),[315] namely, the resolution of tartaric acid by fermenting yeast, an experiment that marks the introduction of stereochemistry into "physiological principles" (i.e., into biochemistry).[316] Consequently, there exists a larger body of data about enzymatic resolutions than about any other kind of kinetic resolution.

An enzymatically catalyzed reaction, such as an ester hydrolysis, may be described by a simplified mechanism (Eq. 7.21)

$$
E + R \xrightleftharpoons[k_{-1}]{k_1} E\text{-}R \xrightleftharpoons[k_{-2}.R'OH]{k_2} E\text{-}R^* \xrightleftharpoons[k_{-3}]{k_3.H_2O} E\text{-}P \xrightleftharpoons[k_{-4}]{k_4} E + P
$$

$$
E + S \xrightleftharpoons[k'_{-1}]{k'_1} E\text{-}S \xrightleftharpoons[k'_{-2}.R'OH]{k'_2} E\text{-}S^* \xrightleftharpoons[k'_{-3}]{k'_3.H_2O} E\text{-}Q \xrightleftharpoons[k'_{-4}]{k'_4} E + Q
$$

$$(7.21)$$

where E = enzyme, R and S represent the fast and slow reacting substrate enantiomers, respectively, and $E\text{-}R$ (and $E\text{-}S$) represent the enzyme–substrate complexes. Further reaction leads to the acyl–enzymes $E\text{-}R^*$ and $E\text{-}S^*$, to the enzyme–product complexes $E\text{-}P$ and $E\text{-}Q$, and finally to the products, P and Q, respectively, with liberation of the enzyme. Application of steady-state kinetics to the rate constants of Eq. 7.21 with the assumption that the steps leading to $E\text{-}R^*$ (and $E\text{-}S^*$) are essentially irreversible (and that products P and Q do not inhibit the reaction) leads to Eq. 7.22[44,317]:

$$
E = \frac{\ln[(1 - C)(1 - ee)]}{\ln[(1 - C)(1 + ee)]} \tag{7.22}
$$

where **E** is the biochemical stereoselectivity factor (corresponding to s). The parameter **E** may also be defined in terms of the rates of reaction of the competing

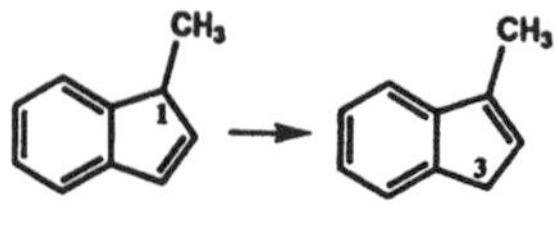

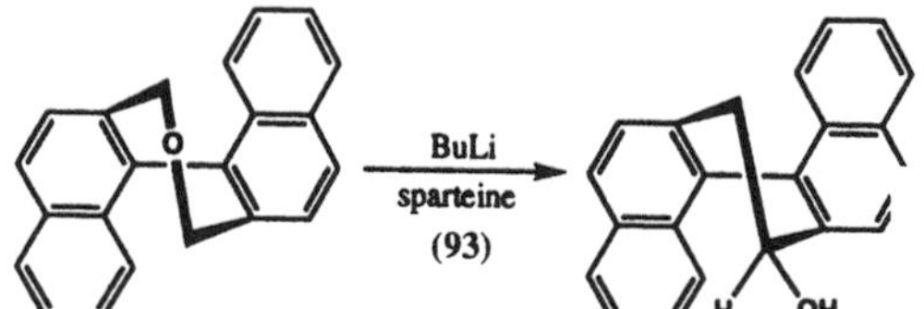

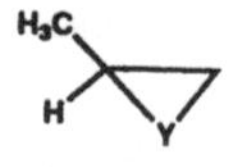

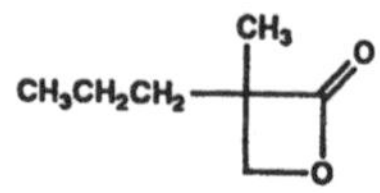

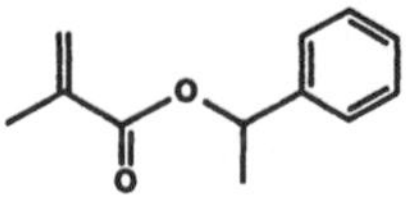

Figure 7.58. Compounds that undergo kinetic resolution with alkaloid and amino acid catalysts.

enantiomer substrates, k_2 and k'_2; for cases in which $k_{-1} \gg k_2$ and $k'_{-1} \gg k'_2$ (Eq. 7.21) the following relation holds:

$$\frac{\left(\dfrac{k_{cat}}{K_m}\right)_R}{\left(\dfrac{k_{cat}}{K_m}\right)_S} = e^{-\Delta\Delta G^{\ddagger}/RT} = \mathbf{E} \qquad (7.23)$$

where k_{cat} and K_m denote the turnover number and Michaelis constant, respectively. While Eqs. 7.15 and 7.22 are formally identical and equally predictive, the latter equation is applicable only to irreversible reactions (see below). Equation 7.22 is also dependent on the ratio of enzymatic specificity constants V/K, where V and K are the maximal velocities and Michaelis constants of the substrate enantiomers, and it is independent of substrate concentrations.[317] Comparison of enzyme efficiencies must be carried out at equal conversions; values of C are conveniently obtained from the enantiomer compositions of unreacted resolution substrates (ee) and of resolution products (ee′) by means of Eq. 7.18. However, under conditions such that the substrate can be racemized in situ during the kinetic resolution, that is, an asymmetric transformation takes place (cf. Section 7-2.d), the enantiomeric purity of the product becomes independent of the extent of conversion; moreover, the reaction (e.g., hydrolysis) is apparently more enantiospecific than when concomitant racemization does not take place.[318]

Enzymatic resolutions that take place in two-phase (aqueous + organic), systems are reversible, and hence are not governed by Eq. 7.22. For such systems, a new expression relating $\mathbf{E}$, C, and ee must be employed that incorporates K, the equilibrium constant for the reactions

$$E + R \rightleftharpoons E + P$$

$$E + S \rightleftharpoons E + Q \qquad (7.24)$$

where $K = k_2/k_1 = k_4/k_3$ and

$$\mathbf{E} = \frac{\ln[1 - (1 + K)(C + ee\{1 - C\})]}{\ln[1 - (1 + K)(C - ee\{1 - C\})]} \qquad (7.25)$$

Note that K, being a thermodynamic parameter unlike $\mathbf{E}$ (a kinetic parameter that varies from enzyme to enzyme), is independent of the nature of the enzyme.[44]

A value of $\mathbf{E} > 100$ is necessary if *both* enantiomers of a racemate are to be obtained with high enantiomeric purity in a single kinetic resolution. With lower $\mathbf{E}$ values, one enantiomer (the unreacted one) may be obtained with high enantiomeric enrichment (at high conversion); to obtain the other one with high enantiomeric purity requires recycling.[319]

It is surprising to find that in quite a few instances only modest $\mathbf{E}$ values (1–10) obtain when commercial enzyme preparations or intact microorganisms are used in kinetic resolutions in vitro. The reasons for this are not easy to establish; the enzyme may lack stereoselectivity or the preparation in use may contain several enzymes, some of which may have countervailing enantioselectivities. In order to raise the enantioselectivity, it is not uncommon to screen enzymes against a given substrate, to modify the substrate, to change reaction conditions, or to recycle the product, that is, to repeat the resolution on the product of an enzymatic resolution.[44]

Treatment with sodium deoxycholate and precipitation with ether and ethanol (1:1) leads to a significant increase in the enantioselectivity of lipase (of *Candida cylindracea*) catalyzed ester hydrolysis (e.g., on arylpropionate esters). This result has been ascribed to a noncovalent modification of the native enzyme protein, that is, to the generation of a more stable conformer.[320] Following a pragmatic and often empirical selection of enzyme, substrate, and reaction conditions, enantiomeric enrichments greater than 90% ee in enzymatic resolutions are not difficult to attain.

An alternative approach to raising the enantioselectivity that is likely to be increasingly useful in the future is the development of reliable active-site models that permit the interpretation and prediction of an enzyme's stereospecificity: That is, will a given enzyme catalyze a reaction on a specific substrate and if so what will be the configuration of the product? An example is the cubic-space section model encompassing the active site region of horse liver alcohol dehydrogenase (HLADH) for predicting the enzyme's specificity in oxidoreductions.[321] This model (inspired by the diamond lattice section model developed by Prelog),[322] is empirical in that it incorporates specific X-ray and kinetic data about the enzyme. The model has features very reminiscent of a chiroptical sector rule (Chapter 12); elaboration of the model by eventual application of computer graphics is evident.

A significant development in enzymatic catalysis that may also reduce some of the extant empiricism in enzymatic resolutions is the discovery that antibody proteins catalyze chemical reactions.[323,324] Since antibodies can be made toward many chemical species, this finding opens up the possibility of "enzymes," that is, catalytic *anti*bodies (so-called abzymes), being designed to catalyze specific chemical reactions, including resolutions. The possibility of designing catalytic antibodies with enzymatic activity may circumvent the need to adjust substrate structures and conditions to overcome the limitations of naturally occurring enzymes.[325,326]

A stable phosphonate ester **162** (Fig. 7.59) that mimics the transition state of a transesterification reaction has (after conjugation with a carrier protein) elicited the formation of a monoclonal antibody that catalyzes a kinetic resolution. Compound **162** acts as a hapten (Section 6-5.h). Synthesis of a single diastereomer of the hapten bearing a suitably placed stereogenic atom (albeit as the racemate) ensured that the antibody would behave in a stereoselective manner. Intramolecular transesterification of *rac*-**163** (Fig. 7.59) in the presence of antibody 24B11 spontaneously terminated at $C = 0.5$, suggesting that the reaction was indeed a kinetic resolution. Confirmation of this fact was obtained by NMR analysis of the product lactone **164** with a chiral shift reagent (ee = 94 ± 8%).[327] In this study, only antibodies preferring one enantiomer of the substrate **163** were found.[328] However, in a later study also using a racemic hapten, two classes of antibodies were found: some antibodies that

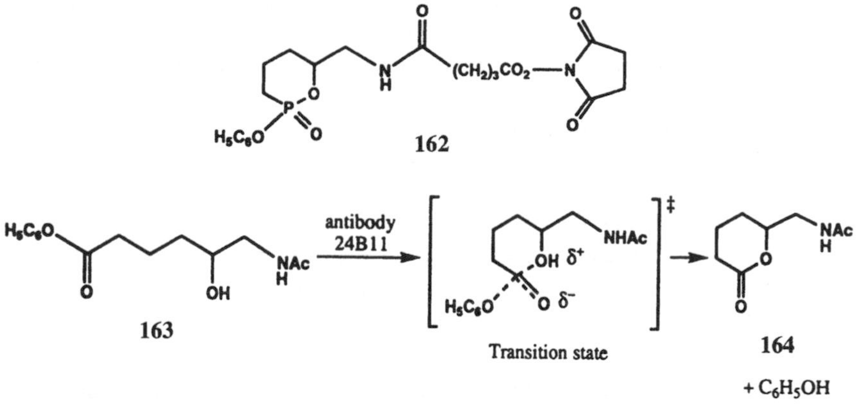

Figure 7.59. Kinetic resolution in a transesterification catalyzed by a monoclonal antibody. Compound **162** is a transition state mimic designed to elicit formation of the antibody.

catalyzed reaction of the *S* enantiomer of the substrate and others that catalyzed reaction of the *R* enantiomer.[329]

In general, enzymes are versatile, highly efficient, and very selective in their activity, that is, selective with respect to functionality, regioselectivity, and stereoselectivity. Two significant limitations in the use of enzymes are that (a) they are easily denatured and (b) they often require stoichiometric amounts of cofactors [coenzymes, such as adenosine triphosphate (ATP)]. These limitations have been addressed, respectively, by (a) immobilization (attachment of the enzyme to a support without affecting its active site); this tactic has been especially important in connection with enzymatic resolutions carried out on a commercial scale (Section 7-5); and (b) incorporation of cofactor regeneration schemes into the enzymatic resolution process.[330,331]

Of the six main classes of enzymes, the first three (oxido-reductases, transferases, and hydrolases) have been the most useful in kinetic resolutions. Both intact cells, for example, microbial preparations or yeast (crude yet effective mixtures containing enzymes) and cell-free enzyme "preparations" have been used in enzymatic resolutions. A significant increase in the application of enzymes to organic synthesis in recent years may be attributed in large measure to new purification methods that have significantly lowered the price of commercial enzyme preparations. In light of the very large number of enzymatic resolutions in the literature and the availability of excellent reviews,[44,262,330,332–337] we limit ourselves to some examples that illustrate significant features of enzymatic resolutions.

Numerous carboxylic acids (including amino acids,[338] e.g., arylglycines),[339] alcohols, and amines have been enzymatically resolved by hydrolysis catalyzed by hydrolases (acylases). These were among the first enzymatic resolutions to be exploited. Substrates for such reactions include esters, amides, carbamates, and hydantoins. An example follows.

On a 2-g scale, hydrolysis of *rac-N*-acetylphenylalanine methyl ester (Fig. 7.60, **165**) in the presence of a commercial serine proteinase (Subilopeptidase A; Alcalase™) is complete in 45 min with the *R* ester being quantitatively recovered by extraction into CH_2Cl_2. The reaction comes to a near standstill at $C \cong 50\%$ as would be expected for a resolution occurring with a high value of **E**. The *S* acid is isolated from the reaction mixture following acidification (96%; 98% ee after one recrystallization).[340]

Microbial resolution of Mosher's acid (Fig. 7.11*a*, **56**), widely used as a reagent for the determination of enantiomer composition (Section 6-5), has been effected by hydrolysis of cyanohydrin acetate (Fig. 7.60, **166**) with dry cells of *B. coagulans*. The recovered (*R*)-**166** (100% ee at $C = 0.7$) was converted to **56** in three steps without loss of enantiomeric integrity.[341]

The discovery that lipases catalyze reactions very well in nearly anhydrous organic solvents has been a stimulus to the application of such enzymes to esterification and transesterification reactions. These are reactions that normally are suppressed when carried out in water as the result of competing hydrolysis.[342–344] Esterification of 2-bromopropionic acid with 1-butanol in hexane in the presence of yeast lipase (the latter is insoluble in organic media), for example, afforded optically active butyl (*R*)-(+)-2-bromopropionate (96% ee at $C = 0.45$) and unreacted (*S*)-(–) acid (99.6% ee at $C = 0.78$). The catalyst could be repeatedly used in resolutions with little loss in activity.[345]

Menthol (Fig. 7.9*c*, **41**) may be enzymatically resolved by ester hydrolysis, by ester formation, or by ester interchange in the presence of lipase from *C. cylindracea*. These reactions are carried out in water (hydrolysis) or in heptane (ester formation or interchange) as appropriate.[341] The aggregation pheromone of ambrosia beetle (a pest that attacks North American timber) is alcohol **167** (Sulcatol, Fig. 7.60). The corresponding racemate has been resolved, for example, with pig pancreatic lipase (PPL) by transesterificaton of laurate from trifluoroethyl laurate in diethyl ether (**E** = 100).[347,348]

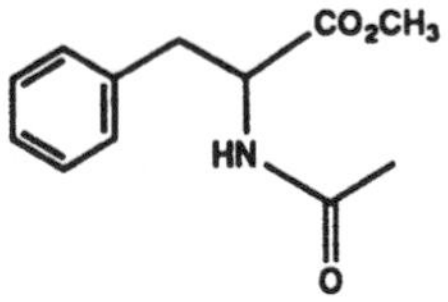

165

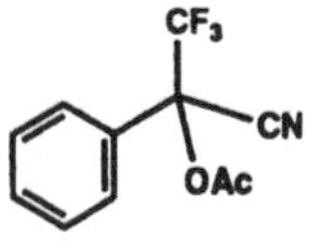

166

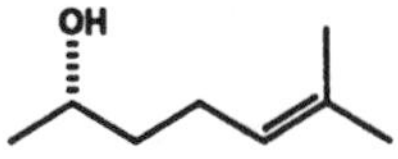

167
(*S*)(+)-Sulcatol

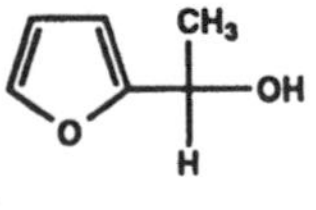

168

2-(2-Furyl)ethanol

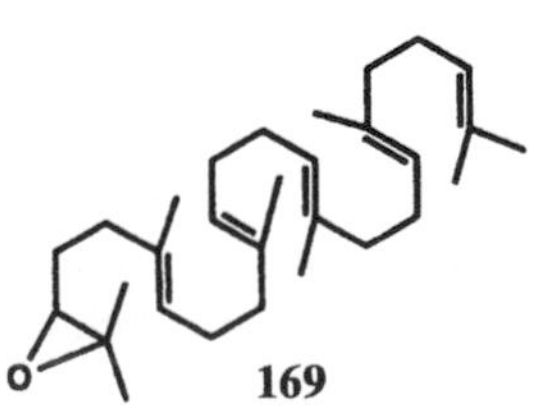

169

Squalene 2,3-epoxide

Figure 7.60. Enzymatic resolution substrates.

In enzymatic esterification or transesterification a given enzyme catalyzes both the forward and the reverse reaction. Consider the hypothetical reaction described in Eq. 7.26 in which $k_R > k_S$. The R acetate accumulates in the reaction along with unreacted

$$R,S \qquad\qquad R \qquad\qquad S \qquad\qquad (7.26)$$
$$+ \ CH_3OH$$

S alcohol (together with minor amounts of the enantiomeric ester and alcohol). As a consequence of microscopic reversibility, it is the major acetate R enantiomer that reacts faster with methanol in the reverse reaction, because [R acetate] > [S acetate] together with the rates k_R (acetate) > k_S (acetate). Consequently, in such reversible reactions, as the conversion increases, the enantiomeric purity of both products gradually decreases. This difficulty has been circumvented, as first described by Degueil-Castaing et al.,[349] by causing the reaction to become irreversible by application of enol esters, for example, vinyl acetate (the by-product is acetaldehyde), as esterification (or transesterification) agents.[350,351]

A two-step enantioselective synthesis of chiral cyanohydrin acetates RCH(OAc)CN, e.g., R = p-CH$_3$C$_6$H$_4$, is effected by (a) base-catalyzed formation of rac-cyanohydrin from aldehydes in the presence of acetone cyanohydrin, (CH$_3$)$_2$C(OH)CN, and (b) enantioselective and irreversible acetylation of the unstable racemic cyanohydrin by isopropenyl acetate, CH$_2$=C(CH$_3$)OAc, in diisopropyl ether in the presence of *P. cepacia*. The yield of a single enantiomer (63–100%; ee up to 94%) exceeds 50% in view of the continuous racemization of the slower reacting cyanohydrin enantiomer that equilibrates with the precursor aldehyde. Thus the entire process incorporates the elements of a kinetic resolution coupled to an asymmetric transformation.[352]

Enzymatic resolution of 2,2′-dihydroxy-1,1′-binaphthyl (binol) by hydrolysis of diesters was effected with several strains of soil microorganisms. With *Bacillus* sp. LP-75, the hydrolysis rate, the efficiency (ee), and the sense of enantioselectivity depended strongly on the nature of the R group of the acid. The diacetate furnished the S diol (50% ee) directly while the (±) dibutyrate preferentially hydrolyzed to the R diol (97% ee).[353] In other cases, enzymatic resolution of diesters furnish monoesters.[354] Subsequently, a much faster resolution of binol (hydrolysis of the divalerate ester) has been described that is catalyzed by an enzyme (PPL) preparation affording (S)-diol (95% ee) at 46% conversion.[355]

The enzymatic resolution of simple 2° alcohols [e.g., C$_6$H$_5$CH(CH$_3$)OH] by lipase catalyzed hydrolysis of the acetate from *Pseudomonas* sp. affords both enantiomers of the alcohol (after hydrolysis of the unreacted one) in nearly quantitative yields (ee >99% for each enantiomer; **E** > 1000) on a molar scale. The high yield, simplicity, and efficiency of this type of optical activation process may lead it to displace resolution methods mediated by covalent diastereomers as well as enantioselective ketone reductions.[351]

7-6. MISCELLANEOUS SEPARATION METHODS

Prelog et al. found that chiral α-amino alcohols (e.g., *rac-erythro*-norephedrine, Fig. 7.61, **170**) can be resolved by partition of their salts with lipophilic anions (e.g., PF_6^-) between an aqueous phase (aq) and an immiscible lipophilic phase (lp) (1,2-dichloroethane) containing an ester of tartaric acid, for example, (−)-di-5-nonyl-(*S,S*)-tartrate.[322] The selectivity rose with decreasing temperature, reaching an enantiomer excess ratio $ee_{lp}/ee_{aq} = 7.1$ [23.3 (1*R*) − 3.3 (1*S*)] at 4°C in the two phases. No selectivity was exhibited by α-methylbenzylamine salts, thus demonstrating again that at least two hydrogen bonds between the substrate and the chiral extractant molecules are necessary for differentiation (three such bonds according to Prelog). This partition experiment was translated into a preparative chromatographic resolution.

Donald J. Cram and co-workers at UCLA have been particularly successful in the design of synthetic chiral "solvents" to effect extraction with high selectivity. In their experiments, the racemic solute and solvent (a solution of an optically active compound diluted by, e.g., $CHCl_3$) are often called guest and host, respectively, and the interaction is so specific and so tight that the diastereomeric complexes thus formed may even be isolated (Section 7-3.c).

Extraction of aqueous *rac*-α-methylbenzylammonium hexafluorophosphate $C_6H_5CH(CH_3)NH_3^+PF_6^-$, with a solution of (−)-crown ether **172** (Fig. 7.62) dissolved in $CHCl_3$, leads to recovery of (+) amine having ee = 24%.[356] Subsequently, partial resolution by extraction (one equilibrium = one plate distribution) was achieved with phenylglycine methyl ester (Fig. 7.61, **171a**) and other amino acid derivatives using host compound **173** (Fig. 7.62). The selectivity was determined by measuring enantiomer distribution constants, EDC = D_A/D_B, where D_A is the distribution constant for $CDCl_3$ versus D_2O of the enantiomer more tightly bound to the chiral crown ether, while D_B is the corresponding coefficient for the less tightly complexed enantiomer. The EDC values are directly related to the diastereomer discrimination energy: $\Delta(\Delta G°) = -RT \ln(EDC)$.[357]

It was found[358] that the highest selectivity was achieved with compounds **171a** and **171b** (Fig. 7.61) for which EDC = 12 and 18, respectively, using host **173** (Fig. 7.62). By modifying the nature and the amount of inorganic salt in the water layer ($LiPF_6$ or $LiClO_4$) and of CH_3CN in the organic layer, the amount of guest transferred and the

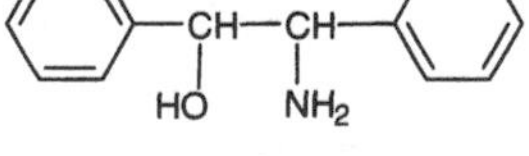

171 a R = H
171 b R = OH $X^- = ClO_4^-, PF_6^-$

170

erythro-Norephedrine   **171**

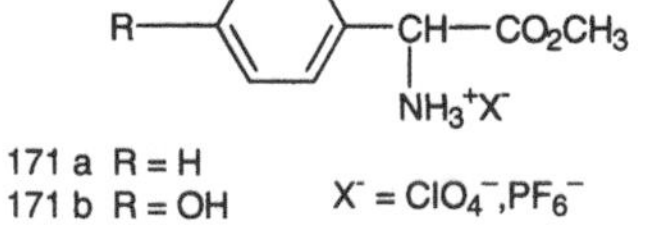

Figure 7.61. Compounds resolved by partition or transport.

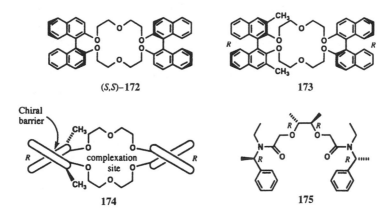

Figure 7.62. Ion-selective ligands (ionophores). [Structure **173** is reprinted with permission from Cram, D. J. and Cram, J. M. *Acc. Chem. Res.* **1978**, *11*, 8. Copyright © 1978 American Chemical Society.]

selectivity could be adjusted. Amino acids, such as phenylglycine, $C_6H_5CH(NH_3^+)CO_2H$ ClO_4^- could be extracted with EDC = 52 leading to the possibility of obtaining guest with an enantiomer purity of 96% ee through just one equilibration.[152]

The fact that compounds such as **172** and **173** (Fig. 7.62) exhibit enantiomer selectivity toward α-methylbenzylamine salts requires that the nature of the hydrogen bonding in these cases differ from that which obtains with the tartrate ester (see above). This selectivity is understandable in light of the rigid shape of the chiral host molecule and the necessity of the substrate to fit *inside* the host (ligand) cavity wherein the hydrogen bonding is manifest. As cogently stated by Cram and Cram,[357] in macrocyclic chiral ligands the binding sites *converge* while those of the substrate molecule (three hydrogen bonds involving $-NH_3^+$) *diverge*. The resultant complementary stereoelectronic arrangement of binding sites and steric fit may account for the greater selectivity of these ligands relative to that possible with open ligands, such as "simple" tartrate esters, although the latter are capable of a greater variety of hydrogen-bonding modes with multifunctional substrates.

The model of guest–host complexation that emerges is pictured in Figure 7.62, **174**. This general model has been elaborated on the basis of ^{1}H NMR evidence as well as from X-ray spectra of both diastereomeric complexes of a given guest–host pair.[359,360] Hosts, such as **172** (called dilocular since they have two stereogenic units), have homotopic faces [C_2 or higher (e.g., D_2) symmetry] such that complexation may take place with equal ease from either face to provide identical diastereomic complexes. Substituent groups, such as CH_3, on the naphthalene rings in **173** (Fig. 7.62) extend the chiral barrier and are responsible for the dramatically higher diastereomer discrimination, for example, to EDC = 52, where $\Delta\Delta G° = -2.15$ kcal mol^{-1} (9.0 kJ mol^{-1}).

Interest in the transport of amino acids, and of other substances, across membranes in biological systems has stimulated experiments in which synthetic lipophilic substances of the type described in Figure 7.62, and others, might serve as carriers, that is, ferry ions across an otherwise repelling barrier.[361] The idea was to create synthetic membranes consisting of water immiscible solvents into which one of the lipophilic chiral compounds might be dissolved. Since these "membranes" are chiral and nonracemic, it was hoped that enantioselective transport might be observed as a consequence of the operation of diastereomer discrimination within the membrane. The results of partition experiments (see above) provided guidance in the design of several types of transport experiments that are summarized schematically in Figure 7.63. Bulk toluene and chloroform liquid phases served as representative membranes. One of the goals of such transport experiments was the development of practical resolution methods.

One of the first reports of enantiomer discrimination in transport is that of Newcomb, Helgeson, and Crarn in 1974.[362] The driving force for transport in this experiment is provided by entropy of dilution and "salting out" of the chiral organic salt from its reservoir by an achiral inorganic salt (Fig. 7.63a, U-tube). The enantiomers of methyl phenylglycinate·HPF$_6$ (Fig. 7.61, **171a**) were shown to migrate (from the α aqueous reservoir toward the β aqueous reservoir) across a chloroform layer containing R,R ligand **175** (Fig. 7.63) (the carrier C) at differing rates (rate ratio $\cong 10$). The faster moving (R)-**171a** (obtained with ee = 78%) is that shown to form a tighter complex with R,R-ligand **175** in one-plate partition experiments. Similar experiments carried out in W-shaped tubes permit simultaneous and competitive

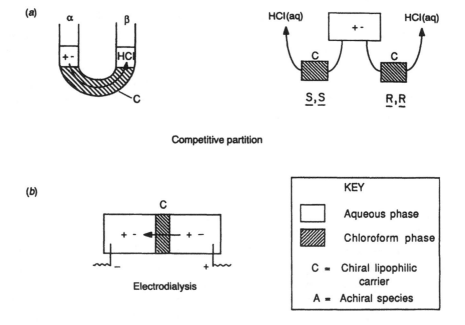

Competitive partition

Electrodialysis

Figure 7.63. Types of enantioselective transport.

transport of both enantiomers of the same glycine ester toward separate reservoirs thereby accomplishing a virtually complete resolution of the racemate. Newcomb, Cram, and co-workers dubbed such a system a *resolving machine*.[363]

In a second type of transport experiment, electrodialysis provides the driving force for migration across the thin (0.2 mm) liquid membrane (Fig. 7.63*b*). The membrane consists of the ionophore **175** (Fig. 7.62) dissolved in *o*-nitrophenyl octyl ether containing PVC. Migration of α-methylbenzylammonium [MBA, $C_6H_5CH(CH_3)NH_3^+$] ions ($10^{-3} M$) under a potential difference affords a selectivity of 8% for the two enantiomers.[364,365] The results were subjected to a chiral cross-check (Section 6-2). In these experiments, the enantiomeric compositions of the very dilute solutions were measured radiochemically by means of a novel double-labeling technique (^3H label for the *R* enantiomer of MBA and ^{14}C label for the *S*) thus avoiding the need for measuring optical rotations (Section 6-5.h).

7-7. RACEMIZATION

Racemization is the formation of a racemate from a pure enantiomer having the same constitution (Eq. 7.27). While the constitution of the compound is unaffected, the

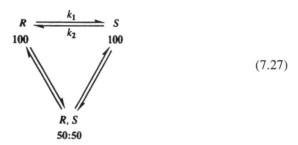

$$(7.27)$$

configurational integrity of the sample is lost; that is, an enantiopure (or nonracemic) sample becomes racemic during the process. Racemization is an irreversible process arising from the reversible interconversion of enantiomers; it is always associated with the disappearance of optical activity.[366]

Epimerization is a related term that we use to describe a (reversible) change in configuration at one or more (but not all) stereocenters in a molecule possessing several stereocenters. Diastereomers differing in configuration at just one stereocenter are often called *epimers*.[27] Because at least one stereocenter remains unaffected in the molecule, epimerization of a nonracemic substance does not lead to the loss of optical activity. However, the magnitude and sign of the sample's rotation may, and generally will, change over time. This change in rotation is called mutarotation.[367]

Racemization is an energetically favored process since the concentration of the starting enantiomer is reduced by one-half during its operation. But, it is favored energetically only to the extent of $\Delta G° = RT \ln \frac{1}{2} = -RT \ln 2$, corresponding to -0.41 kcal mol^{-1} (-1.7 kJ mol^{-1}) at 25°C. Since $\Delta H° = 0$ for the enantiomers undergoing racemization (assuming no differential intermolecular interactions; cf. Chapter 6), it follows that the driving force for racemization is entirely entropic.

Racemization is generally quite slow unless a suitable mechanistic pathway is available. Yet, because numerous mechanisms exist by which the configuration at a stereocenter may be "lost," racemization is a ubiquitous phenomenon which is described and discussed in numerous papers incidental to other work. Racemization has the distinction of being at times something to avoid (as during the synthesis of a polypeptide) and, at other times, something to encourage (to avoid having to discard the unwanted enantiomer after a resolution). There are two contrasting studies of racemization mechanisms that illustrates these opposite goals.[368,369]

For a system undergoing racemization, a useful quantity is the racemization half-life, τ (more accurately, τ is the inversion half-life in view of its definition:

$$\tau = \frac{\ln 2}{k_1 + k_2} = \frac{\ln 2}{(1 + K')k_1} \tag{7.28}$$

for the time when the enantiomeric purity of a chiral sample has been reduced from 100% to 50% ee. In Eq. 7.28, $K' = 1/K = k_2/k_1$ and, when $k_1 = k_2$, as is true in racemizations, Eq. 7.28 reduces to $\tau = \ln 2/2k$.[370] Another quite useful and practical quantity, particularly in the study of racemization of compounds having large rotations, is $t_{1/100}$, the time necessary to lose 1% of the rotation.[371]

a. Racemization Processes

There are three general ways of making racemates. (a) Most frequently, a racemate is formed whenever a synthesis results in the generation of a stereocenter in the absence of a chiral (intramolecular or extramolecular) influence. By way of example, addition of a nucleophilic reagent, such as methyllithium, to either of the heterotopic faces of an achiral but prochiral aldehyde (e.g., benzaldehyde) gives rise to a racemic carbinol. (b) Racemates may be formed as the result of chemical transformations on nonracemic samples. Typically, this requires the cleavage of bonds (usually just one bond) at a stereocenter or the stretching of or rotation about bonds. The cleavage can take place via formation of free radicals, carbocations, carbanions, or excited states. The basis of racemization mediated by these intermediates and transition states is that they are planar (or that they easily attain a planar geometry). Examples of this approach to racemization are given below. (c) The most direct way of making racemates is by mixing equal parts of the two enantiomers of a given compound. While this process may seem trivial, it is occasionally practiced when a sample of a racemate (say of a natural product) is needed that is otherwise unavailable, for example, dimers of camphor,[372] α-pinene,[373] and a protein.[374] A special application of the mixing approach is the preparation of 50:50 mixtures of enantiomers wherein one enantiomer is isotopically labeled (or both enantiomers are labeled, each with a different isotope) so as to permit facile subsequent analysis of the enantiomer composition (e.g., by mass spectrometry; cf. Section 6-5.f).

The second of the general types of racemization processes (see above) may take place by mechanisms involving the intervention of discrete intermediates[375] or of planar (or near-planar) transition states. In this section we limit ourselves to citing some examples of the several types from the large literature of racemization studies. Leading references to the earlier literature may be found in Eliel[27] and in Morrison and Mosher.[268] Racemization of amino acids is dealt with in Section 7-7.b.

Thermal Methods. Racemization without the breaking of covalent bonds may take place if thermal molecular deformation (of bond angles and lengths) can be achieved without demolition of the structure. Attainment of a truly planar transition state in thermal racemization is less important than is the facile interconversion of the enantiomers (Eq. 7.27, top line). The total spontaneous resolution of configuration labile compounds (e.g., TOT **7** and 1,1'-binaphthyl **10**; Figs. 7.4 and 7.5, respectively), implies their easy racemization without breaking of covalent bonds; for a discussion of this point, see Section 7-2.d and Chapter 13.

The pioneering studies of thermal racemizations were those carried out on the biaryls. It is well worth pointing out that the calculation of activation energies for the racemization of biaryls are among the seminal studies leading to the development of the powerful empirical force field calculation (molecular mechanics, MM) method (cf. Section 2-6).[376] In contrast to tetra-*o*-alkylsubstituted biphenyls, single (**176**) and doubly bridged biphenyls (Fig. 7.64, **177**) undergo thermal racemization. The E_a values calculated by the fledgling MM method were in remarkably good accord with the experimental ones (cf. Section 13-5).[377] The measurement of racemization rates of nonracemic **176** and of its dimethyl-d_6 analogue (with the finding that $k_D < k_H$) was the first demonstration of the operation of a steric kinetic isotope effect (Mislow et al.).[377]

The helicenes (e.g., hexahelicene, Fig. 6.6) are prototypes of overcrowded molecules adopting helical configurations. The fact that helicenes as large as [9]helicene (Fig. 7.64, **178**) have been found to racemize thermally [the latter is completely racemized in 10 min at 380°C with little decomposition; $\Delta G^{\ddagger} = 43.5$ kcal

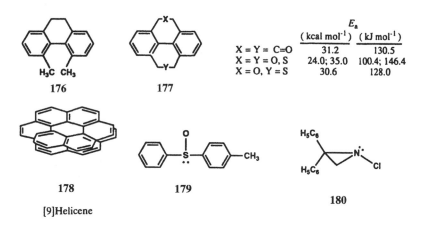

	E_a	
	(kcal mol⁻¹)	(kJ mol⁻¹)
X = Y = C=O	31.2	130.5
X = Y = O, S	24.0; 35.0	100.4; 146.4
X = O, Y = S	30.6	128.0

176   **177**

178   **179**   **180**

[9]Helicene

Figure 7.64. Thermally racemizable compounds.

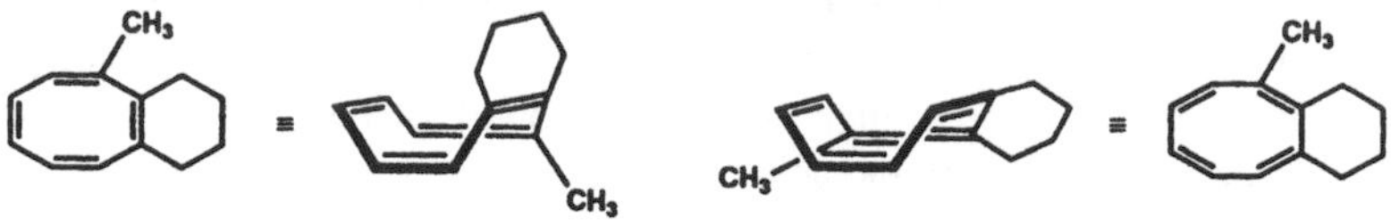

Figure 7.65. Thermal racemization of annulated cyclooctatetraenes.

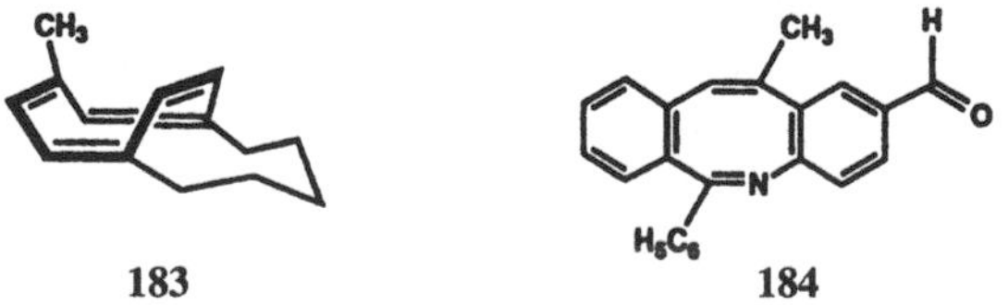

mol^{-1} (182 kJ mol^{-1}) in naphthalene] is remarkable.[378] Even [11]helicene has been found to undergo thermal racemization (at 400–410°C).[379] The ingeneous suggestion by J. Nasielski, that racemization of helicenes might conceivably take place by an internal double Diels–Alder reaction, was discarded by means of an isotope label study. The most reasonable explanation for the racemization remains that it takes place by a "conformational pathway"; that is, the helicenes are significantly more flexible than our prejudices and examination of space-filling models previously led us to believe, with the necessary bond deformations being spread over much of the molecular skeleton.[380]

Chiral substituted cyclooctatetraenes (COTs; e.g., **181**, Fig. 7.65), thermally racemize principally by ring inversion rather than by bond shifting of **181** to isomer **182**. Thermal racemization is completely inhibited in a cyclooctatetraene that is 1,4-annulated (Fig. 7.65, **183**).[381] The racemization rate of COT analogue **184** was measured by following changes in the pitch of the cholesteric mesophase generated when **184** was added to a nematic liquid crystal[382] (cf. Section 12-4.3); see also Section 13-7.

Racemization of the alkaloid (−)-vincadifformine occurs by a sequential retro-Diels–Alder and Diels–Alder cycloaddition process. Racemization is assured by the fact that the intermediate (cycloreversion) product is achiral (Fig. 7.66).[382]

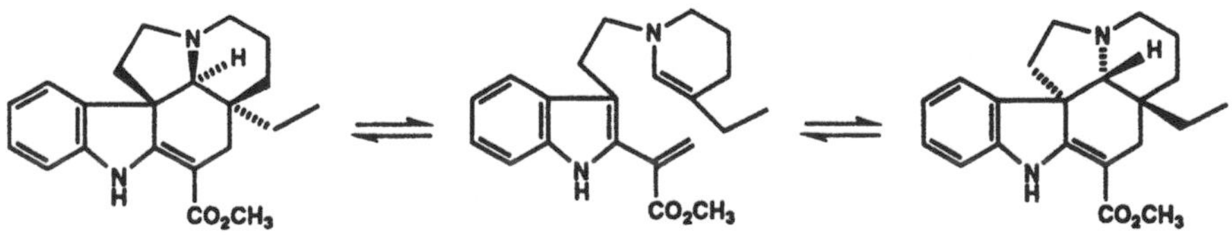

Figure 7.66. Racemization of (−)-vincadifformine.

Perchlorotriphenylamine (Fig. 7.30, **100**), a compound whose chirality is due to its shape as a molecular propeller (the nitrogen is nonstereogenic), undergoes thermal racemization at 120°C [$\Delta G^{\ddagger}$ = 31.4 kcal mol^{-1} (131 kJ mol^{-1})] presumably by a "two-ring flip" mechanism (see Section 13-6.a).[195]

Some chiral compounds with stereogenic heteroatoms are able to racemize by a pyramidal inversion mechanism at the heteroatom (equivalent to quantum mechanical tunneling). Two examples of compounds subject to this type of racemization mechanism are phenyl p-tolyl sulfoxide (Fig. 7.64, **179**) with inversion at sulfur [$\Delta G^{\ddagger}$ = 38.6 kcal mol^{-1} (162 kJ mol^{-1}) at 200°C in p-xylene] and 2,2-diphenyl-N-chloroaziridine (Fig. 7.64, **180**) (with inversion at nitrogen [$\Delta G^{\ddagger}$ = 24.4 kcal mol^{-} (102 kJ mol^{-1}) in CCl$_4$].[383,384] Note the much lower inversion barrier at nitrogen.

Racemization Via Stable Achiral Intermediates, Excited States, and Free Radicals. Partial racemization of **181** (Fig. 7.67) takes place on catalytic hydrogenation in the presence of palladium; the enantiomeric purity is reduced (by > 10% ee) during the reduction. It is believed that the racemization is most likely due to double-bond isomerization to an achiral alkene occurring concurrently with the reduction (cf. also Chapter 9).[385]

In contrast to other biaryls, it has been suggested that the thermal racemization of hetero-biaryls **184** is best accounted for by a ring opening–ring closure reaction (Eq. 7.29).[386]

$$(7.29)$$

184
X = O, S

achiral

Acyl halides with a stereocenter at the α-carbon atom are prone to racemization provided that at least one α-hydrogen atom is present in the structure. There is evidence that elimination of and readdition of HCl to a ketene intermediate is responsible for this racemization.[387]

rac-Cryptone (Fig. 7.67, **182**) is obtained from the more abundant (–)-isomer on ketalization with ethylene glycol. Derivatization is attended by double-bond migration to the achiral-dioxolane Δ^3-dioxolane **183**. Regeneration of the ketone on acid hydrolysis returns the double bond to its original location but leaves the product racemic.[388]

Palladium black catalyzes the racemization and concomitant alkylation of nonracemic α-methylbenzylamine.[389] Deuterium label studies have demonstrated the intermediacy of achiral enamine and/or imine palladium complexes in the racemization (Eq. 7.30):

$$(7.30)$$

PdH$_2$ H$_2$

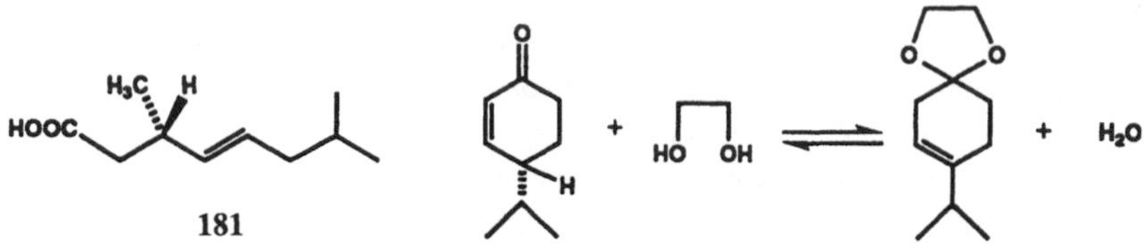

Figure 7.67. Racemization via stable achiral intermediates and radicals.

trans-1,2-Disubstituted cyclopropanes enantiomerize, that is, racemize thermally without competing epimerization to the cis isomers, via ring-opened 1,3-disubstituted trimethylene diradicals. This insight stems from the fact that experimental $\Delta G^{\ddagger}_{\text{enant}}$ values correlate linearly with the sum of substituent radical stabilization energy terms.[390]

Acid-Catalyzed Processes. Acid catalysts have the potential of generating carbocations at chiral centers provided that suitable leaving groups are present there. Since the resulting charged intermediates are typically planar, hence achiral, the topographic criterion would seem to demand that the products of reactions of carbocations generated at chiral centers be invariably racemic. However, due to the intervention of ion pairs (especially in solvolytic reactions), or of solvated (hence potentially chiral) carbocations, matters are significantly less clear-cut than is implied by the preceding statement. (For a summary of results bearing on this point see Eliel, ref. 27, p.372; March, ref. 391, p. 302.)

Carbocations that cannot achieve planarity in the vicinity of the positively charged carbon may be able to maintain their configuration for long periods of time. When ion **186**, prepared by dissolving the (+) alcohol **185** in H_2SO_4 at $-20°C$, is quenched by pouring it into H_2O, (−)-alcohol is recovered (with 91% ee; Fig. 7.68). Since rotation of the α-naphthyl group would lead to interconversion of the chiral cation to its enantiomer, the maintenance of optical activity forces the conclusion that such rotation is prevented. It was suggested that ionization occurs from the syn conformer of the alcohol, and that water attacks the more accessible rear face of the cation (with

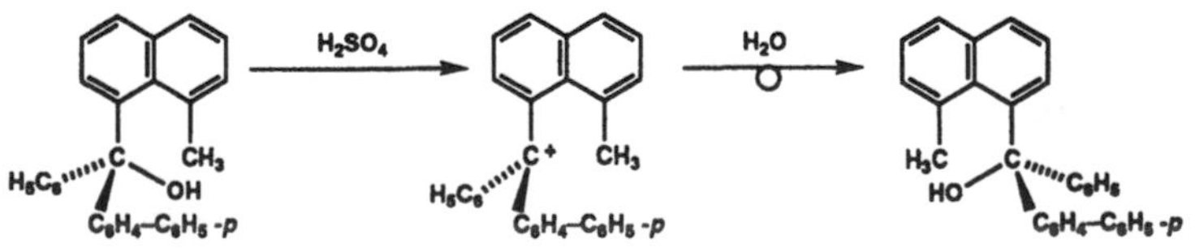

Figure 7.68. Hydration of a chiral carbocation.

inversion). In contrast, ionization of (+)-**185** with BF_3 in CH_2Cl_2 leads to (+)-**185** on quenching. These results were explained by assuming that both ionization and quenching steps involved inversion due to increased steric requirements of the $-OH·BF_3$ leaving group.[392,393]

Cases previously termed *autoracemization* (spontaneous racemization on standing in the absence of added reagents) are now believed to be due to the effect of traces of catalysts, including the glass of the container.[68] α-Methylbenzyl chloride racemizes during chromatography on silica gel (and on acidic, but not basic, alumina); formation of styrene as a minor by-product of the chromatography was noted.[394] Because of the extremely high polarity of silica gel,[395] this racemization may reasonably be attributed to the intervention of the $C_6H_5CH(CH_3)^+$ cation.

Addition of $(CF_3CO_2)_2Hg$ to **187** refluxing in THF or dioxane leads to complete racemization but without the accumulation of the trans isomer. It was concluded that the symmetrical carbocation **189** must account for the racemization (Fig. 7.69). On equilibration with $(CF_3CO_2)_2Hg$, the analogous urethane **188** affords a 1:2 mixture of diastereomers (asymmetric transformation) from which the enantiomerically enriched methyl 5-hydroxy-3-cyclohexenecarboxylate ester may be recovered.[25]

Racemization of Tröger's base (Fig. 7.27, **98**) with acid has been ascribed to the reversible formation of a ring-opened methyleneiminium intermediate.[184] However, a spectroscopic study (UV and ^{13}C NMR) has failed to turn up any evidence for such an intermediate.[396] Nevertheless, formation of a methyleneiminium ion in lower than detectable concentrations remains the most likely acid-catalyzed racemization mechanism for **98**.

Base Catalyzed Processes.   Removal of α-hydrogen atoms from carbonyl compounds and nitriles gives rise to carbanions. This property plays a central role in the methodology of synthetic organic chemistry. If the α-carbon atom is a stereogenic atom, then as a rule, stereochemical integrity is not maintained when carbanion formation initiates with a nonracemic sample, with the rate of racemization equaling that of enolization and of deuterium incorporation (in protonic solvents containing $-OD$).[391]

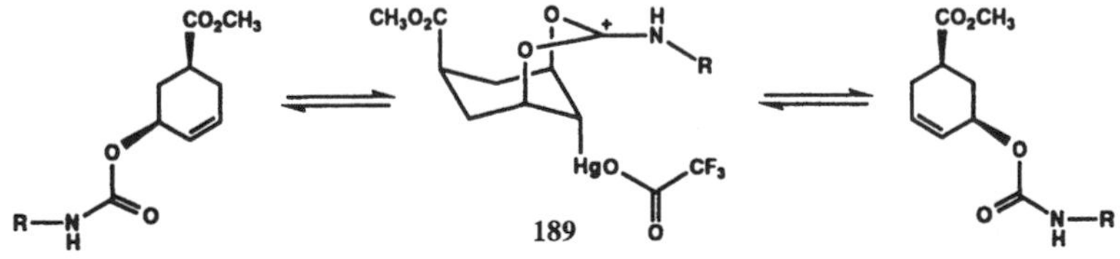

(S,S)-**187**   R = C_6H_5–CH_2
(S,S,S)-**188**   R = α-(1-naphthyl)methyl

(R,R)-**187**
(R,R,S)-**188**

Figure 7.69. Acid-catalyzed racemization via a symmetrical intermediate.

Numerous studies dealing with carbanion racemization have appeared. In general, carbanions have potentially less tendency to racemize quickly than do carbocations. Two instances in which stereochemical integrity is partially or fully retained during carbanion formation are:

1. formation of nonracemic CHBrClF in the haloform reaction of CHBrClC(=O)CH$_3$[397,398] and in the decarboxylation of the strychnine salt of CFClBrCO$_2$H.[399]

2. N-Pivaloylphenylalanine dimethylamide (Fig. 7.70, **190**) has been found to exchange its α-hydrogen atom for deuterium (k_c) at a rate faster than the loss of optical activity (k_a)[k_c/k_a = 2.4 with t-BuOK in t-BuOD at 30°C].[400]

Although simple ketones bearing an α-hydrogen atom are remarkably resistant to racemization on standing and distillation, in the presence of base, rapid racemization ensues. The rate of racemization is highly sensitive to structure; whereas ketone **191** (Fig. 7.70) racemizes with $t_{1/2}$ = 18.4 min, replacement of α-CH$_3$ by α-tert-butyl leads to $t_{1/2}$ = 13,680 min (both with EtONa in EtOH at 25°C.[401]

Carbanionic intermediates have been implicated also in the racemization of mandelic acid by mandelate racemase enzyme from *Pseudomonas putida*.[402] The racemization requires a metal cation (Mg^{2+} is most effective); it is inferred that the cation increases the acidity of the mandelate substrate in the active site.[403] Based on a double-isotope labeling study, it has been demonstrated that proton transfer by the enzyme is strictly intramolecular and heterofacial, that is, the proton abstracted from one face of the substrate is returned to the opposite face of the same molecule.[404]

b. Racemization of Amino Acids

There is a remarkably large literature dealing with the racemization of amino acids and peptides.[405] There are at least four reasons for the interest: (a) efforts to minimize racemization incidental to peptide and protein synthesis; (b) attempts to find ways of promoting racemization in connection with the large scale resolution and asymmetric transformation of amino acids so as to avoid the need to discard 50% of the starting

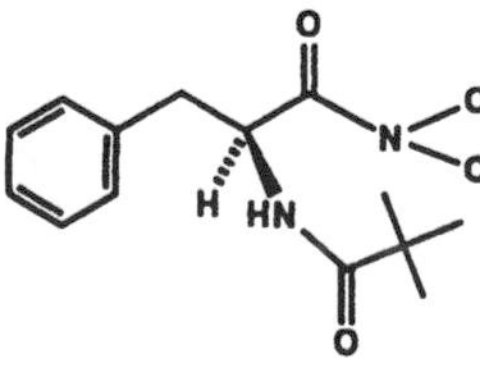

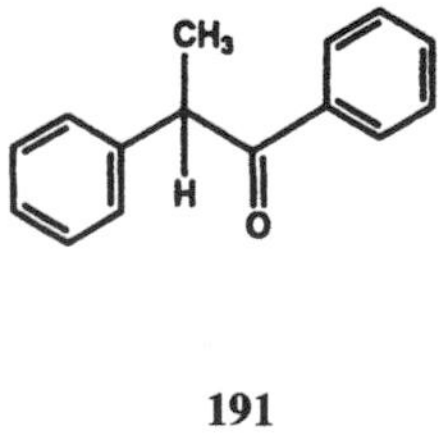

Figure 7.70. Structures **190** and **191**.

material; (c) the relation of amino acid racemization to the age of fossils requires clarification of the mechanism of racemization; and (d) the study of racemization contributes to the search for the origins of the chiral homogeneity of natural products and, by extension, for the origins of life. For reviews of the subject, see Kaneko et al.[5]; Williams and Smith[370]; Bada [406]; Bodanszky.[407] It is worthwhile to emphasize the fact that the modern study of racemization of amino acids has benefited greatly from the very sensitive techniques now available for the determination of their enantiomer composition by nonchiroptical methods, principally those involving GC analysis of volatile derivatives on enantioselective GC columns (cf. Section 6-5.d).[408]

It is well known that free amino acids are relatively difficult to racemize in aqueous medium. Racemization is catalyzed by both acids and bases, especially the latter. A mechanism for racemization was proposed as early as 1910 by Dakin.[409] The proposal of a mechanism at the time is remarkable since it antedated the work of Robert Robinson by some decades.[410] The contemporary accepted mechanism, a modification of that of Dakin (a) in basic medium and (b) in acidic medium, is that of Neuberger (Eq. 7.31).[411]

$$\text{(7.31a)}$$

$$\text{(7.31b)}$$

The racemization rate is enhanced by a number of factors: an increase in the electronegativity of R; a decrease in the negative charge on the carboxylate group; and substitution of hydrogen on the amino group by electronegative atoms or groups, such as acyl and by transition metal ions.[408] A detailed modern study (of the racemization of phenylglycine as a model compound) indicates that the racemization is in accord with the S_E1 mechanism. The mechanism is consistent with the observation that amino acid derivatives (N-acyl, esters) and peptides racemize faster than do free amino acids. In addition, racemization of N-acyl amino acids is facilitated by intramolecular formation of and deprotonation of azlactones (oxazolinones)[251] and of hydantoins.[412] Racemization of dipeptides is influenced by neighboring groups and by formation of diketopiperazines.[413] The role of hydantoins in racemization and the labilization of amino acids and their derivatives to racemization by transition metal ions have been dealt with in Section 7-2.d.

Efforts to suppress racemization during peptide synthesis (coupling activated by dicyclohexylcarbodiimide) have principally taken the form of adding weak acids (e.g., 1-hydroxybenzotriazole) that suppress proton abstraction from the α-carbon atom of amino acid intermediates and have additional benefits as well.[407]

The avoidance of racemization in the context of stereoselective synthesis is subject to both thermodynamic and kinetic control. A strategy for alkylation of amino acids without loss of stereochemical integrity is illustrated by the electrophilic substitution of (S)-proline at the α-carbon atom. Formation of the N,O-acetal **192** of proline and pivalaldehyde (one diastereomer only) takes place under thermodynamic control while the subsequent alkylation of the acetal enolate (e.g., with methyl iodide) affords, after hydrolysis, enantiomerically pure (S)-2-methylproline under kinetic control (steered by the induced chiral center; Eq. 7.32).[414]

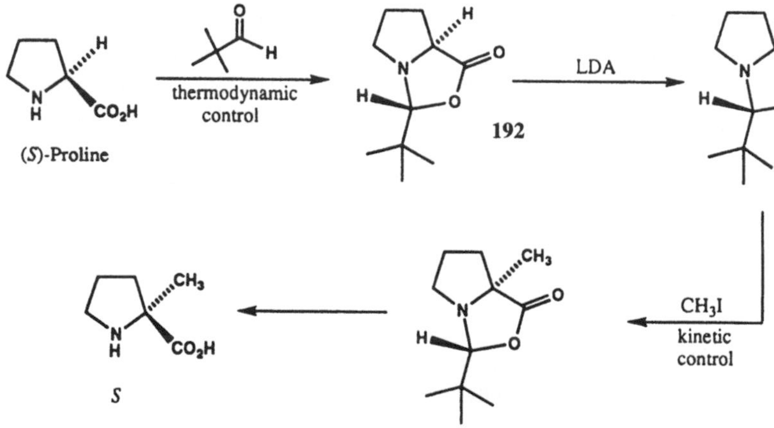

(7.32)

Readers are reminded that the racemization of amino acids has been coupled to the preferential crystallization of derivatives crystallizing as conglomerates giving rise to highly efficient asymmetric transformations (Section 7-2.d). Efforts to simplify their racemization have led to the finding that a wide range of amino acids (15 out of the 17 amino acids studied) are substantially racemized on heating in acetic acid in 1 h (at 80–100°C) in the presence of aldehydes, such as salicylaldehyde.[369] The proposed mechanism of this racemization involves an imine (Schiff's base). Proton abstraction from the α-carbon atom of the protonated imine yields a stabilized zwitterion (Eq. 7.33).

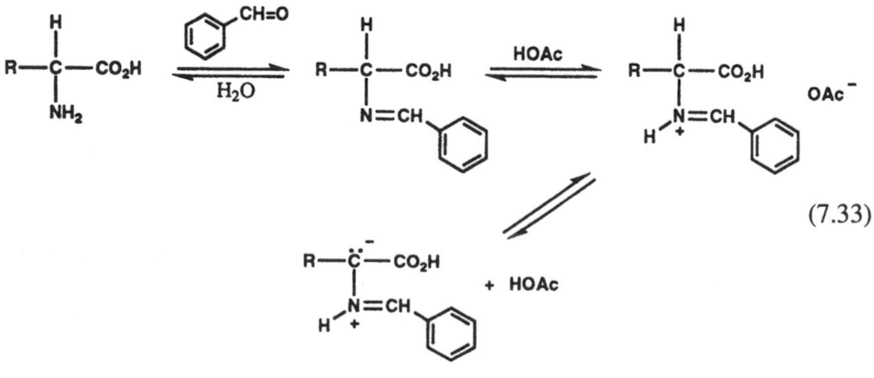

(7.33)

It was also observed that mineral surfaces (silica or clays) increased the tendency of amino acids to undergo radioracemization.[415-417] Radioracemization of amino acids in contact with such materials was intended to more closely approximate conditions under which geologic samples (either terrestrial or extraterrestrial) were irradiated (e.g., by cosmic rays). Radioracemization tends to reduce the significance of the finding that amino acids in meteorites are racemic in regard to their biogenic or abiogenic origin.[418]

Finally, we consider the application of amino acid racemization to the dating of fossil samples. Abelson[419] suggested that the decomposition of protein as a function of time might provide a dating method. After noting that amino acid samples in fossil shell and bone were significantly racemized, a dating method based on measurement of the enantiomer composition was developed.[420] This dating method (often called *aminostratigraphy*) is based on the following premises: (a) Biosynthesis of amino acids in most living cells (bacteria excepted) produces material having the L-configuration. After the death of the individual, replenishment of amino acids ceases and accumulation of increasing amounts of the D-enantiomers takes place as a result of racemization. (b) The racemization rate is governed by first-order kinetics. (c) The rate of racemization is dependent on environmental conditions in which the amino acids are stored, the main variable (along with pH, bound and free metals, and surfaces) being the average temperature of the sample t_0 (called the diagenetic temperature). Application of the method requires the determination of k_1 and K values for the individual amino acids (in the laboratory), and calibration by an independent method of the environment in which a given sample of amino acid is found (e.g., ^{14}C dating). Samples as old as 10^5-10^6 years can be dated by this method, this being a time range complementary to those accessible by other dating methods, for example, ^{14}C dating that is only useful for $(30-70) \times 10^3$ years if activation analysis methods are used. Numerous examples, evaluation of assumptions, and description of the pitfalls of this method are given in the detailed review by Williams and Smith,[370] see also Sykes[421] and Meyer.[422,423]

The amino acid dating method is applicable even to living humans (this might be used, e.g., if their age is not well documented) provided that the analysis is carried out on a sample derived from metabolically inert tissue (e.g., tooth enamel or the eye lens). The analysis focuses on the enantiomer composition of aspartic acid, one of the fastest racemizing amino acids.[406]

REFERENCES

1. Pasteur, L. *Ann. Chim. Phys.* **1848**, *24*, [3] 442.

2. Kauffman, G. B. and Myers, R. D. *J. Chem. Educ.* **1975**, *52*, 777.

3. Piutti, A. *C. R. Hebd. Séances Acad. Sci.* **1886**, *103*, 134.

4. Jacques, J., Collet, A., and Wilen, S. H. *Enantiomers, Racemates and Resolutions,* Wiley, New York, 1981.

5. Kaneko, T., Izumi, Y., Chibata, I., and Itoh, T., eds., *Synthetic Production and Utilization of Amino Acids,* Kodansha, Tokyo, and Wiley, New York.

6. Jacques, J., Leclercq, M., and Brienne, M.-J. *Tetrahedron* **1981**, *37*, 1727.

7. Collet, A., Brienne, M.-J., and Jacques, J. *Chem. Rev.* **1980**, *80*, 215.

8. Collet, A. *Enantiomer* **1999**, *4*, 153. See also Inagake, M. *Chem. Pharm. Bull.* **1977**, *25*, 2497.

9. Kuroda, R. and Mason, S. F. *J. Chem. Soc. Perkin 2* **1981**, 167.

10. Amiard, G. *Bull. Soc. Chim. Fr.* **1956**, 447.

11. Velluz, L. and Amiard, G. *Bull. Soc. Chim. Fr.* **1953**, 903.

12. Velluz, L., Amiard, G., and Joly, R. *Bull. Soc. Chim. Fr.* **1953**, 342.

13. Gernez, D. *C. R. Acad Sci.* **1866**, *63*, 843.

14. Duschinsky, R. *Chem. Ind. (London)* **1934**, 10. Wüest, H. M., ed., *Festschrift Emil Barell*, F. Reinhardt Verlag, Basel, Switzerland, 1936, p. 375.

15. Marckwald, W. and Meth, R. *Ber. Dtsch. Chem. Ges.* **1905**, *38*, 801. Ault, A. in *Organic Syntheses, Coll. Vol. V*, Baumgarten, H. E., ed., Wiley, New York, 1973, p. 932.

16. Nohira, H., Kai, M., Nohira, M., Nishikawa, J., Hoshiko, T., and Saigo, K. *Chem. Lett.* **1981**, 951.

17. Brienne, M.-J., Collet, A., and Jacques, J. *Synthesis* **1983**, 704.

18. Jommi, G. and Teatini, A. *Chim. Ind (Milan)* **1962**, *44*, 29.

19. Coquerel, G., Bouaziz, R., and Brienne, M.-J. *Chem. Lett.* **1988**, 1081.

20. Yamada, S., Yamamoto, M., and Chibata, I. *J. Agric. Food Chem.* **1973**, *21*, 889.

21. Kimoto, H., Saigo, K., Ohashi, Y., and Hasegawa, M. *Bull. Chem. Soc. Jpn.* **1989**, *62*, 2189.

22. Kondepudi, D. K., Kaufman, R. J., and Singh, N. *Science* **1990**, *250*, 975.

23. McBride, J. M. and Carter, R. L. *Angew. Chem. Int. Ed. Engl.* **1991**, *30*, 293.

24. Anon. *Chem. Eng. (N. Y.)* **1965**, *72* (Nov. 8, 1965), 247.

25. Trost, B. M. "Approaches for Asymmetric Synthesis as Directed Toward Natural Products," in Eliel, E. L. and Otsuka. S., eds., *Asymmetric Reactions and Processes in Chemistry*, ACS Symposium Series 185, American Chemical Society, Washington, DC, 1982, p. 3.

26. Arai, K. *Yuki Gosei Kagaku Kyokaishi* **1986**, *44*, 486.

27. Eliel, E. *Stereochemistry of Carbon Compounds*, McGraw-Hill, New York, 1962.

28. Mislow, K., Gust, D., Finocchiaro, P., and Boettcher, R. J. *Top. Curr. Chem.* **1974**, *47*, 1.

29. Craig, D. P. and Mellor, D. P. *Top. Curr. Chem.* **1976**, *63*, 1.

30. Johnson, C. A., Guenzi, A., and Mislow, K. *J. Am. Chem. Soc.* **1981**, *103*, 6240.

31. Casarini, D., Lunazzi, L., Placucci, G., and Macciantelli, D. *J. Org. Chem.* **1987**, *52*, 4721.

32. Inoue, Y. *Chem. Rev.* **1992**, *92*, 741.

33. Khatri, C. A., Andreoloa, C., Peterson, N. C., and Green, M. M. Polymer Preprint, Polymer Division, American Chemical Society, Washington, DC, 1992.

34. Bur, A. J. and Fetters, L. J. *Chem. Rev.* **1976**, *76*, 727.

35. Havinga, E. *Biochim. Biophys. Acta* **1954**, *13*, 171.

36. Baker, W., Gilbert, B., and Ollis, W. D. *J. Chem. Soc.* **1952**, 1443.

37. Newman, A. C. D. and Powell, H. M. *J. Chem. Soc.* **1952**, 3747.

38. Wynberg, H. and Groen, M. B. *J. Am. Chem. Soc.* **1968**, *90*, 5339.

39. Wilson, K. R. and Pincock, R. E. *J. Am. Chem. Soc.* **1975**, *97*, 1474.

40. Chandrasekhar, S. and Ravindranath, M. *Tetrahedron Lett.* **1989**, *30*, 6207.

41. Black, S. N., Williams, L. J., Davey, R. J., Moffat, F., Jones, R. V. H., McEwan, D. M., and Sadler, D. E. *Tetrahedron* **1989**, *45*, 2677.

42. Arai, K., Obara, Y., Takahashi, Y., and Takakuwa, Y. *Jpn. Kokai Tokkyo Koho JP 61 238 734*, Oct. 24, 1986; *Chem. Abstr.* **1987**, *106*, 196063x.

43. Piselli, F. L. *Eur. Patent Appl.* EP 298 395, Jan. 11, 1989; *Chem. Abstr.* **1989**, *111*, 7085a.

44. Sih, C. J. and Wu, S.-H. *Top. Stereochem.* **1989**, *19*, 63.

45. Cecere, F., Galli, G., and Morisis, F. *FEBS Lett.* **1975**, *57*, 192.

46. Olivieri, R., Fascetti, E., Angelini, L., and Degen, L. *Enzyme Microb. Technol.* **1979**, *1*, 201; *Biotechnol. Bioeng.* **1981**, *23*, 2173.

47. Anon. *Jpn. Kokai Tokkyo Koho JP 80 104 890*, Aug. 11, 1980 (to Kanegafuchi Chemical Industry Co.); *Chem. Abstr.* **1980**, *93*, 236942e.

48. Yokozeki, K., Nakamori, S., Eguchi, C., Yamada, K., and Mitsugi, K. *Agric. Biol. Chem.* **1987**, *51*, 355.

49. Schurig, V. and Bürkle, W. *J. Am. Chem. Soc.* **1982**, *104*, 7573.

50. Allenmark, S. G. *Chromatographic Enantioseparation: Methods and Applications,* 2nd ed., Ellis Horwood, Chichester, UK, 1991.

51. Pasteur, L. *C. R. Acad. Sci.* **1853**, *37*, 162.

52. Wilen, S. H. *Top. Stereochem.* **1971**, *6*, 107.

53. Kai, Z. D., Kang, S. Y., Ke, M. J., Jin, Z., and Liang, H. *J. Chem. Soc. Chem. Commun.* **1985**, 168.

54. Rosan, A. M. *J. Chem. Educ.* **1989**, *66*, 608.

55. Marckwald, W. *Ber. Dtsch. Chem. Ges.* **1896**, *29*, 42, 43.

56. ten Hoeve, W. and Wynberg, H. *J. Org. Chem.* **1985**, *50*, 4508.

57. Overby, L. R. and Ingersoll, A. W. *J. Am. Chem. Soc.* **1960**, *82*, 2067.

58. Ingersoll, A. W. *J. Am. Chem. Soc.* **1925**, *47*, 1168.

59. Wong, C.-H. and Wang, K.-T. *Tetrahedron Lett.* **1978**, 3813.

60. Saigo, K., Kubota, N., Takebayashi, S., and Hasegawa, M. *Bull. Chem. Soc. Jpn.* **1986**, *59*, 931.

61. Findlay, A. *The Phase Rule and Its Applications,* 9th ed., by Campbell, A. N. and Smith, N. O., Dover, New York, 1951, p. 409ff.

62. Leclercq, M., Jacques, J., and Cohen-Adad, R. *Bull. Soc. Chim. Fr.* **1982**, I-388. See also Leclercq, M. and Jacques, J. *Nouv. J. Chim.* **1979**, *3*, 629.

63. Pirkle, W. H. and Simmons, K. A. *J. Org. Chem.* **1983**, *48*, 2520.

64. Woodward, R. B., Cava, M. P., Ollis, W. D., Hunger, A., Daeniker, H. V., and Schenker, K. *Tetrahedron,* **1963**, *19*, 247; see, in particular, the footnotes on p. 259.

65. Newman, P. *Optical Resolution Procedures for Chemical Compounds*, Optical Resolution Information Center, Manhattan College, New York, Vols. 1, 2A, 2B, 3, and 4 (Parts 1 and 2), 1978–1993.

66. Wilen, S. H. *Tables of Resolving Agents and Optical Resolutions,* Eliel, E. L., ed., University of Notre Dame Press, Notre Dame, IN, 1971.

67. Boyle, P. H. *Q. Rev. Chem. Soc.* **1971,** *25,* 323.

68. Potapov, V. M. *Stereochemistry,* translated by A. Beknazarov, Mir, Moscow, 1979.

69. Helmchen, G., Nill, G., Flockerzi, D., Schüle, W., and Youssef, M. S. K. *Angew. Chem. Int. Ed. Engl.* **1979,** *18,* 62. Helmchen, G. and Nill, G. *Angew. Chem. Int. Ed. Engl.* **1979,** *18,* 65.

70. Sonnet, P. E. *J. Org. Chem.* **1982,** *47,* 3793; Sonnet, P. E., McGovern, T. P., and Cunningham, R. T. *J. Org. Chem.* **1984,** *49,* 4639.

71. Vercesi, D. and Azzolina, O. *Farmaco Ed. Prat.* **1985,** *40,* 396; *Chem. Abstr.* **1986,** *105,* 114681q.

72. Fizet, C. *Helv. Chem. Acta* **1986,** *69,* 404.

73. Webster, F. X., Millar, J. G., and Silverstein, R. M. *Tetrahedron Lett.* **1986,** *27,* 4941.

74. Wani, M. C., Nicholas, A. W., and Wall, M. E. *J. Med. Chem.* **1987,** *30,* 2317.

75. Levy, J. personal communication to SHW, 1977.

76. Imhof, R., Kyburz, E., and Daly, J. J. *J. Med. Chem.* **1984,** *27,* 165.

77. Rozwadowska, M. D. and Brossi, A. *J. Org. Chem.* **1989,** *54,* 3202.

78. Yamada, M. and Okawa, K. *Bull. Chem. Soc. Jpn.* **1985,** *58,* 2889.

79. Allan, R. D. and Fong, J. *Aust. J. Chem.* **1986,** *39,* 855.

80. Gal, G., Chemerda, J. M., Reinhold, D. F., and Purick, R. M. *J. Org, Chem.* **1977,** *42,* 142.

81. Baldwin, J. E., Adlington, R. M., Rawlings, B. J., and Jones, R. H. *Tetrahedron Lett.* **1985,** *26,* 485.

82. Tashiro, Y. and Aoki, S. *Eur. Patent App.,* EP 133 053, Feb. 13, 1985; *Chem. Abstr.* **1985,** *103,* 37734p.

83. Nohira, H. and Ueda, K. *Eur. Patent Appl.* EP 65 867; *Chem. Abstr.* **1983,** *98,* 161164v.

84. Viret, J., Patzelt, H., and Collet, A. *Tetrahedron Lett.* **1986,** *27,* 5865.

85. Garnier-Suillerot, A., Albertini, J. P., Collet, A., Faury, L., Pastor, J.-M., and Tosi, L. *J. Chem. Soc. Dalton Trans.* **1981,** 2544.

86. Vriesema, B. K., ten Hoeve, W., Wynberg, H., Kellogg, R. M., Boesten, W. H. J., Meijer, E. M., and Shoemaker, H. E. *Tetrahedron Lett.* **1986,** *27,* 2045.

87. Chibata, I., Yamada, S., Hongo, C., and Yoshioka, R. *Eur. Par. Appl.* EP 75 318, Mar. 30, 1983; *Chem. Abstr.* **1983,** *99,* 105702.

88. Yoshioka, R., Tohyama, M., Ohtsuki, O., Yamada, S., and Chibata, I. *Bull. Chem. Soc. Jpn.* **1987,** *60,* 649.

89. Bajgrowicz, J. A., Cossec, B., Pigiére, Ch., Jacquier, R., and Viallefont, P. *Tetrahedron Lett.* **1984,** *25,* 1789.

90. Klyashchitskii, B. A. and Shvets, V. I. *Russ. Chem. Rev.* **1972,** *41,* 592; *Usp. Khim.* **1972,** *41,* 1315.

91. Givens, R. S., Hrinczenko, B., Liu, J. H.-S., Matuszewski, B., and Tholen-Collison, J. *J. Am. Chem. Soc.* **1984,** *106,* 1779.

92. Burns, C. J., Martin, C. A., and Sharpless, K. B. *J. Org. Chem.* **1989,** *54,* 2826.

93. Whitesell, J. K. and Reynolds, D. *J. Org. Chem.* **1983,** *48,* 3548.

94. Corey, E. J., Hopkins, P. B., Kim, S., Yoo, S.-E., Nambiar, K. P., and Falck, J. R. *J. Am. Chem. Soc.* **1979**, *101*, 7131.

95. Smith A. B., III and Konopelski, J. P. *J. Org. Chem.* **1984**, *49*, 4094.

96. Trost, B. M., Belletire, J. L., Godleski, S., McDougal, P. G., Balkovec, J. M., Baldwin, J. J., Christy, M. E., Ponticello, G. S., Varga, S. L., and Springer, J. P. *J. Org. Chem.* **1986**, *51*, 2370.

97. Koreeda, M. and Yoshihara, M. *J. Chem. Soc. Chem. Commun.* **1981**, 974.

98. Canceill, J., Collet, A., Gabard, J., Gottarelli, G., and Spada, G. P. *J. Am. Chem. Soc.* **1985**, *107*, 1299.

99. Desai, T., Fernandez-Mayoralas, A., Gigg, J., Gigg, R., and Payne, S. *Carbohydr. Res.* **1990**, *205*, 105.

100. Desai, T., Fernandez-Mayoralas, A., Gigg, J., Gigg, R., Jaramillo, C., Payne, S., Penades, S., and Schnetz, N. "Preparation of Optically Active *myo*-Inositol Derivatives as Intermediates for the Synthesis of Inositol Phosphates," in Reitz, A. B., ed., *Inositol Phosphates and Derivatives: Synthesis, Biochemistry and Therapeutic Potential*, ACS Symposium Series 463, American Chemical Society, Washington DC, 1991, Chap. 6.

101. Lauricella, R., Kéchayan, J., and Bodot, H. *J. Org. Chem.* **1987**, *52*, 1577.

102. Corey, E. J., Danheiser, R. L., Chandrasekaran, S., Keck, G. E., Gopalan, B., Larsen, S. D., Sizer, P., and Gras, J.-L. *J. Am. Chem. Soc.* **1978**, *100*, 8034.

103. Donaldson, R. E., Saddler, J. C., Byrn, S., McKenzie, A. T., and Fuchs, P. L. *J. Org. Chem.* **1983**, *48*, 2167.

104. Whitesell, J. K., Minton, M. A., and Chen, K.-M. *J. Org. Chem.* **1983**, *48*, 2193.

105. Saito, H., Nishimura, Y., Kondo, S., and Umezawa, H. *Chem. Lett.* **1987**, 799.

106. Fessner, W.-D. and Prinzbach, H. *Tetrahedron,* **1986**, 1797.

107. Arcamone, F., Bernardi, L., Patelli, B., and Di Marco, A. *Ger. Offen.* 2 604, 785, July 29, 1976; *Chem. Abstr.* **1976**, *85*, 142918.

108. Si, Y., Zhou, J., and Huang, L. *Sci. Sin. Ser. B (Engl. Ed.)* **1987**, *30*, 297; *Chem. Abstr.* **1988**, *108*, 5775q.

109. Tyson, R. *Chem. Ind. (London)* **1988**, 118.

110. Enders, D., Fey, P., and Kipphardt, H. *Org. Synth.* **1987**, *65*, 173; *Organic Syntheses, Collective Volume VIII,* Freeman, J. P., ed., Wiley, New York, 1993, p. 26.

110. Pirkle, W. H. and Hoekstra, M. S. *J. Org. Chem.* **1974**, *39*, 3904.

111. Oare, D. A. and Heathcock, C. H. *Top. Stereochem.* **1989**, *19*, 227.

112. Johnson, C. R. *Aldrichim. Acta* **1985**, *18*, 3.

113. (a) Eaton, P. E. and Leipzig, B. *J. Org. Chem.* **1978**, *43*, 2483. (b) Just, G., Luthe, C., and Potvin, P. *Tetrahedron Lett.* **1982**, *23*, 2285.

114. Mangeney, P., Alexakis, A., and Normant, J. F. *Tetrahedron Lett.* **1988**, *29*, 2677.

115. Paquette, L. A. and Doehner, R. F., Jr. *Org. Chem.* **1980**, *102*, 5016.

116. Eschler, B. M., Haynes, R. K., Kremmydas, S., and Ridley, D. D. *J. Chem. Soc. Chem. Commun.* **1988**, 137.

117. Bortolini, O., Di Furia, F., Licini, G., Modena, G., and Rossi, M. *Tetrahedron Lett.* **1986**, *27*, 6257.

118. Fogassy, E., Lopata, A., Faigl, F., Darvis, F., Ács, M., and Toke, L. *Tetrahedron Lett.* **1980**, *24*, 647.

119. Ingersoll, A. W., Babcock, S. H., and Burns, F. B. *J. Am. Chem. Soc.* **1933**, *55*, 411.

120. Leclercq, M. and Jacques, J. *Bull. Soc. Chim. Fr.* **1975**, 2052.

121. Collet, A., "Optical Resolution by Crystallization Methods," in Drstulovic, A. M., ed., *Chiral Separations by HPLC. Applications to Pharmaceutical Compounds,* Ellis Horwood, Chichester, UK, 1989, Chap. 4.

122. Kanoh, S., Muramoto, H., Kobayashi, N., Motoi, M., and Suda, H. *Bull. Chem. Soc. Jpn.* **1987**, *60*, 3659.

123. Shiraiwa, T., Tazoh, H., Sunami, M., Sado, Y., and Kurokawa, H. *Bull. Chem. Soc. Jpn.* **1987**, *60*, 3985.

124. Gould, R. O. and Walkinshaw, M. D. *J. Am. Chem. Soc.* **1984**, *106*, 7840.

125. Fogassy, E., Faigl, F., Ács, M., and Grofcsik, A. *J. Chem. Res. Synop.* **1981,** 346; *Miniprint*, 3981.

126. Arnold, W., Daly, J. J., Imhof, R., and Kyburz, E. *Tetrahedron Lett.* **1983**, *24*, 343.

127. Schwab, J. M. and Lin, D. C. T. *J. Am. Chem. Soc.* **1985**, *107*, 6046.

128. Snatzke, G. and Meese, C. O. *Liebigs Ann. Chem.* **1987**, 81.

129. Pope, W. J. and Peachey, S. J. *J. Chem. Soc.* **1899**, *75*, 1066.

130. Mofaddel, N. and Bouaziz, R. *Bull. Soc. Chim. Fr.* **1991**, 773.

131. Fogassy, E., Lopata, A., Faigl, F., Darvas, F., Ács, M., and Toke, L. *Tetrahedron Lett.* **1980**, *21*, 647.

132. Simon, K., Kozsda, E., Bocskei, Z., Faigl, F., Fogassy, E., and Reek, G. *J. Chem. Soc. Perkin Trans. 2* **1990**, 1395.

133. Bailey, M. E. and Hass, H. B. *J. Am. Chem. Soc.* **1941**, *63*, 1969.

134. Hass, H. B. U.S. Patent 2,388,688, Nov. 13, 1945; *Chem. Abstr.* **1946**, *40*, 1538.

135. Fritz-Langhals, E. *Angew. Chem. Int. Ed. Engl.* **1993**, *32*, 753.

136. Cope, A. C., Ganellin, C. R., and Johnson, H. W., Jr. *J. Am. Chem. Soc.* **1962**, *84*, 3191.

137. Cope, A. C., Ganellin, C. R., Johnson, H. W., Jr., Van Auken, T. V., and Winkler, J. J. S. *J. Am. Chem. Soc.* **1963**, *85*, 3276.

138. Hulshof, L. A., McKervey, M. A., and Wynberg, H. *J. Am. Chem. Soc.* **1974**, *96*, 3906.

139. Lazzaroni, R., Uccelo-Barretta, G., Pini, D., Pucci, S., and Salvadori, P. *J. Chem. Res. Synop.* **1983**, 286.

140. Alcock, N. W., Brown, J. M., and Hulmes, D. I. *Tetrahedron: Asymmetry* **1993**, *4*, 743.

141. Block, P., Jr. and Newman, M. S. *Org. Synth.* **1968**, *48*, 120; *Organic Synthesis, Collective Volume V*, Baumgarten, H. E., ed., Wiley, New York, 1973, p. 1031.

142. Newman, M. S. and Lutz, W. B. *J. Am. Chem. Soc.* **1956**, *78*, 2469. Newman, M. S., Lutz, W. B., and Lednicer, D. *J. Am. Chem. Soc.* **1955**, *77*, 3420.

143. Wynberg, H. and Lammertsma, K. *J. Am. Chem. Soc.* **1973**, *95*, 7913.

144. Schlenk, W., Jr. *Angew. Chem. Int. Ed. Engl.* **1965**, *4*, 139.

145. Cram, D. J. *Science* **1983**, *219*, 1177.

146. Weber, E., ed., *Top Curr. Chem.* **1987**, *140*, (Molecular Inclusion and Molecular Recognition—Clathrates I) Springer, Berlin.

147. Arad-Yellin, R., Green, B. S., Knossow, M., and Tsoucaris, G. "Enantiomeric Selectivity of Host Lattices," in Atwood, J. L., Davies, J. E. D., and MacNicol, D. D., eds., *Inclusion Compounds*, Vol. 3, Academic, London, 1984, Chap. 9.

148. Worsch, D. and Vogtle, F. *Top. Curr. Chem.* **1987**, *140*, 21.

149. Toda, F. *Top. Curr. Chem.* **1987**, *140*, 43.

150. Tsoucaris, G. "Clathrates," in Desiraju, G. R., ed., *Organic Solid State Chemistry,* Elsevier, Amsterdam, The Netherlands, 1987, Chap. 7.

151. Stoddart, J. F. *Top. Stereochem.* **1987**, *17*, 207.

152. Peacock, S. C. and Cram, D. J. *J. Chem. Soc. Chem. Commun.* **1976,** 282.

153. Lingenfelter, D. S., Helgeson, R. C., and Cram, D. J. *J. Org. Chem.,* **1981**, *46*, 393.

154. Cramer, F. and Dietsche, W. *Chem. Ber.* **1959**, *92*, 378.

155. Cramer, F. and Hettler, H. *Naturwissenschaften* **1967**, *54*, 625.

156. Benschop, H. P. and Ven den Berg, G. R. *J. Chem. Soc. D* **1970**, 1431.

157. Addadi, L., Berkovitch-Yellin, Z., Weissbuch, I., van Mil, J., Shimon, L. J. W., Lahav, M., and Leiserowitz, L. *Angew. Chem. Int. Ed. Engl.* **1985**, *24*, 466.

158. Schlenk, W., Jr. *Experientia* **1952**, *8*, 337.

159. Asselineau, C. and Asselineau, J. *Ann. Chem. (Paris)* **1964**, *9*, 461.

160. Tam, W., Eaton, D. F., Calabrese, J. C., Williams, I. D., Wang, Y., and Anderson, A. G. *Chem. Mat.* **1989**, *1*, 128.

161. Gnaim, J. M., Green, B. S., Arad-Yellin, R., and Keehn, P. M. *J. Org. Chem.* **1991**, *56*, 4525.

162. Arad-Yellin, R., Green, B. S., Knossow, M., and Tsoucaris, G. *J. Am. Chem. Soc.* **1983**, *105*, 4561.

163. Gerdil, R. and Allemand, J. *Helv. Chim. Acta* **1980**, *63*, 1750.

164. Toda, F., Tanaka, K., and Ueda, H. *Tetrahedron Lett.* **1981**, *22*, 4669.

165. Toda, F., Tanaka, K., Ueda, H., and Oshima, T. *J. Chem. Soc. Chem. Commun.* **1983**, 743.

166. Wilen, S. H. and Qi, J. Z. 4th Chemical Congress of North America and 202nd ACS National Meeting, New York, August 25,1991, Abstract ORGN 41.

167. Toda, F., and Tanaka, K., *Tetrahedron Lett.* **1988**, *29*, 551.

168. Miyata, M., Shibakami, M., and Takemoto, K. *J. Chem. Soc. Chem. Commun.* **1988,** 655.

169. Toda, F., Tanaka, K., Ootani, M., Hayashi, A., Miyahara, I., and Hirotsu, K. *J. Chem. Soc. Chem. Commun.* **1993**, 1413.

170. Lucas, H. J. and Gould, C. W., Jr. *J. Am. Chem. Soc.* **1942**, *64*, 601.

171. Pavlis, R. R., Skell, P. S., Lewis, D. C., and Shea, K. *J. J. Am. Chem. Soc.* **1973**, *95*, 6735.

172. Pavlis, R. R. and Skell, P. S. *J. Org. Chem.* **1983**, *48*, 1901.

173. Cristol, S. J. *J. Am. Chem. Soc.* **1949**, *71*, 1894.

174. Still, W. C., Kahn, M., and Mitra, A. *J. Org. Chem.* **1978**, *43*, 2923.

175. Comber, R. N. and Brouillette, W. J. *J. Org. Chem.* **1987**, *52*, 2311.

176. Gal, J. "Indirect Chromatographic Methods for Resolution of Drug Enantiomers—Synthesis and Separation of Diastereomeric Derivatives," in Wainer, I. W. and Drayer, D. E., eds., *Drug Stereochemistry. Analytical Methods and Pharmacology,* Marcel Dekker, New York, 1988, Chap. 4. See also, Gal, J. "Indirect Methods for the Chromatographic Resolution of Drug Enantiomers: Synthesis and Separation of Diastereomeric Derivatives," in Wainer, I. W., ed., *Drug Stereochemistry.*

Analytical Methods and Pharmacology, 2nd ed., Marcel Dekker, New York, 1993, Chap. 4.

177. Lindner, W. "Indirect Separation of Enantiomers by Liquid Chromatography," in Zeif, M. and Crane, L. J., eds., *Chromatographic Chiral Separations,* Marcel Dekker, New York, 1988, Chap. 4.

178. Ahnoff, M. and Einarsson, S. "Chiral Derviatization," in Lough, W. J., ed., *Chiral Liquid Chromatography,* Blackie, Glasgow, UK, 1989, Chap. 4.

179. Davankow, V. A. Zolotarev, Y. A., and Kurganov, A. A. *J. Liq. Chromatogr.* **1979**, *2*, 119.

180. Pirkle, W. H. and Hamper, B. C. "The Direct Preparative Resolution of Enantiomers by Liquid Chromatography on Chiral Stationary Phases," in Bidlingmeyer, B. A., ed., *Preparative Liquid Chromatography,* Elsevier, Amsterdam, The Netherlands, 1987, p. 235.

181. Zief, M. "Preparative Enantiomeric Separation," in Zief, M. and Crane, L. J., eds., *Chromatographic Chiral Separations,* Marcel Dekker, New York, 1988, Chap. 13.

182. Taylor, D. R. "Future Trends and Requirements," in Lough, W. J., ed., *Chiral Liquid Chromatography,* Blackie, Glasgow, UK, 1989, p. 287.

183. Francotte, E. and Junker-Buchheit, A. *J. Chromatogr.* **1992**, *576*, 1.

184. Prelog, V. and Wieland, P. *Helv. Chim. Acta* **1944**, *27*, 1127.

185. Hess, H., Burger, G., and Musso, H. *Angew. Chem. Int. Ed. Engl.* **1978**, *17*, 612.

186. Konrad, G. and Musso, H. *Liebigs Ann. Chem.* **1986**, 1956.

187. Pirkle, W. H. and House, D. W. *J. Org. Chem.* **1979**, *44*, 1957.

188. Pirkle, W. H. and Finn, J. M. *J. Org. Chem.* **1981**, *46*, 2935.

189. Pirkle, W. H. and Welch, C. J. *J. Org. Chem.* **1984**, *49*, 138.

190. Pirkle, W. H., Finn, J. M., Hamper, B. C., Schreiner, J., and Pribish, J. R. "A Useful and Conveniently Accessible Chiral Stationary Phase for the Liquid Chromatographic Separation of Enantiomers," in Eliel, E. L. and Otsuka, S., eds., *Asymmetric Reactions and Processes in Chemistry,* ACS Symposium Series 185, American Chemical Society, Washington, DC, 1982, p. 245.

191. Pirkle, W. H. and Finn, J. M. *J. Org. Chem.* **1982**, *47*, 4037.

192. Pirkle, W. H., Tsipouras, A., and Sowin, T. J. *J. Chromatogr.* **1985**, *319*, 392.

193. Pirkle, W. H. "New Developments in Chiral Stationary Phases for HPLC," *11th International Symposium on Column Liquid Chromatography*, Amsterdam, The Netherlands, 1987, cited in ref. 182.

194. Pirkle, W. H., Deming, K. C., and Burke III, J. A. *Chirality* **1991**, *3*, 183.

195. Hayes, K. S., Nagumo, M., Blount, J. F., and Mislow, K. *J. Am. Chem. Soc.* **1980**, *102*, 2773.

196. Roussel, C. and Chemlal, A. *New J. Chem.* **1988**, *12*, 947.

197. Agranat, I., Suissa, M. R., Cohen, S., Isaksson, R., Sandstrom, J., Dale, J., and Grace, D. *J. Chem. Soc. Chem. Commun.* **1987**, 381.

198. Okamoto, Y., Aburatani, R., Kaida, Y., and Hatada, K. *Chem. Lett.* **1988**, 1125.

199. Biali, S. E., Kahr, B., Okamoto, Y., Aburatani, R., and Mislow, K. *J. Am. Chem. Soc.* **1988**, *110*, 1917.

200. Ichida, A. and Shibata, T. "Cellulose Derivatives as Stationary Chiral Phases," in Zeif, M. and Crane, L. J., eds., *Chromatographic Chiral Separations,* Marcel Dekker, New York, 1988, Chap. 9.

201. Johns, D. M. "Binding to Cellulose Derivatives," in Lough, W. J., ed., *Chiral Liquid Chromatography,* Blackie, Glasgow, UK, 1989, p. 166.

202. Francotte, E., Lang, R. W., and Winkler, T. *Chirality* **1991**, *3*, 177.

203. Blaschke, G. *J. Liq. Chromatogr.* **1986**, *9*, 341.

204. Seebach, D., Gysel, U., and Kinkel, J. N. *Chimia* **1991**, *45*, 114.

205. Okamoto, Y., Yashima, E., Hatada, K., and Mislow, K. *J. Org. Chem.* **1984**, *49*, 557.

206. Okamoto, Y. and Hatada, K. "Optically Active Poly(Trophenylmethyl Methacrylate) as a Chiral Stationary Phase," in Zief, M. and Crane, L. J., eds., *Chromatographic Chiral Separations,* Marcel Dekker, New York, 1988, Chap. 8.

207. Yuki, H., Okamoto, Y., and Okamoto, I. *J. Am. Chem. Soc.* **1980**, *102*, 6356.

208. Mason, S. F. *Molecular Optical Activity and the Chiral Discriminations,* Cambridge University Press, Cambridge, UK, 1982.

209. Mislow, K. "Stereoisomerism," in Florkin, M. and Stotz, E. H., eds., *Comprehensive Biochemistry*, Vol. 1, Elsevier, Amsterdam, The Netherlands, p. 223.

210. Reider, P. J., Davis, P., Hughes, D. L., and Grabowski, E. J. J. *J. Org. Chem.* **1987**, *52*, 955.

211. Toda, F. and Tanaka, K. *Chem. Lett.* **1983**, 661.

212. Tanaka, K. and Toda, F. *Nippon Kagaku Kaishi* **1987**, *3*, 456; *Chem. Abstr.* **1987**, *107*, 197525g,

213. Clark, J. C., Phillipps, G. H., Steer, M. R., Stephenson, L., and Cooksey, A. R. *J. Chem. Soc. Perkin 1* **1976**, 471.

214. Clark, J. C., Phillipps, G. H., and Steer, M. R. *J. Chem. Soc. Perkin 1* **1976**, 475.

215. Shiraiwa, T., Chatani, T., Matushita, T., and Kurokawa, H. *Technol. Rep. Kansai Univ.* **1985**, *26*, 103; *Chem. Abstr.* **1986**, *104*, 149365w.

216. Shiraiwa, T., Sakata, S., and Kurokawa, H. *Chem. Express.* **1988,** *3*, 415; *Chem. Abstr.* **1989**, *110*, 95749c.

217. Numata, Y., Okawa, H., and Kida, S. *Chem. Lett.* **1979**, 293.

218. Openshaw, H. T. and Whittaker, N. *J. Chem. Soc.* **1963**, 1461.

219. Oppolzer, W. *Tetrahedron* **1987**, *43*, 1969.

220. Duhamel, L. and Launay, J.-C. *Tetrahedron Lett.* **1983**, *24*, 4209.

221. Hogeveen, H. and Zwart, L. *Tetrahedron Lett.* **1982**, *23*, 105.

222. Duhamel, L. and Plaquevent, J.-C. *Bull. Soc. Chim. Fr.* **1982**, II-69, II-75.

223. Duhamel, L. C. R. *Séances Acad. Sci. Ser. C* **1976**, *282*, 125.

224. Matsushita, H., Tsujino, Y., Noguchi, M., Saburi, M., and Yoshikawa, S. *Bull. Chem. Soc. Jpn.* **1978**, *51*, 862.

225. Pirkle, W. H. and Reno, D. S. *J. Am. Chem. Soc.* **1987**, *109*, 7189.

226. Pirkle, W. H. and Pochapsky, T. C. *J. Am. Chem. Soc.* **1987**, *109*, 5975.

227. Huckel, W., Mentzel, R., Brinkmann, W., and Goth, E. *Justus Liebigs Ann. Chem.* **1925**, *441*, 1. Seyer, W. F. and Walker, R. D. *J. Am. Chem. Soc.* **1938**, *60*, 2125.

228. Kaupp, G. *Angew. Chem. Int. Ed. Engl.* **1994**, *33*, 728.

229. Eliel, E. L. and Biros, F. I. *J. Am. Chem. Soc.* **1966**, *88*, 3334.

230. Pritchard, I. G. and Vollmer, R. L. *J. Org. Chem.* **1963**, *28*, 1545.

231. Bosnich, B. and Harrowfield, J. M. *J. Am. Chem. Soc.* **1972**, *94*, 3425.

232. Uneyama, K., Makio, S., and Nanbu, H. *J. Org. Chem.* **1989**, *54*, 872.

233. Sheehan, J. C. and Whitney, J. G. *J. Am. Chem. Soc.* **1963**, *85*, 3863.

234. Dummel, R. J. and Kun, E. *J. Biol. Chem.* **1969**, *244*, 2966.

235. Sharpless, K. B., Chong, A. O., and Scott, J. A. *J. Org. Chem.* **1975**, *40*, 1252.

236. Still, W. C., Hauck, P., and Kempf, D. *Tetrahedron Lett.* **1987**, *28*, 2817.

237. Baldwin, J. E. and Black, K. A. *J. Am. Chem. Soc.* **1984**, *106*, 1029.

238. Henkel, J. G. and Spector, J. H. *J. Org. Chem.* **1983**, *48*, 3657.

239. Denmark, S. E. and Almstead, N. G. *J. Am. Chem. Soc.* **1991**, *113*, 8089.

240. Elliott, M. L., Urban, F. J., and Bordner, J. *J. Org. Chem.* **1985**, *50*, 1752.

241. Krapcho, A. P. and Dundulis, E. A. *J. Org. Chem.* **1980**, *45*, 3236.

242. Whitesell, J. K., Minton, M. A., and Chen, K.-M. *J. Org. Chem.* **1988**, *53*, 5383.

243. Brown, H. C. and Zweifel, G. *J. Org. Chem.* **1962**, *27*, 4708.

244. Eliel, E. L. and Brett, T. J. *J. Org. Chem.* **1963**, *28*, 1923.

245. Aoyama, Y., Tanaka, Y., and Sugahara, S. *J. Am. Chem. Soc.* **1989**, *111*, 5397.

246. Snatzke, G., Wynberg, H., Feringa, B., Marsman, B. G., Greydanus, B., and Pluim, H. *J. Org. Chem.* **1980**, *45*, 4094.

247. Pluim, H. and Wynberg, H. *Tetrahedron Lett.* **1979**, 1251.

248. Bucciarelli, M., Forni, A., Marcacciolil, S., Moretti, I., and Torre, G. *Tetrahedron* **1983**, *39*, 187.

249. Rossiter, B. E. and Sharpless, K. B. *J. Org. Chem.* **1984**, *49*, 3707.

250. Ács, M., Pokol, G., Faig], F., and Fogassy, E. *J. Thermal Anal.* **1988**, *33*, 1241.

251. Fryzuk, M. D. and Bosnich, B. *J. Am. Chem. Soc.* **1978**, *100*, 5491.

252. Cervinka, O., Fabryova, A., and Sablukova, I. *Collect. Czech. Chem. Commun.* **1986**, *51*, 401.

253. Bir, G. and Kaufmann, D. *Tetrahedron Lett.* **1987**, *28*, 777.

254. Gutzwiller, J., Buchschacher, P., and Furst, A. *Synthesis* **1977**, 167.

255. Hengartner, U., Valentine, D., Jr., Johnson, K. K., Larcscheid, M. E., Pigott, F., Scheidl, F., Scott, J. W., Sun, R. C., Townsend, J. M., and Williams, T. H. *J. Org. Chem.* **1979**, *44*, 3741.

256. Kikukawa, T., Iizuka, Y., Sugimura, T., Harada, T., and Tai, A. *Chem. Lett.* **1987**, 1267.

257. Gabard, J. and Collet, A. *Nouv. J. Chim.* **1986**, *10*, 685.

259. Hanotier-Bridoux, M., Hanotier, J., and De Radzitzky, P. *Nature* (London) **1967**, *215*, 502.

260. Fleming, I. and Ghosh, S. K. *J. Chem. Soc. Chem. Commun.* **1994**, 99.

261. Vigneron, J. P., Dhaenens, M., and Horeau, A. *Tetrahedron* **1973**, *29*, 1055.

262. Kagan, H. B. and Fiaud, J. C. *Top. Stereochem.* **1988**, *18*, 249.

263. Brown, H. C. and Joshi, N. N. *J. Org. Chem.* **1988**, *53*, 4059.

264. Brown, H. C., Jadhav, P. K., and Desai, M. C. *J. Org. Chem.* **1982**, *47*, 4583.

265. Brown, H. C. and Singaram, B. *J. Org. Chem.* **1984**, *49*, 945.

266. Jadhav, P. K., Vara Prasad, J. V. N., and Brown, H. C. *J. Org. Chem.* **1985**, *50*, 3203.

267. Briaucourt, P. and Horeau, A. C. R. *Seances Acad. Sci. Ser. C* **1979**, *289*, 49.

268. Morrison, J. D. and Mosher, H. S. *Asymmetric Organic Reactions*, Prentice-Hall, Englewood Cliffs, NJ, 1971; American Chemical Society, Washington, DC, corrected reprint, 1976.

269. Schoofs, A. R. and Guette, J.-P. "Competitive Reaction Methods for the Determination of Maximum Specific Rotations," in Morrison, J. D., ed., *Asymmetric Synthesis*, Vol. 1, Academic Press, New York, 1983, Chap. 3.

270. Newman, P., Rutkin, P., and Mislow, K. *J. Am. Chem. Soc.* **1958**, *80*, 465.

271. Farina, M. *Top. Stereochem.* **1987**, *17*, 1.

272. Marckwald, W. and McKenzie, A. *Ber. Dtsch. Chem. Ges.* **1899**, *32*, 2130.

273. Martin, V. S., Woodard, S. S., Katsuki, T., Yamada, Y., Ikeda, M., and Sharpless, K. B. *J. Am. Chem. Soc.* **1981**, *103*, 6237.

274. Discordia, R. P. and Dittmer, D. C. *J. Org. Chem.* **1990**, *55*, 1414.

275. Schreiber, S. L., Schreiber, T. S., and Smith, D. B. *J. Am. Chem. Soc.* **1987**, *109*, 1525.

276. Chinchilla, R., Nájera, C., Yus, M., and Heumann, A. *Tetrahedron: Asymmetry* **1990**, *1*, 851.

277. Halpern, B. and Westley, J. W. *Aust. J. Chem.* **1966**, *19*, 1533.

278. Jermyn, M. A. *Aust. J. Chem.* **1967**, *20*, 2283.

279. Bell, K. H. *Aust. J. Chem.* **1979**, *32*, 65.

280. Franck, A. and Rüchardt, C. *Chem. Lett.* **1984**, 1431.

281. Salz, U. and Rüchardt, C. *Chem. Ber.* **1994**, *117*, 3457; see also Rüchardt, C., Gartner, H., and Salz, U. *Angew. Chem. Int. Ed. Engl.* **1984**, *23*, 162.

282. Ellis, M. K., Golding, B. T., and Watson, W. P. *J. Chem. Soc. Chem. Commun.* **1984**, 1600.

283. Wiesner, K., Jay, E. W. K., Tsai, T. Y. R., Demerson, C., Jay, L., Kanno, T., Krepinsky, J., Vilim, A., and Wu, C. S. *Can. J. Chem.* **1972**, *50*, 1925.

284. Hiraki, Y. and Tai, A. *Chem. Lett.* **1982**, 341.

285. Huber, U. A. and Dreiding, A. S. *Helv. Chim. Acta* **1970**, *53*, 495.

286. Mori, A. and Yamamoto, H. *J. Org. Chem.* **1985**, *50*, 5444.

287. Hua, D. H. *J. Am. Chem. Soc.* **1986**, *108*, 3835.

288. Cinquini, M., Colonna, S., and Cozzi, F. *J. Chem. Soc. Perkin Trans. 1* **1988**, 247.

289. Patterson, L., McClure, C. K., and Schumann, R. C. *Tetrahedron Lett.* **1989**, *30*, 1293.

290. Wegener, B., Hansen, M., and Winterfeldt, E. *Tetrahedron: Asymmetry* **1993**, *4*, 345.

291. Brown, H. C., Ayyangar, N. R., and Zweifel, G. *J. Am. Chem. Soc.* **1964**, *86*, 397.

292. Masamune, S., Kim, B., Petersen, J. S., Sato, T., Veenstra, S. J., and Imai, T. *J. Am. Chem. Soc.* **1985**, *107*, 4549.

293. Johnson, C. R. and Meanwell, N. A. *J. Am. Chem. Soc.* **1981**, *103*, 7667.

294. Whitesell, J. K. and Carpenter, J. F. *J. Am. Chem. Soc.* **1987**, *109*, 2839.

295. Brown, J. M. *Chem. Ind London* **1988**, 612.

296. Brown, J. M. *Chem. Br.* **1989**, *25*, 276.

297. Finn, M. G. and Sharpless, K. B. "On the Mechanism of Asymmetric Epoxidation with Titanium-Tartrate Catalysts," Morrison, J. D., ed., *Asymmetric Synthesis*, Vol. 5, Academic Press, New York, 1985, Chap. 8.

298. Rossiter, B. E. "Synthetic Aspects and Applications of Asymmetric Epoxidation" in Morrison, J. D., ed., *Asymmetric Synthesis*, Vol. 5, Academic Press, New York, 1985, Chap. 7.

299. Gao, Y., Hanson, R. M., Klunder, J. M., Ko, S. Y., Masamune, H., and Sharpless, K. B. *J. Am. Chem. Soc.* **1987**, *109*, 5765.

300. Carlier, P. R., Mungall, W. S., Schröder, G., and Sharpless, K. B. *J. Am. Chem. Soc.* **1988**, *110*, 2978.

301. Miyano, S., Lu, L. D.-L., Viti, S. M., and Sharpless, K. B. *J. Org. Chem.* **1985**, *50*, 4350.

302. Kobayashi, Y., Kusakabe, M., Kitano, Y., and Sato, F. *J. Org. Chem.* **1988**, *53*, 1586.

303. Phillips, M. L., Berry, D. M., and Panetta, J. A. *J. Org. Chem.* **1992**, *57*, 4047.

304. Brown, J. M. and Cutting, I. *J. Chem. Soc. Chem. Commun.* **1985**, 578.

305. Kitamura, M., Kasahara, I., Manabe, K., Noyori, R., and Takaya, H. *J. Org. Chem.* **1988**, *53*, 708.

306. Kitamura, M., Manabe, K., Noyori, R., and Takaya, H. *Tetrahedron Lett.* **1987**, *28*, 4719.

307. James, B. R. and Young, C. G. *J. Organomet. Chem.* **1985**, *285*, 321.

308. Hayashi, T., Yamamoto, A., and Ito, Y. *J. Chem. Soc. Chem. Commun.* **1986**, 1090.

309. Meurling, L. *Chem. Scr.* **1974**, *6*, 92.

310. Mazaleyrat, J. P. and Welvart, Z. *Nouv. J. Chim.* **1983**, *7*, 491.

311. Sépulchre, M., Spassky, N., and Sigwalt, P. *Macromolecules* **1972**, *5*, 92.

312. Spassky, N., Leborgne, A., and Sépulchre, M. *Pure Appl. Chem.* **1981**, *53*, 1735.

313. Leborgne, A., Spassky, N., and Sigwalt, P. *Polym. Bull. (Berlin)* **1979**, *1*, 825.

314. Okamoto, Y., Suzuki, K., Kitayama, T., Yuki, H., Kageyama, H., Miki, K., Tanaka, N., and Kasai, N. *J. Am. Chem. Soc.* **1982**, *104*, 4618.

315. Pasteur, L. *C. R. Acad Sci.* **1858**, *46*, 615.

316. Pasteur, L. *Researches on the Molecular Asymmetry [sic] of Natural Organic Products*, Alembic Club Reprint No. 14, W. F. Clay, Edinburgh, UK, 1860, p. 43.

317. Chen, C.-S., Fujimoto, Y., Girdaukas, G., and Sith, C. J. *J. Am. Chem. Soc.* **1982**, *104*, 7294.

318. Fülling, G. and Sih, C. J. *J. Am Chem. Soc.* **1987**, *109*, 2845.

319. Laumen, K., Breitgoff, D., and Schneider, M. P. *J. Chem. Soc. Chem. Commun.* **1988**, 1459.

320. Wu, S.-H., Guo, Z.-W., and Sih, C. J. *J. Am. Chem. Soc.* **1990**, *112*, 1990.

321. Jones, J. B. and Jakovac, I. J. *Can. J. Chem.* **1982**, *60*, 19.

322. Prelog, V., Stojanac, Z., and Kovacević, A. *Helv. Chim. Acta* **1982**, *65*, 377.

323. Pollack, S. J., Jacobs, J. W., and Schultz, P. G. *Science* **1986**, *234*, 1570.

324. Tramontano, A., Janda, K. D., and Lerner, R. A. *Science* **1986**, *234*, 1566.

325. Schultz, P. G. *Acc. Chem. Res.* **1989**, *22*, 287.

326. Schultz, P. G. and Lerner, R. A. *Acc. Chem. Res.* **1993**, 26, 391.

327. Napper, A. D., Benkovic, S. J., Tramontano, A., and Lerner, R. A. *Science* **1987**, *237*, 1041.

328. Benkovic, S. J. personal communication to SHW, 1991.

329. Janda, K. D., Benkovic, S. J., and Lerner, R. A. *Science* **1989**, *244*, 437.

330. Whitesides, G. M. and Wong, C.-H. *Angew. Chem. Int. Ed. Engl.* **1985**, *24*, 617.

331. Wong, C.-H. *Science* **1989**, *244*, 1145.

332. Davies, H. G., Green, R. H., Kelly, D. R., and Roberts, S. N. *Biotransformations in Preparative Organic Chemistry,* Academic Press, San Diego, CA, 1989.

333. Abramowicz, D. A., ed. *Biocatalysis,* Van Nostrand Reinhold, New York, 1990.

334. Jones, J. B. and Beck, J. F. "Asymmetric Syntheses and Resolutions Using Enzymes," in Jones, J. B., Sih, C. J., and Perlman, D., eds., *Applications of Biochemical Systems in Organic Synthesis, Technique of Chemistry,* Vol. 10, Part 1, Wiley-Interscience, New York, 1976, Chap. 4.

335. Fischli, A. "Chiral Building Blocks in Enantiomer Synthesis Using Enzymatic Transformations," in Scheffold, R., ed., *Modern Synthetic Methods,* Vol. 2, Salle and Sauerländer, Frankfurt, Germany, and Aarau, Switzerland, 1980.

336. Svedas, V. and Galaev, I. U. *Usp. Khim.* **1983**, *52*, 2039; *Russ. Chem. Rev.* **1983**, *52*, 1184.

337. Jones, J. B. "Enzymes as Chiral Catalysts," in Morrison, J. D., ed., *Asymmetric Synthesis,* Vol. 5, Academic Press, New York, Chap. 9.

338. Verkhovskaya, M. A. and Yamskov, I. A. *Usp. Khim.* **1991**, 2250; *Russ. Chem. Rev.* **1991**, *60*, 1163.

339. Williams, R. M. and Hendrix, J. A. *Chem. Rev.* **1992**, *92*, 889.

340. Roper, J. M. and Bauer, D. P. *Synthesis* **1983**, 1041.

341. Ohta, H., Miyamae, Y., and Kimura, Y. *Chem. Lett.* **1989**, 379.

342. Zaks, A. and Klibanov, A. M. *Science* **1984**, *224*, 1249.

343. Klibanov, A. M. *CHEMTECH* **1986**, *16*, 354.

344. Klibanov, A. M. *Acc. Chem. Res.* **1990**, *23*, 114.

345. Kirchner, G., Scollar, M. P., and Klibanov, A. M. *J. Am. Chem. Soc.* **1985**, *107*, 7072.

346. Langrand, G., Baratti, J., Buono, G., and Triantaphylides, C. *Tetrahedron Lett.* **1986**, *27*, 29.

347. Stokes, T. M. and Oehlschlager, A. C. *Tetrahedron Lett.* **1987**, *28*, 2091.

348. Belan, A., Bolte, J., Fauve, A., Gourcy, J. G., and Veschambre, H. *J. Org. Chem.* **1987**, *52*, 256.

349. Deagueil-Castaing, M., De Jeso, B., Drouillard, S., and Maillard, B. *Tetrahedron Lett.* **1987**, *28*, 953.

350. Wang, Y.-F., Lalonde, J. J., Momongan, M., Bergbreiter, D. E., and Wong, C.-H. *J. Am. Chem. Soc.* **1988**, *110*, 7200.

351. Laumen, K. and Schneider, M. P. *J. Chem. Soc. Chem. Commun.* **1988**, 598.

352. Inagaki, M., Kiratake, J., Nishioka, T., and Oda, J. *J. Org. Chem.* **1992**, *57*, 5643.

353. Fujimoto, Y., Iwadate, H., and Ikekawa, N. *J. Chem. Soc. Chem. Commun.* **1985**, 1333.

354. Ganey, M. V., Padykula, R. E., Berchtold, G. A., and Braun, A. G. *J. Org. Chem.* **1989**, *54*, 2787.

355. Miyano, S., Kawahara, K., Inoue, Y., and Hashimoto, H. *Chem. Lett.* **1987**, 355.

356. Kyba, E. P., Koga, K., Sousa, L. R., Siegel, M. G., and Cram, D. J. *J. Am. Chem. Soc.* **1973**, *95*, 2692.

357. Cram, D. J. and Cram, J. M. *Acc. Chem. Res.* **1978**, *11*, 8.

358. Helgeson, R. C., Timko, J. M., Moreau, P., Peacock, S. C., Mayer, J. M., and Cram, D. J. *J. Am. Chem. Soc.* **1974**, *96*, 6762.

359. Goldberg, I. *J. Am. Chem. Soc.* **1977**, *99*, 6049.

360. Kyba, E. P., Timko, J. M., Kaplan, L. J., de Jong, F., Gokel, G. W., and Cram, D. J. *J. Am. Chem. Soc.* **1978**, *100*, 4555.

361. Behr, J.-P. and Lehn, J.-M. *J. Am. Chem. Soc.* **1973**, *95*, 6108.

362. Newcomb, M., Helgeson, R. C., and Cram, D. J. *J. Am. Chem. Soc.* **1974**, *96*, 7367.

363. Newcomb, M., Toner, J. L., Helgeson, R. C., and Cram, D. J. *J. Am. Chem. Soc.* **1979**, *101*, 4941.

364. Thoma, A. P., Cimerman, Z., Fiedler, U., Bedeković, D., Güggi, M., Jordan, P., May, K., Pretsch, E., Prelog, V., and Simon, W. *Chimia* **1975**, *29*, 344.

365. Thoma, A. P., Pretsch, E., Horvai, G., and Simon, W., in Semenza, G., and Caforoli, E., eds., *Biochemistry of Membrane Transport,* FEBS Symposium No. 42, Springer, Berlin, p. 116.

366. Mislow, K. *Introduction to Stereochemistry,* Benjamin, New York, 1965.

367. Lowry, T. M. *J. Chem. Soc.* **1899**, *75*, 211.

368. Smith, G. G. and Sivakua, T. *J. Org. Chem.* **1983**, *48*, 627.

369. Yamada, S., Hongo, C., Yoshioka, R., and Chibata, I. *J. Org. Chem.* **1983**, *48*, 843.

370. Williams, K. M. and Smith, G. G. *Orig. Life* **1977**, *8*, 91.

371. Canceill, J., Collet, A., and Gottarelli, G. *J. Am. Chem. Soc.* **1984**, *106*, 5997.

372. Huffman, J. W. and Wallace, R. H. *J. Am. Chem. Soc.* **1989**, *111*, 8691.

373. Weber, L., Imiolczyk, I., Haufe, G., Rehorek, D., and Hennig, H. *J. Chem. Soc. Chem. Commun.* **1992**, 301.

374. Zawadzke, L. E. and Berg, J. M. *J. Am. Chem. Soc.* **1992**, *114*, 4002.

375. Henderson, J. W. *Chem. Soc. Rev.* **1973**, *2*, 397.

376. Westheimer, F. H. "Calculation of the Magnitude of Steric Effects," in Newman, M. S., ed., *Steric Effects in Organic Chemistry,* Wiley, New York, 1956, Chap. 12.

377. Mislow, K., Glass, M. A. W., Hopps, H. B., Simon, E., and Wahl, G. H., Jr. *J. Am. Chem. Soc.* **1964**, *86*, 1710. Mislow, K., Graeve, R., Gordon, A. J., and Wahl, G. H., Jr. *J. Am. Chem. Soc.* **1964**, *86*, 1733.

378. Martin, R. H. and Marchant, M. J. *Tetrahedron* **1974**, *30*, 347.

379. Martin, R. H. and Libert, V. *J. Chem. Res., Synop.* **1980**, 130; *Miniprint* **1980**, 1940.

380. Martin, R. H. *Angew. Chem. Int. Ed. Engl.* **1974**, *13*, 649.

381. Paquette, L. *Acc. Chem. Res.* **1993**, *26*, 476.

382. Ruxer, J.-M., Solladié, G., and Candau, S. *J. Chem. Res. Synop.* **1978**, 82.

382. Takano, S., Kijima, A., Sugihara, T., Satoh, S., and Ogasawara, K. *Chem. Lett.* **1989**, 87.

383. Rayner, D. R., Gordon, A. J., and Mislow, K. *J. Am. Chem. Soc.* **1968**, *90*, 4854.

384. Forni, A. I., Moretti, I., Prosyanik, A. V., and Torre, G. *J. Chem. Soc. Chem. Commun.* **1981**, 588.

385. Chan, K.-K., Cohen, N., De Noble, J. P., Specian, A. C., and Saucy, G. *J. Org. Chem.* **1976**, *41*, 3497.

386. Roussel, C., Adjimi, M., Chemlal, A., and Djafri, A. *J. Org. Chem.* **1988**, *53*, 5076.

387. Sutliff, T. M. S. Thesis, Ohio State University, 1966; cited in Newman, M. S., *An Advanced Organic Laboratory Course*, Macmillan, New York, 1972.

388. Soffer, M. D. and Günay, G. E. *Tetrahedron Lett.* **1965**, 1355.

389. Murahashi, S.-I., Yoshimura, N., Tsumiyama, T., and Kojima, T. *J. Am. Chem. Soc.* **1983**, *105*, 5002.

390. Baldwin, J. E. *J. Chem. Soc. Chem Commun.* **1988**, 31.

391. March, J. *Advanced Organic Chemistry,* 4th ed., Wiley, New York, 1992.

392. Murr, B. L. and Feller, L. W. *J. Am. Chem. Soc.* **1968**, *90*, 2966.

393. Murr, B. L. and Santiago, C. *J. Am. Chem. Soc.* **1968**, *90*, 2964.

394. Denney, D. B. and DiLeone, R. *J. Org. Chem.* **1961**, *26*, 984.

395. Flowers, G. C. and Leffler, J. E. *J. Org. Chem.* **1989**, *54*, 3995.

396. Greenberg, A., Molinaro, N., and Lang, M. *J. Org. Chem.* **1984**, *49*, 1127.

397. Hargreaves, M. K. and Modarai, B. *J. Chem. Soc. C.* **1971**, 1013.

398. Wilen, S. H., Bunding, K. A., Kascheres, C. M., and Wieder, M. J. *J. Am. Chem. Soc.* **1985**, *107*, 6997.

399. Doyle, T. R. and Vogl, O. *J. Am. Chem. Soc.* **1989**, *111*, 8510.

400. Guthrie, R. D. and Nicolas, E. C. *J. Am. Chem. Soc.* **1981**, *103*, 4637.

401. Mills, A. K. and Smith, A. E. W. *Helv. Chim. Acta* **1960**, *43*, 1915.

402. Kenyon, G. L. and Hegeman, G. D. *Biochemistry* **1970**, *9*, 4036.

403. Fee, J. A., Hegeman, G. D., and Kenyon, G. L. *Biochemistry* **1974**, *13*, 2528.

404. Sharp, T. R., Hegeman, G. D., and Kenyon, G. L. *Biochemistry* **1977**, *16*, 1123.

405. Benoiton, N. L. "Quantitation and Sequence Dependence of Racemization in Peptide Synthesis," in Gross, E. and Meienhofer, J., eds., *The Peptides: Analysis, Synthesis, Biology, Vol. 5, Special Methods in Peptide Synthesis,* Part B, Academic Press, New York, 1983, p. 217.

406. Bada, J. L. "Racemization of Amino Acids," in Barrett, G. C., ed., *Chemistry and Biochemistry of Amino Acids,* Chapman and Hall, London, 1985, Chap. 13.

407. Bodanszky, M. *Peptide Chemistry,* Springer, Berlin, 1988, Chap. 8.

408. Smith, G. G., Khatib, A., and Reddy, G. S. *J. Am. Chem. Soc.* **1983**, *105*, 293.

409. Dakin, H. D. *Am. Chem. J.* **1910**, *44*, 48.

410. Robinson, R. *Outline of an Electrochemical (Electronic) Theory of the Course of Organic Reactions,* The Institute of Chemistry of Great Britain and Ireland, London, 1932.

411. Neuberger, A. *Adv. Protein Chem.* **1948**, *4*, 297.

412. Lazarus, R. A. *J. Org. Chem.* **1990**, *55*, 4755.

413. Smith, G. G. and Baum, R. *J. Org. Chem.* **1987**, *52*, 2248.

414. Seebach, D., Boes, M., Naef, R., and Schweizer, W. B. *J. Am. Chem. Soc.* **1983**, *105*, 5390.

415. Bonner, W. A. and Lemmon, R. M. *Bioorg. Chem.* **1978**, *7*, 175.

416. Bonner, W. A. and Lemmon, R. M. *Orig. Life Evol. Biosphere*, **1981**, *11*, 321.

417. Bonner, W. A., Hall, H., Chow, G., Liang, Y., and Lemmon, R. M. *Orig. Life Evol. Biosphere* **1985**, *15*, 103.

418. Bonner, W. A. *Top. Stereochem.* **1988,** *18*, 1.

419. Abelson, P. H. *Carnegie Inst. Wash. Yearb.* **1955**, *54*, 107.

420. Hare, P. E. and Mitterer, R. M. *Carnegie Inst. Wash. Yearb.* **1967**, *67*, 205.

421. Sykes, G. A. *Chem. Br.* **1988**, *24*, 235.

422. Meyer, V. R. "Amino Acid Racemization. A Tool for Dating," in Ahuja, S., ed., *Chiral Separations by Liquid Chromatography,* ACS Symposium Series No. 471, American Chemical Society, Washington, DC, 1991, Chap. 13.

423. Meyer, V. R. *CHEMTECH* **1992**, *22*, 412.

8

HETEROTOPIC LIGANDS AND FACES: PROSTEREOISOMERISM AND PROCHIRALITY

8-1. INTRODUCTION AND TERMINOLOGY

Often (e.g., in stereoselective synthesis) one is interested in the fact that in certain molecules, such as propionic acid (Fig. 8.1, **1**), a nonstereogenic center (here C_α) can be transformed into a stereogenic center by replacement of one or other of two apparently identical ligands by a different one. Such ligands are called[1a] "homomorphic" from Greek *homos* meaning same and *morphe* meaning form; they are identical only when separated from the rest of the molecule. Thus the replacement of H_A at C_α in propionic acid by OH generates the chiral center of (S)-lactic acid (Fig. 8.1, **2**), whereas the analogous replacement of H_B gives rise to the enantiomeric (R)-lactic acid. The C_α center in propionic acid has therefore been called a "prochiral center"[2] (but see ref. 3). H_A and H_B at such a center are called "heterotopic ligands"[1b,1c,4,5] from Greek *heteros* meaning different and *topos* meaning place (see also Section 8-3). Prochiral axes and planes may similarly be defined in relation to chiral axes and planes (see below).

Substitution is one of the common ways of interconverting organic molecules, another is addition. The chiral center in lactic acid (Fig. 8.1, **2**) can also be generated by the addition of hydride (e.g., from sodium borohydride) to the carbonyl group of pyruvic acid (Fig. 8.1, **3**). Depending on which face of the keto acid the hydride adds to, either (S)- or (R)-lactic acid is obtained. Addition to the rear face of the keto acid as depicted in Figure 8.1 will give rise to (S)-lactic acid **2**, whereas (R)-lactic acid is

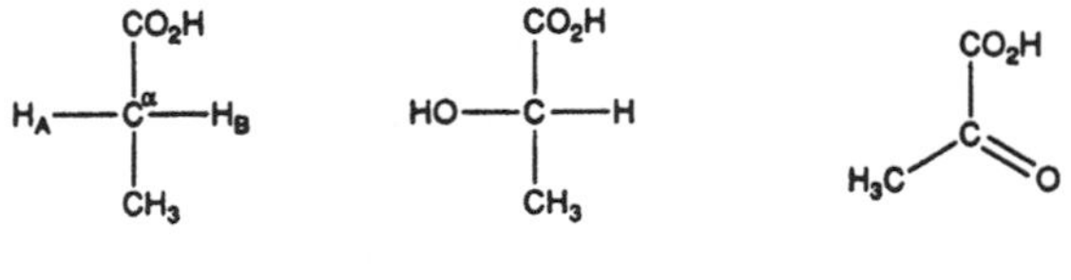

Figure 8.1. Chiral and prochiral molecules.

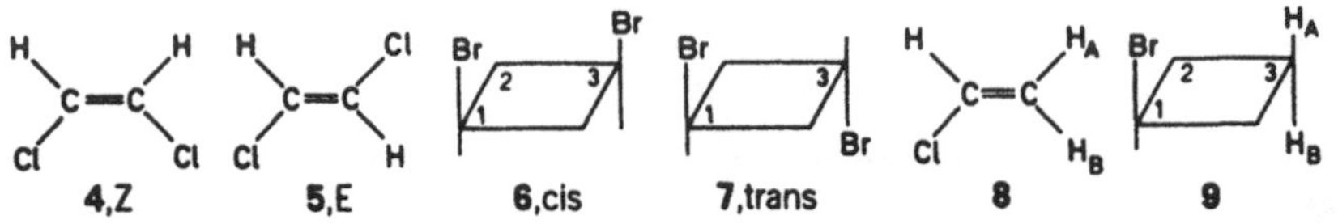

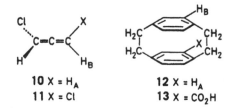

Figure 8.2. Stereogenic and prostereogenic elements.

obtained by addition to the front face. Thus the carbonyl group in pyruvic acid is also said to be prochiral and to present two heterotopic faces.

Although the term prochirality is properly used in relation to prochiral centers, faces, axes, and so on,[6] it suffers from a limitation that arises from a corresponding limitation in the definition of chirality. We have already seen that molecules, such as cis–trans isomers of alkenes and certain cis–trans isomers of cyclanes, may display stereochemical differences without being chiral. Thus (Z)- and (E)-1,2-dichloroethylene (**4**, **5**) are achiral diastereomers, as are *cis*- and *trans*-1,3-dibromocyclobutanes (**6**, **7**) (Fig. 8.2). Thus, just as it is inappropriate to associate stereoisomerism only with the occurrence of chiral elements (cf. Chapter 3), the concept of prochirality needs to be generalized to one of prostereoisomerism.[1a] It is exemplified by chloroethylene (Fig. 8.2, **8**) and bromocyclobutane (**9**); these molecules display prostereoisomerism inasmuch as replacement of the homomorphic atoms H_A and H_B in **8** by chlorine gives rise to achiral stereoisomers **5** and **4**, respectively. Similarly, replacement of H_A and H_B, respectively, in **9** by bromine gives rise to the diastereoisomers **6** and **7**. Thus **9** has a prostereogenic (but not prochiral) center (center of prostereoisomerism) at C(3) and **8** may be said to have a prostereogenic axis (axis of prostereoisomerism) coinciding with the axis of the double bond. The atoms H_A and H_B in both **8** and **9** are heterotopic.

Cases of a prochiral axis (in allene **10**, convertible by replacement of H_A by Cl into chiral allene **11**) and a prochiral plane (in paracyclophane **12**, which can be converted into the chiral structure **13** by replacement of H_A by CO_2H) are shown in Figure 8.3 (see also Chapter 13).

Figure 8.3. Chiral and prochiral axes and planes.

8-2. SIGNIFICANCE AND HISTORY

The most significant aspect of the present subject of prostereoisomerism lies in the possibility of differentiating heterotopic ligands or heterotopic faces. The concept of heterotopic ligands and the possibility of their differentiation, in suitable instances, by NMR spectroscopy was first presented in a pioneering article by Mislow and Raban.[4] Differentiation of heterotopic ligands or faces may be chemical or biochemical (as in stereoselective synthesis, including transformations by enzymes) or spectroscopic (notably by NMR spectroscopy). Before entering upon these topics in detail, we pose here a challenge to illustrate the utility of the concepts: In citric acid (Fig. 8.4, **14**), can the four methylene hydrogen atoms H_A, H_B, H_C, and H_D be distinguished by NMR spectroscopy, or by virtue of their involvement in the enzymatic dehydration of citric acid to *cis*-aconitic acid **15**, or both? This question can easily be answered once the tenets of prostereoisomerism are understood: All the hydrogen atoms can be distinguished by appropriate enzymatic reactions and H_A and H_B (as well as H_C and H_D) can give rise to distinct signals in the ^{1}H NMR spectrum, whereas H_A and H_C (or H_B and H_D) give rise to coincident signals, except in chiral media, where all four protons differ in chemical shift.[7].

Let us consider citric acid (Fig. 8.4, **14**). It was long known[8,9] that when oxaloacetic acid **16** labeled at C(4) (Fig 8.5) is taken through the Krebs cycle, the α-ketoglutaric

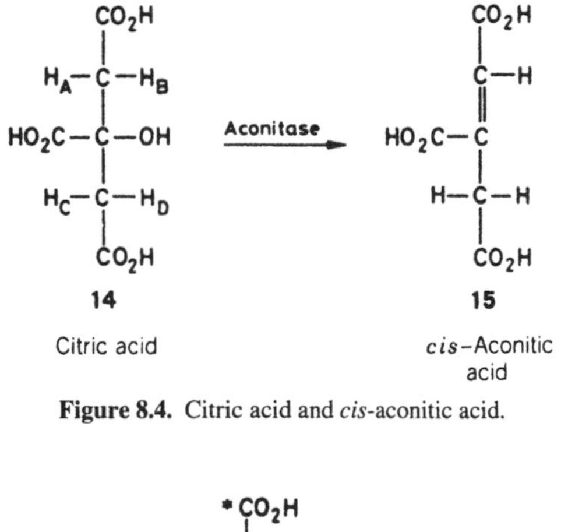

14 Citric acid — **15** *cis*-Aconitic acid

Figure 8.4. Citric acid and *cis*-aconitic acid.

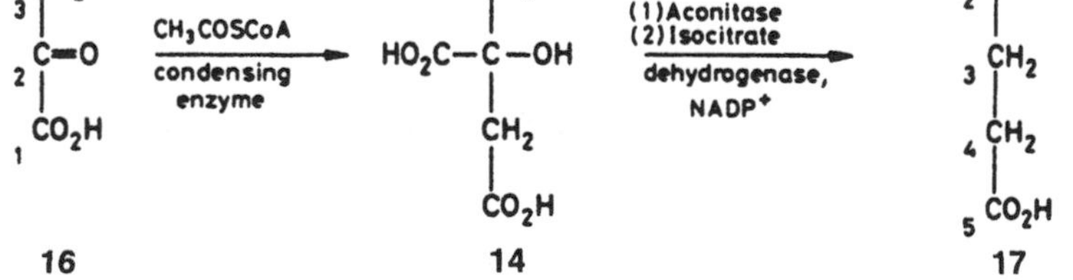

16 **14** **17**

Figure 8.5. Part of citric acid cycle.

acid **17** formed is labeled exclusively at C(1) (next to the keto group), and not at all at C(5). This finding seemed to throw doubt on the theretofore assumed intermediacy of citric acid **14** in the cycle since, it was argued, the two ends of citric acid — CH_2CO_2H are "equivalent" and, therefore, the α-ketoglutaric acid formed through this intermediate should be labeled equally at C(1) and C(5). However, it is now clear (see Sections 8-3 and 8-5) that the experiment in no way eliminates citric acid as a potential intermediate in the oxaloacetic acid–α-ketoglutaric acid transformation, since the two CH_2CO_2H branches are, in fact, distinct and distinguishable by enzymes because they are heterotopic (enantiotopic, see below). Similarly, since phosphorylation of glycerol **18** with adenosine triphosphate (ATP) in the presence of the enzyme glycerokinase gives exclusively (*R*)-(−)-(glycerol I-phosphate) (Fig. 8.6, **19**),[10] it is clear that the enzyme can distinguish between the two enantiotopic primary alcohol groups of glycerol.

The first glimpse of understanding of this type of differentiation came when Ogston[11] (see Ref. 12 for a historical review) pointed out that an attachment of a substrate Caa'bc (a = a') to an enzyme at three sites (so-called three-point contact) could lead to the observed distinction between the homomorphic (as we would now say) groups a and a', as shown in Figure 8.7. If A is a catalytically active site on the enzyme and B and C are binding sites, Figure 8.7 shows that only a but not a' can be brought into juxtaposition with the active site A when b and c are bound to B and C. Therefore a but not a' may be enzymatically transformed; a and a' are clearly distinguishable.

The first mention of a three-point contact (between a chiral drug and its receptor) is actually found in an article by Easson and Stedman published in 1933,[13] and a year later, Max Bergmann[14] postulated a three-point contact (involving CO_2H, H_2N, and the dipeptide linkage) between peptidases and the dipeptides hydrolyzed by them.

At this point, we must pick up one other, at the beginning apparently unrelated, historical thread. In 1957 two groups of investigators[15,16] discovered that in molecules of the type $CX_2Y\overset{*}{C}abc$ [e.g., $CF_2Br\overset{*}{C}HBrC_6H_5$ or $CH_2Br\overset{*}{C}(CH_3)BrCO_2CH_3$] the X nuclei (F in the first example, H in the second) display distinct NMR signals. Although the phenomenon was not clearly understood until some time later,[17,18] it is now clear that the nonequivalence of such X nuclei in NMR rests on the same symmetry

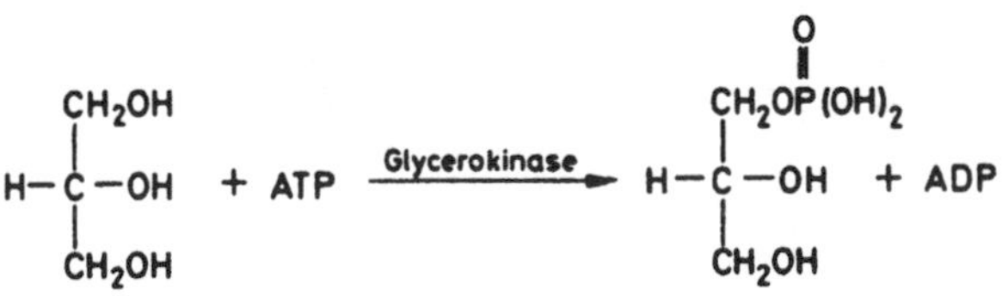

Figure 8.6. Enzymatic phosphorylation of glycerol.

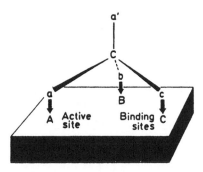

Figure 8.7. Ogston's three-point contact model. [From Florkin, M. and Stotz. G., eds. *Comprehensive Biochemistry*, Vol. 12, Elsevier, Amsterdam, 1964, p. 237.]

principles[4] as the earlier-mentioned nonequivalence in enzymatic reactions and other reactions involving chiral reagents. Sections 8-3 to 8-5 will deal with the explanation of these nonequivalencies and their chemical and spectral consequences.

8-3. HOMOTOPIC AND HETEROTOPIC LIGANDS AND FACES

Ligands and faces may be homotopic or heterotopic.[4,5,19] Heterotopic ligands and faces may be either enantiotopic or diastereotopic. In the following sections we shall define these terms and explain how the pertinent nature of ligands or faces is recognized. (For a preview of the terminology, see Fig. 8.16.)

a. Homotopic Ligands and Faces

We indicated in Section 8-2 that some apparently alike (homomorphic) ligands are, in fact, not equivalent toward enzymes or in their NMR signals. How does one decide, then, whether such ligands are equivalent (homotopic) or not? There are two alternative criteria: a substitution and a symmetry criterion.[4] Similar criteria (addition or symmetry) serve to test the equivalency (homotopicity) of faces.

Substitution and Addition Criteria.   Two homomorphic ligands (see p. 303) are homotopic if replacement of first one and then the other by a different ligand leads to the same structure. (The replacement ligand must be different not only from the original one but also from all other ligands attached to the same atom.) Thus, as shown in Figure 8.8, the two hydrogen atoms in methylene chloride **20** are homotopic because replacement of either by, say, bromine gives the same $CHBrCl_2$ molecule **21**; the three methyl hydrogen atoms in acetic acid **22** are homotopic because replacement of any one of them by, say, chlorine gives one and the same chloroacetic acid **23** (fast rotation about the C–C bond being assumed; see below); the two methine hydrogen

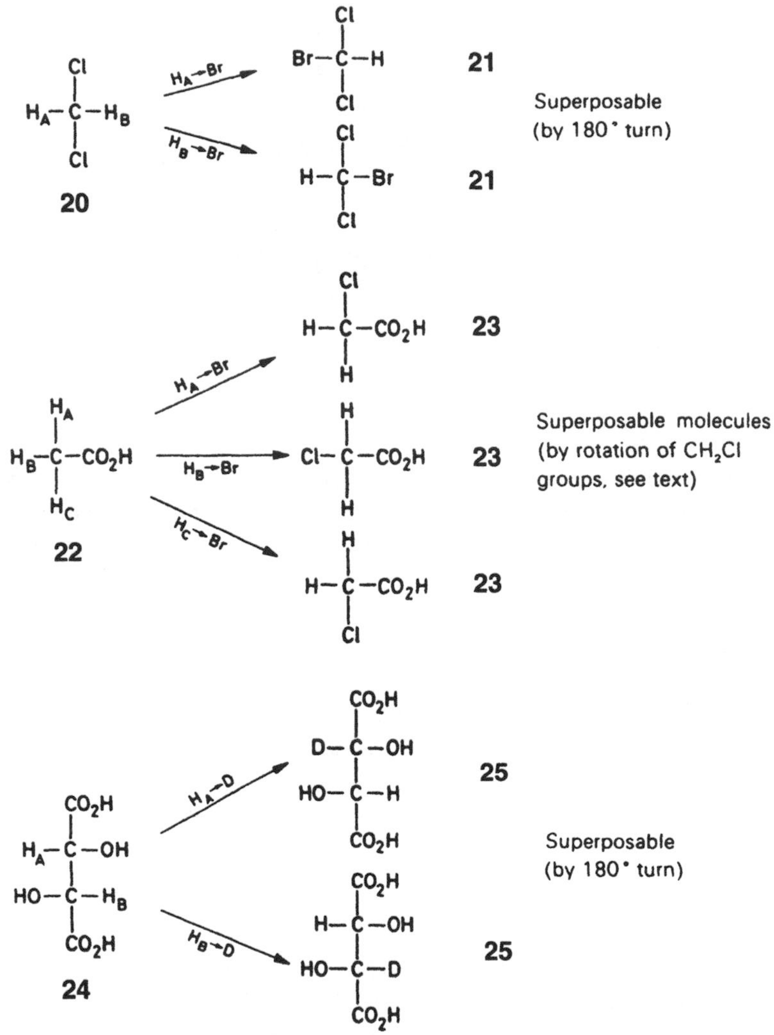

Figure 8.8. Homotopic ligands.

atoms in (R)-(+)-tartaric acid **24** are homotopic because replacement of either of them, for example by deuterium, gives the same (2R,3R)-tartaric-2-d acid **25**.

Two corresponding faces of a molecule (usually, but not invariably, faces of a double bond) are homotopic when addition of the same reagent to either face gives the same product. For example, addition of HCN to acetone **26** will give the same cyanohydrin **27**, no matter to which face addition occurs (Fig. 8.9) and addition of bromine to ethylene similarly gives $BrCH_2CH_2Br$ regardless of the face of approach. The two faces of the C=O double bond of acetone and of the C=C double bond of ethylene are thus homotopic.

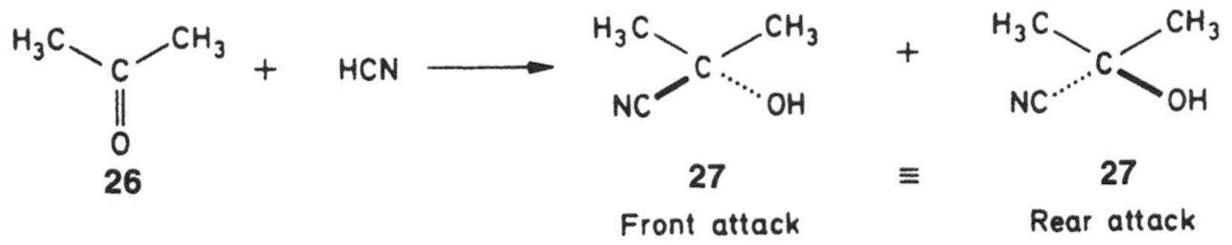

Figure 8.9. Homotopic faces: addition of HCN to acetone.

Symmetry Criterion.   Ligands are homotopic if they can interchange places through operation of a C_n symmetry axis (cf. Chapter 4). Thus the chorine atoms in methylene chloride (Fig. 8.8, **20**; symmetry point group C_{2v}) are homotopic since they exchange places through a 180° turn around the C_2 axis (C_2^1). Similarly, the methine hydrogen atoms of (+)-tartaric acid (Fig. 8.8, **24**) are interchanged by operation of the C_2 axis (the molecule belongs to point group C_2). It is essential that the operation of the symmetry axis in fact interchanges the homotopic ligands: H_A and H_B in Fig. 8.10 are *not* homotopic. The situation in acetic acid is somewhat more complicated. If we depict this molecule as stationary in one of its eclipsed conformations, we see (Fig. 8.11) that the hydrogen atoms are heterotopic. However, rotation around the H_3C—CO_2H axis is rapid on the time scale of most experiments. We are therefore dealing with a case of averaged symmetry leading to interchange of the three methyl hydrogen atoms of CH_3CO_2H, which are thus homotopic when rotation is fast on the time scale of whatever experiment is being considered.

Faces of double bonds are similarly homotopic when they can be interchanged by operation of a symmetry axis. (Since there are only two such faces, the pertinent axis must, of necessity, be of even multiplicity so as to contain C_2.) Thus the two faces of acetone (Fig. 8.9) are interchanged by the operation of the C_2 axis (the molecule is of symmetry C_{2v}); the two faces of ethylene (D_{2h}) are interchanged by operation of two of the three C_2 axes (either the one containing the C=C segment or the axis at right angles to the first one and in the plane of the double bond).

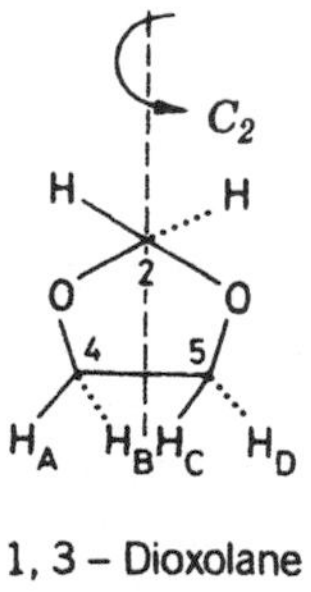

1, 3 – Dioxolane

Figure 8.10. 1,3-Dioxolane.

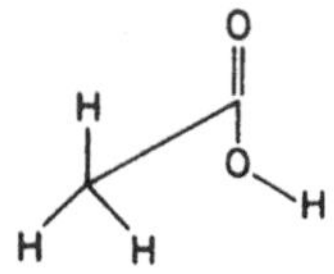

Acetic acid

Figure 8.11. Eclipsed conformation of acetic acid.

b. Enantiotopic Ligands and Faces

Just as one divides stereoisomers into two sets, enantiomers and diastereomers, it is convenient to divide heterotopic (nonequivalent) ligands or faces into enantiotopic and diastereotopic moieties (cf. Fig. 8.16). Enantiotopic ligands are ligands in mirror-image positions, whereas diastereotopic ligands are in stereochemically distinct positions not related in mirror-image fashion; similar considerations relate to faces of double bonds.

Substitution–Addition Criterion.   Two ligands are enantiotopic if replacement of either one of them by a different achiral ligand (which also must be different from other ligands attached to the prochiral element) gives rise to enantiomeric products. Examples are shown in Figure 8.12. The subscripted hydrogen atoms in CH_2ClBr (**28**), *meso*-tartaric acid (**30**), cyclobutanone (**32**) [at C(2) and C(4) but not at C(3)], and chloroallene (**34**) [at C(3)] are enantiotopic as are the methyl carbon atoms in isopropyl alcohol (**36**). *meso*-Tartaric acid exemplifies one of the rare instances of a molecule with heterotopic ligands but no discernible prochiral atom or other element of prochirality.

Similar criteria, but of addition, can be established for enantiotopic faces. Faces are enantiotopic if addition of the same achiral reagent to either one or the other will give rise to enantiomeric products. Thus (Fig. 8.13) addition of HCN to the two enantiotopic faces of acetaldehyde gives rise to the two enantiomers of lactonitrile. [Here, as in the case of substitution, the added group must be different from any group already there. Thus we cannot test the enantiotopic nature of the two faces of the C=O function of acetaldehyde **38** by addition of CH_3MgI, since the group added ($CH_3—$) is the same as one of the existing groups.]

Symmetry Criterion.   Enantiotopic ligands and faces are not interchangeable by operation of a symmetry element of the first kind (C_n, simple axis of symmetry) but must be interchangeable by operation of a symmetry element of the second kind (σ, plane of symmetry; i, center of symmetry; or S_n, alternating axis of symmetry). It follows that, since chiral molecules cannot contain a symmetry element of the second kind, there can be no enantiotopic ligands or faces in chiral molecules. (Nor, for different reasons, can such ligands or faces occur in linear molecules, $C_{\infty v}$, or $D_{\infty h}$.)

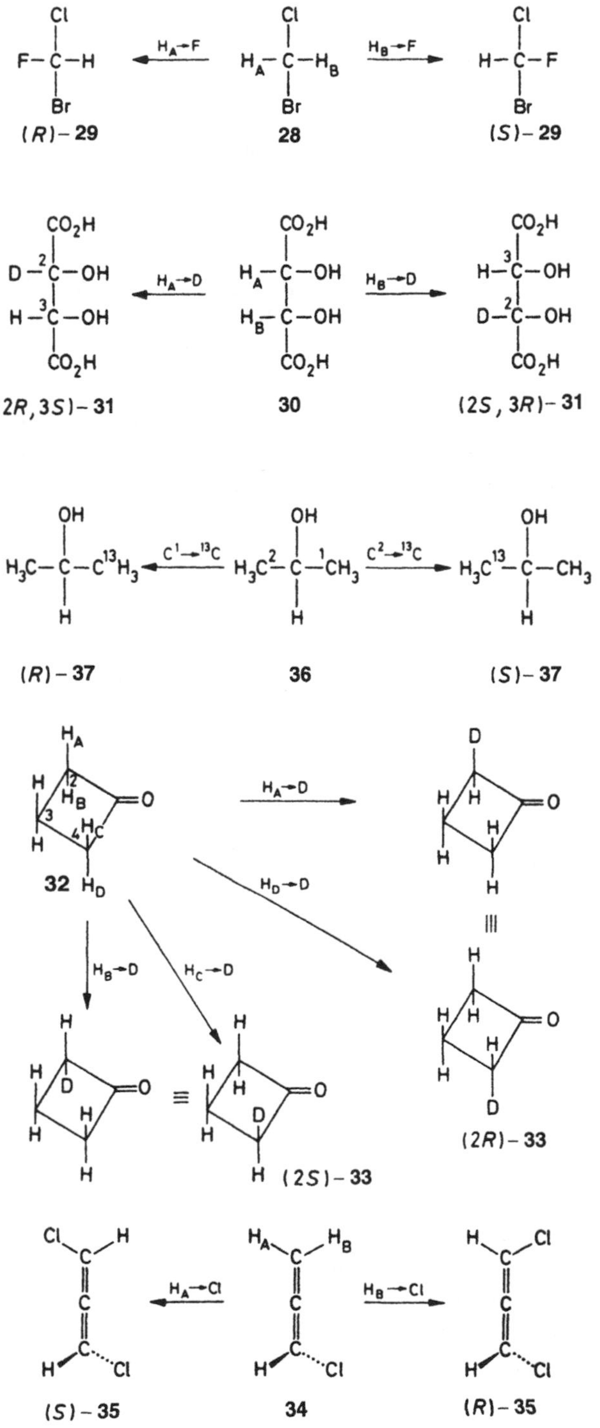

Figure 8.12. Enantiotopic ligands.

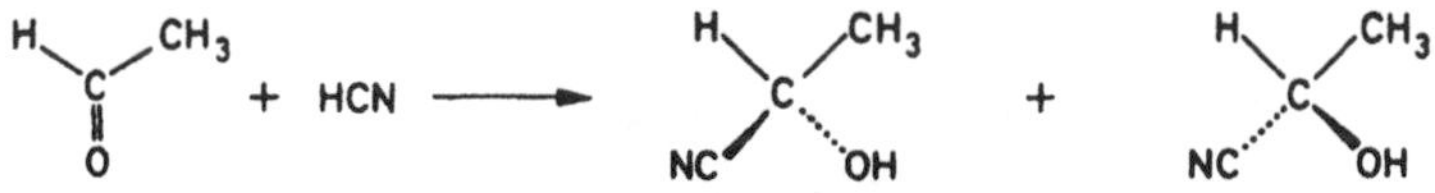

Figure 8.13. Addition of HCN to acetaldehyde.

The symmetry planes σ in molecules **28, 30, 32, 34**, and **36** in Figure 8.12 should be readily evident. It is possible to have both homotopic and enantiotopic ligands in the same set, as exemplified by the case of cyclobutanone **32**: H_A and H_D are homotopic as are H_B and H_C. The ligand H_A is enantiotopic with H_B and H_C; H_D is similarly enantiotopic with H_C and H_B. The sets $H_{A,B}$ and $H_{C,D}$ may be called equivalent (or homotopic) sets of enantiotopic hydrogen atoms. The unlabeled hydrogen atoms at position 3, constitutionally distinct (see Section 3.4) from those at C(2, 4), are homotopic with respect to each other. Enantiotopic ligands need not be attached to the same atom, as seen in the case of *meso*-tartaric acid **30**, and also the just-mentioned pair H_A, H_C (or H_B, H_D) in cyclobutanone.

Enantiotopic faces (Fig. 8.13) are also related by a symmetry plane (e.g., the plane of the double bond in **38**). The faces must not be interchangeable by operation of a symmetry axis, otherwise they are homotopic rather than enantiotopic.

Just as enantiomeric molecules cannot be distinguished in achiral environments, neither can enantiotopic ligands. Such ligands can, however, be distinguished by NMR spectroscopy in nonracemic chiral media[20] or in the presence of chiral shift reagents (discussed in Chapter 6), in synthetic transformations involving either chiral reagents or other types of chiral environment, asymmetric syntheses, and, above all, in enzymatic reactions, since the enzyme catalysts are chiral (cf. Section 8-5). It is because of these potential distinctions between enantiotopic ligands and faces that it is important to be able to recognize them.

c. Diastereotopic Ligands and Faces

The earlier mentioned criteria may also be employed to recognize diastereotopic ligands, that is, ligands that are located in a stereochemically distinct but nonmirror-image environment.

Substitution–Addition Criterion.   Figure 8.14 shows a number of cases where replacement of first one and then another of two homomorphic ligands by a different achiral test ligand gives rise to diastereomeric products.

Such ligands are called diastereotopic and are generally distinct both spectroscopically and chemically. Their NMR signals will generally be different (cf. Section 8-4) and their reactivity will, in general, be unequal.

The case of **40** (Fig. 8.14) is straightforward: since C(2) in 2-bromobutane is chiral, H_A and H_B cannot be enantiotopic and the replacement criterion discloses that they are

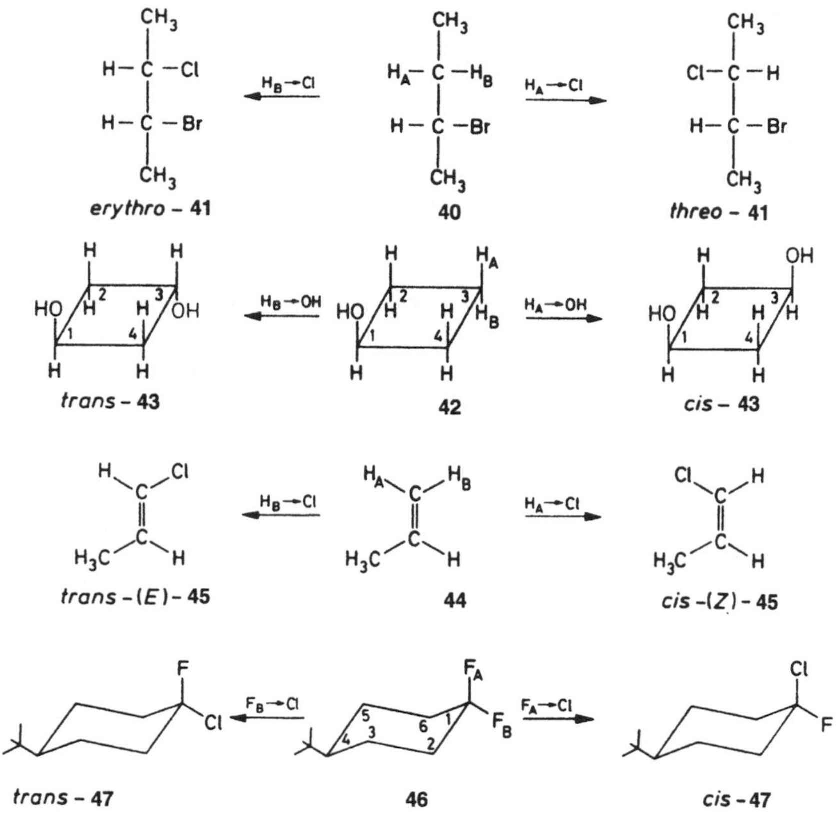

Figure 8.14. Diastereotopic ligands.

diastereotopic rather than homotopic. The examples of cyclobutanol **42** and 4-*tert*-butyl-1,1-difluorocyclohexane **46** show (cf. Section 8-1) that presence of a chiral center is not required for the existence of diastereotopic nuclei. The ligands H_A and H_B in **42** and F_A and F_B in **46** are diastereotopic because they are cis and trans, respectively, to the hydroxyl group at C(1) in **42** or the *tert*-butyl group at C(4) in **46**. It might be noted that, after replacement, C(3) in **43** or C(1) in **47** is not a chiral center but is a stereogenic center; the corresponding atoms in **42** and **46** are prostereogenic. In the case of propene (**44**) replacement of H_A and H_B generates a cis–trans pair of (diastereomeric) alkenes, again making H_A and H_B diastereotopic. (One is cis to the methyl group at the distal carbon atom, the other is trans.)

The addition criterion may similarly be applied to recognize diastereotopic faces. Methyl α-phenethyl ketone (**48**, Fig. 8.15) has a chiral center, thus HCN addition gives rise to diastereomers **49a** and **49b**; hence the faces of the carbonyl carbon are diasteoreotopic. This case is of importance in conjunction with Cram's rule (Chapter 5). Compounds **50** and **52** also display diastereotopic faces even though the products

51 and **53** are not chiral; these are cases of prostereogenicity but not prochirality. α-Phenethyl methyl sulfide **54** displays diastereotopic sides of a molecular plane not due to a double bond and may alternatively be considered a case of diastereotopic ligands (unshared pairs on sulfur); attachment of oxygen to one or other of the diastereotopic pairs gives diastereomeric sulfoxides.

Symmetry Criterion. The symmetry criteria of diastereotopic ligands or faces are simple: Such ligands or faces must not be related either by a symmetry element of the first kind (axis) or by one of the second kind (plane, center, or alternating axis). The reader should become convinced that the even-numbered molecules depicted in Figures 8.14 and 8.15 (middle column) are either devoid of such symmetry elements or that, when such elements (e.g., σ) are present, their operation does not serve to interchange the ligands or faces designated as being diastereotopic.

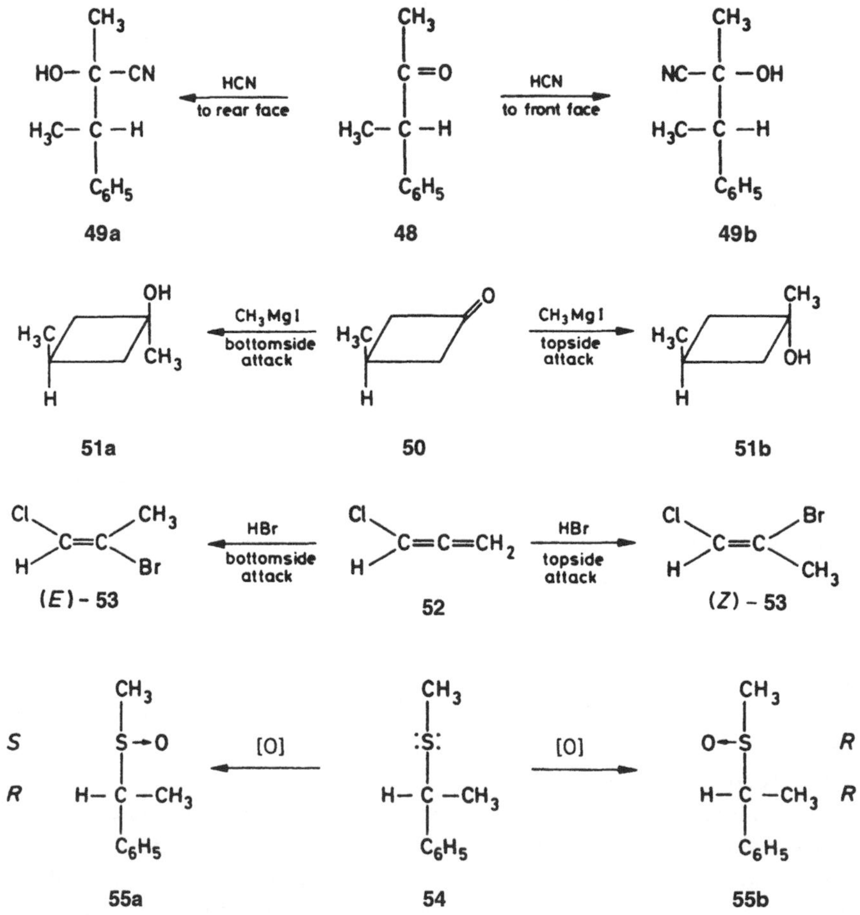

Figure 8.15. Diastereotopic faces of double bonds.

Ligands may be diastereotopic by external as well as by internal comparison. Corresponding ligands in diastereomers are diastereotopic under any circumstances; corresponding ligands in enantiomers are diastereotopic when viewed in a chiral environment (e.g., a chiral solvent; cf ref. 20b; see also Section 6-5.c, p. 153).

d. Concepts and Nomenclature

Concepts. First it may be well to review the symmetry criteria for homotopic, enantiotopic, and diastereotopic ligands or faces.[4] Ligands, or faces, are equivalent or homotopic when they can be brought into coincidence by operation of a proper (C_n) symmetry axis. If this condition is not fulfilled but the ligands or faces can be brought into coincidence by operation of an improper (S_n) axis of symmetry, including a plane σ, or center i of symmetry, the ligands or faces are enantiotopic. If neither symmetry operation C_n nor S_n brings the ligands or faces into coincidence, they are diastereotopic or (see below) constitutionally heterotopic.

It is illuminating to make a comparison between isomeric and "homomeric" (identitical) compounds on the one hand, and heterotopic or homotopic ligands or faces on the other. There is logic to such a comparison since it was explained earlier that stereoisomers are generated by appropriate replacement of heterotopic ligands or addition to heterotopic faces. Figure 8.16 displays such a comparison.[1b] It is convenient, in conjunction with the diagram of homotopic and heterotopic ligands (Fig. 8.16) to introduce an additional term: If homomorphic ligands (e.g., the hydrogen atoms in a methylene group) occur in constitutionally distinct portions of a

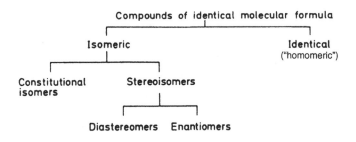

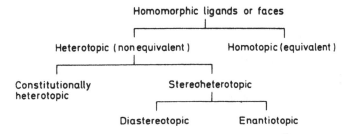

Figure 8.16. Classification of compounds and ligands.[1b]

molecule, we call them constitutionally heterotopic. Examples would be the methylene hydrogen atoms at C(2) and those at C(3) in cyclobutanol (Fig. 8.14, **42**).

Constitutionally heterotopic ligands are in principle always distinguishable, just as constitutional isomers are. (The same is true of diastereotopic ligands.) Diastereotopic and enantiotopic ligands or faces may be lumped together under the term "stereoheterotopic" just as diastereomers and enantiomers are both called stereoisomers.

Nomenclature.   Just as it is convenient to distinguish enantiomers and diastereomers by appropriate descriptors (*R, S, E, Z*, etc.) it is desirable to provide descriptors for stereoheterotopic ligands or faces. The basic nomenclature to this end has been provided by Hanson[1c,2] and is closely related to the nomenclature of stereoisomers.

If, in a prostereogenic assembly (e.g., a prochiral center Caabc) a hypothetical priority, in the sense of the sequence rules (cf. Chapter 5), is given to one of the identical ligands a over the other a′, that ligand a will be called "*pro-R*" if the newly created "chiral center" Caa′bc (sequence a > a′) has the *R* configuration, but it will be called "*pro-S*" if the newly created "chiral center" has the *S* configuration. Let us take ethanol (Fig. 8.17, **56**) as an example. The hydrogen atoms H_A and H_B are enantiotopic. If preference is given to H_A over H_B in the sequence rule, the sequence is OH, CH_3, H_A, H_B and the (hypothetical) configurational symbol for **56** would be *R*, hence H_A is *pro-R*; by default, H_B, is *pro-S*. The answer would have come out the same if H_B had been given precedence over H_A; in that case the sequence would have been OH, CH_3, H_B, H_A and the hypothetical configurational symbol for **56** is then *S*, hence H_B is *pro-S*. It might be noted (cf. Fig. 8.17) that the same result would have been obtained by replacing first one hydrogen and then the other by deuterium since deuterium has priority over hydrogen (Chapter 5); replacement of H_A by D gives (*R*)-ethanol-1-*d* [(*R*)-**57**], and hence H_A is *pro-R*; similarly, replacement of H_B by deuterium gives (*S*)-ethanol-1-*d* [(*S*)-**57**], and hence H_B is *pro-S*.

In a formula, the "*pro-R* group X" is sometimes written as X_R (and similarly X_S for the *pro-S* group). It is important, however, to read X_R as "the *pro-R* group X" and not as the "*R* group X," since heterotopicity or prochirality, not chirality, is implied. Indeed, Figure 8.18 shows a case (type of CGGXY) where both CH_3CHOH (G) ligands have the *S*-configuration but the upper one is *pro-R*, whereas the lower one is *pro-S*.

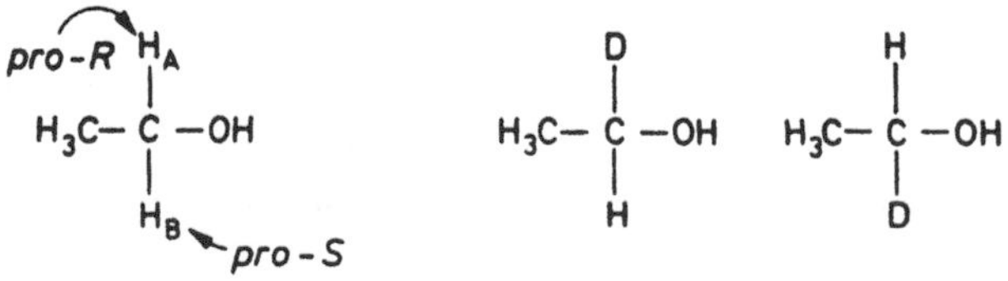

Figure 8.17. Ethanol, (*R*)-ethanol-1-*d* and (*S*)-ethanol-1-*d*.

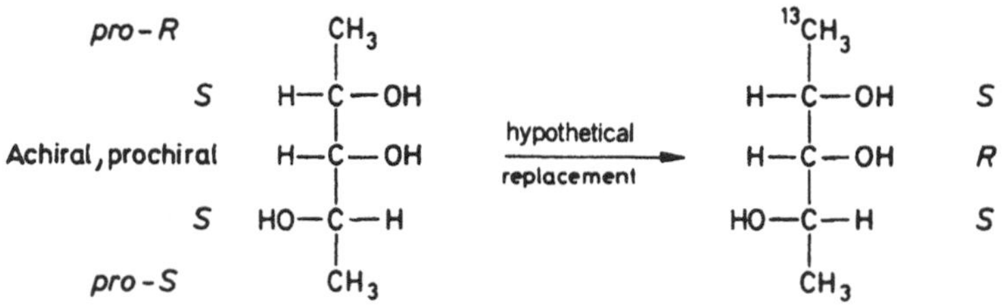

Figure 8.18. Molecule in which a ligand of *S*-configuration is *pro-R*.

In the original structure, the central atom C(3) is achiral (cf. Chapter 3). When priority is given to the upper ligand (indicated in Fig. 8.18 by replacement of CH_3 by $^{13}CH_3$), however, C(3) becomes chiral, and since its configuration is then *R*, the upper CH_3CHOH ligand is *pro-R*.[21]

Hanson also devised a specification of heterotopic faces. Thus if one looks at the plane of the face and the CIP sequence of the attached atoms or ligands is clockwise, one calls it *Re*; if the sequence is counterclockwise, one calls it *Si*, these being the first two letters of *Rectus* and *Sinister*, respectively. Thus the face of acetaldehyde turned toward the reader in Figure 8.13 is *Si* (O, CH_3, H are in counterclockwise order) whereas the front face in **48** (Figure 8.15) is *Re*. (See also ref. 22.)

We have already mentioned (cf. Fig. 8.3) that heterotopicity may also be found to exist in cases where replacement of one of two homomorphic ligands gives rise to molecules of axial or planar chirality. Compounds **10** (Fig. 8.3) and **34** (Fig. 8.12) are examples of axial prochirality giving rise to enantiotopic ligands; compound **12** in Figure 8.3 is an example of planar prochirality giving rise to such ligands. Figure 8.19 shows examples of axial prochirality giving rise to diastereotopic ligands[23] in **58** (ref. 24) and **59**.

Although systematic nomenclature is generally to be preferred, there are some instances (e.g., in steroids) where a local system of nomenclature is still generally used. Thus in 3-cholestanone (Fig. 8.20) the hydrogen atoms above the plane of the paper, which itself represents a projection of the three-dimensional (3D) molecule, are called β and those below the plane α (cf. Chapter 11). Since the geminal hydrogen atoms at each methylene carbon form a diastereotopic pair, it is clear that diastereotopic hydrogen atoms in such pairs may be distinguished by calling them H_α and H_β and this is commonly done. There is obviously no 1:1 connection of such

$$C_2H_5CHBr\text{--}CO\text{--}C(CH_3)\text{=}C\text{=}CH_2$$

58

$$(CH_3)_2CHCR = C = CR'R''$$

59

Figure 8.19. Prochiral axes and planes.

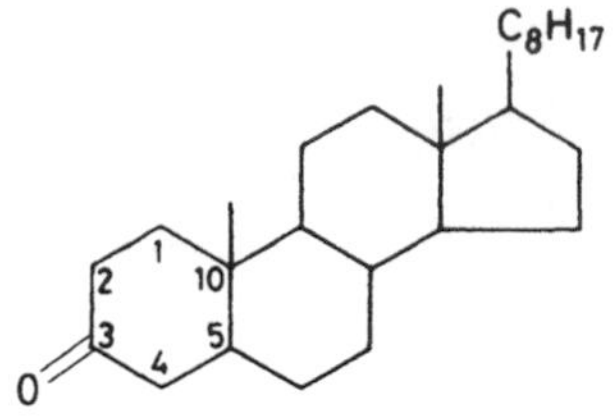

Figure 8.20. 3-Cholestanone.

common with systematic nomenclature; for example, the β-hydrogen atoms at C(2) is *pro-S* but that at C(4) is *pro-R*. Not surprisingly this lack of correlation parallels that between α/β and *R/S* (Chapter 11) when one looks at chiral centers in steroids (as in 2- and 4-cholestanol). The α and β designation may also be used for heterotopic faces; thus the front (*Si*) face of the keto function at C(3) is β and the rear (*Re*) face is α.

8-4. HETEROTOPICITY AND NUCLEAR MAGNETIC RESONANCE

a. General Principles. Anisochrony

Nuclei that are diastereotopic will, in principle, differ in chemical shift (see the reviews in refs. 26 and 27) that is, they will be "anisochronous" (cf. ref. 3, p. 23; the term was coined by G. Binsch following the use of "isochronous" for chemical-shift equivalent).[28] Although such chemical shift differences are often seen, sometimes they are so small that the signals can be resolved only at quite high fields, or not at all. In the latter situation one speaks of "accidental isochrony," meaning that while the nuclei are in principle anisochronous, they are not, in fact, resolved.

Anisochrony for diastereotopic ligands is seen with a number of different nuclei. We have already mentioned that $CH_2BrC(CH_3)BrCO_2CH_3$ displays different signals for the diastereotopic protons (italicized)[16] and that $CF_2BrCHBrC_6H_5$ displays different resonances for the diastereotopic fluorine nuclei.[15] The diastereotopic methyl groups in the ferrocenyl cation **60** (Fig. 8.21) are distinct both in their ^1H and ^{13}C signals.[29]

The immediate cause for anisochrony is, of course, the unequal magnetic field sensed by the diastereotopic nuclei. It follows that, as the source of the diastereotopic environment is removed further and further from the test nuclei, the anisochrony is expected to diminish. This prediction has been tested[30] with the results shown in Table 8.1; the anomaly seen in entry 4 may be due to the molecule "coiling back on itself" so that the methyl groups sense the shielding or deshielding effect of the benzene ring differentially. Table 8.1 also shows a solvent effect; it is thus desirable, in looking for potential anisochronies, to record the spectrum in several different solvents, for example, CCl_4, $CDCl_3$, benzene-d_6, or pyridine-d_5.[27,31,32] Another way of enhancing (or manifesting) anisochronies is to use lanthanide shift reagents.[31]

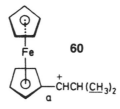

Figure 8.21. Example of diastereotopic methyl groups.

Since NMR is a scalar probe, enantiotopic nuclei are isochronous (i.e., they have the same chemical shift) in achiral media. Such nuclei, however, become diastereotopic in chiral media and thus in principle (though not necessarily in practice) anisochronous. Among many example[20b] are the enantiotopic methyl protons of dimethyl sulfoxide (DMSO), CH_3SOCH_3, which are shifted with respect to each other by[33] 0.02 ppm in nonracemic solvent $C_6H_5CHOHCF_3$. Surprisingly, the ^{13}C signals of the two methyl groups are not resolved under these conditions; this is an exception to the rule that ^{13}C signals of diastereotopic methyl groups generally show larger relative shifts than their ^1H signals.[34,35] Similarly, the methyl protons of DMSO[36] are anisochronous in the presence of chiral lanthanide shift reagents (see Chapter 6) as are the enantiotopic carbinol protons of alcohols RCH_2OH.[37]

The detection of diastereotopic nuclei by NMR is possible only if the diastereotopic nature of such nuclei is maintained on the time scale of the NMR experiment. Thus the equatorial and axial fluorine atoms in 1,1-difluorocyclohexane (Fig. 8.22), though diastereotopic, give rise to a single NMR signal because the rate of interchange of these nuclei by ring reversal at room temperature (ca. 100,000 s^{-1}) is much greater than the shift between the fluorine nuclei (884 Hz: at 56.4 MHz or 884 s^{-1}; cf. Chapter 11; see also ref. 38, p. 158). However, the fluorine atoms F^1 and F^2 become anisochronous below $-46°C$ when interconversion between the two chair forms (Fig. 8.22, **A, B**) is slowed to a rate less than the separation of the fluorine signals. This situation will be further discussed in Section 8-4.d and in Chapter 11.

TABLE 8.1. Observed Anisochrony of Diastereotopic CH_3 Protons in $(CH_3)_2CH—X—CH(CH_3)C_6H_5$

| Entry | X | Shift Difference | |
		in CCl_4 (ppm)	in C_6H_6 (ppm)
1	None	0.182	0.133
2	O	0.067	0.013
3	OCH_2	0.005	0.008
4	OCH_2CH_2	0.042	0.030
5	OCH_2CH_2O	0.000	0.013
6	$OCH_2CH_2OCH_2$	0.000	0.000

A B

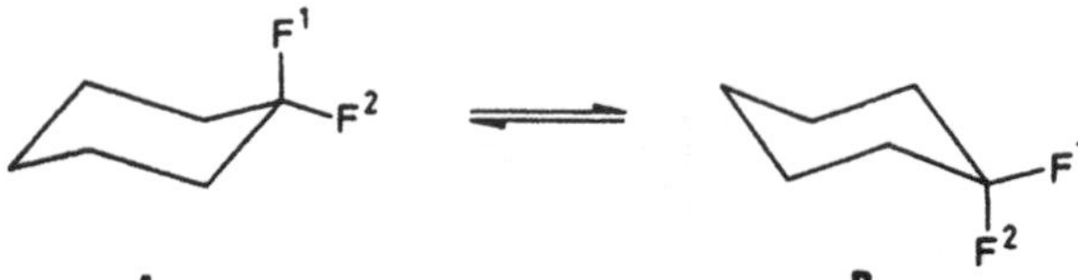

Figure 8.22. 1,1-Difluorocyclohexane.

b. NMR in Assignment of Configuration and of Descriptors of Prostereoisomerism

Determination of Configuration. This section describes the use of the concept of heterotopicity in assignment of stereochemical configuration[39] (see also Chapter 5), usually relative (especially meso vs. chiral) rather than absolute configuration, as well as assignment of the appropriate symbols to heterotopic ligands (i.e., experimental recognition as to which ligand is *pro-R* and which is *pro-S* at a prochiral center). Recognition of heterotopic faces as *Re* and *Si* is usually obvious from the configuration of the addition products thereto and will not be discussed here; examples are found in Section 8.5.b.

In favorable circumstances, chiral and meso stereoisomers may be distinguished directly; an acyclic and a cyclic example are shown in Figure 8.23 (**61**, **62**). In both cases the methylene protons H_C in the chiral species (**61b**, **62b**) are related by a C_2 axis and are therefore homotopic and isochronous, whereas the corresponding protons H_A and H_B in the meso forms (**61a**, **62a**) are not related by either C_2 or σ and thus are diastereotopic and anisochronous. The situation is not altered when the racemic mixture rather than an individual enantiomer is compared with the meso isomer: The (internally homotopic) methylene protons of the two enantiomers are externally enantiotopic and so remain isochronous.

When no suitable probes for distinction of meso forms and racemic pairs are present in the molecule, such probes may sometimes be introduced by combination

meso 61a | chiral 61b | meso 62a | chiral 62b

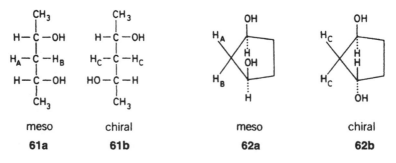

Figure 8.23. Distinction of active and meso forms by NMR.

with appropriate reagents. An example[40] is depicted in Figure 8.24. Benzylation of the amines **63** and **64** gives the *N*-benzyl derivatives **65** and **66**. In **65**, which is derived from the meso isomer **63**, H_A and H_B are enantiotopic, and hence isochronous; the protons constitute a single (A_2) signal. In contrast, in **68**, which is derived from the chiral isomer (whether or not racemic), the benzylic protons are diastereotopic, and hence anisochronous, and constitute an AB system.

Use of this methodology is risky when only one stereoisomer is available. If the benzyl derivative displays a single signal, it may be because one deals with a species of type **65** or because accidental isochrony is encountered in a species of type **66**. If the latter is the case, the method falls even when both stereoisomers are available.

The interplay of external and internal diastereotopicity may sometimes foil attempts to distinguish racemic from meso isomers.[5b] However, this difficulty can be alleviated in cases where internally diastereotopic nuclei couple with each other (of course externally diastereotopic ones cannot do so). An example[33] is the distinction of meso and racemic 2-butylene oxide (*cis*- and *trans*-2,3-dimethyloxirane, **67**, **68**) by means of a chiral shift reagent (Eu,* Fig. 8.25). Upon complexation with such a reagent, the internally enantiotopic C—H protons of the meso isomer **67** become

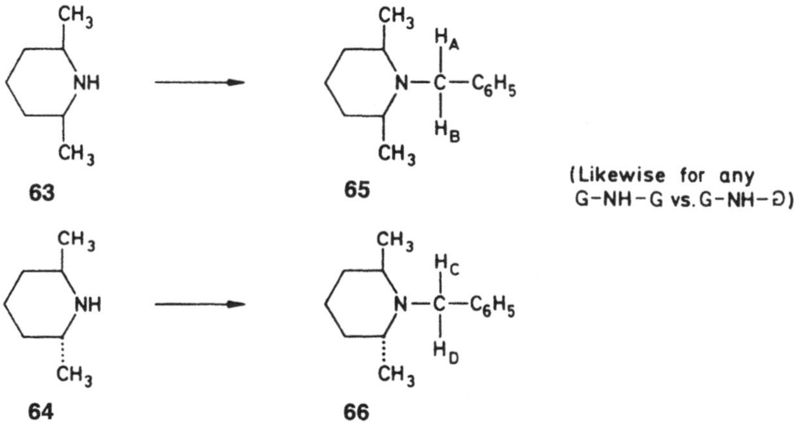

Figure 8.24. Achiral probe to distinguish racemic and meso forms.

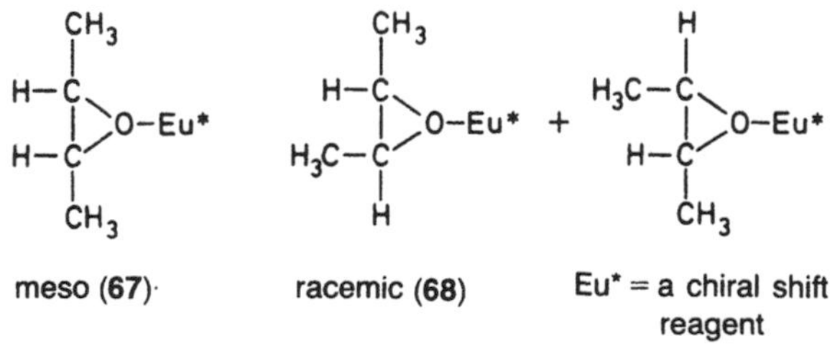

Figure 8.25. Distinction of meso- and racemic 2,3-epoxybutane by chiral shift reagent.

internally diasterotopic and thus anisochronous. The corresponding internally homotopic protons of the chiral isomer **68** remain homotopic and isochronous. But, if **68** is a racemate, the two enantiomers are converted into diastereomers by complexation with the chiral shift reagent and the C–H protons in *rac-***68** thus become diasterotopic and anisochronous as well. So far the situation appears stalemated. However, in *rac-***68** the C–H protons in each individual enantiomer are isochronous and hence do not display coupling. Thus, in the methyl-decoupled proton spectrum, they appear as two singlets. In contrast, the methine protons of the meso form, being anisochronous, do couple and appear as an AB system when the methyl protons are decoupled: the two cases are thus distinguishable.

Assignment of Descriptors to Heterotopic Ligands.   So far in this section we have discussed the use of stereoheterotopic probes in configurational assignment. We now come to the problem of assigning the stereochemical placement (*pro-R* or *pro-S*) of the stereoheterotopic groups themselves. One way of doing this involves replacement of the prochiral by a chiral center, for example, to replace $RR'CH_2$ by $RR'CHD$ or $RR'C(CH_3)_2$ by $RR'C^{12}CH_3^{13}CH_3$ or by $RR'C(CH_3)CD_3$. The groups at the chiral center may then be distinguished and the configuration of that center determined by any of the classical methods described in Chapter 5. Finally, these groups are correlated (usually by NMR) with the corresponding groups at the prochiral center.

If the groups in question are enantiotopic, the correlation of the chiral with the prochiral center is, in most cases, effected through enzymatic reactions (cf. Section 8-5). For example, if an enzyme abstracts the deuterium rather than the hydrogen atoms from (*R*)-RR′CHD it will abstract the *pro-R* rather than the *pro-S* hydrogen in $RR'CH_2$. Another approach would be to observe an enantiomer of RR′CHD of known configuration by NMR in a chiral solvent or in the presence of a chiral shift reagent. If, under these circumstances, the position of the C*H*D proton in one enantiomer is different from that of the corresponding proton in the other (or of the other corresponding proton in the racemate) then the position of this proton, say in (*S*)-RR′CHD (Fig. 8.26, **69**) will correspond, save for small isotope effects, to the position of the *pro-R* proton in RR′CHH, **70**. A case of this type (except that it involves covalent bond formation) has been described by Raban and Mislow[41] and is shown in Figure 8.27. It was found that in the (*R*)-O-methylmandelate of (*S*)-(+)-2-propanol- 1,1,1-d_3 (**71**) of known configuration,[42] the (sole) proton doublet

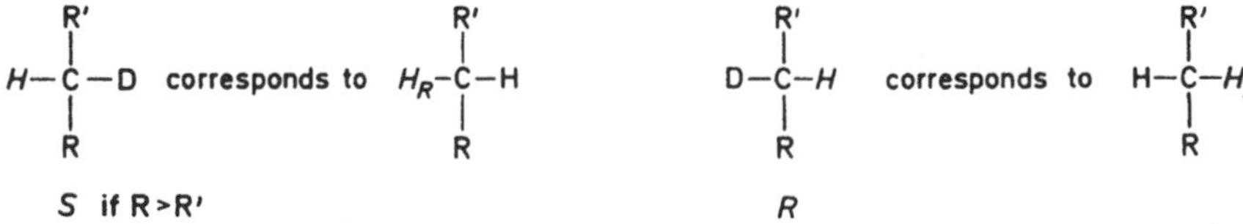

Figure 8.26. Assignment of enantiotopic nuclei in chiral environment.

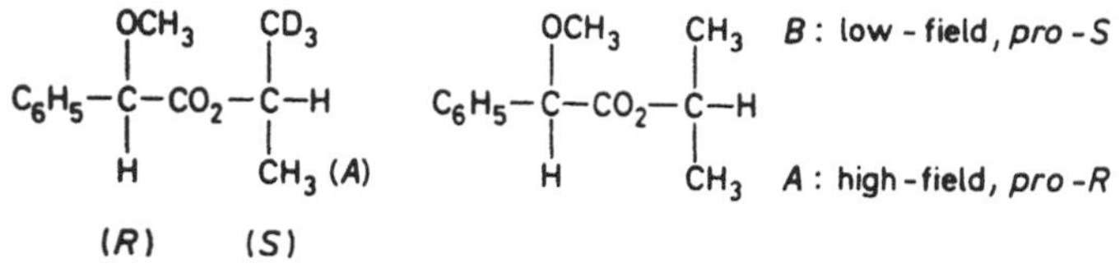

$$OCH_3 \quad CD_3$$
$$C_6H_5-\overset{|}{\underset{|}{C}}-CO_2-\overset{|}{\underset{|}{C}}-H$$
$$H \qquad CH_3 \ (A)$$
$$(R) \qquad (S)$$
$$71$$

$$OCH_3 \quad CH_3 \quad B: \text{low - field}, pro\text{-}S$$
$$C_6H_5-\overset{|}{\underset{|}{C}}-CO_2-\overset{|}{\underset{|}{C}}-H$$
$$H \qquad CH_3 \quad A: \text{high - field}, pro\text{-}R$$
$$72$$

Figure 8.27. Prochirality assignment of C methyl groups in isopropyl O-methylmandelate.

of the CH_3 group (A) of the alcohol corresponds to the higher field doublet A of the corresponding (R)-O-methylmandelate of unlabeled 2-propanol (**72**). The lower field CH_3 doublet B in the unlabeled material disappears in the trideuterated species. It may thus be concluded that the lower field signal is due to the *pro-S* methyl group B and the higher field one to the *pro-R* group A.

The configurational assignment of the isotopically labeled analogue required in all these cases may, of course, be achieved by synthesis from a chiral precursor. A case in point, but relating to diastereotopic nuclei, is shown in Figure 8.28.[43,44] Valine (**73**) has diastereotopic methyl groups resonating at 1.38 and 1.43 ppm (proton spectrum). In connection with an enzymatic transformation of the molecule, it became of importance to determine which group was which. The methyl groups were introduced stereospecifically starting from methyl (S)-lactate via (S)-(+)-2-propanol-d_3 and the two diastereomers ultimately obtained were separated by enzymatic resolution (at the conventional chiral center). Corresponding assignments with ^{13}C labeled methyl groups have also been described.[45,46]

c. Origin of Anisochrony

The early history of the anisochrony of diastereotopic groups is turbid because there was uncertainty as to whether the cause for the anisochrony was conformational, intrinsic, or both. The problem was finally analyzed clearly by Gutowsky[18] whose treatment we present here (Fig. 8.29). The compound chosen for illustration is CxxyCabc in which the x nuclei are diastereotopic and anisochronous. For simplicity's sake we shall consider only the three staggered conformations shown (for one enantiomer) in Figure 8.29, assuming that the populations of all other conformations are negligible. The chemical shift of x_1 in conformers **A**, **B**, and **C** may be denoted as $\delta_{a/b}$, $\delta_{a/c}$, and $\delta_{b/c}$, respectively, according to the groups at the adjacent carbon that are gauche to x_1. If n_a, n_b, and n_c are the mole fractions of **A**, **B**, and **C**, respectively, it follows that the average chemical shift of nucleus x_1 is

$$\delta_1 = n_A\delta_{a/b} + n_B\delta_{a/c} + n_C\delta_{b/c} \tag{8.1}$$

By the same token, the average chemical shift of x_2 is

$$\delta_2 = n_A\delta_{b/c} + n_B\delta_{a/b} + n_C\delta_{a/c} \tag{8.2}$$

Inspection of Eqs. 8.1 and 8.2 immediately discloses that since, ordinarily, $n_A \neq n_B \neq n_C$, $\delta_1 \neq \delta_2$, that is, x_1 and x_2 are anisochronous. Contrary to some misstatements in the literature, this conclusion is independent of the rate of rotation of the CxxyCabc system about the carbon–carbon bond; this rotation is assumed, throughout, to be fast on the NMR scale (see below for what happens in the limit of slow rotation).

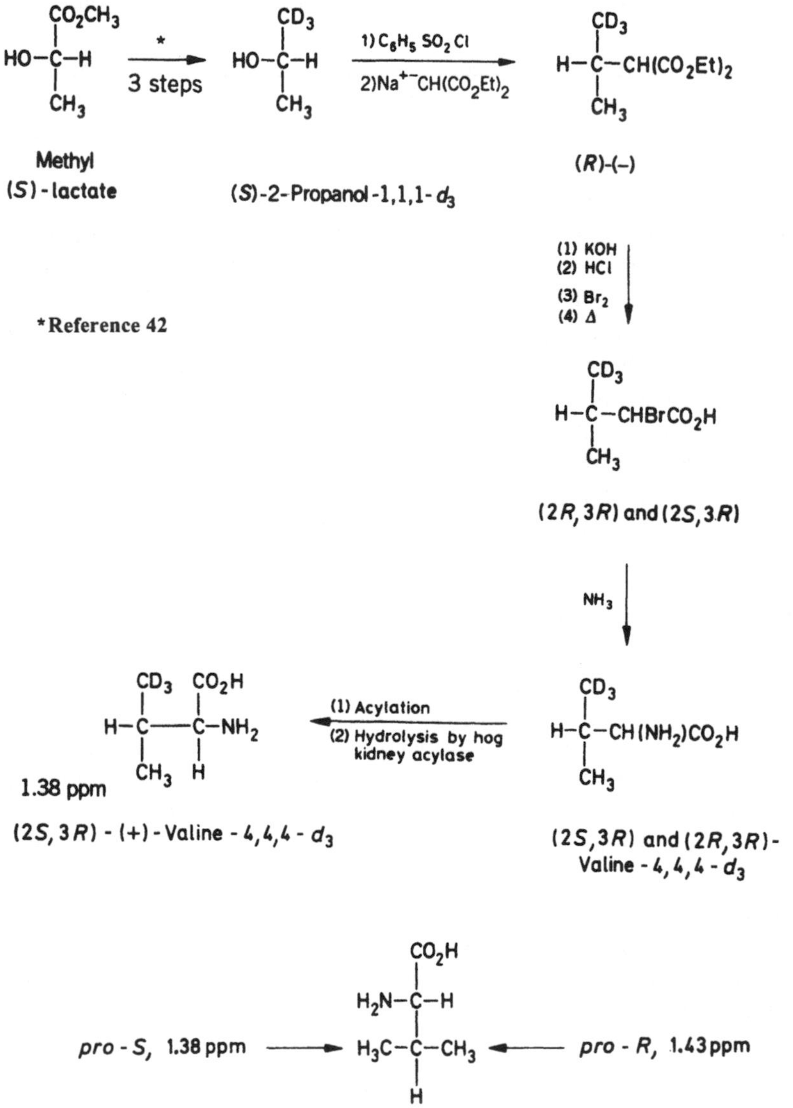

Figure 8.28. Assignment of diastereotopic methyl groups in L-valine.

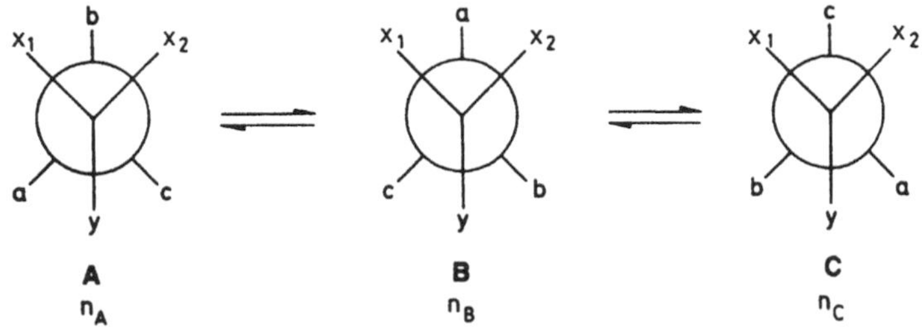

Figure 8.29. Anisochronous nuclei x in mobile systems.

It might thus appear, at this point, that the anisochrony is due to the unequal population of the three conformations depicted in Figure 8.29. Let us therefore consider the case (hypothetical or otherwise), where $n_A = n_B = n_C (= \frac{1}{3})$. Inspection of Eqs. 8.1 and 8.2 might, at first glance, imply that, in that case, δ_1 and δ_2 are equal. But, on more careful consideration, it turns out that this is a fallacy spawned by the inadequacy of the notation. The assumption leading to this fallacious result is that $\delta_{a/b}$ in Eq. 8.1 is the same as $\delta_{a/b}$ in Eq. 8.2 (and likewise for $\delta_{b/c}$, $\delta_{a/c}$). In fact, however, the neighborhood of x_1 in **A** is not the same as the neighborhood of x_2 in **B**. For example, in the former case, passing beyond a from x_1 one reaches y. In the latter case **B**, proceeding from x_2 beyond a one reaches x_1. Hence the environment a/b of x_1 and the shift $\delta_{a/b}$ in **A** is not the same as the environment a/b of x_2 and the shift $\delta_{a/b}$ in **B**: There is thus an intrinsic shift difference so that even if $n_A = n_B = n_C$, $\delta_1 \neq \delta_2$. The conclusion, then, is that both the conformation population difference and the intrinsic difference in chemical shift within each conformer contribute to the observed anisochrony of diastereotopic nuclei in conformationally mobile systems.

An elegant way of demonstrating intrinsic nonequivalence, based on symmetry principles, was suggested by Mislow and Raban[4] and reduced to practice by Binsch and Franzen[47] and subsequently by McKenna et al.[48] Two of the molecules studied, a bicyclic trisulfoxide **74** (ref. 47) and a quinuclidine derivative **75** (ref. 48) are shown in Figure 8.30. In both cases the presence of a threefold symmetry axis in one of the ligands (the bicyclic trisulfoxide moiety in **74**, the quinuclidine moiety in **75**) assures that the three conformers possible by virtue of rotation about the C–C or N–C bond indicated in heavy type are equally populated. The difference in chemical shifts of the CH_3 of CF_3 groups in **74** and the H_A and H_B methylene protons in **75** must therefore be intrinsic in nature. Additional examples of type **75** (general formula **76**; the general type formula for **74** is **77**) have been adduced.[49-51]

d. Conformationally Mobile Systems

This section deals briefly with the problem of averaging of heterotopic nuclei. In general, the symmetry properties of a given species are dependent on the time scale of

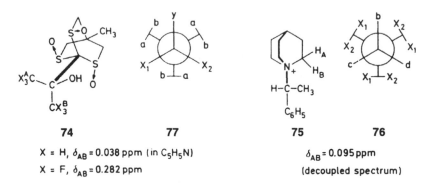

74   **77**   **75**   **76**

X = H, δ_{AB} = 0.038 ppm (in C_5H_5N) δ_{AB} = 0.095 ppm

X = F, δ_{AB} = 0.282 ppm   (decoupled spectrum)

Figure 8.30. Molecules displaying intrinsically anisochronous nuclei.

observation in that the symmetry of structures averaged by site or ligand exchanges may be higher than the symmetry in the absence of such exchanges (Section 4-4). For the present purpose it is significant that structures lacking C_n or S_n axes may acquire such axes as a result of averaging. It follows that diastereotopic nuclei may become enantiotopic, on the average, through operation of S_n, or they may become homotopic through development of a C_n; in other words, averaging may convert anisochronous nuclei into isochronous ones.

To explore the full potential of what is sometimes called "dynamic NMR" or "DNMR,"[38] that is, NMR studies involving site and ligand exchange, is beyond the scope of this chapter and the reader is referred to numerous reviews.[26,27,38,52–55]

An example of ring inversion has already been presented: 1,1-difluorocyclohexane (Fig. 8.22). At room temperature the two chair forms average, and the average symmetry is that of a planar molecule, C_{2v},[56] in which the fluorine atoms are related by the C_2 axis and hence are equivalent. Thus the room temperature spectrum of 1,1-difluorocyclohexane displays a single (except for proton splitting) chemical shift for the two fluorine atoms, as shown in Figure 8.31.[57,58] In contrast, at –110°C the spectrum shows the expected AB pattern for the diastereotopic fluorine atoms expected from the individual structures shown in Figure 8.31. As the temperature is gradually raised, the two doublets broaden and merge into two broad, touching peaks, which on further warming eventually coalesce into one broad single peak at what is called the "coalescence temperature" (in this case –46°C, at 56.4 MHz). The spectrum just above coalescence is also shown in Figure 8.31; the broad single peak sharpens as the temperature is raised further.

A simple formula for determining the rate of site exchange between two equally populated sites[38,59,60] is

$$k_{coal} = \frac{1}{2}\pi \, \Delta \nu \sqrt{2} = 2.221 \, \Delta \nu \tag{8.3}$$

where $\Delta \nu$ is the chemical shift difference (in hertz) between the two exchanging nuclei at a temperature well below coalescence. This equation is valid only for a

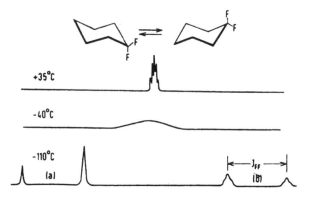

Figure 8.31. The ^{19}F NMR signals, at 56.4 MHz, of 1,1-difluorocyclohexane at various temperatures; signals of (*a*) equatorial and (*b*) axial fluorine atoms. [Reprinted with permission from Roberts, J. D. *Angew. Chem. Int. Ed. Engl.* **1963**, *2*, 58. Copyright © 1963 VCH, Weinheim, Germany.]

(noncoupled) single site-exchanging nucleus, for example, the proton in cyclohexane-d_{11} (observed with deuterium decoupling). In the case of geminal exchanging nuclei, as in 1,1-difluorocyclohexane (proton decoupled), Eq. 8.4 should be used[59–61] ($\Delta\nu = \nu_1 - \nu_2$):

$$k_{coal} = \frac{1}{2}\pi\sqrt{2}\ \sqrt{(\nu_1 - \nu_2)^2 + 6J^2} \tag{8.4}$$

J being the coupling constant of the two nuclei in question. Both ν and J are measured at temperatures well below the coalescence temperature. For 1,1-difluorocyclohexane (see above) $\nu_1 = 1522$ Hz, $\nu_2 = 638$ Hz, $J = 237$ Hz, and thus $k = 2349$ s^{-1} at the coalescence temperature of $-46°$C. From this information, in turn, one can calculate[38,60] the free energy of activation for the site exchange from the Eyring equation (8.5):

$$k = \kappa(k_B T/h)e^{-\Delta G^{\ddagger}/RT} = \kappa(k_B T/h)e^{-\Delta H^{\ddagger}/RT}\ e^{\Delta S^{\ddagger}/R} \tag{8.5}$$

where k is the rate constant for site exchange, κ is the transmission coefficient (usually taken as unity), k_B is Boltzmann's constant, h is Planck's constant, T is the coalescence temperature, $\Delta G^{\ddagger}$ is the free energy of activation, $\Delta H^{\ddagger}$ is the enthalpy of activation, and $\Delta S^{\ddagger}$ is the entropy of activation. From the data of 1,1-difluorocyclohexane, $\Delta G^{\ddagger} = 9.7$ kcal mol^{-1}.[58]

A more general method of measuring site exchange rates is the method of line-shape analysis.[38,59] In this method one compares the shape of the broadened lines some 10°C or 20°C above and below the coalescence temperature, as well as in the fast and slow exchange limit, with the lineshape computed by means of formulas that include the rate of exchange. This method permits one to determine k over a range of temperatures and thus, through a plot of $\Delta G^{\ddagger}$ versus $1/T$, one can arrive at the values of $\Delta H^{\ddagger}$ and $\Delta S^{\ddagger}$ (though the accuracy of determining $\Delta H^{\ddagger}$ and $\Delta S^{\ddagger}$ is often low). The method is applicable to relatively complex spin systems, not just to singlet or AB exchange. It is considered to be the method of choice in determination of rate constants by NMR. A typical comparison of experimental and computed lineshapes, referring to the site exchange in furfural, is shown in Figure 8.32.[62]

Because each chemical shift difference between exchanging sites, as well as each spin coupling constant, gives rise to a coalescence of its own when $\Delta\nu$ or $J \approx k$, where k is the rate of site exchange, a system having many such parameters will, because of the presence of a multitude of "internal clocks," be more sensitive in the response of its NMR spectrum to temperature changes. Thus, within the limits of feasibility of computer treatment, the more shifts and coupled spins, the better.

It is convenient to have terms for structures such as those in Figures 8.22 and 8.32, **78**, which differ only in the position of designated nuclei, and for the process of exchange of such heterotopic nuclei. The term "topomers" has been proposed for the interconverting structures, "topomerization" being the process of interchange.[63] An older term, "degenerate isomerization," seems inappropriate since the two structures shown Figure 8.22 are homomers not isomers. "Automerization" has also been used;[64] it properly denotes the identity of the two interconverting structures but does not address itself to the significance of the process of their interconversion.

Analysis of topomerization rates can also be effected by "dynamic enantioselective chromatography."[55] The accessible barriers, 17–30 kcal mol^{-1} (71–126 kJ mol^{-1}), are in a higher range than those analyzable by DNMR and include, at the higher end, rates approachable by classical equilibration.

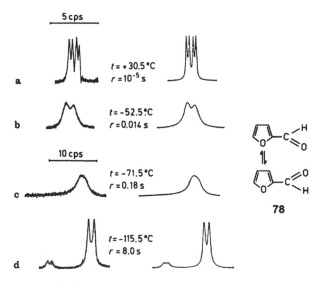

Figure 8.32. Experimental and calculated DNMR spectra for the aldehyde proton of 2-furaldehyde. [Reprinted with permission from Dahlquist. K.-I. and Forsén. S. *J. Phys. Chem.* **1965,** *69,* 4068. Copyright © 1965 American Chemical Society, Washington, DC.]

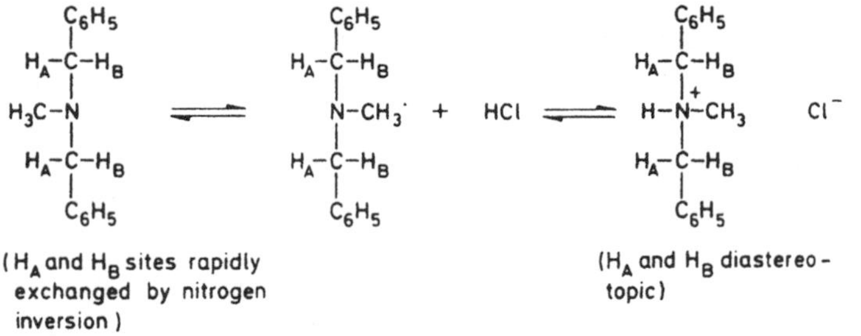

$$\begin{array}{ccc}
\text{C}_6\text{H}_5 & \text{C}_6\text{H}_5 & \text{C}_6\text{H}_5 \\
| & | & | \\
\text{H}_A\text{-C-H}_B & \text{H}_A\text{-C-H}_B & \text{H}_A\text{-C-H}_B \\
| & | & | \\
\text{H}_3\text{C-N} \rightleftharpoons & \text{N-CH}_3 \quad + \quad \text{HCl} \rightleftharpoons & \text{H-N-CH}_3 \quad \text{Cl}^- \\
| & | & | \\
\text{H}_A\text{-C-H}_B & \text{H}_A\text{-C-H}_B & \text{H}_A\text{-C-H}_B \\
| & | & | \\
\text{C}_6\text{H}_5 & \text{C}_6\text{H}_5 & \text{C}_6\text{H}_5
\end{array}$$

(H_A and H_B sites rapidly exchanged by nitrogen inversion)

(H_A and H_B diastereo-topic)

79

Figure 8.33. Inversion of dibenzylmethylamine.

We conclude this section with a discussion of inversion of amines of the type $NR_1R_2R_3$.[65,66] In general, this process is too rapid to be studied except in special circumstances.[67] Nevertheless, Saunders and Yamada[68] were able to determine the very high rate of inversion of dibenzylmethylamine (Fig. 8.33, **79**, $k = 2 \times 10^5$ s^{-1} at 25°C) by the elegant trick of partially neutralizing the amine with hydrochloric acid. Since the hydrochloride cannot invert, the benzylic protons in it are diastereotopic and are hence anisochronous. Only the small amount of free amine in equilibrium with the salt at a given pH (the measurements were carried out on the acid side) inverts at the rate indicated and it can easily be shown[68] that $k_{obs} = k[\text{amine}]/[\text{salt} + \text{amine}]$, where k_{obs} is the observed rate of site exchange of the diastereotopic protons at a given pH, k is the rate constant for the amine inversion to be determined, and the quantity of free amine in the fraction can be ascertained from the measurement of pH and the known basicity of the amine.

8-5. HETEROTOPIC LIGANDS AND FACES IN ENZYME-CATALYZED REACTIONS

a. Heterotopicity and Stereoelective Synthesis

In Section 8-3 we saw that replacement of stereoheterotopic groups or addition to stereoheterotopic faces gives rise to stereoisomers. The rates of such replacements of one or the other of two ligands or additions to one or the other of two faces are frequently not the same. In particular, replacements of diastereotopic ligands or additions to diastereotopic faces usually proceed at different rates because the transition states for such replacements or additions are diastereomeric and therefore are unequal in energy. Thus the reactions shown in Figures 8.14 and 8.15 not only give rise to diastereomeric products, depending on which ligand or face is involved, but they give these products in unequal, sometimes quite unequal amounts; that is, they display diastereoselectivity[69] or "diastereodifferentiation."[70] Replacement of enantiotopic ligands or addition to enantiotopic faces gives rise to enantiomeric products, but here replacement of the two ligands or addition to the two faces ordinarily occurs at the same rate, because the pertinent transition states are enantiomeric and therefore are equal in energy. This situation changes, however, when the reagent (or other entity participating in the transition state, such as the solvent or a catalyst) is chiral. In that circumstance, the two transition states will, once again, become diastereomeric and the two enantiomeric products will be formed at unequal rates and in unequal amounts: The reaction will be enantioselective[69] or "enantiodifferentiating."[70] In this case, where prochiral starting materials give rise to nonracemic chiral products, one speaks of enantioselective (or, less appropriately, asymmetric) syntheses. Here we shall deal only with a few applications in enzyme chemistry.

b. Heterotopicity and Enzyme-Catalyzed Reactions

The answer to the question as to which of two heterotopic ligands or faces is involved in an enzyme-catalyzed reaction depends on, and conversely may be used to elucidate, the fit between the substrate and the active site of the enzyme. (However, it is only one of several techniques used to fathom this relationship.) The literature in this area is extensive and only the principles involved and one or two representative examples can be presented here; for more detailed and extensive information, the reader is referred to pertinent books.[6,71,72]

We start the discussion with a classical experiment related to the stereochemistry of the oxidation of ethanol and reduction of acetaldehyde mediated by the enzyme yeast alcohol dehydrogenase in the presence of the oxidized (NAD^+) and reduced (NADH) forms, respectively, of the coenzyme nicotinamide adenine dinucleotide (Fig. 8.34). The stereochemically significant feature of this reaction stems from the

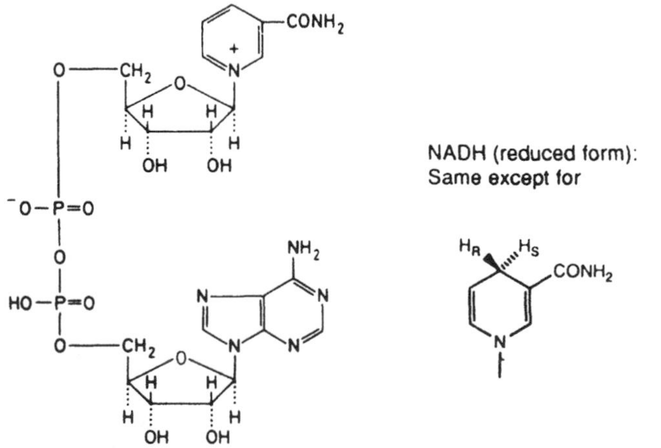

Figure 8.34. Nicotinamide adenine dinucleotide (NAD^+).

fact that the methylene hydrogen atoms in CH_3CH_2OH and the faces of the carbonyl in $CH_3CH=O$ are enantiotopic. The question thus arises as to which of the CH_2 hydrogen atoms is removed in the oxidation and to which of the $C=O$ faces the hydrogen attaches itself in the reduction as mediated by the enzyme and coenzyme.

Loewus, Westheimer, and Vennesland[73] found that reduction of ethanal-1-d with NADH in the presence of yeast alcohol dehydrogenase gave ethanol-1-d, which, upon enzymatic reoxidation by NAD^+, returned ethanal-1-d without loss of deuterium (Fig. 8.35). There is thus a "stereochemical memory effect" involved in this reaction: The H and D of the CH_3CHDOH do not get scrambled but the same H that is introduced in the reduction is the one removed in the oxidation. The reason, as we now know, is that the two methylene hydrogen atoms are not identical, but distinguishable by bearing an enantiotopic relationship.

When the configuration of the ethanol-1-d is inverted by conversion to the tosylate followed by treatment with hydroxide and the inverted ethanol-1-d is then oxidized with yeast alcohol dehydrogenase and NAD^+, the deuterium atom, which has taken the stereochemical position of the original hydrogen, is now removed and the product is largely unlabeled $CH_3CH=O$. The sequence of events is summarized in Figure 8.35.

Later experiments on a larger scale[74] established that the ethanol-1-d from $CH_3CD=O$ and NADH (Fig. 8.35, **80**) is levorotatory, $[\alpha]_D^{28} - 0.28 \pm 0.03$ (neat) and this finding, coupled with the elucidation of configuration of (−)-ethanol-1-d as S[75] leads to the stereochemical picture summarized in Figure 8.35. It follows that the hydrogen transferred from the NADH in the enzymatic reduction attaches itself to the *Re* face of the aldehyde and that this hydrogen thus becomes H_R in the ethanol; it is

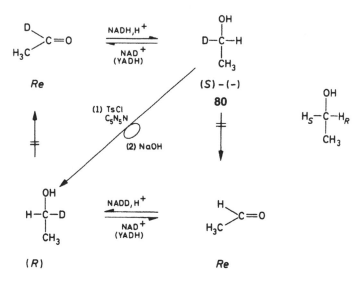

Figure 8.35. Oxidation of ethanol and reduction of acetaldehyde by NAD–NADH in the presence of yeast alcohol dehydrogenase (YADH).

H_R (the *pro-R* carbinol hydrogen), in turn, which is abstracted by NAD^+ in the oxidative step.

It is clear that ethanol (and acetaldehyde) must fit into the active site of YADH in such a way as to conform to these stereochemical findings. A model for the reduction of a very similar substrate, pyruvic acid, which is reduced by NADH in the presence of liver alcohol dehydrogenase (LADH) to (S)-lactic acid,[76,77] is shown in Figure 8.36. Here one can discern Ogston's picture of the three-point contact (cf. Fig. 8.7), one contact being established by the salt bond pyruvate–arginine-H^+, the second by the hydrogen bond (histidine) N–H $\cdots$ O=C (pyruvate), and the third one involving delivery of the hydrogen of NADH (bound to the enzyme) to the *Re* face of the C=O of pyruvate. Thus only (S)-lactate [not (R)-lactate] is formed in the reduction. Similarly, the model explains why, in the reverse reaction, the enzyme is substrate stereoselective for (S)-lactate: (R)-Lactate, if locked into the enzyme cavity, would have CH_3 rather than C–H juxtaposed with the NAD^+ and could thus not be oxidized.

The reduction of acetaldehyde is probably similar though the absence of the COO^- group requires the contact at the third site to be established in a different manner. It is not certain that covalent or ionic bonding is actually involved in this contact; the shape of the enzyme cavity itself, and the attendant hydrophilic and hydrophobic interactions between certain parts of the enzyme and parts of the substrate, may contribute to the required orientation of the substrate.

The study of the stereochemistry of ethanol oxidation and acetaldehyde reduction and the information relating to the topography of the enzyme derived from this study are typical of a large number of other investigations of this type. For example, the transfer of the hydrogen to and from the coenzyme involves a stereochemical problem of its own (Fig. 8.37): In the reductive step, is it H_R or H_S of the dihydronicotinamide moiety that is transferred from the coenzyme to the substrate; correspondingly, in the oxidation step, is the hydrogen abstracted from the substrate added to the *Re* or *Si* face of the pyridinium moiety of the coenzyme? This question was answered as summarized in Figure 8.37.[78,79]

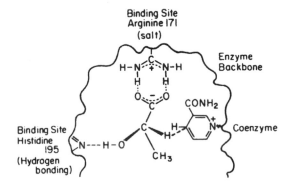

Figure 8.36. Reduction of pyruvate by NADH in the presence of LADH.

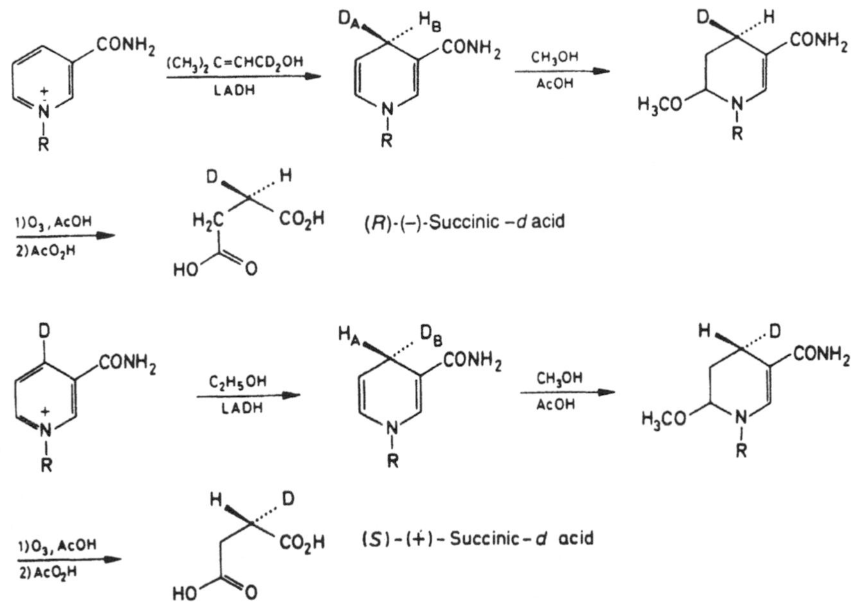

Figure 8.37. Prostereoisomerism of hydrogen transferred to the C(4) of NADH from alcohol in the presence of LADH.

Deuterium was transferred from a dideuterated alcohol RCD_2OH to NAD^+ in the presence of LADH. This transfer creates a chiral center at C(4) in the NAD^2H formed. Degradation of this material in the manner shown (Fig. 8.37, top half) yielded (R)-(−)-succinic-d acid, which is recognized by its known optical rotatory dispersion (ORD) spectrum. It follows that the configuration of the NAD^2H formed was R and it is therefore H_R that is transferred from (and to) the alcohol; the attachment of hydride to NAD^+ thus occurs from the Re face. (This result is not general; that is, it does not apply to all oxidation–reduction reactions mediated by NAD^+-NADH.[80]) To confirm this finding and to avoid any remote possibility that the β,β-dimethylallyl-d_2 alcohol used as the source of deuterium would behave differently from ethanol, the experiment was repeated with NAD-4-d^+ and ethanol (shown in Fig. 8.37, bottom part). In this case, of course, the ultimate degradation product is (S)-(+)-succinic-d-acid.

Next, we take up the stereochemistry of an enzymatic addition to a C=C double bond: the hydration of fumaric to (S)-malic acid[81] and the amination of fumaric to (S)-aspartic acid.[82] Both of these reactions are of industrial importance.[83] They are summarized in Figure 8.38. The absolute configurations of both (−)-malic and (−)-aspartic acids are well known and erythro- and threo-malic-3-d acids have been identified by NMR spectroscopy (being diastereomers, they differ in NMR spectrum) and their configurations have been unambiguously assigned by a synthesis of controlled stereochemistry, (Fig. 8.39).[84,85] In the dianion of the threo acid (81) in D_2O

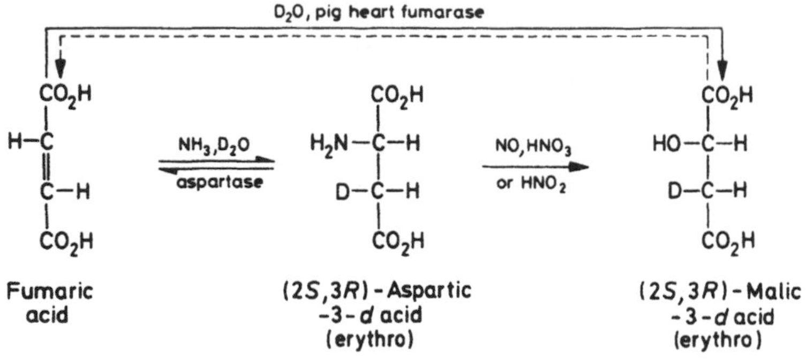

D₂O, pig heart fumarase

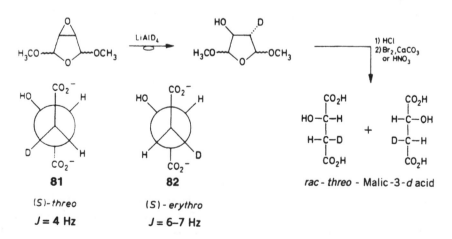

Figure 8.38. Stereochemistry of fumarase and aspartase reactions.

solution, the carboxylate groups are predominantly anti (because of electrostatic repulsion) and it follows that the hydrogen atoms are gauche; the coupling constant of these protons is therefore small ($J = 4$ Hz). In contrast, the erythro isomer **82** obtained by biosynthesis (see below) has its hydrogen atoms predominantly anti to each other and their coupling constant is thus larger ($J = 6$–7 Hz).

Though addition of D₂O to fumaric acid is reversible, it yields only mono-deuterated malic acid and involves recovery of undeuterated fumaric acid. This indicates that the addition and elimination steps are stereospecific and that their stereochemical course (syn or anti) is the same. The formation of the erythro isomer of (S)-(−)-malic-3-d acid in the fumarase-mediated D₂O addition (Fig. 8.38) relates the absolute stereochemistry of C(3) to that at C(2) and proves that the configuration at C(3) is R. Since the (S)-(−)-aspartic-3-d acid formed in aspartase-mediated ammonia addition to fumaric acid (Fig. 8.38) is converted to the same (2S,3R)-(−)-malic-3-d acid by nitrous acid deanimation, and since the latter reaction

Figure 8.39. Synthesis of racemic threo-malic-3d acid.

Figure 8.40. Steric course of addition of D₂O and ND₃ (or NH₃/D₂O) to fumaric acid.

does not affect the configuration at C(3), the aspartic acid must also be 3*R*. [The fact that the steric course of the nitrous acid deanimation at C(2) involves retention was already well known in the literature.] Then it follows that the hydrogen added (or abstracted, in the reverse reaction) at C(3) in the conversion of fumaric to malic or aspartic acid is the *pro-R* hydrogen and that addition to the fumaric acid of the proton from either water or ammonia proceeds from the *Re* face at C(3). (The *Re* face is the front face of the double bond in Fig. 8.38.) On the other hand, since the configuration at C(2) in both cases is *S*, the addition at C(2) in both cases must be from the rear face in Figure 8.38 (i.e., from the *Si* face). The overall picture, then, is one of anti addition giving the 2*S*,3*R* isomer (Fig. 8.40). An analogous steric course is observed in the addition of water to maleic, citraconic (α-methylmaleic), and mesaconic (α-methylfumaric) acids.[75]

Many additional examples of the elucidation of prostereoisomerism in biochemical reactions could be given, for example, the elegant elucidation by Cornforth and co-workers[86] (cf. ref. 75) of the biosynthesis of squalene, which was recognized by the Nobel Prize in Chemistry in 1975, or the more recent studies of the enzymatic decarboxylation of tyrosin,[87a] histidine,[87b] and 5-hydroxytryptophan,[87c] as well as the elucidation of the stereochemistry of pyrrolizidine alkaloid biosynthesis.[88] The stereochemistry of metabolic reactions of amino acids has been reviewed.[89]

Finally, it should be mentioned that isotopes of carbon (^{12}C, ^{13}C) and of oxygen (^{16}O, ^{18}O) have been used, in addition to hydrogen isotopes, in the elucidation of enzyme stereochemistry.[5b]

Further details on the topics discussed in this section may be found in ref. 90.

REFERENCES

1. Hirschmann, H. and Hanson, K. R. (a) *Tetrahedron* **1974**, *30*, 3649. (b) *Eur. J. Biochem.* **1971**, *22*, 301. (c) *J. Org. Chem.* **1971**, *36*, 3293.

2. Hanson, K. R. *J. Am. Chem. Soc.* **1966**, *88*, 2731.

3. Mislow, K. and Siegel, J. *J. Am. Chem. Soc.* **1984**, *106*, 3319.

4. Mislow, K. and Raban, M. *Top. Stereochem.* **1967**, *1*, 1.

5. Eliel, E. L. (a) *J. Chem. Educ.* **1980**, *57*, 52. (b) *Top. Curr. Chem.* **1982**, *105*, 1.

6. Bentley, R. *Molecular Asymmetry in Biology*, Academic Press, New York, Vol. 1, 1969, Vol. 2, 1970.

7. Anet, F. A. L. and Park, J. *J. A Chem. Soc.* **1992**, *114*, 411.

8. Evans, E. A. and Slotin, L. *J. Biol Chem.* **1941**, *141*, 439.

9. Wood, H. G., Werkman, C. H., Hemingway, A., and Nier, A. O. *J. Biol. Chem.* **1942**, *142*, 31.

10. Bublitz, C. and Kennedy, E. P. *J. Biol. Chem.* **1954**, *211*, 951.

11. Ogston, A. G. *Nature (London)* **1948**, *162*, 963.

12. Bentley, R. *Nature (London)* **1978**, *276*, 673.

13. Easson, L. H. and Steadman E. *Biochem. J.* **1933**, *27*, 1257.

14. Bergman, M. *Science* **1934**, *79*, 439.

15. Drysdale, J. J. and Phillips, W. D. *J. Am. Chem. Soc.* **1957**, *79*, 319.

16. Nair, P. M. and Roberts, J. D. *J. Am. Chem. Soc.* **1957**, *79*, 4565.

17. Waugh, J. S. and Cotton, F. A. *J. Phys. Chem.* **1961**, *65*, 562.

18. Gutowsky, H. S. *J. Chem. Phys.* **1962**, *37*, 2196.

19. Hirschmann, H. "Newer Aspects of Enzyme Stereochemistry," in Florkin, M. and Stotz, G., eds., *Comprehensive Biochemistry*, Vol. 12, Elsevier, New York, 1964, p. 236.

20. (a) Pirkle, W. H. *J. Am. Chem. Soc.* **1966**, *88*, 1837. (b) Pirkle, W. H. and Hoover, D. J. *Top. Stereochem.* **1992**, *13*, 263.

21. Eliel, E. L. *J. Chem. Educ.* **1971**, *48*, 163.

22. Hanson, K. R. *Annu. Rev. Biochem.* **1976**, *45*, 307.

23. Martin, M. L., Mantione, R., and Martin, G. J. *Tetrahedron Lett.* **1965**, 3185.

24. Martin, M. L., Martin, G. J., and Coufignal, R. *J. Chem. Soc. B* **1971**, 1282.

25. Beaulieu, P. L., Morriset, V. M., and Garratt, D. G. *Can. J. Chem.* **1980**, *58*, 928.

26. Siddall, T. H. and Stewart, W. E. *Proc. Nucl. Mag. Reson. Spectrosc.* **1969**, *5*, 33.

27. Jennings, W. B. *Chem. Rev.* **1975**, *75*, 307.

28. Abragam, A. *The Principles of Nuclear Magnetism*, Oxford University Press, Oxford, 1961, p. 480.

29. Sokolov, V. I., Petrovskii, P. V., and Reutov, O. A. *J. Organometal. Chem.* **1973**, *59*, C27.

30. Whitesides, G. M., Holtz, D., and Roberts, J. D. *J. Am. Chem. Soc.* **1964**, *86*, 2628.

31. Schiemenz G. P. and Rast, H. *Tetrahedron Lett.* **1971**, 4685.

32. Martin, M. L. and Martin, G. J. *Bull. Soc. Chim. Fr.* **1966**, 317.

33. Kainosho, M., Ajisaka, K., Pirkle, W. H., and Beare, S. D. *J. Am. Chem. Soc.* **1972**, *94*, 5942.

34. Devriese, G., Ottinger, R., Zimmerman, D., Reisse, J., and Mislow, K. *Bull Soc. Chim. Belg.* **1976**, *85*, 167.

35. Wilson, N. K. and Stothers, J. B. *Top. Stereochem.* **1974**, *8*, 1; especially p. 17.

36. Goering, H. L., Eikenberry, J. N., Koermer, G. S., and Lattimer, C. J. *J. Am. Chem. Soc.* **1974**, *96*, 1493.

37. Fraser, R. R., Schuber, F. J., and Wigfield, Y. Y. *J. Am. Chem. Soc.* **1972**, *94*, 8795.

38. Binsch, G. *Top. Stereochem.* **1968**, *3*, 97.

39. Gaudemer, A. "Determination of Configuration by NMR Spectroscopy," in Kagan, H. B., ed., *Stereochemistry, Fundamentals and Methods*, Vol. 1, Thieme, Stuttgart, Germany, 1977, p. 73.

40. Hill, R. K. and Chan, T.-H. *Tetrahedron* **1965**, *21*, 2015.

41. Raban, M. and Mislow, K. *Tetrahedron Lett.* **1966**, 3961.

42. Mislow, K., O'Brien, R. E., and Schaefer, H. *J. Am. Chem. Soc.* **1960**, *82*, 5512.

43. Hill, R. K., Yan, S., and Arfin, S. M. *J. Am. Chem. Soc.* **1973**, *95*, 7857.

44. Aberhart, D. J. and Lin, L. J. *J. Am. Chem. Soc.* **1973**, *95*, 7859; *J. Chem. Soc. Perkin I*, **1974**, 2320.

45. Baldwin, J. E., Loliger, J., Rastetter, W., Neuss, N., Huckstep, L. L., and De La Higuera, N. *J. Am. Chem. Soc.* **1973**, *95*, 3796.

46. Kluender, H., Bradley, C. H., Sih, C. J., Fawcett, P., and Abraham, E. P. *J. Am. Chem. Soc.* **1973**, *95*, 6149.

47. Binsch, G. and Franzen, G. R. *J. Am. Chem. Soc.* **1969**, *91*, 3999.

48. McKenna, J., McKenna, J. M., and Wesby, B. A. *J. Chem. Soc. Chem. Commun.* **1970**, 867.

49. Franzen, G. R. and Binsch, G. *J. Am. Chem. Soc.* **1973**, *95*, 175.

50. Morris, D. G., Murray, A. M., Mullock, E. B., Plews, R. M., and Thorpe, J. E. *Tetrahedron Lett.* **1973**, 3179.

51. Gielen, M., Close, V., and de Poorter, B. *Bull. Soc. Chim. Belg.* **1974**, *83*, 339.

52. Kessler, H. *Angew Chem. Int. Ed. Engl.* **1970**, *9*, 219.

53. Jackman, L. M. and Cotton, F. A., eds., *Dynamic Nuclear Magnetic Resonance Spectroscopy*, Academic Press, New York, 1975.

54. Roberts, J. D. *Pure Appl. Chem.* **1979**, *51*, 1037; Binsch, G. and Kessler, H. *Angew. Chem. Int. Ed. Engl.* **1980**, *19*, 411.

55. Gasparrini, F., Lunazzi, L., Misti, D., and Villani, C. *Acc. Chem. Res.* **1995**, *28*, 163.

56. Leonard, J. E., Hammond, G. S., and Simmons, H. E. *J. Am. Chem. Soc.* **1975**, *97*, 5052.

57. Roberts, J. D. *Angew. Chem. Int. Ed. Engl.* **1963**, *2*, 53.

58. Spassov, S. L., Griffith, D. L., Glazer, E. S., Nagarajan, K., and Roberts, J. D. *J. Am. Chem. Soc.* **1967**, *89*, 88.

59. Pople, J. A., Schneider, W. G., and Bernstein, H. J. *High Resolution Nuclear Magnetic Resonance Spectroscopy*, McGraw-Hill, New York, 1959.

60. Binsch, G. "Band-Shape Analysis," Jackman, L. M. and Cotton F. A., eds., *Dynamic Nuclear Magnetic Resonance Spectroscopy*, Academic Press, New York, 1975, p. 45.

61. Kurland, R. J., Rubin, M. B., and Wise, W. B. *J. Chem. Phys.* **1964**, *40*, 2426.

62. Dahlqvist, K.-I. and Forsén, S. *J. Phys. Chem.* **1965**, *69*, 4062.

63. Binsch, G., Eliel, E. L., and Kessler, H. *Angew. Chem. Int. Ed. Engl.* **1971**, *10*, 570.

64. Balaban, A. T. and Farcasiu, D. *J. Am. Chem. Soc.* **1967**, *89*, 1958.

65. Lehn, J.-M. *Top. Curr. Chem.* **1970**, *15*, 311.

66. Lambert, J. B. *Top. Stereochem.* **1971**, *6*, 19.

67. Buschweller, C. H., Anderson, W. G., Stevenson, P. E., and O'Neil, J. W. *J. Am. Chem. Soc.* **1975**, *97*, 4338.

68. Saunders, M. and Yamada, F. *J. Am. Chem. Soc.* **1963**, *85*, 1882.

69. Izumi, Y. *Angew. Chem. Int. Ed. Engl.* **1971**, *10*, 871.

70. Izumi, Y. and Tai, A., *Stereo-Differentiating Reactions: The Nature of Asymmetric Reactions*, Academic Press, New York, 1977.

71. Alworth, W. L. *Stereochemistry and Its Applications in Biochemistry*, Wiley, New York, 1972.

72. Frey, P. A., ed., *Mechanism of Enzymatic Reactions. Stereochemistry*, Elsevier, New York, 1986.

73. Loewus, F. A., Westheimer, F. H., and Vennesland, B. *J. Am. Chem. Soc.* **1953**, *75*, 5018.

74. Levy, H. R., Loewus, F. A., and Vennesland, B. *J. Am. Chem. Soc.* **1957**, *79*, 2949.

75. See Arigoni, D. and Eliel, E. L. *Top. Stereochem.* **1969**, *4*, 127.

76. Adams, M. J., Rossman, M. G., Kaplan, N. O., et al. *Proc. Natl. Acad Sci. USA* **1973**, *70*, 1968.

77. Vennesland, B. *Top. Curr. Chem.* **1974**, *48*, 39.

78. Cornforth, J. W., Ryback, G., Popjak, G., Donninger, C., and Schroepfer, G. *Biochem. Biophys. Res. Commun.* **1962**, *9*, 371.

79. Oppenheimer, N. J., Marschner, T. M., Malver, O., and Kam, B. L. "Stereochemical Aspects of Coenzyme–Dehydrogenase Interactions," in Frey, F. A., ed., *Mechanism of Enzymatic Reactions. Stereochemistry*, Elsevier, New York, 1986, p. 15.

80. You, K.-S., Arnold, L. J., Allison, W. S., and Kaplan, N. O. *Trends Biochem. Sci.* **1978**, *3*, 265.

81. Englard, S. and Colowick, S. P. *J. Biol. Chem.* **1956**, 221, 1019.

82. Krasna, A. I. *J. Biol. Chem.* **1958**, *233*, 1010.

83. Chibata, S. "Applications of Immobilized Enzymes for Asymmetric Reactions," in Eliel, E. L. and Otsuka, S., eds., *Asymmetric Reactions and Processes in Chemistry*, ACS Symposium Series 185, American Chemical Society, Washington, DC, 1982, p. 195.

84. (a) Gawron, O. and Fondy, T. P. *J. Am. Chem. Soc.* **1959**, *81*, 6333; (b) Gawron, O., Glaid, A. J., and Fondy, T. P. *J. Am. Chem. Soc.* **1961**, *83*, 3634.

85. Anet, F. A. L. *J. Am. Chem. Soc.* **1960**, 82, 994.

86. Cornforth, J. W. Q. *Rev. Chem. Soc.* **1969**, *23*, 125; *Chem. Soc. Rev.* **1973**, *2*, 1; *J. Mol. Catalysis* **1976**, *1*, 145.

87. (a) Battersby, A. R., Chrystal, E. J. T., and Staunton, J. *J. Chem. Soc. Perkin I* **1980**, 31; (b) Battersby, A. R., Nicoletti, M., Staunton, J., and Vleggaar, R. *J. Chem. Soc. Perkin I* **1980**, 43; (c) Battersby, A. R., Scott, A., and Staunton, J. *Tetrahedron* **1990**, *46*, 4685.

88. Kunec, E. K. and Robins, D. J. *J. Chem. Soc. Chem. Commun.* **1985**, 1450.

89. Young, D. W. *Top. Stereochem.* **1994**, *21*, 381.

90. Eliel, E. L. and Wilen , S. H. *Stereochemistry of Organic Compounds*, Wiley, New York, 1994, pp. 513–532.

9

STEREOCHEMISTRY OF ALKENES

9-1. STRUCTURE OF ALKENES AND NATURE OF CIS–TRANS ISOMERISM

a. General

In most alkenes (olefins), the two double-bonded carbon atoms and the four additional ligands attached to them are coplanar (Fig. 9.1). The generally accepted orbital description involves sp^2 hybridized carbon atoms.

The carbon atoms are linked to each other and to the attached ligands (a, b and c, d, respectively) by sp^2 hybridized σ bonds, and they are further linked by a π bond formed by lateral overlap of the remaining p orbitals of the two carbon atoms. Whereas the C–C σ bond strength is about 83 kcal mol^{-1} (347 kJ mol^{-1}), the strength of the π bond, with its less favorable lateral overlap, is only 62 kcal mol^{-1} [259 kJ mol^{-1} the two numbers add up to the generally accepted total energy of a C=C double bond, 145 kcal mol^{-1} (607 kJ mol^{-1})].

The activation barrier to the thermal isomerization of 2-butene[1] (Fig. 9.1, $Z \rightleftharpoons E$, a = c = CH$_3$, b = d = H), E_a = 62 ± 1 kcal mol^{-1} (259 ± 5 kJ mol^{-1}), provides a direct measure of the strength of the π bond, since, in the process of rotation of the Z to the E isomer, the p orbitals of the two olefinic carbon atoms become perpendicular, with no overlap. Thereby the π bond is completely broken in the transition state. The barrier in CHD=CHD,[2] 65 kcal mol^{-1} (272 kJ mol^{-1}), is similar to the 2-butene value.

The C=C bond length in unstrained, unconjugated ethenes ranges from 133.5 to 135 pm (1.335 to 1.35 Å)[3] but is lengthened in conjugated ethenes and other alkenes in which the C=C bond is weakened (Section 9-1.d). Although elementary textbooks often state that the

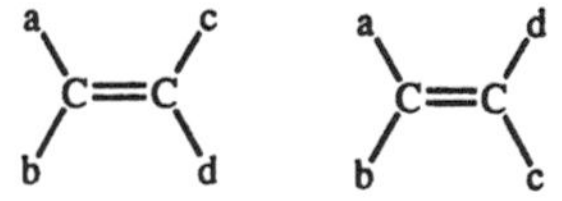

Figure 9.1. Z- and E-alkenes. Cahn–Ingold–Prelog priority is a > b and c > d.

$$\underset{R'}{\overset{R}{>}}C{=}C$$

bond angles are 120°, this is only approximately true; since a moiety of this type cannot possibly have local C_{3h} symmetry, there is no reason why the three bond angles (R–C=C, R'–C=C, and R–C–R') should be equal. In fact, in ethene (ethylene) itself the H–C–H angle has been measured by electron diffraction as[4] 116.6° or[5] 117.8°. In propene, the C=C–C angle is 124.3° and the C=C–H angle[6,7a] is 119° and in *cis*-2-butene the C–C=C angle[8] is 125.8°. The CH_3–C–CH_3 angle in isobutylene (2-methylpropene) is[7b,9] 115.3°. It appears, thus, that the R–C–R' angle in the above moiety is generally smaller than 120° and the R–C=C angle is larger.

cis–trans Isomerism (sometimes called geometric isomerism, a term we shall not use in this book) is a type of diastereomerism: The cis and trans isomers are (with rare exceptions, see below) not mirror-image stereoisomers. The necessary and sufficient condition for cis–trans isomers to exist is that one substituent at each end of the double bond differ from the other; referring to Figure 9.1 this means a ≠ b and c ≠ d. There are no restrictions as to the identity of a, b with c, d; thus abC=Cab displays cis–trans isomerism. The other conditions for cis–trans isomerism are implicit: One condition is that the torsion angles a–C–C–c, a–C–C–d, b–C–C–c, and b–C–C–d be near 0° or near 180° (i.e., that the alkene be planar or near-planar; deviations of these torsion angles from 0° or 180° by a few degrees are common) and the second condition is that the barrier for interconversion of the cis–trans isomers be high enough for these isomers to be distinguishable entities. Both of these conditions are generally fulfilled with alkenes; as indicated above, the rotational barriers in alkenes are very much higher than those in alkanes [e.g., 3.6 kcal mol^{-1} (15.1 kJ mol^{-1}) for the central C–C bond in butane]. Later we shall discuss exceptional cases where the barriers are low and/or the alkenes are appreciably nonplanar.

b. Nomenclature

The two arrangements shown in Figure 9.1 are called "Z" (from German *zusammen* meaning together) and "E" (from German *entgegen* meaning opposite) depending on whether the atoms of highest priority in the Cahn–Ingold–Prelog sequence (cf. Section 5-2), which are assumed to be a and c in Figure 9.1, are on the same side or on opposite sides.[10,11] Examples are shown in Figure 9.2, including cases of cis–trans

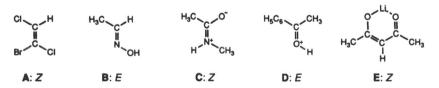

Figure 9.2. Examples of *E–Z* nomenclature.

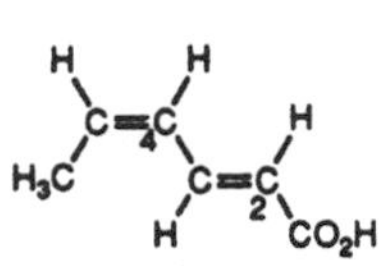

(2E,4Z)-2,4-
Hexadienoic acid

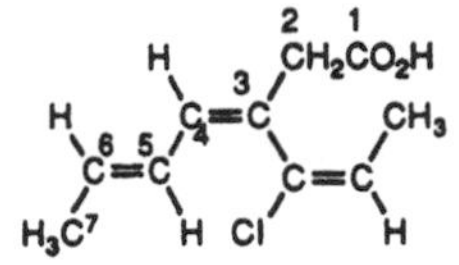

3-[E-1-chloropropenyl]-
(3Z,5E)-3,5-
heptadienoic acid

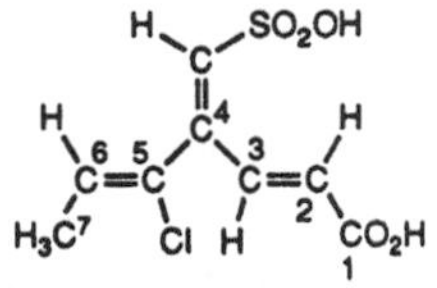

(2E,5Z)-5-Chloro-4-
[(E)-hydroxysulfonylmethylene]-
2,5-heptadienoic acid [Cl(5) > H(3)
decides sequence at C(4)]

Figure 9.3. Additional cases of Z–E nomenclature.

isomerism of C=N and C=O double bonds or partial double bonds. When the descriptor (E or Z) is part of a name, it is placed in parentheses in front of the name; thus structures **A** and **B** in Figure 9.2 are called (Z)-1-bromo-1,2-dichloroethene and (E)-(ethanal oxime), respectively. The descriptors E and Z are always italicized and, in the case of the oxime, since the descriptor refers to the entire name, the two parts of the name are enclosed in parentheses. Figure 9.3 indicates situations where locants must be used in conjunction with the descriptors.

Before 1968, the prefixes cis and trans were used for alkenes and the prefixes syn and anti for oximes and other aldehyde and ketone derivatives. As shown in Figure 9.4 the old nomenclature ranges from straightforward (as in the butene **F**) to cumbersome (as in the acetophenone oxime **G**) to confusing (as in the 3-phenylbutenoic acid **H**) to being inapplicable (as in many tetrasubstituted ethylenes, such as **I**). The E–Z nomenclature can always be applied and is always unequivocal. It is important to realize that in cases where cis and trans (or syn and anti) can be used, Z does not always correspond to cis or syn (and E does not necessarily correspond to trans or anti). Thus compound **A** in Figure 9.2 is trans but Z; acetaldoxime **B** is (by convention) syn but E.

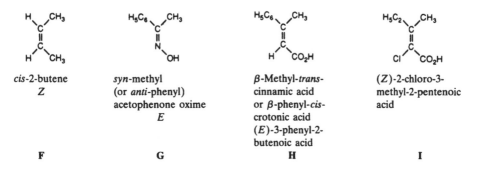

cis-2-butene
Z

syn-methyl
(or anti-phenyl)
acetophenone oxime
E

β-Methyl-trans-
cinnamic acid
or β-phenyl-cis-
crotonic acid
(E)-3-phenyl-2-
butenoic acid

(Z)-2-chloro-3-
methyl-2-pentenoic
acid

F   **G**   **H**   **I**

Figure 9.4. Old and new nomenclature.

c. Cumulenes

As van't Hoff[12] already recognized, cis–trans isomerism exists not only in monoenes but also in polyenes of the cumulene type with an odd number of double bonds [ab(C=)$_n$Ccd, where n is odd]. (When n is even, as in allenes, etc., there is enantiomerism; this will be discussed in Chapter 13.) This isomerism is due to the fact that successive planes of π bonds are orthogonal to each other, as shown in Figure 9.5. The cis–trans isomers in a butatriene (Fig. 9.6) were first observed in 1959.[13] The barrier between the diastereomers is relatively low: $\Delta H^{\ddagger} = 31.0$ kcal mol^{-1} (129.7 kJ mol^{-1}) for $CH_3CH=C=C=CHCH_3$.[14] The low barrier is presumably due to zwitterionic or biradical resonance (cf. Fig. 9.5). It is therefore not surprising that the cis–trans isomers interconvert readily, either photochemically (in diffuse daylight; see the later discussion regarding photochemical cis–trans isomerization of alkenes) or thermally, at 160°C. In the case of the substituted hexapentaenes, [5]cumulenes, the barrier measured by NMR, is considerably lower: 19.1 kcal mol^{-1} (79.9 kJ mol^{-1}) for

$$C_6H_5CH_2(CH_3)_2C(C_6H_5)C=C=C=C=C=C(C_6H_5)C(CH_3)_2CH_2CH_5$$

and 19.1 kcal mol^{-1} (79.9 kJ mol^{-1}) for t-$BuC_6H_5C=C=C=C=C=CC_6H_5t$-Bu.[15]

d. Alkenes with Low Rotational Barriers and Nonplanar Alkenes

There are two ways in which the barriers in alkenes can be lowered: by raising the ground-state energy or by lowering the transition-state energy (or by a combination of the two).[16,17] Steric factors sometimes destabilize the ground state; two examples are shown in Figure 9.7. In the substituted ethylene A,[18] as the size of the R group increases from hydrogen to *tert*-butyl, the rotational free energy barrier drops from about 27.7 kcal mol^{-1} (116 kJ mol^{-1}) to 18.3 kcal mol^{-1} (76.6 kJ mol^{-1}). In the fulvene

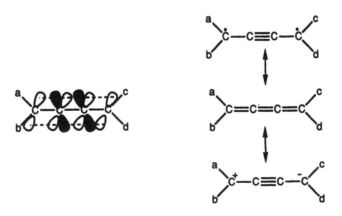

Figure 9.5. Stereoisomerism in cumulenes.

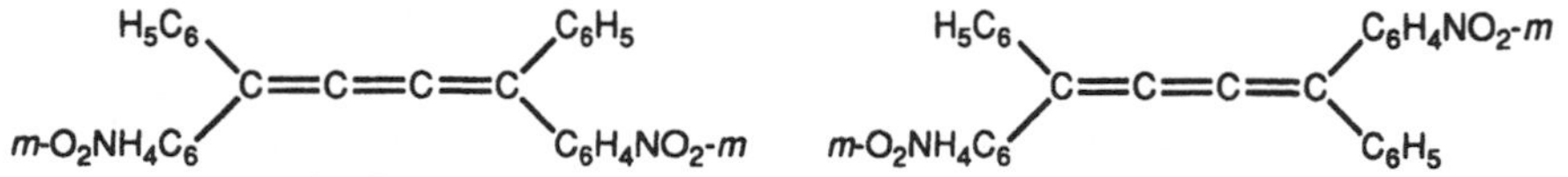

Figure 9.6.   The cis–trans isomerism in a butatriene.

B, the simple change of the R group from hydrogen to methyl lowers the barrier[19] by 5.7 kcal mol^{-1} (23.8 kJ mol^{-1}); this effect is probably not *entirely* steric.

Even in compound **A**, R = H (Fig. 9.7), the barrier is about 35 kcal mol^{-1} (146 kJ mol^{-1}) lower than that in 2-butene. The relatively mild repulsion of the methoxy and cis-located carbomethoxy group can in no way account for this difference. Rather, compounds **A** are representatives of "push–pull" or "capto–dative" alkenes[17,20] in which there is extensive delocalization of the π electrons, as shown in Figure 9.7, **A'**. One might consider that the central C=C bond has considerable single-bond character and that the barrier is therefore unusually low. Perhaps a better way of looking at the situation[17] is to say that what is shown as canonical form **A'** in Figure 9.7 actually represents the structure of the transition state in which, since the torsion angle at the central bond is now 90°, there is no double-bond character left in that bond. The structure written for **A'** now implies a high degree of stabilization of the zwitterionic transition state with a concomitant sizable lowering of the activation energy for rotation. Numerous other cases of low rotational barriers in push–pull ethylenes are known.[16,17]

Delocalization of a π bond by resonance does not necessarily require a push–pull situation. Thus the barrier in stilbene, $C_6H_5CH=CHC_6H_5$ (cis → trans) is 42.8 kcal

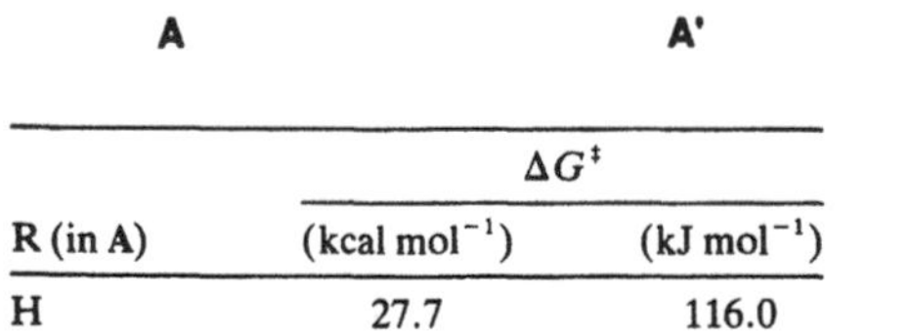

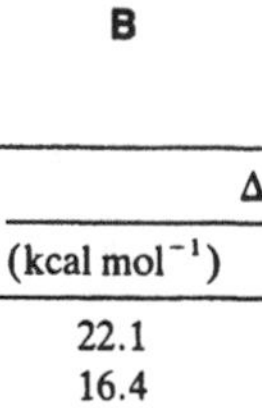

R (in A)	$\Delta G^{\ddagger}$		R (in B)	$\Delta G^{\ddagger}$	
	(kcal mol^{-1})	(kJ mol^{-1})		(kcal mol^{-1})	(kJ mol^{-1})
H	27.7	116.0	H	22.1	92.5
CH_3	25.7	107.5	CH_3	16.4	68.6
C_2H_5	24.7	103.3			
i-C_3H_7	23.3	97.5			
t-C_4H_9	18.3	76.6			

Figure 9.7.   Steric effect on barriers in alkenes.

mol^{-1} (179 kJ mol^{-1}), about 20 kcal mol^{-1} (83.7 kJ mol^{-1}) less than that in 2-butene.[21,22] Steric strain in the ground state can account for only a small part of this difference, inasmuch as the strained *cis*-stilbene is only 3.7–4.2 kcal mol^{-1} (15.5–17.6 kJ mol^{-1}) less stable than the presumably unstrained *trans*-stilbene.[23,24] There must, therefore, also be stabilization of the transition state, presumably (see Section 9-3.b) as a low-lying resonance-delocalized triplet biradical, $H_5C_6\dot{C}HCHC_6H_5$ formed by intersystem crossing.[25]

In some alkenes, steric crowding in the planar state may be so severe that the latter is no longer an energy minimum but becomes an energy barrier with the ground state becoming twisted.[3] In this context the compounds in Figure 9.8 are of interest. The drop in rotational barrier from stilbene (42.8 kcal mol^{-1}, 179 kJ mol^{-1}) to compound **A**[26] (21.1 kcal mol^{-1}, 88.3 kJ mol^{-1}) to compound **B** ("small barrier", cf. ref 16) may be explained by increasing stabilization of the triplet biradical transition state. However, the "negative" barrier in **C**, that is, the fact that the stable ground state of **C** is nonplanar and that the planar conformation represents the transition state for rotation,[27,28] requires a different explanation; the difference between **B** and **C** is presumably due to the steric interaction of the four chlorine substituents in the planar conformation of **C**. This situation resembles that in *o,o'*-tetrasubstituted biphenyls to be discussed in Chapter 13.

Remarkably, the polycyclic species shown in Figure 9.9 is chiral, by virtue of a twist of the terminal benzene moieties induced by the half-chair conformation of the seven-membered rings. X-ray structure determination revealed that the two CH_2CH_2 bridges in the seven-membered rings are syn, so the symmetry point group of the molecule is C_2 (not C_i). Thus the molecule is chiral and can, in fact, be resolved by chromatography on swollen microcrystalline triacetylcellulose (cf. Section 7-3.d). The stiffness of the CH_2CH_2 bridges is indicated by the ^1H NMR spectrum that indicates an ABCD system. Both from exchange broadening of this spin system and from measurement of the rate of racemization by circular dichroism (CD) (Chapter

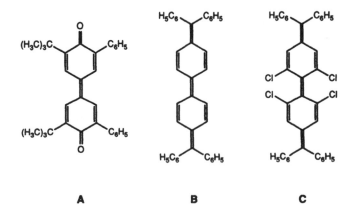

A   **B**   **C**

Figure 9.8. Alkenes with low or "negative" barriers.

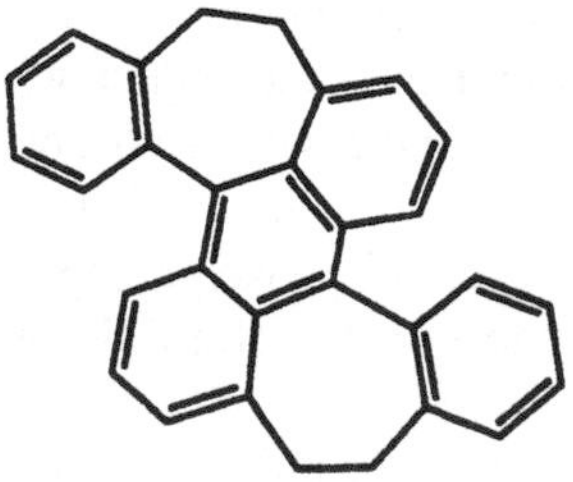

Figure 9.9. Chiral bridged diphenylanthracene.

12), the barrier to "flipping" of the seven-membered rings was found to be[29] 23.0–23.2 kcal mol^{-1} (96.2–97.1 kJ mol^{-1}).

Push–pull ethylenes can also have twisted ground states. An example[30] is the dithiodiketone **A** shown in Figure 9.10. The angle of twist (81°) and very long inter-ring bond distance [148 pm (1.48 Å)] suggest that this molecule had better be looked at as existing largely in canonical form **A′**. Related twisted push–pull ethylenes, shown in the general formula **B** in Figure 9.10,[31] are actually chiral, have been resolved, and racemize with an activation energy of 29.9–30.3 kcal mol^{-1} (125–127 kJ mol^{-1}).

We conclude this section with mention of one of the challenges in the area of twisted ethylenes, namely, tetra-*tert*-butylethylene (Fig. 9.11, **C**, (CH$_3$)$_3$C in place of

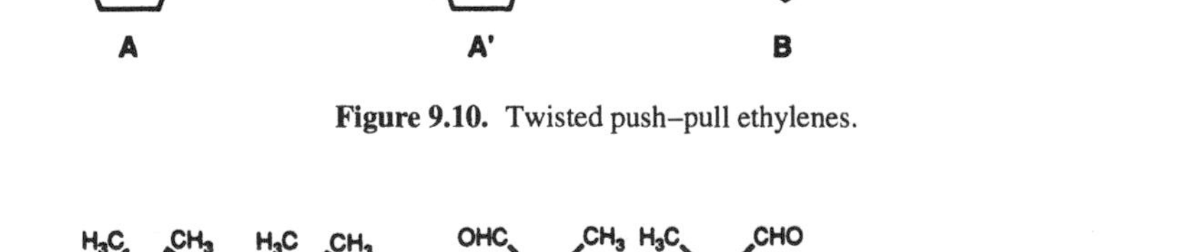

Figure 9.10. Twisted push–pull ethylenes.

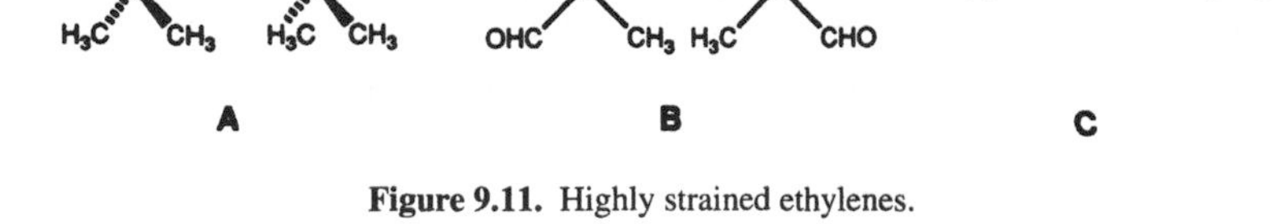

Figure 9.11. Highly strained ethylenes.

C_2H_5). It has been calculated[32] that this molecule should exist in two different (diastereomeric) twisted conformations: one with a twist of 45° and a strain energy of 82.3 kcal mol^{-1} (344 kJ mol^{-1}), the other with a twist of 13° and a strain energy of 86.3 kcal mol^{-1} (361 kJ mol^{-1}). Although the molecule has not as yet been synthesized,[33] cyclic analogues have been prepared.[34a] Two examples are shown in Figure 9.11, **A**, $n = 1$ or 2; the compound with $n = 2$ shows little of the ordinary reactivity of an alkene.[34b]

An even closer approach to the as yet unknown tetra-*tert*-butylethylene[34c] is shown in structures **B** and **C** in Figure 9.11. The tetraaldehyde **B** has a long [136 pm (1.36 Å)] central double bond and is twisted by an angle of 28.6°, as evidenced by X-ray diffraction analysis; the CH_3–C–CHO bond angle is compressed to 96 ± 1° (see also ref. 35). The C=C Raman stretching frequency is at an unusually low 1461 cm^{-1}; the UV spectrum is anomalous also and rotation about the alkyl bonds is hindered by a barrier of 12.2 kcal mol^{-1} (51.0 kJ mol^{-1}) as shown by low-temperature NMR spectroscopy. Unfortunately, various efforts to reduce the aldehyde groups terminated in formation of a very stable, double six-membered hemiacetal [cf. Fig. 9.11, structure **A**, CH_2–O–CHOH in place of $(CH_2)_n$] or other cyclic species. A more extensive discussion of strained and nonplanar alkenes may be found elsewhere.[3]

e. The C=N and N=N Double Bonds

The cis–trans isomerism about C=N double bonds is of importance in oximes, imines, hydrazones, and so on, and that about N=N double bonds is of interest in azo compounds.[16,36] Partial double bonds also are found in amides and thioamides (pertinent canonical structure

$$R-\overset{+}{C}=NR_2 \text{ and } R-\overset{+}{C}=NR_2 \text{), esters (} R-\overset{+}{C}=OR' \text{), enolates } \left(\begin{array}{c}\diagup\\\diagdown\end{array}C=\overset{|}{C}-O^-\right), \text{ and so on)}$$
with the lower substituents being O^-, S^-, O^- respectively.

Barriers about various aldehyde and ketone derivatives with amines (imines and hydrazones) are shown in Table 9.1; included are barriers about N=N double bonds. Some of these barriers, which were determined by NMR spectroscopy, involve topomerization rather than isomerization; nonetheless it is clear that they are, in most cases, low enough as to make isolation of stable cis–trans isomers in these series difficult. Notable exceptions are the oximes and oxime ethers[38] whose E–Z (or syn–anti; cf. Section 9-1.b) isomers are quite stable with barriers above 39 kcal mol^{-1} (163 kJ mol^{-1}). Here, as in the case of F–N=N–F (Table 9.1), the presence of electron-withdrawing substituents on nitrogen greatly increases the barriers for reasons we shall explore shortly. Most of the other (minor) trends in Table 9.1 can readily be explained on the basis of resonance effects (delocalization of the π electrons in C=N lowers the barrier). The low barrier in the hexafluoroacetone phenylimine is an exception; it appears that the normal $\overset{+}{C}-\overset{-}{N}$ polarization of the imine bond here is reversed with the negatively charged carbon being stabilized by the electron-withdrawing CF_3 groups:

TABLE 9.1. Barriers to E–Z Isomerization or Topomerization about C=N and N=N Bonds[a]

	$\Delta G^{\ddagger}$	
Compound	(kcal mol^{-1})	(kJ mol^{-1})
$(CH_3)_2C=NC_6H_5$[b]	20.3	84.9
$(CH_3)_2C=NCN$[b]	18.9	79.1
$(CH_3)_2C=N-N(CH_3)C_6H_5$[b]	21.1[c]	88.3[c]
$(CF_3)_2C=NC_6H_5$[b]	15.45	64.6
$p-ClC_6H_4(C_6H_5)C=NCH_3$	25[c]	105[d]
$C_6H_5CH=NC_6H_5$	16.5[d,e]	69.0[d,e]
$C_6H_5N=NC_6H_5$	23.7	99.2
F–N=N–F	35.2	147
$(CH_3)_2\overset{+}{N}=N-\bar{O}$[b]	23.3	97.5
$p-ClC_6H_4(C_6H_5)C=N-O-CH_3$	>39	>163

[a]Data taken from ref. 16 refer to isomerization unless otherwise indicated.
[b]Topomerization, cf. Chapter 8.
[c]In hexachlorobutadiene. A much higher barrier was found in diphenyl ether.
[d]E_a.
[e]Data from ref. 37.

$$\overset{\displaystyle CF_3}{\overset{\uparrow}{F_3C \leftarrow \underset{}{\underline{C}}-\overset{+}{N}-C_6H_5}}$$

Accordingly electron-donating substituents in the aryl group lower the barrier still further.[39]

As mentioned in Chapter 2, amides and thioamides,

$$\overset{\displaystyle X}{\overset{\|}{R'-C-NR_2}} \quad (X = O \text{ or } S)$$

though generally written with C=X double and C–N single bonds, in fact have partial C=N double bonds by virtue of the contribution of the

$$\overset{\displaystyle R'C=\overset{+}{N}R_2}{\underset{\displaystyle X^-}{|}}$$

canonical form.

Thioamides have a higher barrier than amides presumably because the canonical form with C=S double bond is less important than that with a C=O double bond; correspondingly, the canonical form with the $\bar{S}-C\overset{+}{=}N$ double bond is more important for the thioamide than for the amide. Thus the barrier in N,N-dimethylthioacetamide (27.8 kcal mol^{-1}, 116.4 kJ mol^{-1}) is substantially higher than that in N,N-dimethylacetamide (17.3 kcal mol^{-1}, 72.4 kJ mol^{-1}).[36,40] In L-alanyl–L-proline,

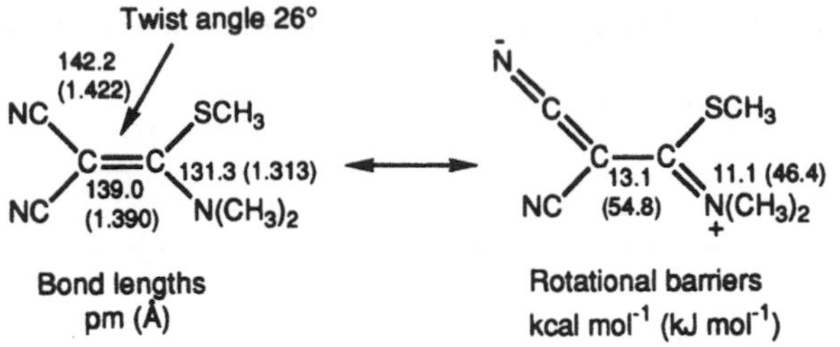

Figure 9.12. Energy barriers, bond lengths, and torsion angle in a push–pull ethylene.[17]

cis–trans isomers are sufficiently stable to appear as separate peaks in a reverse-phase liquid chromatogram;[41] cis–trans isomerism in proline-containing polypeptides is an important determining factor in protein structure.

Returning, for the moment, to the push–pull ethylenes discussed earlier, we note (see Fig. 9.12) that such compounds have not only diminished C=C bond rotation barriers but also enhanced barriers to rotation about the single bonds of the attached donor and acceptor atoms (in the case shown, only the C–N barrier is measurable). Along with this goes a C=C distance considerably longer than the normal 133 pm (1.33 Å) and foreshortened bonds to the donor and acceptor atoms [normal C–C bond length is 153 pm (1.53 Å); C–N, 147 pm (1.47 Å)]. Perhaps surprisingly, in view of the small size of the CN substituents, the molecule shown in Figure 9.12 is also twisted by 26° around the C–C bond.

9-2. DETERMINATION OF CONFIGURATION OF CIS–TRANS ISOMERS

The configuration of cis–trans isomers (*E* or *Z*) can be determined by either physical or chemical methods.[42,43] Chemical methods were the first to be applied; these methods often rest on a very firm basis, and the assignments made on chemical grounds have served as underpinnings for the later developed physical (mainly spectroscopic) methods. However, they are now largely of historical interest; physical methods are simpler and easier to employ and are therefore used almost exclusively in contemporary chemistry. Our discussion of chemical methods will therefore be limited to but a few cases illustrating the principles involved.

a. Chemical Methods

These methods are essentially of three types: absolute, correlative not affecting the configuration of the double bond, and mechanistic.

Absolute methods rest on the fact that isomers in which functional or reactive groups are located cis to each other can sometimes be converted to cyclic lactones, anhydrides, amides, and so on, whereas the corresponding trans isomers cannot.

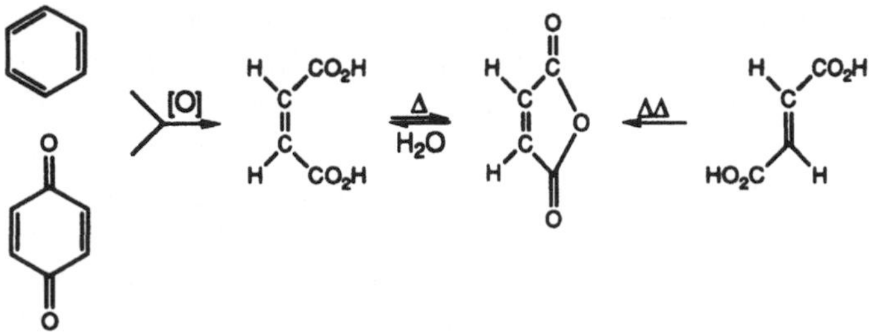

Figure 9.13. Configurations of maleic and fumaric acids.

Alternatively, the cis, but not the trans, isomers can be synthesized from small ring alkenes. Thus maleic, (Z)-butenedioic, acid is converted, by gentle heating, into its cyclic anhydride (Fig. 9.13) from which the acid can be regenerated by hydration; on the basis of this long-known fact van't Hoff[44] assigned the cis or Z configuration to maleic acid. The trans isomer, fumaric or (E)-butenedioic acid, is converted into the same anhydride only at a much higher temperature, presumably as a result of thermal isomerization. The formation of maleic acid by oxidation of benzene or p-quinone (Fig. 9.13) supports the Z assignment. Similar principles have been applied to configurational assignments of oximes.[43]

As in the case of determination of configuration in chiral molecules (Section 5-5), once the configuration of a few alkenes was known with certainty, that of others could be determined by correlative methods. The reliability of such methods is limited on the one hand by the possibility of cis–trans isomerization in the course of the transformations used in the correlations and, on the other hand, in cases where reactions at an olefinic carbon are involved, by potential uncertainty concerning the steric course of such reactions.

An example not involving reaction at the olefinic carbon atoms is shown in Figure 9.14, the correlation of the configuration of the higher melting (trans) crotonic acid with fumaric acid via the relay of trichlorocrotonic acid.[45] Although the assignment of the E configuration to the crotonic acid melting at 72°C is now known to be correct, the argument embodied in Figure 9.14 is somewhat weak, since isomerization (E → Z) in the course of the transformation was not convincingly excluded. It is preferable in such correlations to make them for both the E and the Z isomer to assure that there

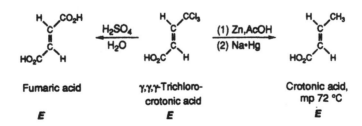

Fumaric acid
E

γ,γ,γ-Trichloro-
crotonic acid
E

Crotonic acid,
mp 72 °C
E

Figure 9.14. Configurational correlation of crotonic acid (mp 72°C) with fumaric acid.

is no stereoconvergence, or at least to check (by physical means) that a configurationally pure starting material gives a configurationally pure product. The assumption underlying the latter check is that configurational change, if it occurred, would not be complete but would lead to a cis–trans mixture; this assumption is more certain if the correlation is effected with the less stable (cis?) isomer.

A third way of assigning E or Z configuration to an alkene is by a directed stereospecific synthesis. The terms "stereospecific" and "stereoselective" are defined below; "directed" means that the stereochemical outcome of the reaction chosen can be confidently predicted; the principle involved is the same as that discussed in Section 5-5.f for configurational assignment at chiral centers. According to Zimmerman et al.,[46] a *stereoselective* reaction is one in which only (or largely) one of two or more possible stereoisomers is formed. Thus the reaction shown in Figure 9.15 is stereoselective: Of the two possible stereoisomers (E, Z) of the addition product, only the Z isomer is formed. In contrast, the term *stereospecific* denotes a defined configurational relationship between starting material and product. Thus, for example, the dehydrobromination (by KOH) on the left side of Figure 9.16 is stereospecific: The meso dibromide gives the cis (E) bromostilbene whereas the chiral stereoisomer gives the trans (Z) isomer. The term "stereoselective" may be modulated ("highly stereoselective", "moderately stereoselective"). All stereospecific reactions must also be stereoselective but the converse is not true (the reaction shown in Fig. 9.15 is not stereospecific since phenylacetylene has a unique structure without stereoisomers).

There are three approaches to this type of assignment: addition to alkynes (acetylenes), synthesis from saturated compounds of known configuration, and "other" methods, of which the Wittig reaction is probably the most important.

Of the three approaches, addition to alkynes is the simplest, since alkynes have no stereochemistry of their own and the stereochemical outcome (formation of cis or trans alkene) depends only on the stereochemical course (syn or anti) of the addition. Thus, in contrast to electrophilic additions,[47] nucleophilic additions, exemplified in Figure 9.15, proceed reliably in anti fashion,[43,48] giving rise to Z enol (or thioenol) ethers. It is believed that the incoming electron pair of the nucleophile and the p electrons of the π bond tend to stay as far away from each other as possible; subsequent protonation of the anion with retention of configuration then leads to anti addition.[49]

In the second approach, alkene configuration is mechanistically linked to the configuration (assumed known) of a saturated precursor. Only such reactions are useful whose stereochemical course is known "with confidence." Ionic E2 eliminations and pyrolytic elimination reactions[50] best fill this bill. Bimolecular ionic elimination reactions in unconstrained (i.e., noncyclic) systems almost invariably

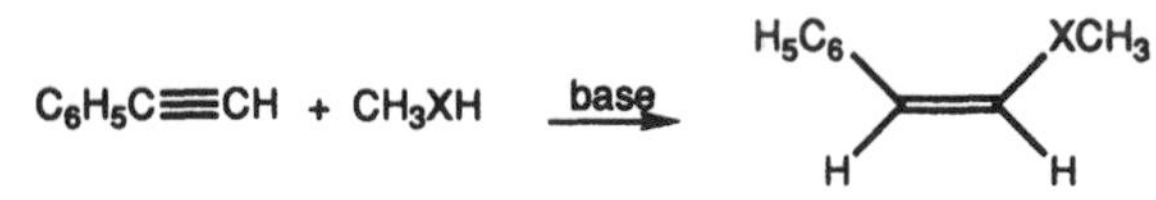

$$C_6H_5C\equiv CH \ + \ CH_3XH \ \xrightarrow{\text{base}}$$

Figure 9.15. Nucleophilic addition to acetylenes, where X = O or S.

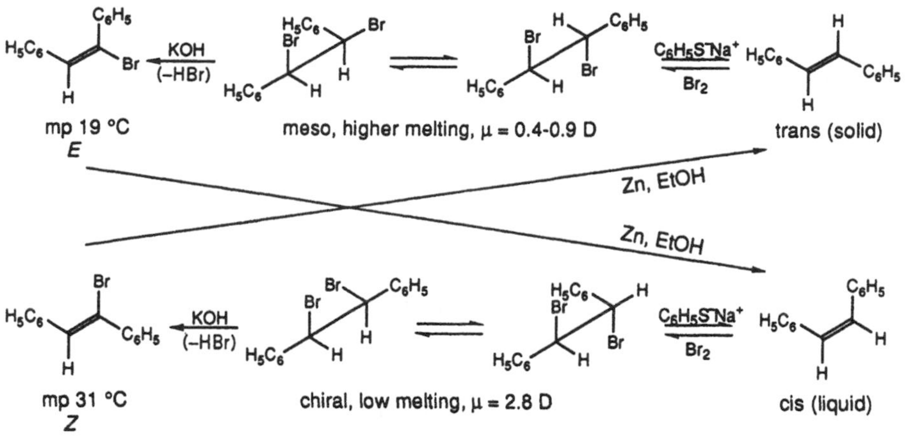

Figure 9.16. Interconversion of stilbene dibromides, stilbenes, and bromostilbenes.

proceed in antiperiplanar fashion (the few exceptions seem to be confined to elimination of quaternary ammonium salts with hindered bases, such as tert-alkoxides). The necessary condition for success of the method, then, is that the configuration of the saturated starting material be known. A classical example[51] is shown in Figure 9.16; the configuration of the precursor "dibromostilbenes" (1,2-dibromo-1,2-diphenylethanes) is readily inferred from their dipole moments[52]: Since the phenyl groups tend to be antiperiplanar, the dipole moment of the meso isomer (antiperiplanar bromines) will be much lower than that of the chiral isomers (gauche bromines). If it is assumed that the steric course of both KOH and sodium thiophenolate (PhS⁻Na⁺) elimination is anti, the configurations of the stilbenes (1,2-diphenylethenes) and α-bromostilbenes shown in Figure 9.16 follow; moreover it follows that the reduction of the bromostilbenes to stilbenes with zinc and ethanol proceeds with retention of configuration.

Perhaps surprisingly, substantial stereoselectivity, and thereby prediction of the configuration of the product, may also be attained in the Wittig reaction[53,54] in which the alkene is, so to speak, put together from its two halves, one in the form of alkylidenephosphorane, $RCH=P(C_6H_5)_3$ (or, in general, $RCH=PR_3$), and the other in the form of an aldehyde or ketone, $O=CRR'$. When the second reagent is an aliphatic aldehyde, $O=CHR$ and the reaction is carried out under carefully circumscribed conditions, the cis or Z alkene will generally be the near-exclusive product, formed via a cis-oxaphosphetane intermediate (Fig. 9.17).

Interestingly, when the Wittig reagent is derived from dibenzophosphole (Fig. 9.18), trans alkenes are obtained in better than 6:1 predominance in a kinetically controlled reaction even from aliphatic aldehydes.[55] In contrast, triarylphosphine-derived Wittig reagents in which the aryl groups bear o-methyl or o,o'-difluoro substituents lead to an enhancement in the predominance of cis alkenes in cases where stereoselectivity is otherwise not satisfactory.[56]

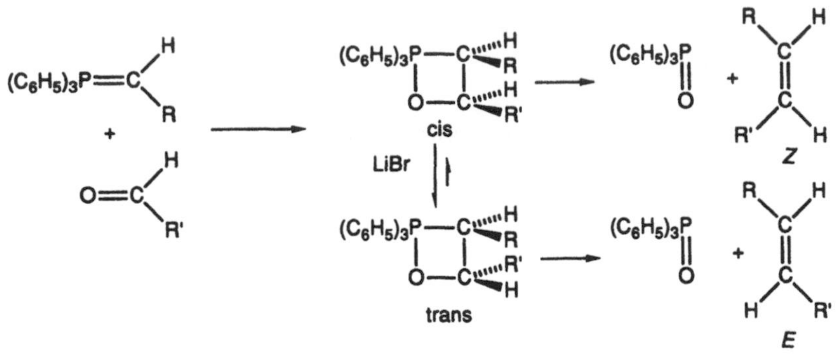

Figure 9.17. Wittig reaction leading to Z alkene (in the absence of Li salt). (Li salts promote equilibration of the intermediate oxaphosphetane furthering formation of the E product.)

An interesting possibility is to use reaction *rate* rather than reaction product in configurational assignment. This device (exploitation of kinetics) can be used both in alkene-forming reactions and in alkene-consuming reactions. Two examples must suffice to illustrate the principle (see also ref. 57).

In the dehydrohalogenation of 2-bromobutane, the 2-butene isomers are formed in a ratio E:Z of about 6:1.[58] The reaction profile is shown in Figure 9.19. On the basis of the following argument, one can predict that the major product formed is the E (trans) isomer: Formation of the trans and cis isomers involves transition states with antiperiplanar Br and diastereotopic hydrogen atoms H_1 or H_2. Transition state $A^{\ddagger}$ (elimination of H_2) is preferred over $B^{\ddagger}$, since it lies between starting conformation A and product P_A, whereas $B^{\ddagger}$ lies between B and P_B, and since the energy level of A is below that of B (absence of CH_3/CH_3 gauche interaction) and that of P_A lies below that of P_B (trans product more stable than cis; cf. Sec. 9-3a). Therefore, barring the most unlikely possibility of a double crossing of the energy surfaces, transition state $B^{\ddagger}$ lies above $A^{\ddagger}$ (Fig. 9.19) and the major product must thus be the trans isomer.

One more rather important case where the constitution of a reaction product is used to assign configuration to its precursor is the Beckmann rearrangement of oximes.[59,60] We have already mentioned (Section 5-5.f) that the migrating group in the Beckmann rearrangement retains its configuration. However, even when the migrating group is achiral, there is an important stereochemical aspect to the rearrangement: The group that migrates is the group trans to the OH moiety of the oxime (Fig. 9.20). This fact allows one to infer the stereochemistry of the oxime from the nature of the amide

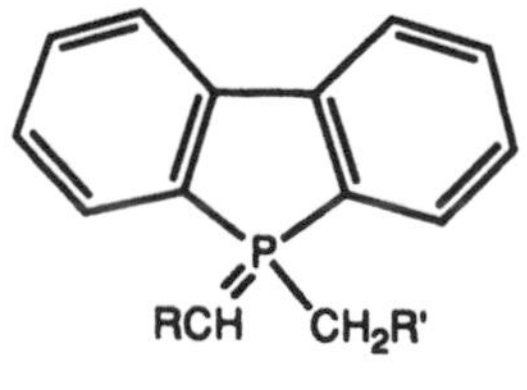

Figure 9.18. Dibenzophosphole-derived alkylidenephosphorane.

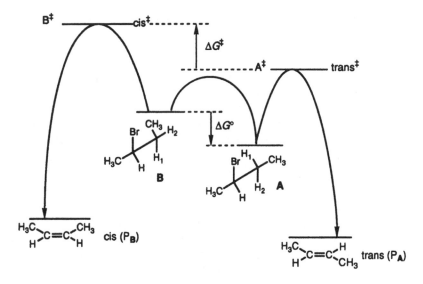

Figure 9.19. Dehydrohalogenation of 2-bromobutane.

formed from it. Thus acid treatment of the (*E*)-(acetophenone oxime) (Fig. 9.20, **A**, where R = C_6H_5, R′ = CH_3) would lead to acetanilide ($CH_3CONHC_6H_5$) through phenyl migration, whereas the corresponding *Z* oxime **B** would give *N*-methylbenzamide ($C_6H_5CONHCH_3$) by methyl migration.

b. Physical Methods

Since cis–trans isomers are diastereomers, they will generally differ in physical properties. If the differences are well understood in terms of configuration (*E* or *Z*), they can be used for configurational assignment. Only the more easily measurable and often clearcut differences in properties—dipole moments, boiling point, density, refractive index, vibrational (IR–Raman) spectra, and NMR spectra, both ^{1}H and [13]C—will be discussed here. Among other techniques, the earlier mentioned X-ray and electron diffraction methods, as well as microwave spectroscopy, also lead to

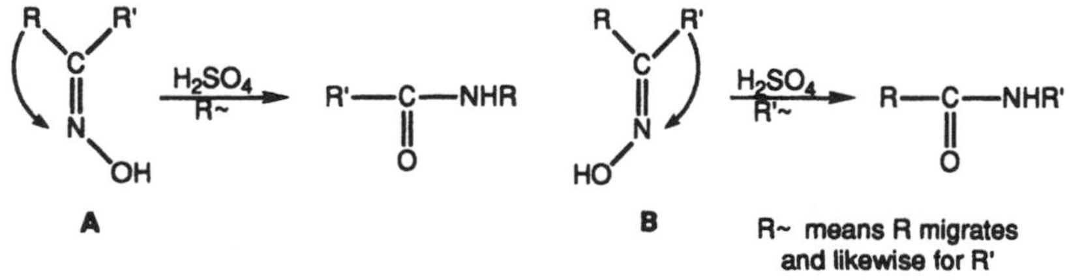

Figure 9.20. Course of the Beckmann rearrangement.

configurational assignment, though the latter two only in the case of relatively small molecules.[61]

Dipole Moments.   As shown in Figure 9.21, the relation of dipole moment (cf. ref. 62) to configuration is quite direct. If, in a 1,2-disubstituted alkene (XCH=CHY), X and Y are both electron donating or both electron withdrawing, the dipole moment of the cis isomer will generally be sizable, whereas that of the trans isomer will be small or zero (i.e. $\mu_{cis} > \mu_{trans}$). If, on the other hand, X is electron donating and Y is electron withdrawing, or vice versa, $\mu_{trans} > \mu_{cis}$. In trisubstituted alkenes XCH=CYZ the situation is less clear cut, though if Z is alkyl and X and Y are halogens or other strongly electron withdrawing groups, the disposition (cis or trans) of X and Y will be decisive. Some salient dipole moments are shown in Table 9.2, column 3. Except for the case of cyclooctene, configuration can be inferred from dipole moments where known, though in some cases the differences are perhaps uncomfortably small. In the case of 1-chloro-2-iodoethylene the original, seemingly unreasonable order of dipole moments (cf. ref. 43) was later reversed.[65] In *trans*-cyclooctene, the normal C_{2h} symmetry of *trans*-XCH=CHX is reduced to at most C_2 (in the prevalent twist conformation, see Chapter 13); therefore (cf. Section 4-5.b) the compound can have a dipole moment, whereas compounds of C_{2h} symmetry cannot. That the moment is so large may be related to the highly twisted nature of the double bond (torsion angle between 136° and 157°; cf. ref. 17, p. 166 and Chapter 13).

Density, Refractive Index, and Boiling Point.   Equipment to measure dipole moments, while not complex, is perhaps not widely available. Fortunately, the "dipole rule"[67] predicts that the isomer of higher dipole will have the lower molar volume (presumably because of the enhanced tendency of dipolar molecules to self-associate through electrostatic attraction of the oppositely charged ends), and hence the higher density, refractive index, and boiling point, as well as retention volume in gas chromatography (GC) on a nonpolar column. As shown in Table 9.2, columns 4–6, the rule is generally obeyed and so the simple measurements just mentioned can be used for configurational assignments, provided both diastereomers are available and provided that the measured quantities differ appreciably for the two.

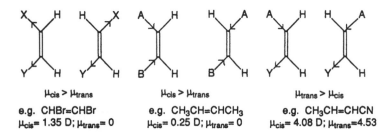

Figure 9.21. Dipole moment and cis–trans configuration.

TABLE 9.2. Dipole Moments and the Dipole Rule[a]

Compound		$\mu(D)$[b]	bp (°C; 760 mm)	n^{20}	D^{20}(g mL^{-1})
CHCl=CHCl	cis	1.84	60.3	1.4486	1.2835
	trans	0	47.4	1.4454	1.2583
CHCl=CHI	cis	1.27[c,d]	116–117	1.5829	2.2080 (15°C)
	trans	0.55[c]	113–114	1.5715	2.1048 (15°C)
CH$_3$CH=CHCl	cis	1.64[e]	32.8	1.4060	0.9347
	trans	1.97[e]	37.4	1.4058	0.935
CH$_3$CCl=CHCl	cis	2.20[f]	93	1.4549	1.1870 (25°C)
	trans	0.84[f]	76	1.4498	1.1704 (25°C)
CH$_3$CH=CHCH$_3$	cis	0.25[e]	3.7	1.3931 (−25°C)	0.6213
	trans	0	0.9	1.3848 (−25°C)	0.6044
(CH$_3$)$_3$CCH=CHC(CH$_3$)$_3$	cis	nr[g]	143	1.4266	0.7439
	trans	nr[g]	125.0	1.4115	0.7167
CH$_3$CH=CHCN	cis	4.08[e]	108	1.4182	0.8244
	trans	4.53[e]	122	1.4216	0.8239
EtO$_2$CCH=CHCO$_2$Et	cis	2.59[h]	223	1.4413	1.067
	trans	2.40[h]	218	1.4411	1.052
Cyclooctene	cis	0.43[i]	74–75[j]	1.4682 (25°C)	0.8443
	trans	0.82[i]	75[k]	1.4741 (25°C)	0.8483
Cyclodecene	cis	0.44[i]	194–195[l]	1.4858	0.8770
	trans	0.15[i]	194[l]	1.4821	0.8672

[a]Data from ref. 43, Beilstein, 3rd and 4th supplement, and McClellan[63] unless otherwise indicated.
[b]In benzene at 25°C unless otherwise indicated.
[c]Data of Errera[64] as corrected (see footnote d).
[d]Reference 65.
[e]In the gas phase.[66]
[f]At 30°C.
[g]Not reported.
[h]In carbon tetrachloride.
[i]In heptane.
[j]At 84 mm.
[k]At 78 mm.
[l]At 740 mm.

Acid Strength.   Bjerrum's law[68] (cf. Section 11-2.c):

$$\Delta pK_a = pK_2 - pK_1 = 0.60 + 2.3 \ Ne^2/RT\varepsilon r$$

is useful in the assignment of configuration of unsaturated dicarboxylic acids. Here ΔpK_a is the difference between the two pK_a values of such an acid, N is Avogrado's number, e is the electronic charge, R is the gas constant, T is the absolute temperature, ε is the effective dielectric constant of the medium, and r is the distance between the acid functions. Since r is generally smaller for a cis dicarboxylic acid than for its trans isomer, ΔpK_a will be larger, a fact that may be used for configurational assignment. For example, ΔpK_a in maleic acid is 4.19, whereas in fumaric acid it is only 1.36 (Fig. 9.22). For unsaturated monocarboxylic acids (Fig. 9.22) the difference in pK_a between E and Z isomers is much smaller though still palpable, with the Z acid being the stronger one. This is probably due to steric inhibition of resonance in the undissociated acid,

$$RHC=CH-C\overset{O}{\underset{OH}{\diagup}} \leftrightarrow R\overset{+}{H}C-CH=C\overset{O^-}{\underset{OH}{\diagup}}$$

This type of resonance is less important in the anion than in the acid itself, since in the anion it would place two negative charges on geminal oxygen atoms. Thus its net effect is to stabilize the anion less than the acid, resulting in a weakening of the latter (Fig. 9.23). For such resonance to be at its maximum, the conjugated C=C–C=O system must be coplanar. In the cis isomer, the β substituent (Fig. 9.22) interferes with coplanarity; therefore resonance is somewhat inhibited sterically, and since resonance was originally acid weakening, the cis acid becomes stronger, since the energy required to ionize it is less.

Vibrational (Infrared–Raman) Spectra.   Differences   in   IR   and   Raman spectra[43,72,73] of cis–trans isomers are found both in the C=C stretching region (ca. 1650 cm^{-1}) and the =C–H out-of-plane vibration region (970–690 cm^{-1}). For a molecular vibration to give rise to IR absorption, it must produce a change in dipole

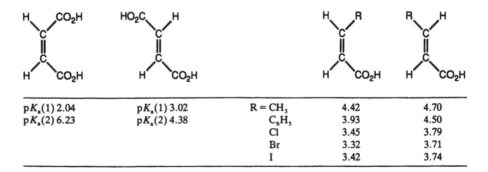

$pK_a(1)$ 2.04	$pK_a(1)$ 3.02	$R = CH_3$	4.42	4.70
$pK_a(2)$ 6.23	$pK_a(2)$ 4.38	C_6H_5	3.93	4.50
		Cl	3.45	3.79
		Br	3.32	3.71
		I	3.42	3.74

Figure 9.22. pK_a values of E and Z β-substituted propenoic acids. Data from ref. 69 except those for maleic and fumaric acid, which are from refs. 70 and 71.

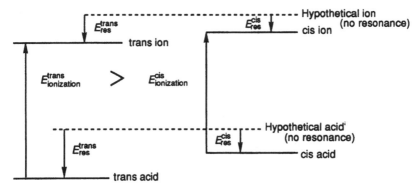

Figure 9.23. Resonance weakening of conjugated carboxylic acids. (It is assumed that effects other than resonance affect the acids and ions equally and thus cancel in the difference.)

moment of the molecule; no such condition applies to Raman absorption, which, however, requires that the vibration produce a change in polarizability. Thus *trans*-1,2-dichloroethylene, (*E*)-CHCl=CHCl, shows no IR absorption due to C=C stretching, since the dipole moment is zero and remains so during the stretching motion. However, there is a strong Raman absorption at 1577 cm^{-1} corresponding to this vibration. *cis*-Dichloroethylene, (*Z*)-CHCl=CHCl, shows a strong IR C=C stretching vibration at 1590 cm^{-1}, since the molecule has a dipole moment that is altered in the course of the stretching. Similar differences appear in the IR spectra of fumaric and maleic acids, (*E*)- and (*Z*)-HO$_2$CCH=CHCO$_2$H and of *trans*- and *cis*-3-hexenes, (*E*)- and (*Z*)-C$_2$H$_5$CH=CHC$_2$H$_5$. The differences are less marked when the substitution at the olefinic double bond is not symmetrical, for in that case even the trans isomer has a small dipole moment that changes during the C=C stretching. Thus *trans*-2-hexene, (*E*)-CH$_3$CH=CHC$_3$H$_7$, shows a C=C stretching frequency at 1670 cm^{-1}, though much less intense than that of the cis (*Z*) isomer at 1656 cm^{-1} and the *Z* and *E* isomers of 1,2-dichloropropene, ClCH=CClCH$_3$, both show absorptions in the C=C stretching region (at 1614 and 1615 cm^{-1}, respectively), the absorption of the *Z* isomer being considerably more intense.[74] As one might expect, this criterion fails in an alkene with an electron-donating group at one end and an electron-withdrawing group at the other (cf. Fig. 9.21); in fact, there is little difference in C=C stretching frequency in the isomeric crotonic acids, CH$_3$CH=CHCO$_2$H, and the situation is actually reversed (trans more strongly absorbing than cis) in the corresponding esters.[75] In cases where both isomers display the C=C stretching frequency, that of the trans isomer is often at higher wavenumber (1665–1675 cm^{-1}) than that of the cis (1650–1660 cm^{-1}).[73]

Infrared spectroscopy is also useful to distinguish s-cis from s-trans α,β-unsaturated ketones (cf. Fig. 10.20). The s-cis compounds show a higher intensity of the C=C stretching vibration (near 1625 cm^{-1}) and a lower intensity of the C=O stretching mode (near 1700 cm^{-1}) than the s-trans compounds. As a result, the intensity ratio of the two bands (C=O over C=C) is less (0.7–2.5) for the s-cis than for the s-trans compounds (6–9).

Nuclear Magnetic Resonance Spectroscopy. NMR[76,77] is by far the most useful and versatile technique for distinguishing cis and trans isomers. Both ^{1}H and ^{13}C NMR can be used and both chemical shifts and coupling constants are useful for inferring configuration. Moreover, the method is not generally confined to disubstituted alkenes (RCH=CHR'), though some of its variants are.

The chemical shift of the olefinic proton in RCH=CR'R", where R" may be H, can be used for configuration assignment by use of a formula first proposed by Pascual et al.[78] and later perfected.[79] The formula depends on the fact that the chemical shift increments due to the substituents R, R', and R" are additive and, moreover, for R' and R" depend on whether these groups are cis or trans to the proton in question: Referring to Figure 9.24 $\delta_{C=CH} = 5.25 + Z_{gem} + Z_{cis} + Z_{trans}$, where Z_{gem}, Z_{cis}, and Z_{trans} are parameters characteristic of substituents R_{gem}, R_{cis}, and R_{trans} (Fig. 9.24), given in Table 9.3. The precision of the formula has been assessed through a statistical treatment,[79] which indicated that 81% of the 4298 cases considered yielded calculated shifts within 0.20 ppm of those actually found and 94% yielded shifts within 0.30 ppm of the experimental. Since the calculated differences in shift between a cis and trans disubstituted alkene will be the differences between Z_{cis} and Z_{trans} in Table 9.3, one can easily figure out which substituents allow one to compute the result at the 81% or 94% level of confidence and make a prediction of configuration accordingly. By way of illustration, let us choose two trisubstituted alkenes (since methods of configuration assignment are more limited in this series than in disubstituted alkenes), namely, $CH_3CH=CBrCH_3$ and $CH_3CH=CClCO_2H$. For the 2-bromo-2-butenes, the calculated shifts are $5.25 + 0.45 - 0.28 + 0.45 = 5.87$ ppm for the *E* (cis) isomer and $5.25 + 0.45 - 0.25 + 0.55 = 6.00$ ppm for the *Z* (trans). The difference of 0.13 ppm is insufficient to allow the *Z* or *E* configuration to be assigned to individual isomers at a reasonable level of confidence; although, if shifts for both stereoisomers are available, it may be tentatively concluded that the isomer with the downfield olefinic proton has the *Z* configuration. In contrast, for the 2-chloro-2-butenoic acids, the calculated vinyl proton shift for the *E* isomer is $5.25 + 0.45 + 0.32 + 0.18 = 6.20$ ppm and that for the *Z* $5.25 + 0.45 + 0.98 + 0.13 = 6.81$ ppm. The difference of 0.61 ppm is sufficient to make the assignment from the NMR data with virtually complete confidence even if only one isomer is available. (For a shift difference of 0.50 ppm, the confidence level of assignment is 99.7%. We are assuming here that the shift actually measured is close to one of the two calculated ones.)

In alkenes of type RCH=CHR" or RCH=CFR' or RCF=CFR', proton–proton, proton–fluorine, or fluorine–fluorine coupling constants can be used to assign configuration. Table 9.4 shows the pertinent coupling constants for cis and trans disposed nuclei, respectively (corresponding to cis–trans arrangements of the R and

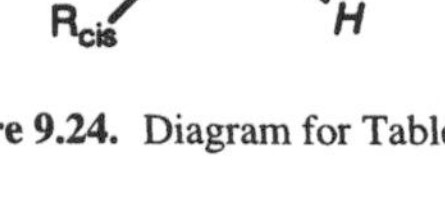

Figure 9.24. Diagram for Table 9.3.

TABLE 9.3. Parameters for Calculation of Proton Shift in the Alkene[a] Shown in Figure 9.24

Substituent R	Z_i for R (ppm)		
	Z_{gem}	Z_{cis}	Z_{trans}
–H	0	0	0
–Alkyl	0.45	–0.22	–0.28
–Alkyl-Ring	0.69	–0.25	–0.28
–CH_2O	0.64	–0.01	–0.02
–CH_2S	0.71	–0.13	–0.22
–CH_2X; X = F, Cl, Br	0.70	0.11	–0.04
–CH_2N	0.58	–0.10	–0.08
–C=C isolated	1.00	–0.09	–0.23
–C=C conjugated	1.24	0.02	–0.05
–C≡N	0.27	0.75	0.55
–C≡C	0.47	0.38	0.12
–C=O isolated	1.10	1.12	0.87
–C=O conjugated	1.06	0.91	0.74
–CO_2H isolated	0.97	1.41	0.71
–CO_2H conjugated	0.80	0.98	0.32
–CO_2R isolated	0.80	1.18	0.55
–CO_2R conjugated	0.78	1.01	0.46
H –C=O	1.02	0.95	1.17
N –C=O	1.37	0.98	0.46
Cl –C=O	1.11	1.46	1.01
–OR, R: aliphatic	1.22	–1.07	–1.21
–OR, R: conjugated	1.21	–0.60	–1.00
–OCOR	2.11	–0.35	–0.64
–CH_2–C=O; –CH_2–C≡N	0.69	–0.08	–0.06
–CH_2-Aromatic ring	1.05	–0.29	–0.32
–Cl	1.08	0.18	0.13
–Br	1.07	0.45	0.55
–I	1.14	0.81	0.88
–N–R, R: aliphatic	0.80	–1.26	–1.21
–N–R, R: conjugated	1.17	–0.53	–0.99
–N–C=O	2.08	–0.57	–0.72
–Aromatic	1.38	0.36	–0.07
–Aromatic o-subst	1.65	0.19	0.09
–SR	1.11	–0.29	–0.13
–SO_2	1.55	1.16	0.93

[a]From ref. 79. [Adapted, with permission, from Matter, E. V. et al. *Tetrahedron* **1969**, *25*, 693/4. Copyright © 1969 Pergamon Press, Headington Hill, Oxford, UK.]

TABLE 9.4. Proton–Proton, Proton–Fluorine, and Fluorine–Fluorine Coupling Constants[a]

Coupled Nuclei (X,Y)	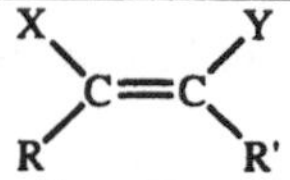 $^3J_{cis}$ (Hz)	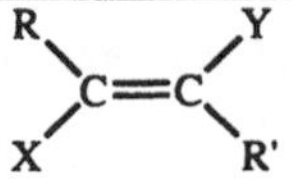 $^3J_{trans}$ (Hz)
H, H	+4 to +12	+12 to +19
H, F[b]	−4 to +20	+10 to +50
F, F[b]	+15 to +35	−115 to −134

[a]In part, from ref. 80, where X, Y = H or F.
[b]See also ref. 81.

R′ groups). The proton–proton constants follow the Karplus equation (cf. Fig. 10.26) with the cis coupling constant (torsion angle 0°) being smaller than the trans (torsion angle 180°). There is little overlap of range. For the proton–fluorine couplings, although the range overlaps somewhat, J_{trans} is said to be always larger than J_{cis} and the same is true for the absolute values of the fluorine–fluorine vicinal couplings.[81] Some examples are shown in Figure 9.25. A complication arises in that for a symmetrical alkene (RCH=CHR or RCF=CFR) the proton (or fluorine) nuclei are chemical shift equivalent so that the coupling constant cannot be observed directly. However, it is possible to obtain the coupling constant indirectly from the ^{13}C satellite spectrum (since X—^{13}C—^{12}C—X, where X = F or H, gives rise to an ABX pattern) or, in the case of protons, from the H—D coupling constant in the spectrum of the deuterated analogue, RCH=CDR ($J_{HH} = 6.49\ J_{HD}$).

Since neither IR spectra nor proton coupling constants in NMR are useful for configurational assignment in trisubstituted alkenes (RR′C=CHR″) and since proton chemical shift criteria may also fail in some such cases (see above), it is fortunate that ^{13}C NMR spectroscopy is a particularly valuable tool for configurational assignment

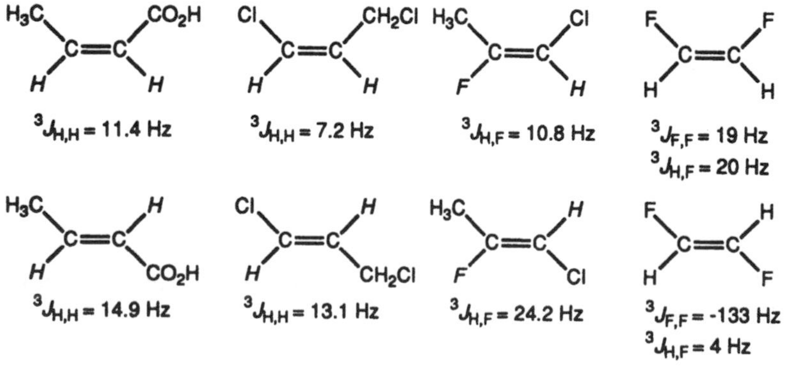

Figure 9.25. Examples of H/H, H/F, and F/F vicinal couplings.

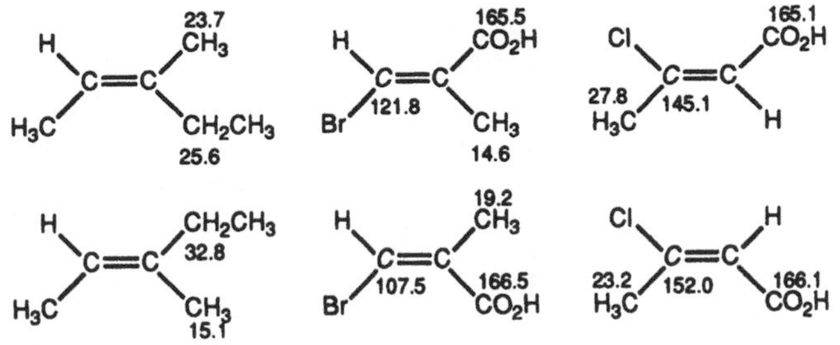

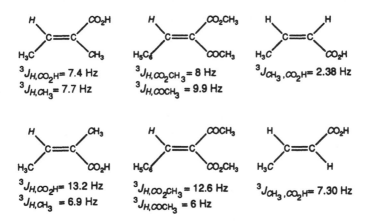

Figure 9.26. The ^{13}C shifts (ppm) of cis–trans isomers.

in this instance. The only condition is that either R or R′ (or both) must have a carbon atom at the point of attachment to the C=C moiety. The principle involved is that the carbon nucleus cis to the R″ group will be upfield shifted, through a "γ compression effect," relative to the same carbon nucleus positioned cis to H. Both the principle and examples are displayed in Figure 9.26 (data from ref. 82). It is seen that the CO_2H group does not manifest this upfield compression effect in its shift and that substantial (but apparently not as yet systematized) differences in shift at the β-carbon atom in acrylic acids are observed in some cases.

Recent improvements in NMR techniques, notably 2D NMR and double-quantum coherence experiments, have facilitated the determination of $^3J_{C/H}$ and even $^3J_{C/C}$ coupling constants.[83] Since the Karplus relationship applies to such couplings just as it does to proton–proton couplings (see above), $J_{trans} > J_{cis}$. Examples are shown in Figure 9.27.

We conclude this section on the use of NMR in cis–trans configurational assignment with a discussion of the nuclear Overhauser effect (NOE).[84,85] The NOE implies that irradiation of a nucleus (e.g., a proton) in any molecule will facilitate the

Figure 9.27. Use of $^3J_{C/H}$ and $^3J_{C/C}$ in configurational assignments of cis–trans isomers (The italicized C's represent ^{13}C-enriched ligands).

relaxation of a nearby nucleus (^{13}C, proton, etc.) and thereby enhance its signal, since the more rapidly a nucleus that has been excited to its upper spin state by absorption of energy returns to the ground state, the more energy it can absorb. The NOE, also, is a through-space effect that drops off with the sixth power of internuclear distance. Thus it will generally be important only between cis located nuclei, not between those located trans to each other. Indeed, the first application of the NOE in organic chemistry[86] involved assignment of the methyl proton signals in β,β-dimethylacrylic acid, $(CH_3)_2C{=}CHCO_2H$. Irradiation of the protons in one of the methyl groups leads to a substantial enhancement in the signal of the olefinic ($=CH$) proton, whereas irradiation of the other methyl group produced, if anything, a slight diminution in the olefinic proton signal. It was concluded that the methyl group whose irradiation produced the enhancement was located cis to the olefinic hydrogen, its proton signal thus being assigned. In this case, there is only one species with two diastereotopic methyl groups, but determination of the NOE has also been used to assign configuration[87] to the two isomers of citral (E)- and (Z)-$(CH_3)_2C{=}CHCH_2CH_2C^\beta(CH_3){=}C^\alpha HCHO$. In one isomer there is no enhancement of the α-hydrogen when the β methyl group is irradiated, whereas in the other case there is an 18% enhancement. The isomer showing this enhancement must have $CH_3(\beta)$ cis to $H(\alpha)$ and is therefore the Z isomer (chain cis to CHO), whereas the isomer not displaying the NOE is E. It is important to recognize that in conformationally mobile systems, such as N,N-dimethylformamide (DMF), $HCON(CH_3)_2$, the NOE (unlike the chemical shift or coupling constant) is *not* averaged and the equalization of NOEs may occur at lower temperatures than coalescence of chemical shifts.

The value of the NOE technique has been enhanced by the possibility of carrying it out for several nuclei at a time in a two-dimensional nuclear Overhauser and exchange spectroscopy (2D NOESY) experiment.[88,89]

9-3. INTERCONVERSION OF CIS–TRANS ISOMERS: POSITION OF EQUILIBRIUM AND METHODS OF ISOMERIZATION

a. Position of cis–trans Equilibria

When such equilibria (Fig. 9.28) can be established experimentally (see below), the equilibrium constant K (usually obtained chromatographically or spectroscopically from the composition of the equilibrated products) leads directly to the free energy difference between isomers, $\Delta G° = -RT \ln K$. If the measurements are carried out at

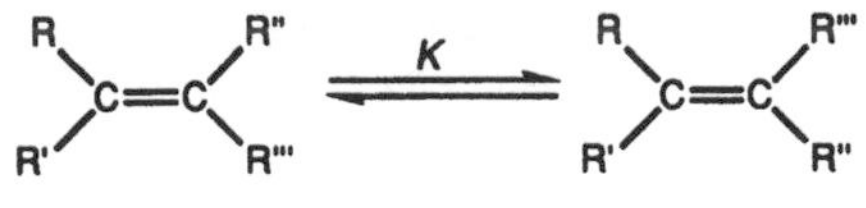

Figure 9.28. cis–trans Equilibrium.

more than one temperature, $\Delta H°$ can also be obtained as the slope of the plot of ln K versus $1/T$. Some of the energy values mentioned in this section have been obtained in this fashion, notably those for the 1,2-dihaloethylenes, CHX=CHY. Unfortunately, the method has several drawbacks. One is that cis–trans isomerization is often not clean but is accompanied by double-bond migration. As long as the latter process is not excessive at equilibrium, it is still possible to obtain the equilibrium ratio of the desired cis–trans isomers. Clearly, however, the formation of a number of positional isomers would be a serious problem in equilibration of higher alkenes. With acid catalysts there is also the danger of skeletal rearrangements. In addition, if the equilibrium is quite one-sided, it becomes difficult to measure. An example is the equilibration of the stilbenes (Fig. 9.28, R, R″ = C_6H_5, R′, R‴ = H) by means of iodine activated by visible light[23,24] (cf. the following section); the equilibrium mixture contains only 0.09–0.21% of cis-stilbene at room temperature, corresponding to $\Delta G°_{300}$ = 3.7–4.2 kcal mol^{-1} (15.5–17.6 kJ mol^{-1}).

Another method of determining $\Delta H°$ for the equilibrium shown in Figure 9.28 is to determine the heats of combustion (or of formation) of the two isomers and take their difference. The general principle of this method is shown in Figure 9.29. The cis–trans isomers are converted to a common product and the enthalpy change for the process involved is determined. In the case of heats of combustion, the common products are CO_2 and H_2O (formed in the same molar amount for the two stereoisomers), whereas in the case of heats of formation, which must be determined indirectly, they would be hydrogen and carbon in their standard states. Since the products are identical for the two starting isomers, the enthalpy difference found ($\Delta H_{cis} - \Delta H_{trans}$) is equal to $\Delta H°$ between the isomers (cf. Fig. 9.29). Some pertinent data for alkenes are shown in Table 9.5. In disubstituted alkenes, the trans isomer is more stable than the cis. (In the trisubstituted alkenes the enthalpy differences are small or nil.) For unhindered cases the enthalpy difference is close to 1 kcal mol^{-1} (4.2 kJ mol^{-1}). Interestingly, $\Delta G°$ is smaller than $\Delta H°$ because the cis isomer has a slightly larger entropy (for an explanation, see ref. 43, p. 339). When the group at one end of the double bond is *tert*-butyl, it will interact sterically with a cis substituent at the other end (more so if

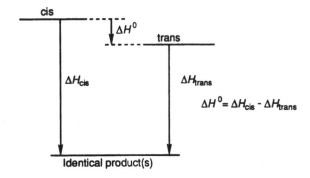

Figure 9.29. Energy picture for conversion of cis–trans isomers to a common product.

TABLE 9.5   Differences in Enthalpies of Combustion (or Formation) and in Free Energies for cis (Z) and trans (E) Isomers[a,b]

Compounds	$\Delta H°$		$\Delta G°$	
	(kcal mol^{-1})	(kJ mol^{-1})	(kcal mol^{-1})	(kJ mol^{-1})
CH$_3$CH=CHCH$_3$	0.86[c]	3.6[c]	0.5	2.12
C$_2$H$_5$C=CHC$_2$H$_5$	1.19	5.0	1.22	5.1
(CH$_3$)$_2$CHCH=CHCH$_3$	0.96	4.0	0.60	2.5
(CH$_3$)$_3$CCH=CHCH$_3$	3.85	16.1	e	
(CH$_3$)$_3$CCH=CHC$_2$H$_5$	4.42	18.5	e	
(CH$_3$)$_3$CCH=CHC(CH$_3$)$_3$[d]	10.5	44.0	e	
C$_6$H$_5$CH=CHCH$_3$	0.98	4.1	0.62	2.6
CH$_3$CH$_2$(CH$_3$)C=CHCH$_3$	0.12	0.5	0.07	0.3
CH$_3$CH$_2$(CH$_3$)C=CHC$_2$H$_5$	0.76	3.2	e	
C$_6$H$_5$CH=CHC$_6$H$_5$[f]	4.59	19.2	3.7-4.2	15.5-17.6

[a]At 25°C; gas phase.
[b]From *TRC Thermodynamic Tables, Hydrocarbons, Vol. VIII* (1983, 1988) unless otherwise indicated: pp. p-2600, p-2630, p-2650, p-2651, p-2672, p-4490, Thermodynamic Research Center, College Station, TX.
[c]Reference 90 reports $\Delta H° = 1.2$ kcal mol^{-1} (5.0 kJ mol^{-1}), $\Delta S° = 1.2$ cal mol^{-1} deg^{-1} (5.0 J mol^{-1} deg^{-1}) at 400°C by direct Equilibration. The $\Delta G°$ values in the tables are computed as $\Delta G° = \Delta H° - T\Delta S°$, the entropy difference, $\Delta S°$, being obtained experimentally.
[d]Liquid phase.[91]
[e]Not available.
[f]See text.

that substituent is ethyl than when it is methyl) and the cis–trans energy difference increases. Further examples of this type will be discussed below.

It is important to indicate whether differences in heat of combustion refer to the gas or liquid phase; in the latter case they include the negative heats of vaporization of the two isomers, which may not be the same.

Yet another means of determining heats of isomerization involves calorimetric measurement of heats of hydrogenation. Since the hydrogenation product of a pair of cis–trans alkenes is one and the same alkane, the energy diagram in Figure 9.29 applies here as well, except that the vertical arrows now represent heats of hydrogenation and $\Delta H°$ is computed as the difference in heats of hydrogenation of the cis and trans isomers. The advantage of this method over that of determining heats of combustion is that heats of hydrogenation are much smaller, of the order of 30 kcal mol^{-1} (126 kJ mol^{-1}), and can generally be measured to ±0.1 kcal mol^{-1} (0.4 kJ mol^{-1}) without much difficulty; the minimum precision in $\Delta H°$, which may amount to no more than 1 kcal mol^{-1} (4.2 kJ mol^{-1}) (cf. Table 9.5) will then be of the same magnitude, (i.e., ca. ±10%). [In contrast, heats of combustion tend to amount to well over 1000 kcal mol^{-1} (4200 kJ mol^{-1}) and it is difficult to measure them accurately enough to get meaningful differences of the order of a kcal mol^{-1} (4.2 kJ mol^{-1}) or so.]

Heats of hydrogenation and their differences are shown in Table 9.6.[92] Here, again, phase must be considered: Data measured in acetic acid include differences in heat of

TABLE 9.6. Heats of Hydrogenation for cis–trans Isomers and Their Differences[a]

Entry	Compounds	ΔH_{hyd} (kcal mol^{-1})			ΔH_{hyd} (kJ mol^{-1})		
		cis or Z	trans or E	$\Delta\Delta H°$	cis or Z	trans or E	$\Delta\Delta H°$
1	$CH_3CH{=}CHCH_3$	28.6[b]	27.6[b]	1.0	120[b]	116[b]	4.2
2	$(CH_3)_2CHCH{=}CHCH_3$	27.3	26.4	0.9	114	110	3.8
3	$(CH_3)_3CCH{=}CHCH_3$	30.8	26.5	4.3	129	111	18
4	$(CH_3)_2CHCH{=}CHCH(CH_3)_2$	28.7	26.8	1.9	120	112	7.9
5	$(CH_3)_3CCH{=}CHC(CH_3)_3$	36.2	26.9	9.3	151	112	39
6	$(CH_3)_3CCH_2CH{=}CHCH_2C(CH_3)_3$	26.9	26.0	0.9	113	109	3.8
7	$H_5C_2O_2CCH{=}CHCO_2C_2H_5$	33.2	29.0	4.2	139	121	18
8	$C_6H_5CH{=}CHCO_2CH_3$	27.8	19.5	8.3	116	81	35
9	$C_6H_5CH{=}CHC_6H_5$	25.8	20.1	5.7	108	84	24

[a]Reference 92.
[b]These are gas-phase values.[93] Most of the other values were obtained in acetic acid solvent by the group of Turner.[94]

solution. Considering this complication, the data agree quite well with those in Table 9.5 where comparison is possible (entries 1–3, 5, 9) except in the case of the stilbenes (entry 9) where the heat of hydrogenation difference seems to be too large. In unconjugated alkenes large differences in heats of hydrogenation (and therefore stability) are seen if the group at one end of the double bond or both is *tert*-butyl (entries 3 and 5); the case with isopropyl at both ends (entry 4) shows only a marginally enhanced difference and that with two neopentyl groups (entry 6) shows none. Conjugated systems (entries 7–9) show larger differences than unconjugated ones, presumably because of steric inhibition of resonance stabilization in the cis isomer as discussed earlier (p. 356).

An exception to the rule that trans isomers are more stable than cis isomers occurs with the 1,2-dihaloethylenes. This apparent anomaly was originally found in 1,2-dichloroethylene[95] and was ascribed to the halogens being both donors and acceptors of electronic charge (Fig. 9.30). In the meantime it has been found that the cis isomer is more stable than the trans in most 1,2-dihaloethylenes[96a] the only exceptions being the bromo-iodo and diodo compounds,[96b] where the steric interference of the halogens in the cis isomer is apparently large enough to make it less stable than the trans. Pertinent data, mostly obtained by thermal equilibration of the cis–trans isomers, are shown in Table 9.7. In the case of the dichloro compound, heat

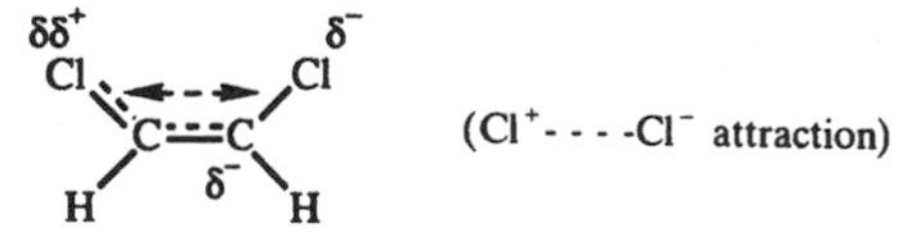

(Cl$^+$- - - -Cl$^-$ attraction)

Figure 9.30. Assumed reason for the high stability of *cis*- 1,2-dichloroethylene.

TABLE 9.7. Relative Stability of *cis*- and *trans*- 1,2-Dihaloethylenes[a]

Compound	$\Delta G°$ (trans $\rightleftharpoons$ cis)	
	(kcal mol^{-1})	(kJ mol^{-1})
FCH=CHF	−0.5; −0.89[d]	−2.1; −3.72[d]
FCH=CHCl	−0.8; −0.72[e]	−3.3; −3.01[e]
FCH=CHBr	−0.8	−3.3
FCH=CHI	−0.77	−3.21
ClCH=CHCl[b]	−0.5; −0.56[e]	−2.1; −2.34[e]
BrCH=CHBr	−0.1; −0.15[e]	−0.4; −0.63[e]
BrCH=CHI	+0.2	+0.8
ICH=CHI	+0.7; +0.32[e]	+2.9; +1.34[e]
BrCH=CH−CH$_3$[c]	−0.85	−3.55

[a]From ref. 96.
[b]Reference 95.
[c]Gas phase.[98] For the liquid phase, see ref. 99.
[d]At 25°C.[100a]
[e]At 25°C.[100b]

of combustion data are also available;[97] they point to the cis isomer being lower by 0.25 kcal mol^{-1} (1.05 kJ mol^{-1}) in enthalpy.

b. Methods of Equilibration

There are two conceptually distinct ways of interconverting cis and trans isomers. One is to use thermal, catalytic, or photochemical means. These will lead to an equilibrium mixture or a photostationary state, respectively. Unless one isomer predominates greatly over the other at equilibrium (e.g., in the case of the stilbenes, Table 9.5, where equilibration produces virtually pure trans isomer), such procedures give rise to mixtures and, while of mechanistic interest (see above), are thus of limited synthetic usefulness. A different approach is to convert an *E* isomer into the *Z* or vice versa in directed fashion. Such interconversion is valuable synthetically in cases where the most readily accessible isomer is not the one wanted as a synthetic target. Both approaches have been reviewed.[101] We shall first deal with equilibration processes.

Thermal Equilibration. Barriers to rotation have already been discussed in Section 9-1. In simple alkenes, such as 2-butene, the barrier (62 kcal mol^{-1}, 259 kJ mol^{-1}) is too high to make thermal isomerization feasible in the absence of a catalyst.[90] Even a somewhat lower barrier of 40 kcal mol^{-1} (167 kJ mol^{-1}) implies a half-life ($t_{1/2}$) of over 58 h at 200°C and though such compounds as fumaric acid have been isomerized "thermally" as mentioned earlier, these must, in fact, be acid-autocatalyzed reactions. In push–pull ethylenes (Fig. 9.7) thermal isomerization may, of course, be facile or may even occur spontaneously. An example of thermal isomerization is the conversion of 15,15′-*cis*-β-carotene to the all-trans isomer[102] by heating at 80°C (Fig. 9.31).

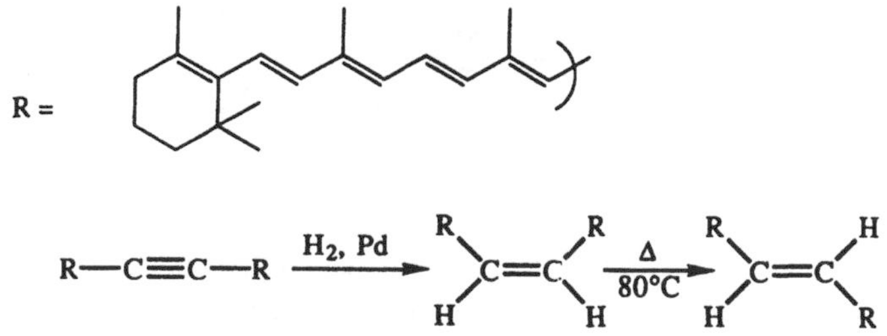

Figure 9.31. Carotene isomerization.

Catalyzed Equilibration.   A wide variety of catalysts has been employed to bring about cis–trans interconversion,[43,101,103] among them free radicals and free radical generators, such as oxides of nitrogen, halogens in the presence of light, iodine, thioglycolic acid, and diphenyl disulfide in the presence of light.[104] Also employed have been acids, such as halogen acids, sulfuric acid, and boron trifluoride; alkali metals, such as sodium; and hydrogenation–dehydrogenation catalysts and reagents, such as selenium and platinum. Radicals $\dot{X}$ probably add reversibly to the alkene RR′C=CR″R‴ to give RR′CX–ĊR″R‴ radicals in which the central bond is now single and rotation around it therefore facile. Subsequent departure of $\dot{X}$ may then give either the original alkene or its stereoisomer. Light is often required to produce the radicals ($\dot{I}$ from I_2, Br from Br_2, and $C_6H_5\dot{S}$ from C_6H_5—SC_6H_5); however, a photochemical isomerization process (cf. the following section) is not involved, inasmuch as the light is not absorbed by the double bond and there is no triplet species formed to produce photosensitization (alkene triplet formation; see below). Moreover the product composition in such cases is different from that produced by triplet excitation.[101]

Acid catalysis may similarly involve addition of the Lewis acid or proton to the double bond generating a cation that may then rotate, similarly as the radical, and subsequently lose the Lewis acid or proton to return the alkene—either the original one or the isomerized one. The α,β-unsaturated carbonyl compounds are particularly easily isomerized by acid, presumably through involvement of an enolic resonance hybrid: H–$\overset{+}{O}$=C–C=C ↔ H–O–C=C–$\overset{+}{C}$ in which the original C=C double bond is weakened and rotation around it thus facilitated. However, in some instances, although acid catalysis appears to be operating, this is actually not the case. Thus the isomerization of (Z)-stilbene to E by HBr requires oxygen and is stopped by a radical inhibitor, such as hydroquinone.[105] Moreover, HCl does not bring about isomerization. Thus it appears that the catalyst is a bromine atom formed from the HBr, not the acid itself.

Nucleophilic catalysis of cis–trans isomerization has been observed in the case of diethyl maleate and fumarate, (E)- and (Z)-$EtO_2CCH=CHCO_2Et$, which are interconverted by secondary amines.[106] This process presumably involves a reversible 1,4 addition of the amine to the unsaturated ester.

The general mechanism of catalyzed isomerization is shown in Figure 9.32, where the asterisk denotes a positive or negative charge or unshared electron. For

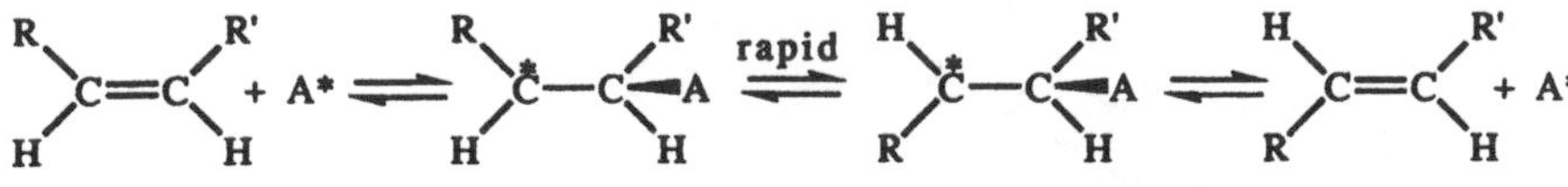

$$R \bigg\rangle C = C \bigg\langle {R' \atop H} + A^* \rightleftharpoons R \bigg\rangle \dot{C} - C \bigg\langle {R' \atop A} \xrightarrow{\text{rapid}} {H \atop R} \bigg\rangle \dot{C} - C \bigg\langle {R' \atop A} \rightleftharpoons {H \atop R} \bigg\rangle C = C \bigg\langle {R' \atop H} + A^*$$

Figure 9.32. cis–trans Isomerization. A superscript * represents a cation, radical, or anion.

isomerization to take place, it is essential that A^* add to the alkene (e.g., nucleophiles do not catalyze isomerization of alkenes unless the latter bear electron-withdrawing groups) but that the addition be readily reversible. Species that add exoergically to alkenes (i.e., for which $\Delta G^\circ_{\text{addition}} \ll 0$) will generally add irreversibly rather than induce isomerization, thus I_2 is a much better catalyst for alkene isomerization than Cl_2 which will add to the alkene instead.

Photochemical Isomerization. The cis–trans isomerization of alkenes occurs under irradiation with light that they absorb.[25,107] Absorption of light leads to a singlet excited state whose stable geometry is one in which the planes defined by the three atoms attached to each olefinic carbon are at right angles to each other (Fig. 9.33; for an introduction to photochemistry, see. ref. 108). With this geometry the mutual repulsion of the unpaired electrons is minimized and their interaction with the σ-electrons of the bonds to the adjacent carbon (hyperconjugation) is maximized. The initial vertically excited singlet thus relaxes to the twisted minimum and from there to the twisted ground state, which subsequently partitions itself between cis and trans products (Fig. 9.34). The process is difficult to study with simple alkenes because of the very short wavelength of absorption and the intervention of side reactions, such as $2\pi_s + 2\pi_s$ cyclodimerization;[109] however, it has been accomplished with (*E*)-1,2-di-*tert*-butylethylene.[110] The ratio of trans/cis isomer in various solvents ranged from 5 to 8; however, there were other products and it is not certain that a photostationary state was reached. Though an *E/Z* ratio of 5–8 is much less than the equilibrium ratio ($\Delta\Delta H^\circ = 9.3$ kcal mol^{-1}, 39 kJ mol^{-1}, cf. Table 9.6; $\Delta\Delta G^\circ$ must be similar), it does show that the twisted state relaxes preferentially to the trans isomer, presumably for steric reasons.

c. Directed cis–trans Interconversion

Unless equilibria between cis–trans isomers are very one-sided, equilibration methods are of limited synthetic usefulness, since they ordinarily require subsequent diastereomer separation. It should be noted, however, that trans alkenes are often

Figure 9.33. Excited state of alkenes abC=Ccd.

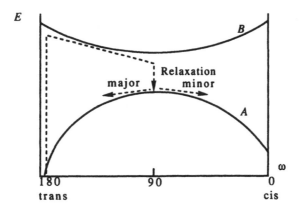

Figure 9.34. Energy profile of ground state and singlet excited states of alkenes. E is the potential energy. A is the ground-state energy as a function of torsion angle, B is the excited-state energy as a function of torsion angle, and the dotted line represents the reaction path for trans excitation.

substantially more stable than their cis isomers (Table 9.5), so that quite trans-rich alkene mixtures can frequently be generated by chemical equilibration or by thermodynamically controlled syntheses. An example is the already mentioned synthesis of the all-trans carotene by heating of the 15,15′-cis isomer (Fig. 9.31). In contrast, photostationary states (see above) are frequently rich in cis isomers that can thus be produced from their more stable trans diastereomers either by irradiation at a wavelength where the trans isomer absorbs much more strongly than the cis or by photosensitization with a sensitizer whose triplet energy is below that of the cis but above that of the trans triplet. An example of the latter type is the photochemical conversions of *trans*- to *cis*-β-ionol (Fig. 9.35), which is nearly quantitative with photosensitizers of triplet energies below 65 kcal mol^{-1} (272 kJ mol^{-1}), such as 2-acetonaphthone, E_t, 59 kcal mol^{-1} (247 kJ mol^{-1}), whereas sensitizers of much higher energy [e.g., acetone, E_t, 78 kcal mol^{-1} (326 kJ mol^{-1})] produce a cis–trans mixture containing only about 65% of the cis because of indiscriminate pumping of both stereoisomers to the triplet excited state.[111]

A *directed* interconversion of a trans alkene to its cis isomer can be effected in principle by an anti addition followed by syn elimination, by a syn addition followed by an anti elimination, by a syn elimination (to an alkyne) followed by an anti addition, or by an anti elimination followed by a syn addition. Anti addition–anti elimination (or syn addition–syn elimination) or their corresponding

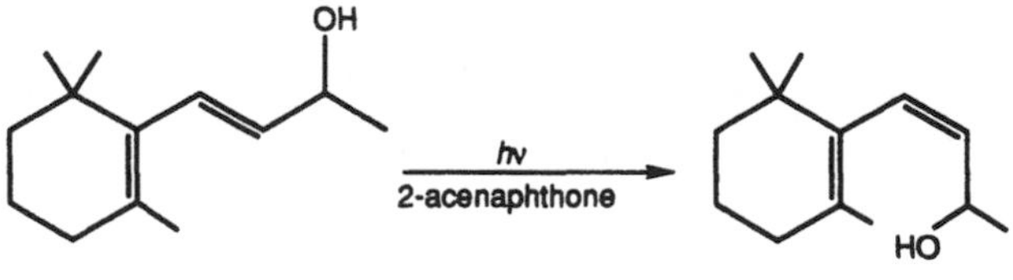

Figure 9.35. Photochemical conversion of *trans*- to *cis*-β-ionol.

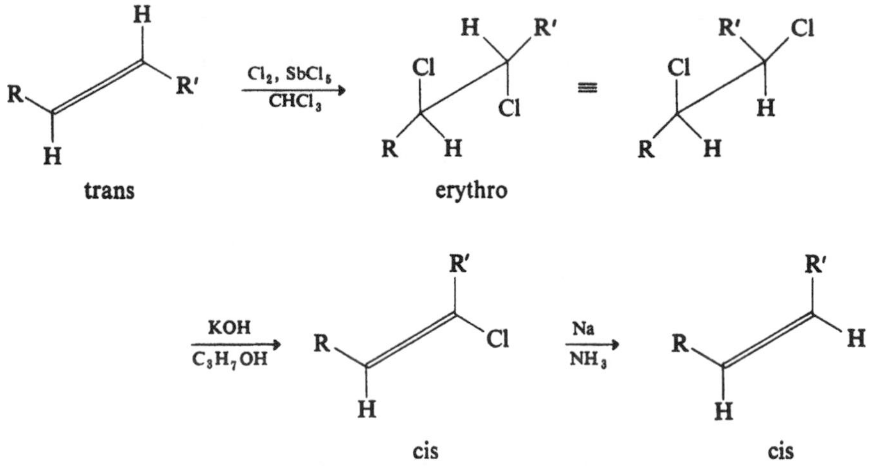

Figure 9.36. Conversion of trans to cis alkenes.

elimination–addition sequences, in contrast, normally return alkene of unchanged configuration. However, when the elements added are different from those eliminated, such may not be the case and the sequence may serve for ultimate trans–cis interconversion. An example[112] is shown in Figure 9.36 (see also Fig. 9.16). Here the addition is of Cl_2 (anti) but the elimination (also anti) is of HCl; the original constitution is regenerated by reducing the chloroalkene to an alkene with retention of configuration; the overall result is trans–cis interconversion.

A sequence of anti addition followed by apparent syn elimination[113] is shown in Figure 9.37 [R' = $(CH_3)_2CH(CH_2)_4$—; R = —$(CH_2)_9CH_3$]. Evidently the direct anti elimination of BrCl by iodide is so slow that S_N2 displacement of Br by I (with inversion of configuration) occurs instead;[114] subsequent anti elimination of ICl leads to the cis alkene in high purity, the sequence of S_N2 inversion plus anti elimination being formally equivalent to syn elimination. Similar apparent syn elimination had

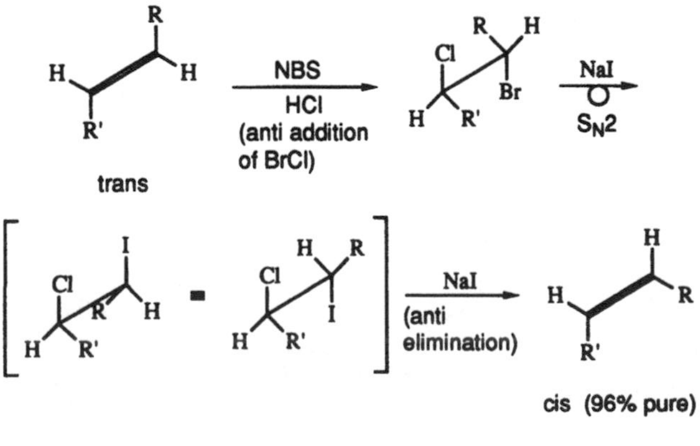

Figure 9.37. A trans–cis interconversion. NBS = N-bromosuccinemide.

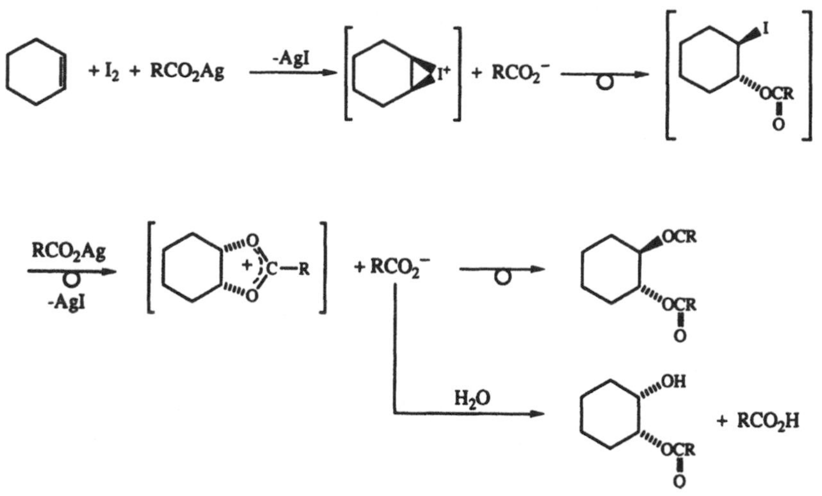

Figure 9.38. Dry and wet Prévost reactions.

been observed earlier in the iodide induced conversion of *meso*-CHDBrCHDBr to *cis*–CHD=CHD.[115] These results are in contrast to the normal anti elimination of nonterminal dibromides with iodide.[43]

The Prévost (iodine–silver acetate or benzoate) reaction shown in Fig. 9.38 involves syn addition of iodine to form an iodonium ion, followed by inversion to form an iodoacetate or iodobenzoate, followed by a second inversion with neighboring group participation[116] to give a five-membered cyclic intermediate that is then opened by a third inversion to give a diester.[117a] Since the syn addition is followed by an odd number of inversions, the overall steric course of addition of the two alkanoate moieties is anti (Fig. 9.38, the example is that of a cyclic alkene). However, when water is present, the cyclic cationic intermediate is hydrolyzed to a cis monoester and the overall addition is syn.[117b]

REFERENCES

1. Rabinovitch, B. S. and Michel, K.-W. *J. Am. Chem. Soc.* **1959**, *81*, 5065; Jeffers, P. M. and Shaub, W. *J. Am. Chem. Soc.* **1969**, *91*, 7706; Jeffers, P. M. *J. Phys. Chem.* **1974**, *78*, 1469.

2. Douglas, J. E., Rabinovitch, B. S., and Looney, F. S. *J. Chem. Phys.* **1955**, *23*, 315.

3. Luef, W. and Keese, R. *Top. Stereochem.* **1991**, *20*, 231.

4. Kuchitsu, K. *J. Chem. Phys.* **1966**, *44*, 906.

5. Duncan, J. L., Wright, I. J., and Van Lerberghe, D. *J. Mol. Spectrosc.* **1972**, *42*, 463.

6. Lide, D. R. and Christensen, D. *J. Chem. Phys.* **1961**, *35*, 1374.

7. Tokue, I., Fukuyama, T., and and Kuchitsu, K. (a) *J. Mol. Struct.* **1973**, *17*, 207; (b) *ibid.* **1974**, *23*, 33.

8. Kondo, S., Sakurai, Y., Hirota, E., and Morino, Y. *J. Mol. Spectrosc.* **1970**, *34*, 231.

9. Scharpen, L. H. and Laurie, V. W. *J. Chem. Phys.* **1963**, *39*, 1732.

10. Blackwood, J. E., Gladys, C. L., Loening, K. L., Petrarca, A. E., and Rush, J. E. *J. Am. Chem. Soc.* **1968**, *90*, 509.

11. (a) Cross, L. C. and Klyne, W. *Pure Appl. Chem.* **1976**, *45*, 11; (b) *J. Org. Chem.* **1970**, *38*, 2849. (c) Moss, G. P. *Pure Appl. Chem.* **1996**, *68*, 2193.

12. van't Hoff, J. H. *Die Lagerung der Atome in Raume*, Viehweg & Sohn, Braunschweig, Germany, 1877, p. 14.

13. Kuhn, R. and Blum, D. *Chem. Ber.* **1959**, *92*, 1483.

14. Roth, W. R. and Exner, H.-D. *Chem. Ber.* **1976**, *109*, 1158.

15. Bertsch, K. and Jochims, J. C. *Tetrahedron Lett.* **1977**, 4379.

16. Kalinowski, H.-O. and Kessler, H. *Top. Stereochem.* **1973**, *7*, 295.

17. Sandström, J. *Top. Stereochem.* **1983**, *14*, 83.

18. Shvo, Y. *Tetrahedron Lett.* **1968**, 5923.

19. Downing, A. P., Ollis, W. D., and Sutherland, I. O. *J. Chem. Soc. B* **1969**, 111.

20. Viehe, H. G., Janousek, Z., Merenyi, R., and Stella, L. *Acc. Chem. Res.* **1985**, *18*, 148.

21. Kistiakowsky, G. B. and Smith, W. R. *J. Am. Chem. Soc.* **1934**, *56*, 638.

22. Schmiegel, W. W., Litt, F. A., and Cowan, D. O. *J. Org. Chem.* **1968**, *33*, 3344.

23. Fischer, G., Muszkat, K. A., and Fischer, E. *J. Chem. Soc. B* **1968**, 1156.

24. Saltiel, J., Ganapathy, S., and Werking, C. *J. Phys. Chem.* **1987**, *91*, 2755.

25. Saltiel, J. and Charlton, J. L. "cis–trans Isomerization of Olefins," in de Mayo, P., ed., *Rearrangement in Ground and Excited States*, Vol. 3, Academic Press, New York, 1980.

26. Rieker, A. and Kessler, H. *Chem. Ber.* **1969**, *102*, 2147.

27. Müller, E. and Neuhoff, H. *Ber. Dtsch. Chem. Ges.* **1939**, *72*, 2063.

28. Müller, E. and Tietz, E. *Ber. Dtsch. Chem. Ges.* **1941**, *74B*, 807.

29. Agranat, I., Suissa, M. R., Cohen, S., Isaksson, R., Sandström, J., Dale, J., and Grace, D. *J. Chem. Soc. Chem. Commun.* **1987**, 381.

30. Sandström, J., Stenvall, K., Sen, N., and Venkatesan, K. *J. Chem. Soc. Perkin 2* **1985**, 1939.

31. Khan, A. Z.-Q., Isaksson, R., and Sandström, J. *J. Chem. Soc. Perkin 2* **1987**, 491.

32. Favini, G., Simonetta, M., and Todeschini, R. *J. Comp. Chem.* **1981**, *2*, 149.

33. Dannheim, J., Grahn, W., Hopf, H., and Parrodi, C. *Chem. Ber.* **1987**, *120*, 871.

34. (a) Krebs, A., Rüger, W., Ziegenhagen, B., Hebold, M., Hardtke, I., Müller, R., Schütz, M., Wietzke, M., and Wilke, M. *Chem. Ber.* **1984**, *117*, 277. (b) Krebs, A., Rüger, W., Nickel, W.-U., Wilke, M., and Burkert, U. *Chem. Ber.* **1984**, *117*, 310. (c) Krebs, A., Kaletta, B., Nickel, W.-U., Rüger, W., and Tikwe, L. *Tetrahedron* **1986**, *42*, 1693.

35. Deuter, J., Rodewald, H., Irmgartinger, H., Loerzer, T., and Lüttke, W. *Tetrahedron Lett.* **1985**, *26*, 1031.

36. Jackman, L. M. "Rotation About Partial Double Bonds in Organic Molecules," in Jackman, L. M. and Cotton, F. A., eds., *Dynamic Nuclear Magnetic Resonance Spectroscopy*, Academic Press, New York, 1075, p. 203.

37. Anderson, D. G. and Wettermark, G. *J. Am. Chem. Soc.* **1965**, *87*, 1433.

38. Meisenheimer, J. and Theilacker W., in Freudenberg, K., ed., *Stereochemie*, Franz Deuticke, Leipzig, Germany, 1933.

39. Hall, G. E., Middleton, W, J., and Roberts, J. D. *J. Am. Chem. Soc.* **1971**, *93*, 4778.

40. Feigl, M. *J. Phys. Chem.* **1983**, *87*, 3054.

41. Melander, W. R., Jacobson, J., and Horvath, C. *J. Chromatogr.* **1982**, *234*, 269.

42. Brewster, J. H. "Assignment of Stereochemical Configurations by Chemical Methods," in Bentley, K. W. and Kirby, G. W., eds., *Techniques of Chemistry*, Vol. IV, Part 3, 2nd ed., Interscience-Wiley, New York, 1972.

43. Eliel, E. L. *Stereochemistry of Carbon Compounds*, McGraw-Hill, New York, 1962, Chap. 12.

44. van't Hoff, J. H. Letter to Buys Ballot, 1875, cited in ref. 42.

45. von Auwers, K. and Wissebach, H. *Ber. Dtsch. Chem. Ges.* **1923**, *56*, 715.

46. Zimmerman, H. E., Singer, L., and Thygarajan, B. S. *J. Am. Chem. Soc.* **1959**, *81*, 108; see also ref. 43, p. 436 and ref. 11c.

47. Fahey, R. C. *Top. Stereochem.* **1968**, *3*, 237.

48. Winterfeldt, E. "Ionic Addition to Acetylenes," in Viehe, H. G., ed., *Chemistry of Acetylenes*, Marcel Dekker, New York, 1969, Chap. 4.

49. Miller, S. I. *J. Am. Chem. Soc.* **1956**, *78*, 6091.

50. Saunders, W. H. and Cockerill, A. F. *Mechanism of Elimination Reactions*, Wiley, New York, 1973.

51. Pfeiffer, P. *Z. Phys. Chem.* **1904**, *48*, 40.

52. Weissberger, A. *J. Am. Chem. Soc.* **1945**, *67*, 778.

53. Schlosser, M. *Top. Stereochem.* **1970**, *5*, 1.

54. Vedejs, E. and Peterson, M. J. *Top. Stereochem.* **1994**, *21*, 1.

55. Vedejs, E. and Marth, C. F. *Tetrahedron Lett.* **1987**, *30*, 3445.

56. Schaub, B., Jeganathan, S., and Schlosser, M. *Chimia* **1986**, *40*, 246.

57. Curtin, D. Y. *Rec. Chem. Prog.* **1954**, *15*, 11.

58. Lucas, H. J., Simpson, T. P., and Carter, J. M. *J. Am. Chem. Soc.* **1925**, *47*, 1462.

59. Donaruma, I. G. and Heldt, W. *Z. Org. React.* **1960**, *11*, 1.

60. McCarty, C. G. "syn–anti Isomerizations and Rearrangements," in Patai, S., ed., *The Chemistry of the Carbon–Nitrogen Double Bond*, Wiley-Interscience, New York, 1970, Chap. 9.

61. Landolt-Börnstein, *Structure Data of Free Polyatomic Molecules*, Vol. 7, Hellwege, K.-H. and Hellwege, A. M., eds., Springer Verlag, New York, 1976.

62. Minkin, V. I. "Dipole Moments and Stereochemistry of Organic Compounds. Selected Applications," in Kagan, H. B., ed. *Stereochemistry, Fundamentals and Methods*, Vol. 2, Thieme, Stuttgart, Germany, 1977.

63. McClellan, A. L. *Tables of Experimental Dipole Moments*, Vol. 1, Freeman, San Francisco, 1963; Vol. 2, 1974, Vol. 3, 1989, Rahara Enterprises, El Cerrito, CA.

64. Errera, J. *Phys. Z.* **1928**, *29*, 689.

65. Henderson, G. and Gajjar, A. *J. Org. Chem.* **1971**, *36*, 3834.

66. Kondo, S., Sakurai, Y., Hirota, R. E., and Morino, Y. *J. Mol. Spectrosc.* **1970**, *34*, 231.

67. van Arkel, A. E. *Recl. Trav. Chim. Pays-Bas* **1932**, *51*, 1081; *ibid.* **1933**, *52*, 1013.

68. Bjerrum, N. *Z. Phys. Chem.* **1923**, *106*, 219.

69. Serjeant, E. P. and Dempsey, B. *Ionization Constants of Organic Acids in Aqueous Solution*, Pergamon Press, New York, 1979.

70. Kortüm, G., Vogel, W., and Andrussow, K. *Pure Appl. Chem.* **1960**, *1*, 187.

71. Lowe, B. M. and Smith, D. G. *J. Chem. Soc. Faraday Trans. 1* **1974**, 362.

72. Bellamy, L. J. *The Infrared Spectra of Complex Molecules*, 3rd. ed., Vol. 1, Chapman and Hall, London, 1975.

73. Golfier, M. "Determination of Configuration by Infrared Spectroscopy," in Kagan, H. B., ed., *Stereochemistry, Fundamentals and Methods*, Vol. 1, Thieme, Stuttgart, Germany, 1977, Chap. 1.

74. Bernstein, H. J. and Powling, J. *J. Am. Chem. Soc.* **1951**, *73*, 1843.

75. Allan J. L. H., Meakins, G. D., and Whiting, M. C. *J. Chem. Soc.* **1955**, 1874.

76. Watts, V. S. and Goldstein, J. H. "Nuclear Magnetic Resonance Spectra of Alkenes," in Zabicky, J., ed., *The Chemistry of Alkenes*, Vol. 2, Wiley-Interscience, New York, 1970, p. 1.

77. Gaudemer, A. "Determination of Configurations by NMR Spectroscopy," in Kagan, H. B., ed., *Stereochemistry, Fundamentals and Methods*, Vol. 1, Thieme, Stuttgart, Germany, 1977, Chap. 2.

78. Pascual, C., Meier, J., and Simon, W. *Helv. Chim. Acta* **1966**, *49*, 164.

79. Matter, U. E., Pascual, C., Pretsch, E., Pross, A., Simon, W., and Sternhell, S. *Tetrahedron* **1969**, *25*, 691.

80. Phillips, L. "Application of ^{19}F Nuclear Magnetic Resonance," in Bentley, K. W. and Kirby, G. W., eds., *Elucidation of Organic Structures by Physical and Chemical Methods*, Vol. IV, Part. 1 of *Techniques of Chemistry*, 2nd ed., Wiley-Interscience, New York, 1972, p. 323.

81. Emsley, J. W., Phillips, L., and Wray, V. *Prog. Nucl. Magn. Reson. Spectrosc.* **1976**, *10*, 83.

82. Breitmaier, E., Hass, G., and Voelter, W. *Atlas of Carbon-13 NMR Data*, IFI/Plenum Data Co., New York, 1979.

83. Marshall, J. L. *Carbon–Carbon and Carbon–Proton NMR Couplings*, Verlag Chemie, Deerfield Park, FL, 1983.

84. Noggle, J. H. and Schirmer, R. E. *The Nuclear Overhauser Effect, Chemical Applications*, Academic Press, New York, 1971.

85. Neuhaus, D. *The Nuclear Overhauser Effect in Structural Conformational Analysis*, VCH Publishers, New York, 1989.

86. Anet, F. A. L. and Bourn, A. J. R. *J. Am. Chem. Soc.* **1965**, *87*, 5250.

87. Ohtsuru, M., Teraoka, M., Tori, K., and Takeda, K. *J. Chem. Soc. B* **1967**, 1033.

88. Sanders, J. K. M. and Hunter, B. K. *Modern NMR Spectroscopy, A Guide for Chemists*, 2nd ed., Oxford University Press, New York, 1993.

89. Wüthrich, K. *NMR of Proteins and Nucleic Acids*, Wiley, New York, 1986.

90. Golden, D. M., Egger, K. W., and Benson, S. W. *J. Am. Chem. Soc.* **1964**, *86*, 5416.

91. Rockenfeller, J. D. and Rossini, F. D. *J. Phys. Chem.* **1961**, *65*, 267.

92. Jensen, J. L. *Prog. Phys. Org. Chem.* **1976**, *12*, 189.

93. Kistiakowsky, G. B., Ruhoff, J. R., Smith, H. A., and Vaughan, W. E. *J. Am. Chem. Soc.* **1935**, *57*, 876.

94. Turner, R. B., Nettleton, D. E., and Perelman, N. *J. Am. Chem. Soc.* **1958**, *80*, 1430.

95. Pitzer, K. S. and Hollenberg, J. L. *J. Am. Chem. Soc.* **1954**, *76*, 1493.

96. (a) Viehe, H. G. *Chem. Ber.* **1960**, *93*, 1697; Viehe, H. G., Dale, J., and Franchimont, E. *Chem. Ber.* **1964**, *97*, 244. (b) Viehe, H. G. and Franchimont, E. *Chem. Ber.* **1963**, *96*, 3153.

97. Smith, L., Bjellerup, L., Krook, S., and Westermark, H. *Acta Chem. Scand.* **1953**, *7*, 65.

98. Skell, P. S. and Allen, R. G. *J. Am. Chem. Soc.* **1958**, *80*, 5997.

99. Harwell, K. E. and Hatch, L. F. *J. Am. Chem. Soc.* **1955**, *77*, 1682.

100. (a) Craig, N. C. and Entemann, E. A. *J. Am. Chem. Soc.* **1961**, *83*, 3047. (b) Craig, N. C., Piper, L. G., and Wheeler, V. L. *J. Phys. Chem.* **1971**, *75*, 1453.

101. Sonnet, P. E. *Tetrahedron* **1980**, *36*, 557.

102. Isler, O., Lindlar, H., Montavon, M., Rüegg, R., and Zeller, P. *Helv. Chim. Acta* **1956**, *39*, 249.

103. Crombie, L. *Q. Rev. Chem. Soc.* **1952**, *6*, 101.

104. Rokach, J., Young, R. N., and Kakushima, M. *Tetrahedron Lett.* **1981**, *22*, 979.

105. Kharasch, M. S., Mansfield, J. V., and Mayo, F. R. *J. Am. Chem. Soc.* **1937**, *59*, 1155.

106. Clemo, G. R. and Graham, S. B. *J. Chem. Soc.* **1930**, 213.

107. Cowan, D. O. and Drisko, R. L. *Elements of Organic Photochemistry*, Plenum Press, New York, 1976, Chap. 9.

108. Turro, N. J. *Modern Molecular Photochemistry*, Benjamin/Cummings, Menlo Park, CA, 1978.

109. Yamazaki, H., Cvetanović, R. J., and Irwin, R. S. *J. Am. Chem. Soc.* **1976**, *98*, 2198.

110. Kropp, P. J. and Tise, F. P. *J. Am. Chem. Soc.* **1981**, *103*, 7293.

111. Ramamurthy, V. and Liu, R. S. H. *J. Am. Chem. Soc.* **1976**, *98*, 2935.

112. Hoff, M. C., Greenle, K. W., and Boord, C. E. *J. Am. Chem. Soc.* **1951**, *73*, 3329.

113. Sonnet, P. E. and Oliver, J. E. *J. Org. Chem.* **1976**, *41*, 3284.

114. Hine, J. and Brader, W. H. *J. Am. Chem. Soc.* **1955**, *77*, 361.

115. Schubert, W. M., Steadly, H., and Rabinovitch, B. S. *J. Am. Chem. Soc.* **1955**, *77*, 5755.

116. Capon, B. and McManus, S. P. *Neighboring Group Participation*, Plenum Press, New York, 1976.

117. Winstein, S. and Buckles, R. E. *J. Am. Chem. Soc.* **1942**, *64*, a) 2780; b) 2787.

10

CONFORMATION OF ACYCLIC MOLECULES

10-1. CONFORMATION OF ETHANE, BUTANE, AND OTHER SIMPLE SATURATED ACYCLIC MOLECULES

a. Alkanes

As indicated in Section 2-4, the term "conformation" relates to the different spatial arrangements of the atoms in a molecule that arise through rotation about the bonds linking such atoms. Different conformations thus differ in torsion angles about one or more bonds. Since it requires four atoms to define a torsion angle, a molecule must have at least four atoms to display conformational variability. Thus a tetraatomic molecule in which the valency angles differ from 180° and in which the atoms are linked in a row (as in H–O–O–H) will display an infinite number of conformations (Fig. 10.1), since the torsion angle ω around the central O–O bond can vary continuously between 0° and + and −180° (cf. Section 2-4 and Table 2.2).

Most of these conformations are unstable (like the extreme or intermediate positions in the swing of a pendulum); stable conformations that are located at energy minima are called "conformational isomers" or "conformers." Thus, while ethane ($H_3C–CH_3$) has an infinite number of conformations by virtue of rotation about the C–C bond, there are only three minima (as shown in Fig. 10.2); that is, ethane has only three conformers. Since they are indistinguishable, they are "degenerate."

Conformational isomers, even if nondegenerate as in the case of the butanes (see below), are generally not isolable because of the small energy barriers that separate them (cf. Section 2-1). It was therefore assumed, in the early days of stereochemistry, that rotation about single bonds was "free."

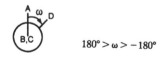

$$180° > \omega > -180°$$

Figure 10.1. Conformations of A–B–C–D.

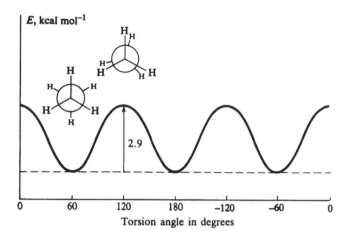

Figure 10.2. Potential energy of ethane as a function of the torsion angle.

The first surmise that this might not be true appears in work by Bischoff[1] concerned with the reactivity of substituted succinic acids; in the 1891 paper the staggered and eclipsed conformations of ethane are shown with the surmise that the staggered form is the more stable. (Bischoff's formulas are exemplified in Fig. 5.21.) Little attention seems to have been paid to Bischoff's work or to conformational analysis in acyclic systems in general until the 1920s when Hermans[2] explained the difference in ease of acetal formation of racemic and *meso*-hydrobenzoin ($C_6H_5CHOHCHOHC_6H_5$) in conformational terms. (Hermans' formulas are exemplified in Fig. 5.21 also.) The field became more active in the 1930s[3-6] and the idea that rotation around the C–C bond in ethane and related compounds was not free had hardened by the middle of the decade (for an historical perspective, see ref. 7). In 1936, Kemp and Pitzer[8] not only asserted, on the basis of lack of agreement of the calculated (spectroscopic) and experimental (calorimetric) entropy of ethane, that rotation was not free but calculated the barrier to be 3.15 kcal mol^{-1} (13.2 kJ mol^{-1}), later revised[9] to 2.88 kcal mol^{-1} (12.05 kJ mol^{-1}), which is practically identical with the most recent experimental values of 2.89–2.93 kcal mol^{-1} (12.09–12.26 kJ mol^{-1}).[10,11] The potential energy curve of ethane is shown in Figure 10.2.

The torsional potential (sometimes called *Pitzer potential*) is approximated by the equation $E = \frac{1}{2}E_0(1 + \cos 3\omega)$,[8] where ω is the torsion angle (taken as zero for the eclipsed conformation) and E_0 is the energy barrier. Three indistinguishable minima, corresponding to staggered conformations, appear at 60°, 180°, and –60° (see Section 2-4 for the sign convention of torsion angles) with maxima of $E_0 =$ 2.9 kcal mol^{-1} (12.1 kJ mol^{-1}) at 0°, 120°, and –120°. Because of the occurrence of three maxima and three minima in the potential function (Fig. 10.2) this is sometimes called a "V_3 potential" (see also Section 2-6). (The symbols V and E for energy are used interchangeably.)

The torsional barrier in ethane[12] is *not* mainly due to steric causes, since the hydrogen atoms of the methyl groups are barely within van der Waals distance (cf. Section 2-6). Thus steric (van der Waals) repulsion accounts for less than 10% of the experimental barrier; electrostatic interactions of the weakly polarized C–H bonds are not of importance either. The major contribution to the barrier has been ascribed to the unfavorable overlap interaction between the bond orbitals in the eclipsed conformation (Pauli's exclusion principle[13]); but other calculations suggest it results from a favorable interaction (bonding–antibonding orbitals) in the staggered conformation.[14] Recent calculated values of the barrier[15,16] range from 2.7 to 3.07 kcal mol^{-1} (11.3 to 12.8 kJ mol^{-1}) and thus bracket the experimental value.

The conformational situation in propane is very similar to that of ethane: There are three indistinguishable energy minima (staggered conformations) and three indistinguishable barriers (eclipsed conformations) in the potential energy curve. However, since the barrier now involves the eclipsing of a methyl group (rather than a hydrogen atom) with a hydrogen atom, it is somewhat higher[17,18] than in ethane, about 3.4 kcal mol^{-1} (14.2 kJ mol^{-1}). Presumably the additional 0.5 kcal mol^{-1} (2.1 kJ mol^{-1}) is due to CH_3/H van der Waals repulsive interaction.

A different conformational situation is encountered in molecules of the type $XCH_2–CH_2Y$, such as butane (X = Y = CH_3). The potential curve for butane is shown in Figure 10.3. There are now three different energy minima, two corresponding to the enantiomeric (and therefore equienergetic) gauche forms of C_2 symmetry and one corresponding to the achiral anti form of C_{2v} symmetry, which is of lower energy because it is devoid of the repulsive CH_3/CH_3 van der Waals interactions of the gauche forms. There are also three barriers between the anti and gauche conformers, two low ones involving CH_3/H eclipsing only (these two barriers are enantiomeric, and therefore equienergetic) and one high one involving CH_3/CH_3 eclipsing.

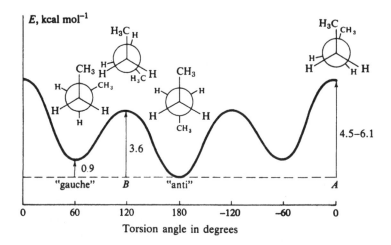

Figure 10.3. Potential energy of butane as a function of torsion angle.

The energy difference between the gauche and anti conformers of butane has been determined numerous times (for summaries see refs. 17 and 19) and it has become clear that the resulting value is quite phase dependent, being of the order of 0.89–0.97 kcal mol^{-1} (3.7–4.1 kJ mol^{-1}) in the gas phase and 0.54–0.57 kcal mol^{-1} (2.3–2.4 kJ mol^{-1}) in the liquid. This phase dependence may be taken as a manifestation of the von Auwers–Skita or conformational rule,[6,20] which states that the isomer of higher enthalpy has the lower molecular volume, and therefore the higher density, refractive index, boiling point, and heat of vaporization. In the case of the butane conformers, the gauche isomer is more compact; therefore it has the lower molecular volume and greater *inter*molecular van der Waals interactions (as a result of a high surface/volume ratio). Since the latter are predominantly attractive rather than repulsive, the heat of vaporization is larger for the gauche conformer with the resulting situation shown in Figure 10.4. Clearly, the gauche–anti enthalpy difference is greater in the gas than in the liquid (or solution).

If one wants to estimate the population of gauche and anti conformers in butane, one has to keep in mind that the equilibrium constant $K = \%\text{anti}/\%\text{gauche}$ depends on $\Delta G°$ ($K = e^{-\Delta G°/RT}$), rather than on $\Delta H°$, so that the entropy difference between the two conformers also comes into play ($\Delta G° = \Delta H° - T\Delta S°$). Although, the rotational and vibrational entropies of the two conformers are not the same (the two structures differ in moments of inertia and in normal vibrations), the difference from this source is apt to be small (but see ref. 21); moreover, there is no difference in translational entropy. However, there are two additional sources of entropy differences, entropy of symmetry and entropy of mixing, which often differ substantially between conformers. The entropy of symmetry is given as $S_{\text{sym}} = -R \ln \sigma$, where σ is the symmetry number characteristic of the symmetry point group of the conformer in question (cf. Section 4-5.c). The entropy of mixing of i components is $S_{\text{mix}} = -R \Sigma_i n_i \ln n_i$. Thus, for two components $S_{\text{mix}} = -R(n_1 \ln n_1 + n_2 \ln n_2)$. In the case of butane, both the anti and gauche conformers (C_{2v} and C_2, respectively) have $\sigma = 2$; hence $S_{\text{sym}} = -R \ln 2$ will cancel out in the difference between the two. On the other hand, there is no entropy of mixing for the achiral anti isomer, but the chiral gauche isomer has an entropy of mixing of $-R(0.5 \ln 0.5 + 0.5 \ln 0.5) = -R \ln 0.5 = R \ln 2$. This is the

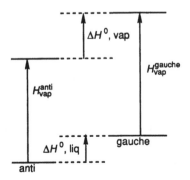

Figure 10.4. Enthalpy difference of butane conformers in the liquid and in the vapor state.

entropy of mixing of any pair of enantiomers. Thus the entropy of the gauche isomers exceeds that of the anti by $R \ln 2$ or 1.38 cal mol^{-1} K^{-1} (5.76 J mol^{-1} K^{-1}). Hence

$$\Delta G° = \Delta H° - 1.38T \text{ (cal mol}^{-1}) \text{ (anti } \rightleftharpoons \text{ gauche)}$$

or

$$\Delta G° = \Delta H° - 5.76T \text{ (J mol}^{-1})$$

For gaseous butane, taking $\Delta H° = 900$ cal mol^{-1} (3800 J mol^{-1}), $\Delta G°$ at room temperature (298.16 K) is $900 - 410 = 490$ cal mol^{-1} (2050 J mol^{-1}); hence $K = 2.3$, corresponding to 30% gauche (g) and 70% anti (a) isomer. The corresponding percentage in the liquid phase [$\Delta H° = 550$ cal mol^{-1} (2300 J mol^{-1})] is 44% gauche and 56% anti ($n_g = 0.44$, $n_a = 0.56$).

Whereas the torsion angle in the anti form of butane is exactly 180°, that in the gauche form is larger than the 60° expected for a perfectly staggered conformation.[22,23] This result obtains because the van der Waals repulsion of the methyl groups in the gauche form makes these groups rotate away from each other, even at the expense of increasing the torsional energy. Energy minimization (cf. Section 2-6) occurs near a 65° torsion angle.

Figure 10.3 displays two barriers, one due to the CH$_3$/CH$_3$ eclipsing at $\omega = 0°$ and one due to twofold CH$_3$/H eclipsing at $\omega = \pm 120°$. As expected, the value of the latter[17,23,24] of about 3.6 ± 0.2 kcal mol^{-1} (15.1 ± 0.8 kJ mol^{-1}) is higher than that in propane (because of the double eclipsing). It must be noted that this is the energy difference between the lower lying anti conformer and the top of the barrier; the difference between the higher energy gauche isomers and the top is about 0.9 kcal mol^{-1} (3.8 kJ mol^{-1}) less (in the gas phase), this being the enthalpy difference between the two conformers. [The height of the barrier always depends on whether it is measured from the side of the more stable or that of the less stable conformer. The barrier height above the less stable conformer is always less, namely, by the ground-state energy difference of the conformers (cf. Fig. 10.3).]

The higher (CH$_3$/CH$_3$) barrier, derived from spectroscopic experiments,[24] amounts to 4.5–4.9 kcal mol^{-1} (18.8–20.5 kJ mol^{-1}); recent values calculated[25] by either molecular mechanics or *ab initio* methods appear to agree quite well with the experimental ones.

The gauche–anti enthalpy difference and the rotational barrier in *n*-pentane seem to be similar to corresponding values in butane.[17,19,26] Nonetheless, the conformational situation in *n*-pentane is more complex than that in butane, since rotation around *two* CH$_2$–CH$_2$ bonds is now possible, giving rise to nine staggered conformations that are depicted in Figure 10.5. Conformations about the C(2, 3) and C(3, 4) bonds are denoted as anti (a) or gauche (g); for the gauche conformations it is also specified whether the torsion angle is near +60° (g$^+$) or near –60° (g$^-$). Conformations above the horizontal line are indistinguishable from (degenerate with) those below, and conformers to the right of the vertical line are mirror images of those on the left. Thus ag$^+$ and g$^+$a are interconvertible by a rotation of the molecule as a

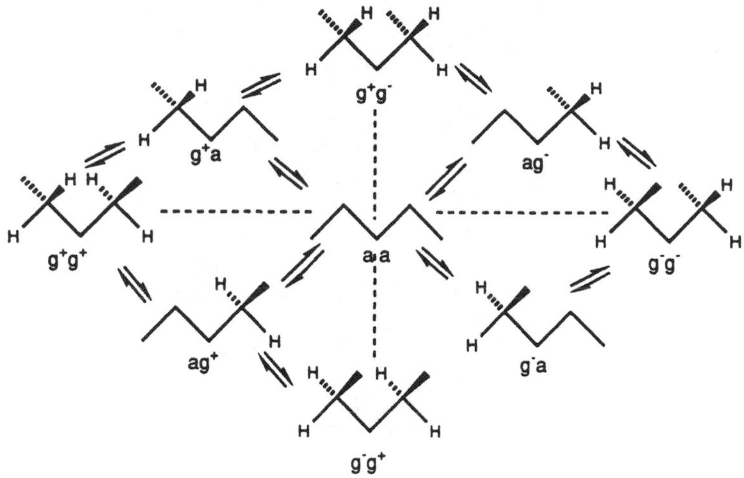

Figure 10.5. Pentane conformations.

whole (C_2) and the same is true of ag⁻ and g⁻a and of g⁺g⁻ and g⁻g⁺; there are only six distinct conformers. Moreover, g⁺a (or ag⁺) is the mirror image of ag⁻ (or g⁻a) and these conformers are thus equal in energy. The same is true of g⁺g⁺ and g⁻g⁻. This leaves four conformations of different enthalpy, the low-lying aa, the intermediate ag, the high-lying (gg)$^{\pm}$ and the extremely unstable g⁺g⁻. (Regarding the exclusion of the very crowded g⁺g⁻ conformation, see refs. 27 and 28). For an assumed gauche–anti enthalpy difference in the liquid phase of 0.55 kcal mol⁻¹ (2.3 kJ mol⁻¹) and taking into account that the ga conformers have a statistical weight of 4 and the gg$^{\pm}$ of 2, the calculated conformer distribution at 298.16 K in liquid pentane is aa, 34.6%; ag, 54.6%; gg, 10.8%; g⁺g⁻, less than 0.4% (see below for its energy level).

Situations such as that in pentane, where two independent variables [here the C(2, 3) and C(3, 4) torsion angles (ω)] affect the enthalpy of the system, are quite common, for example, in polypeptides. It is convenient to depict the situation in the form of a contour diagram (Fig. 10.6) in which the abscissa and ordinate represent $\omega_{2,3}$ and $\omega_{3,4}$, respectively, and the contours represent energy (the dots at the centers of the contours are energy minima). The deepest minimum occurs exactly at $\omega_{2,3} = \omega_{3,4} = 180°$ (aa conformer). Traversal of barriers (high-energy contours) leads to conformers of higher energy, where $\omega_{2,3} = 65°$, $\omega_{3,4} = 180°$, or vice versa; these are the ag conformers. From these conformers, by further rotation over a barrier, one goes to the still higher lying g⁺g⁺ and g⁻g⁻ conformers at torsion angles of +65°, +65° or −65°, −65°. Direct transition to these minima from the aa conformer is not favored in that it leads over very high ground (shaded). Minima are not found at +60°, −60° (g⁺g⁻) or −60°, +60° (g⁻g⁺) but there are some high shallow minima on both sides of the g⁺g⁻ conformation representing somewhat deformed g⁺g⁻ conformers between which the g⁺g⁻ conformation constitutes a saddle point. The energy of this conformation has been calculated[19] to lie over 3.3 kcal mol⁻¹ (13.8 kJ mol⁻¹) above that of the aa conformer.

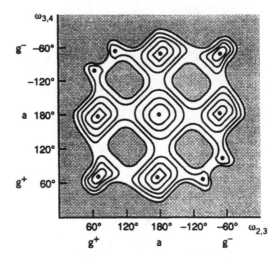

Figure 10.6. Energy contour diagram of *n*-pentane. [Reprinted with permission from Dale, J. *Stereochemistry and Conformational Analysis*. Copyright © 1978 Universitetsforlaget, Oslo, Norway, p. 98.]

Of the branched alkanes, we mention here only 2,3-dimethylbutane (Fig. 10.7) because of the rather unusual finding[29–31] that the anti and gauche conformations are nearly equal in enthalpy (and that the gauche conformation therefore predominates by nearly a factor of 2 at equilibrium, since it is a pair of enantiomers and has an entropy advantage of $R \ln 2$, or, statistically, since there are two gauche conformations but only one anti).

The explanation for the perhaps unexpectedly high stability of the gauche form relative to the anti in 2,3-dimethylbutane and related hydrocarbons of the RR′CH–CHRR′ type[32–34] (see also ref. 35) seems to be as follows: In butane, the CH_3–C–H bond angle is close to tetrahedral. However, in 2,3-dimethylbutane, the CH_3–C–CH_3 angle opens up to near 114°.[30,34] As a result, the ordinary Newman projections (Fig. 10.7, **A** and **G**) no longer apply; rather the anti conformer in 2,3-dimethylbutane tends toward a CH_3/CH_3 eclipsed conformation (Fig. 10.7, **A′**) and the gauche toward a CH_3/CH_3 perpendicular one (Fig. 10.7, **G′**). This deformation will enhance van der Waals repulsion in the anti conformer (though the enlargement

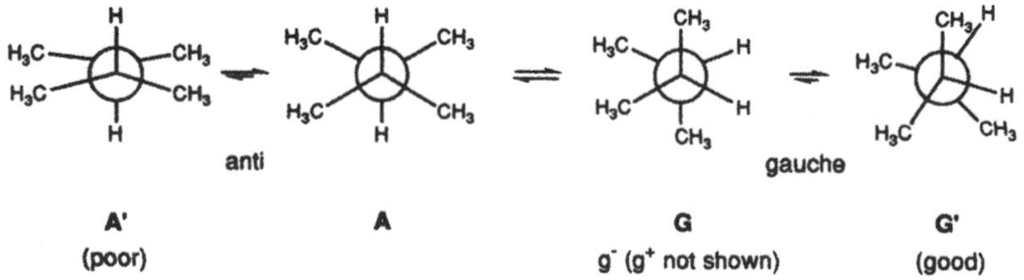

Figure 10.7. Conformational isomers of 2.3-dimethylbutane.

of the bond angle in that conformer may actually be resisted) but will diminish it in the gauche isomer, which is thereby stabilized. The experimental barrier[31] to rotation in 2,3-dimethylbutane, 4.3 kcal mol^{-1} (18.0 kJ mol^{-1}) is remarkably small if considered to represent simultaneous eclipsing of CH_3/CH_3 plus twice CH_3/H; presumably the angle deformations indicated in Figure 10.7 imply[33] that the eclipsing of the various ligands that have to pass each other is *not* simultaneous, the barrier thus being lowered. Similar considerations apply to other molecules of the RR′CHCHRR′ type[36] (see also Section 13-5.d). An extensive review of the conformations of larger branched hydrocarbons has been published by Anderson[37]; see also ref. 38.

b. Saturated Acyclic Molecules with Polar Substituents or Chains and the Anomeric Effect

The molecules discussed so far are hydrocarbons, devoid of polar groups; therefore electrostatic interaction between groups (V_E, Eq. 2.1, Chapter 2, p. 22) is nonexistent and solvation energy, V_S (except for phase changes on dissolution, as in the case of butane discussed earlier) is negligible. The situation changes in molecules with polar groups. Such molecules possess substantial dipoles (both local and overall). Since dipole interactions, which are responsible for the electrostatic term V_E in Eq. 2.1, vary from one conformer to another, they affect the conformational energy differences. The magnitude of these interactions may be solvent dependent. Moreover, differences in overall dipole moment between conformers may lead to differences in solvation energy (V_s in Eq. 2.1) between them.

Before entering upon these matters, we shall discuss briefly the history of the conformational analysis of the 1,2-dihaloethanes, XCH_2CH_2X (X = Cl or Br), since these molecules played an important part in the development of the subject[5] (for a historical retrospective, see ref. 39). It was early recognized, on the basis of dipole and Raman spectral measurements,[40] that 1,2-dihaloethanes (cf. Fig. 10.8) could not exist as pure anti isomers: Their dipole moments differ from zero and their Raman spectra, which show a centrosymmetric (anti) structure in the solid, display a number of additional lines in the liquid state. Although it was recognized that both observations could be explained by librational motion (torsional vibrations) in the fluid state, it was considered more likely that a second conformation with a finite dipole moment and a distinct Raman spectrum contributed in the liquid state. Perhaps the best evidence for the latter hypothesis is that molecules of the type CCl_3CHCl_2

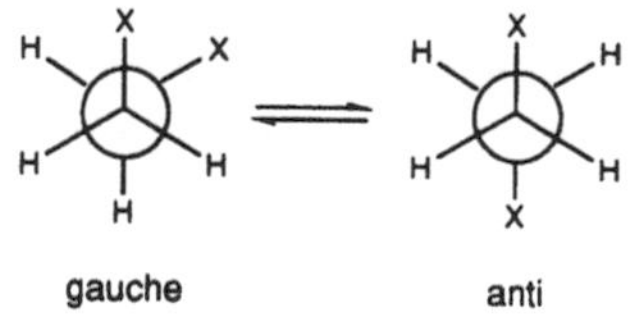

gauche   anti

Figure 10.8. Conformers of 1,2-dihaloethanes.

or CCl_3CH_2Cl, which can have only one kind of staggered conformation, do not show additional Raman lines on melting.

That the second conformer corresponds to the gauche and not the syn conformation was shown elegantly by Neu and Gwinn[41] (see also ref. 42) and is discussed in detail elsewhere.[43]

The anti–gauche enthalpy difference in gaseous 1,2-dichloroethane (Fig. 10.8, X = Cl) is in the 0.9–1.3 kcal mol^{-1} (3.8–5.4 kJ mol^{-1}) range[5,17,44] and determinations for 1,2-dibromoethane (Fig. 10.8, X = Br) range from 1.4 to 1.8 kcal mol^{-1} (5.9 to 7.5 kJ mol^{-1}).[5,17,45–47] Both differences are appreciably larger than those for butane (see above), not for steric reasons but because of the strong dipole–dipole repulsion of the C–X dipoles in the gauche conformation (cf. Fig. 10.8). This dipole effect, which can be simulated by molecular mechanics calculations,[44,48] is less important in polar solvents with a consequent increase in the population of the gauche conformation (cf. Table 10.1). This results in part from a diminution of the coulombic interaction of the dipoles in the more polar solvent; a more important factor is that the conformer of higher dipole moment gains more energy by solvation in polar solvents.

The situation in 1,2-difluoroethane (Fig. 10.8, X = F), surprisingly, is quite different from that in the dichloro and dibromo analogues in that the gauche form is preferred even in the gas phase (not just in terms of free energy, where it is favored by the statistical factor of 2, but in enthalpy). The $\Delta H°$ preference seems to amount to 0.6–0.9 kcal mol^{-1} (2.5–3.8 kJ mol^{-1}).[46,51–53] The reason for this preference for the gauche form, despite the repulsive dipole–dipole interaction (a so-called V_1 potential, since there is only one maximum and one minimum of energy in the course of a 360° rotation about the C–C bond), probably lies in a combination of a relatively small van der Waals repulsion (affecting the normal V_3 potential) thanks to the small size of fluorine, plus the intervention of a V_2 potential, which displays energy minima when the C–F bonds are at 90° (or –90°) to each other and maxima when they are at 0° or 180° torsion angles.[53–55] The V_2 (twofold) potential has been pictorially described as being due to a hyperconjugative interaction of the type $FCH_2–CH_2F \leftrightarrow F^- CH_2{=}CHF\ H^+ \leftrightarrow H^+CHF{=}CH_2F^-$, which is triggered by the high electron demand of the fluorine. In order to involve both fluorine atoms simultaneously in this interaction, the two C–F bonds must be orthogonal. The combination of this V_2 potential, which is optimal at 90°, and the normal V_3 potential, which is optimal at 60° or 180°, gives rise to the near-gauche energy minimum; the actual F–C–C–F torsion angle is 71°.[56] Smaller V_2 potentials may occur in other molecules, such as butane. An alternative explanation of the preferred gauche conformation of FCH_2CH_2F is in terms of the so-called "gauche effect,"[57] which implies that a chain segment A–B–C–D will prefer the gauche conformation when A and D either are very electronegative relative to B and C or are unshared electron pairs (see also below). The two explanations are related in their origin; both imply a stabilization of the gauche conformer. A different explanation, involving destabilization of the anti conformer, has been given by Wiberg et al.[58] These authors postulate that the bonds in FCH_2CH_2F are bent appreciably; this leads to diminution of the carbon–carbon (sigma) bond overlap,

TABLE 10.1. Solvent Dependence of Conformational (Enthalpy) Differences[a] **of 1,2-Dichloro- and 1,2-Dibromoethane**[b]

Solvent	ε^c	$ClCH_2CH_2Cl$		$BrCH_2CH_2Br$	
		Experimental[d]	Calculated[e]	Experimental[f]	Calculated[f]
Vapor	1.5[d]	0.9–1.3	0.91[d]	1.4–1.8	NA
		(3.8–5.4)	(3.8)	(5.9–7.4)	
$CCl_2{=}CCl_2$	2.5	0.89	0.82, 0.92	1.24	1.38
		(3.7)	(3.4, 3.8)	(5.2)	(5.9)
Et_2O	4.3	0.69	0.69, 0.68		
		(2.9)	(2.9, 2.8)	NA	NA
EtOAc	6.0	0.42	0.61, 0.57		
		(1.8)	(2.6, 2.4)	NA	NA
C_6H_6 or C_6D_6	2.3	0.60	0.56[g], 0.92[h]	0.69	1.38[h]
		(2.5)	(2.3, 3.8)	(2.9)	(5.8)[h]
Liquid (neat)	i	0.31[j]	0.48, 0.43	0.86[j]	1.10
		(1.3)	(2.0, 1.8)	(3.6)[j]	(4.6)
Acetone	20.7	0.18	0.25, 0.26		
		(0.75)	(1.05, 1.1)	NA	NA
CH_3CN	36.2	0.15	0.04, 0.12	0.66	0.57
		(0.63)	(0.17, 0.50)	(2.8)	(2.4)

[a] $\Delta H°$ in kcal mol^{-1}, values in parentheses are in kJ mol^{-1}. Data not available are indicated as NA.
[b] See Figure 10.8.
[c] Dielectric constant, rounded off to one decimal, from ref. 49 unless otherwise noted.
[d] References 44 and 46.
[e] When there are two figures, the first refers to ref. 48 and the second to ref. 47.
[f] Reference 47.
[g] This value was computed assuming a dielectric constant of benzene of 7.5; see text.
[h] This value was computed using the real dielectric constant of benzene of 2.3.
[i] 10.1 for ClCH₂CH₂Cl, 4.8 for BrCH₂CH₂Br.[49]
[j] These values are corrected for the variation of $\Delta H°$ with temperature (resulting, in turn, from the variation of solvent dielectric ε with temperature) and therefore differ from the raw values of 0.0 kcal mol^{-1} (0.0 kJ mol^{-1}) for ClCH₂CH₂Cl and 0.74 kcal mol^{-1} (3.1 kJ mol^{-1}) for BrCH₂CH₂Br reported in ref. 17. The actual range of values for the dibromide is large, 0.65–1.3 kcal mol^{-1} (2.7–5.4 kJ mol^{-1}): See ref. 50.

and hence weakening of the C–C bond, more so in the anti than in the gauche conformer.

In a more polar medium, such as the pure liquid (dielectric constant $\varepsilon = 34.4$) the gauche conformer of FCH₂CH₂F is, as expected, even more strongly preferred, to the near exclusion of the anti; $\Delta G° = 2.0–2.6$ kcal mol^{-1} (8.4–10.9 kJ mol^{-1}).[51,59] The parameter $\Delta S°$ is 1.36 cal mol^{-1} K^{-1} (5.69 J mol^{-1} K^{-1}) favoring the gauche isomer,[60] very close to the calculated $R \ln 2$.

A gauche preference similar to that in 1,2-difluoroethane is found in liquid succinonitrile (NCCH₂–CH₂CN), the β-halopropionitriles (XCH₂–CH₂CN),[61] in 1,2-dimethoxyethane (glyme, CH₃O–CH₂–CH₂–OCH₃), and probably also in CH₃O–CH₂CH₂–X (X = halogen).[62] Polyoxyethylene (OCH₂CH₂)ₙ also prefers the gauche conformation about the C–C bond and therefore, in contrast to polyethylene, assumes a helical conformation overall[63] (see also ref. 64).

We pass, now, to 1-propanol and the 1-halopropanes, $CH_3CH_2CH_2X$ (X = OH, F, Cl, Br, or I). In all these molecules the gauche–anti energy difference is small and in the first four the gauche conformation is enthalphically slightly preferred over the anti in the vapor state.[17,65–67] Stabilization of the gauche conformation would seem to be due to an attractive electrostatic interaction between X (negative) and the opposite (positive) end of the carbon chain.[68] An extreme case of this type of attraction is found in acetylcholine, $XCH_2CH_2N(CH_3)_3$ Y^- (X = OAc), choline (X = OH),[69] and fluorocholine (X = F),[70] which exist almost exclusively in the gauche conformation; this is not true of chlorocholine (X = Cl)[69] nor of thio- and selenocholine (X = SH or SeH),[70] where steric factors in the gauche conformers are more important.

Yet another potential cause for predominance of the gauche conformation (in addition to the V_2 potential, see above) is found in ethylene glycol, in its monoether and in the haloethanols $X–CH_2CH_2OH$ (X = OH, OCH_3, F, Cl, or Br),[17,71,72] where hydrogen bonding contributes to stabilization of the gauche conformer.[73]

We end this rather brief discussion of the conformation of saturated acyclic molecules by considering the conformation of molecules of the type $CH_3O–CH_2X$ (X = halogen, OR, or SR). Such species are important because of their relation to sugars, glycosides, and sugar halides (cf. Chapter 11). In dimethoxymethane ($CH_3OCH_2OCH_3$) one might anticipate, on steric grounds, that gauche conformations would be rather unstable because of the short C–O bond distances (see above). In fact, however, the molecule exists largely in the gg conformation.[74] From polymer studies of polyoxymethylene, $(–CH_2O–)_n$, which has a helical conformation,[64,75] it had been estimated[63,76] that the gauche preference is 1.1 kcal mol^{-1} (4.6 kJ mol^{-1}), but calculations for $CH_3OCH_2OCH_3$ referring to the gas phase suggest a value of 2.2 kcal mol^{-1} (9.2 kJ mol^{-1})[77] or 2.5 kcal mol^{-1} (10.5 kJ mol^{-1}).[78] The discrepancy may well be due to differences in the medium, since calculation also indicates that solvent effects are very important in this type of equilibrium.[79] In dimethoxymethane, as in pentane (Fig. 10.5), there are, in all, four different conformations of minimum energy engendered by rotation about two C–O bonds: g^+g^+, ag, aa, and g^+g^-, with the lowest minimum at gg, higher minima at ag, and a still higher and shallow minimum at aa; the situation is very different from that in pentane (Fig. 10.6).

The difference between the enthalpy difference (gauche–anti) one would expect on "purely steric" grounds and that actually found is called the "anomeric effect"[79–82] or "generalized anomeric effect"[64c,83]; the origin of the term (which comes from sugar chemistry), its history, and its extensive application in cyclic systems will be discussed in Chapter 11.

Phenomenologically, the generalized anomeric effect implies that the preferred conformation of a $R–O–CH_2–X$ fragment (where X is an electronegative group, such as halogen, OH, OCOR', OR', SR', or NR'R'') is gauche or, in other words, that an unshared pair of electrons on oxygen will be antiperiplanar to X, as shown in Figure 10.9. A more general statement about the anomeric and related stereoelectronic effects[80,84] is: "There is a stereoelectric preference for conformations in which the best donor lone pair or bond is antiperiplanar to the best acceptor bond." Originally, the effect was ascribed[85] to dipole–dipole repulsion of the C–X dipole and the dipole of

gauche   anti

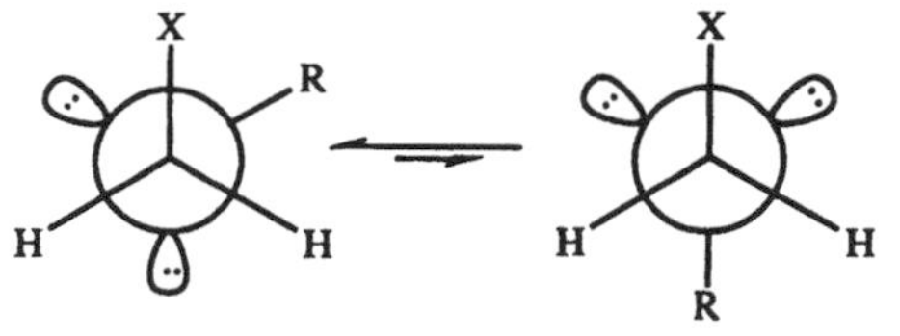

Figure 10.9. Anomeric preferences.

the C–O–C moiety (see Fig. 10.9; the C–O–C dipole is directed along the bisector of the lone pair axes; the lone pairs are considered to be sp^3 hybridized). The solvent dependence of the effect supports this idea; moreover, conformations in which the lone pairs on O and X (which may be considered the generators of the dipoles) are parallel are particularly unstable.[64a] Later, it was pointed out[86] that the calculated electrostatic repulsion is inadequate to account for the magnitude of the anomeric effect. At the same time, attention was drawn to certain anomalies in bond lengths in molecules displaying the anomeric effect. These anomalies are shown for CH_3–O–CH_2–X (X = F or Cl, ref. 87; X = OCH_3; ref. 74b) in Figure 10.10; the normal bond distances (cf. Chapter 2) are 137.9 pm (1.379 Å) for C–F, 176.7 pm (1.767 Å) for C–Cl, and 142.6 pm (1.426 Å) for C–O. The data in Figure 10.10 suggest that, whereas the CH_3–O bond length is normal, the O–CH_2 bond is considerably foreshortened; more so when X = halogen than when X = OCH_3. The C–X bond is markedly lengthened when X = Cl, less so when X = F; but it is shortened when X = OCH_3 (in which case, for reason of symmetry, the two CH_2–O bonds must be of equal length). The suggestion was made[86] that at least part of the anomeric effect is due to double-bond-no-bond resonance of the type R–O–CH_2–X $\leftrightarrow$ R–$\overset{+}{O}$=CH_2 $\overset{-}{X}$ or its molecular orbital equivalent, which implies overlap of the n orbital of oxygen with the σ^* orbital of the C–X bond (Fig. 10.11). For this overlap to be maximal, one of the lone pairs on oxygen must be antiperiplanar to the C–X bond, as shown in Figure 10.9. When X (Figs. 10.10 and 10.11) is halogen, the O–CH_2 bond order is increased and the bond therefore shortened, whereas the CH_2X bond order is decreased and the bond

	CH_3———O———CH_2———X		
X = F	142.4	136.2	138.5
	(1.424)	(1.362)	(1.385)
X = Cl	142.1	136.2	182.2
	(1.421)	(1.362)	(1.822)
X = OCH_3	143.2	138.2	138.2
	(1.432)	(1.382)	(1.382)

Figure 10.10. Bond lengths (in picometers; parenthesized values in angströms) in compounds displaying anomeric effect.

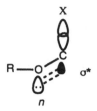

Figure 10.11. Anomeric effect: molecular orbital interpretation.

thereby lengthened as shown. But when X = OCH_3 the hyperconjugative resonance can work either way:

$$CH_3-\overset{+}{O}=CH_2-\overset{-}{O}-CH_3 \leftrightarrow CH_3-O-CH_2-O-CH_3 \leftrightarrow CH_3-\overset{-}{O}-CH_2=\overset{+}{O}-CH_3$$

and the effect on the CH_2–O bond length is equivocal; in actual fact, the bond shortening due to partial double-bond character is more important than the bond lengthening resulting from partial no-bond character.

10-2. CONFORMATION OF UNSATURATED ACYCLIC AND MISCELLANEOUS COMPOUNDS

a. Unsaturated Acyclic Compounds

Ethylene is planar and, because of the high barrier to rotation about the double bond (cf. Chapter 9), exists in a single conformation. Propylene, on the other hand, can undergo conformational changes by rotation about the $H_3C-CH=CH_2$ single bond. The two achiral conformations (in which the plane of the double bond is the symmetry plane) are shown in Figure 10.12. Interestingly, the most stable conformation is that in which the double bond is eclipsed with one of the C–H bonds of the methyl group (cis or eclipsed, Fig. 10.12).[88] The barrier to rotation in propylene is nearly 2.0 kcal

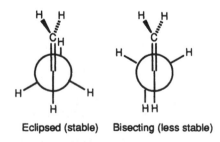

Eclipsed (stable)   Bisecting (less stable)

Figure 10.12. Conformations of propylene.

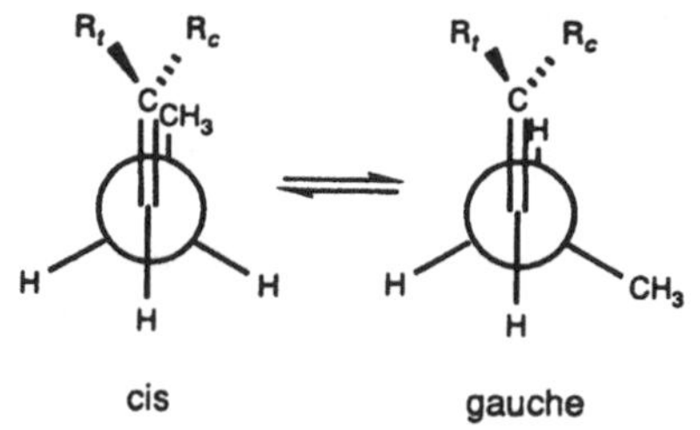

Figure 10.13. Stable conformers of l-butene, where $R_t = R_c = H$.

mol^{-1} (8.4 kJ mol^{-1})[17] and it would appear that the $H_2C=C/C-H$ staggered (synclinal or, *bisecting*) conformation corresponds to the energy maximum. Since the H/H eclipsing that occurs in this conformation only accounts for about 1 kcal mol^{-1} (4.2 kJ mol^{-1}), that is, one-half of the total barrier (cf. the earlier discussion on ethane, Fig. 10.2), there seems to be an additional destabilization, of about the same magnitude, due to unfavorable interaction of the other two C–H bonds with the $C=C$ π orbitals in this conformation.[89] An alternative would be a corresponding attractive interaction in the eclipsed conformation.

In l-butene (Fig. 10.13) $R_t = R_c = H$ there are two possible eclipsed conformations, one "cis," with the $C=CH_2$ group eclipsed with CH_3, and the other "gauche" (really anticlinal), with the $C=CH_2$ group eclipsed with H. Different investigators have found either the cis or the gauche conformer to be lower in enthalpy, but the difference is small, 0.53 kcal mol^{-1} (2.2 kJ mol^{-1}) or less.[90,91] In some substituted propylenes of the type $R-CH_2-CH=CH_2$, where R is a small electronegative substituent, such as F, CH_3O, or CN, the cis conformer clearly predominates.[92] The barrier to interconversion of the l-butene conformers is 1.74 kcal mol^{-1} (7.3 kJ mol^{-1}) above the more stable gauche form.

The situation in aldehydes is similar to that in alkenes. Thus acetaldehyde is more stable in the eclipsed conformation (actually the stable conformation has the methyl group rotated about 9° away from perfect eclipsing) than in the bisecting one by 1.17 kcal mol^{-1} (4.9 kJ mol^{-1}) (Fig. 10.14).[93] That the energy advantage of the cis over the bisecting form is smaller in acetaldehyde than in propylene may again strike one as unexpected, since O is smaller than CH_2 and, in addition, one might have expected an electrostatic attraction between O and eclipsed H due to hyperconjugation:

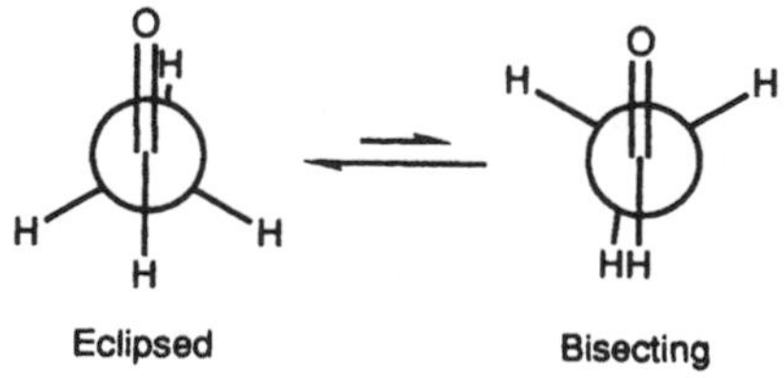

Eclipsed   Bisecting

Figure 10.14. Conformations of acetaldehyde.

$O=C-CH_2H \leftrightarrow {}^-O-C=CH_2H^+$. However, the dominating factor seems to be the repulsive $C-H/C=X(\pi)$ interaction in the bisecting conformation that is less when X $= O$ than when $X = CH_2$ because of the lesser π character of the (polar) $C=O \leftrightarrow C-O$ bond. In propionaldehyde (unlike in 1-butene) the conformation in which CH_3 is eclipsed with O is clearly preferred [by 0.7–1.2 kcal mol^{-1} (2.9–5.0 kJ mol^{-1})[94–96]; cf. Fig. 10.15, **A**]. That this energy difference represents a net stabilization of the cis conformer is confirmed by the value of the rotational barrier, 2.3 kcal mol^{-1} (9.6 kJ mol^{-1})[95] which is 1.1 kcal mol^{-1} (4.6 kJ mol^{-1}) above that in acetaldehyde, presumably because the cis form in propionaldehyde is stabilized by that amount of energy. That steric factors can vitiate this stabilization is demonstrated in that *tert*-butylacetaldehyde $(CH_3)_3C-CH_2-CHO$, (Fig. 10.15, **A**, *t*-Bu instead of CH_3) prefers the gauche conformation over the cis by 0.25 kcal mol^{-1} (1.05 kJ mol^{-1}).[92]

The preferred conformation of acetone corresponds to that of acetaldehyde in that one hydrogen in each methyl group is eclipsed with the carbonyl oxygen; similarly diethyl ketone resembles propionaldehyde in having as its preferred conformation the one with both terminal methyl groups similarly eclipsed.[92] In 2-butanone (Fig. 10.15, **B** and **C**) the most populated conformation (**B**, $\omega = 0°$) corresponds to that in propionaldehyde; however, an interesting new feature is a long, flat region in the potential energy curve (as calculated by molecular mechanics) near $\omega = 120°$ (**C**). It appears that in the $\omega = 120° - 90°$ region, relief of nonbonded interaction between the terminal methyl groups very nearly balances the energy increase incurred by the loss of the (optimal) eclipsing between one of the methylene hydrogen atoms and the carbonyl oxygen.[97]

We turn now to acid derivatives: acid halides, RCOX, esters, RCO_2R', and amides $RCONH_2$. The situation in acid halides is very similar to that in aldehydes in that the oxygen will eclipse the carbon chain (cis conformation Fig. 10.15, halogen instead of H at the lowermost position).[92,98] The situation with esters is more interesting, since in principle, these esters might exist in either cis or trans conformations (Fig. 10.16). It has been shown that even for methyl[99] and ethyl[100] formate (Fig. 10.16, R = H, R' $= CH_3$ or CH_3CH_2), where the competition is between CH_2/O and CH_2/H eclipsing, the conformer with CH_2 trans to R (Fig. 10.16, **Z**) wins out. The hydrogen atoms of the methyl group are staggered with the O–C bond, with a barrier to methyl rotation of 1.19 kcal mol^{-1} (5.0 kJ mol^{-1}) in methyl formate; the corresponding lowest barrier for the methylene group in ethyl formate is 1.10 kcal mol^{-1} (4.6 kJ mol^{-1}). Apart from

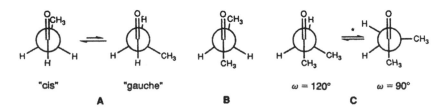

Figure 10.15. Eclipsed conformers of propionaldehyde **A** and 2-butanone **B**, **C**. The asterisk (*) designates nearly free rotation in this region of ω.

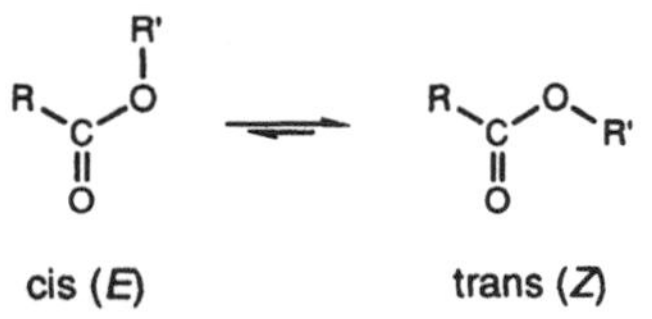

Figure 10.16. Ester conformations.

that, methyl formate is planar; but ethyl formate has two almost equally stable conformations, one planar zigzag and another, only 0.19 kcal mol^{-1} (0.79 kJ mol^{-1}) above the first, in which the terminal methyl group is nearly at right angles (95°) to the plane of the rest of the molecule.[100]

The $E–Z$ energy difference in methyl formate (Fig. 10.16, R = H, R′ CH$_3$) is so large as to be difficult to measure, there being only a negligible amount of the E isomer at room temperature. Nonetheless, an estimate of ΔH_{E-Z} has been obtained[101] by rapidly cooling hot beams ($T = 286, 662$, and 803 K) of the ester in an argon matrix so as to "freeze in" the high temperature equilibrium and then analyzing the stable mixture by IR spectroscopy at liquid helium temperatures (ca. 4 K). From the temperature dependence of the absorbance ratio (cf. Section 10-3.c), $\Delta H°$ was found to be 4.75 ± 0.19 kcal mol^{-1} (19.9 ± 0.8 kJ mol^{-1}) favoring the Z conformer.[101,102]

The geometry of methyl formate[99] is shown in Figure 10.17. The C–O–C angle (114.8°) is considerably larger than that in ethers (111.5°), whereas the H–C=O angle is remarkably small. The acyl oxygen single bond, 133.4 pm (1.334 Å) is considerably shorter than the alkyl oxygen bond, 143.7 pm (1.437 Å). This result suggests considerable overlap of one of the p orbitals on the alkyl oxygen with the C=O bond (Fig. 10.17b shows the corresponding resonance form). As a result, the H–C–O–C framework is planar and the barrier to rotation about the C–O "single" bond (see above) is high. Surprisingly, the C=O bond is not lengthened; its length of 120.0 pm (1.200 Å) is actually slightly shorter than the 122.0 pm (1.22 Å) usual in aldehydes and ketones. The reasons why the Z conformer of methyl formate is more stable than E are complex. The following factors may contribute: (a) H/CH$_3$ steric repulsion in the E conformer (Fig. 10.16). This factor is probably small in formates but important in methyl esters of higher acids where the repulsion is of the R/CH$_3$ type. (b) van der

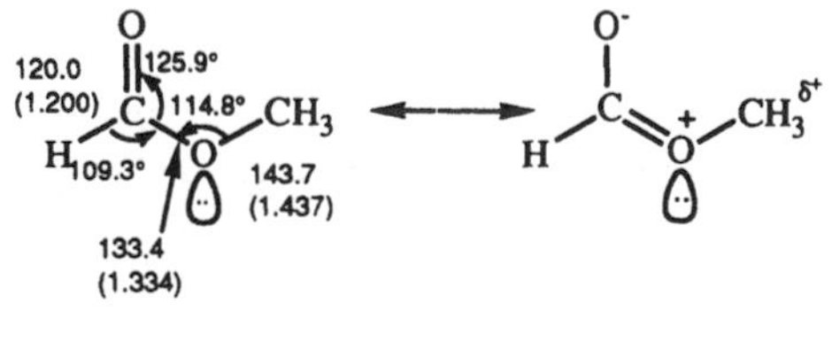

(a) (b)

Figure 10.17. Geometry of methyl formate (bond lengths in picometers; values in parentheses are in angströms).

Waals attraction between CH_3 and $=O$ in the Z conformer. (c) Electrostatic attraction between $O^{\delta-}$ and $H_3C^{\delta+}$ in the Z conformer. (This factor, equivalent to a dipole–dipole repulsion in the E isomer, is probably the most important one.[103]) (d) n–σ^* overlap in the Z conformer (Fig. 10.17).[80] One of the unshared pairs on the alkyl oxygen is in the R–C–O–R′ plane of the ester function and antiperiplanar to the $C=O$ bond. While this pair is not involved in the p–π overlap, which contributes to the planarity of the ester function, it is properly disposed for overlap with the σ^* orbital of the $C=O$ bond, thus stabilizing the Z conformation in a manner similar to that involved in the anomeric effect discussed earlier.

From the conformational aspect, amides are probably the most significant acyl derivatives, inasmuch as they serve as prototypes for the important class of peptides. Both formamide[104,105a] and acetamide[105b] are planar or nearly planar, as a result of extensive resonance of the ^-O–$CR=NH_2$ type (cf. Section 2-4), with barriers to rotation in the 18–22 kcal mol^{-1} (75–92 kJ mol^{-1}) range.[106] The structures of N-methylformamide[105c] and N-methyl- acetamide[105d] as determined by electron diffraction (gas phase) are shown in Figure 10.18. The extensive foreshortening of the OC–$NHCH_3$ bond suggests that this linkage has much double-bond character (the bond lengths in N-methylacetamide are similar to those in N-methylformamide). This point is underscored by the nearly 120° C–N–C angle. The barrier to OC–NC rotation is 20.6 kcal mol^{-1} (86 kJ mol^{-1}) for $HCONHCH_3$ and 21.3 kcal mol^{-1} (89 kJ mol^{-1}) for $CH_3CONHCH_3$ [these are free energies of activation; the enthalpies of activation are about 2 kcal mol^{-1} (8 kJ mol^{-1}) higher, the entropy of activation being positive].[107] The Z conformation is more stable than the E by 1.4–1.6 kcal mol^{-1} (5.9–6.7 kJ mol^{-1}) in the case of the formamide and by 2.1–2.5 kcal mol^{-1} (8.8–10.5 kJ mol^{-1}) in the case of the acetamide[107] corresponding to the conformer composition (at room temperature) shown in Figure 10.18. Ab initio quantum mechanical calculations[108] are

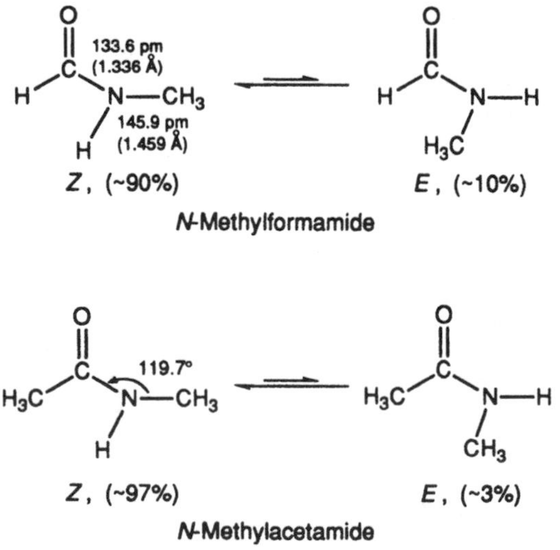

Figure 10.18. Structures of N-methylformamide and N-methylacetamide.

in agreement with the experimental findings. The explanation for the preferred Z conformation for esters in terms of $n-\sigma^*$ overlap cannot apply to amides, since there are no free electron pairs in the amide plane (the pair involved with C=O is at right angles to that plane). Therefore the marked preference for the Z conformation in amides must be due to a combination of steric (attractive in Z, repulsive in E) and charge interaction factors. Because of strong hydrogen bonding, the conformational energies of N-alkylamides and peptides might be expected to be considerably affected by solvation and self-association.

We next consider unsaturated systems with conjugated double bonds, the simplest of which is 1,3-butadiene (Fig. 10.19). The most stable conformation of this molecule is the antiperiplanar or so-called s-trans conformer ($\omega = 180°$; s refers to the fact that one is dealing with conformation about single bonds rather than an alkene configuration), which has a maximum of conjugation and a minimum of steric interaction. There is a second, higher energy conformer that may be either gauche ($\omega = 60°$) or s-cis (synperiplanar, $\omega = 0°$), the gauche form suffering from some steric interaction and less than optimal orbital overlap, whereas the s-cis form has much better overlap but also greater steric repulsions. Calculations[109,110] suggest either an s-cis or a gauche conformer 1.5–2.6 kcal mol^{-1} (6.3–10.9 kJ mol^{-1}) above the s-trans with a barrier to conformational inversion of 5.5–7.3 kcal mol^{-1} (23.0–30.5 kJ mol^{-1}), the decision as to whether this is an s-cis or a gauche conformer being made difficult by the shallow nature of the potential function in the 0°–60° region. Despite some doubts that had been raised as to the reality of this higher energy conformer it has actually been isolated by a matrix isolation technique.[111,112] A mixture of 1,3-butadiene and argon at 400–900°C was rapidly cooled to –243°C; the UV spectrum of the matrix so obtained displays a band not seen when the same experiment is carried out with vapor cooled from room temperature to –243°C. The nature of the UV absorption suggests that the band is due to the planar s-cis rather than the gauche isomer (but see below) and the rate of its disappearance when the matrix is warmed to –213°C leads to an activation energy of 3.9 kcal mol^{-1} (16.3 kJ mol^{-1}) for the s-cis $\rightleftharpoons$ s-trans interconversion. Raman spectroscopic studies[113,114] imply a barrier in the opposite direction of 6.6–7.2 kcal mol^{-1} (27.7–30.1 kJ mol^{-1}) and a ground-state energy difference of 2.5–3.2 kcal mol^{-1} (10.5–13.4 kJ mol^{-1}) (see also refs. 112 and 115). Overall, these data

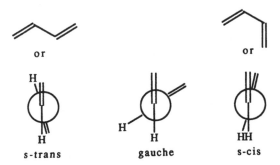

Figure 10.19. Conformations of 1,3-butadiene.

are compatible with an s-cis conformer lying 2.9 ± 0.4 kcal mol^{-1} (12.1 ± 1.7 kJ mol^{-1}) above the s-trans with a barrier from the s-trans side of 6.8 ± 0.4 kcal mol^{-1} (28.5 ± 1.7 kJ mol^{-1}). Calculations suggest that the s-cis form is actually a saddle point between two nearby high minima[116–118] with a torsion angle of about 38° for these minima.[118] The rotational barrier in styrene, 3.13–3.27 kcal mol^{-1} (13.1–13.7 kJ mol^{-1})[117,119] is considerably smaller than that in 1,3-butadiene, presumably because of less effective π orbital overlap in the (thereby less stabilized) ground state.

Conformational preferences for other conjugated unsaturated molecules have been tabulated.[120,121] We mention here only three examples: acrolein, acrylic acid, and methyl acrylate. Acrolein (propenal, Fig. 10.20, X = H) seems to exist largely in the sterically less crowded s-trans conformation with an energy difference between it and the less stable s-cis conformer of 1.7 kcal mol^{-1} (7.1 kJ mol^{-1}) and an energy barrier of 5.0–6.4 kcal mol^{-1} (20.9–26.8 kJ mol^{-1}).[113b,122,123a] The structure of both conformers has been determined accurately by microwave spectroscopy: Both structures are planar.[123b] In the case of acrylic acid (propenoic acid, Fig. 10.20, X = OH) and methyl acrylate (Fig. 10.20, X = OCH$_3$) there is evidently little steric difference between the s-cis and s-trans conformations; accordingly the energy difference is also small, 0.17 kcal mol^{-1} (0.71 kJ mol^{-1}) in acrylic acid[124] (with a barrier of 3.8 kcal mol^{-1} or 16 kJ mol^{-1}) and 0.31 kcal mol^{-1} (1.3 kJ mol^{-1}) in methyl acrylate.[125] Here, also, it is not clear whether the less stable conformer is s-cis or gauche. A similar situation pertains in acryloyl (propenoyl) chloride.[126] That the s-trans conformer is favored so little in these compounds suggests that it has little stereoelectronic advantage and that its substantial predominance in the acrolein case is due to steric reasons, that is, an unfavorable =CH$_2$/O= interaction in the s-cis or gauche conformation (see also Section 9-2.b). The fact that the barrier (from the s-trans side) in acrylic acid is lower than that in acrolein is in accord with the assumption of a sterically more congested ground state (=CH$_2$/OH eclipsed) in the acid. In benzaldehyde, as in acrolein, the planar conjugated conformation is preferred; measurements of the barrier to rotation out of the plane have given results of 7.6–7.9 kcal mol^{-1} (31.8–33.1 kJ mol^{-1}).

Conjugation also occurs with a cyclopropane ring. The barrier in methylcyclopropane, 2.86 kcal mol (12.0 kJ mol^{-1}), is quite similar to the ethane barrier, but that in cyclopropanecarboxaldehyde [Fig. 10.21; $V_2 = 4.39$ kcal mol^{-1} (18.4 kJ mol^{-1})[127]] is closer to that in acrolein than to that in propionaldehyde (see above). The two stable conformations, nearly equal in energy, have the aldehyde hydrogen either synperiplanar or antiperiplanar to the cyclopropyl hydrogen on the adjacent carbon. Similarly, in cyclopropylbenzene (Fig. 10.21) the tertiary C–H bond

s-cis   s-trans

Figure 10.20. Conformation of conjugated unsaturated compounds.

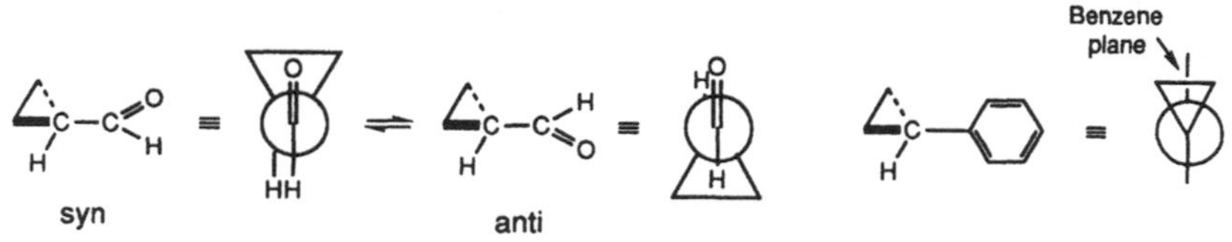

Figure 10.21. Cyclopropanecarboxaldehyde and cyclopropylbenzene.

is eclipsed with the phenyl plane, the barrier to rotation being of the order of 2.0 kcal mol^{-1} (8.4 kJ mol^{-1}).[128] The explanation may be found in the Walsh picture of cyclopropyl bonding,[129] which ascribes near sp^2 character to the cyclopropyl hydrogen atoms and places p orbitals in the plane of the cyclopropyl ring (contributing to the ring C–C bonds) and at right angles to the plane of the H–C–H or (exo-C)–C–H bonds. The stable conformations of the molecules shown in Figure 10.21 are such that this p-orbital plane extends in the direction of the filled C=O or aromatic orbitals for maximum overlap, thus explaining both the height and the mainly twofold nature of the barrier.

b. Alkylbenzenes

We now turn to hydrocarbons with aromatic substituents. One might perhaps have expected the barrier to methyl rotation in toluene to be similar to that in propene. However, this is not the case: Methyl rotation in toluene is nearly free, the barrier being only 14 cal mol^{-1} (59 J mol^{-1}).[130–132] Unlike the twofold V_2 and threefold V_3 barriers encountered so far, that in toluene (Fig. 10.22) is sixfold, as a result of the presence of the local C_2 axis in the plane of the aromatic ring. Sixfold barriers, in general, tend to be very low; for example, that in nitromethane, which also has a local C_2 axis along the C–N bond of the nitro group, is about 6 cal mol^{-1} (25 J mol^{-1}).[17]

The barriers in o-xylene, in higher alkylbenzenes, and in other substituted aromatics are often of the V_3 type (compounds of type ArCX$_3$, such as t-butylbenzene, are exceptions) and the barriers are therefore substantially higher than in toluene [e.g., 1.49 kcal mol^{-1} (6.23 kJ mol^{-1}) in o-xylene],[132] though not high enough to study by low-temperature NMR spectroscopy. As we shall see in Section 10-3.a, whether spectral transitions from distinct conformations of a molecule can be seen separately or whether experimental data from (mole fraction weighted) averaged conformations

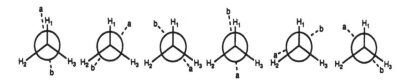

Figure 10.22. Toluene conformations. (a---b stands for an "oriented" benzene ring plane.)

are obtained, depends on the relation between the rate of interconversion of the various conformations and the time scale of the observation. It is therefore fortunate that supersonic jet mass resolved excitation spectroscopy (MRES) has been applied to the study of a number of substituted aromatics.[133] MRES uses optical absorption processes ($S_1 \leftarrow S_0$ transitions), which occur much more rapidly than nuclear motion (Franck–Condon principle). In addition, the expansion of the molecules through the supersonic jet nozzle leads to very substantial cooling (molecules at near zero K); thus the resolution of the experiment is excellent [<3 cal mol^{-1} (12.5 J mol^{-1})]. In principle, each conformation of a molecule has its own spectral properties, and the spectroscopic transitions can be seen for each component of a multicomponent (multiconformational) system.

Thus MRES has been applied to ethylbenzene[134] (Fig. 10.23). In principle, the molecule may be most stable in the "gauche" **A**, "perpendicular" **B**, or "eclipsed" **C** conformation. An assignment was reached on the basis of the MRES spectra. For ethylbenzene itself, a spectrum consistent with a single conformation was observed, indicating that only one of the **A–C** conformations is present. For 1,3-diethylbenzene, conformation **B** would give rise to two conformers, one (meso) with both methyl groups up (or down, the two possibilities are identical) or the other (chiral) with one methyl group up, the other down (a diastereomeric situation). Conformation **C**, on the other hand, can give rise to three conformers, one with both methyl groups pointing inward, a second with both pointing outward, and the third with one in and the other out. Conformation **A** of 1,3-diethylbenzene can exist in six stereoisomeric forms. As the MRES showed two conformations, the perpendicular conformer **B** is in accordance with the spectral evidence for ethylbenzenes. This is also in agreement with calculations,[135,136] which further suggested that conformation **C** corresponds to the energy maximum [barrier 1.16 kcal mol^{-1} (4.85 kJ mol^{-1}).[137]

The situation in anisole is different from that in alkylbenzenes in that the methyl group is in the plane of the aromatic ring (presumably because this favors overlap of the p electrons on oxygen with the π electrons of the ring).[138] Thus p-dimethoxybenzene (in contrast to p-diethylbenzene) shows the presence of *two* conformers (Fig. 10.24 **A** and **B**) both in the vapor phase (by MRES)[139] and in the liquid phase (by Raman spectroscopy),[140] whereas the crystalline solid is entirely in conformation **A**.[141]

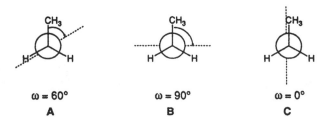

$\omega = 60°$ $\omega = 90°$ $\omega = 0°$
  A   **B**   **C**

Figure 10.23. Ethylbenzene conformations (benzene plane dashed).

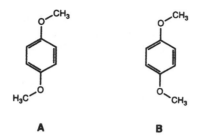

A   **B**

Figure 10.24. Stable conformations of *p*-dimethoxybenzene.

Styrene, like anisole, has a planar conformation for similar reasons (optimal π–π electron overlap). This fact has been confirmed by MRES[142a] through studying of *p*-ethylstyrene (one conformer), *m*-methylstyrene (two conformers), and *p*-methyl-*trans*-β-methylstyrene (two conformers). In contrast, data for α-methylstyrene[142b] suggest that in this molecule the alkene plane is inclined by about 30° with respect to the benzene plane, presumably in order to avoid steric interaction of the α-methyl group with one of the ortho hydrogen atoms. This subject will be returned to in Chapter 13, where it will be shown that replacing the ortho hydrogen atoms by methyl groups increases the steric hindrance sufficiently to make isolation of the (nonplanar) enantiomers possible.

c. Miscellaneous Compounds

The barrier in methanol (Table 10.2)[143] is only about one-third of that in ethane and that in methylamine is about two-thirds. These results are most easily rationalized by assuming that there is no bond eclipsing (Pitzer) strain for the opposition of a bond with a lone pair; thus in methanol there is only one set of H/H eclipsed bonds and in

TABLE 10.2. Rotational Barriers

Compound	Barrier (kcal mol^{-1})	(kJ mol^{-1})	Source
H_3C-CH_3	2.91	12.2	Ref. 12
H_3C-OH	1.07	4.48	Ref. 143
H_3C-NH_2	1.96	8.20	Ref. 144
$H_3C-C_2H_5$	3.4	14.2	Refs. 17, 18
H_3C-OCH_3	2.7	11.3	Ref. 17
$H_3C-OC_2H_5$	2.61	10.9	Ref. 145a
$H_3C-NHCH_3$	3.62	15.1	Ref. 17
$H_3C-NHC_2H_5$	3.12	13.1	Ref. 145b
$H_3C-SC_2H_5$	2.05	8.57	Ref. 145c
$HO-OH$	1.1	4.6	Ref. 17
H_2N-NH_2	3.15	13.2	Ref. 17

methylamine there are two, compared to three in ethane. In contrast, the barrier to methyl rotation (Table 10.2) in dimethyl ether, in ethyl methyl ether ($H_2C–O$ rotor), in dimethylamine and in ethylmethylamine ($H_3C–N$ rotor) are not much lower than the propane barrier; at first sight, this might seem to contradict the assumption that the eclipsing of hydrogen atoms with lone pairs does not cost energy. However, it must be remembered that in ethers and amines, because of the relatively short C–N and C–O distances (relative to C–C), the *H–C–X–C* (X = O or NR) *H–C* distance in the eclipsed form is considerably less than in propane (X = CH_2). There may thus be a rather important steric component to the barrier in this H/lone pair eclipsing energy, which fortuitously makes the barriers in the amines and ethers similar to those in propane. The barrier in ethyl methyl sulfide (long C–S bond!) is, in fact, lower than that in the corresponding ether and amine.

The last two molecules to be discussed here (see also Table 10.2) are hydrogen peroxide and hydrazine. In hydrogen peroxide[146,147] (H–O–O–H) the dihedral angle between the H–O–O planes is near 120°. In hydrazine,[148] the preferred conformation has an :–N–N–: torsion angle of 91°. On the other hand, dialkyl peroxides, such as $CH_3O–OCH_3$, (ref. 149) and *t*-BuO–OBu-*t* (ref. 150), have torsion angles of 166–180°; that is, the alkyl groups are antiperiplanar. The reason generally advanced for the anticlinal (cf. Fig. 2.10) preference in H_2O_2 and H_2NNH_2 is that in the antiperiplanar conformation there would be an unfavorable orbital interaction of the antiparallel unshared electron pairs (see also refs. 57 and 151). However, this does not explain the variability of the torsion angle nor the fact that RO–OR does, in fact, seem to be nearly antiperiplanar when R = alkyl [when R = $(CH_3)_3Si$ the angle is 143.5°].[150] An attractive interaction between the (partially positive) hydrogen atoms and the lone pairs may be a contributing factor,[147] at least in H_2O_2 but does not seem to explain the observed conformation of $H_2N–NH_2$.

There are many molecules other than the few representative ones mentioned here whose conformation has been determined; extensive tabulations have been made.[17,120,152]

10-3. PHYSICAL AND SPECTRAL PROPERTIES OF DIASTEREOMERS AND CONFORMERS

a. General

Before entering into the subject matter of the physical and spectral properties of acyclic diastereomers, we must recall the discussion concerning isomers given in Chapter 3. With acyclic compounds, in comparing diastereomers, one is actually comparing conformer mixtures. If, as frequently happens, the time scale of the experiment is slow relative to that of conformational change, no individual conformers are detected and the measurement yields only the ensemble average for all conformers. This is true of virtually all chemical experiments, plus measurements of such bulk properties as vapor pressure, boiling point, density, refractive index, dipole moment, viscosity, optical rotation, optical rotatory dispersion, and many others. But if the time scale of the experiment is short relative to that of conformer interconversion, individual conformers will be detected. Many of the measurements discussed in Sections 10-1 and 10-2 are of this type, notably measurements of microwave spectra, IR spectra, Raman spectra, UV–vis spectra, and MRES, but excluding electron diffraction. The NMR spectral measurements are on the edge; depending on the frequency separation of the signals observed and the exact rate of the conformational exchange, they may either involve individual conformers or ensemble averages.

b. Dipole Moments

In Section 4-5.b we have seen that molecules in symmetry point groups C_n (including C_1), C_s, and C_{nv} can (and normally will) have dipole moments. If the compounds are hydrocarbons, the dipole moment will ordinarily be small, but for molecules with polar substituents (halogens, OR, SR, NR_2, etc.) the dipole moment is often sizeable.

Dipole moments depend on molecular geometry. This fact is particularly evident in molecules that have two (or perhaps three) polar substituents, such as the 1,2-dihaloethanes. From Figure 10-8, it is obvious that the anti conformer of XCH_2CH_2X (X = halogen) belongs to point group C_{2h} and must therefore have zero dipole moment, whereas the moment of the gauche conformers (C_2) is sizeable.

There are two major methods of measuring dipole moments that give different results. One is the classical method of measuring dielectric constant along with density or refractive index in solution as a function of concentration. This method gives an ensemble average dipole moment μ, which is given by the equation

$$\mu = \left(\sum_i n_i \mu_i^2 \right)^{1/2} \tag{10.1}$$

or, for the particular case under consideration

$$\mu = (n_g \mu_g^2 + n_a \mu_a^2)^{1/2} \tag{10.2}$$

where n_g and n_a are the mole fractions of the gauche and anti conformers, respectively, and μ_g and μ_a are their respective dipole moments. (It must be noted that the dipole moment of the ensemble average is *not* the weighted average of the individual conformer dipoles, but that it is the squares of the individual dipole moments that are averaged to give the square of the ensemble dipole moment. This is so because the actual averaging is of the polarizations, which are proportional to the squares of the dipole moments.) The other method is by microwave spectroscopy, which yields the dipole moments of the individual conformers whose spectra are being observed (microwave spectra can, in fact, be obtained only of molecules that have dipole moments).

c. Infrared Spectra

Of the various techniques mentioned in Sections 10-1 and 10-2 for the determination of conformer population, IR and NMR spectroscopy are the most accessible. The use of these techniques is therefore described in some detail.

For the reasons discussed earlier, IR spectra of conformationally heterogeneous substances will usually display absorption bands due to individual conformers. The intensity I of these absorption bands will be proportional to the conformer population[61]; thus for anti (a) and gauche (g) conformers, $I_a = \alpha_a c_a \ell$ and $I_g = \alpha_g c_g \ell$, where the α values are integrated absorption coefficients, the c values are concentrations, and ℓ is the cell length. It follows that the equilibrium constant is

$$K = c_a/c_g = \alpha_g I_a/\alpha_a I_g \tag{10.3}$$

The ratio I_a/I_g can be measured from the spectrum, provided (and this may cause a problem) that absorption bands can be unequivocally assigned to one conformer or the other. However, the ratio of absorption coefficients, α_a/α_g is generally not known and usually differs from unity.

To circumvent this difficulty, the ratio I_a/I_g is measured at two or more temperatures. Then, according to the van't Hoff isochore

$$\ln (K_{T_2}/K_{T_1}) = (\Delta H°/R)(1/T_1 - 1/T_2) \tag{10.4}$$

where $\Delta H°$ is the enthalpy difference between two conformers, T_1 and T_2 are two temperatures, and K_{T_1} and K_{T_2} are the equilibrium constants at those temperatures. Insertion of Eq. 10.3 into Eq. 10.4 yields

$$\Delta H° = [RT_1 T_2/(T_2 - T_1)][\ln(I_a/I_g)_{T_2} - \ln(I_a/I_g)_{T_1}] \tag{10.5}$$

on the assumption that the absorption coefficients α do not vary with temperature.

If the measurement is carried out at more than two temperatures, $\ln(I_a/I_g)$ is plotted versus $1/T$; the slope of the straight line so obtained gives $\Delta H°/R$. Thus the IR method in this form yields $\Delta H°$, not $\Delta G°$.

Infrared spectroscopy is particularly suitable for probing intramolecular hydrogen bonding[73] of the type $-O-H \cdots :X$, $-S-H \cdots :X$, $-N-H \cdots :X$, and so on. Thus the presence of an IR absorption peak at 3612 cm^{-1} in addition to a peak at 3644 cm^{-1} (due to free hydroxyl) for ethylene glycol in dilute CCl$_4$ solution indicates the presence of intramolecularly bonded $-O-H \cdots O-H$ as well as free hydroxyl, and thereby points to the existence of at least part of the material in the gauche conformation (cf. Fig. 10.8, X = OH).[153a]

Manifestation of intramolecular hydrogen bonding can be used for configurational as well as conformational assignment. For example, both meso and racemic 2,3-butanediol (Fig. 10.25) display intramolecular hydrogen bonding.[153b] However, the meso isomer must take up a relatively unfavorable conformation, with methyl groups gauche, to achieve the necessary proximity of hydroxyl groups (Fig. 10.25, **A**

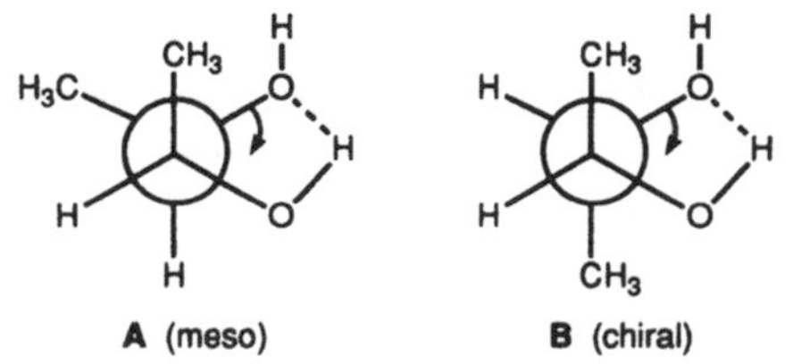

Figure 10.25. Hydrogen-bonded conformations of meso and racemic 2,3-butanediol.

or enantiomer), whereas the chiral isomer can exist in conformation **B** in which the OH groups are gauche but the methyl groups are anti. It is therefore not surprising that the ratio of bonded to unbonded OH [measured, though not accurately (see above), by the area ratio of the corresponding bands] is higher for the chiral than for the meso isomer. In addition, the difference (in wavenumbers) between intramolecularly bonded and nonbonded OH is larger in the chiral isomer (3632 − 3583 = 49 cm^{-1}) than in the meso (3633 − 3591 = 42 cm^{-1}), pointing to a stronger hydrogen bond in the former. This result may be explained by noting (Fig. 10.25) that it is easy to rotate the OH groups together (thus strengthening the bond) in the chiral isomer **B**, whereas the same rotation in the meso isomer **A** requires (unfavorable) increased eclipsing of the methyl groups.

d. NMR Spectroscopy

As explained in Section 10-3.a, the time scale of NMR spectroscopic measurements may be either fast or slow relative to the time scale of rotation around single bonds. At room temperature, except for bonds having partial double bond character, such as the OC–N bond in amides (cf. Section 10-2.a), rotation around single bonds is generally fast on the NMR time scale, and therefore NMR chemical shifts are averaged over all conformations,[154] as are coupling constants:

$$\delta = \sum_i n_i \delta_i \qquad (10.6)$$

and

$$J = \sum_i n_i J_i \qquad (10.7)$$

where δ and J are the observed chemical shifts and coupling constants, respectively, δ_i and J_i are the shifts and coupling constants of the ith conformer and n_i the corresponding mole fractions. At low temperatures, however, if rotation is "slow" (cf. Section 10-3.a) the conformations may be observed individually. In that case, on the assumption that signal intensity is directly proportional to the number of nuclei (or, in the parlance of IR, Section 10-3.c, that all extinction coefficients are equal; for justification see refs. 155 and 156; it is essential to compare corresponding nuclei in the conformers in question), equilibrium constants at the temperature of observation are obtained directly from ratios of signal areas.

In fluoroethanes containing other halogens, for example, $CBr_2ClCHBrF$, barriers are indeed high, in the 9–10 kcal mol^{-1} (38–42 kJ mol^{-1}) range when there are four other halogens (and one hydrogen), and even higher, 13–15 kcal mol^{-1} (54–63 kJ mol^{-1}) when there are five other halogens.[157] Thus it was possible to undertake a complete conformational analysis (i.e., to assess the population of all three rotamers) in a series of fluorinated haloethanes.[158]

Under circumstances of slow exchange (i.e., well below coalescence), both the chemical shifts for salient nuclei and the coupling constants for appropriate pairs of nuclei can be ascertained for all three conformers of appropriately substituted ethanes. Above coalescence, in principle at least, Eqs. 10.6 and 10.7 can be used to assess the conformational composition at temperatures where interconversion among conformers is fast on the NMR time scale. However, there are two practical difficulties. One is that chemical shifts are temperature dependent; therefore Eq. 10.6 cannot safely be used because the contributing shifts δ_i at, say, room temperature will not be the same as the actual shifts measured at low temperature. Attempting a temperature extrapolation of the shifts is rarely successful, since the temperature range over which the shifts of individual conformers can be observed is too small, being limited on the upper bound by exchange and coalescence phenomena (cf. Section 8-4.d) and on the lower bound by difficulties due to freezing or excessive viscosity of the solvent, crystallization of the solute, and so on. On the other hand, coupling constants are nearly temperature independent and therefore Eq. 10.7 should be useful to determine room temperature conformer populations.

Table 10.3 gives a correlation between coalescence temperature (cf. Section 8-4.d) and energy barrier between conformers for two signals originating from two different conformers separated by $\delta = 100$ and 500 Hz. (The former separation would correspond to two protons separated by 0.5 ppm in a 200-MHz instrument or two carbon nuclei separated by 2 ppm in the same instrument, the latter to two protons separated by 1 ppm in a 500-MHz instrument or two carbon nuclei separated in such an instrument by 4 ppm.) The data apply strictly to two conformers of equal population only but may be used as an approximation for other cases where the populations do not differ greatly.

Temperatures around $-100°C$ (ca. 175 K) can be reached routinely and coalescence can thus be observed for signals of conformers separated by barriers of about 8 kcal mol^{-1} (33.5 kJ mol^{-1}). However, since complete signal separation and reliable integration requires going well below the coalescence temperature, the barriers required must actually be somewhat higher or the temperature somewhat lower. A few investigators have been able to reach much lower temperatures, down to 100 K $(-173°C)$[159,160] and have thus been able to measure barriers in the 4.2–5.2 kcal mol^{-1} (17.6–21.8 kJ mol^{-1}) range.

TABLE 10.3. Relation of Coalescence Temperature (T_c) to Barrier ($\Delta G^{\ddagger}$)^a for Signals Separated by 100 and 500 Hz, Respectively

	$\Delta G^{\ddagger}$			
	$\delta = 100$ Hz		$\delta = 500$ Hz	
T_c (K)	(kcal mol^{-1})	(kJ mol^{-1})	(kcal mol^{-1})	(kJ mol^{-1})
100	4.6	19.1	4.2	17.8
110	5.0	21.1	4.7	19.6
120	5.5	23.1	5.1	21.5
130	6.0	25.1	5.6	23.4
140	6.5	27.1	6.0	25.2
150	7.0	29.1	6.5	27.1
175	8.2	34.2	7.6	31.9
200	9.4	39.3	8.7	36.7
225	10.6	44.5	9.9	41.5

$^a\Delta G^{\ddagger} = 1.987\ T_c\ (23.76 + \ln T_c/k_c)$ cal mol^{-1}, where $k_c = 1/2\sqrt{2}\,\pi\delta$.

Vicinal ^{1}H/^{1}H coupling constants ($^3J_{HH}$) as a function of torsion angle ω can be obtained from the Karplus equation[161] (Eq. 10.8; see also ref. 162), which is plotted in Figure 10.26. Because of the uncertainty of the coefficients in the equation

$$J = A\cos^2\omega - B\cos\omega + C \qquad (10.8)$$

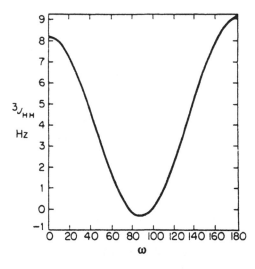

Figure 10.26. Karplus relationship.

as well as the uncertainty concerning the exact torsion angle in the acyclic conformers, a set of empirical parameters is often used to solve Eq. 10.8. Even so the variability of J is a problem: Typically, J_{gauche} varies between 1.5 and 5 Hz and J_{anti} between 10 and 14 Hz. The parameter J_{syn} is typically around 9–9.5 Hz and $J_{90°}$ is near zero. In a later paper[163] the Karplus equation is given in the form $J = 7.76 \cos^2 \omega - 1.10 \cos \omega + 1.40$. This gives values of $J_0 = 8.08$, $J_{60°} = 2.79$, $J_{90°} = 1.40$, $J_{180} = 10.26$ Hz; the values for $J_{0°}$ and $J_{180°}$ appear somewhat small. It is important to recognize that the coupling constants in a segment H^1–CX–CY–H^2 depend not only on the torsion angle but also on several other factors[161b] of which the electronegativity of X and Y[165,166] and the torsion angles between Y and H^1 and X and H^2 [167] are the most important.[164] Taking these factors into account, an empirical relationships (Eq. 10.9) has been developed[163] (see also ref. 164) which appears to be considerably more accurate and more general than the simple Karplus relationship, though at the expense of introducing several additional parameters.

$$J = A \cos^2 \omega + B \cos \omega + \sum_i \Delta\chi_i[D + E \cos^2 (\zeta_i \omega + F|\Delta\chi_i|)] \qquad (10.9)$$

The first two terms in Eq. 10.9 are Karplus terms; the constant term C in the Karplus equation (10.8) is taken to be zero. The parameter χ_i is the difference in Huggins electronegativity[168] between the ith substituent and hydrogen and ζ_i, is either +1 or −1: it is +1 if the X/H^2 or Y/H^1 torsion angle (looking along the C–C bond) is +60° and −1 if it is −60°. The parameters for general use are $A = 13.70$, $B = -0.73$, $D = 0.56$, $E = -2.47$, $F = 16.9°$. An additional parameter G is needed if the substituents X and Y are not single atoms but polyatomic groups (–U–V); in that case $\Delta\chi_i = \Delta\chi_i^U - G\Sigma_j\Delta\chi_j^V$, the subscript j referring to all the substituents attached to atom U

$$(\text{V in } -U-V,V + V \text{ in } -UV_2,V + W \text{ in } U\overset{\displaystyle W}{\underset{\displaystyle V}{\diagup}}_{\diagdown} \text{, and so on})$$

with $G = 0.14$. Equation 10.9 correlates 315 coupling constants from the literature with a correlation coefficient of 0.992 and a root mean square deviation of 0.479 Hz,[163] even better correlations have been developed since.[164]

In the assessment of chemical shift differences in the ^{13}C spectra of diastereomers the upfield shifting γ effect[169] is particularly revealing. It must be recalled, however, that in an ensemble averaged spectrum at room temperature the observed effect depends on the conformational composition of the sample; that is, it is conformation that dictates the nature of spectral differences, not configuration as such. Some data from the literature,[170–172] summarized in Table 10.4 will illustrate this point.

In most of the molecules of the $CH_3CHXCHXCH_3$ type, the methyl shift of the chiral isomer is upfield of that of the meso isomer. The reason[170] is implicit in Figure 10.27. The meso isomer is dominated by conformer **A**, whereas the chiral isomer contains substantial amounts of conformer **D** in which there is a shielding

TABLE 10.4. The ^{13}C NMR Spectra of Diastereomer Pairs

Entry No.	Compound	Solvent	CH$_3^a$		CHX^a		J_{HH}(Hz)^b	
			meso	chiral	meso	chiral	meso	chiral
1	CH$_3$CHOHCHOHCH$_3^c$	CCl$_4$	17.0	19.4	70.7	72.3		
2	CH$_3$CHOHCHOHCH$_3^c$	DMSO	19.9	19.2	71.6	71.4		
3	CH$_3$CH(OAc)CH(OAc)CH$_3^d$	CFCl$_3$	15.1	16.0	71.0	71.1	3.58	5.08
4	CH$_3$CH(OCH$_3$)CH(OCH$_3$)CH$_3^c$	^{e}	15.3	13.7	80.3	78.9		
5	CH$_3$CHClCHClCH$_3^c$	neat	21.9	19.8	61.3	60.2	7.39	3.45
6	CH$_3$CHBrCHBrCH$_3^c$	neat	25.2	20.5	53.7	52.1	8.81	3.11
7	HO$_2$CCH(CH$_3$)CH(CH$_3$)CO$_2$H^d	CH$_3$OH	15.6	13.9	43.8	42.8		
8	NaO$_2$CCH(CH$_3$)CH(CH$_3$)CO$_2$Na^d	D$_2$O	18.0	14.5	48.5	46.4		
9	CH$_3$CHClCH$_2$CHClCH$_3^f$	neat	24.7	25.5	54.5	55.6		
10	CH$_3$CHBrCH$_2$CHBrCH$_3^g$	CDCl$_3$	25.4	26.4	46.8	49.4		

^{a}Chemical shifts in parts per million from tetramethylsilane (TMS).
^{b}Proton (CHX–CHX) coupling constants.[173]
^{c}See ref. 171.
^{d}See ref. 172.
^{e}Solvent CH$_3$CHOHCH(OCH$_3$)CH$_3$.
^{f}See ref. 174.
^{g}See ref. 175.

405

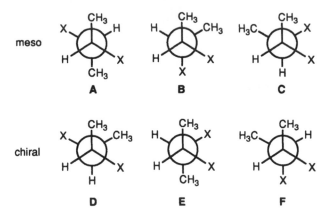

Figure 10.27. Conformers of meso and chiral 2,3-dihalobutanes.

CH_3/CH_3 gauche interaction. The compound $CH_3CHOHCHOHCH_3$ in carbon tetrachloride provides a plausible exception: As mentioned earlier, hydrogen bonding leads to a favoring of the OH/OH gauche conformer: **B–C** for the meso isomer and **E** or **F** for the chiral one. Since both **B** and **C** but only **F** (and not **E**) have CH_3/CH_3 gauche interactions, the upfield methyl shift in the meso isomer is explained. Again plausibly, in DMSO where intramolecular hydrogen bonding is disrupted, the normal order (chiral upfield of meso) is restored. The only other exception in Table 10.4 is ethylene glycol diacetate; this compound is apparently dominated by an attractive gauche effect of the type discussed in Section 10-1.b for FCH_2CH_2F. This is also evidenced by the larger coupling constant in the chiral isomer, which suggests that conformer **F** of the chiral isomer contributes more than conformer **A** of the meso. In $CH_3CHClCH_2CHClCH_3$, Figure 10.28 indicates more gauche interaction in the meso isomer **B** than in the chiral isomer (predominant conformation **A**); therefore the methyl groups in the meso isomer, both in the dichloride and dibromide, are more shielded.

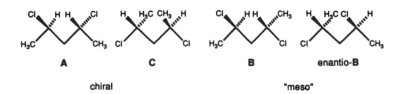

Figure 10.28. Significant conformations for $CH_3CHClCH_2CHClCH_3$.

10-4. CONFORMATION AND REACTIVITY: THE
WINSTEIN–HOLNESS EQUATION AND THE CURTIN–HAMMETT
PRINCIPLE

Although, from a practical point of view, one might want to focus on the different reactivity, in a given reaction, of configurational isomers, the primary relationship of reactivity in conformationally mobile systems is to conformation. Both steric and stereoelectronic factors on reactivity and product composition depend primarily on conformational factors in such systems, in contrast to the situation in (rigid) alkenes (Chapter 9) where reactivity and product composition correlate directly with configuration.

Since the stereochemical aspects of many basic chemical reactions, such as elimination, addition, or substitution, are dealt with in most elementary textbooks, and, to some extent, elsewhere in this text, we shall not take them up here in detail. We must, however, deal with the fact that both reaction rates and reaction products may depend on the conformational composition of starting materials, as well as the conformation of preferred transition states. The dependence of reaction rate on conformational composition is most simply described by the Winstein–Holness equation[176–179] and the related dependence of product composition is governed by the Curtin–Hammett principle.[178–180].

Both the Winstein–Holness equation and the Curtin–Hammett principle refer to the kinetic scheme shown in Eq. 10.10. While the scheme is general and has been so treated,[178] the Curtin–Hammett principle and the Winstein–Holness equation

$$ \overset{k_C}{C \leftarrow} A \underset{k_B}{\overset{k_A}{\rightleftharpoons}} B \overset{k_D}{\rightarrow} D . \qquad (10.10) $$

apply to situations in which A and B are in rapid equilibrium (i.e., k_A, $k_B >> k_C$, k_D), such as when A and B are tautomers or conformational isomers; it is this last situation with which we shall deal in what follows. Products C and D may be different substances, or they may be conformers of the same substance, or (in rare cases) they may even be identical. The energy diagram for the situation where C and D are different is shown in Figure 10.29. For the general kinetic treatment, three different situations must be distinguished.[178]

1. *Case 1:* k_A, $k_B >> k_C$, k_D. (In practice, this condition may be considered fulfilled when k_A, k_B are at least 10 times as large as the larger of k_C and k_D, provided also the A–B equilibrium is not highly one-sided. i.e., k_A and k_B are of the same order of magnitude). In this case the conformational equilibrium $A \rightleftharpoons B$ is maintained throughout the reaction: $[B]/[A] = K = k_A/k_B$. Both the Winstein–Holness equation and the Curtin–Hammett treatment (see below) apply to this case and to this case only.
2. *Case 2:* k_C, k_D, $>> k_A$, k_B. In this case the A/B ratio will not change during the reaction and the product ratio C/D will simply reflect the conformer ratio A/B. This

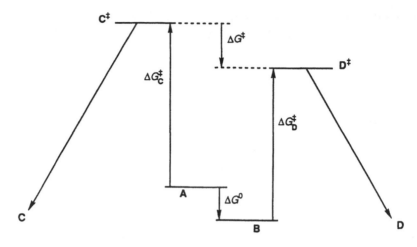

Figure 10.29. Energy diagram for Winstein–Holness and Curtin–Hammett kinetics.

case is called one of "kinetic quenching" since the $A \rightleftharpoons B$ interconversion is stopped; that is, A and B are quenched, by the reaction, in their equilibrium concentrations.

3. *Case 3*: k_C, k_D are of the same order of magnitude as k_A, k_B. In this case, equilibrium is not maintained during the reaction and a kinetic treatment more general than that given below must be applied;[181] the Winstein–Holness equation and the Curtin–Hammett principle in their usual form do not apply.

We proceed, now, to derive the Winstein–Holness equation for Case 1 above. The equation refers to the total empirical rate of reaction of an equilibrating system of the type shown in Eq. 10.10. If the reaction under consideration is first order or pseudo first order, we may write the overall rate as

$$\text{Rate} = \frac{d[\mathbf{A} + \mathbf{B}]}{dt} = k_{\text{WH}}[\mathbf{A} + \mathbf{B}] \tag{10.11}$$

where $[\mathbf{A} + \mathbf{B}]$ is the stoichiometric concentration of the substrate and k_{WH} (the Winstein–Holness rate constant) is the experimentally observed rate constant. (We may consider $\mathbf{A}$ and $\mathbf{B}$ to be two conformers of one and the same substance; we shall see later that one need not confine oneself to two reacting conformations.) But, from Eq. 10.10,

$$\text{Rate} = \frac{d[\mathbf{A}]}{dt} + \frac{d[\mathbf{B}]}{dt} = k_C[\mathbf{A}] + k_D[\mathbf{B}] \tag{10.12}$$

and

$$[\mathbf{B}]/[\mathbf{A}] = K \tag{10.13}$$

Combining Eqs. 10.11 and 10.12 and replacing [**B**] by K[**A**] from Eq. 10.13, one obtains

$$k_{\mathrm{C}}[\mathbf{A}] + k_{\mathrm{D}}K[\mathbf{A}] = k_{\mathrm{WH}}\{[\mathbf{A}] + K[\mathbf{A}]\}$$

or, dividing by [**A**],

$$k_{\mathrm{C}} + k_{\mathrm{D}}K = k_{\mathrm{WH}}(1 + K)$$

whence

$$k_{\mathrm{WH}} = (k_{\mathrm{C}} + k_{\mathrm{D}}K)/(1 + K) \tag{10.14}$$

Equation 10.14 is the Winstein–Holness equation as derived by Eliel and co-workers.[177] It relates the observed rate constant (k_{WH}) to the individual rate constants (k_{C}, k_{D}) at which the conformers **A** and **B** react and to the equilibrium constant K between them. To transform Eq. 10.14 into the form originally given by Winstein and Holness,[176] we set $K = n_{\mathrm{B}}/n_{\mathrm{A}}$, where n_{A} and n_{B} are the mole fractions of **A** and **B** at equilibrium, disregarding other components ($n_{\mathrm{A}} + n_{\mathrm{B}} = 1$). Then

$$k_{\mathrm{WH}} = (k_{\mathrm{C}} + n_{\mathrm{B}}k_{\mathrm{D}}/n_{\mathrm{A}})/(1 + n_{\mathrm{B}}/n_{\mathrm{A}})$$

Multiplying the numerator and denominator on the right-hand side by n_{A} gives

$$k_{\mathrm{WH}} = (n_{\mathrm{A}}k_{\mathrm{C}} + n_{\mathrm{B}}k_{\mathrm{D}})/(n_{\mathrm{A}} + n_{\mathrm{B}})$$

hence, since $n_{\mathrm{A}} + n_{\mathrm{B}} = 1$,

$$k_{\mathrm{WH}} = n_{\mathrm{A}}k_{\mathrm{C}} + n_{\mathrm{B}}k_{\mathrm{D}} \tag{10.15}$$

Equation 10.15 states that the observed rate constant is the average of the rate constants at which the individual conformational isomers (or other rapidly interconverting isomers) react, weighted by their respective mole fractions. This is true, however, only if the rates of interconversion of **A** and **B** are fast relative to their transformation into **C** and **D**, that is, only if equilibrium between **A** and **B** is maintained throughout the reaction.

Equation 10.15 can be generalized for any number of contributing conformations:

$$k_{\mathrm{WH}} = \sum_i n_i k_i \tag{10.16}$$

where n_i is the population (mole fraction) of the ith conformer and k_i is the rate constant for its conversion to product or products. In other words, the observed rate constant for a conformationally heterogeneous system, under conditions of fast conformational interchange, is the weighted average of all the individual rate constants for the contributing conformers (weighted by the respective mole fractions). This result is already familiar from the averaging of enthalpies, polarizations, chemical shifts, coupling constants, and so on, which were discussed earlier.

By way of an example of Winstein–Holness kinetics we shall consider the debromination of meso and chiral 2,3-dibromobutane with iodide ion.[182] These reactions are quite highly stereospecific (see also p. 370) in that the meso dibromide gives largely (96%) *trans*-2-butene, whereas the chiral dibromide gives largely (91%) *cis*-2-butene.[183] These reactions have been interpreted in terms of an E2 mechanism in which the groups to be eliminated must, for maximum reactivity, be antiperiplanar (Fig. 10.30) in the starting material. Referring to Figure 10.27 this means that the meso dibromide needs to be in conformation **A** and the chiral dibromide needs to be in conformation **D**. If the intrinsic reaction rates for iodide-promoted debromination of **A** and **D** are the same, which is not certain, and if it is assumed that no elimination occurs from the other conformations, one has, for the meso isomer (cf. Eq. 10.16)

$$k_{meso} = n_A k_{anti}$$

and for the racemic one

$$k_{rac} = n_D k_{anti}$$

on the assumption that $k_{gauche} = 0$ for conformers **B**, **C**, **E**, and **F**. Here k_{meso} and k_{rac} are the observed reaction rates (k_{WH}) and k_{anti} is the intrinsic rate of reaction for antiperiplanar bromine atoms. It follows, then, that $k_{meso}/k_{rac} = n_A/n_D$, a ratio that is in the 1.3–1.6 region. The experimentally found ratio at 59.7° was 1.93, in reasonable agreement considering the uncertainties of some of the assumptions (for a more detailed treatment of halogen eliminations, see ref. 184).

Additional applications and examples of the Winstein–Holness equation will be discussed in Chapter 11. Whereas this equation deals with reaction *rates*, the conceptually related Curtin–Hammett principle[178,180] deals with reaction *products* derived from two starting materials where $k_A, k_B \gg k_C, k_D$; that is, **A** and **B** (Eq. 10.10) remain in equilibrium throughout the reaction. Let us direct attention once more to Eq. 10.10 and Figure 10.29 (kinetics), but this time focusing on the product ratio **D/C**.

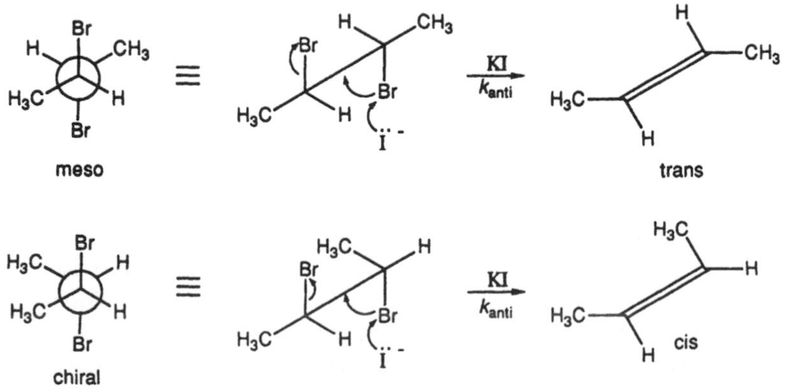

Figure 10.30. Eliminaiion Reaction of $CH_3CHBrCHBrCH_3$ with K1. See also Figure 10.27.

How will this ratio depend on k_A and k_B (or the equilibrium constant $K = k_A/k_B$)? And how will it depend on the rate constants k_C and k_D? In terms of the energy diagram, these questions relate to the role of the ground-state energy difference $\Delta G°$, the activation energies $\Delta G_C^\ddagger$ and $\Delta G_D^\ddagger$, and the transition-state energy difference $\Delta G^\ddagger$.

In the derivation of the Curtin–Hammett principle, we shall again assume Condition 1 above; that is, k_A, $k_B >> k_C$, k_D. In that case equilibrium is maintained, that is, $[B]/[A] = K$ (Eq. 10.10), $dC/dt = k_C[A]$, and $dD/dt = k_D[B]$, the reaction, as before, being taken to be first order or pseudo first order. It follows that $dD/dC = k_D[B]/k_C[A] = (k_D/k_C)$ $([B]/[A]) = k_D K/k_C$ and since, with the assumption made, the right-hand side is constant, one can integrate to

$$[D]/[C] = k_D K/k_C \qquad (10.17)$$

which equation gives the product ratio at the completion of reaction (or at any intermediate point), assuming the initial concentrations of C and D are zero.

Indeed, that $[D]/[C]$ is constant during the course of the reaction is an experimental verification of the Curtin–Hammett assumption.

For a treatment of the Curtin–Hammett and Winstein–Holness kinetics involving second-order reactions, that is, $C \leftarrow R + A \rightleftharpoons B + R \rightarrow D$ (R = second reagent), see ref. 189.

Equation 10.17 states that one can calculate the product ratio if one knows the individual rate constants k_C and k_D for the contributing conformers (e.g., from model studies) and the equilibrium constant K between the conformers. Conversely, if one measures the product ratio D/C and knows K, one can calculate the ratio of rate constants, k_D/k_C.

An application of Curtin–Hammett kinetics (to a cyclic substrate) is the determination of the ratio of rate constants k_C and k_D (Fig. 10.31) in the N-oxidation of N-methy-4-tert-butylpiperidine by hydrogen peroxide in acetone.[185] The product ratio D/C is 95:5 in favor of the axial N-oxide D. The ratio of equatorial to axial N-methyl compound (B/A) is found, by quenching kinetics[186] to be about 60.

This determination of the equilibrium constant K is interesting in itself, since it involves the unusual kinetic case 2 (pp. 407, 408), where the rate of reaction of conformers A and B, in this case with strong acid, is much faster than the rate of equilibration of A and B. The latter can be estimated from the barrier of nitrogen inversion in piperidine [6.1 kcal mol^{-1} (25.5 kJ mol^{-1})][159] to be about 10^8 s^{-1} at room temperature, whereas the rate of protonation, assuming it to be diffusion controlled, is about 10^{10} s^{-1}. By kinetic quenching of N-methyl-4-tert-

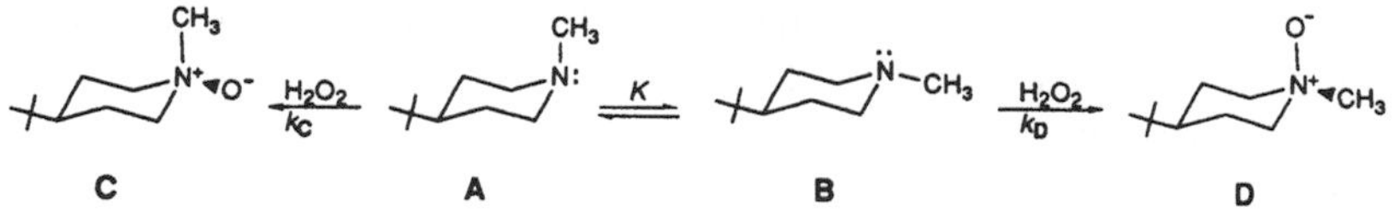

Figure 10.31. Nitrogen oxidation of Nmethyl-4-tert-butylpiperidine.

butylpiperidine (Fig. 10.31, **A** and **B**) with acid under conditions where the result was not affected by mixing control,[187] the conformational energy difference between **A** and **B** was found[186] to range from 2.4 to 3.15 kcal mol^{-1} (10.1–13.2 kJ mol^{-1}), depending on solvent and phase. The lower value refers to the most hydrogen-bonding solvent chloroform and we have assumed that a similar value applies in aqueous acetone, thus $K \sim 60$.

Applying Eq. 10.17, and assuming pseudo-first-order kinetics (since H_2O_2 is in large excess), one obtains $95:5 = 60k_D/k_C$ or $k_D/k_C = 0.3$. This result is reasonable, since N oxidation into the more crowded axial position should be slower than equatorial oxidation.

Returning now, to the derivation of the Curtin–Hammett principle, we note that Eq. 10.17 can be expanded in terms of the free energy and activation free energy parameters displayed in Figure 10.29

$$K = [\mathbf{B}]/[\mathbf{A}] = e^{-\Delta G^\circ/RT}$$

$$k_C = e^{-\Delta G_C^\ddagger/RT}$$

$$k_D = e^{-\Delta G_D^\ddagger/RT}$$

Inserting in Eq. 10.17,

$$[\mathbf{D}]/[\mathbf{C}] = e^{-\Delta G^\circ/RT} \times e^{-\Delta G_D^\ddagger/RT}/e^{-\Delta G_C^\ddagger/RT}$$

or

$$[\mathbf{D}]/[\mathbf{C}] = e^{(-\Delta G^\circ + \Delta G_C^\ddagger - \Delta G_D^\ddagger)/RT}$$

But, from Figure 10.29, $-\Delta G^\circ + \Delta G_C^\ddagger - \Delta G_D^\ddagger = -\Delta G^\ddagger$, where $\Delta G^\ddagger$ is the difference in activation energies for the formation of the two products.

It follows that

$$[\mathbf{D}]/[\mathbf{C}] = e^{-\Delta G^\ddagger/RT} \tag{10.18}$$

(It should be noted that ΔG° and $\Delta G^\ddagger$ in Figure 10.29 are vectors pointing downward; that is, ΔG° and $\Delta G^\ddagger$ are negative. The corresponding upward pointing vectors $\Delta G_C^\ddagger$ and $\Delta G_D^\ddagger$ are positive, as are $-\Delta G^\circ$ and $-\Delta G^\ddagger$.)

Equation 10.18 seems to imply that the product ratio depends only on the free energy difference of the transition states and not, per se, on the energy difference of the ground states. Curtin[180] phrased the Curtin–Hammett principle in approximately the following form (see also ref. 188a): "The relative amounts of products formed from two pertinent conformers are completely independent of the relative populations of the conformers and depend only on the difference in free energy of the transition states, provided the rates of reaction are slower than the rates of conformational interconversion." The principle was interpreted in this way for many years and served mainly as a warning *not* to interpret product ratios ([**D**]/[**C**]) in terms of ground-state conformational equilibrium constants K. However, this interpretation does not allow one to extract quantitative insight, since the difference in energy levels of the transition states ($\Delta G^{\ddagger}$) is not an experimentally determinable quantity. A restated form of the principle (cf. ref. 188b) is more satisfactory: "*The Curtin–Hammett principle implies that in a chemical reaction that yields one product from one conformer and a different product from another conformer (and provided these two conformers are rapidly interconvertible relative to the rate of product formation, whereas the products do not interconvert), the product composition is not solely dependent on the relative proportions of the conformers in the substrate; it is controlled by the difference in standard Gibbs free energies of the representitive transition states. It is also true that the product composition is related to the relative concentrations of the conformers, that is, the conformational equilibrium constant, and the respective rate constants of their reactions; these parameters are, however, often unknown.*"

Yet another way of stating the principle is to say that the product ratio under Curtin–Hammett conditions (conformer equilibration rate faster than rate of product formation) must take into account not only the equilibrium constant of the starting conformers but also the rates of their respective reactions; the sum of the corresponding three free energy terms (ΔG°, $\Delta G_{C}^{\ddagger}$, and $\Delta G_{D}^{\ddagger}$) is expressed in the difference in activation energies of the transition states ($\Delta G^{\ddagger}$), which is directly related to the product ratio.

From the deeper insight gained by a thorough analysis of the scheme in Figure 10.29 it became clear[178] that maximum information about the kinetic behavior of a system is gained when one focuses not only on the product ratio **C/D** but also on the equilibrium constant K and the individual rate constants k_C and k_D. In fact, k_C and k_D can be formulated by Eqs. 10.19 and 10.20, derived from Eqs. 10.14 and 10.17 and setting $P = $ [**D**]/[**C**]. The treatment embodied in Figure 10.31 amounts to just that.

$$k_{C} = k_{WH}[(K + 1)/(P + 1)] \tag{10.19}$$

$$k_{D} = k_{WH}[(K + 1)/K][P/(P + 1)] \tag{10.20}$$

Another example[189] is shown in Figure 10.32. In this example, additionally, the total rate constant, k_{WH}, was determined and thus, knowing **D/C** $= P$ and K, k_C, and k_D could be determined individually by a combination of the Winstein–Holness and Curtin–Hammett kinetic schemes.

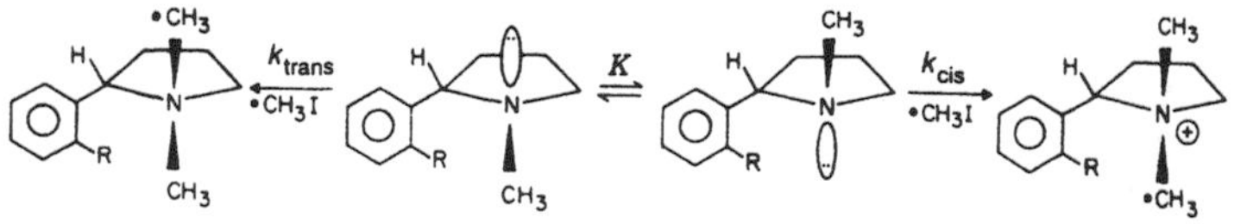

N'$_{trans}$(**C**)   **A**   **B**   N'$_{cis}$(**D**)

Figure 10.32. Quaternization of *N*-methyl-2arylpyrrolidines, $k_{cis} = k_D$; $k_{trans} = k_C$. The dot indicates ^{13}C.

The reaction studied was the quaternization of 1-methyl-2-arylpyrrolidines (Fig. 10.32) in which the 2-aryl group was phenyl or an ortho-substituted phenyl, with the substituent being CH_3, Et, i-Pr, or t-Bu. The (unlabeled) substrates were methylated with $^{13}CH_3I$ and the product ratio (**D/C**) of the isotopic diastereomers was determined by ^{13}C NMR spectroscopy. The overall rate constant k_{WH} was determined by conductometric titration of the methylation product (quaternary salt) formed and the equilibrium constant K was determined by acid quenching of the starting amines (**A**, **B**) and analysis of the product ratio of the ammonium salts so formed by NMR spectroscopy. The results are summarized in Table 10.5.

The following findings from such a complete kinetic analysis are of interest: (a) k_{trans} is about 10–20 times as large as k_{cis}. The aryl group clearly hinders approach from the side of the pyrrolidine ring it occupies. The ratio k_{trans}/k_{cis} increases from about 10 to about 20 when an ortho alkyl group smaller than $tert$-butyl is introduced; the factor of 2 may suggest that the ring turns with the o-alkyl group away from the site of approach of the methyl (the unsubstituted aryl ring, because of its twofold symmetry axis along the aryl–pyrrolidine bond, has twice as many favorable orientations as the substituted ones). With $tert$-butyl the ratio increases to about 15, thus further hindrance clearly comes into play. (b) $K = [\mathbf{B}]/[\mathbf{A}]$ increases (though not sharply) as the size of the ortho substituent increases. This is reasonable: The larger the group, the less comfortable it is in being located cis to the N-methyl substituent. Again this ratio for the substituted aryl groups to the unsubstituted one appears to be not much in excess of two, perhaps for reasons similar to those just discussed. (c) The product ratio is nearly constant for the first four compounds; it varies, but only by a factor of 6, for the $tert$-butyl substituted species. This type of constancy in quaternizations is often observed and, in terms of Eq. 10.17, reflects a compensating factor: As the 2-substituent becomes bulkier, K in the expression $[\mathbf{D}]/[\mathbf{C}] = k_{cis}K/k_{trans}$ becomes larger but k_{cis}/k_{trans} becomes smaller.

We shall return to further examples of the Winstein–Holness equation in Chapter 11.

TABLE 10.5. Experimental Data and Calculated Rate Constant for Quaternization of 1-Methyl-2-arylpyrrolidines[a] with Methyl-^{13}C Iodide[b]

Ortho Substituent R	D/C[c]	$k_{WH}{}^c \times 10^4$	K^c	$k_{cis}{}^d \times 10^4$	$k_{trans}{}^d \times 10^3$
H	1.7	30.6	17	20	20
CH_3	1.4	7.61	>30	4.6	9.8
C_2H_5	1.3	6.17	>30	3.6	8.0
$(CH_3)_2CH$	1.3	5.32	>30	3.0	6.9
$(CH_3)_3C$	0.28	1.25	>40	0.28	4.0

[a]See Figure 10.32.
[b]Reference 189.
[c]Determined experimentally.
[d]$k_{cis} = k_D$; $k_{trans} = k_C$. $k_D/k_C = [\mathbf{D}]/K[\mathbf{C}]$ from Eq 10.17; with this ratio known, k_C and k_D can be calculated from Eq. 10.14.

REFERENCES

1. Bischoff, C. A. *Ber. Dtsch. Chem. Ges.* **1890**, *23*, 623; *ibid.* **1891**, *24*, 1085. See also Bykov, G. V. "The Conceptual Premises of Conformational Analysis in the Work of C. A. Bischoff," in Ramsey, O. B., ed., *van't Hoff–Le Bel Centennial*, American Chemical Society Symposium Series 12, American Chemical Society, Washington, DC, 1975, p. 114.

2. Hermans, P. H *Z. Phys. Chem.* **1924**, *113*, 337.

3. Weissberger, A. and Sängewald, R. *Z. Phys. Chem.* **1930**, *B9*, 133.

4. Wolf, K. L. *Trans. Faraday Soc.* **1930**, *26*, 315.

5. Mizushima, S.-I. *Structure of Molecules and Internal Rotation*, Academic Press, New York, 1954.

6. Eliel, E. L., Allinger, N. L., Angyal, S. J., and Morrison, G. A. *Conformational Analysis*, Interscience-Wiley, New York, 1965; reprinted by American Chemical Society, Washington, DC, 1981.

7. Long, D. A. *J. Mol. Struct.* **1985**, *126*.

8. Kemp, J. D. and Pitzer, K. S. *J. Chem. Phys.* **1936**, *4*, 749.

9. Pitzer, K. S. *Discuss. Faraday Soc.* **1951**, *10*, 66.

10. Weiss, S. and Leroi, G. E. *J. Chem. Phys.* **1968**, *48*, 962.

11. Fantoni, R., van Helroot, K., Knippers, W., and Reuss, J. *J. Chem. Phys.* **1986**, *110*, 1.

12. Pitzer, R. M. *Acc. Chem. Res.* **1983**, *16*, 307; Goodman, L., Porphristic, V., and Weinhold, F. *Acc. Chem. Res.* **1999**, *32*, 983.

13. Sovers, O. J., Kern, C. W., Pitzer, R. M., and Karplus, M. *J. Chem. Phys.* **1968**, *49*, 2592.

14. Bader, R. F. W., Cheeseman, J. R., Laidig, K. E., Wiberg, K. B., and Breneman, C. *J. Am. Chem. Soc.* **1990**, *112*, 6530.

15. Luke, B. T., Pople, J. A., Krogh-Jesperson, M. B., Apeloig, Y., Chandrasekhar, J., and Schleyer, P. v. R. *J. Am. Chem. Soc.* **1986**, *108*, 260.

16. Császár, A. G., Allen, W. D., and Schaefer, H. F. *J. Chem. Phys.* **1998**, *108*, 9751.

17. Lowe, J. P. *Prog. Phys. Org.* **1968**, *6*, 1.

18. Owen, N. L. "Studies of Internal Rotation by Microwave Spectroscopy," in Orville-Thomas, W., ed., *Internal Rotation in Molecules*, Wiley, New York, 1974, p. 157.

19. Wiberg, K. B. and Murcko, M. A. *J. Am. Chem. Soc.* **1988**, *110*, 8029.

20. Allinger, N. L. *J. Am. Chem. Soc.* **1957**, *79*, 3443.

21. Reisse, J. "Quantitative Conformational Analysis of Cyclohexane Systems," in Chiordoglu, G., ed., *Conformational Analysis*, Academic Press, New York, 1971, p. 219.

22. Bradford, W. F., Fitzwater, S., and Bartell, L. S. *J. Mol. Struct.* **1977**, *38*, 85; Heenan, R. K. and Bartell, L. S. *J. Chem. Phys.* **1983**, *78*, 1270.

23. Compton, D. A. C., Montero, S., and Murphy, W. F. *J. Phys. Chem.* **1980**, *84*, 3587.

24. Stidham, H. D. and Durig, J. R. *Spectrochim. Acta* **1986**, *42A*, 105.

25. Allinger, N. L., Grew, R. S., Yates, B. F., and Schaefer, H. F. *J. Am. Chem. Soc.* **1990**, *112*, 114.

26. Bartell, L. F. and Kohl, D. A. *J. Chem. Phys.* **1963**, *39*, 3097.

27. Pitzer, K. S. *Chem. Rev.* **1940**, *27*, 39.

28. Buschweller, C. H., Whalon, M. R., and Laurenzi, B. J. *Tetrahedron Lett.* **1981**, *22*, 2945.

29. Verma, A. L., Murphy, W. F., and Bernstein, H. J. *J. Chem. Phys.* **1974**, *60*, 1540.

30. Bartell, L. F. and Boates, T. L. *J. Mol. Struct.* **1976**, *32*, 379.

31. Lunazzi, L., Macciantelli, D., Bernardi, F., and Ingold, K. U. *J. Am. Chem. Soc.* **1977**, *99*, 4573.

32. Boyd, R. H. *J. Am. Chem. Soc.* **1975**, *97*, 5353.

33. Hounshell, W. D., Dougherty, D. A., and Mislow, K. *J. Am. Chem. Soc.* **1978**, *100*, 3149.

34. Rüchardt, C. and Beckhaus, H.-D. *Angew. Chem. Int. Ed. Engl.* **1980**, *19*, 429.

35. Eliel, E. L. *J. Mol. Struct.* **1985**, *126*, 385.

36. Ōsawa, E., Shirahama, H., and Matsumoto, T. *J. Am. Chem. Soc.* **1979**, *101*, 4824.

37. Anderson, J. E. "Conformational Analysis of Acyclic and Alicyclic Saturated Hydrocarbons," in Patai, S. and Rappoport, Z., eds., *The Chemistry of Alkanes and Cycloalkanes*, Wiley, New York, 1992, p. 95.

38. Berg, U. and Sandström, J. *Adv. Phys. Org. Chem.* **1989**, *25*, 1.

39. Morino, Y. *J. Mol. Struct.* **1985**, *126*, 1.

40. Mizushima, S., Morino, Y., and Higasi, K. *Sci. Pap. Inst. Phys. Chem. Res. Tokyo* **1934**, *25*, 159; see also ref. 5.

41. Neu, J. T. and Gwinn, W. D. *J. Chem. Phys.* **1950**, *18*, 1642.

42. Bose, P. K., Henderson, D. O., Ewig, C. S., and Polavarapu, P. L. *J. Phys. Chem.* **1989**, *93*, 5070.

43. Eliel, E. L. and Wilen, S. H. *Stereochemistry of Organic Compounds*, Wiley, New York, 1994, Chap. 10.

44. Dosĕn-Mićović, L., Jeremić, D., and Allinger, N. L. *J. Am. Chem. Soc.* **1983**, *105*, 1723.

45. Seki, W. and Choi, P. K. *Seisan Kenkyu* **1982**, *34*, 437; *Chem. Abstr.* **1983**, *98*, 125043w.

46. Hammarström, L.-G., Liljefors, T., and Gasteiger, J. *J. Comp. Chem.* **1988**, *9*, 424.

47. Abraham, R. J. and Brettschneider, E. "Medium Effects on Rotational and Conformational Equilibria," in Orville-Thomas, W. J., ed., *Internal Rotation in Molecules*, Wiley, New York, 1976, p. 481.

48. Dosĕn-Mićović, L. and Zigman, V. *J. Chem. Soc. Perkin 2* **1985**, 625.

49. Gordon, A. J. and Ford, R. A. *The Chemist's Companion*, Wiley, New York, 1972, p. 4.

50. Takagi, K., Choi, P.-K., and Seki, W. *J. Chem. Phys.* **1983**, *79*, 964. Tanabe, K., Hiraishi, J., and Tamura, T. *J. Mol. Struct.* **1976**, *33*, 19.

51. Abraham, R. J. and Kemp, R. H. *J. Chem. Soc. B* **1971**, 1240.

52. Radom L., Latham, W. A., Hehre, W. J., and Pople, J. A. *J. Am. Chem. Soc.* **1973**, *95*, 693.

53. Durig, J. R., Liu, J., Little, T. S., and Kalasinsky, V. F. *J. Phys. Chem.* **1992**, *96*, 8224.

54. Bartell, L. S. *J. Am. Chem. Soc.* **1977**, *99*, 3279.

55. Allinger, N. L. and Chang, S. H. M. *Tetrahedron* **1977**, *33*, 1561.

56. Takeo, H., Matsumura, C., and Morino, Y. *J. Chem. Phys.* **1986**, *84*, 4205.

57. Wolfe, S. *Acc. Chem. Res.* **1972**, *5*, 102.

58. Wiberg, K. B., Murcko, M. A., Laidig, K. E., and MacDougall, P. J. *J. Phys. Chem.* **1990**, *94*, 9656.

59. Harris, W. C., Holtzclaw, J. R., and Kalasinsky, V. F. *J. Chem. Phys.* **1977**, *67*, 3330.

60. Felder, P. and Günthard, H. H. *Chem. Phys.* **1984**, *85*, 1.

61. Park, P. J. D., Pethrick, R. A., and Thomas, B. N. "Infrared and Raman Band Intensities and Conformational Change," in Orville-Thomas, W. J., ed., *Internal Rotation in Molecules*, Wiley, New York, 1974, p. 57.

62. Matsuura, H., Miyauchi, N., Murata, H., and Sakakibara, M. *Bull. Chem. Soc. Jpn.* **1977**, *52*, 344.

63. Abe, A. and Mark, J. E. *J. Am. Chem. Soc.* **1976**, *98*, 6468.

64. Eliel, E. L. (a) *Kem. Tidskr.* **1969**, *81*, 6/7, 22; (b) *Acc. Chem. Res.* **1970**, *3*, 1; (c) *Angew. Chem. Int. Ed. Engl.* **1972**, *11*, 739.

65. Meyer, A. Y. *J. Mol. Struct.* **1983**, *94*, 95.

66. Durig, J. R., Godbey, S. E., and Sullivan, J. F. *J. Chem. Phys.* **1984**, *80*, 5983.

67. Yamanouchi, K., Sugi, M., Takeo, H., Matsumura, C., and Kuchitsu, K. *J. Phys. Chem.* **1984**, *88*, 2315.

68. Szasz, G. J. *J. Chem. Phys.* **1955**, *23*, 2449.

69. Terui, Y., Ueyama, M., Satoh S., and Tori, K. *Tetrahedron* **1974**, *30*, 1465.

70. Birdsall, N. J. M., Partington, P., Datta, N., Mondal, P., and Pauling, P. J. *J. Chem. Soc. Perkin 2* **1980**, 1415.

71. Snyder, E. I. *J. Am. Chem. Soc.* **1966**, *88*, 1165.

72. van Duin, M., Baas, J. M. A., and van de Graaf, B. *J. Org. Chem.* **1986**, *51*, 1298.

73. Tichý, M. "The Determination of Intramolecular Hydrogen Bonding by Infrared Spectroscopy and Its Applications in Stereochemistry," in Raphael, R. A., Taylor, E. C., and Wynberg, H., eds., *Advances in Organic Chemistry, Methods and Results*, Vol. 5, Wiley, New York, 1965, p. 115.

74. Astrup, E. E. (a) *Acta Chem. Scand.* **1971**, *25*, 1994; (b) *ibid.* **1973**, *27*, 3271.

75. Miyasaka, T., Kinai Y., and Imamura, Y. *Makromol. Chem.* **1981**, *182*, 3533.

76. Abe, A. *J. Am. Chem. Soc.* **1976**, *98*, 6477.

77. Jeffrey, G. A., Pople, J. A., Binkley, J. S., and Vishveshwara, S. *J. Am. Chem. Soc.* **1978**, *100*, 373.

78. Abe, A., Inomata, K., Tanisawa, E., and Ando, I. *J. Mol. Struct.* **1990**, *238*, 315.

79. Tvaroska, I. and Bleha, T. *Coll. Czech. Chem. Commun.* **1980**, *45*, 1883.

80. Kirby, A. J. *The Anomeric Effect and Related Stereoelectronic Effects at Oxygen*, Springer, New York, 1983.

81. Juaristi, E., and Cuevas, G. *The Anomeric Effect*, CRC Press, Boca Raton, FL, 1995.

82. Graczyk, P. P. and Mikolajczyk, M. *Top. Stereochem.* **1994**, *21*, 159.

83. Lemieux, R. U. *Pure Appl. Chem.* **1971**, *25*, 527.

84. Epiotis, N. D., Yates, R. L., Larson, J. R., Kirmaier, C. R., and Bernardi, F. *J. Am. Chem. Soc.* **1977**, *99*, 8379.

85. Edward, J. T. *Chem. Ind. (London)* **1959**, 568.

86. Romers, C., Altona, C., Buys, H. R., and Havinga, E. *Top. Stereochem.* **1969**, *4*, 39.

87. Hayashi, M. and Kato, H. *Bull Chem. Soc. Jpn.* **1980**, *53*, 2701.

88. Herschbach, D. R. and Krisher, L. C. *J. Chem. Phys.* **1958**, *28*, 728.

89. Wiberg, K. B. and Martin, E. *J. Am. Chem. Soc.* **1985**, *107*, 5035.

90. Van Hemelrijk, D., Van den Enden, L., Geise, H. J., Sellers, H. L., and Schäfer, L. *J. Am. Chem. Soc.* **1980**, *102*, 2189.

91. Durig, J. R. and Compton, D. A. C. *J. Phys. Chem.* **1980**, *84*, 773.

92. Karabatsos, G. J. and Fenoglio, D. J. *Top. Stereochem.* **1970**, *5*, 167.

93. Kilb, R. W., Lin, C. C., and Wlson, E. B. *J. Chem. Phys.* **1957**, *26*, 1695.

94. Abraham, R. J. and Pople, J. A. *Mol. Phys.* **1960**, *3*, 609.

95. Butcher, S. S. and Wilson, E. B. *J. Chem. Phys.* **1964**, *40*, 1671.

96. Durig, J. R., Compton, D. A. C., and McArver, A. Q. *J. Chem. Phys.* **1980**, *73*, 719.

97. Bowen, J. P., Pathiaseril, A., Profeta, S., and Allinger, N. L. *J. Org. Chem.* **1987**, *52*, 5162.

98. Stiefvater, O. L. and Wilson, E. B. *J. Chem. Phys.* **1969**, *50*, 5385.

99. Curl, R. F. *J. Chem. Phys.* **1959**, *30*, 1529.

100. Riveros, J. M. and Wilson, E. B. *J. Chem. Phys.* **1967**, *46*, 4605.

101. Blom, C. E. and Günthard, H. H. *Chem. Phys. Lett.* **1981**, *84*, 267.

102. Ruschin, S. and Bauer, S. H. *J. Phys. Chem.* **1980**, *84*, 3061.

103. Wennerström, H., Forsén, S., and Roos, B. *J. Phys. Chem.* **1972**, *76*, 2430.

104. Hirota, E., Sugisaki, R., Nielsen, C. J., and Sørensen, G. O. *J. Mol. Spectrosc.* **1974**, *49*, 251.

105. Kitano, M. and Kuchitsu, K. (a) *Bull. Chem. Soc. Jpn.* **1974**, *47*, 67; (b) *ibid.* **1973**, *46*, 3048; (c) *ibid.* **1974**, *47*, 631; (d) Kitano, M., Fukuyama, T., and Kuchitsu, K. *Bull. Chem. Soc. Jpn.* **1973**, *46*, 384.

106. Yoder, C. H. and Gardner, R. D. *J. Org. Chem.* **1981**, *46*, 64.

107. Drakenberg, T. and Forsén, S. *J. Chem. Soc. D* **1971**, 1404.

108. Perricaudet, M. and Pullman, A. *Int. J Peptide Protein Res.* **1973**, *5*, 99.

109. Tai, J. C. and Allinger, N. L. *J. Am. Chem. Soc.* **1976**, *98*, 7928.

110. Momicchioli, F., Baraldi, I., and Bruni, M. C. *Chem. Phys.* **1982**, *70*, 161.

111. Squillacote, M. E., Sheridan, R. S., Chapman, O. L., and Anet, F. A. L. *J. Am. Chem. Soc.* **1979**, *101*, 3657.

112. Huber-Wälchli, P. *Ber. Busen-Ges. Phys. Chem.* **1978**, *82*, 10.

113. Carreira, L. A. (a) *J. Chem. Phys.* **1975**, *62*, 3851; (b) *J. Phys. Chem.* **1976**, *80*, 1149.

114. Durig, J. R., Bucy, W. E., and Cole, A. R. H. *Can. J Phys.* **1975**, *53*, 1832.

115. Mui, P. W. and Grunwald, E. *J. Am. Chem. Soc.* **1982**, *104*, 6562.

116. Furukawa, Y., Takeuchi, H., Harada, I., and Tasumi, M. *Bull. Chem. Soc. Jpn.* **1983**, *56*, 392.

117. Bock, C. W., Trachtman, M., and George, P. *Chem. Phys.* **1985**, *93*, 431.

118. Breulet, J., Lee, T. J., and Schaefer, H. F. *J. Am. Chem. Soc.* **1984**, *106*, 6250.

119. Hollas, J. M. and Ridley, T. *Chem. Phys. Lett.* **1980**, *75*, 94.

120. Wilson, E. B. *Chem. Soc. Rev.* **1972**, *1*, 293; Kiss, A. I. and Lukovits, I. *Chem. Phys. Lett.* **1979**, *65*, 169.

121. Bastiansen, O., Sep, H. M., and Boggs, J. E. "Conformational Equilibria in the Gas Phase," in Dunitz, J. D., and Ibers, J. A., eds., *Perspectives in Structural Chemistry*, Vol. 4, Wiley, New York, 1971, p. 60.

122. Courtieu, J., Gounelle, Y., Gonord, P., and Kan, S. K *Org. Magn. Reson.* **1974**, *6*, 151.

123. (a) Blom, C. E. and Bauder, A. *Chem. Phys. Lett.* **1982**, *88*, 55. (b) Blom, C. E., Grassi, G., and Bauder, A. *J. Am. Chem. Soc.* **1984**, *106*, 4727.

124. Bolton, K., Lister, D. G., and Sheridan, J. *J. Chem. Soc. Faraday Trans. 2* **1974**, *70*, 113.

125. George, W. O., Hassid, D. V., and Maddams, W. F. *J. Chem Soc. Perkin 2* **1972**, 400.

126. Hagen, K. and Hedberg, K. *J. Am. Chem. Soc.* **1984**, *106*, 6150.

127. Volltrauer, H. N. and Schwendeman, R. H. *J. Chem. Phys.* **1971**, *54*, 260.

128. Parr, W. J. E. and Schaefer, T. *J. Am. Chem. Soc.* **1977**, *99*, 1033.

129. Walsh, A. D. *Trans. Faraday Soc.* **1949**, *45*, 179.

130. Pitzer, K. S. and Scott, D. W. *J. Am. Chem. Soc.* **1943**, *65*, 803.

131. Lambert, J. B., Nienhuis, R. L., and Finzel, R. B. *J. Phys. Chem.* **1981**, *85*, 1170.

132. Rudolph H. D., Walzer, K., and Krutzik, I. *J. Mol. Spectrosc.* **1973**, *47*, 314.

133. Seeman, J. I., Secor, H. V., Breen, P. J., Grassian, V. H., and Bernstein, E. R. *J. Am. Chem. Soc.* **1989**, *111*, 3140.

134. Breen, P. J., Bernstein, E. R., and Seeman, J. I. *J. Chem. Phys.* **1987**, *87*, 3269.

135. Kříž, J. and Jakeš, J. *J. Mol. Struct.* **1972**, *12*, 367.

136. Umeyama, H. and Nakagawa, S. *Chem. Pharm. Bull.* **1979**, *27*, 2227.

137. Miller, A. and Scott, D. W. *J. Chem. Phys.* **1978**, *68*, 1317.

138. Radom, L., Lathan, W. A., Hehre, W. J., and Pople, J. A. *Austr. J. Chem.* **1972**, *25*, 1601.

139. Breen, P. J., Bernstein, E. R., Secor, H. V., and Seeman, J. I. *J. Am. Chem. Soc.* **1989**, *111*, 1958.

140. Tylli, H., Konschin, H., and Fogerström, B. *J. Mol. Struct.* **1985**, *128*, 297.

141. Goodwin, T. H., Przybylska, M., and Robertson, J. M. *Acta. Crystallogr.* **1950**, *3*, 279.

142. Grassian, V. H., Bernstein, E. R., Secor, H. V., and Seeman, J. I. (a) *J. Phys. Chem.* **1989**, *93*, 3470; (b) *ibid.* **1990**, *94*, 6691.

143. Lees, R. M. and Baker, J. G. *J. Chem. Phys.* **1968**, *48*, 5299.

144. Tagaki, K. and Kojima, T. *J. Phys. Soc. Jpn.* **1971**, *30*, 1145.

145. (a) Durig, J. R. and Compton, D. A. C. *J. Chem. Phys.* **1978**, *69*, 4713. (b) Durig, J. R. and Compton, D. A. C. *J. Phys. Chem.* **1979**, *83*, 2873. (c) Durig, J. R., Compton, D. A., and Jalilian, M.-R. *J. Phys. Chem.* **1979**, *83*, 511.

146. Oelfke, W. C. and Gordy, W. *J. Chem. Phys.* **1969**, *51*, 5336.

147. Bair, R. A. and Goddard, W. A. *J. Am. Chem. Soc.* **1982**, *104*, 2719.

148. Kohata, K., Fukuyama, T., and Kuchitsu, K. *J. Phys. Chem.* **1982**, *86*, 602.

149. Kimura, K. and Osafune, K. *Bull. Chem. Soc. Jpn.* **1975**, *48*, 2421.

150. Käss, D., Oberhammer, H., Brandes, D., and Blaschette A. *J. Mol. Struct.* **1977**, *40*, 65.

151. Fink, W. H. and Allen, L. C. *J. Chem. Phys.* **1967**, *46*, 2261, 2276.

152. George, W. O. and Goodfield, J. E. "Vibrational Spectra at Variable Temperature and the Determination of Energies Between Conformers," in Durig, J. R., ed., *Analytical Applications of FT-IR to Molecular and Biological Systems, NATO Advanced Study Series,* Series C, Vol. 57, Reidel, Boston, p. 293.

153. Kuhn, L. P. (a) *J. Am. Chem. Soc.* **1952**, *74*, 2492; (b) *ibid.* **1958**, *80*, 5950.

154. Eliel, E. L. *Chem. Ind. (London)* **1959**, 568.

155. Booth, H. and Josefowicz, M. L. *J. Chem. Soc. Perkin 2* **1976**, 895.

156. Eliel, E. L., Kandasamy, D., Yen, C.-Y., and Hargrave, K. D. *J. Am. Chem. Soc.* **1980**, *102*, 3698.

157. Weigert, F. J., Winstead, M. B., Garrels, J. I., and Roberts, J. D. *J. Am. Chem. Soc.* **1970**, *92*, 7359.

158. Norris, R. D. and Binsch, G. *J. Am. Chem. Soc.* **1973**, *95*, 182.

159. Anet, F. A. L. and Yavari, I. *J. Am. Chem. Soc.* **1977**, *99*, 2794, 6752.

160. Bushweller, C. H., Fleischman, S. H., Grady, G. L., McGoff, P., Rithner, C. D., Whalon, M. R., Brennan, J. G., Marcantonio, R. P., and Laurenzi, B. J. *J. Am. Chem. Soc.* **1982**, *104*, 6224.

161. Karplus, M. (a) *J. Chem. Phys.* **1959**, *30*, 11; (b) *J. Am. Chem. Soc.* **1963**, *85*, 2870.

162. Conroy, H. *Adv. Org. Chem.* **1960**, *2*, 265.

163. Haasnoot, C. A. G., de Leeuw, F. A. A. M., and Altona, C. *Tetrahedron* **1980**, *36*, 2783; but see ref 164.

164. Altona, C., Ippel, J. H., Westra Hoekzema, A. J. A., Erkelens, C., Groesbeek, M., and Donders, L. A. *Magn. Reson. Chem.* **1989**, *27*, 564; Altona, C., Francke, R., de Haan, R., Ippel, J. H., Daalmans, G. J., Westra Hoekzema, A. J. A., and van Wijk, J. *Magn Reson. Chem.* **1994**, *32*, 670.

165. Glick, R. E. and Bothner-By, A. A. *J. Chem. Phys.* **1956**, *25*, 362.

166. Schaefer, T., Hruska, F., and Kotowycz, G. *Can. J. Chem.* **1965**, *43*, 75.

167. Booth, H. *Tetrahedron Lett.* **1965**, 411.

168. Huggins, M. L. *J. Am. Chem. Soc.* **1953**, *75*, 4123.

169. Duddeck, H. *Top. Stereochem.* **1986**, *16*, 219.

170. Ernst, L. and Trowitzsch, W. *Chem. Ber.* **1974**, *107*, 3771.

171. Levy, G. C., Pehk, T., and Lippmaa, E. *Org. Magn Reson.* **1980**, *14*, 214.

172. Schneider, H.-J. and Lonsdorfer, M. *Org. Magn. Reson.* **1981**, *16*, 133.

173. Bothner-By, A. A. and Naar-Cohn, C. *J. Am. Chem. Soc.* **1962**, *84*, 743.

174. Carman, C. J., Tapley, A. R., and Goldstein, J. H. *J. Am. Chem. Soc.* **1971**, *93*, 2864.

175. Cais, R. E. and Brown, W. L. *Macromolecules* **1980**, *13*, 801.

176. Winstein, S. and Holnes, N. J. *J. Am. Chem. Soc.* **1955**, *77*, 5562.

177. Eliel, E. L. and Ro, R. S. *Chem. Ind. (London)* **1956**, 251. Eliel, E. L. and Lukach, C. A. *J. Am. Chem. Soc.* **1957**, *79*, 5986.

178. Seeman, J. I. *Chem. Rev.* **1983**, *83*, 83.

179. Seeman, J. I. *J. Chem. Educ.* **1986**, *63*, 42.

180. Curtin, D. Y. *Rec. Chem. Prog.* **1954**, *15*, 111.

181. Seeman, J. I. and Farone, W. A. *J. Org. Chem.* **1978**, *43*, 1854.

182. Young, W. G., Pressman, D., and Coryell, C. D. *J. Am..Chem. Soc.* **1939**, *61*, 1640.

183. Winstein, S., Pressman, D., and Young W. G. *J. Am. Chem. Soc.* **1939**, *61*, 1645.

184. Saunders, W. H. and Cockerill, A. F. *Mechanisms of Elimination Reactions*, Wiley, New York, 1973, pp. 332–376.

185. Shvo, Y. and Kaufman, E. D. *Tetrahedron* **1972**, *28*, 573.

186. Crowley, P. J., Robinson, M. J. T., and Ward, M. G. *Tetrahedron* **1977**, *33*, 915.

187. Rys, P. *Acc. Chem. Res.* **1976**, *9*, 345.

188. Gold, V. (a) *Pure Appl. Chem.* **1979**, *51*, 1725; (b) *ibid.* **1983**, *55*, 1281.

189. Seeman, J. I., Secor, H. V., Hartung, H., and Galzerano, R. *J. Am. Chem. Soc.* **1980**, *102*, 7741.

11

CONFIGURATION AND CONFORMATION OF CYCLIC MOLECULES

11-1. STEREOISOMERISM AND CONFIGURATIONAL NOMENCLATURE OF RING COMPOUNDS

2,2-Dimethylcyclopropanecarboxylic acid (Fig. 11.1, **A**), a derivative of the smallest cyclane (cyclopropane), has a chiral center at C(1) and exists as a pair of enantiomers; its stereoisomerism in no way differs from that of acyclic chiral molecules. The same might be said of cyclopropane-1,2-dicarboxylic acid, which has three stereoisomers: a meso form (Fig. 11.1, **B**) and a pair of enantiomers (**C, D**) diastereomeric with the meso form; the situation is the same as in tartaric acid. However, an additional feature results from the rigidity of the cyclic framework: In the meso diacid **B** the carboxyl groups are on the same side of the ring, whereas in the pair of enantiomers **C** and **D** they are on opposite sides. Therefore one may call the meso form cis and the two (isometric) chiral isomers trans. The Cahn–Ingold–Prelog (CIP)[1] system may, of course, be applied to cyclanes; thus enantiomer **A** is *R*, the meso form **B** is 1*R*,2*S* (equivalent to 1*S*,2*R*), and the enantiomers **C** and are 1*S*,2*S* and 1*R*,2*R*, respectively. The *E–Z* descriptors (Chapter 9) should not be used for cyclanes.

The CIP system is always unequivocal; nonetheless, when one deals with diastereomers (whether chiral or meso) the cis–trans nomenclature is easier to grasp. Unfortunately it has a built-in ambiguity analogous to the earlier discussed problem with the same nomenclature in alkenes (p. 341). Is compound **A** in Figure 11.2 to be called *cis*- or *trans*-2-hydroxy-2-phenylcyclopropanecarboxylic acid? The rule here[2] is that the "fiducial" (reference) substituent is to be marked by the prefix *r*- and that

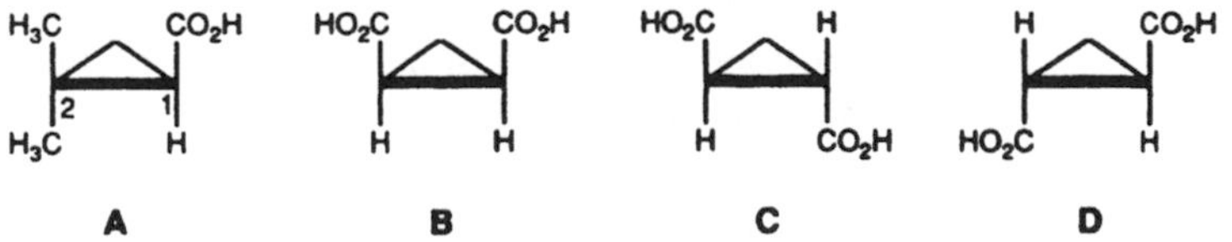

Figure 11.1. Stereoisomerism in cyclopropanes.

421

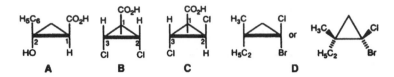

Figure 11.2.   Trisubstituted and tetrasubstituted cyclopropanes.

the positions of other substituents relative to it are to be denoted as *c*- (cis) or *t*- (trans). Since compound **A** is a cyclopropanecarboxylic acid, the carbon to which the CO_2H group is attached is the reference carbon and the compound is *t*-2-hydroxy-2-phenylcyclopropane-*r*-1-carboxylic acid or *t*-2-hydroxy-2-phenyl-*r*-1-cyclopropanecarboxylic acid. (The symbol *c* for the phenyl group may be omitted since it necessarily follows from that of the geminal hydroxyl group being *t*.) It should be noted that even if the name and numbering are assigned incorrectly, for example, if the compound were named *r*-1-hydroxy-1-phenyl-*t*-2-carboxycyclopropane, use of the *r*- symbol ensures that the correct stereochemistry can be derived unequivocally from the name given.

The *r* prefix is also useful when there are more than two stereogenic centers in the ring, as in Figure 11.2 **B**. Confusion might result here because the chlorine substituents are trans to the carboxyl group but cis to each other. However, the name *t*-2,*t*-3-dichlorocyclopropane-*r*-1-carboxylic acid is unequivocal: The reference point is the carboxylic acid group and both chlorine substituents are trans to it. In stereoisomer **C** in Figure 11.2, there is a question as to which way around the ring should be numbered: The rule used here is that cis precedes trans, so the compound is properly called *c*-2,*t*-3-dichlorocyclopropane-*r*-1-carboxylic acid (or *c*-2,*t*-3-dichloro-*r*-1-cyclopropanecarboxylic acid). A compound in which there might be a question both about the numbering system and about the reference group is **D** (Fig. 11.2), since there is no suffix (such as -carboxylic acid, -carboxaldehvde, -carbinol, or -ol) in its name; in such cases the reference carbon is the one substituted with the highest priority (CIP) group (here Br), thus the proper name is *r*-1-bromo-1-chloro-*c*-2-ethyl-2-methylcyclopropane.

If it is not known whether a substituent in a cyclane is cis or trans relative to other substituents, this is indicated by a wiggly line from the ring to the substituent (in lieu of a solid line or wedge for a substituent located in front of or above the plane of the ring and a dotted, cross-hatched, or dashed line for one behind or below that plane as in Fig. 11.2, **D**). The symbol used for such a substituent is ξ (Greek xi) in lieu of *c* or *t*.

In a four-membered ring (Fig. 11.3) a new feature emerges: While the stereoisomerism in 1,2-disubstituted cyclobutanes (e.g., Fig. 11.3, **A**) is analogous to that in 1,2-disubstituted cyclopropanes, a different situation arises in 1,3-disubstituted cyclopropanes (Fig. 11.3, **B, C**). Although these compounds exist as cis (**B**) and trans (**C**) diastereomers, both **B** and **C** are achiral; C(1) and C(3) are stereogenic but they are not chiral centers. One should *not* call such diastereomers meso forms, in as much as the set does not contain any chiral isomers.

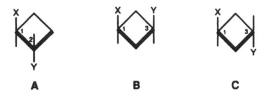

Figure 11.3.   1,2- and 1,3-Disubstituted cyclobutanes.

cis–trans Isomerism without accompanying chirality occurs in all saturated carboxylic rings of n members when n is even and the substituents are at positions 1 and $1 + (n/2)$. The IUPAC (International Union of Pure and Applied Chemistry) definition[2] specifies that "rings are to be considered in their most extended form; reentrant angles are not permitted," but there is the difficulty that, while planar graphs are easily written in the "most extended form," molecular models are often not easily forced into that form (the reader might experiment with a model of cyclooctane) and real molecules are, as we shall see later, not likely to exist in such a conformation at all.[3]

Parochial (local) systems of nomenclature have been used with sugars, steroids, and certain other polycyclic systems and will be introduced as we deal with these systems.

11-2. DETERMINATION OF CONFIGURATION OF SUBSTITUTED RING COMPOUNDS

a. Introduction

Just as with acyclic molecules, methods of determining relative (cis or trans) configuration frequently also depend on conformation. Most notably this is true of spectral (e.g., NMR) methods, whose consideration will therefore be postponed to later parts of this chapter (e.g. Section 11-4.d). Cyclopropane derivatives, in which the ring must necessarily be planar, since three points define a plane, will best serve to illustrate methods that are not dependent on conformational considerations. Only methods for assigning the (relative) configuration of diastereomers will be discussed here; determination of the absolute configuration of enantiomers in cyclic systems is no different, in its principles, from similar determination in acyclic systems (Section 5-3).

Six methods for the determination of cis–trans configuration in cyclic molecules may be recognized: (a) testing resolvability, (b) determining the number of isomers obtained in certain chemical transformations, (c) establishing the ease of bridge formation, (d) drawing conclusions from physical, including spectral, data, (e) drawing conclusions from considerations of reaction mechanism, and (f) correlating one compound chemically or physically with another.

b. Symmetry-Based Methods

The first two methods (a and b) depend on symmetry considerations (see also Section 5-5.c and Chapter 8) and are therefore unequivocal. Among the cyclopropane-1,2-dicarboxylic acids (Fig. 11.1, **B–D**) only the trans isomer can be resolved, since the cis (meso) isomer has a symmetry plane; successful resolution will therefore lead to assignment of (trans) configuration. The converse is not true: Nonresolvability does not guarantee that one is dealing with the meso isomer, since resolution may have failed for technical reasons. This point is particularly vexing if only the achiral meso isomer is at hand: Positive proof of its configuration by the resolvability criterion is evidently impossible. A useful trick here is to convert the achiral compound into a chiral one, resolve the latter, and then reconvert it to its achiral precursor. Thus conversion of the meso acid (Fig, 11.1, **B**) into a monomethyl ester will make it chiral; of course the monomethyl ester of the chiral acid (**C, D**) is chiral also. If one can resolve the monomethyl ester, then, regardless of which diacid was at hand, the configurational problem can be solved: The resolved monomethyl ester of the meso acid **B** will give an optically inactive acid upon hydrolysis, whereas the resolved ester of the chiral acid **C** or **D** will give an optically active acid when hydrolyzed.

A number of symmetry-based methods of distinguishing meso and chiral isomers on the basis of the spectroscopically observable heterotopicity (or lack thereof) of appropriate nuclei, either in the parent compound itself or in a derivative, have been described in Section 8-4-b. Similar *chemical* methods have been known for many years; for example, Wislicenus[4] one of the pioneers of stereochemistry, determined the configuration of the 2,5-dimethylcyclopentane-1,1-dicarboxylic acids by decarboxylating them to the corresponding monocarboxylic acids (Fig. 11.4): The

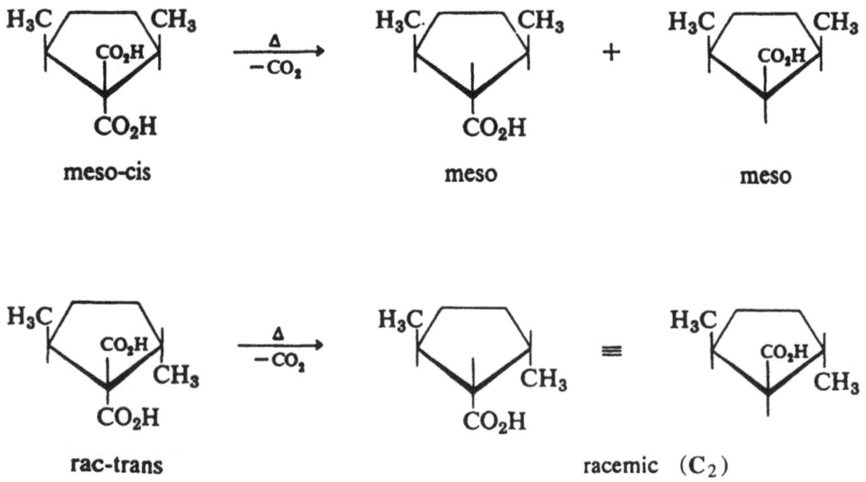

Figure 11.4.   Proof of configuration of 2,5-dimethylcyclopentane-1,1-dicarboxylic acids.

meso (cis) acid, which has diastereotopic CO_2H groups, gives rise to two different meso products, whereas the racemic trans acid, in which the CO_2H groups are homotopic, yields a single racemic product (Fig. 11.4, only one enantiomer shown).

Thus, while symmetry-based methods are in principle unequivocal, they may yet meet with practical difficulties. The situation is even less safe, however, in the case of methods discussed below, which are based on chemical reactivity, physical properties, or mechanistic considerations. Since such methods are usually based on spatial proximity, or on a definitive orientation in space of salient substituents, they are basically indicators of conformation rather than configuration. Thus they can be used for the determination of configuration only if conformational factors are well understood.

c. Methods Based on Physical and Chemical Properties

In *cis*- and *trans*-cyclopropane-1,2-dicarboxylic acids (Fig. 11.1, **B** and **C** or **B** and **D**), the carboxyl groups are close to each other in the cis isomer but not in the trans. Therefore one might expect the cis but not the trans isomer to form a cyclic anhydride; such is, in fact, the case. Also one might expect the cis isomer, because of its eclipsed carboxyl groups, to be less stable; indeed, the cis isomer is converted, by heating with mineral acid, to the trans isomer. Furthermore, passing from chemical to physical properties, one would expect the cis isomer to be more acidic in its first ionization constant (since the monoanion may be stabilized by a through-space effect, either electrostatic or hydrogen bonding, with the free CO_2H), and this is the case: $pK_1^{cis} = 3.56$; $pK_1^{trans} = 3.80$.[5] In addition, one might expect the acidity of the second carboxyl group to be higher for the trans acid than for the cis, since double ionization in the latter case leads to CO_2^- groups that are proximal in space and highly repulsive. Again the expectation is fulfilled:[5] $pK_2^{cis} = 6.65$; $pK_2^{trans} = 5.08$.[5] The ratio K_1/K_2 is 1210 for the cis isomer and 19.4 for the trans. This ratio is a much more reliable indicator of proximity than are K_1 and K_2 individually; there is a relationship, called Bjerrum's law,[6] which correlates the distance of the carboxyl groups with the K_1/K_2 ratio:

$$\ln(K_1/4K_2) = Ne^2/\varepsilon r \, RT \qquad \text{Bjerrum's law}$$

where N is Avogadro's number, e is the charge of an electron, ε is the effective dielectric constant of the medium, R is the gas constant, T is the absolute temperature, and r is the distance between the carboxylic acid groups.

In the formulation of Bjerrum's law the free energy difference between the first and second ionization processes, corrected by a statistical factor of 4, is set equal to the electrostatic interaction energy of the two carboxylate charges as given by Coulomb's law. The statistical factor of 4 comes from the fact that there are two possible monoanions but only one dicarboxylic acid and one dianion; thus the dissociation constant K_1 of the dicarboxylic acid is favored by a statistical factor of 2 but that, K_2, of the monoanion is disfavored by a factor of 2; the factor of 4 in the denominator compensates both for the statistically enhanced K_1 and the similarly diminished K_2.

Rings larger than three membered are, in general, not planar (see individual discussions in Sections 11-5.b–d) and so the criteria just discussed (anhydride or other ring formation from the cis isomer, greater stability of the trans isomer, larger K_1/K_2 ratio for the cis isomer) tend to become less certain as ring size increases from four to seven and more. Thus Perkin, Jr.[7] found that both in the cyclobutane-1,2-dicarboxylic and the cyclopentane-1,2-dicarboxylic acids, the cis but not the trans isomer forms an anhydride on heating or mild treatment with acetyl chloride. that the anhydride is readily reconverted to the cis acid on treatment with water, and that the cis acid is converted to the trans acid on heating with mineral acid. However, Baeyer[8] encountered a less clear-cut situation with cyclohexane-1,2-dicarboxylic (hexahydrophthalic) acid: The trans isomer is indeed more stable than the cis (from which it is formed by treatment with hot mineral acid) but *both* acids now form (different) anhydrides, though that from the cis acid is formed more easily, whereas that formed from the trans acid is less stable (apparently more strained) and converted to the cis anhydride on prolonged treatment with acetyl chloride. By the same token, the ratio K_1/K_2 (Table 11.1) for the cis versus trans acids shows much more pronounced differences in the cyclopropane series than for the larger ring diacids. Perhaps somewhat unexpectedly, this ratio diverges again in the seven- and eight-membered rings, with K_1/K_2 being extremely large for the cis diacids. It would thus appear that conformational factors lead back to eclipsing of the cis carboxyl groups in such rings.

Intramolecular hydrogen bonding in vicinal diols, already discussed for acyclic diastereomers in Chapter 10, is also useful for configurational assignment in the cyclic analogues, as shown in Table 11.2. In the case of cyclobutane-1,2-diol and cyclopentane-1,2-diol, consideration of models indicates that only the cis but not the trans isomer is capable of intramolecular hydrogen bonding and assignment of configuration can readily be made on this basis. In the case of the cyclohexane-, cycloheptane-, and cyclooctane-1,2-diols the situation is not so extreme: Both the cis

TABLE 11.1. Dissociation Constants of Cycloalkane-1,2-dicarboxylic Acids

1,2-Dicarboxylic Derived from	Acid	pK_1	pK_2	K_1/K_2	References
Cyclopropane	cis	3.56	6.65	1210	5
	trans	3.80	5.08	19.4	5
Cyclobutane	cis	4.16	6.23	130	9
	trans	3.94	5.55	41	9
Cyclopentane	cis	4.42	6.57	138	10
	trans	4.14	5.99	70	10
Cyclohexane	cis	4.44	6.89	282	11, 12
	trans	4.30	6.06	58	11, 12
Cycloheptane	cis	3.87	7.60	5370	11, 12
	trans	4.30	6.18	76	11, 12
Cyclooctane	cis	3.99	7.34	2240	11, 12
	trans	4.37	6.24	74	11, 12

TABLE 11.2.   Hydroxyl Stretching Frequencies ν in the Infrared[a]

Cycloalkane-1,2-Diol Derived from		ν_{free}	ν_{bonded}	$\Delta\nu$
Cyclobutane[b]	cis	3640	3580	60[c]
	trans	3610		c
Cyclopentane	cis	3633	3572	61
	trans	3620		c
Cyclohexane	cis	3626	3588	38
	trans	3633	3600	33
Cycloheptane	cis	3632	3588	44
	trans	3626	3589	37
Cyclooctane	cis	3635	3584	51
	trans	3631	3588	43

[a]In reciprocal centimeters (cm^{-1}). Reference 13 unless otherwise noted.
[b]Reference 14.
[c]No intramolecular hydrogen bond formed.

and the trans diol are capable of forming intramolecular hydrogen bonds, but the bond in the cis diol is stronger. The cis isomer therefore shows the greater difference in IR stretching frequency between unbonded and intramolecularly bonded OH. However, these generalizations break down for rings larger than eight-membered because of their flexibility.

^{13}C NMR spectra of disubstituted cyclanes (Table 11.3) are useful in permitting the following generalizations (at least up to and including seven-membered rings): In the 1,2-disubstituted series, the resonances of the cis isomers are upfield of corresponding compounds for the trans isomers, whereas the reverse is true for 1,3-disubstituted compounds.

While physical and spectral properties are often suggestive of relative configuration in rings, they are rarely conclusive.

d. Correlation Methods

In some instances it is convenient to establish the configuration of cyclic molecules by correlative methods, similar to those described in connection with the determination of absolute configuration in Section 5-5.b. For example, the configurations of *cis*- and *trans*-3-hydroxycyclohexanecarboxylic acids are readily established in as much as the cis acid (Fig. 11.5) forms a five-membered lactone but the trans acid does not. Chemical correlation of these acids with the 3-methylcyclohexanols (shown for the cis isomer in Fig. 11.5) establishes the configuration of the latter.[20] One must, of course, make sure that there is no epimerization (change of configuration at one or other of the chiral centers) in the course of the transformation used for the correlation. In the case depicted in Figure 11.5, this was assured by carrying out the correlation with both the cis and the trans acid; but even if only one of the stereoisomers is available for correlation, the absence

TABLE 11.3. The ^{13}C Resonances of Dimethylcycloalkanes^a

Compound		C_1	C_β^b	C_4	C_5	CH_3	References
Cyclopropane	cis-1,2	9.8	13.6			13.0	15, 16
	trans-1,2	14.2	14.6			19.0	15, 16
Cyclobutane	cis-1,2	32.2	26.6			15.4	17
	trans-1,2	39.2	26.8			20.5	17
	cis-1,3	26.9	38.5			22.5	17
	trans-1,3	26.1	36.4			22.0	17
Cyclopentane	cis-1,2	37.9	33.5	23.5		15.4	18a
	trans-1,2	43.0	35.3	23.6		19.0	18a
	cis-1,3	35.7	45.3	34.6^c		21.4^c	18a
	trans-1,3	33.8	43.4	35.5^c		21.7^c	18a
Cyclohexane	cis-1,2	34.6	31.7	23.9		15.9	19a
	trans-1,2	39.7	36.2	27.0		20.4	19a
	cis-1,3	33.0	44.9	35.6	26.6	23.0	19a
	trans-1,3	27.2	41.6	34.1	20.9	20.7	19a
Cycloheptane	cis-1,2	37.5	34.0	26.6	29.2	17.9	18b
	trans-1,2	41.3	35.8	26.7	29.7	22.6	18b
	cis-1,3	34.2	46.9	37.4^c	26.5^c	24.9	18b
	trans-1,3	31.1	44.7	37.5^c	29.1^c	24.3	18b

^{a}In parts per million from tetramethylsilane (TMS).
^{b}The C(3) atom in a 1,2-disubstituted compound and the C(2) atom in a 1,3-disubstituted compound.
^{c}These entries represent reversals from the expected order.

428

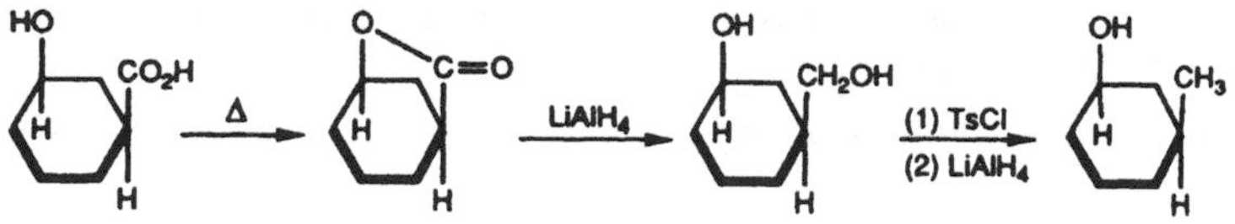

Figure 11.5.   Correlation of *cis*-3-hydroxycyclohexanecarboxylic acid with *cis*-3-methylcyclohexanol.

of epimerization (leading to formation of a second diastereomer) can nowadays be monitored by chromatographic or spectroscopic (especially ^{13}C NMR) methods.

Conclusions concerning configuration deduced from evidence of reaction mechanism usually involve consideration of conformation and so will be discussed later (e.g., Section 11-4.e).

11-3. STABILITY OF CYCLIC MOLECULES

a. Strain

It was von Baeyer[21] who first pointed out that the construction of a small ring compound involves strain (*Spannung*). For example, in cyclopropane the angle between the carbon–carbon bonds must, for geometrical reasons, be 60°. (This is the angle between lines connecting the nuclei, i.e., the internuclear, not the interorbital angle, see Section 11-5.a). Since the "normal" bond angle at carbon is 109°28′ (tetrahedral), there is a deviation of 49°28′ from the norm. Baeyer apportioned this strain equally between the two ring bonds flanking the 60° angle and thus called the strain 24°44′ (one-half of 49°28′). In a similar fashion, he reported strain values for other carbocyclic rings as shown in Figure 11.6. This picture, and the very concept of "angle strain," was evidently based on the Kekulé models used at the time of Baeyer.[22]

As discussed in Chapter 2 (eq. 2.1), strain introduced in a molecule in any fashion tends to be minimized by becoming distributed among several modes, such as bond strain, angle strain, torsional strain, and van der Waals compression. Thus the strain in cyclanes is actually not purely angle strain; and so it becomes desirable to define strain in an entirely different manner, in terms of energy. Strain is the excess of observed over "calculated" heat of formation (or more conveniently from the practical

24°44′	9°44′	0°44′	−5°16′	"angle strain"
9.17	6.58	1.24	0.02	strain per CH$_2$ group^a in kcal mol^{-1}
38.4	27.5	5.19	0.09	strain per CH$_2$ group^a in kJ mol^{-1}

^{a}See Table 11.4.

Figure 11.6.   "Angle strain" (Baeyer) and actual strain per methylene group.

point of view, heat of combustion). The question, then, becomes what is meant by the "calculated" enthalpy value. For a cycloalkane, this is conveniently, though somewhat arbitrarily, taken as the heat of formation or heat of combustion of a CH_2 group multiplied by the number of carbon atoms in the ring. (See also refs. 23–27.) The value for a CH_2 group, in turn, is obtained by taking the difference between a large straight-chain hydrocarbon, $CH_3(CH_2)_nCH_3$, where $n > 5$, and its next lower homologue. This difference is quite constant, 157.44 kcal mol^{-1} (658.73 kJ mol^{-1}) for the heat of combustion or 4.93 kcal mol^{-1} (20.6 kJ mol^{-1}) for the heat of formation in the vapor state[27,28]; the liquid-phase value is 6.09 kcal mol^{-1} (25.5 kJ mol^{-1}).[27] The strain thus calculated (Table 11.4), expressed either as total strain or as strain per CH_2 group, is high for cyclopropane, drops to near zero as one proceeds to cyclohexane, increases again to a maximum in the cyclooctane to cycloundecane region and then drops again, reaching values near zero from 14-membered rings on up. The strain in three-, four-, and five-membered rings indeed runs parallel with Baeyer's prediction (Fig. 11.6) but the six-membered ring is nearly strainless. The Kekulé models that Baeyer used apparently concealed the fact that a puckered (chair or boat shaped) form of cyclohexane can be constructed that is free from angle strain. (The bond angles in a nonplanar ring are always smaller than those in a planar one; thus puckering reduces the angle in cyclohexane from 120° to near tetrahedral.) No more than five years after the publication of Baeyer's paper, however, Sachse[31] had the insight to recognize that nonplanar cyclohexane can be strainless, or at least has no angle strain. We shall return to this point later.

The data in Table 11.4 suggest that ring compounds might be divided into four families. The three- and four-membered rings are obviously highly strained; they are classified as "small rings." There is relatively little strain in the five-, six-, and seven-membered rings (so-called "common rings"), which abound in natural as well as synthetic products. Strain increases again in the C_8–C_{11} family; rings of this size are named "medium rings" and the larger, nearly strainless ones (C_{12} and larger) are called "large rings." (For the nomenclature, see ref. 32.)

One might question why the medium rings are so strained, even though it is easy to construct puckered models for them that are free of angle strain (Baeyer strain). The answer, in light of what is now known about molecular mechanics (Section 2-6) is as follows: Models of medium rings that are free of angle strain tend to have large numbers of pairs of eclipsed hydrogen atoms on adjacent CH_2 groups; moreover, with rings in the C_8–C_{11} range, such hydrogen atoms will tend to "bump" each other across the ring, leading to so-called "transannular strain"[33]; this strain arises from the van der Waals compression (nonbonded energy) term V_{nb} in Eq. 2.1. Energy minimization (Section 2-6) will, of course, tend to minimize the total strain; therefore angle deformation will occur, with some development of angle strain that is more than compensated for by a concomitant reduction in eclipsing (Pitzer strain) and transannular (van der Waals) strain. The picture of a puckered ring with normal bond angles suggested by mechanical molecular models is therefore incorrect (just as Baeyer's picture of planar rings was incorrect); in fact, bond angles in cyclononanes and cyclodecanes are as large as 124°.[34,35] Models tend to lead to overestimation of

TABLE 11.4. Heat of Combustion and Ring Strain for Cyclanes

Size of Ring (n)	Heat of Combustion[a]		Total Strain		Strain per CH_2[d]	
	(kcal mol⁻¹)	(kJ mol⁻¹)	(kcal mol⁻¹)[b]	(kJ mol⁻¹)[c]	(kcal mol⁻¹)	(kJ mol⁻¹)
3	499.83	2091.3	27.5	115.1	9.17	38.4
4	656.07	2745.0	26.3	110.1	6.58	27.5
5	793.40	3319.6	6.2	26.0	1.24	5.19
6	944.77	3952.9	0.1	0.5	0.02	0.09
7	1108.3	4637.3	6.2	26.2	0.89	3.74
8	1269.2	5310.3	9.7	40.5	1.21	5.06
9	1429.6	5981.3	12.6	52.7	1.40	5.86
10	1586.8	6639.1	12.4	51.8	1.24	5.18
11	1743.1	7293.3	11.3	47.3	1.02	4.30
12	1893.4	7921.9	4.1	17.2	0.34	1.43
13	2051.9	8585.0	5.2	21.5	0.40	1.66
14	2206.1[e]	9230.9[e]	1.9	8.0	0.14	0.57
15	2363.5	9888.7	1.9	7.8	0.13	0.51
16	2521.0	10547.7	2.0	8.0	0.12	0.50
17	2673.2	11184.5	-3.3	-13.9	-0.19	-0.82

[a]Gas phase values. From ref. 29, p. n-1960.
[b]Heat of combustion minus 157.44n.
[c]Heat of combustion minus 658.7n.
[d]Total strain divided by n.
[e]Corrected values: see ref. 30.

angle and underestimation of torsional strain and, in the case of non space-filling models, van der Waals strain. Rings of specific sizes will be discussed later in Sections 11-4 and 11-5.

b. Ease of Cyclization as a Function of Ring Size

The relative thermodynamic instability of medium-size rings has made their derivatives relatively difficult to obtain, the first successful general synthesis (acyloin reaction) having been described only in the 1940s by Prelog and by Stoll[36] although, since then, numerous preparations have been developed. Large rings, despite their low strain, present synthetic difficulties as well. It was recognized many years ago[37] that a complicating factor is the difficulty of getting the ends of a long chain to approach each other: The conformational entropy of a chain compound is greater than that of a ring. A competing reaction in ring closure tends to be dimerization or oligomerization, since, if in X-$(CH_2)_n$Y the functional groups X and Y are capable of interacting to form a ring, there is also the possibility of similar interaction between two molecules to form a dimer. These competing reactions are not much of a problem in small and common ring compounds where the loss of translational entropy in a dimerization or oligomerization reaction tends to be much more severe than the loss of conformational entropy in cyclization. But this situation is reversed in medium and especially in large rings, where the possibility of rotation about a large number of bonds leads to a high conformational entropy in the open-chain precursor as well as the linear dimer or oligomer, which is largely lost in the cyclic product. Cyclizations leading to such rings must therefore frequently be carried out in high dilution, where bimolecular reactions tend to be suppressed (but, unfortunately, operating conditions make for long reaction times).[38,39] Here again a number of new synthetic methods have been developed,[40] principally with the impetus of synthesizing large-ring lactones that occur in a number of natural products known as macrolides.[41]

Both ring strain (3 > 4 > 5 > 6) of the cyclic product, an enthalpy factor, and ease of approach of the ends of a chain to close a ring (3 > 4 > 5 > 6), an entropy factor, affect facility of cyclization. As a result of the counterplay of these factors ease of ring closure is usually in the order three-, five-membered greater than four- or six-membered rings. Since the high rate of ring closure of three- and five-membered rings is based on different factors, predictions as to which closure will occur more rapidly in a given case are unsafe. Six-membered rings usually cyclize less rapidly than five-membered rings, but a firm prediction on this point is risky since (Table 11.4) the strain in the six-membered ring is appreciably less; however, this is usually more than outweighed by the much greater loss of entropy in the formation of the six-membered ring. A factor that must be considered is thermodynamic versus kinetic control; since six-membered rings tend to be more stable than five-membered rings, their formation may be favored thermodynamically, even if not kinetically. Moreover, since enthalpy and entropy factors work in opposite directions in the 6:5 equilibrium, such equilibria might be expected to be strongly temperature dependent: Since $K = e^{-\Delta H^\circ/RT} e^{\Delta S^\circ/R}$, as ΔH° favors the six-membered ring and ΔS° favors the five-membered ring, it is clear that lowering the temperature should shift the equilibrium toward the six-membered ring. An example is the acetalization of glycerol with isobutyraldehyde; the reaction scheme is shown in Figure 11.7.

The activation energy is less for formation of the five-membered ring, hence this ring is formed faster (i.e., it is the product of kinetic control: $k_5 > k_6$). However, the reaction is reversible and the six-membered ring is slightly more stable ($K < 1$), so after some time it will become the major product (product of thermodynamic control; both five- and six-membered ring products will exist as a mixture of cis and trans isomers). In the equilibrium shown in Figure 11.7 (bottom), since enthalpy favors the nearly strain-free six-membered ring (whereas entropy favors the more flexible five-membered ring), lowering the temperature should shift the equilibrium toward the six-membered ring, as was indeed found to be the case.[42]

c. Ease of Ring Closure as a Function of the Ring Atoms and Substituents: The Thorpe–Ingold Effect

One of the major sources of strain in medium rings is the transannular repulsion of the hydrogen atoms of the CH_2 groups. Therefore replacement of such groups by heteroatoms (O, S, or NH) or sp^2 hybridized carbon atoms (C=O, –CH=CH–) would be expected to reduce the strain. There is a dearth of thermochemical information in this regard, but it is known that the presence of several (not just one) such elements facilitates ring closure. Thus the rates of cyclization to phenolic ethers of the compounds shown in Figure 11.8 do not show the customary minimum in the region of the medium rings, but rather display a nearly monotonic decrease with ring size, coming to a plateau with the large rings.[43]

Another interesting manifestation of the effect of changing the nature of the groups in the ring is the "Thorpe–Ingold effect" or "gem dialkyl effect."[44] In the formation of small rings, there must occur a substantial reduction in C–C–C bond angle (see above). However, the reductions in angle postulated by Baeyer (Fig. 11.6) for these

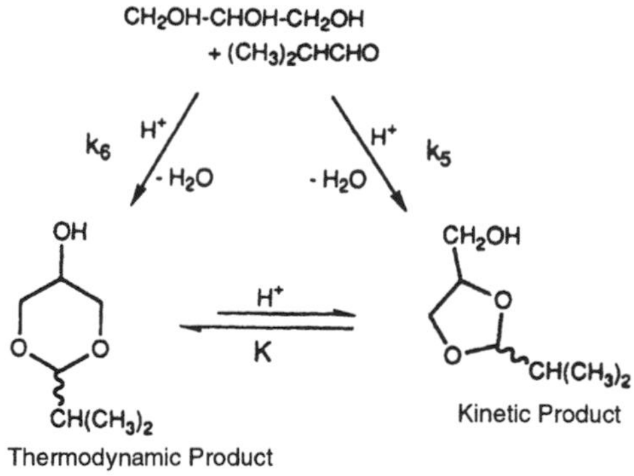

Figure 11.7.   Reaction of glycerol with isobutyraldehyde.

$X = O$ or CH_2

Figure 11.8. Cyclization of phenolic mono- and di-ethers.

rings are actually not quite correct, since the normal bond angle in propane is 112.5°, not 109.5°; thus the strain is larger than calculated by Baeyer. This situation changes when the fragment that cyclizes, $-CH_2CR_2CH_2-$ or $-CH_2CR_2CH_2CH_2-$, bears geminal methyl ($R = CH_3$) or higher alkyl groups instead of hydrogen ($R = H$). Here the carbon that is geminally substituted resembles the central carbon in neopentane at which the bond angle (for symmetry reasons) is reduced to the tetrahedral angle, 109°28′. This decreases the angle deformation (i.e., strain) incurred upon cyclization; in other words, the formation of rings bearing gem-dialkyl groups should be easier than it would be in the absence of such groups. Although the effect might appear small, it does, in fact, manifest itself, as shown in Table 11.5 for halohydrin cyclization[45]; it is also seen in other contexts.

d. Baldwin's Rules

In 1970, Eschenmoser et al.[46] observed that a compound of type $\overline{Y}(CH_2)_n\overset{+}{X}-CH_3$ would not undergo an intramolecular S_N2 displacement [to give $H_3C-Y(CH_2)_n-X$] because of the virtual impossibility of approaching a collinear arrangement of the nucleophile Y^- and the leaving group X^+ (cf. Fig. 5.38). In a generalization and expansion of this observation, Baldwin in 1976 enunciated rules of ring closure bearing his name.[47] For the purpose of classifying ring closures, Baldwin distinguishes "exo" and "endo" cases, as shown in Figure 11.9.[48] In endo ring closures, the nucleophile at one end of the cyclizing chain attaches itself to the other end of that chain, whereas in exo ring closures, attachment is at the penultimate atom and the terminal (end) atom remains outside of the newly formed ring. Also, the reactions are classified as "tet," "trig," or "dig" depending on whether ring closure

TABLE 11.5.   Relative Rates of Ring Closures of Chlorohydrins

Compound	Relative Rate
$HOCH_2CH_2Cl$	1
$HOCH_2CHClCH_3$	5.5
$CH_3CHOHCH_2Cl$	21
$HOCH_2CCl(CH_3)_2$	248
$(CH_3)_2COHCH_2Cl$	252
$(CH_3)_2COHCHClCH_3$	1360
$CH_3CHOHCCl(CH_3)_2$	2040
$(CH_3)_2COHCCl(CH_3)_2$	11,600

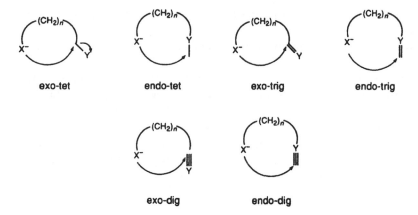

Figure 11.9. Exampies of exo- and endo-tet, -trig, and -dig ring closures.

involves displacement at a tetrahedral atom or addition to a trigonal (sp^2 hybridized) or digonal (sp hybridized) moiety. Finally, the size of the ring to be formed is indicated by an appropriate leading numeral.

The Eschenmoser case indicates, then, that a 6-endo-tet reaction is disfavored (note that this would not actually be a ring closure) and this is true for other n-endo-tet reactions, where $n \leq 6$ because of the great difficulty in achieving the appropriate collinearity of the incoming and outgoing groups (see above). On the other hand, the n-exo-tet reactions are favorable even for small n values (3–7).

The situation for trig reactions is similar, though for different reasons. According to the Bürgi–Dunitz trajectory[49] (see also p. 469), approach of a nucleophile to a double bond does not occur orthogonally but at an angle of about 109°. This angle of approach again makes the exo approach more favorable than the endo, and it is therefore not surprising that, whereas endo-trig reactions are disfavored for rings five-membered and smaller (but not necessarily in formation of six- and

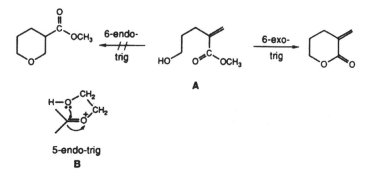

Figure 11.10. Limitations of Baldwin's rules.

seven-membered rings), exo-trig reactions are favored.[47b,50] An example is shown in Figure 11.10, **A**.

In digonal (*sp*) systems, however, the angle of nucleophilic approach appears to be less than 90°, rather than 109° (*sp²*) or 180° (*sp³*).[47a] Therefore, in this case, the exo-dig approach is disfavored for small rings (three- or four-membered) though it seems to be possible for larger rings; the endo-dig process is generally favored.

Baldwin's rules have been applied to a variety of reactions: Conjugate additions of oxygen nucleophiles,[51] endocyclic alkylations of ketone enolates,[52] and intramolecular aldol condensations.[53] However, the rules probably do not apply to nonconcerted reactions, such as the second step in the ring closure of dioxolanes (Fig. 11.10, **B**), where an oxycarbenium ion intermediate is involved.[48]

11-4. CONFORMATIONAL ASPECTS OF THE CHEMISTRY OF SIX-MEMBERED RING COMPOUNDS

a. Cyclohexane

Three-membered rings are of necessity flat, but all other rings (from four-membered rings on up) are nonplanar, and thereby present important conformational aspects affecting physical and chemical properties. We consider the six-membered ring first, not only because of its wide occurrence in natural, as well as purely synthetic substances, but also because its conformation is easier to study than that of either smaller or larger rings. This is so because the well-known chair conformer of cyclohexane lies in a deep energy valley, such that chemical changes at the periphery of the ring are unlikely to change the conformation of the ring itself [except (see below) possibly to invert the chair into an inverted chair]. Indeed, it is because of the rigidity of the chair conformer that the entropy of cyclohexane is much lower than one would calculate on the basis of constant entropy increments in the homologous series of cyclanes.

As we have already seen, Baeyer considered cyclohexane a strained planar molecule, but Sachse, nearly 100 years ago, recognized that it might be chair or boat shaped (Fig. 11.11) and unstrained.[31] Sachse died (in 1893) soon after publication of his pioneering papers and his ideas died with him, only to be resurrected a quarter

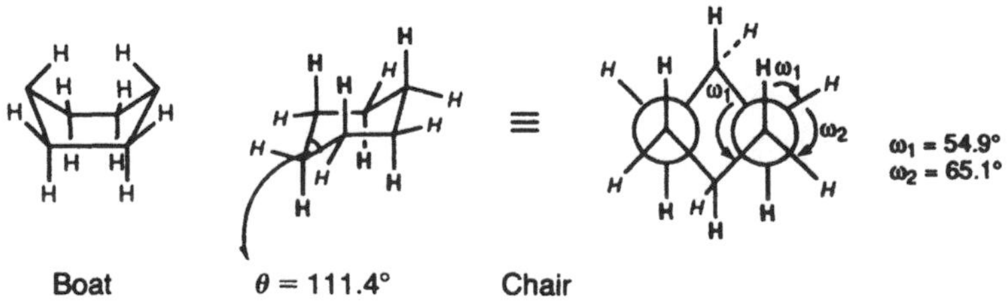

$\omega_1 = 54.9°$
$\omega_2 = 65.1°$

Boat $\theta = 111.4°$ Chair

Figure 11.11. Cyclohexane. **H**, axial hydrogen atoms; *H*, equatorial hydrogen atoms in chair conformer.

century later by Mohr.[54] The difficulty with Sachse's proposal was that, in light of what was known in 1890, it had to be interpreted in terms of a *rigid* chair model, which should have given rise to two different monosubstituted cyclohexyl derivatives, such as hexahydrobenzoic acid: one (to use present terminology) equatorially substituted, the other axially substituted. For now obvious reasons (to be detailed below) such isomers were, however, never found, and in a widely used textbook of the time[55] it was stated that, "The non-existence of two forms of hexahydrobenzoic acid makes Sachse's notions untenable." It was recognized only many years later that rotation around single bonds (cyclohexane inversion, which interconverts the two forms of hexahydrobenzoic acid, being a process of this type) could be rapid though not instantaneous. In fact, a well-known textbook of a much later time[56] errs in the opposite direction: "It seems probable that there is an equilibrium between the two forms and that the two models vibrate from one to the other so rapidly that the net average is a planar molecule." However, as pointed out in Section 2-1, the axially and equatorially monosubstituted cyclohexanes are spectroscopically distinguishable molecules; in fact, in the case of chlorocyclohexane, the equatorial isomer has been obtained in pure crystalline form and the axial isomer has been enriched in the mother liquor of crystallization; the two conformers are quite stable at $-150°C$.[57]

The idea that cyclohexane was, in fact, a chair-shaped molecule gained much ground based on a variety of physical and chemical experiments carried out in the 1920s, 1930s, and 1940s, but it was not until the appearance of a pioneering paper by Barton[58] that the physical and chemical consequences of the chair conformations were fully realized. It is with these consequences that this section is mainly concerned; the detailed history of earlier findings that led up to Barton's insight has been dealt with elsewhere.[22,59-61]

Prominent among the approaches to establishing the chair shape of the cyclohexane ring has been Hassel's work using electron diffraction.[62] Application of this technique to cyclohexane itself[63,64] established the molecule to be a slightly flattened chair with bond angles (1971 determination) of $111.4 \pm 0.2°$ and torsion (C–C–C–C) angles in the ring of 54.9 ± 0.4.[65] The bond angle is larger than tetrahedral but smaller than the "normal" (i.e., optimal) bond angle in propane, $112.4°$; the torsion angle deviates from the optimum of $60°$. One might interpret this deviation from optimal values as follows: In propane, bond angle deformation and torsion are independent, so the bond angle can be expanded beyond the tetrahedral with the torsion angle still remaining at or near the $60°$ optimum. In cyclohexane, because of the constraint of the ring,[66] this is not possible: If the bond angle expands beyond the tetrahedral, the intraannular torsion angle must necessarily decrease below $60°$. Such a decrease causes torsional strain and will be resisted; the total strain is minimized when a compromise is reached with the bond angle being slightly "too small" ($111.4°$ instead of $112.4°$) and the torsion angle slightly "too small" also ($54.9°$ instead of $60°$).

There is a general relationship[67] between bond angles θ and adjacent torsion angles ω in cyclohexane: $\cos \omega = -\cos \theta/(1 + \cos \theta)$. Thus, for a bond angle near the tetrahedral, a variation in bond angle of $1°$ causes a variation in torsion angle of $2.5°$.

Closing the intraannular torsion angle to 54.9° brings with it a *decrease* in the (external) H–C–C–H torsion angle of cis located hydrogen atoms from 60° to 54.9° and of trans diaxially located hydrogen atoms from 180° to 174.9° and an *increase* for the corresponding trans diequatorially located hydrogen atoms from –60° to –65.1° (cf. Fig. 11.11). In view of the Karplus relationship (Fig. 10.26) this has an important impact on vicinal coupling constants in cyclohexane, which will be discussed later.

Cyclohexane has two geometrically different sets of hydrogen atoms, six extending up and down along the S_6 axis of the molecule (whose symmetry point group is $\mathbf{D_{3d}}$), called axial (a) hydrogen atoms and six alternating about an "equatorial" plane at right angles to this axis (not a symmetry plane!), which are called equatorial (e)[68] (Fig. 11.11). The axial hydrogen atoms are homotopic relative to each other since they can be brought into coincidence by operation of the symmetry axes (C_3, C_2) of the molecule; the same is true of the equatorial hydrogen atoms relative to each other. However, the axial set is diastereotopic with the equatorial set, the two types being related neither by a symmetry axis nor by a symmetry plane.

Sachse already recognized that a cyclohexane ring could invert (*snap over* as Hassel later called it or *flip*) from one chair form to another (Fig. 11.12). Although he did not explicitly appreciate the ease of this process, he called this motion "version." Today the terms "ring inversion" or "ring reversal" are preferred. When ring reversal occurs, the set of axial hydrogen atoms becomes equatorial and the equatorial set becomes axial; that is, the two diastereotopic sets exchange places: topomerization occurs (cf. Chapter 8). As explained in Section 8-4.d, processes of this type can be detected and the barrier to inversion measured, by means of dynamic (variable-temperature) NMR. The barrier to ring reversal in cyclohexane has been determined a number of times by this method,[69,70] the most accurate results coming from determination in cyclohexane-d_{11}: The "best overall values" estimated for the kinetic parameters are $\Delta G^{\ddagger} = 10.25$ kcal mol^{-1} (42.9 kJ mol^{-1}) at –50°C to –60°C, $\Delta H^{\ddagger}$: $= 10.7$ kcal mol^{-1} (44.8 kJ mol^{-1}), $\Delta S^{\ddagger} = 2.2$ cal mol^{-1} (9.2 J mol^{-1} K^{-1}); that is, $\Delta G^{\ddagger} = 10.$ 1 kcal mol^{-1} (42.3 kJ mol^{-1}) at 25°C.[71] Calculations of the barrier by force field methods[69,72] span the same range of $\Delta H^{\ddagger}$ as the experimental values. The activated complex for the ring reversal is probably close to the half-chair depicted in Figure 11.13 but is actually quite flexible, as evidenced by the fairly large positive entropy of activation.

There are probably three factors contributing to the sizeable positive $\Delta S^{\ddagger}$. (a) The transition state is much less symmetrical than the starting cyclohexane, which has a symmetry number $\sigma = 6$ and hence an entropy of symmetry of $-R$ ln 6. (b) The

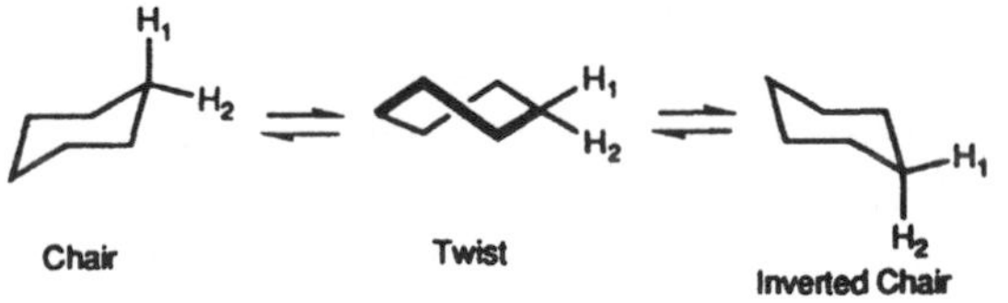

Chair   Twist   Inverted Chair

Figure 11.12. Cyclohexane inversion (reversal).

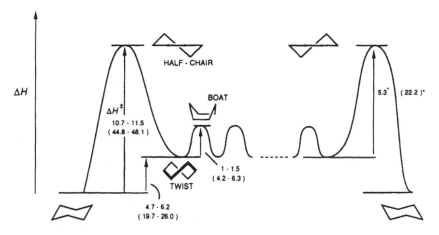

Figure 11.13. Energy profile [$\Delta H^{\ddagger}$ kcal mol^{-1} (kJ mol^{-1})] for cyclohexane ring reversal. The asterisk (*) indicates that $\Delta S^{\ddagger} = 0$ is assumed.

activated complex has a number of nearly equienergetic states and therefore a sizeable entropy of mixing, including the entropy that results because the half-chair is a racemic mixture. (c) The half-chair has a number of low-lying vibrational modes that contribute to its vibrational entropy (this point is related to the previous one).

As shown in Figure 11.12, the inversion of the cyclohexane chair[73] is not a one-step process. This inversion first produces an intermediate which, in a second step equienergetic with the first, leads to the inverted chair. Manipulation of models might suggest that the intermediate is Sachse's boat form, but force field calculations make this highly unlikely; rather, the intermediate is almost certainly the twist form depicted in Figures 11.12 and 11.13.[74] This conformer, which is obtained from the boat by slight deformations, is less stable than the chair conformer by 4.7–6.2 kcal mol^{-1} (19.7–26.0 kJ/mol^{-1}) in $\Delta H°$ according to various indirect experimental determinations; most of the force field calculations also fall into this range.[25,72,74,75] The calculations also suggest that the true boat form is about 1–1.5 kcal mol^{-1} (4.2–6.3 kJ mol^{-1}) in energy above the twist form; it is apparently the energy maximum in the interconversion of two distinct twist conformers.

b. Monosubstituted Cyclohexanes

In monosubstituted cyclohexanes, the topomerization process shown in Figure 11.12 becomes an isomerization process (Fig. 11.14) involving the interconversion of two diastereomers.[76] Like the inversion of cyclohexane itself, this process is very rapid, its rate in most cases being close to that in cyclohexane (ca. 2×10^5 s^{-1} at room temperature). This result explains why temperatures near −150°C were required to isolate conformational isomers of the type shown in Figure 11.14[57] and why the notion of Sachse's contemporaries that such isomers might be observable under ambient conditions was erroneous. In fact, no conformational isomers of monosubstituted

$$\Delta G^\circ = -RT \ln K$$

Figure 11.14. Conformational inversion of monosubstituted cyclohexane.

cyclohexanes have ever been isolated at room temperature, several claims to the contrary having been subsequently disproved (but see p. 236). Nonetheless the equilibrium depicted in Figure 11.14 is very real, with most monosubstituted cyclohexanes being mixtures of two conformers with the equatorial one generally predominating.[62] This phenomenon can be readily demonstrated by IR spectroscopy at room temperature [e.g., cyclohexyl bromide (Fig. 11.14, X = Br) shows a C–Br stretching vibration[77] for the equatorial conformer at 685 cm^{-1} and for the axial conformer at 658 cm^{-1} and by NMR spectroscopy at temperatures below about $-50°C$, under which conditions distinct signals of nuclei due to the two conformers are seen.[78,79] In fact, as already described in Chapter 10, both methods lend themselves to a *quantitative* determination of the equilibrium depicted in Figure 11.14: IR spectroscopy[80] for determination of $\Delta H°$ by observation of the change in band area with temperature (cf. Eq. 10.5), NMR spectroscopy through direct measurement of the intensity of the signals due to the two species shown in Figure 11.14. The ratio of these signal intensities gives K, and hence $\Delta G°$. Either ^1H or ^{13}C (or, in the case of fluorinated compounds, ^{19}F) NMR spectroscopy may be used for this purpose. Application of ^1H NMR spectroscopy is limited because the signals of the two conformers, with the possible exception of that for the proton geminal to X (CHX), are often broad and poorly resolved both within one conformer and between the two. ^{13}C NMR spectroscopy[81] is therefore much to be preferred; whereas at room temperature $C_6H_{11}X$ (Fig. 11.14) generally shows four distinct ^{13}C signals [C(2,6) and C(3,5) being enantiotopic, and hence isochronous], at low temperature ($-80°C$ or lower) eight signals (for the two individual conformers) are often seen. This finding allows one to compute four different values for K, the degree of whose agreement reflects the accuracy of the determination.

The NMR method usually yields $\Delta G°$, in the $-80°C$ to $-100°C$ region, since area measurements of signals must be made well below the coalescence temperature. It does not, however, lend itself to determination of $\Delta H°$ and $\Delta S°$, which requires accurate measurement of K over a sizeable range of temperature. Unfortunately, the accessible temperature range is quite limited, at the upper bound by incipient coalescence, at the lower by crystallization of the solute or solvent or both. The ΔG^0 values measured at low temperature are therefore often reported as if they were the same at 25°C, which implies $\Delta S^0 = 0$ and ΔH^0 independent of temperature. Neither of these assumptions is likely to be correct, however.

In general it would be preferable to measure $\Delta G°$ at 25°C, which is close to the temperature at which one frequently deals with physical, spectral, and pharmacological properties, as well as synthesis, reaction mechanism, and so on. In principle, conformational composition can be measured at any temperature by use of the Winstein–Holness equation (Eq. 10.15) which, applied to a cyclohexyl system (Fig. 11.14) with two conformations and generalized to any appropriate property P, becomes

$$P = n_E P_E + n_A P_A \qquad (11.1)$$

or

$$K = n_E/n_A = (P_A - P)/(P - P_A) \qquad (11.2)$$

where n denotes mole fraction and the subscripts E and A refer to equatorial and axial conformers, respectively. The first application going beyond kinetics[82] and dipole measurements (where $P = \mu^2$ is polarization, μ being the dipole moment) was, in fact, in NMR.[83] Application to chemical shifts, δ, and coupling constants, J, gives $\delta = n_E \delta_E + n_A \delta_A$ and $J = n_E J_E + n_A J_A$; that is, both observed chemical shifts and observed coupling constants in an equilibrating system, such as shown in Figure 11.14, are the weighted averages of the corresponding constants in the individual conformers. Using the form of Eq. 11.2 (cf. p. 401), this becomes

$$K = (\delta_A - \delta)/(\delta - \delta_E) = (J_A - J)/(J - J_E) \qquad (11.3)$$

Unfortunately, the room temperature shifts of the pure conformers δ_A and δ_E are not readily accessible; because of sizable variations of chemical shift with temperature, they cannot be taken to be equal to the low-temperature shifts. The situation is much more favorable with coupling constants which are very nearly temperature independent and can thus be used to estimate K at room temperature from J, if J_E and J_A are known from low-temperature measurement.[71b] In the past, this method has been limited to proton spectra, but with the increasing availability of ^1H–^{13}C coupling constants,[84] the method may find extension to ^{13}C NMR spectroscopy.

More popular, but less reliable, have been methods to determine ΔG^0 through the use of model compounds. These are compounds confined to a single conformation in which the substituent X is either strictly equatorial or strictly axial. Such confinement may be brought about either by "conformational locking," such as in the *trans*-decalin-2-ols (Fig. 11.15 A, R or R′ = OH), or by "conformational biasing," such as in the *cis*- and *trans*-4-*tert*-butylcyclohexanols (Fig. 11.15, **B**, R or R′ = OH, respectively). Conformational locking was present in many of the compounds cited in

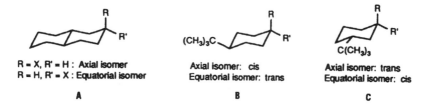

R = X, R′ = H : Axial isomer
R = H, R′ = X : Equatorial isomer

A

Axial isomer: *cis*
Equatorial isomer: *trans*

B

Axial isomer: *trans*
Equatorial isomer: *cis*

C

Figure 11.15. Conformationally locked (**A**) and conformationally biased (*anancomeric*: **B** and **C**) compounds.

Barton's classical paper[58] dealing with physical and chemical properties of axially and equatorially substituted cyclohexyl compounds. However, it has not been used much in quantitative work, perhaps because of concern that the double substitution only two and three positions away from the carbon bearing the substituent X might affect reactivity. Prototypes of the biased model[82] are the 4-*tert*-butylsubstituted compounds (Fig. 11.15, **B**); compounds of this type have also been called[85] "anancomeric," meaning "fixed in one conformation" (the word is derived from Greek *anankein* meaning to fix by some fate or law). Models of this type can be used to ascertain appropriate axial or equatorial parameters, such as chemical shifts (of various nuclei), coupling constants, dipole moments, pK_a values of acids, and extinction coefficients for IR determinations (for a review, see ref. 59).

Two methods other than low-temperature NMR studies have been used widely to determine conformational equilibria (Fig. 11.14). One is chemical equilibration of conformationally locked models[86a] (Fig. 11.16), based on Barton's recognition[58] that equatorially substituted isomers are generally more stable than axially substituted ones. Since this equilibration is one of cis–trans diastereomers, it requires chemical intervention and one obvious condition for the method to succeed is that equilibrium must be established cleanly, that is, without appreciable side reactions. The second condition is that the equilibria in Figures 11.16 and 11.14 correspond, which again depends on the innocuousness of the holding group. In this situation, experience suggests that a 4-*tert*-butyl substituent does seem to be satisfactory.

"Conformational energies," $-\Delta G^0$ or "A values"[82] (cf. Fig. 11.14) for a variety of common substituents are summarized in Table 11.6. Since almost all ΔG^0 values are negative, we find it convenient to tabulate $-\Delta G^0$. More detailed tabulations have been compiled elsewhere.[76,87–89] Perusal of the halogen values indicates that ΔG^0 is not solely a function of substituent size. As the halogens get larger (i.e., their van der Waals radius increases), the C–X bond becomes longer (i.e., the bond length increases also), and thus X becomes more distant from the carbon and hydrogen atoms at C(3,5) including, especially, the synaxial hydrogen atoms, which are mainly responsible for crowding of the axial substituent. This result leads to a form of compensation that is reinforced because the substituent with the longer bond also benefits more from the outward bending caused by the flattening of the cyclohexane ring (lever principle). The fact that the larger atoms (in the lower part of the periodic table) are also the softer or more polarizable ones, and therefore the atoms for which the attractive part of the van der Waals potential (London force) is more important (cf. Section 2-6) may be a contributing factor to their apparently low conformational energy values.

Conformational energies for OX [X = H, CH_3, Ac, Ts, $Si(CH_3)_3$, or $C(CH_3)_3$] vary little with X, presumably because the group X can be turned so as to point away from

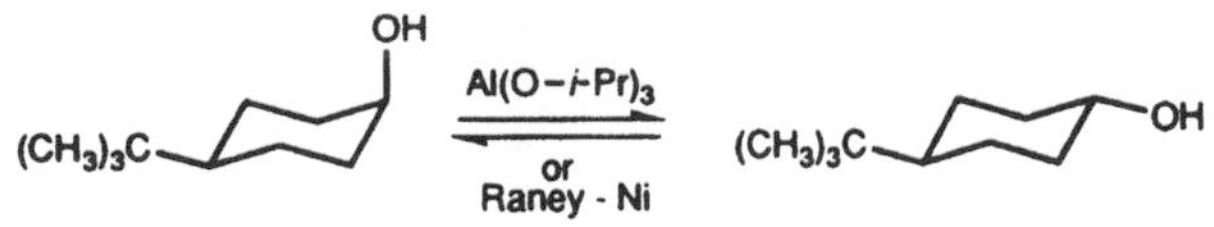

Figure 11.16. Chemical equilibration to determine conformational energies.

TABLE 11.6.   Conformational Energies[a]

Group[b,c]	Conformational Energies (kcal mol⁻¹)	(kJ mol⁻¹)	t (°C)
D	0.006	0.025	25
T	0.011	0.046	−88
F*	0.25–0.42	1.05–1.75	−86 to −93, −25
Cl*	0.53–0.64	2.22–2.68	−80 to −93, 25–27
Br*	0.48–0.67	2.01–2.80	−81, 25–27
I*	0.47–0.61	1.97–2.55	−78, −93, 25
OH(C₆H₁₂)*	0.60	2.51	25
OH(CS₂)	1.04[d,e]	4.35[d,e]	−83
OH(CH₃CHOHCH₃)	0.95	3.97	25
OCD₃*, OCH₃	0.55, 0.58, 0.63, 0.75	2.30, 2.43, 2.64, 3.14	−82, −93
OC(CH₃)₃	0.75	3.14	36
OC₆H₅	0.65	2.72	−93
OC₆H₄NO₂-p*	0.62	2.59	−93
OC₆H₄Cl-p*	0.65	2.72	−93
OC₆H₄OCH₃-p*	0.70	2.92	−93
OCHO	0.27, 0.60[d]	1.13, 2.51[d]	25, −80 to −93
OCOCH₃*	0.68, 0.71, 0.79, 0.87	2.85, 2.97, 3.31, 3.64	25, −90 ± 3
OCOCF₃	0.68, 0.56	2.85, 2.34	25, −88 to −93
OCOC₆H₅*	0.5[d]	2.09[d]	−92 ± 1
OCONHC₆H₅*	0.77	3.22	−91
OSO₂C₆H₄CH₃-p	0.50[p]	2.09[d]	−80 to −83
OSO₂CH₃	0.56	2.34	−88
ONO₂	0.59, 0.62[d]	2.47, 2.59[d]	25, −101
OSi(CH₃)₃	0.74	3.10	−103
SH	1.21[d]	5.06[d]	−80
SCD₃, SCH₃	1.04[d]	4.35[d]	−79 to −100
SC₆H₅	1.10–1.24	4.60–5.19	−80
SOCH₃	1.20	5.02	−90 to −100
SO₂CH₃	2.50	10.5	−90 to −100
SCN	1.23	5.15	−79
SeC₆H₅	1.0–1.2	4.2–5.0	−50
SeOC₆H₅	1.25	5.23	−60
TeC₆H₅	0.9	3.7	−30
NH₂ (toluene-d_8: CFCl₃)	1.23, 1.47[d]	5.15, 6.15[d]	−80 to −100
NH₂(CH₃OCH₂CH₂OH /H₂O)	1.7	7.1	20
NH₃⁺	1.7–2.0	7.1–8.4	20–25
NHCH₃(CFCl₃-CDCl₃)	1.29	5.40	−80
N(CH₃)₂(CFCl₃-CDCl₃)	1.53	6.40	−90
N(CH₃)₂(CH₃OCH₂CH₂OH /H₂O)	2.1	8.8	20
NH(CH₃)₂⁺	2.4	10.0	20
NHCOC₆H₅	1.6	6.7	−90
NC	0.20[p]	0.84[p]	−80 to −93
NCO	0.44, 0.51	1.84, 2.13	−70 to −80

TABLE 11.6. (continued)

	Conformational Energies		
Group[b,c]	(kcal mol^{-1})	(kJ mol^{-1})	t (°C)
N$_3^*$	0.45–0.62	1.88–2.59	−183, −93
NCS	0.25^d	1.05^d	−79 to −93
N=CHCH(CH$_3$)$_2$	0.75	3.14	32
N=C=NC$_6$H$_{11}$	0.96	4.02	−80
NO$_2^*$	1.1^d	4.8^d	−80 to −90, 25
PH$_2$	1.6	6.7	−90, 27
P(CH$_3$)$_2$	1.5, 1.6	6.3, 6.7	−90, 27
P(C$_6$H$_5$)$_2$	1.8	7.5	37
PCl$_2$	1.9, 2.0	7.9, 8.4	−90, 27
P(OCH$_3$)$_2$	1.9; 1.5	7.9; 6.3	−90, 27
O=P(C$_6$H$_5$)$_2$	2.46	10.3	−80
S=P(C$_6$H$_5$)$_2$	3.13	13.1	−102
CHO	0.56–0.73, 0.8	2.34–3.05, 3.35	25
COCH$_3$	1.02, 1.21, 1.52	4.27, 5.06, 6.36	−100, 25
CO$_2$H	1.4	5.9	25
CO$_2^-$	2.0	8.4	25
CO$_2$CH$_3$	1.2–1.3	5.0–5.4	25, −78
CO$_2$Et	1.1–1.2	4.6–5.0	25
COF	1.4–1.7	5.9–7.1	25
COCl*	1.3	5.4	25
CN*	0.2	0.84	−79 to −95
C≡CH	0.41–0.52	1.71–2.18	−91
CH=CH$_2$	1.49, 1.68	6.23, 7.0	−100
CH=C=CH$_2$	1.53	6.40	−80
CH$_3$*	1.74	7.28	27
CD$_3$	0.0115^f	0.048^f	25–27
CH$_2$CH$_3$*	1.79	7.49	27
CH(CH$_3$)$_2$*	2.21	9.25	27
C(CH$_3$)$_3$	4.7; 4.9	19.7; 20.5	−120
CH$_2$Br	1.79	7.49	27
CH$_2$OH	1.76	7.36	27
CH$_2$OCH$_3$	1.72	7.20	27
CH$_2$CN	1.77	7.41	27
CH$_2$Si(CH$_3$)$_3$	1.65	6.90	27
CH$_2$Sn(CH$_3$)$_3$	1.79	7.49	27
CH$_2$Pb(CH$_3$)$_3$	1.81	7.57	27
CH$_2$HgOAc	2.05	8.57	27
CF$_3$	2.4–2.5	10.0–10.5	27
C$_6$H$_5$	2.8^d	11.71^d	−100, 700
CH$_2$C$_6$H$_5$	1.68	7.03	−71
C$_6$H$_{11}$	2.2	9.2	36
SiH$_3$	1.45; 1.52	6.07; 6.36	−85, 75
Si(CH$_3$)$_3$	2.5	10.5	33
SiCl$_3$	0.61	2.55	−80
Ge(CH$_3$)$_3$	2.1	8.8	−70
Ge(C$_6$H$_5$)$_3$	2.90	12.1	not given
Sn(CH$_3$)$_3$	1.0^d	4.2^d	−69 to −90

TABLE 11.6.   (continued)

| Group[b,c] | Conformational Energies | | t (°C) |
	(kcal mol⁻¹)	(kJ mol⁻¹)	
Sn(i-Pr)$_3$	1.10	4.6	not given
Sn(CH$_3$)$_2$C$_6$H$_5$	1.08	4.5	not given
SnCH$_3$(C$_6$H$_5$)$_2$	1.20	5.02	not given
Sn(C$_6$H$_5$)$_3$	1.44	6.0	not given
Pb(CH$_3$)$_3$	0.67	2.80	−69
HgOAc	0, −0.3	0, −1.3	−79, −90
HgCl	−0.25	−1.05	−90
HgBr	0[d]	0[d]	−79
MgBr(Et$_2$O)	0.78	3.26	−75
MgC$_6$H$_{11}$(Et$_2$O)	0.53	2.22	−82

[a]Complete references to all tabulated energy values are listed in ref. 90, p. 697.
[b]Starred values mean that ΔH^0 and ΔS^0 also available in the original reference.
[c]The solvent is in parentheses in cases where large solvent dependence is observed.
[d]Averaged value; all values given are within experimental error of each other.
[e]The alcohol may have been self-associated (oligomeric).
[f]Difference between CH$_3$ and CD$_3$; CD$_3$ has lesser equatorial preference than CH$_3$.

the ring when OX is axial. At worst this will decrease the number of rotamers (rotational conformers) for the axial OX, and thereby produce a slight drop in its entropy. The OH group itself shows a large solvent effect, $-\Delta G_{OH}^0$ being appreciably larger in hydrogen-bond forming solvents (e.g., isopropyl alcohol) than in those not forming hydrogen bonds (e.g., cyclohexane). The value for OH obtained by low-temperature NMR in CS$_2$ seems to be out of line; it must be strongly suspected that solutions about 0.2 M in cyclohexanol (concentrations appropriate for an NMR experiment) at −80°C are subject to extensive oligomerization of the solute by intermolecular hydrogen bonding, so that the measured $-\Delta G^0$ is not that of the monomeric alcohol.

The value of $-\Delta G^0$ for SH is slightly larger than that for OH. That for SCH$_3$ is somewhat smaller; that of SOCH$_3$ is similar to SCH$_3$, but that of SO$_2$CH$_3$ is considerably larger. The former three groups presumably confront the ring with their lone electron pairs, which are evidently not greatly repulsive, but SO$_2$CH$_3$ must confront the ring with an O or CH$_3$ moiety (probably the former), which is sterically much more demanding. The progression of $-\Delta G^0$ in the series NH$_2$, NHCH$_3$, N(CH$_3$)$_2$ is also slight, for the same reasons adduced for OH versus OCH$_3$, but (CH$_3$)$_3\overset{+}{N}$, as expected, has a very large $-\Delta G^0$ value (too large to be measured). The values for PR$_2$ are of the same magnitude as those for NR$_2$; the diminution in $-\Delta G^0$ seen for Group 14(IVA) and Group 17(VIIA) elements as one goes down the periodic table is not evident in Group 15(VA) and Group 16(VIA), though in the latter groups there are data for only the first two members. Linear substituents, such as −NC, −NCO, N$_3$, CN, and C≡CH have expectedly small conformational energy values and those of planar groups, such as COR, CO_2R, $CH{=}CH_2$ are intermediate between those of linear and those of tetrahedral groups, such as CH_3. The vinyl group has the largest conformational energy in this series; apparently, when it is axial, one of its β (methylene) hydrogen atoms interferes seriously with one of the equatorial ring hydrogen atoms of the cyclohexane.

The sp^2 hybridized groups orient themselves so as to confront the ring with their flat sides, in other words the plane of the substituent is perpendicular or nearly perpendicular to the bisector plane of the cyclohexane ring. In the case of axial phenyl, this rotational conformation, though optimal, imposes steric crowding of the ortho hydrogen atoms (*o*-H) of the phenyl with *both* adjacent equatorial hydrogen atoms (e-H) of the cyclohexane chair[91] (Fig. 11.17*a*); this explains the high conformational energy of the phenyl group (Table 11.6). Equatorial phenyl, in contrast, is most stable in the bisector plane of the cyclohexane chair (Fig. 11.17*b*), where the unfavorable *o*-H/e-H interaction is avoided (see also Section 2-6, Fig. 2.15).

The conformational energy of methyl in methylcyclohexane (a key datum; cf. ref. 81) has been determined by low-temperature ^{13}C NMR spectorscopy.[92] Since the contribution of the axial conformer in the temperature range of the experiment (140–195 K) is only about 1%, it was necessary to work with ^{13}C-enriched material to see the methyl peak of the minor conformer. (Data from a careful reexamination[93] of the methyl parameter are included in Table 11.7). With the value for methyl in hand, values for ethyl and isopropyl were determined[92b] by use of the "counterpoise method,"[94] employing *cis*-1-alkyl-4-methylcyclohexanes as objectives of low-temperature NMR investigation (Fig. 11.18). Making the reasonable assumption that the conformational energies of CH_3 and R are additive, one has $\Delta G^0 = \Delta G_R - \Delta G_{CH_3}$, where ΔG^0 is the free energy change for the process shown in Figure 11.18; hence $\Delta G_R = \Delta G^0 + \Delta G_{CH_3}$.

Thermodynamic parameters (ΔH^0 and ΔS^0) have been determined for some of the substituents in Table 11.6; ΔG^0 values in such cases have been starred: ΔH^0 and ΔS^0 may be found in the original references, cited in ref. 90.

The difference between CO_2^- and CO_2H and between NH_3^+ and NH_2 is of note; in both cases $-\Delta G^0$ for the ion is considerably larger than that for the uncharged species. One way of explaining this is to say that the axial group, when ionic, is swelled by solvation, and therefore more subject to steric repulsion than the neutral ligand. Another complementary explanation implies that the axial substituent, because of crowding, is less readily solvated than the equatorial one, and therefore benefits less

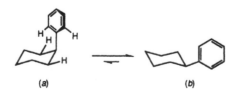

(a)   (b)

Figure 11.17.   Equatorial and axial conformers of phenylcyclohexane; $\Delta G^0 = -2.87$ kcal mol^{-1} (12.0 kJ mol^{-1}).

TABLE 11.7. Conformational Thermodynamic Parameters for Alkyl Groups

Alkyl	$-\Delta H^{0\,a}$		$-\Delta S^{0\,b}$		$-\Delta G_{25}^{0\,a}$	
	Found	Calculated	Found	Calculated	Found	Calculated
CH_3	1.75 (7.32)	1.77 (7.41)	−0.03 (0.13)	0 (0)	1.74 (7.28)	1.77 (7.41)
C_2H_5	1.60 (6.69)	1.69 (7.07)	0.64 (2.68)	0.61 (2.55)	1.79 (7.49)	1.87 (7.82)
$(CH_3)_2CH$	1.52 (6.36)	1.40 (5.86)	2.31 (9.67)	2.18 (9.12)	2.21 (9.25)	2.05 (8.58)

[a]In kilocalories per mole (kcal mol^{-1}); values in parentheses are in kilojoules per mole (kJ mol^{-1}).
[b]In calories per mole per kelvin (cal mol^{-1}K^{-1}); values in parentheses are in joules per mole per kelvin (J mol^{-1} K^{-1}).

from the diminution in free energy than any charged species experiences when it is solvated. (In the case of NH_3^+ vs. NH_2, the steric effect of the extra hydrogen may, of course, also contribute to the larger $-\Delta G^0$ of the former.) The true answer probably lies somewhere in between.

c. Disubstituted and Polysubstituted Cyclohexanes

1,2-, 1,3-, and 1,4-Disubstituted cyclohexanes each exist as cis and trans isomers (Section 11-1). When the two substituents are identical, the cis-1,2 and cis-1,3 isomers are meso forms, whereas the corresponding trans isomers are chiral. In the 1,4-disubstituted series, both the cis and the trans isomers are achiral, regardless of whether the substituents are the same or not.

When one considers conformational factors, the situation becomes somewhat more complex. The 1,4-dimethylcyclohexanes are shown in Figure 11.19. The cis isomer exists as an equimolar mixture of two indistinguishable conformers. Its steric energy (cf. Section 2-6, p. 21) is that of the axial methyl group or 1.74 kcal mol^{-1} (7.28 kJ mol^{-1}). It has no entropy of symmetry (symmetry point group C_s; $\sigma = 1$) and no entropy of mixing, since the two conformers are superposable. The trans isomer consists of two conformers, the predominant e,e and the much less abundant a,a whose energy level is 2×1.74 or 3.48 kcal mol^{-1} (14.56 kJ mol^{-1}) above the e,e, since it has two axial methyl groups. The Boltzmann distribution thus corresponds to 99.7% e,e conformer and 0.3% a,a at 25°C (the amount of the diaxial conformer is less at lower temperatures and more at elevated ones). Use of Eq. 11.1 (P = H) thus leads to an overall conformational enthalpy of $0.997 \times 0 + 0.003 \times 3.48$ or 0.01 kcal mol^{-1} (0.04 kJ mol^{-1}). The entropy of symmetry is $-R \ln 2$ (symmetry point group C_{2h}; $\sigma = 2$) or −1.38 cal mol^{-1} K^{-1} (−5.76 J mol^{-1} K^{-1}) and the entropy of mixing of the two

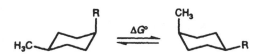

Figure 11.18. Counterpoise method.

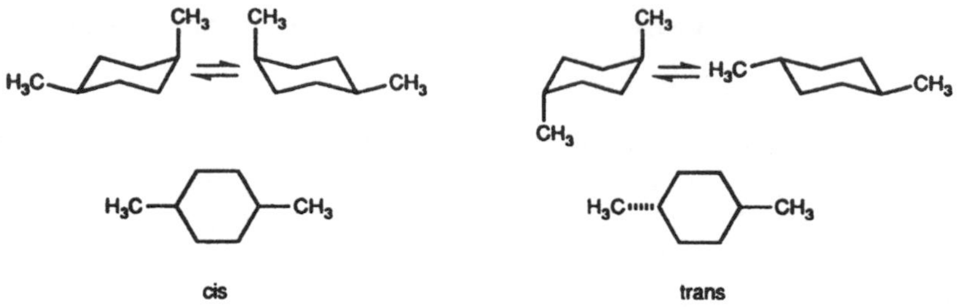

Figure 11.19.   1,4-Dimethylcyclohexanes.

conformers is $-R(0.997 \times \ln 0.997 + 0.003 \times \ln 0.003)$ or 0.04 cal mol^{-1} K^{-1} (0.17 J mol^{-1} K^{-1}). One thus calculates an enthalpy difference between the two diastereomers of 1.73 (1.74 – 0.01) kcal mol^{-1} (7.24 kJ mol^{-1}) and an entropy difference of –0.04 + 1.38 or 1.34 cal mol^{-1} K^{-1} (5.59 J mol^{-1} K^{-1}), the enthalpy favoring the trans isomer and the entropy the cis. The experimental data, along with the calculations, are shown in Table 11.8. It must be kept in mind that the 1.74 kcal mol^{-1} (7.28 kJ mol^{-1}) value for the conformational energy of methyl is a liquid-phase value and that, as already explained in the case of the butane conformers (p. 379), gas-phase enthalpy differences (Table 11.8) differ from liquid-phase ones because of differences in heats of vaporization.

The situation in 1,3-dimethylcyclohexane (Table 11.8) is simpler because both diastereomers exist in single conformations.

A more complex situation arises in 1,2-dimethylcyclohexane (Fig. 11.20). The (more stable) trans isomer (Fig. 11.20b) has two conformers, the more stable e,e and the less stable a,a. The steric energy of the a,a conformer amounts to 3.48 kcal mol^{-1} (14.56 kJ mol^{-1}) because it has two axial methyl groups. However, the e,e conformer also has non zero steric energy since its two methyl groups are gauche to each other. One could take its steric energy as equal to that of the gauche form of butane (Section 10-1), but, in fact, a better value is that determined experimentally[96,97] as 0.73–0.90 kcal mol^{-1} (3.05–3.77 kJ mol^{-1}), inter alia by direct determination of the a,a $\rightleftharpoons$ e,e equilibrium (Fig. 11.20b). Using 0.74 kcal mol^{-1} (3.10 kJ mol^{-1}) for this interaction yields a difference of 2.74 kcal mol^{-1} (11.5 kJ mol^{-1}) between the e,e and a,a trans conformers, leading, at room temperature, to a Boltzmann distribution of 99% e,e, 1% a,a. This yields a steric enthalpy of 0.77 kcal mol^{-1} ($0.74 \times 0.99 + 3.48 \times 0.01$) or 3.21 kJ mol^{-1} for the ensemble; the steric entropy is composed of an entropy of mixing of the two conformers of 0.11 cal mol^{-1} K^{-1} (0.46 J mol^{-1} K^{-1}), $-R(0.99 \ln 0.99 + 0.01 \ln 0.01)$, an entropy of symmetry of $-R \ln 2$ (symmetry point group C_2; $\sigma = 2$), and an entropy of mixing for the enantiomer pair of $R \ln 2$, total 0.11 cal mol^{-1} K^{-1} (0.46 J mol^{-1} K^{-1}).

The cis isomer (Fig. 11.20a) presents a puzzling aspect: In the planar representation **A″** it has a symmetry plane and appears to be a meso form. In contrast, the chair representation (Fig. 11.20, **A** and **A′**) is chiral in either (equilibrating) conformation. However, if one takes the **A′** chair and turns it 120° around a vertical axis, one realizes

TABLE 11.8. Enthalpy, Entropy, and Free Energy Differences Between Diastereomeric Dimethylcyclohexanes[a]

Energy or Entropy Difference		1,2(liq)	1,2(vap)	1,3(liq)	1,3(vap)	1,4(liq)	1,4(vap)
$-\Delta\Delta H^0$	calcd[b]	1.71		1.74		1.73	
	found[b]	1.5	1.9	1.7	1.94	1.63	1.89
	calcd[c]	7.17		7.28		7.24	
	found[c]	6.4	7.8	7.2	8.1	6.8	7.9
$-\Delta\Delta S^0$	calcd[d]		1.27		1.38		1.34
	found[d]	0.22	0.74	0.88	1.05	0.74	1.10
	calcd[e]		5.31		5.77		5.59
	found[e]	0.9	3.1	3.7	4.4	3.1	4.6
$-\Delta\Delta G^0_{25}$	calcd[b]						
	found[b]	1.48 1.46f	1.65 1.47f	1.46	1.63	1.41 1.43f	1.58
	calcd[c]						
	found[c]	6.2 6.11f	6.9 6.15f	6.1	6.8	5.9 5.98f	6.6

[a]Experimental values for $\Delta\Delta H^0$, $\Delta\Delta S^0$, and $\Delta\Delta G^0$ from ref. 29, p. n-2100 (differences in enthalpies, entropies, and free energies of formation: 1,2 and 1,4: cis $\rightleftharpoons$ trans; 1,3: trans $\rightleftharpoons$ cis).

[b]In kilocalories per mole (kcal mol^{-1}).

[c]In kilojoules per mole (kJ mol^{-1}).

[d]In gibbs, that is, calories per mole per kelvin (cal mol^{-1} K^{-1}).

[e]In joules per mole per kelvin (J mol^{-1} K^{-1}).

[f]Experimental values for $\Delta\Delta G^0$ at 298 K from ref. 95. Since these latter values were obtained by equilibration in the 480–600 K range (ΔG_{25} being computed from the experimental values of $\Delta\Delta H^\circ$ and $\Delta\Delta S^\circ$ with corrections for the presence of some twist form at the elevated temperature), they are probably in between liquid and vapor data, since, while most of the material was in the liquid form, some undoubtedly existed as a vapor.

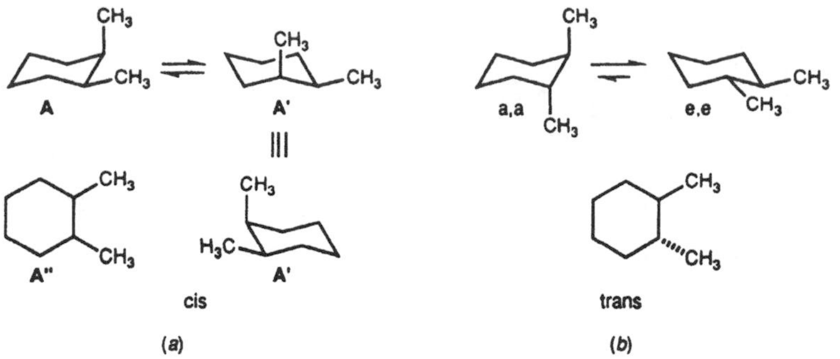

Figure 11.20.   1,2-Dimethylcyclohexane.

that it is the mirror image of the **A** chair; in other words, conformational inversion converts one enantiomer of *cis*-1,2-dimethylcyclohexane into the other. The compound is therefore a racemate in which the enantiomers interconvert rapidly at room temperature, and therefore cannot be individually isolated. Thus *cis*-1,2-dimethylcyclohexane is an example of a molecule whose averaged symmetry (C_{2v}) is higher than the symmetry of the contributing conformers (C_1). This situation has been discussed in general in Section 4-4; in the specific case of *cis*-1,2-disubstituted cyclohexanes (identical substituents) the planar representation properly represents the average symmetry of the two rapidly interconverting chairs.[98] On the assumption that the gauche interaction of adjacent equatorial and axial methyl groups is the same as that of two equatorial methyl groups, the steric enthalpy of *cis*-1,2-dimethylcyclohexane is $1.74 + 0.74 = 2.48$ kcal mol^{-1} (10.4 kJ mol^{-1}) and the steric entropy is the entropy of mixing of the two enantiomeric conformers (*not* superposable in this case) or $R \ln 2$ (1.38 cal mol^{-1} K^{-1}, 5.76 J mol^{-1} K^{-1}); the cis–trans differences in Table 11.8 are derived from these numbers. The agreement between calculated and experimental $\Delta\Delta G^0$ values in Table 11.8 is quite good, especially if one takes into account that the high temperature at which the experiment had to be carried out may produce a variety of complications.

We have already seen that conformational energies of substituents located 1,4 to each other are close to being additive (Fig. 11.18); this has been confirmed in several cases,[94,96a] where $-\Delta G^0$ values have been determined both by direct low-temperature equilibration (Fig. 11.14) and by the counterpoise method (Fig. 11.18). Additivity is not as good for substituents located 1,3 to each other because of buttressing problems.[25,99] Perhaps not unexpectedly, additivity breaks down altogether for most vicinal 1,2 substituents because of potential interactions between the substituent, for example, in 1,2-dimethylcyclohexane where agreement between calculated and experimental energy values (Table 11. 8) is relatively poor.

Additivity of conformational energies in geminally (1,1) disubstituted cyclohexanes (for tabulations see refs. 100–102) tends to break down for reasons similar to those operative in vicinally disubstituted ones: One substituent interferes with the otherwise optimal rotational conformation of the other in one of the two possible conformations (but not both; or at least the interference is not the same in the two ring-inverted conformations).

Large steric interactions occur when two groups are placed synaxially (Fig. 11.21). The synaxial interaction energies are tabulated[103] in Table 11.9; most of these interactions are so large that counterpoise methods had to be used to determine them.

Among hexasubstituted cyclohexanes, the inositols[105] (Fig. 11.22) are important. All the inositols are known, either as natural or as synthetic products.[106] Of the eight diastereomers, seven are achiral; the eighth (*chiro*-inositol) is chiral. The symmetry planes are easy to spot except in the allo isomers in which individual chair forms are chiral, but chair inversion leads to the enantiomer, much as in *cis*-1,2-dimethylcyclohexane (see p. 448).

Figure 11.21. The synaxial interaction. $\Delta G^0 = \Delta G_{\text{synaxial}}^{xy} + \frac{1}{2}(\Delta G_x + \Delta G_y)$.

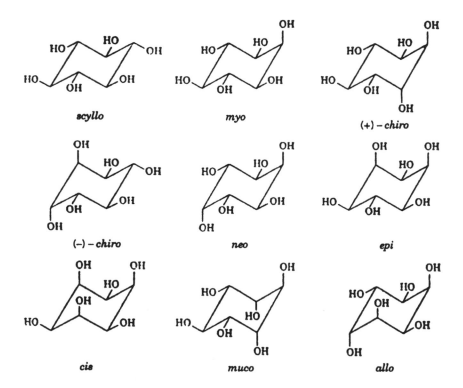

scyllo   *myo*   (+) – *chiro*

(–) – *chiro*   *neo*   *epi*

cis   *muco*   *allo*

Figure 11.22. The inositols. [Reproduced with permission from Eliel, E. L., Allinger, N. L., Angyal, S. J., and Morrison, G. A. *Conformational Analysis.* Copyright © 1981 American Chemical Society, Washington, DC, p. 353.]

TABLE 11.9. Synaxial Interactions[a]

Groups	CH$_3$/CH$_3$	OH/OH	CH/OH	OAc/OAc	Cl/Cl	CH$_3$/F	CH$_3$/Br	CN/CN	CH$_3$/CN	CH$_3$/C$_6$H$_5$
$-\Delta G^0$ (kcal mol^{-1})[b]	3.7	1.9	2.4[c]	2.0	5.5	0.4	2.2	3.0	2.7	3.4[d]
$-\Delta G^0$ (kJ mol^{-1})[b]	15.5	7.95	10.0[c]	8.4	23.0	1.67	9.2	12.5	11.1	14.2[d]

Groups	CH$_3$/CO$_2$Et	CH$_3$/CO$_2^-$	CO$_2$CH$_3$/CO$_2$CH$_3$	CO$_2$H/CO$_2$H	CO$_2^-$/CO$_2^-$	CO$_2$H/NH$_3^+$	CO$_2^-$/NH$_3^+$
$-\Delta G^0$ (kcal mol^{-1})[b]	2.8–3.2	3.4	1.5[e]	1.1	~4.2	0.5	-1.8
$-\Delta G^0$ (kJ mol^{-1})[b]	11.7–13.4	14.1	6.3[e]	4.6	~17.6	2.1	-7.5

[a]From ref. 103 unless otherwise indicated.
[b]Synaxial interaction.
[c]Reference 104.
[d]Reference 96.
[e]Revised value.

d. Conformation and Physical Properties in Cyclohexane Derivatives

Many of the differences in physical and spectral properties between conformational and configurational isomers in cyclohexane have been discussed in detail elsewhere.[59] The most important properties in this category are NMR spectral properties (relating to both ^{1}H and ^{13}C spectra) and chiroptical properties [optical rotatory dispersion (ORD) and circular dichroism (CD)]. Most of this section will therefore be concerned with NMR spectra; chiroptical properties are dealt with in Chapter 12.

Before dealing with NMR spectroscopy, we mention briefly some other salient properties of cyclohexane stereoisomers. Relative boiling points, refractive indices, and densities of stereoisomers can often be predicted on the basis of the conformational rule, a modification of the classical von Auwers–Skita rule[107]: "The isomer with the smaller molar volume has the greater heat content." Since smaller molar volume implies greater density, refractive index, and boiling point, another way to state the rule is to say the isomer of greater enthalpy (*not* free energy) has the higher boiling point, refractive index, density, and also heat of vaporization (Trouton's rule). This rule, incidentally, is the reason why in Table 11.8, $-\Delta\Delta H_{vap}^0 > -\Delta\Delta H_{liq}^0$ (cf. Fig. 10.4). The pertinent physical properties of the dimethylcyclohexanes are given in Table 11.10 (for enthalpy data see Table 11.8).

The conformational rule applies only to nonpolar compounds[108]; for molecules with dipole moments the dipole (Van Arkel) rule applies, but with less reliability: The isomer with the higher dipole moment has the higher boiling point, density, and refractive index.[107,109]

In connection with Table 11.6, we mentioned that because of solvation or ion pairing, ionic ligands (such as NH_3^+ and CO_2^-) are more bulky than the corresponding neutral ligands (NH_2 and CO_2H), and therefore less comfortable in axial positions. It is thus not surprising that axial amines and axial carboxylic acids are weaker than their equatorial counterparts: for example, the pK_a, 5.55, of *cis*-4-*tert*-butylcyclohexanecarboxylic acid (Fig. 11.23, X = CO_2H) is higher than that of the trans acid with pK_a 5.10, and the cis amine (X = NH_2) with pK_b 3.50 is weaker than

TABLE 11.10.   Boiling Points, Refractive Indexes, and Densities of Dimethylcyclohexanes

Isomer	Major Conformation	bp (°C)	n_D^{25}	d_4^{25}
cis-1.2	e,a	129.7	1.4336	0.7922
trans-1,2[a]	e,e	123.4	1.4247	0.7720
cis-1,3[a]	e,e	120.1	1.4206	0.7620
trans-1,3	e,a	124.5	1.4284	0.7806
cis-1,4	e,a	124.3	1.4273	0.7787
trans-1,4[a]	e,e	119.4	1.4185	0.7584

[a]The isomer of lower heat content in each diastereomeric pair.

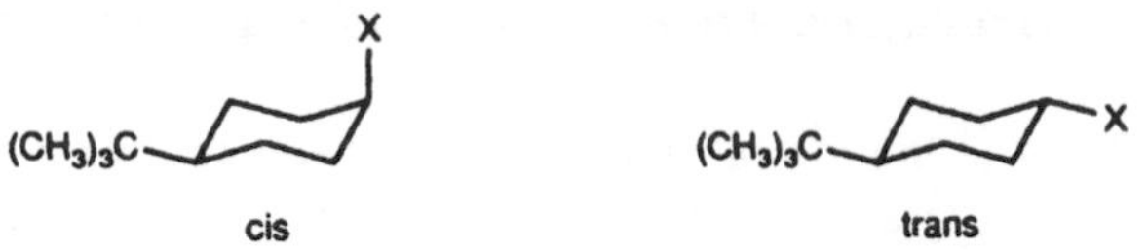
X
$(CH_3)_3C$
cis
$(CH_3)_3C$
X
trans

Figure 11.23. The *cis*- and *trans*-4-*tert*-butylcyclohexanecarboxylic acids (X = CO_2H) and butylcyclohexylamines (X = NH_2).

the trans amine with pK_b 3.40[110] [data in H_2O extrapolated from aqueous dimethyl sulfoxide (DMSO)].

Nuclear magnetic resonance spectroscopy is undoubtedly the most powerful method for the elucidation of stereochemistry of cyclohexyl derivatives. The most important generalities in ^1H NMR spectroscopy[111] are that axial protons generally resonate upfield of equatorial protons and that, because of the operation of the Karplus relationship (Fig. 10.26) and the fact that axial protons generally have other axial protons antiperiplanar to them, axial protons show larger splitting (or bandwidth, if the splitting is not resolved) than equatorial protons. Thus, since the torsion angles (cf. Fig. 11.11) are $\omega_{aa} = 175°$, $\omega_{ea} = 55°$ and $\omega_{ee} = 65°$, J_{aa} (9–13 Hz) > J_{ea} (3–5 Hz) > J_{ee} (2–4 Hz).

These ranges exclude protons that are antiperiplanar to electronegative atoms or groups, such as the halogens or OR. If one of the two coupled protons is antiperiplanar to such a group, the coupling constant is diminished by 1–2 Hz below the normal value. For example, in the low-temperature spectrum of the axial conformer of cyclohexyl-d_8 methyl-d_3 ether (Fig. 11.24), $J_{ea} < J_{ee}$ because the axial proton at C(2) (but not the equatorial one) is antiperiplanar to electronegative oxygen.[71b] J_{ea} in the equatorial conformer is much larger.

In the case of equatorial or axial protons for which the splitting is poorly resolved, width at half height W may be taken as a qualitative conformational criterion: Equatorial protons generally have $W < 12$ Hz, while axial protons have $W > 15$ Hz.

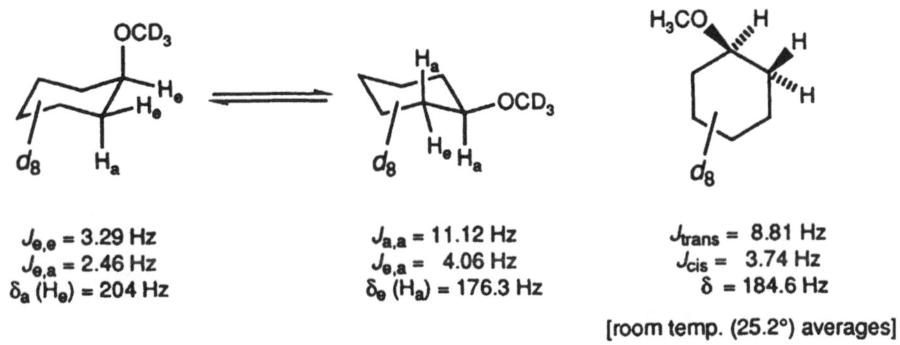
OCD_3
H_e
H_e
d_8
H_a
H_a
OCD_3
d_8
H_e H_a
H_3CO
H
H
H
d_8
$J_{e,e}$ = 3.29 Hz
$J_{e,a}$ = 2.46 Hz
$\delta_a(H_e)$ = 204 Hz
$J_{a,a}$ = 11.12 Hz
$J_{e,a}$ = 4.06 Hz
$\delta_e(H_a)$ = 176.3 Hz
J_{trans} = 8.81 Hz
J_{cis} = 3.74 Hz
δ = 184.6 Hz
[room temp. (25.2°) averages]

Figure 11.24. Conformational equilibrium in cyclohexyl-d_8 methyl-$d_3$3 ether; δ at 60 MHz.

The differences in chemical shifts between axial and equatorial protons in cyclohexane have been rationalized in terms of the diamagnetic anisotropy of the C(2)–C(3) bonds (Fig. 11.25). If, in oversimplified fashion, one ascribes the shielding effect of the C–C bond to a magnetic point dipole located at the electrical center of gravity of the bond causing the shielding, the shielding (σ) may be expressed by the McConnell equation as

$$\sigma = \Delta_\chi \, (1 - 3 \cos^2\theta)/3r^3 \tag{11.4}$$

where θ is the angle between the bond causing the shielding and a line drawn from the electric center of gravity (G) of this bond to the shielded proton; r is the distance of the proton from G and χ is the diamagnetic anisotropy of the shielding bond (C–C in Fig. 11.25).[112] Equation 11.4 describes the angular dependence of the shielding: If $\theta < 54.8°$, it gives a negative value for σ (assuming positive χ); that is, there is deshielding (this will be the case for the equatorial proton). But for the axial proton, $\theta > 54.8°$, so the effect of the C(2)–C(3) bonds on this proton is one of shielding. Thus a C–C bond will deshield a proton antiperiplanar to it but will shield a proton gauche to it. The effect of a vicinal methyl group can be explained in the same way, as shown in Figure 11.26[113a]: The equatorial carbinol proton in **A** is shielded by the methyl groups gauche to it in **B** and **C**; the axial carbinol proton in **D** (upfield of the equatorial one) is shielded by the methyl group gauche to it in **E** but deshielded by the methyl group anti to it in **F**. Compound **G** is included to illustrate another effect, namely, a deshielding caused by van der Waals compression of the axial carbinol proton by the synaxial methyl group.

An application to configurational assignment of 4-*tert*-butylcyclohexyl *p*-toluenesulfonates and 4-*tert*-butylcyclohexyl phenyl thioethers[113b] is shown in Figure 11.27. The broad (*b*) upfield signals are due to the axial hydrogen atoms next to sulfur or oxygen, respectively in the trans isomers, whereas the narrow (*n*), downfield signals originate from the corresponding equatorial protons in the cis isomers. The assignment is in agreement with that made earlier on chemical grounds (cf. Fig. 5.39).

Carbon-13 NMR spectroscopy[114–117] is at least as useful in identifying configuration and conformation as ¹H NMR. Commonly, each peak in the broad-band decoupled ¹³C spectrum is sharp and well resolved so that a number of signals can be

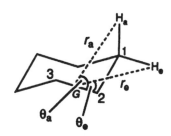

Figure 11.25.   Shielding of axial and equatorial protons by C(2)–C(3) bond.

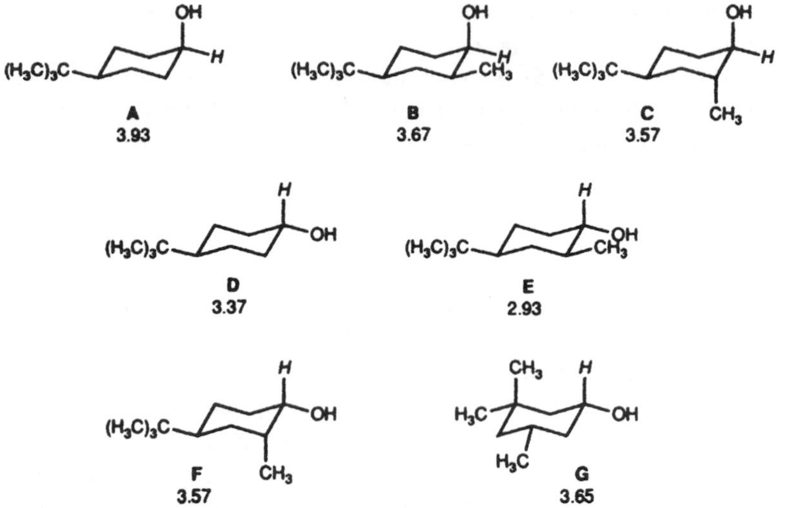

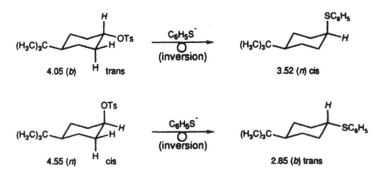

Figure 11.26. Shielding and deshielding effects of vicinal and synaxial methyl substituents (shift in parts per million). (**A** and **D** are reference compounds.)

used for configurational assignment. Diastereomers, with very few exceptions, will differ in the position of at least some of their signals, so that the differences can be used for identification as well as assignment of configuration and quantitative analysis. Most useful in signal assignment and configurational and conformational identification are the Grant parameters,[19] which indicate the effect of a methyl substituent (either equatorial or axial) on the chemical shifts of the ring carbon atoms in cyclohexane. These parameters are summarized in Table 11.11 (see also ref. 118); they are to be added to the chemical shift of the appropriate ring carbon in the absence of the substituent. For cyclohexane itself the basic shift [downfield from tetramethylsilane (TMS) in CDCl$_3$ solvent] is 27.3 ppm. The α-carbon atom is the one

Figure 11.27. The ^{1}H NMR signals of starting 4-*tert*-butylcyclohexyl tosylates and product 4-*tert*-butylphenyl thioethers in thiophenolate displacement reaction. Shifts are in parts per million; *b*, broad, *n*, narrow signal.

TABLE 11.11.   Methyl Substitution Parameters

α_e	β_e	γ_e	δ_e	ϵ_e	α_a	β_a	γ_a	δ_a	ϵ_a
+6.0	+9.0	−0.3	−0.5	−0.4	+1.4	+5.4	−6.4	+0.2	−0.1

to which the substituent (methyl) is attached, the β-carbon atom is next to the α, and so on, as shown in Figure 11.28. [The most remote carbon within the ring is δ. The ε parameter refers to the effect of a ring substituent on the attached carbon of another (equatorial) substituent located 1,4 to it.] By way of an example, in methylcyclohexane (equatorial CH_3), the calculated parts per million shift at C(1) is 33.3 ppm (27.3 + α_e) and that at C(2) is 36.3 ppm (27.3 + β_e); the experimental values[118] are 33.0 and 35.6 ppm. The agreement is better than appears at first sight: If one takes into account that methylcyclohexane is 95% equatorial and 5% axial at 25°C, and adds the approximately weighted shifts for the two conformers (cf. Eq. 11.1; P, chemical shift) one calculates 33.1 ppm for C(1), 36.1 ppm for C(2).

One attractive feature of the shift parameters is that, except for disturbances that usually occur when substituents are geminal or vicinal, they tend to be additive. Thus in cis-1,3-dimethylcyclohexane (Fig. 11.29) the observed shift at C(2) is 45.3 ppm, and is found to be (perhaps fortuitously) exactly that calculated for a cyclohexane (basic shift 27.3 ppm) with two equatorial β_e effects of 9.0 ppm each.

Shift parameters for a large number of groups other than methyl have been tabulated[89] and are shown in Table 11.12. Except for I and SH, the α effects of heteroatom-linked groups are much larger than those for methyl. On the other hand, the β effects are of the same order of magnitude, ranging from 1.7 to 13.8 ppm for β_e and from −0.4 to 9.5 ppm for β_a. In general, $\alpha_e > \alpha_a$ and $\beta_e > \beta_a$, except for bromine and iodine in the former case and $N(CH_3)_2$ in the latter. (The global α effect for a substituent −X−Y includes the γ effect of Y and the global β effect of such a combination includes the γ effect of Y.) All γ_a effects are upfield shifting and of a magnitude similar to that for methyl (Table 11.11). However, most of the substituents also have sizeable negative γ_e effects.[119]

e.   Conformation and Reactivity in Cyclohexanes

We mentioned earlier that Barton[58] recognized the greater stability of equatorially as compared with axially substituted cyclohexanes. But probably the most important

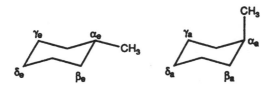

Figure 11.28. Denomination of substituent effects (parameters) in a (hypothetical) conformationally homogeneous species. The substituent effect (ppm) is equal to the observed shift −27.3 (cyclohexane shift).

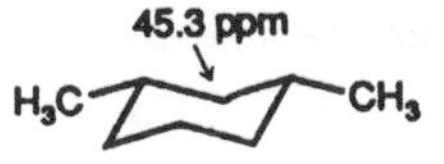

Figure 11.29. Additivity of parameters for C(2) in *cis*-1,3-dimethylcyclohexane.

insight gained from his pioneering work on conformational analysis concerns the effects of conformation on chemical reactivity. It is convenient to divide these effects into two kinds, steric and stereoelectronic effects.

By "steric effects" we mean effects due to close approach of two groups in a molecule (or between molecules) such that appreciable van der Waals forces (either attractive, at relative long distances, or repulsive, at short distances) are called into play. Such effects may occur in the ground state of a molecule, or in the transition state for a given reaction, or both. For the sake of simplicity, we shall, in the discussion that follows, assume that the effects are repulsive. While it must be kept in mind that van der Waals interactions can also be attractive, the attractive potential is always quite small (cf. Fig. 2.13), whereas the repulsive potential can become quite large if the nonbonded distances are sufficiently short.

Two kinds of situations are depicted in Figure 11.30. In the first, more familiar situation (*a*) the repulsion is substantial in the transition state (TS) and small or absent in the ground state (GS). Compared to a reference case, the activation energy in this

TABLE 11.12. Miscellaneous Substituent Shifts[a]

Substituent	Shift Parameter							
	α_e	α_a	β_e	β_a	γ_e	γ_a	δ_e	δ_e
F	+64.5	+61.1	+5.6	+3.1	−3.4	−7.2	−2.5	−2.0
Cl	+32.7	+32.3	+10.5	+7.2	−0.5	−6.9	−1.9	−0.9
Br	+25.0	+27.5	+8.1	+8.1	+0.7	−6.3	−2.0	−1.1
I	+2.1	+9.5	+13.8	+9.5	+2.4	−4.5	−2.4	−0.8
OH	+44.1	+38.9	+8.5	+6.0	−2.3	−6.9	−1.5	−0.6
OCH$_3$	+52.9	+47.7	+5.2	+3.1	−2.0	−6.3	−0.7	−0.1
OSi(CH$_3$)$_3$	+43.5	+39.1	+9.0	+6.1	−2.3	−7.2	−2.0	−2.0
OCOCH$_3$	+46.5	+42.3	+ 4.8	+3.2	−2.3	−6.1	−1.5	−1.1
OCOCF$_3$	+51.8	+48.1	+4.2	+2.8	−2.4	−6.3	−1.6	−1.2
OTs	+55.5	+52.2	+5.5	+3.9	−2.2	−6.7	−2.0	−1.4
SH	+11.1	+8.9	+10.7	+6.1	−0.6	−7.6	−2.2	−1.3
NH$_2$	+23.9	+18.1	+10.0	+6.5	−1.6	−7.2	−1.3	−0.3
NHCH$_3$	+32.1	+26.8	+6.3	+3.2	−1.8	−6.6	−0.7	−0.1
N(CH$_3$)$_2$	+37.1	+33.7	+1.7	+2.6	−1.1	−6.2	−0.6	+0.3
N$_3$	+32.5	+29.8	+4.5	+2.0	−2.5	−6.9	−2.5	−1.8
NO$_2$	+58.0	+53.9	+4.0	+1.7	−2.4	−5.6	−2.0	−1.1
−C≡CH	+1.7	+1.0	+5.1	+3.0	−1.8	−5.8	−2.1	−1.3
−CN	+0.7	−0.6	+2.2	−0.4	−2.6	−5.1	−2.6	−2.0
−NC	+24.9	+23.3	+6.7	+3.5	−2.6	−6.9	−1.8	−1.8

[a]Data from ref. 89.

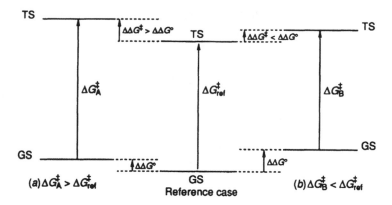

Figure 11.30. Steric hindrance and steric assistance.

situation is increased, because the energy level of the transition state is elevated more than that of the ground state: The reaction is slowed down relative to the reference case. This situation is termed one of "steric hindrance." In the other situation (*b*) the steric repulsion is more important in the ground state than in the transition state. As a result, the energy level of the ground state is elevated more than that of the transition state; thus the activation energy for the reaction is decreased and the reaction is accelerated relative to the reference case. This situation is called one of "steric assistance."

We shall limit ourselves to one example each of these two situations. Saponification of anancomeric ethyl cyclohexanecarboxylates (Fig. 11.31)[120,121] illustrates steric hindrance. The carbonyl group in the ground state is sp^2 hybridized; the rate-determining transition state, in which an HO^- moiety becomes attached to the CO_2Et group, is sp^3 hybridized. As we have already seen in Table 11.6,

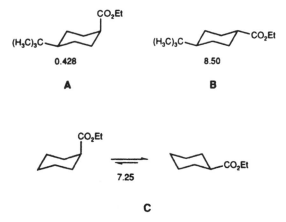

Figure 11.31. Specific saponification rates of ethyl cyclohexanecarboxylates (L mol^{-1} s^{-1} × 10^4).

conformational energies of sp^3 hybridized groups are generally larger than those of sp^2 hybridized ones: the latter (in contrast to the former) can escape crowding by turning their flat sides to the ring. In the present case there is the additional factor that the ground state is neutral (at least as far as the organic moiety is concerned) but the transition state is negatively charged, and therefore more solvated. As already pointed out in connection with the conformational energy of CO_2^- versus CO_2H, solvation leads to additional bias against the axial position. The saponification rate of the cis-4-tert-butyl (axial-CO_2Et, **A**) isomer is thus about 20 times less than that of the trans (equatorial-CO_2Et, **B**) isomer. [The saponification rate for the unsubstituted, conformationally heterogenous compound **C** is intermediate; cf. Eq. 11.1: $k_c = n_e k_e + n_a k_a$, where n_e and n_a refer to the mole fractions of the equatorial and axial conformers, respectively, and k_e and k_a refer to their specific saponification rates, approximately equal (see above) to k_B and k_A.]

A (presumed) case of steric assistance is shown in Figure 11.32; it concerns the oxidation rates of anancomeric cyclohexanols.[122] It has been pointed out[123] that the rates parallel the degree of crowding of the hydroxyl group, or, perhaps more concisely, the degree of strain relief that occurs when the sp^2 hybridized alcohol (or the corresponding chromate, which is an intermediate in the oxidation) is converted, in the rate-determining step, to the sp^2 hybridized ketone, with resulting relief of synaxial strain. Thus **A** reacts faster than **B** (relief of two synaxial OH/H interactions); **C** also reacts faster than **B** (relief of a synaxial CH_3/H interaction), and **D** reacts much faster than any of the others (relief of a severe CH_3/OH synaxial interaction, cf. Table 11.9). Again the rate of the conformationally mobile species **E** is intermediate.

We turn now to "stereoelectronic effects"[124] though at this point we shall confine the discussion to cyclohexane systems. Deslongchamps defines stereoelectronic effects as effects on reactivity of the spatial disposition of particular electron pairs, bonded or nonbonded. In many cases these are electron pairs in bonds that are formed, broken, or otherwise dislocated in the reaction in question; in other examples they are unshared electrons on exocyclic or endocyclic hetero atoms (cf. the discussion of the anomeric effect, Section 10-1.b and p. 387).

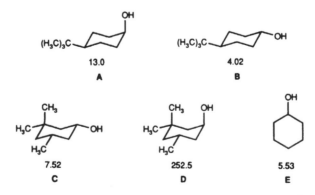

Figure 11.32.   Oxidation rates of cyclohexanols (L mol^{-1} s^{-1} × 10^3).

A well-known example is the S_N2 displacement reaction; it has long been known (cf. Chapter 5) that this reaction proceeds with inversion of configuration, that is, that the incoming nucleophile enters from the back of the leaving one. The stereoelectronic situation is that the σ^* orbital of the bond joining the reacting carbon to the leaving group is the one that gives rise to the new σ bond to the incoming nucleophile. The S_N2 reaction of cyclohexane p-toluenesulfonates with thiophenolate is depicted in Figure 11.27. Clearly, the reaction proceeds with inversion. Moreover, the axial (cis) tosylate ($k_a = 3.61 \times 10^{-4}$ L mol^{-1} s^{-1}) reacts about 31 times faster than the trans (equatorial) tosylate ($k_e = 0.116 \times 10^{-4}$ L mol^{-1} s^{-1}) at 25°C.[86b] This finding would appear to be due to a combination of steric and stereoelectronic effects: There is ground-state compression in the axial compound, giving rise to steric assistance [but since, according to Table 11.6 $-\Delta G_{OTs} = 0.50$ kcal mol^{-1} (2.1 kJ mol^{-1}), ground-state compression can only give rise to a rate factor of 2.3 at 25°C] and there is also steric hindrance to backside attack (from the axial side) in the equatorial compound, that is, the stereoelectronic requirement of backside attack produces an adverse steric factor.

A number of bimolecular elimination (E2) reactions in cyclohexyl systems have the stereoelectronic requirement that the ligands to be eliminated be antiperiplanar. This requirement may be an example of the principle of "least motion": After departure of the ligands to be eliminated, two antiperiplanar, that is, antiparallel, orbitals are optimally disposed to overlap and form a π bond. The next best orientation, from the viewpoint of optimal orbital overlap, is a synperiplanar one (which does not normally occur in cyclohexyl systems) whereas a synclinal orientation of the ligands to be eliminated is unfavorable. This condition leads to the conclusion that E2 eliminations in cyclohexanoid systems should be trans–diaxial, but not trans–diequatorial or cis–equatorial–axial, since the former array is antiperiplanar, whereas the latter two are synclinal. Numerous examples are found in the literature,[59,125] of which only one can be detailed here.

Of the *cis*- and *trans*-4-*tert*-butylcyclohexyl p-toluenesulfonates (Fig. 11.27), only the cis isomer undergoes bimolecular elimination with ethoxide in ethanol; the trans isomer is constricted to E1 (and S_N) pathways. Here, again, in the cis isomer OTs and H are antiperiplanar (diaxial), whereas in the trans isomer they are synclinal, either e,e or e,a. Cyclohexyl tosylate itself, though predominantly in the conformation with equatorial OTs, does undergo bimolecular elimination with ethoxide at a rate of 0.26 that of the *cis*-4-*tert*-butyl isomer[82] and this result can be readily explained in terms of the Winstein–Holness equation (Section 10-4): $k = n_e k_e + n_a k_a$; if $k_e = 0$, $k = n_a k_a$. According to Table 11.6 the conformational energy of the tosylate group is 0.50 kcal mol^{-1} (2.1 kJ mol^{-1}) from which one may calculate $n_a = 0.326$, at 75.2°C, in reasonable agreement with the experimentally found 0.26.

A substrate may react in a minor conformation, provided the reaction rate for that conformation is reasonably large. This statement applies to reactions of biochemical and pharmacological import (enzyme–substrate and drug–receptor interactions) just as well as to simpler chemical ones. For example (Fig. 11.33), although the diaxial mole fraction of *cis*-3-hydroxycyclohexanecarboxylic acid is very small, [estimated ΔG^0 ca. 3 kcal mol^{-1} (12.6 kJ mol^{-1}) corresponding to 0.6% diaxial conformer at room

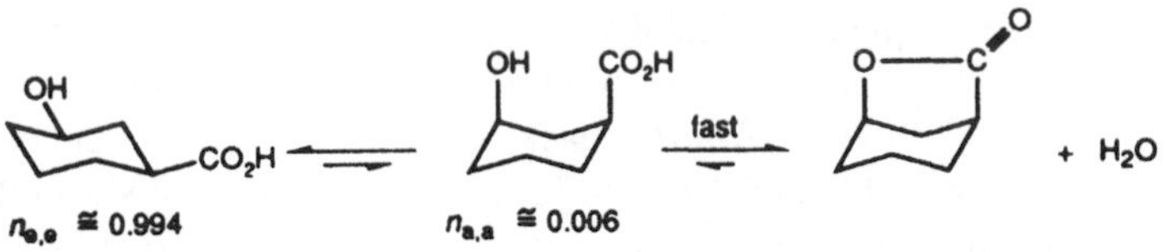

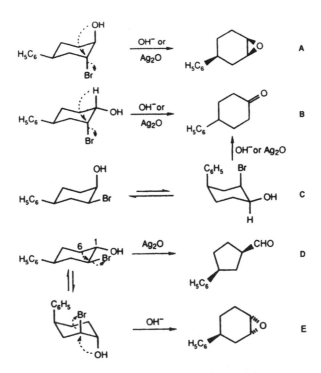

Figure 11.33.   Lactonization of *cis*-3-hydroxycyclohexanecarboxylic acid.

temperature], the acid lactonizes readily: Once the carboxyl and hydroxyl group are juxtaposed in the synaxial disposition, the rate of lactonization is, for entropic reasons, very high and the equilibrium very favorable to the lactone.

Ring formation, rearrangement, neighboring group participation, and fragmentation constitute another (interrelated) set of reactions with the stereoelectronic requirement that the groups involved must be antiperiplanar. Normally, this involves diaxial disposition; however, since an element of the ring may be antiperiplanar to an equatorial leaving group, ring contractions can involve equatorial substituents. This situation applies also to fragmentation reactions.[126] The reactions of bromohydrins (derived from cyclohexanes) with base and with silver ions illustrate these principles. These reactions are summarized in Figure 11.34 for the four diastereomeric 2-bromo-4-phenylcyclohexanols.[127] [The phenyl group, $-\Delta G^0_{confo} =$

Figure 11.34.   Reactions of 2-bromo-4-phenylcyclohexanols with base and silver oxide.

2.9 kcal mol^{-1} (12.0 kJ mol^{-1}) serves to bias the conformational equilibria of the various stereoisomers although the bias is clearly not compelling.] Case **A** exemplifies epoxide ring formation: The entering (OH or O$^-$) and leaving (Br) groups are antiperiplanar (a,a). Case **B** exemplifies ketone formation by either hydride shift or enolate formation (HBr elimination): The hydrogen involved (but not the hydroxyl group) is antiperiplanar to the leaving bromine. In Case **C**, where neither proper stereoelectronic situation is attained in the most stable starting conformation, the molecule apparently reacts in the alternate conformation, even though this involves synaxial phenyl and bromine. The fourth diastereomer (Case **D**) is particularly interesting. With Ag$_2$O, departure of the equatorial bromine is induced and the ring bond antiperiplanar to the departing equatorial bromine [C(1)/C(6)] migrates to produce a ring contraction to *cis*-3-phenylcyclopentanecarboxyaldehyde. With base (hydroxide) the driving force for ring contraction in the secondary halide seems to be insufficient (although such a contraction can occur with tertiary halide) and the molecule reacts in the alternate conformation to produce an epoxide (Case **E**).

Two other reactions in which stereoelectronic factors are crucial are electrophilic addition to alkenes and epoxide ring opening; these reactions will be discussed in the next section. Some additional examples for the operation of stereoelectronic effects will be illustrated later in the steroid system (Section 11-6.a).

f. *sp*^{2} Hybridized Cyclohexyl Systems

Various such systems have been discussed in a collection of reviews.[117]

Cyclohexene.   Cyclohexene has long been assumed to exist in the half-chair form depicted in Figure 11.35, **A**[128] and this has been confirmed by both microwave spectroscopy[129,130] and electron diffraction.[131,132] The torsion angles in cyclohexene are shown in Figure 11.35, **B** (see also refs. 133). Carbon atoms 3 and 6 lie in the plane of the double bond. Although this plane does not exactly bisect the H–C–H angles at C(3) and C(6) (this would require $\omega_{2,3} = \omega_{1,6}$ to be 0^0 (see Fig. 11.35, **C**; the actual value is around 15^0), the methylene hydrogen atoms at these positions are quite far from being in truly axial or equatorial positions; such ligands are called "pseudoaxial" (Ψ-a) and "pseudoequatorial" (Ψ-e). On the other hand, $\omega_{4,5}$ is very close to $60°$ and the hydrogen atoms at these positions are therefore truly axial or equatorial. (For a summary of torsion angles in cyclohexene and cyclohexadienes, see ref. 133.)

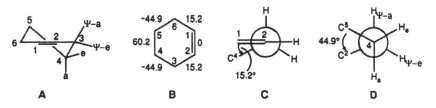

Figure 11.35. Cyclohexene half-chair and its torsion angle.

The inversion barrier in cyclohexene has been determined as 5.3 kcal mol^{-1} (22.2 kJ mol^{-1})[57b,134] using low-temperature NMR of deuterated analogues. Molecular mechanics and *ab initio* quantum mechanical calculations are in agreement with a barrier in this range; the transition state corresponds to the boat form[135,136] (see also refs. 137–139).

In 4-substituted cyclohexenes[140] the equatorial conformer is more stable than the axial one, though $-\Delta G^0$ is less than in correspondingly substituted cyclohexanes, presumably because there is only one synaxial X/H interaction in the axially substituted cyclohexene, whereas there are two such interactions in axially substituted cyclohexane. The conformational free energy values are compiled in Table 11.13; except for OH, where intramolecular hydrogen bonding may stabilize the axial conformer, these values are close to one-half the values in cyclohexane (Table 11.6). This finding supports the hypothesis that the steric interaction with the olefinic carbon C(2) is small or absent. Some of the variability of the data in Table 11.13, which were obtained in different solvents, may be due to solvent dependence. Polar solvents stabilize the axial conformer when the substituent is itself polar.[141]

A different situation arises for 3-substituted (allylically substituted) cyclohexenes. Although a 3-methyl group prefers the pseudoequatorial position by 0.97 kcal mol^{-1} (4.1 kJ mol^{-1}),[142] electronegative groups, such as OH, OCH$_3$, OAc, Cl, and Br are predominantly pseudoaxial,[143] with the preference for OH being 0.45 kcal mol^{-1} (1.9 kJ mol^{-1})[142] and that for Cl 0.13 kcal mol^{-1} (0.54 kJ mol^{-1}) by electron diffraction.[144] Values of 0.64 kcal mol^{-1} (2.68 kJ mol 1^{-1}) for Cl and 0.70 kcal mol^{-1} (2.93 kJ mol^{-1}) for Br have also been reported based on IR.[145] An effect akin to the anomeric effect (Section 10-1.b), that is, double-bond/no-bond resonance, may be responsible for the greater stability of the pseudoaxial groups (cf. Fig. 11.36).

The ΔG^0 values for conformational equilibria in di- and trisubstituted cyclohexenes have been tabulated.[128]

An important reaction of cyclohexenes is electrophilic addition. In the case of bromine addition, which normally involves a bromonium ion intermediate, this addition is generally antiperiplanar. It would appear that in cyclohexenes there are two possible modes of antiperiplanar attack (cf. Fig. 11.37) leading to diaxially or

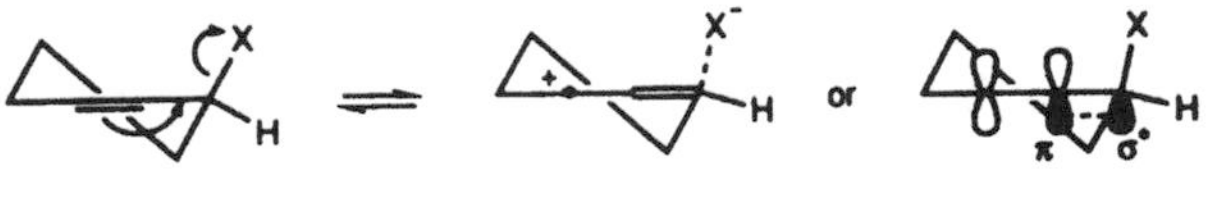

Figure 11.36.   Cyclohexenes with electronegative substituents at C(3).

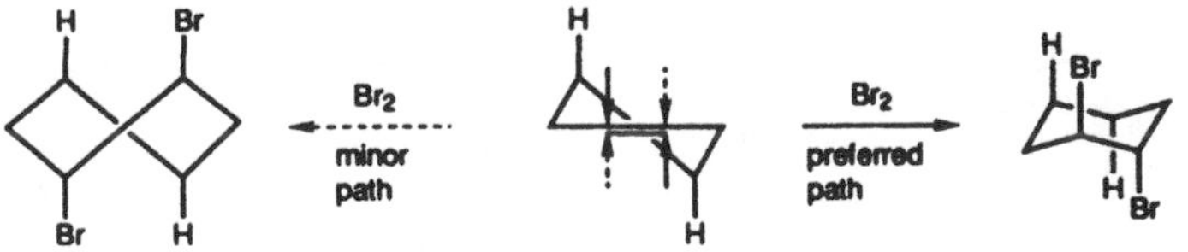

Figure 11.37.   Diaxial electrophilic addition to cyclohexenes.

TABLE 11.13. Conformational Energies in 4-Substituted Cyclohexenes[a]

Substituent	CH₃	C₆H₅	CO₂H	CHO	CO₂CH₃	COC₆H₅	CN	NO₂	OH
$-\Delta G^0$ (kcal mol⁻¹)	ca. 1; 0.86	0.99	1.0	0.45	0.85	0.45	0.1; 0.15	0.25	0; 0.22
$-\Delta G^0$ (kJ mol⁻¹)	ca. 4.2; 3.6	4.1	4.2	1.9	3.6	1.9	0.1; 0.63	1.05	0; 0.92

[a]References to the origins of these data may be found in ref. 90.

diequatorially 1,2-disubstituted cyclohexanes. However, a careful study of models (cf. Fig. 11.37) shows that antiperiplanar attack from the sides of the proximal axial hydrogen atoms gives rise to a diaxially disubstituted chair, whereas attack from the opposite sides gives rise to a diaxially disubstituted twist form, which must then subsequently invert to the diequatorially disubstituted chair. Since the twist form is over 5 kcal mol^{-1} (21 kJ mol^{-1}) less stable than the chair, the corresponding transition state leading to it is also quite destabilized and it is therefore not surprising that, under conditions of kinetic control, the predominant or exclusive product is that of diaxial addition.[146] In fact, it has been shown[147] that bromine addition to an anancomeric cyclohexene, in this case 4-*tert*-butylcyclohexene, gives the diaxial dibromide in at least 94:6 preference over the diequatorial. However, the same is not true for 3-*tert*-butylcyclohexene, which gives predominantly the product of diequatorial addition. Models show that the otherwise preferred mode of addition is impeded, at C(2), by the steric bulk of the *tert*-butyl group. Either this factor here overwhelms the normal preference for diaxial attack, or the molecule actually reacts in the alternate conformation with pseudoaxial *tert*-butyl.

Not all electrophilic additions are antiperiplanar; depending on mechanisms, they may, alternatively, by synperiplanar or nonstereoselective.[148]

Cyclohexene Oxide. Epoxidation of cyclohexenes, for example, with peracids give rise to cyclohexene oxides. These molecules, also, exist as half-chairs.[149,150] Ring opening of epoxides, either in acidic or in basic medium, usually proceeds with inversion of configuration.[151] In addition, however, opening of cyclohexene oxides generally proceeds in such fashion that the diaxial rather than the diequatorial reaction product is obtained (Fürst–Plattner rule[152]). Thus reduction of *cis*-4-*tert*-butylcyclohexene oxide (Fig. 11.38, **A**) with lithium aluminum deuteride leads to *c*-4-*t*-butylcyclohexan-*r*-ol-*t*-2-*d* (**B**).[153]

Cyclohexadienes. In the cyclohexadiene series, the conjugated 1,3 isomer[154] is nonplanar, as evidenced by microwave,[155] electron diffraction,[156] and NMR[133] techniques (see also ref. 139). In contrast, the 1,4 isomer 1,4-dihydrobenzene is planar[133,157–159] (there is also low-temperature X-ray evidence[160]) although its energy minimum is quite shallow. We shall see more of these two molecules later: of 1,3-cyclohexadiene as an inherently dissymmetric chromophore of interest in ORD and CD studies (Chapter 12), and of the 1,4 isomer in connection with its dibenzo derivative 9,10-dihydroanthracene,[161] which is nonplanar and presents interesting conformational features (p. 499).

Cyclohexanone. The geometry and conformation of cyclohexanone are summarized in a paper by Dillon and Geise.[163a] The most stable conformation of cyclohexanone is the chair; the C–C(O)–C angle is somewhat less than usual (115° with a corresponding expansion of the O–C–C angle to 122°) but the chair is nevertheless appreciably flattened at the carbonyl site, values for the H$_{eq}$–C–C–O torsion angle ranging from[164] 3.3° to 12.7°[165,166] The structure of 4,4-diphenylcyclohexanone has been determined by X-ray diffraction[137]; the ring

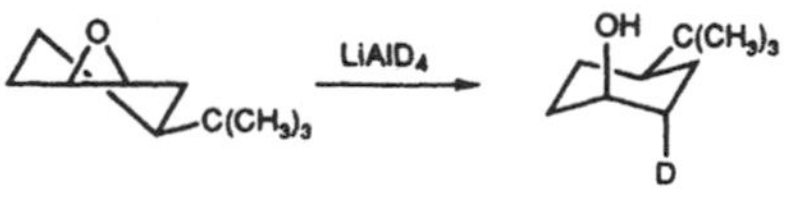

Figure 11.38.   Ring opening of *cis*-4-*tert*-butylcyclohexene oxide with lithium aluminum deuteride.

torsion angles are $\omega_{1,2}$ and $\omega_{6,1}$ (average) = 42°, $\omega_{2,3}$, and $\omega_{5,6}$ = 52°, $\omega_{3,4}$ and $\omega_{4,5}$ = 59°; that is, the ring is flattened at the carbonyl end and puckered at C(4) (the latter possibly because of the phenyl substituents). The twist form of cyclohexanone is calculated to be only 2.72 kcal mol^{-1} (11.4 kJ mol^{-1}) above the chair and the activation energy for chair inversion is found[168] to be 4.0 kcal mol^{-1} (16.7 kJ mol^{-1}); these values are considerably lower than corresponding values for cyclohexane (Section 11-4.a). The strain in cyclohexanone exceeds the strain in cyclohexane by about 3 kcal mol^{-1} (12.5 kJ mol^{-1}).[169] Cyclohexanone and cyclobutanone (and presumably also cyclopropanone, for which calculations are not available) are the only cyclanones for which this is the case; cyclopentanone and medium ring ketones are less strained than the corresponding hydrocarbons and the strain in simple aliphatic ketones is near zero.[164] Angle strain is clearly responsible for the excess strain in the small-ring ketones (deformation of the near-120° angle corresponding to sp^2 hybridization of the carbonyl to 90° or 60° is more difficult than corresponding deformation of the near-tetrahedral 109.5° angle in the hydrocarbon. The relief of strain occurring in addition reactions of cyclohexanone makes such reactions particularly facile both with respect to rate and with respect to equilibrium. Thus the reduction of cyclohexanone with sodium borohydride is 355 times as fast as that of di-*n* hexyl ketone and the equilibrium constant for HCN addition (ketone + HCN ⇌ cyanohydrin) is 70 times as large for cyclohexanone as for di-*n*-octyl ketone.

The stereochemistry of addition of nucleophiles to cyclohexanones is complex. Both the addition of organometallics[170] and of metal hydrides[171,172] have been studied extensively; equilibria in cyanohydrin formation have also been elucidated[173] (see also ref. 59, p. 116).

In nucleophilic additions to conformationally locked cyclohexanones, either equatorial attack (to give the axial alcohol) or axial attack (to give the equatorial alcohol) may occur (Fig. 11.39). The observed facts are as follows: Nucleophiles of small steric requirement add to unhindered cyclohexanones from the axial side to give equatorial alcohol **B**. In contrast, bulky nucleophiles add from the equatorial side to give axial alcohols **A**; predominant equatorial attack is also observed in ketones where the axial side is screened, for example, by synaxial substituents (Fig. 11.39, **C** and **D**). Thus reduction of the sterically open 4-*tert*-butylcyclohexanone with the unhindered LiAlH$_4$ in tetrahydrofuran (THF) gives 90% equatorial alcohol, whereas corresponding reduction of 3,3,5-trimethylcyclohexanone gives only 23% equatorial alcohol, hydride approach from the axial side being sterically hindered. The somewhat more bulky

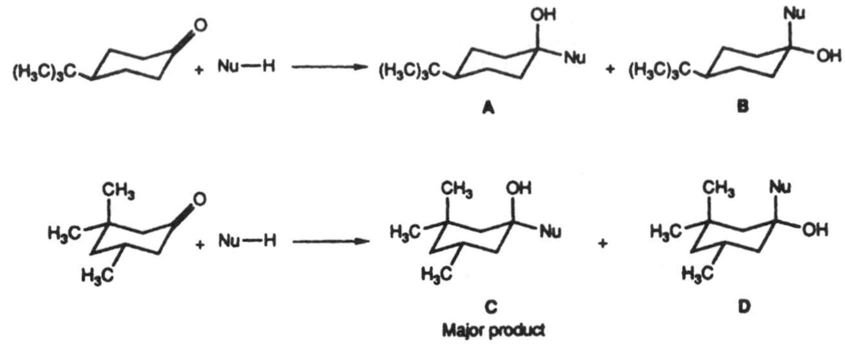

Figure 11.39.   Stereochemistry of addition of nucleophiles to cyclohexanones.

LiAlH(t-BuO)$_3$ still gives 90% equatorial alcohol with the former ketone but only 4–12% with the latter. To reduce 4-*tert*-butylcyclohexanone to the axial alcohol requires a very bulky hydride, such as L-Selectride™, LiBH(sec-Bu)$_3$, which gives 93% axial alcohol at room temperature and 96.5% at –78°C.[174a] (3,3,5-Tri-methylcyclohexanone with the same reagent gives 99.8% axial alcohol.) Even more stereoselective is lithium trisiamylborohydride, LiBH[CH$_3$CHCH(CH$_3$)$_2$]$_3$] which gives almost exclusively the axial alcohol even with the unhindered 4-*tert*-butylcyclohexanone.[174b] To get pure equatorial alcohol (**B**, Nu = H), one may equilibrate the aluminum complex of the axial alcohol or alcohol mixture using AlHCl$_2$ and excess ketone.[175]

Steric factors are also important in the reaction of cyclohexanones with organometallics. Thus the following percentages of equatorial attack (to give **A**, Fig. 11.39) on 4-*tert*-butylcyclohexanone are reported[170]: HC≡CNa, 12% (i.e., 88% axial attack); CH$_3$Li, 65%; CH$_3$MgI, 53%; C$_2$H$_5$MgI, 71%; (CH$_3$)$_2$CHMgBr, 82%; and (CH$_3$)$_3$CMgCl, 100%; that is, the more bulky the reagent, the more equatorial and the less axial attack and vice versa. A high degree of equatorial attack (82–99%) is observed with organoiron reagents.[176] An interesting situation arises with (CH$_3$)$_3$Al: when the reagent is used in a 1:1 ratio to ketone, methylation is mostly equatorial (via a four-centered transition state), whereas with a 2:1 ratio it is mostly axial (via a six-membered ring transition state). With 3,3,5-trimethylcyclohexanone, all the reagents [save (CH$_3$)$_3$Al in excess] give exclusively equatorial attack with formation of the axial alcohol **C**. High stereoselectivity in reactions with both hydrides and organometallics is also seen with camphor and norcamphor (Fig. 11.40): Camphor reacts nearly exclusively from the endo side (the exo side is screened by the geminal methyl groups), but the opposite is true for norcamphor in which the nucleophile approaches from the less hindered exo side (see also p. 503). There seems to be general agreement that equatorial attack with bulky reagents or hindered ketones is due to steric

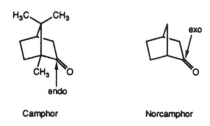

Figure 11.40. Nucleophilic approach to camphor and norcamphor.

interference to approach from the axial side ("steric approach control").[177] The reason why small nucleophiles add to unhindered cyclohexanones from the axial side is still controversial.[90] The trajectory of the incoming nucleophile (Nu) is not perpendicular to the C=O σ bond but at an angle of 109° (Nu–C–O), in the π plane.[49,178]

Methylenecyclohexane.   The structure of methylenecyclohexane[162] has not been determined experimentally, but force field calculations[138,179] suggest that it is quite similar to that of cyclohexanone. The barrier to inversion, however, is considerably higher, between 7.7 ± 0.5 kcal mol^{-1} (32.2 ± 2.1 kJ mol^{-1}) and 9.0 ± 0.6 kcal mol^{-1} (37.7 ± 2.5 kJ mol^{-1}).[143,180,181] About two-thirds of this barrier derives from angle strain in the transition state.[138,182] This finding lends credence to the hypothesis that the much higher inversion barrier in methylenecyclohexane compared to cyclohexanone results from the greater torsional potential of the X=C–C single bonds when X = CH$_2$ compared to X = O. A similar situation is seen in the isobutylene versus acetone barriers [2.2 kcal mol^{-1} (9.2 kJ mol^{-1}) vs. 0.78 kcal mol^{-1} (3.26 kJ mol^{-1})]. Methylenecyclohexane is thermodynamically unstable with respect to 1-methylcyclohexene; the equilibrium, which is easily established in acid, lies far on the side of the endocyclic alkene. The enthalpy difference of 1.71–1.74 kcal mol^{-1} (7.2–7.3 kJ mol^{-1})[183,184] is somewhat larger than that between 2-methyl-1-pentene and 2-methyl-2-pentene (1.34 kcal mol^{-1}, 5.6 kJ mol^{-1}); in addition, the entropy is higher for methylcyclohexene so that the free energy advantage of the latter at 25°C is 2.84–2.87 kcal mol^{-1} (11.9–12.0 kJ mol^{-1}). The conformational energy for a methyl group in the 3-position of methylenecyclohexane, 0.8 kcal mol^{-1} (3.3 kJ mol^{-1}),[185] on the other hand, is substantially less, not only than the value in methylcyclohexane (1.74 kcal mol^{-1}, 7.28 kJ mol^{-1}), but also than the value in 3-methylcyclohexanone (1.36 kcal mol^{-1}, 5.69 kJ mol^{-1}); again it is not obvious why. For polar substituents in the 3-position, however, the conformational energies tend to be larger than in corresponding cyclohexanes in nonpolar solvents but smaller in polar solvents (for a summary, see ref. 162).

   The conformational energy of a methyl group at C(2) methylenecyclohexane is also relatively small, 1.0 kcal mol^{-1} (4.2 kJ mol^{-1}).[143] Polar substituents at C(2), such as CH$_3$O, in this case actually prefer the axial position by 0.4 kcal mol^{-1} (1.7

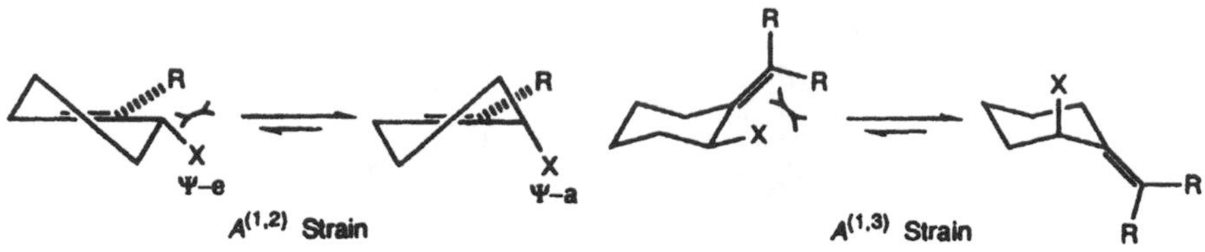

Figure 11.41. Examples of $A^{(1,2)}$ and $A^{(1,3)}$ strain.

kJ mol^{-1}). The reason may, in part, be similar to that shown for 3-substituted cyclohexenes in Figure 11.36 (hyperconjugative orbital overlap). However, there is also a steric factor that is quite important in cyclohexenes and very important in methylenecyclohexanes, especially when the terminal substituents on the alkene (R in Fig. 11.41) are larger than hydrogen. This factor is called $A^{(1,2)}$ strain in a 1,6-disubstituted cyclohexene and $A^{(1,3)}$ strain in a 2-substituted methylenecyclohexane[162,186a]; the two types of strain are shown in Figure 11.41. In both instances the strain leads to a disfavoring of the pseudoequatorial position.

The $A^{(1,3)}$ strain is of particular interest in enamines and other ketone derivatives, where it has found useful synthetic applications. Examples are the synthesis of trans-1,3-dimethylcyclohexane[186b] and trans-2,6-dimethylcyclohexanone[187] shown in Figure 11.42. The pyrrolidine enamine shows both regio- and stereoselectivity in the following way: In the enamine itself, after alkylation the double bond will be on the unsubstituted side in order to avoid a cis interaction between the 2-alkyl group and the enaminic nitrogen. Similarly, $A^{(1,3)}$ strain forces the 2-alkyl substituent, whether from the incoming nucleophile (CH$_3$I) or already present in the ketone, into the axial position. Of course it requires a mild method of hydrolysis, as in the case of the trans-2,6-dimethylcyclohexanone or avoidance of hydrolysis altogether, as in the case of trans-1,3-dimethylcyclohexane to prevent epimerization of the initial product to the more stable diequatorial epimer. In the case of the alkylation of the 2-methylcyclohexanone N',N'-dimethylhydrazone, avoidance of $A^{(1,3)}$ strain leads to formation of the anti or E isomer of the hydrazone, which then (again to avoid $A^{(1,3)}$ strain) must be alkylated axially. (Geminal dialkylation is presumably shunned because the tertiary CH$_3$CH site is kinetically disfavored in proton abstraction because of lesser acidity.) In the third example, methyl groups at C(2) and C(6) were both introduced axially, even though this causes a severe synaxial interaction; unfortunately partial epimerization during hydrolysis leads to a mixture of only 60% 2a,6a, and 40% 2e,6a dimethyl ketone. One surprising and as yet not well-understood feature is that the first alkylation even in the imine occurs syn to the N-alkyl substituent.[188]

An interesting application of $A^{(1,2)}$ strain is seen in the synthesis of solenopsin A (Fig. 11.43).[189] Ordinary hydride reduction of the precursor imine (a tetrahydropyricline) gives largely the undesired cis (diequatorial) product.

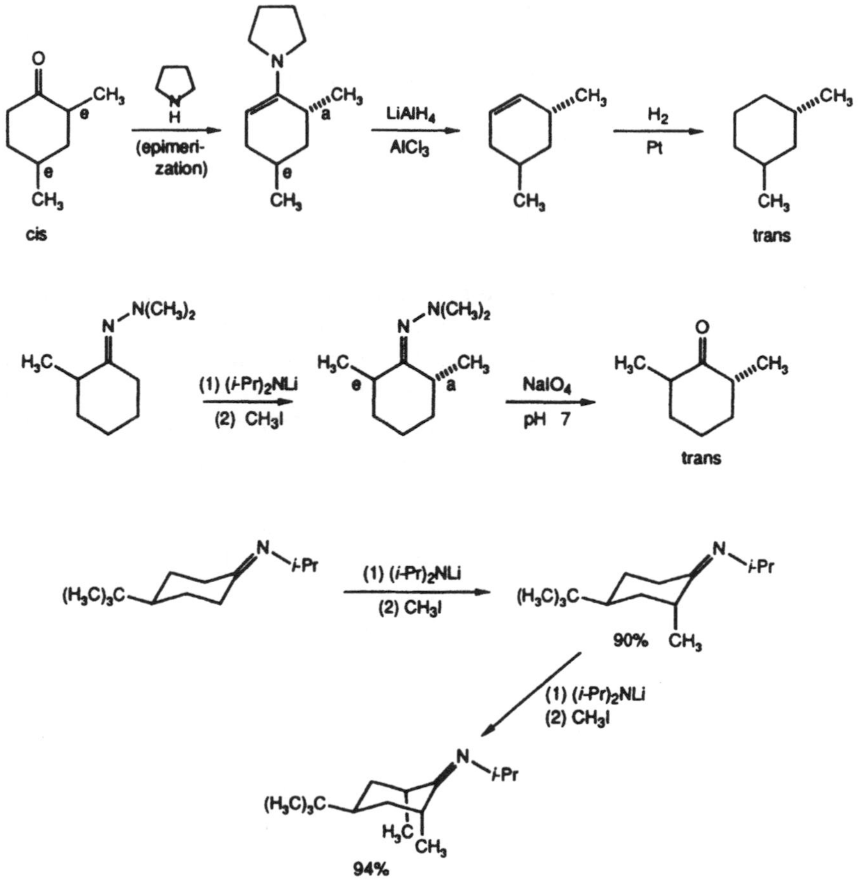

Figure 11.42. Synthetic applications of $A^{(1,3)}$ strain.

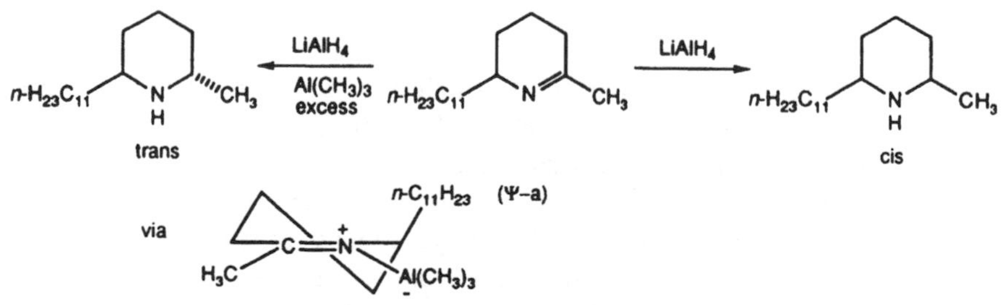

Figure 11.43. Synthesis of solenopsin A.

However, when $LiAlH_4$ reduction is carried out in the presence of an excess of $Al(CH_3)_3$, complexation of this Lewis acid to the ring nitrogen engenders $A^{(1,2)}$ strain and presumably forces the *n*-undecyl group into the pseudoaxial position; reduction of the C=N bond leading to equatorial CH_3 as before then produces axial–equatorial, that is, trans stereochemistry.

g. Six-Membered Saturated Heterocycles

In a book[59] on conformational analysis published in 1965, the conformational analysis of saturated heterocycles was dealt with in 12 pages. The thinking at that time was that the conformational analysis of saturated heterocycles was, with some minor modifications, very similar to that of cyclohexane. In the intervening years, however, attention has been focused on the differences and by 1980 the subject was worthy of a 152-page monograph[190] (see also refs. 191–193). Only a brief account of the subject can be given here.

The simple saturated heterocycles all exist as chairs in their most stable conformations, with the inversion barriers given in Table 11.14.

It is seen from Table 11.14 that the inversion barriers for cyclohexane, piperidine, and oxane (tetrahydropyran) are essentially identical; as one goes down the periodic table (O, S, also, Se, Te — not shown) the barriers become lower, probably because the torsional potentials diminish.[194] Substitution on the heteroatoms seems to raise the barrier slightly: $S < SO < SO_2$; $NH < NCH_3$, perhaps because the torsional barriers increase.

Chair–twist energy differences seem not to be known for the simplest (mono-hetero) systems; a comparison between cyclohexane and the dihetero systems 1,3-dioxane and 1,3-dithiane[74,196] is given in Table 11.15. The twist form in 1,3-dioxane is of higher energy, relative to the chair, than that in cyclohexane, but the opposite is true for 1,3-dithiane. These differences have been explained in terms of molecular dimensions; since the bond lengths are C–O < C–C < C–S, steric interactions of carbon and hydrogen atoms across the ring in the twist form should be most serious in 1,3-dioxane and least so in 1,3-dithiane. A contributing factor may be the somewhat lesser torsional potential in 1,3-dithiane. A ring with four sulfur atoms, 3,3,6,6-tetramethyl-1,2,4,5-tetrathiane, in fact exists in a readily measurable chair–twist equilibrium.[197]

Conformational equilibria of methyl groups in six-membered saturated heterocycles are tabulated in Table 11.16, which lists the ground-state free energy difference between a ring with an axial methyl substituent at the position indicated and one with the corresponding equatorial substituent.

Little variation in $-\Delta G^0$ between different heterocycles and cyclohexane is seen at the 4 position; equatorial and axial methyl groups at C(4) are evidently "cyclohexane-like" except for possible small variations in the exact ring shape, as one would expect. Rather larger variations are seen at C(2) and C(3), for reasons that become obvious from Figure 11.44. In the 3-substituted systems, one of the CH_3/H synaxial interactions (in axially substituted cyclohexanes) is replaced by a CH_3/: (lone

TABLE 11.14. Inversion Barriers in $C_5H_{10}X$

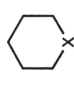

X:	CH_2	O	S	SO	SO_2	NH	NCH_3
$\Delta G^{\ddagger}$ (kcal mol^{-1})	10.25	10.3	9.4	10.1	10.3	10.1	11.9
$\Delta G^{\ddagger}$ (kJ mol^{-1})	42.9	43.1	39.3	42.3	43.1	42.3	49.8
t (°C); ref.	-60^a	-61^b	-81^b	-70^b	-63^b	-63^c	-29^c

[a]See p. 438.
[b]Reference 194.
[c]Reference 195. $\Delta G^{\ddagger}$ values calculated from chemical shifts and coalescence temperatures.

TABLE 11.15.   Chair–Twist (c–t) Energy and Entropy Differences in

Compound	X	ΔG^0_{c-t} [a]	ΔH^0_{c-t} [a]	ΔS^0_{c-t} [b]
Cyclohexane[c]	CH_2	4.9 (20.5)	5.9 (24.7)	3.5 (14.6)
1,3-Dioxane	O	5.7 (23.8)	7.1 (29.7)[d]	4.8 (20.1)
1,3-Dithiane[e]	S	2.9 (12.1)	4.3 (17.9)	4.7 (19.5)

[a]In kilocalories per mol (kcal mol^{-1}) at 25°C; values in parentheses are in kilojoules per mole (kJ mol^{-1}).
[b]In gibbs or calories per degree per mole (cal deg^{-1} mol^{-1}); values in parentheses are in joules per degree per mole (J deg^{-1} mol^{-1}).
[c]See also ref. 74.
[d]The true value may be even higher.[74]
[e]Reference 198.

pair) interaction that is known (see below) to be less severe, leading to a drop in $-\Delta G^0$, albeit not a very large one (the difference is especially small for piperidine and N-methylpiperidine).

In the 2-substituted systems, the nitrogen and especially oxygen heterocycles display a larger $-\Delta G^0$ for methyl than does cyclohexane, whereas the value in 2-methylthiane is smaller. This appears to be a consequence of changes in molecular dimensions: Since bond lengths are in the order C–O < C–N < C–C < C–S, the

TABLE 11.16.   Conformational Free Energies[a] ($-\Delta G^0$) of Methyl Substituents in Saturated Heterocycles

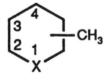

	Position				
Group X	1	2	3	4	Reference
CH_2	1.74 (7.28)	1.74 (7.28)	1.74 (7.28)	1.74 (7.28)	92b, 93
O		2.86 (12.0)	1.43 (5.98)	1.95 (8.16)	199
S	[0.28 (1.17)][b]	1.42 (5.94)	1.40 (5.86)	1.80 (7.53)	200b
NH	[0.36 (1.51)][c]	2.5 (10.5)	1.6 (6.7)	1.9 (7.95)	203
NCH_3	[3.0 (12.6)][d]	1.7 (7.1)	1.6 (6.7)	1.8 (7.5)	203
NCH_3H^+	2.1 (8.8)	1.4 (5.9)	2.2 (9.2)	1.6 (6.7)	203

[a]In kilocalories per mole (kcal mol^{-1}); values in parentheses are in kilojoules per mole (kJ mol^{-1}).
[b]S-Methylsulfonium salt: ref. 200a.
[c]This is the value for the N–H (equatorial vs. axial H) equilibrium: refs. 201 and 202.
[d]From ref. 204.

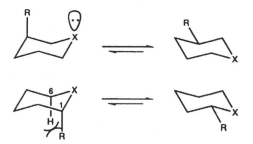

Figure 11.44. Conformational inversion in 3- and 2-substituted heterocyclohexanes.

distance between an axial methyl at C(2) and the synaxial hydrogen at C(6) increases as one passes from oxane to piperidine to cyclohexane to thiane and $-G^0_{2\text{-}CH_3}$ decreases accordingly (cf. Fig. 11.44). The 2-methyl groups in N-methylpiperidine and its salt are clearly exceptions from the norm, for reasons not yet understood. The N-methyl group in N-methylpiperidine has a very large $-\Delta G^0$ value.[204]

We shall deal only briefly with rings having more than one hetero atom. Study of the 1,3-dioxane and 1,3-dithiane systems (Fig. 11.45), preceded that of the simpler monoheterosubstituted species by several years, for whereas the oxane, thiane, and piperidine systems in general have to be studied by low-temperature ^{13}C NMR,[205] the conformational equilibria in the disubstituted systems shown in Figure 11.45 could be approached through configurational changes by acid-catalyzed equilibration, as shown. Since these equilibria, in general, present no surprises and have been extensively reviewed,[190,206,207] we shall refer here to only one case: 2-methyl-5-*tert*-butyl-1,3-dioxane [Fig. 11.45, X = O, R = CH$_3$, R' = C(CH$_3$)$_3$]. This case is remarkable because NMR spectra unequivocally show[208,209] that the preferred conformer has the 2-methyl group in the equatorial position and the 5-*tert*-butyl one in the axial; moreover, $-\Delta G^0$ is only 1.4 kcal mol^{-1} (5.9 kJ mol^{-1}) (see also ref. 210). Clearly, the steric compression of the axial 5-*tert*-butyl group by the 1,3-oxygen atoms with their lone pairs is very small compared to the compression by synaxial hydrogen atoms in axial *tert*-butylcyclohexane (cf. Table 11.6).

The N-substituted 1,3-diazane system (Fig. 11.45, X = N–CH$_3$) is of interest because cis–trans equilibrium is established spontaneously at room temperature

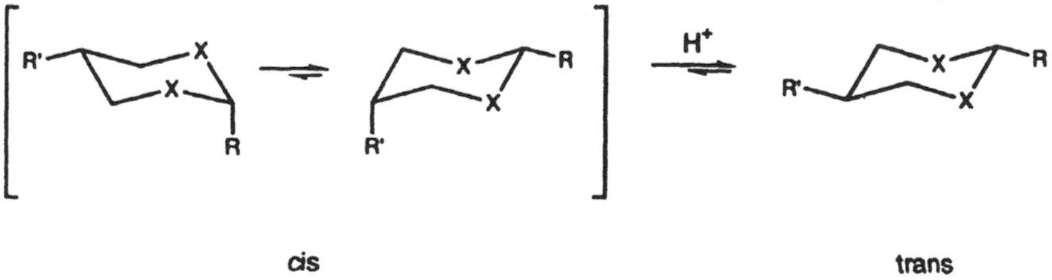

Figure 11.45. Equilibria in 1,3-dioxanes and 1,3-dithianes, where X = O or S.

though equilibration is slow on the NMR time scale: Both isomers can be seen spectrally.[211] Similar observations have been made in oxazanes.[212]

In cyclohexyl systems (cf. Table 11.6) the equatorial position of a substituent is almost invariably preferred, the lowest observed conformational energy value (for HgX) being about zero. As we shall see presently, this is not necessarily true in heterocyclic systems where there are many cases of axial preference. A case in point is thiane sulfoxide (Fig. 11.46) in which axial SO is preferred by 0.18 kcal mol^{-1} (0.73 kJ mol^{-1})[213]; even in the thiane sulfonium salt the equatorial preference is only 0.28 kcal mol^{-1} (1.15 kJ mol^{-1}) (cf. Table 11.16).

The most important conformational differences between heterocyclic and carbocyclic systems are due to polar factors. In cyclohexanes, polar factors only come into play when there are at least *two* polar substituents in the ring. In heterocyclic systems, on the other hand, at least one polar group (and sometimes more than one) is already in the ring and the parent compounds usually have dipole moments (except when, as in the case of 1,4-dioxane, there is a compensation of dipoles). Thus a single polar exocyclic substituent will give rise to dipolar interactions (either attractive or repulsive), which depend on the location of the substituent relative to the endocyclic heteroatom or heteroatoms, as well as on the nature of these atoms. However, dipolar interactions are not the only ones operative; other, stereoelectronic factors are often called into play as well, of these the most important is the anomeric effect.[124,214–217] Since recognition of this effect has its origin in carbohydrate chemistry, a brief excursion into the stereochemistry of the aldopyranoses is in order here.[59,218]

Representative stereoformulas of aldohexoses were shown in Fig. 3.17. The sugars were depicted as open-chain polyhydroxyaldehydes which are, however, in equilibrium with the cyclic hemiacetals shown for D-glucose in Figure 11.47, equilibrium being largely on the side of the latter. And, as also shown in Figure 11.47, the actual shape of the hemiacetals is that of an (oxane) chair, the more stable of the two chair forms being that with the CH$_2$OH group (and, as it happens, also the hydroxyl groups) equatorial. While it is recognized that cyclization can also lead to five-membered (THF) rings, as in the furanose forms of sugars, only the pyranose forms (six membered) will be discussed here.

We have seen earlier that the aldohexoses, having four chiral centers of the CHOH type, exist as 2^4 or 16 stereoisomers (8 diastereomeric pairs of enantiomers). While, in the modern definition of the term, any pair of diastereomeric aldohexoses differing only in the configuration of one chiral center may be called a pair of epimers, in classic sugar chemistry the term "epimers" referred to pairs of sugars specifically differing in configuration at C(2), such as glucose and mannose (Fig. 3.17). Contemplation of Figure 11.47 indicates that cyclization to a six-membered ring (hemiacetal formation)

Figure 11.46. Thiane sulfoxide equilibrium; $\Delta G^0 = +0.18$ kcal mol^{-1} (+0.75 kJ mol^{-1}).

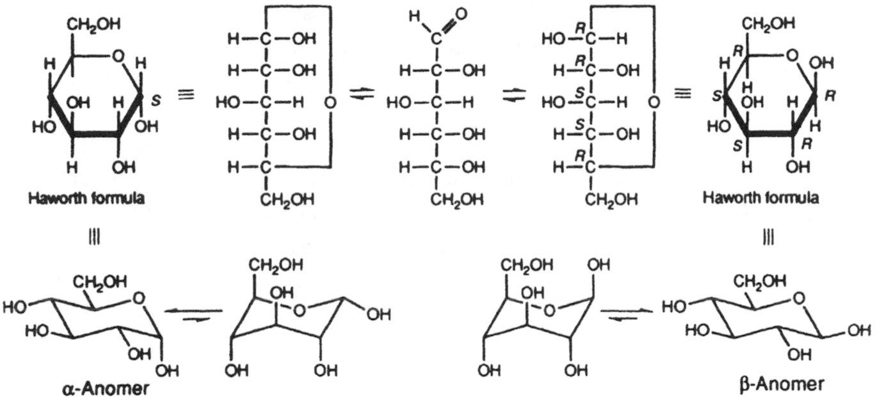

Figure 11.47. Open-chain and various cyclic formulas for D-glucose.

in any one of the 16 stereoisomeric aldohexoses will lead to two possible stereoisomeric pyranoses differing now in configuration at C(1). Such diastereomers are called "anomers" and C(1) is sometimes called the "anomeric carbon" or "anomeric center." Since equilibrium between the hemiacetal and open-chain forms of sugars is readily established, and since ring closure can lead to either anomer, the anomers are in equilibrium in solution. It is, however, possible to purify them by crystallization; thus crystallization of glucose from water below 50°C in the ordinary manner or from ethanol produces α-D-glucose, $[\alpha]_D^{20}$ + 112.2 (H_2O), whereas crystallization from pyridine or by vacuum evaporation of the glucose syrup at 115°C produces the β anomer, $[\alpha]_D^{20}$ + 17.5. When either anomer is dissolved in water, equilibrium is established in a matter of hours (the exact half-life is dependent on pH) and the initial rotation gradually changes to the equilibrium value of $[\alpha]_D^{20}$ 20 = +52.7; this phenomenon is called "mutarotation."

Anomeric sugars are given the symbols α or β, as shown in Figure 11.47. When a D sugar (pyranose) is written in such a way that the six-membered ring is oriented with the ring oxygen at the rear and the anomeric carbon on the right (so-called *Haworth formula* as in Fig. 11.47), the α form is the one with the anomeric hydroxyl (or other functional group) below the plane of the ring, whereas the β form has it above that plane. The opposite (α: OH up; β: OH down) applies in the L series: the mirror image of the α-D isomer is the α-L isomer, not the β-L isomer. In both cases the hydroxyl or other anomeric group in the β series is on the same side as the CH_2OH group in aldohexopyranoses. In the major chair conformer of most aldohexoses (Fig. 11.47) the α anomer has an axial OH and the β has an equatorial OH.

From the initial and equilibrium values of the rotations of the anomers of D-glucose one can compute the equilibrium mixture to be made up of about 36% of the axial α anomer and 64% of the equatorial β, hence $K = 1.78$ favoring the equatorial isomer and $\Delta G_{25}^0 = -0.34$ kcal mol^{-1} (−1.43 kJ mol^{-1}) (solvent water). This value should be compared with the conformational energy of hydroxyl in cyclohexane (Table 11.6): $\Delta G^0 = -0.95$ kcal mol^{-1} (−3.97 kJ mol^{-1}) in a hydroxylic solvent; evidently the preference for the β (equatorial) anomer of glucose is considerably less than anticipated. The discrepancy is even larger than this calculation might imply, since, as exemplified in Table 11.16, substituents at the 2 position in oxanes ordinarily have *larger* conformational energies than corresponding substituents in cyclohexane (see also ref. 219). Moreover, in the corresponding acetal, methyl D-glucoside [CH_3O instead of HO at C(1)] the α (axial) isomer predominates over the equatorial at equilibrium by about 2:1. This tendency for the axial anomer to predominate was first explained by Edward[221] and has been termed "anomeric effect" by Lemieux.[222,223] The effect is not limited to sugars but may be seen in simple 2-alkoxytetrahydropyrans as well as the corresponding 2-alkylthio analogues and analogues with sulfur in the ring (2-alkoxy- and 2-alkylthiothians), as shown in Figure 11.48.[214]

The origin of the anomeric effect, which has long been a matter of controversy, has now been well established by both experiment and quantum mechanical calculations (some of which were alluded to in Chapter 10). It is clear from Figure 11.48 that the axial isomer **A** has a lower dipole moment than the equatorial isomers **B**; therefore it should be favored by the electrostatic term V_E in Eq. 2.1, but, as explained earlier, that advantage should be less in solvents of increasing dielectric constant, both because V_E decreases (Coulomb's law) and because the solvation term V_S becomes more favorable to the higher dipole conformation. There is, however, a second factor, as explained earlier (see Fig. 10.11), namely, the overlap of the p orbital on X with the σ^* orbital of the C(1)–Y bond, shown as the equivalent double-bond/no-bond resonance in Figure 11.48, **A'**. This kind of overlap is not possible for the equatorial CH_3Y group.

However, the reverse motion of electrons, in which the exocyclic CH_3Y is the donor and the endocyclic X the acceptor of electrons, is possible in both isomers, as shown in Figure 11.49 (antiperiplanar pairs and bonds in heavy print). This "reverse" electron donation is called the "exo-anomeric effect."[214,224–227] Figure 11.49 also shows the C–O bond distances in glycosides.[228] The O(endo)–C(5) (143.3–143.5 pm) distance may serve as the standard of comparison. The O(endo)–C(1) distance 141.6 pm is appreciably shorter in the axial glycosides than in the equatorial ones because only the axial glycoside can partake of the electron overlap shown in Figure 11.49, **A ↔ A'**, which leads to a foreshortening of the O(endo)–C(1) bond. However,

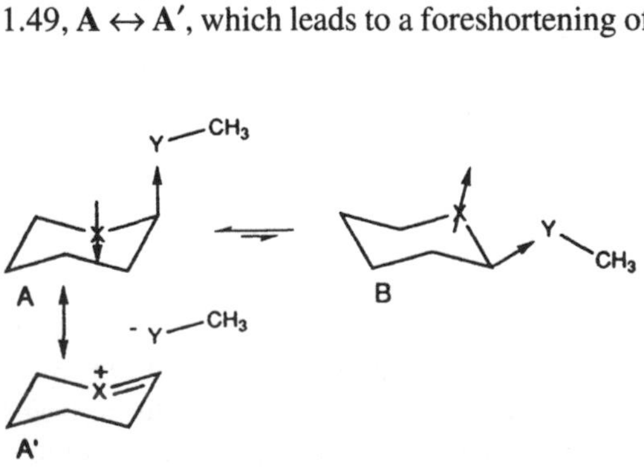

	$+\Delta G^0$ kcal/mol (kJ/mol)	
	CCl₄	CH₃CN
X=Y=O	0.89 (3.72)	0.37 (1.55)
X=O, Y=S	0.48 (2.01)	0.00 (0.00)
X=S, Y=O	1.53 (6.40)	
X=Y=S	0.42 (1.76)	

Figure 11.48. Anomeric effect. For X = Y = O see ref. 220a; for X = O, Y = S see ref. 220b. Data for X = S, Y = O; X = Y = S are from ref. 214.

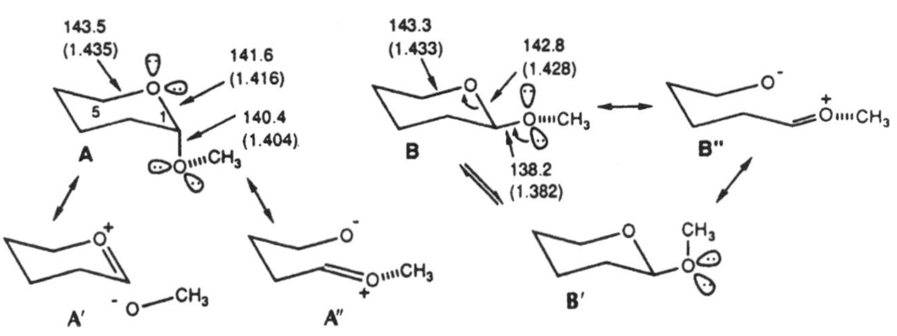

Figure 11.49. The C–O bond distances (Å) in glycosides.

both the axial and equatorial glycosides are capable of the reverse electron motion shown in Figure 11.49; **A ↔ A″, B ↔ B″**, or **B′ ↔ B″**; therefore the C(1)–O(exo) bond is foreshortened in both. This effect (foreshortening to 138.2 pm) is more important in the equatorial glycoside because it represents the only orbital overlap possible there, whereas in the axial glycoside the exocyclic OCH_3 can be either donor **A″** or acceptor **A′**; the foreshortening of the exocyclic C(1)–O(exo) bond (to 140.4 pm) is therefore less pronounced in the latter.

For the kind of overlap shown in Figure 11.49, **A″** and **B″**, to occur, one of the unshared pairs on the exocyclic oxygen (shown in bold in **A**, **B** and **B′**) must be antiperiplanar to the C(1)–O(endo) bond. In the axial isomer (and granting that the methoxyl methyl group can, for steric reasons, not point into the ring) this requires the exocyclic methyl group to be gauche to the ring oxygen as shown. In the equatorial isomer, it must also be gauche to the ring oxygen, but there are two possible conformations (**B** and **B′**) of which the first one is preferred on steric grounds. In either case there is a definitive preference for the conformation in which the methoxy methyl is gauche to the ring oxygen rather than the ring $CH_2(2)$ group.

A simpler situation is encountered when the group at C(1) in a sugar, or C(2) in a tetrahydropyran or 1,4-dioxan, is a halogen, which is a good acceptor but a poor donor for electrons. In that case there is a strong preference for the axial conformation of the halogen and a substantial shortening of the endocyclic C–O bond next to it. An example[214] is shown in Figure 11.50. The fact that the anomeric effect is more important for Br than for Cl[214] and for SCH_3 than for OCH_3 (cf. Fig. 11.48) is a

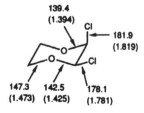

Figure 11.50. Crystal structure of *cis*-2,3-dichloro-1,4-dioxane; distances in pm and (in parentheses) in Å.

manifestation of the lower energy level of the σ^* orbital of the C–X bond as X moves down the periodic table. (For studies of the anomeric effect in second- and third-row elements see, for example, refs. 229 and 230.

11-5. CHEMISTRY OF RING COMPOUNDS OTHER THAN SIX-MEMBERED ONES

a. Three-Membered Rings

Three-membered rings are necessarily planar, and therefore quite highly strained (cf. Fig. 11.6), even though the interorbital angle in cyclopropane is considerably larger than the internuclear angle of 60° inasmuch as the orbitals do not overlap end-over-end (see ref. 231 for a survey of heats of formation of three-membered ring systems).

The total strain in cyclopropane is only slightly greater than that in cyclobutane (cf. Table 11.4), which implies that angle strain (Fig. 11.6) cannot exclusively account for the difference. There must be an electronic (orbital) stabilization of cyclopropane that partly offsets its greater apparent strain (see Chapter 2 in ref. 24; and ref. 232).

Introduction of double bonds (either endocyclic or exocyclic) into the cyclopropane ring increases the strain further, as expected; the strain in cyclopropene is 52 kcal mol^{-1} (218 kJ mol^{-1})[233] and that in methylenecyclopropane 41.0 kcal mol^{-1} (171.5 kJ mol^{-1}); the isomerization of methylcyclopropene to methylenecyclopropane is exothermic by 10.3 kcal mol^{-1} (43.1 kJ mol^{-1}).[234] Even though cyclopropene avoids the H/H bond eclipsing that exists in cyclopropane and (to a lesser extent) in methylenecyclopropane, angle strain is maximized when two sp^2 hybridized carbon atoms are in the ring.

Despite the high strain in cyclopropene, two cyclopropene derivatives, sterculic acid (Fig. 11.51, $n = 7$) and malvalic acid (Fig. 11.51, $n = 6$), occur in nature in seed oils.[235] Derivatives of cyclopropane also are found in nature, for example, the pyrethrins.

Cyclopropanone[236] is also very strained and quite unstable; it easily adds water or alcohols to give geminal diols or hemiketals. Its C=O stretching frequency is unusually high at 1813 cm^{-1}, suggesting a high degree of s character in the C=O bond resulting from the C–C bonds in the ring having a high degree of p character to minimize angle deformation. (The normal angle between p bonds is 90°, that between sp^3 hybridized bonds is 109°28′.) On the other hand, cyclopropenone is

$H_3C (H_2C)_7$ $(CH_2)_nCO_2H$

Figure 11.51.   Sterculic and malvalic acids.

Figure 11.52. Aromatic character of cyclopropenone and the cyclopropenium cation.

quite stable (its preparation has been described in *Organic Syntheses*[237] as is the cyclopropenium cation (Fig. 11.52); both of these systems are aromatic.[238a] In contrast, anions derived from cyclopropane, which tend to be highly pyramidalized[239] are "antiaromatic."[238b]

b. Four-Membered Rings

A four-membered ring can be either planar or puckered. The planar ring has minimal angle strain (cf. Fig. 11.6) but will have maximal torsional strain because of its eight pairs of eclipsing hydrogen atoms. The torsional strain can be reduced by puckering (at the expense of some increase in angle strain) and this is exactly what happens in cyclobutane (cf. ref. 240, from which much of the following information is taken). Cyclobutane is therefore a wing-shaped molecule with an "angle of pucker" ϕ (cf. Fig. 11.53) of 28°[241] and a barrier to ring inversion of 1.45 kcal mol^{-1} (6.06 kJ mol^{-1})[241]; the angle between the planes defined by atoms 1,2,3 and 3,4,1 is 180 − ϕ.

The inversion barrier in cyclobutane is high enough for the molecule to display two distinct energy minima, that is, two (superposable) conformations can be defined; this is also true for most heteracyclobutanes. The barrier in thietane (trimethylene sulfide) amounts to 0.75 kcal mol^{-1} (3.1 kJ mol^{-1}). However, oxetane (trimethylene oxide) has a barrier of only 0.1 kcal mol^{-1} (0.4 kJ mol^{-1}) and may be considered effectively planar since a single molecular vibration will carry the molecule over the barrier. A similar situation occurs in 1,3-dithiacyclobutane which, while showing puckering in the vibrational spectrum, crystallizes in the average planar conformation (according to X-ray structure analysis).[242]

As a result of the puckering in cyclobutane, monosubstituted derivatives, such as cyclobutyl bromide, can exist as axial or equatorial conformers (cf. Fig. 11.53). As in cyclohexyl compounds the equatorial conformer is, for steric reasons, the more stable one. In the case of bromocyclobutane the energy difference between the two conformers, about 0.85 kcal mol^{-1} (3.6 kJ mol^{-1}),[243] appears to be greater than the

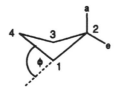

Figure 11.53. Geometry of cyclobutane.

corresponding difference [0.49 kcal mol^{-1} (2.05 kJ mol^{-1})] in bromocyclohexane. The axial conformer is less puckered ($\phi_a = 14°$) than the equatorial one ($\phi_e = 20°$).[243]

In 1,2-disubstituted cyclobutanes the trans isomer is diequatorial and more stable than the equatorial–axial cis isomer; the equilibrium ratio (trans–cis) of 9 for the cyclobutane-1,2-dicarboxylic acids is nearly the same as for the cyclohexane-1,2-dicarboxylic acids and the ratio of dissociation constants (Table 11.1) is also similar. X-ray structure determination of the trans diacid has shown it to be the diequatorial conformer. In 1,3-disubstituted cyclobutanes the cis (e,e) isomer is predictably the more stable; for example, in the 1,3-dibromocyclobutanes the trans (a,e) isomer is less stable than the cis by 0.58 kcal mol^{-1} (2.43 kJ mol^{-1}); electron diffraction study shows the ring to be puckered in both stereoisomers, which indeed have the expected e,e and a,e conformations. Dipole moments are in agreement with this assessment but suggest that the rings in the trans (e,a) isomers are somewhat flatter than in the cis (e,e); the difference is particularly marked in 1.3-diiodocyclobutane for which ϕ (cf. Fig. 11.53) is 48° for the cis and 24° for the trans isomer.

The strain in cyclobutene (29 kcal mol^{-1}, 121 kJ mol^{-1}) and in methylene cyclobutane (27.9 kcal mol^{-1}, 116.7 kJ mol^{-1}).[28,233,243,244] exceeds that in cyclobutane by less than 3 kcal mol^{-1} (12.6 kJ mol^{-1}) in contradistinction to the situation in the corresponding three-membered rings mentioned earlier. Although the C=C–C angles in cyclobutene are reduced from 124.3° (in propene) to 94.0°,[245] the increase in angle strain (less than in the three-membered ring!) seems to be nearly offset by a decrease in torsional strain due to the reduction in H–H eclipsing.

c. Five-Membered Rings

Five-membered rings are common in natural products: carbocyclic rings in the D ring of steroids[246,247] and in prostaglandins; oxygen containing rings in furanose sugars, nucleosides, nucleotides,[248] and nucleic acids[249]; nitrogen-containing rings in the amino acids proline and hydroxyproline,[250] and in alkaloids, such as nicotine[251] among others. A carbocyclic planar five-membered ring would have valence angles of 108° and thus be nearly free of angle strain. On the other hand, the torsional strain in such a conformation would be sizeable [estimated at 10 kcal mol^{-1} (42 kJ mol^{-1}) for 10 pairs of H–H eclipsings]. To minimize this strain, cyclopentane becomes nonplanar,[252] thus reducing the sum of residual torsional plus angular (Baeyer) strain to about 60% of the value in a planar ring (cf. Table 11.4). The most stable conformations of cyclopentane are the envelope (or C_s, naming it after its symmetry

Figure 11.54.   Envelope (C_s) and half-chair (C_2) conformations of cyclopentane derivatives.

point group, cf. Section 4-3) and the half-chair (or C_2) conformations shown in Figure 11.54. For cyclopentane itself the difference in energy between these conformations is slight and the barriers between the two conformations are also very low (the molecule never passes through the high-energy planar form). It is thus not surprising that cyclopentane is in a rapid "conformational flux" among various C_2 and C_s as well as intermediate conformations. It was early recognized[253] that this conformation change can be brought about by successive oscillatory motions of the five carbon atoms of cyclopentane in a direction perpendicular to the plane of the ring. The apparent effect of this motion is to create a "bulge" (out-of-plane atom), which appears to rotate around the ring, even though there is in fact no motion of the atoms in that direction. This process has been called "pseudorotation." In cyclopentane itself it is so rapid that it is probably better to consider it a molecular vibration rather than a conformational change. However, in substituted cyclopentanes the barrier is higher, for example 3.40 kcal mol^{-1} (14.2 kJ mol^{-1}) in methylcyclopentane in which the most stable conformation (by 0.9 kcal mol^{-1} or 3.8 kJ mol^{-1}) is that with the methyl group "equatorial" at the flap of the envelope (cf. ref. 254 for these and many of the following data) (Fig. 11.54). In cyclopentanone the barrier is 1.15 kcal mol^{-1} (4.81 kJ mol^{-1}) (above the less stable conformation) but the preferred conformation, by 2.4 kcal mol^{-1} (10.0 kJ mol^{-1}), is the half-chair with the carbonyl group in the least puckered region (Figs. 11.54 and 11.55).

The implications in the conformational analysis of cyclopentanes are twofold. One stems from the low barrier which, in general, makes it impossible to "freeze out," by low-temperature NMR spectroscopy, individual conformers in substituted cyclopentanes in the manner described earlier for cyclohexanes. In this regard cyclopentane resembles cyclobutane. However, the pseudorotation causes an additional complication, in that in an unsymmetrically substituted cyclopentane (say cyclopentane-d) there are a total of 20 conformational minima (10 half-chairs and 10 envelopes) on the pseudorotational circuit, plus an infinite number of additional conformations of nearly the same energy as the unique ones just mentioned. Thus, since there is little preference for any particular conformation of the cyclopentane

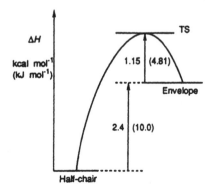

Figure 11.55.   Energy diagram for cyclopentanone.

framework itself, substituted cyclopentanes will take up conformations such that the interactions of the substituents with the framework (plus, if there is more than one substituent, the interactions of the substituents with each other) are minimized. As a result, the conformation of any one substituted cyclopentane (or heterocyclopentane) is likely to be different from that of any other. In this regard, cyclopentane is totally different from cyclohexane in which the chair is located in a deep energy valley or mold, the next higher energy minimum (corresponding to the twist form) being about 5 kcal mol^{-1} (20.9 kJ mol^{-1}) above the chair with the energy barrier to interconversion being more than 10 kcal mol^{-1} (42 kJ mol^{-1}) (Fig. 11.13). It takes two very large substituents, for example, two *tert*-butyl groups located such that one would have to be axial in the chair conformation, to force the cyclohexane ring out of its "natural" chair conformation: In contrast, there is no "natural" conformation of the five-membered ring!

In the course of a pseudorotation circuit (i.e., one 360° turn of the puckering) one encounters 10 envelope forms (5 with one of the carbon atoms in turn pointing up and 5 with one of the carbon atoms pointing down) and 10 half-chair forms lying in between two envelope forms on the circuit. (There are five pairs of adjacent carbon atoms that may be located at the site of maximum pucker of the half-chair: 1,2; 2,3; 3,4; 4,5 and 5,1, and the puckering may be either up or down, since the half-chair is chiral. Thus there are 10 nonidentical half-chairs.) The puckering may thus be described by two parameters: an angle ψ that describes how far puckering has moved around the five-membered ring starting at an arbitrarily defined origin and a "puckering amplitude" q that indicates how far a given atom moves above and below the average plane of the ring. In cyclopentane itself q is 43 pm (0.43 Å).[255] If one starts the pseudorotation circuit with a $\mathbf{C_s}$ form, one will encounter a $\mathbf{C_2}$ form at $\psi = 18°$, another $\mathbf{C_s}$ form at $\psi = 36°$, a $\mathbf{C_2}$ form again at $\psi = 54°$ and so on, with $\mathbf{C_s}$ and $\mathbf{C_2}$ forms alternating at 18° intervals. All 20 forms (10 $\mathbf{C_s}$, 10 $\mathbf{C_2}$) thus make an appearance as ψ changes by 360° (i.e., as the puckering moves once around the ring; cf. ref. 254).

For the reasons already indicated, the conformations of monosubstituted cyclopentanes (e.g., chlorocyclopentane[256]) are much less predictable than those of the corresponding cyclohexanes. Oversimplifications may have been made in the literature by assuming that the ring is either an envelope or a half-chair when, in fact, it may assume a conformation somewhere in between, depending on the substituent. On the other hand, clear-cut experimental data are available for configurational (as distinct from conformational) equilibria in disubstituted cyclopentanes. For 1,2-disubstituted cyclopentanes, the trans isomer is more stable than the cis, the energy difference for the dimethyl compounds. 1.73–1.94 kcal mol^{-1} (7.24–8.12 kJ mol^{-1}), being about the same as for the dimethylcyclohexanes (cf. Table 11.8). However, the agreement may be fortuitous: whereas in the cyclohexane series the major difference is between two equatorial versus one equatorial and one axial methyl (the mutual gauche interaction of the methyl groups nearly cancels), the major factor destabilizing the cis isomer in the cyclopentane series may be the direct interaction of the substituents that are probably at a torsion angle ($CH_3-C-C-CH_3$) of no more than 50°, since an angle larger than that would involve a serious increase in energy of the cyclopentane framework.[257] Greater stability for the trans isomer was also found in the 1,2-dicarbomethoxy- and 1,2-diphenylcyclopentanes.

The strain in cyclopentene is nearly the same as that in cyclopentane[258] and that in cyclopentadiene is actually considerably less.[233,244] Evidently, the diminution in eclipsing strain more than compensates for the increase in angle strain in these compounds. The strain in methylenecyclopentane is somewhat less than that in cyclopentene,[23] yet the transformation of methylenecyclopentane to methylcyclopentene is exothermic, by 3.5 kcal mol^{-1} (14.6 kJ mol^{-1}),[259] that is, more so than the corresponding process in the six-membered ring (p. 469). The explanation may be that in methylenecyclohexane (but not in methylenecyclopentane) the favored C=C/H eclipsed conformation (cf. Section 10-2.a) is achieved.

The conformational situation in cyclopentanone has already been discussed. Addition reactions to cyclopentanone are less favorable than those to cyclohexanone since they lead to an increase in bond eclipsing. Accordingly, the equilibrium constant for addition of HCN to cyclopentanone to give the cyanohydrin is only 3.33 as against 70 in cyclohexanone[260] and the reduction of cyclohexanone with sodium borohydride is 23 times as fast as the corresponding reduction of cyclopentanone.[261]

d. Rings Larger Than Six-Membered

Several excellent reviews of the conformation of seven-membered, medium-sized, and larger rings are available.[35,69,262,263] By way of introduction to this subject, let us consider that cyclopentane has basically but a single, pseudorotating conformation. In cyclohexane there are two families of conformations: the chair, which happens to be rigid and thus constitutes a single-membered family (for an explanation see ref. 66) and the twist–boat family. As we proceed to higher cyclanes, we shall find an increasing number of conformational families, each with several members. Within each family there are several conformations that are easily interconverted by a process of "pseudorotation" (see above) which involve a very low barrier, generally inaccessible by NMR study. Between families, however, the barriers approach the 10.3 kcal mol^{-1} (43.1 kJ mol^{-1}) value in cyclohexane and the separation into families can therefore often be studied by low-temperature NMR, with ^{1}H and ^{13}C spectra yielding complementary results.

In the case of cycloheptane and cyclooctane the number of possible conformations possessing a symmetry element is of manageable size (cf. Figs. 11.57 and 11.59) and common names have been attached to these conformations as shown. However, for larger rings both this form of representation and the naming become too complex and either the torsion angle notation of Bucourt[137] or the wedge notation of Stoddart and Szarek[265] (cf. ref. 262) are best to use (Fig. 11.56). In the torsion angle notation, the symbol + against a bond indicates that the torsion angle, with that bond (2–3) in the middle of the four-atom (1–2–3–4) segment, is positive, that is, the sequence 1–2–3–4 describes a right-handed helix. The opposite arrangement (left-handed helix) is described by the symbol –. In the wedge notation, 1 ◁ 2 means that atom 2 is

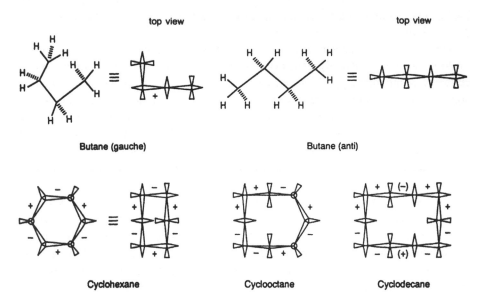

Figure 11.56. Examples of torsion angle notation (Bucourt) and wedge notation (Dale). [From Allinger, N. L. and Eliel, E. L. *Top. Stereochem.* **1976**, *9*, 204–205. Copyright © 1976 John Wiley & Sons, with permission of the publisher.]

in front of atom 1 or that atom 1 is behind atom 2; a combination of such wedges, as shown in Figure 11.56, will reveal the conformation.

The principal methods for conformational study in seven-membered and larger rings have been NMR, force field calculations, and, to a lesser extent, study of vibrational spectra, electron diffraction, and X-ray study of derivatives.

The conformational situation in cycloheptane[163b,266] was first unraveled by the classical study of Hendrickson,[257,267] which represents the first application of the

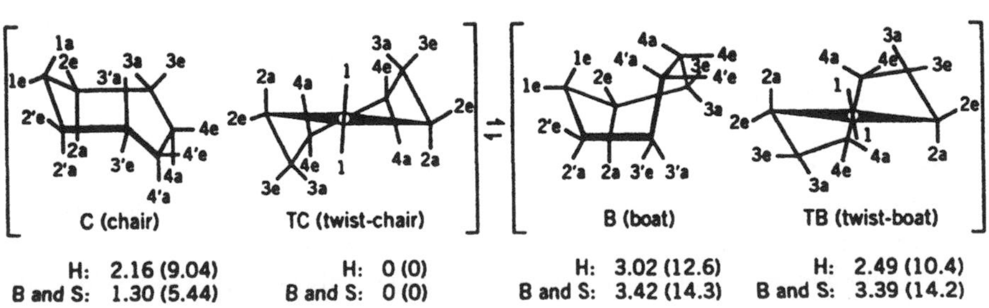

C (chair)	TC (twist-chair)	B (boat)	TB (twist-boat)
H: 2.16 (9.04)	H: 0 (0)	H: 3.02 (12.6)	H: 2.49 (10.4)
B and S: 1.30 (5.44)	B and S: 0 (0)	B and S: 3.42 (14.3)	B and S: 3.39 (14.2)

Figure 11.57. Conformations of cycloheptane. Relative potential energy values are in kilocalories per mole (kcal mol⁻¹) and (in parentheses) in kilojoules per mol (kJ mol⁻¹). H refers to values given by Hendrickson[257] and B and S to values given by Bocian, D. F. and Strauss, H. L. *J. Am. Chem. Soc.* **1977**, *99*, 2876. [Adapted, with permission, from Hendrickson, J. B. *J. Am. Chem. Soc.* **1961**, *83*, 4543. Copyright © 1961 American Chemical Society, Washington, DC.]

computer to molecular mechanics calculation. There are two families of conformations in cycloheptane, shown in Figure 11.57. One comprises the chair and twist–chair, the other the boat and twist–boat. The situation is thus quite similar to that in cyclohexane, except that the chair is part of a family of flexible conformations and, because of the severe eclipsing at the "flat" end, lies about 2.16 kcal mol^{-1} (9.0 kJ mol^{-1}) above the twist–chair into which it can be readily pseudorotated and which represents the most stable conformation of cycloheptane. In the second family, comprising the boat and twist–boat, the twist–boat is more stable [by 0.53 kcal mol^{-1} (2.22 kJ mol^{-1})] than the true boat because of the more severe eclipsing in the latter, similarly as in cyclohexane. The difference between the twist–boat and the twist–chair is 2.49 kcal mol^{-1} (10.4 kJ mol^{-1}), that is, only about one-half the twist/chair difference in cyclohexane: whereas in cyclohexane the chair is nearly strain-free, in cycloheptane the more stable twist–chair cannot completely escape eclipsing interactions. The chair/twist–chair and boat/twist–boat families in cycloheptane, like their cyclohexane analogues, can be interconverted only by passing over a relatively high barrier of about 8.5 kcal mol^{-1} (35.6 kJ mol^{-1}).

The cycloheptene system seems to be largely in the chair form, the C=C moiety taking the place of one of the ring carbon atoms in the cyclohexane chair.[268,269] Benzocycloheptene derivatives[270] and heterocyclic analogues[271] have been studied especially in the laboratories of St.-Jacques with interesting conformational results.

We shall consider only one other compound in the seven-membered ring series: cycloheptatriene. There is a valence bond tautomerism (cf. Fig. 2.5) between this compound and the isomeric norcaradiene (see also Fig 11.58) with the position of equilibrium depending on the substitutent(s) at the saturated carbon. In addition, however, the boat-shaped[272,273] cycloheptatriene moiety can undergo a conformational inversion (methylene flip) with a barrier of 6.1 kcal mol^{-1} (25.5 kJ mol^{-1}).[274,275] In the case of a 7-substituted 1,3,5-cycloheptatriene, this involves two different conformers that may be called axial and equatorial. Studies involving a variety of substituents[266] suggest that the equatorial conformer is more stable, except when there is also a substituent at C(1) that causes eclipsing. In that case, for example, in 1-methyl-7-tert-butylcycloheptatriene, the axial conformer is the more stable (see also ref. 139).

Cyclooctane exhibits no fewer than 10 symmetrical conformations (Fig. 11.59),[276] which fall into four families. Studies of cyclooctane,[277–279a] as well as X-ray study of a number of derivatives,[276] indicate that the most stable family **I** is that of the boat–chair (BC) and twist–boat–chair (TBC) in which the boat–chair represents the

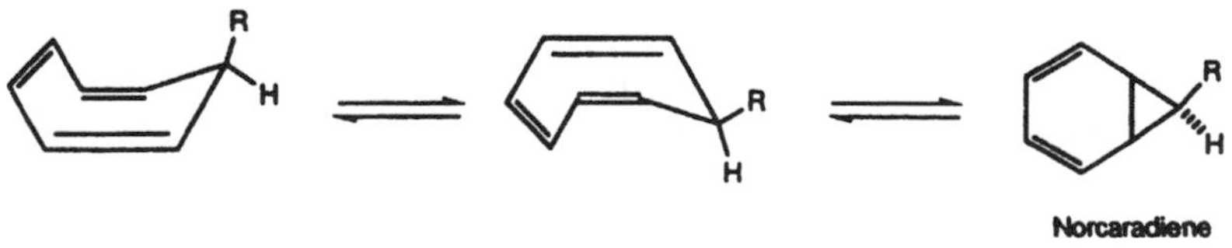

Figure 11.58. Conformational equilibrium in 7-substituted cycloheptatrienes.

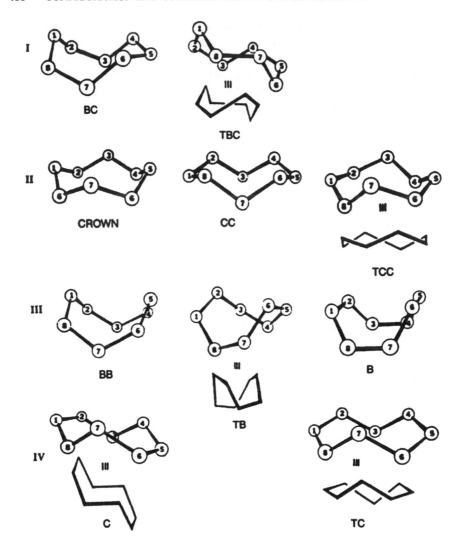

Figure 11.59. Cyclooctane conformations. [Adapted, with permission, from Anet, F. A. L. *Top. Curr. Chem.* **1974**, *45*, 178. Copyright © 1974 Springer Verlag, Heidelberg, Germany.]

energy minimum. Interconversion of this family into the one of next-higher energy, the crown/chair–chair (CC) twist–chair–chair (TCC) family **II**, requires an activation energy calculated to be 11.4 kcal mol^{-1} (47.7 kJ mol^{-1}), that is, somewhat higher than the chair–twist barrier in cyclohexane. The crown family lies only slightly above the boat–chair family [ca. 1.0–1.6 kcal mol^{-1} (4.2–6.7 kJ mol^{-1})] and it has been estimated that cyclooctane contains about 6% of the crown conformer at room temperature.[277,279] The main problem in the crown form, in addition to high symmetry, and hence low entropy, is eclipsing strain, which is reduced in heteracyclooctanes; thus 1,3,5,7-tetraoxcyclooctane is most stable in the crown form. The third family **III** of cyclooctane conformations comprises the boat (B), twist–boat (TB), and boat–boat (BB) conformations. It is easily seen that these conformations suffer from severe eclipsing strain; they do not appear to be appreciably populated in cyclooctane, even though the barrier between this family and the boat–chair family is calculated to be only 9.4 kcal mol^{-1} (39.3 kJ mol^{-1}). The remaining conformations **IV**, chair (C), and twist–chair (TC) are calculated to be very high in energy [over 8 kcal mol^{-1} (33.5 kJ mol^{-1}) above the boat–chair] and are not viable conformations of cyclooctane (the same applies to the boat member of the boat/twist–boat/boat–boat family).

The conformations of cyclooctadienes and cyclooctatrienes have been discussed by Anet.[263] Cyclooctatetraene[280] (Fig. 11.60) with its eight ($4n$ with $n = 2$) π electrons is not aromatic and therefore derives no advantage from being planar; in fact, the molecule is tub shaped[281] and thereby relatively unstrained. Similar to the cyclohexane chair, the cyclooctatetraene tub can undergo a ring inversion (**A** ⇌ **B**) to an alternate tub; the energy barrier to this process in variously substituted cyclooctatetraenes is 14.7 kcal mol^{-1} (61.5 kJ mol^{-1}),[282] whereas the energy barrier for double-bond migration (**B** ⇌ **C**) is appreciably higher, somewhat over 17 kcal mol^{-1} (71 kJ mol^{-1}). The latter process, unlike the former which allows the double bonds to remain localized, may involve an antiaromatic 8π-electron transition state **D** (Fig. 11.60; see also p. 280).

Both cyclononane and cyclodecane[263,283] present interesting stereochemical aspects. The four minimum-energy conformations of cyclononane, twist–boat–chair (TBC or [333]) (the latter referring to the fact that this conformation has three "straight" segments of three bonds each), twist–chair–boat (TCB or [225]), twist–chair–chair (TCC or [144]), and twist–boat–boat (TBB or [234]) are depicted in Figure 11.61, along with their calculated potential energies. From the calculated energies, it is clear that cyclononane should exist largely as the [333] form, especially at low temperatures, and this is borne out by NMR study.[284] It was therefore rather surprising that several cyclononane derivatives, such as cyclononylamine hydrobromide,[285] crystallize in the [225] conformation. Detailed NMR reinvestigation of cyclononane[286] not only disclosed existence of two of the minor conformers, [225] and [144], but also indicated that these conformers have a substantially higher entropy than the symmetrical [333] conformer, something that was not evident from the calculations. Thus it was estimated that, at room temperature,

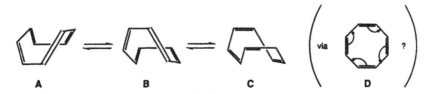

Figure 11.60.   Ring inversion **A** ⇌ **B** and bond migration (**B** ⇌ **C**) in cyclooctatetraene.

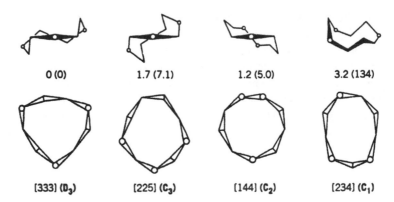

0 (0)	1.7 (7.1)	1.2 (5.0)	3.2 (134)
[333] (**D$_3$**)	[225] (**C$_3$**)	[144] (**C$_2$**)	[234] (**C$_1$**)

Figure 11.61. Minimum-energy conformations of cyclononane. Relative potential energies are in kilocalories per mole (kcal mol^{-1}); values in parentheses are in kilojoules per mole (kJ mol^{-1}). [Adapted, with permission, from Glass, R. S. *Conformational Analysis of Medium-Sized Heterocycles.* Copyright © 1988 VCH, New York, p. 53.]

cyclononane consists of 40% of the [333], 50% of the [225], and 10% of the [144] conformer; that is, the [225] conformer is predominant at room temperature. This may explain why the [225] conformer is found in crystals of cyclononane derivatives; it also points up one of the limitations of force field calculations prior to the advent of MM3 (see Section 2-6; see also ref. 279b): They estimate potential energy, which is closely related to enthalpy, but in cases where entropy differences are sizeable, not to free energy. It should be noted also that, since entropy contributions are small at low temperature, calculations may be borne out by low-temperature NMR experiments; yet, unless a careful determination of temperature dependence of conformational population is made, the composition at room temperature may be appreciably different from both the measured and the calculated values.

Cyclodecane derivatives have provided two interesting conformational findings. The lowest energy conformation calculated is "rectangular" [2323]; it fits into a diamond lattice and may be called boat–chair–boat (BCB) (Fig. 11.62). Most cyclodecane derivatives that have been studied by X-ray diffraction fall into this conformation,[283] with one notable exception.[287] Indeed, when the electron diffraction pattern of cyclodecane itself was studied,[288] the experimental intensity and radial distribution curves agreed very well with those calculated for the BCB conformer. However, the bond lengths, bond angles, and torsion angles found agreed poorly with

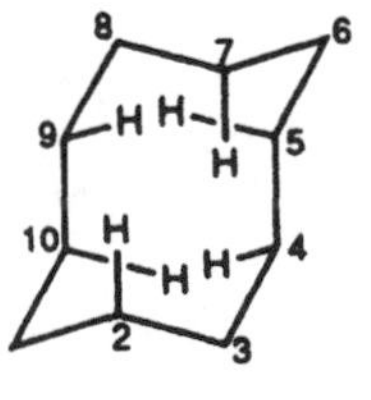

Figure 11.62. Boat–chair–boat (BCB) conformation of cyclodecane.

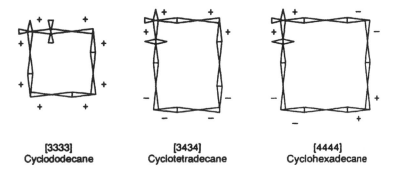

[3333]
Cyclododecane

[3434]
Cyclotetradecane

[4444]
Cyclohexadecane

Figure 11.63. Most stable conformers of cyclododecane, cyclotetradecane, and cyciohexadecane.

those from X-ray studies of various cyclodecane derivatives. It was therefore decided to combine the electron diffraction (ED) studies with force field calculations (this is a known method for enhancing the accuracy of electron diffraction studies). When this was done, it was found that the BCB conformation alone no longer gave a good fit for the ED data, but that a better fit was obtained by assuming cyclodecane to be a conformer mixture of about one-half the BCB conformer, one-third the TBC conformer, and the remainder about equal parts of TBCC and BCC conformers (see also ref. 289).

Only limited information is available for 11-, 13-, and 15-membered rings.[262,290a] The 12-membered ring has been studied by X-ray diffraction,[291] by NMR spectroscopy,[263,289,290b] and by a combination of electron diffraction and molecular mechanics[292]; the stable conformation is the "square" [3333] one (D_4, Fig. 11.63), and the barrier for site exchange is 7.3 kcal mol^{-1} (30.5 kJ mol^{-1}). For monosubstituted cyclododecanes, there is a conformational equilibrium between "corner substituted" and "noncorner substituted" conformers.[293] For the larger even-membered rings the most stable conformation corresponds to the diamond lattice type: "rectangular" [3434] in cyclotetradecane (Fig. 11.63)[289] and "square" [4444] in cyclohexadecane (Fig. 11.63)[294] with a barrier to site exchange of 6.7 kcal mol^{-1} (28.0 kJ mol^{-1}). A cyclotridecane derivative has been investigated by a combination of X-ray diffraction and molecular mechanics.[295]

The conformation of rings larger than six-membered containing oxygen, nitrogen, sulfur, and phosphorus has been reviewed[264] as has the conformation and stereochemistry of phosphorus-containing rings of all sizes.[296]

11-6. STEREOCHEMISTRY OF FUSED, BRIDGED, AND CAGED RING SYSTEMS

The chemistry of ring systems of the above types has been extensively studied[23,24,297–300] and only a few representative cases will be discussed here. We shall pick examples either because they are of particular interest (high strain, extreme cases, high symmetry, etc.) or because the ring systems in question are particularly widespread in occurrence.

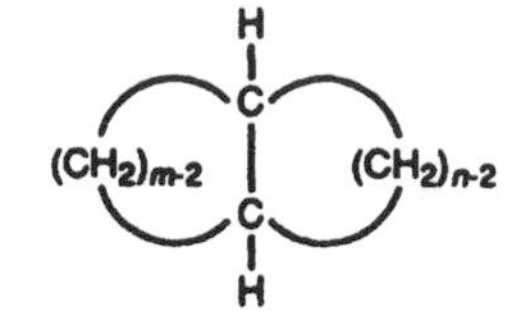

Figure 11.64.   Fused rings.

a.   Fused Rings

The general formula for fused rings is given in Figure 11.64. The smallest, and most strained example is bicyclo[1.1.0]butane, where $m = n = 3$ (Fig. 11.65, **A**), with a strain of 66.5 kcal mol^{-1} (278 kJ mol^{-1}). (The strain energy here and elsewhere is taken from ref. 23 unless otherwise indicated.) Its structure has been determined[301]; its angle of pucker, 58°, is much larger than that of cyclobutane and its C–C bonds are unusually short (149.8 pm, 1.498 Å).

The strain in bicyclo[1.1.0]butane (Fig. 11.65, **A**) exceeds the sum of the strains in two cyclopropane moieties (Table 11.4) by about 11.5 kcal mol^{-1} (48 kJ mol^{-1}). In contrast, the strain in the next higher fused system, bicyclo[2.1.0]pentane (Fig. 11.65, **B**), 57.3 kcal mol^{-1} (240 kJ mol^{-1}) exceeds that of the sum of cyclopropane and cyclobutane by less than 4 kcal mol^{-1} (16.7 kJ mol^{-1}) and the strain in higher bicyclo[n.1.0]alkanes is close to that of the sum of the strain in the two fused rings.

The hydrocarbons shown in Figure 11.65 are cis fused. What is the smallest bicyclic hydrocarbon of the type shown in Figure 11.64 with trans fused rings? *trans*-Bicyclo[5.1.0]octane (Fig. 11.64, $m = 7$, $n = 3$) has been prepared[302,303] as have ketone and alcohol derivatives of *trans*-bicyclo[4.1.0]heptane (Fig. 11.64, $m = 6$, $n = 3$), the smallest known trans fused bicyclic hydrocarbon with a three-membered ring[304]; for calculations on even smaller trans fused systems, see ref. 305. In the [5.1.0] structure, the trans isomer is about 9 kcal mol^{-1} (37.6 kJ mol^{-1}) less stable than the cis[306] but this difference vanishes in the [6.1.0] compound.[307] *trans*-Bicyclo[3.2.0]heptane (Fig. 11.64, $m = 5$, $n = 4$)[308] and its ketone precursor[309] are also known. The structures of the next higher homologues, *cis*- and *trans*-bicyclo[4.2.0]octane (Fig. 11.64, $m = 6$, $n = 4$) have been determined by electron diffraction.[310] The angle of pucker ϕ of the four-membered ring (Fig. 11.53) in the cis isomer (23°) is within the normal range and the cyclohexane ring bears most of the

A   **B**

Figure 11.65.   Bicyclo[1.1.0]butane and bicyclo[2.1.0]pentane.

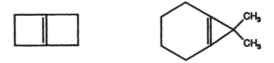

Figure 11.66. Bicyclo[2.2.0]hex-1(4)-ene and 7,7-dimethylbicyclo[4.1.0]hept-1(6)-ene.

strain, with its endocyclic torsion angle at the ring juncture reduced from the normal 55° to 32.8°. In the trans isomer, the pucker of the four-ring (45°) is abnormally high; at the same time the cyclohexane ring is also exceedingly puckered with an endocyclic torsion angle of 69.8° (so as to reduce the exocyclic e,e torsion angle). Only calculations are available for *trans*-bicyclo[2.2.0]hexane (Fig. 11.64, $m = n = 4$).[311] The bridged alkenes bicyclo[2.2.0]hex-1(4)-ene[312] and 7,7-dimethyl-bicyclo[4.1.0]hept-1(6)-ene[313] (Fig. 11.66) have also been synthesized, but though theoretical work has been done on their lower homologues,[314,315a] bicyclo[3.1.0]hex-1(5)-ene[315b] and systems containing the bicyclo[2.1.0]-pent-1(4)-ene,[316] and bicyclo[1.1.0]but-1(3)-ene[317] exist at best as fleeting intermediates.

We pass now to the more common fused ring systems. The cis and trans isomers of hydrindane (Fig. 11.67) have long been known. The trans (racemic) isomer has a lower heat of combustion than the cis (meso) but the difference is only 1.065 kcal mol^{-1} (4.46 kJ mol^{-1}),[318] appreciably less than that (1.74 kcal mol^{-1}, 7.28 kJ mol^{-1}) between *cis*- and *trans*-1,2-dimethylcyclohexane. Presumably the smaller difference is due to the strain of ring fusion in the trans isomer being larger than in the cis. This is not surprising; it can be seen in models and is a consequence of the normal maximum value of the endocyclic torsion angle in a five-membered ring being about 45°, whereas the normal exocyclic 1,2-trans torsion angle in cyclohexane is of the order of 64°. There must therefore be a sizable and energetically unfavorable puckering of the six-membered ring and perhaps of the five-membered ring as well. Electron diffraction data of *trans*-hydrindane[319] indicate that the endocyclic torsion angle in the six-membered ring at the ring juncture is indeed increased from the normal 55° to 61.1° with a corresponding decrease of the exocyclic (e,e) angle. There is less problem in the cis isomer; although the maximum desirable torsion angle in cyclopentane (45°) is appreciably less than the exocyclic cis torsion angle in

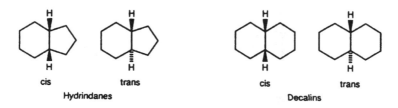

Figure 11.67. Hydrindanes and decalins.

cyclohexane (56°), flattening of the six-membered ring to reduce the latter angle is relatively facile (calculated geometry: ref. 319).

Entropy favors the cis isomer in the hydrindanes as it does in the 1,2-dimethylcyclohexanes (cf. Table 11.8 and explanation thereto) but the difference is somewhat larger (2.04 cal mol^{-1} K^{-1}, 8.54 J mol^{-1} K^{-1})[318]; the resulting ΔG^0 at 25°C is 0.50 kcal mol^{-1} (2.09 kJ mol^{-1}) in favor of the trans isomer. Because of the countertrend of ΔH^0 (favoring trans) and ΔS^0 (favoring cis) there is a cross-over of ΔG^0 at about 200°C; above that temperature the cis isomer is more stable.[320a]

The lower homologue of hydrindane, bicyclo[3.3.0] octane (Fig. 11.68), was obtained as cis and trans isomers as early as 1936.[321] The considerations mentioned in connection with *trans*-hydrindane apply to an even greater extent to the trans isomer of this lower homologue: The maximum normal pucker in cyclopentane (45°) is ill adjusted to the exocyclic trans torsion angle in cyclopentane (75°) and considerable distortion of both rings must result. Indeed the strain in this isomer is 6.4 kcal mol^{-1} (26.8 kJ mol^{-1})[322] [a very similar difference, 6.0 kcal mol^{-1} (25.1 kJ mol^{-1}) was measured earlier[321]]. Suprisingly, the trans isomers of hetero analogues of *trans*-bicyclo[3.3.0]nonane, with O or S in lieu of CH$_2$(2)[323] or O and NR in lieu of CH$_2$ (1 and 3),[324] are readily prepared by ordinary synthetic methods; it is not known whether these systems are less strained or whether they can be readily prepared despite being strained.

Decalin (Fig. 11.67) is a molecule of historical importance because it served to confirm, experimentally, the Sachse–Mohr theory of the chair (and boat) shaped six-membered ring (cf. Section 11-4.a). Two planar six-membered rings (as pictured by Baeyer) can only be fused cis; Mohr[54] predicted, in contrast, that decalin with puckered six-membered rings should exist as both cis and trans isomers. This hypothesis was confirmed through the isolation of *trans*-decalin.[325] At that time it was believed that *cis*-decalin existed as a double boat, but it is now quite certain, and has, in fact, been demonstrated by electron diffraction measurements[326] that both isomers exist as double chairs, as undistorted as cyclohexane itself (Fig. 11.69).

The framework of *trans*-decalin is locked: It can invert to a chair–boat or even boat–boat system of considerably higher energy, but it cannot invert to an alternative chair–chair system (as cyclohexane can), since this would require spanning two axial positions on opposite sides of the chair with four methylene groups, a geometric impossibility. The *trans*-decalin system, like the anancomeric 4-*tert*-butylcyclohexyl system discussed earlier, thus serves as a conformational anchor (Fig. 11.15); we shall

cis   trans

Figure 11.68.   The structures of *cis*- and *trans*-bicyclo[3.3.0] octane.

Figure 11.69.   Chair forms of *cis*- and *trans*-decalin.

see evidence of this later. In contrast, *cis*-decalin has two interconverting chair–chair combinations (both chairs have to be inverted to effect the interconversion). Just as in *cis*-1,2-dimethylcyclohexane, these two conformers (**A, A′**, Fig. 11.69) are not identical but enantiomeric.

If one considers each ring in decalin as being a substituent of the other ring, one notes that in *trans*-decalin all pertinent methylene substituents are equatorial, and therefore antiperiplanar. There are three extraannular gauche interactions in *cis*-decalin indicated in Figure 11.69, **A** either by heavy bonds or by an extra dotted line. It would appear at first that there are four such interactions, since one methylene in each ring is axial to the other, giving rise to two gauche interactions, or four in the two rings. However, if one marks the pertinent interactions, as done in Figure 11.69, one notices that one of them (dotted) is common to the two axial methylenes, so there remain only three. If each such interaction is worth 0.87 kcal mol^{-1} (3.64 kJ mol^{-1}) in the liquid phase (one-half the axial methylcyclohexane interaction of 1.74 kcal mol^{-1} or 7.28 kJ mol^{-1}; cf. Table 11.6) the total should be 2.61 kcal mol^{-1} (10.9 kJ mol^{-1}); the experimental value (heat of isomerization[320b,327]; difference in heats of combustion,[328] 2.7 kcal mol^{-1} (11.3 kJ mol^{-1}) is in excellent agreement.

The symmetry point groups of *cis*- and *trans*-decalin are **C$_2$** and **C$_{2h}$**, respectively; both have a symmetry number of 2 and an entropy of symmetry of $-R \ln 2$. However, the cis isomer exists as a pair of enantiomers (albeit not isolable at room temperature since ring inversion converts one enantiomer into the other), and therefore has an entropy of mixing of $R \ln 2$; the trans isomer is achiral, so the entropy advantage of the cis should be $R \ln 2$ or 1.38 cal mol^{-1} K^{-1} (5.76 J mol^{-1} K^{-1}). The experimental value[320b,327] of 0.55–0.60 cal mol^{-1} K^{-1} (2.30–2.51 J mol^{-1} K^{-1}) is appreciably less than that; thus there must be additional factors affecting the entropy difference between the cis and trans isomers. The barrier to ring inversion $\Delta G^{\ddagger}$ in *cis*-decalin is 12.3–12.6 kcal mol^{-1} (51.5–52.7 kJ mol^{-1}) at room temperature,[329,330] appreciably higher than that in cyclohexane.

Since *trans*-decalin is a conformationally rigid chair–chair system, substituents in *trans*-decalin will occupy well-defined axial or equatorial positions (cf. Fig. 11.15). Examples are shown in Figure 11.70: In the 1-hydroxy-*trans*-decalin shown the hydroxyl substituent is equatorial, whereas in the 2-methyl-*trans*-decalin, the methyl substituent is axial.

The parochial descriptors α and β are used to denote relative configuration in the decalin system. The descriptor β is used for a ligand on the same side as the proximal

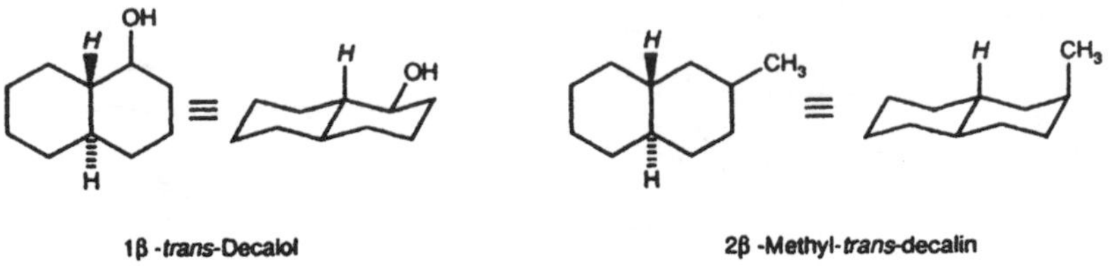

1β -*trans*-Decalol   2β -Methyl-*trans*-decalin

Figure 11.70. Monosubstituted *trans*-decalins.

ring-juncture hydrogen, and α is used for a ligand on the opposite side. (The proximal hydrogen atoms are italicized in Figs. 11.70 and 11.71.)

Since the *cis*-decalin system is mobile, substituents in it can occupy either the equatorial or the axial position, and since the framework equilibrium is unbiassed (1:1, inasmuch as the two conformers are enantiomeric, see above), the preferred conformation will be that with the substituent equatorial. Thus equilibrium **A** in Figure 11.71 will be shifted to the right (equatorial OH) to the extent of the ΔG^0 value of the OH group (Table 11.6), whereas equilibrium **B** will be shifted to the right as the result of a sizable $NH_2/CH_2/H$ interaction in the other conformer (cf. Table 11.9).

Among fused systems containing three rings, the perhydrophenanthrenes and perhydroanthracenes have been most extensively studied. The perhydro-phenanthrenes (Fig. 11.72) constitute an ABBA system and there are four pairs of enantiomers and two meso forms (cf. the hexaric acids, Fig. 3.22). The compounds were synthesized and their stereochemistry elucidated in an elegant piece of work many years ago,[331] which has been summarized elsewhere[332] and even today presents a challenging exercise in stereochemical reasoning.

The relative stabilities of the six isomers were assigned[333] on the basis of conformational arguments: **A** has no axial substituents, **B** and **C** have one each, **D** has two, **E** has a pair of synaxial substituents and the configuration of **F** is such that the central ring must be in the twist form. (If it were a chair, the junctions to the outer rings would be e, e, a, a, and, as we have already discussed, the a,a fusion in the

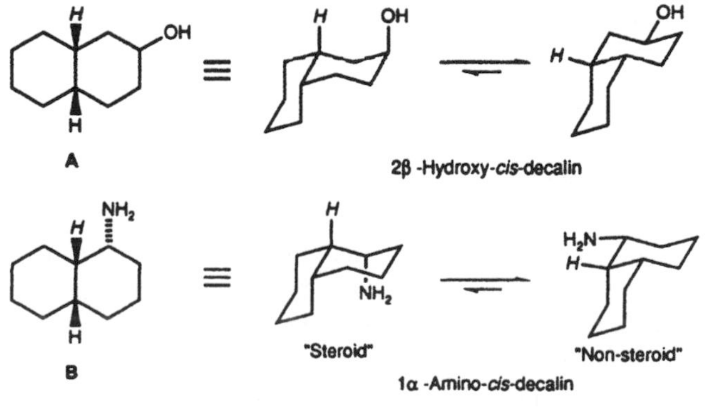

A

2β -Hydroxy-*cis*-decalin

B

"Steroid"   "Non-steroid"

1α -Amino-*cis*-decalin

Figure 11.71. Monosubstituted *cis*-decalins.

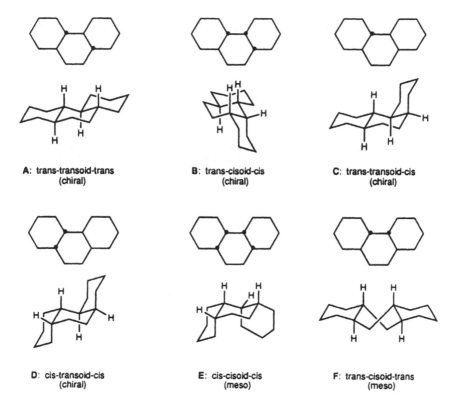

A: trans-transoid-trans
(chiral)

B: trans-cisoid-cis
(chiral)

C: trans-transoid-cis
(chiral)

D: cis-transoid-cis
(chiral)

E: cis-cisoid-cis
(meso)

F: trans-cisoid-trans
(meso)

Figure 11.72. The perhydrophenanthrenes.

trans-decalin moiety is sterically prohibited.) The twist structure has been verified in a derivative by X-ray structure analysis.[334] The relative conformational energies of the stereoisomers were computed by molecular mechanics calculations[335] and confirmed by experiment.[336]

A note on nomenclature and notation is in order here. A dot at a ring juncture indicates an upward or forward hydrogen, the absence of a dot indicates a downward or backward hydrogen. The descriptors "trans" and "cis" are used for the decalin-like ring junctures (cf. Figs. 11.67 and 11.69), whereas the descriptors "cisoid" and "transoid" are used for two adjacent ring junctures to the central ring involving the two peripheral rings. (In the older literature, "syn" and "anti" were used instead of cisoid and transoid.)

The perhydroanthracenes are shown in Figure 11.73. There are three meso isomers and two enantiomer pairs (the cis–transoid–cis isomer has a center of symmetry). The relative stability of the isomers, originally predicted qualitatively,[333] has been both calculated by molecular mechanics and determined experimentally[337]; the calculated and experimental data are summarized in Table 11.17. The instability of isomer **E** is due to the enforced boat (not twist!) conformation of the central ring, that of isomer **D** results from the presence of the synaxial methylene interaction. It is interesting that

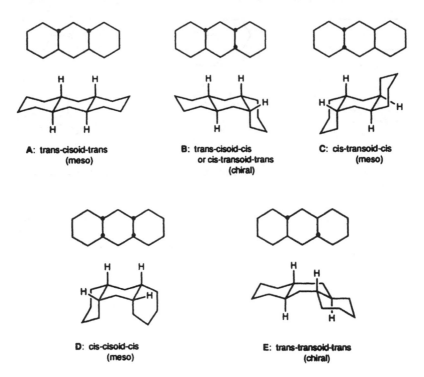

Figure 11.73. The perhydroanthracenes.

TABLE 11.17. Calculated and Experimental Conformational Enthalpies and Entropies of Perhydroanthracenes[a]

	ΔH^0		ΔS^0	
	Calculated	Experimental	Calculated	Experimental
Isomer	(kcal mol^{-1})[b]	(kcal mol^{-1})[b]	(cal mol^{-1} K^{-1})[c]	(cal mol^{-1} K^{-1})[c]
A	0 (0)	0 (0)	0 (0)	0 (0)
B	2.62 (11.0)	2.76 (11.5)	+2.8 (+11.7)	+2.1 (+8.8)
C	5.56 (23.3)	5.58 (23.3)	+1.4 (+5.9)	+0.3 (+1.3)
D	8.13 (34.0)[d]	8.74 (36.6)	+2.2 (+9.2)[d]	+4.0 (+16.7)
E	5.86 (24.5)	4.15 (17.4)	0 (0)	−1.6 (−6.7)

[a]Relative to isomer **A**. Ref. 337.
[b]Values in parentheses are in kilojoules per mole (kJ mol^{-1}).
[c]Values in parentheses are in joules per mole per kelvin (J mol^{-1} K^{-1}).
[d]Includes contribution of 13% boat form.

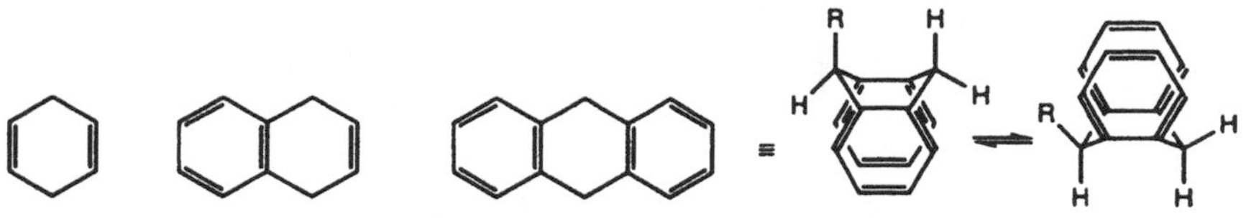

A   **B**   **C**   **C'**

Figure 11.74. Structures of 1,4-dihydrobenzene, 1,4-dihydronaphthalene, and 9,10-dihydroanthracene.

an alternate conformer of **D** with the central ring as a boat is of only slightly higher energy than the chair and contributes to the extent of 13% at 271°C.

We now digress briefly into a highly unsaturated system, 9,10-dihydroanthrcene (Fig. 11.74, **C**).[117,157,161] We had mentioned earlier that 1,4-dihydrobenzene (Fig. 11.74, **A**) is a planar molecule with a rather shallow energy minimum. Introduction of bulky substituents at C(1) deplanarizes both this molecule[338,339] and the related 1,4-dihydronaphthalene (Fig. 11.74, **B**).[159,161] The substituents will occupy a pseudoaxial position, presumably because of $A^{(1,2)}$ strain (Fig. 11.41) involving the peri position in the dihydronaphthalene. When there are two peri positions, as in 9,10-dihydroanthracene, the nonplanar geometry (Fig. 11.74, **C'**) is assumed even by the parent molecule, R = H (at least in the solid state).[340,341]

Study of the conformations of 9-substituted 9,10-dihydroanthracenes in almost all cases shows that the substituent is predominantly pseudoaxial.[157,161] The situation here is quite different from that in cyclohexane: There are no synaxial hydrogen atoms and the only interference for a pseudoaxial substituent comes from the transannular axial hydrogen at C(10). This interference can be minimized by some flattening of the ring. In contrast, in the planar or pseudoequatorial conformation, there would be severe steric interaction of the substituent with the peri hydrogen atoms of the benzene rings.

We return now to the saturated fused ring systems and in particular to steroids, a group of compounds of wide natural occurrence. The constitution and relative as well as absolute configuration of these compounds has been determined both by chemical methods[342,343] and by X-ray crystallography.[247,344] Virtually all natural steroids fall into two stereochemical categories: the A/B trans (trans–transoid–trans–transoid–trans) and A/B cis (cis–transoid–trans–transoid–trans) series, as shown in Figure 11.75. The conformational formulas shown in Figure 11.76 were proposed by Barton[58] and account for many if not most physical and chemical properties of steroids. The conformational principles are the same as those earlier enunciated (Section 11-4.a) for simpler cyclohexanes. Before discussing them, we need to explain the local stereochemical nomenclature in steroids: Positions on the same side as the angular methyl groups are denoted as β, those on the opposite side are denoted as α. Figure 11.76 shows these positions for the A/B trans fused system (left) and, for the A ring only, in the A/B cis fused system (right; the descriptors in rings B, C, and D are the same in the two systems). There is no 1:1 relationship between α and β and axial and equatorial, but Figure 11.76 allows one to infer this relationship for each position of the A/B trans or A/B cis fused steroid nucleus.

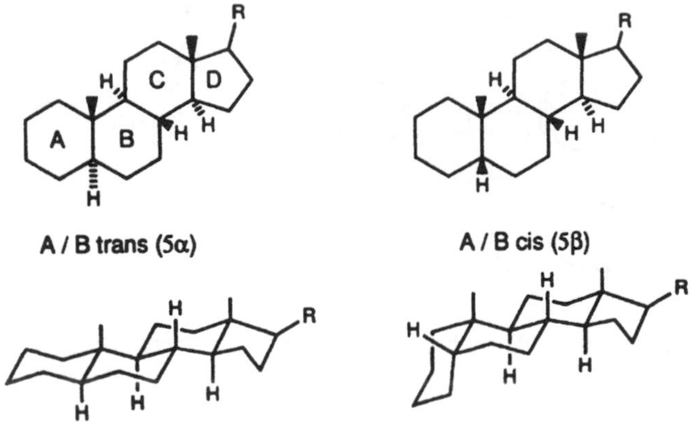

A / B trans (5α)   A / B cis (5β)

Figure 11.75. Steroids.

Numerous examples of the conformational rationale of the reactivity and stability of stereoisomers in steroids as well as triterpenoids and alkaloids have been given elsewhere.[59] Equatorial substituents are more stable than axial substituents and they react more rapidly in reactions governed by steric hindrance, such as acetylation of alcohols and saponification of esters; they react more slowly in reactions governed by steric assistance, such as chromic acid oxidation (cf. Figs. 11.30 to 11.32). Addition and elimination reactions proceed via the diaxial mode; thus bromine addition to 2-cholestene (A/B trans, double bond in 2,3) gives initially the diaxial 2β,3α-dibromocholestane which, on heating, is partially converted to the diequatorial 2α,3β isomer (thermodynamic control; both isomers are present at equilibrium because of the earlier discussed counterplay of steric and polar factors). Iodide-induced elimination of the 2β,3α-dibromide (diaxial) is very much faster than that of the 2α,3β-dibromide (diequatorial), and so on. The axial 2β,4β,6β, and 8β positions are more hindered than other axial positions by the presence of synaxial methyl groups; the 11β position, which has two synaxial methyl groups, is the most hindered of all. The five-membered ring is less puckered than the six-membered ring, thus the axial and equatorial nature of the substituents in this ring is not as definitive.

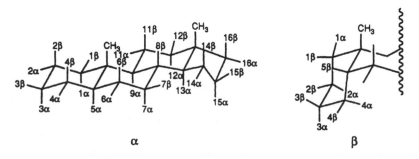

Figure 11.76. Stereochemical descriptors in steroids.

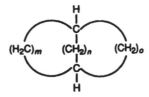

Figure 11.77. Bridged rings.

The two angular methyl groups inhibit approach of reagents from the top (β) side. As a result, additions to double bonds (e.g., epoxidations and carbene additions), which in simple cyclohexenes have an equal chance of occurring on either face of the double bond, in steroids usually proceed by α attack.

b. Bridged Rings

Figure 11.77 shows the general formula of simply bridged rings; m, n, and $o \neq 0$ (contrast the fused ring system shown in Fig. 11.64). The smallest possible bridged ring system, bicyclo[1.1.1]pentane (Fig. 11.78, **A**) has been synthesized and is reasonably stable. Its structure has been determined by electron diffraction[131b,346] and displays a remarkably short transannular nonbonded C(1)—C(3) distance of 184.5–187.4 pm (1.845–1.874 Å); in fact this is the shortest distance known for nonbonded carbon atoms. There is also a remarkably large H(1)—H(3) coupling in the ^{1}H NMR: $^4J = 18$ Hz, suggesting a strong transannular interaction. However, this interaction is not manifest in other properties of the bicyclo[1.1.1]pentane system, such as the effect of 3-substituents on the strength of the 1-carboxylic acid.[347]

The strain energy of the bicyclo[1.1.1]pentane system has not been determined experimentally but has been cumputed by *ab initio* calculations to be 60–68 kcal mol^{-1} (251–284 kJ mol^{-1}).[311,348,349] The systems with larger m, n, and o (Fig. 11.77) ranging from 1, 1, and 2 to 3, 3, and 3 are also all known[23,24] and their strains[349] have been computed by force field[350,351] or *ab initio* calculations[311]; they are, of course, less strained than the [1.1.1] system. Thus the strain in the very common norbornane system (Fig. 11.78, **B**, $m = n = 2$, $o = 1$) is 17.0 kcal mol^{-1}1 (71.1 kJ mol^{-1}); the main source of this strain is a serious deformation of the angle at the one-carbon bridge to 93°–96°,[352–354] plus the eclipsing of the hydrogen atoms at the two-carbon bridges.

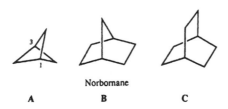

Norbornane

A   **B**   **C**

Figure 11.78. Bicyclo[1.1.1]pentane, norbornane, and bicyclo[2.2.2]octane.

Eclipsing must be a major source of strain in bicyclo[2.2.2] octane (Fig. 11.78, **C**). There has been some controversy[355] whether the molecule is a triple boat, as shown in Figure 11.78 (**D$_{3h}$**) or whether it is twisted out of this conformation of highest symmetry into a **D$_3$** shape to relieve eclipsing strain. In the solid state, the compound seems to be **D$_{3h}$**; but the electron diffraction pattern in the vapor suggests some twisting (about 10°–12° on either side) with a double-minimum potential curve with a slight hump [ca. 0.1 kcal mol^{-1} (0.4 kJ mol^{-1})] at the **D$_{3h}$** conformation, that is, in the middle[356]; yet, since the energy barrier is very low the molecule will rapidly librate from one twist form to the other.

We shall dwell here briefly on the norbornane structure (Fig. 11.78, **B**), both because of its importance in physical organic chemistry and because of its relation to at least one important class of monoterpenes of which camphor (1R,4R)-1,7,7-trimethylbicyclo[2.2.1]heptan-2-one (Fig. 11.79) is an important representative. Camphor is, incidentally, one of the few chemicals readily available as both (+) and (−) enantiomers (from different natural sources as well as synthetically, from natural, nonracemic precursors) and as the racemate (by synthesis).

The two diastereotopic faces of the carbonyl group of camphor (exo and endo, cf. Fig. 11.79) are of quite different accessibility to nucleophilic reagents: Whereas approach to the bottom face is somewhat impeded by the U-shaped cavity of the molecule, that to the top face is strongly hindered by the overhanging methyl group at C(7). Thus Grignard reagents react with camphor exclusively by approach from the endo side[170] and hydrides attach themselves predominantly to that side, the more so, the more bulky the hydride.[171] These facts are summarized in Figure 11.80. An entirely different picture is seen with trisnorcamphor, commonly called "norcamphor" (bicyclo[2.2.1]heptan-2-one), also shown in Figure 11.80. Here there is no hindrance from the exo side, but the impedance to approach from the endo side remains, so Grignard reagents in almost all cases approach exclusively from the exo side and hydrides approach mainly from that side.

Bicyclo[2.2.2]octatriene (barrelene; Fig. 11.81) has also been synthesized[357]; the molecule has **D$_{3h}$** symmetry and its bond lengths and angles are only slightly abnormal.[358]

Among the higher bicycloalkanes we mention here bicyclo[3.3.1]nonane (for a review, see ref. 359) as an example where primitive model considerations led to an erroneous result. Inspection of a model of the compound (cf. Fig. 11.82) led one of us[332] to state that the molecule could not exist as a double chair (Fig. 11.82) because

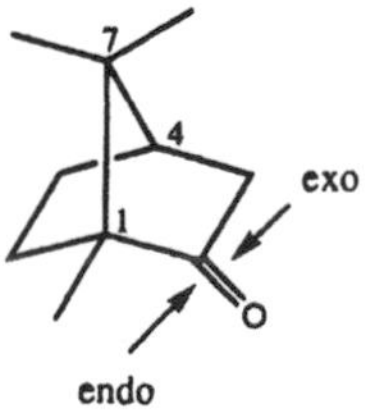

Figure 11.79.   (+)-Camphor.

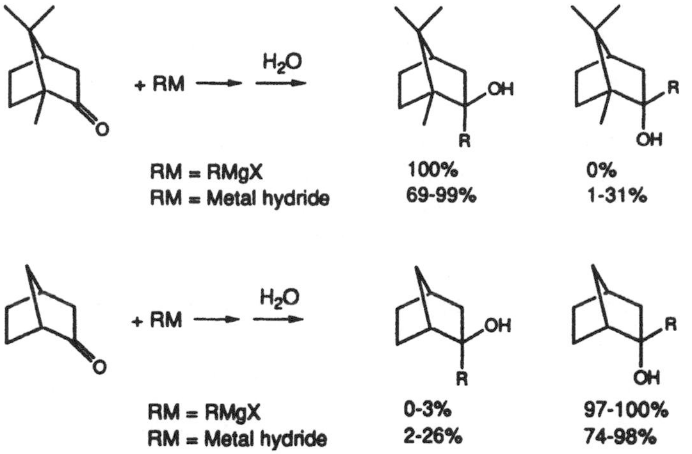

RM = RMgX	100%	0%
RM = Metal hydride	69-99%	1-31%

RM = RMgX	0-3%	97-100%
RM = Metal hydride	2-26%	74-98%

Figure 11.80.   Reaction of nucleophiles with camphor and norcamphor.

of excessive steric compression of the endo hydrogen atoms at C(3) and C(7). However, this compression can be relieved by moderate flattening of the two chairs and it was soon found by X-ray crystallography[360,361] that bicyclo[3.3.1]nonanes do exist in the double-chair conformation. Recent electron diffraction work at different temperatures confirms this conclusion; the ratio of chair–chair to chair–boat conformers at room temperature is 95:5, the chair–chair conformation being favored by 2.5 kcal mol^{-1} (10.5 kJ mol^{-1}) in enthalpy but disfavored by 1.5 cal mol^{-1} K^{-1} (6.3 J mol^{-1} K^{-1}) in entropy.[362] The transannular H $\cdots$ H distance has been calculated to be a very close 195–196 pm (1.95–1.96 Å)[363,364] with a resulting strain of 12.3 kcal mol^{-1} (51.5 kJ mol^{-1}).[365]

Although, because of limitations of space, we cannot expand on the subject of unsaturated bridged ring systems, we do want to touch on one aspect of such systems, namely, Bredt's rule[366,367] (see also ref. 332). Bredt's rule states that in a small bridged

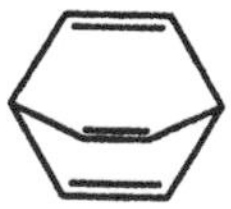

Figure 11.81.   Structure of barrelene.

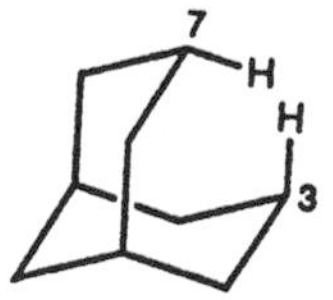

Figure 11.82.   Structure of bicyclo[3.3.1]nonane.

system one cannot, for reasons of excessive strain, have a double bond at the bridgehead position. Thus, for example, the rule accounts for the fact that bicyclo[2.2.2]octane-2,6-dione (Fig. 11.83, **A**) lacks the normal acidic properties of a 1,3-diketone: The corresponding enolate would have a bridgehead double bond.

The evident reason for Bredt's rule is that a bridgehead double bond implies that one of the rings containing this bond must be an (excessively strained?) (*E*)-cycloalkene. However, as will be seen in Section 13-8.c, (*E*)-cycloalkenes are isolable with a ring as small as seven membered, and there is good evidence for the fleeting existence of 1-phenyl-*trans*-cyclohexene.[368]

It is therefore not suprising that [3.3.1]bicyclo-1-nonene (Fig. 11.83, **B**) is, in fact, a stable substance,[369–371] as are other bridgehead alkenes (paradoxically called Bredt olefins) containing (*E*)-cyclooctene rings.[368] More suprising is the high stability of 4-(1-adamantyl)homoadamant-3-ene (Fig. 11.83, **C**) containing an (*E*)-cycloheptene ring.[372] Apparently, the attached bulky adamantyl substituent sterically prevents the otherwise facile dimerization of the adamantene moiety. Smaller bicycloalkenes containing (*E*)-cyclohexene rings are predictably unstable and their existence can, at best, be inferred from the isolation of trapping products.[368]

Although camphor and norcamphor (Fig. 11.80) have two chiral centers (the two bridgehead atoms), in both cases there is only one pair of enantiomers rather than the two expected. The reason for this is that the configurations of the bridgehead atoms are not independently variable: In "small" bridged bicyclic compounds, the "bridge" must be cis with respect to the outer ring. In norbornane and bicyclo[2.2.2]octane (Fig. 11.78) the "outer ring" may be considered six-membered with the "bridge" being either CH_2 or $(CH_2)_2$; it is clear from models that attempting to bridge a six-membered ring 1,4-trans would incur excessive strain, unless the bridge is quite large, for example, $(CH_2)_5$, in which case the trans (in–out) configuration becomes possible (e.g., ref. 373).

There are many other fascinating aspects to the chemistry of bridgehead bicyclic diamines, especially those containing oxygen atoms in the bridge (cryptands), which show high selectivity in the formation of metal ion complexes (cryptates).[374–378] Detailed discussion of these compounds (Fig. 11.84) and the related crown ethers or coronands[379,380] (Fig. 11.84), podands (Fig. 11.84), spherands[381] (Fig. 11.84), hemispherands, carcerands[382] and hemicarcerands,[383] and calixarenes (Fig. 11.85)[384–388] is, unfortunately, beyond the scope of this book (see also refs. 387, 389, and 390).

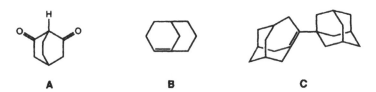

A	B	C

Figure 11.83.   Bredt's rule and bridgehead alkenes.

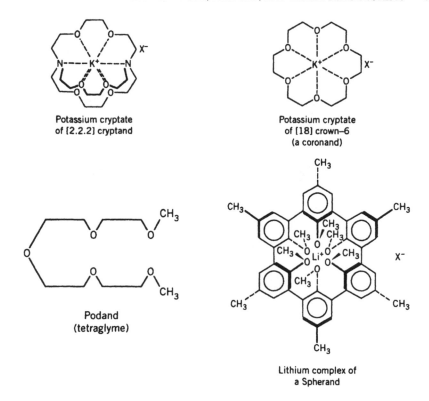

Potassium cryptate
of [2.2.2] cryptand

Potassium cryptate
of [18] crown–6
(a coronand)

Podand
(tetraglyme)

Lithium complex of
a Spherand

Figure 11.84. Cryptands, crown ethers (coronands), podands, and spherands.

c. Propellanes

The propellanes (Fig. 11.86), so called because of their propeller shape, are fascinating molecules, the first example of which was synthesized by Ginsburg and co-workers in 1966.[391] Since the fusion of two small rings already involves strain beyond that of the individual rings themselves (cf. Figs. 11.64 and 11.65 and accompanying text), it is surprising that propellanes with $m + n + o < 6$ can be obtained, and even the compound with $m = n = o = 1$ has been synthesized[392] and is modestly stable. A detailed discussion of the extensive work in this field is beyond the scope of this text, but a number of salient references are available.[90,346b,393–395]

d. Catenanes, Rotaxanes, Knots, and Möbius Strips

This section (for reviews see refs. 396–400) deals with some molecules of unusual topology. Catenanes (Fig. 11.87) (from Latin *catena* meaning chain; the name was coined by Wasserman[401]) are molecules containing two or more intertwined rings. Rotaxanes[402] (from Latin *rota* meaning wheel and axis; Fig. 11.87) are molecules in which a linear molecule (axle) is threaded through a cyclic one (wheel) and prevented

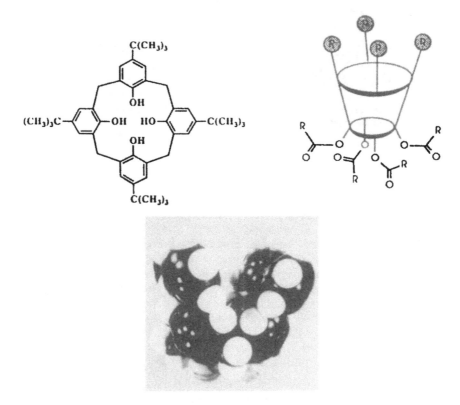

Figure 11.85.   Calix|4|arenes.

from slipping out by large, bulky end-groups.[403] The trefoil knot[404] (one example of the family of knots) is depicted in Figure 11.87. A Möbius strip molecule is a two-dimensional structure the ends of which are connected after a twist[397]; a macroscopic analogue is shown in Figure 11.87.

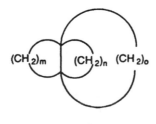

Propellane

Figure 11.86.   Propellane.

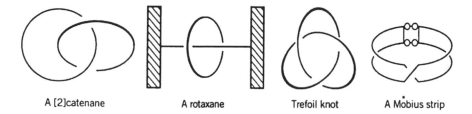

A [2]catenane   A rotaxane   Trefoil knot   A Möbius strip

Figure 11.87. Catenanes, rotaxanes, trefoil knots, and Möbius strips. [Adapted, with permission, from Walba, D. M., Richards, R. M., and Haltiwanger, R. C. *J. Am. Chem. Soc.* **1982,** *104,* 3220. Copyright © American Chemical Society, Washington, DC.]

Catenanes and rotaxanes have been synthesized in two different ways: statistical syntheses and directed syntheses. In a statistical synthesis of a catenane, a large ring (say 34-membered or larger) is formed in the presence of another large ring (Fig. 11.88). A finite number of chain molecules will, on a statistical basis, have become threaded through the ring before cyclization. These threaded molecules, upon cyclization of the chain, give rise to a catenane. This method has been used by Wasserman[401,405] as shown in Figure 11.88. However, the yield is very small [$10^{-4}\%$, based on the ultimate starting material, sebacic (decanedioic) acid.[405]]

In the statistical synthesis of rotaxanes,[406] where, in an equilibrium assembly of a large ring and a straight-chain molecule, the latter is capped at both ends with a bulky group so as to "fix" the rotaxane (Fig. 11.89), yields depend on (and increase with) the size of the ring and the length of the chain and may be as high as 11.3%.[407] Indeed, an ingenious modification of this method constitutes the first published synthesis of a rotaxane[408]: The large-ring molecule (a C_{30} acyloin) was connected chemically to a polystyrene resin of the Merrifield[409] type, the open-chain molecule (1,10-decanediol) was then threaded and capped [with $(C_6H_5)_3CCl$], and any remaining monomeric material (ditrityl ether of decanediol) washed away. Although the threading is statistically inefficient and only a small fraction of the (bound) ring was converted to a (bound) rotaxane, the fact that the unconverted starting ring compound remains attached to the resin allowed the authors to repeat the threading and capping process 70 times after which the product was chemically detached from the resin and separated into starting material and rotaxane.

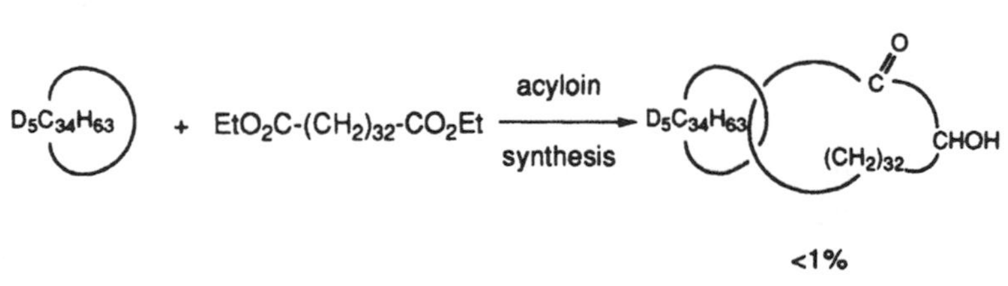

$$D_5C_{34}H_{63} \quad + \quad EtO_2C\text{-}(CH_2)_{32}\text{-}CO_2Et \quad \xrightarrow[\text{synthesis}]{\text{acyloin}} \quad D_5C_{34}H_{63}(CH_2)_{32}\ \overset{O}{\underset{CHOH}{\overset{\|}{C}}}$$

<1%

Figure 11.88. Statistical synthesis of a catenane.

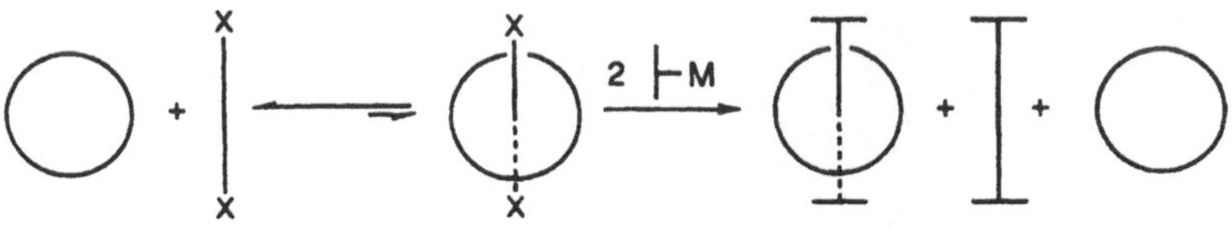

Figure 11.89. Statistical synthesis of a rotaxane.

It is also possible to increase the chances of threading by chemical affinity. An early effort[410] involves the threading of a polyethylene glycol through a large crown ether followed by capping with trityl chloride. The threading is enhanced by the affinity of the crown ether and linear polyether functions and the yield is 15%. In yet another synthesis of similar type, a terminal diamine (1,10-diaminodecane or 1,12-diaminododecane) was threaded in α- or β-cyclodextrin and then capped as a complex with $CoCl(en)^{2+}$ end groups (en = ethylenediamine).[411]

The threading method is now the preferred approach to rotaxane synthesis. The compounds constituting the wheel and axle moieties, respectively, are chosen so as to promote noncovalent (e.g., hydrophobic or multiple hydrogen bonding) interaction. When they are brought together, they form a complex in which the axle moiety is reversibly inserted into the wheel to form what is called a "pseudorotaxane."[412] The ends of the molecule constituting the axle are then capped by covalent chemical bonding by "stoppers," which are large enough to prevent slippage out of the wheel, thus freezing the equilibrium previously established on the side of the rotaxane (Fig. 11.89). Numerous examples of this process have been studied[400,413]; a few particularly interesting ones will be presented below.

The rotaxane prepared from crown ether and polyethylene glycol (see above) was also converted to a catenane[414] in the following way: In lieu of trityl chloride, p-BrCH$_2$C$_6$H$_4$CCl(C$_6$H$_5$)$_2$, mono-p-bromomethyltrityl chloride, was used as a cap. This reactive cap then allowed cyclization of the linear part of the rotaxane by zinc–copper coupling of the benzylic bromide to a bibenzyl, with formation of a catenane in 14% yield.

We return now to directed syntheses of catenanes and rotaxanes. In these syntheses the thread, or the second ring, is initially linked covalently to its ring partner, but in a way that scission of certain covalent bonds (designed to be broken readily) will lead to the desired catenane or rotaxane. The final stages of a directed synthesis of a rotaxane of this type[415] are depicted in Figure 11.90. Directed syntheses of [2]catenanes, that is, catenanes with two interlocked rings[416] have been reviewed.[396]

A very interesting synthesis of catenanes involves the use of a metal as a template.[417] The initial product is a catenane complex (catenate), which may subsequently be demetalated to give the catenane. The sequence is illustrated here by the synthesis of a trefoil knot[418] (Fig. 11.91).

Template syntheses of catenanes without intervention of a metal have been described by Stoddart's group[419]; an example, shown in Fig. 11.92, relies on complexation, prior to ring closure, of biparaphenylene-34-crown-10 with a chain precursor containing two 4,4'-bispyridine units that are then capped with p-xylyl

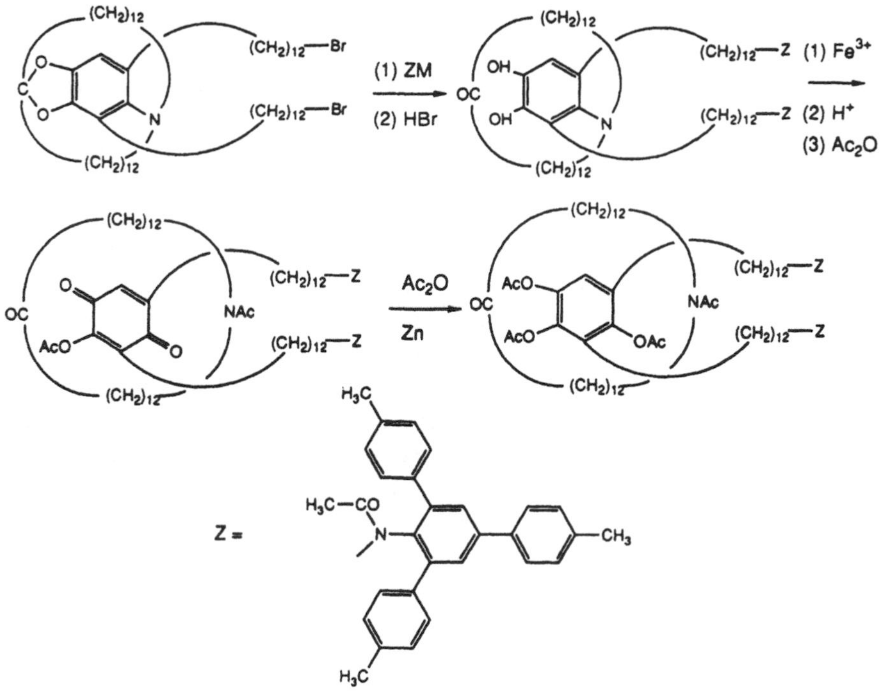

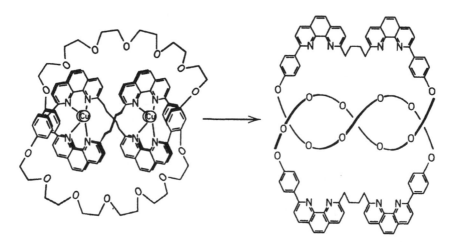

Figure 11.90. Directed synthesis of a rotaxane.

Figure 11.91. Template synthesis of trefoil knot. [Reproduced, with permission, from Dietrich-Buchecker, C. O. and Sauvage. J.-P. *Angew. Chem. Int. Ed. Engl.*, **1989**, *28*, 190. Copyright © 1989 VCH Publishers, Weinheim, Germany.]

dibromide; the yield of catenane is an amazing 70%. In this case there are two translational arrangements: Whereas one of the bipyridine moieties is complexed within the crown ether (shown), the other is wedged between two crown ether moieties in a polymolecular stacked arrangement. Exchange between the two topomers can be observed on the NMR time scale: Depending on which ring moves, the activation energy for the exchange is 12.2 or 14.0 kcal/mol^{-1} (51.0 or 58.6 kJ mol^{-1}).[420]

In a somewhat similiar fashion, a "template" synthesis of a rotaxane in 32% yield was accomplished.[421] The difference between this synthesis and that of the catenane shown in Figure 11.92 is that the "thread" of the rotaxane (Fig. 11.93) was introduced as a silyl-capped podand-type (cf. Fig. 11.84) species. Here, also, either of the two phenyl rings of the "thread" may be included in the bisbipyridine-dixylylene ring, and here again a motion of the thread within the ring (interchange of two topomers) can be observed on the NMR time scale with an activation energy of about 13 kcal mol^{-1} (54.4 kJ mol^{-1}). This system (Figure 11.93) has been called a "molecular shuttle"[421] (See also ref. 422). Much additional exciting work has been done and is being done in the rotaxane field. We mention here only the work on "daisy chains" in which the

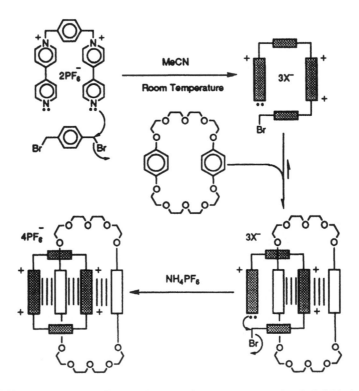

Figure 11.92. Template and self-assembly synthesis of a catenane. The shaded blocks represent 4,4-bispyridine units, the open blocks stand for p-phenylene. [Adapted with permission from Ashton, P. R. et al. *Angew. Chem. Int. Ed. Engl.* **1989**, *28*, 1396. Copyright © 1989 VCH Publishers, Weinheim, Germany.]

wheel and axle are linked covalently in a single molecule; this can lead to cyclic[423] or linear olligomers or even polymers;[413,425] (Fig. 11.94) and the work on distereromeric 3-rotaxanes[426] (the number 3 referring to the fact that there are three molecules involved, one axle and two wheels. In the case cited, the disteromerism results from the fact that the sequence of links in the two wheels is such as to impart directionality and either a meso or a chiral isomer can result, depending on whether the direction is the same or opposite (Fig. 11.95). Several interesting studies are underway to adapt the concept of a molecular shuttle (Fig. 11.93) to a molecular machine or molecular calculator by effecting the shuttling process in catenanes or rotaxanes photochemically or electrochemically in a reversible manner.[424,427]

We now come to the synthesis of a Möbius strip (named after the nineteenth century mathematician Möbius). The Möbius (**A**) and cyclinder (**B**) molecules shown schematically in Figure 11.96 were synthesized, from the linear array of crown ethers shown,[428a] in 24 and 22% yields, respectively.[428b] The cylindrical molecule **B** was crystalline and its structure was confirmed by X-ray crystallography. The Möbius strip molecule **A** is chiral and its chirality could be demonstrated by a doubling of the olefinic ^{13}C signals in the presence of the chiral solvating agent (+)-2,2,2-trifluoro-1-(9-anthryl)-ethanol. [It might appear at first sight that molecule **A** contains diastereotopic olefinic carbon atoms even in the absence of a chiral solvating agent. However, this is in fact not the case because the "twist" moves rapidly (on the NMR time scale) around the ring. thereby leading to averaging of the signals of all olefinic (and also of all saturated) carbon atoms.]

The synthesis of the topologically interesting trefoil knot has been shown in Figure 11.91. Subsequently, a molecule in the shape of a trefoil knot has been

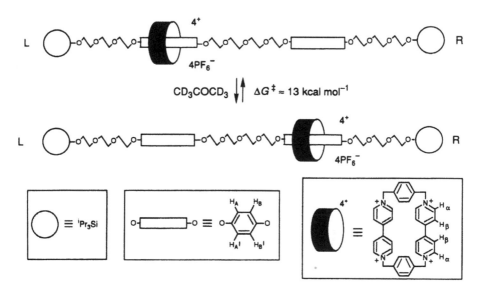

Figure 11.93. "Molecular shuttle." [Adapted, with permission, from Anelli, P. L., Spencer, N., and Stoddart, J. F. *J. Am. Chem. Soc.* **1991**, *113*, 5132. Copyright © 1991 American Chemical Society, Washington, DC.]

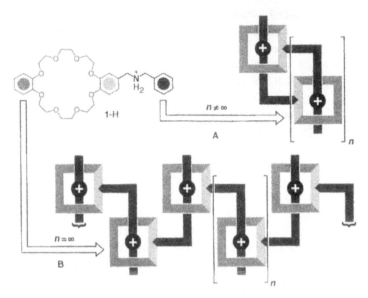

Figure 11.94. Cartoon representations of the viable types of self-assembled superarchitectures that can be created fromthe plerotopic **1-H⁻** cation. (**A**) The supramolecular association of a finite number of **1-H⁻** monomer units in a cyclic fashion generates oligomeric daisy-chain-like macrocycles. (**B**) The noncovalent polymerization of an infinite number of **1-H⁻** monomer units furnishes a supramolecular array that is commensurate with a macromolecular daisy chain. Reproduced with permission from Ashton, P. R. et al. *Angew. Chem. Int. Ed. Engl.* **1998**, *37*, 1294. Copyright © 1998 Wiley-VCH, Weinheim, Germany.

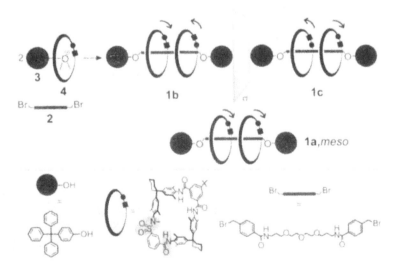

Figure 11.95. Schematic representation of the synthesis of the cyclodiastereomeric [3]rotaxane as a pair of enantiomers (**1b, 1c**) and as the *meso* form (**1a**). Reproduced with permission from Schmieder, R. et al. *Angew. Chem. Int. Ed. Engl.* **1999**, *38*, 3528. Copyright © 1999 Wiley-VCH, Weinheim, Germany.

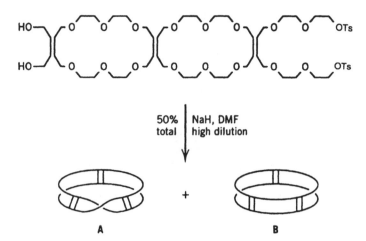

Figure 11.96. Synthesis of a Möbius strip molecule. DMF = N,N-Dimethylformamide. [Reproduced with permission from Walba, D. M., Richards, R. M., and Haltiwanger, R. C. *J. Am. Chem. Soc.* **1982**, *104*, 3220. Copyright © 1982 American Chemical Society, Washington, DC.]

synthesized in quantity sufficient to allow resolution and determination of absolute configuration.[429]

e. Cubane, Tetrahedrane, Dodecahedrane, Adamantane, and Buckminsterfullerene

We shall conclude this chapter with an esthetically pleasing subject, namely, the Platonic solids. The Platonic solids are polyhedra of high symmetry: the tetrahedron ($\mathbf{T_d}$), the octahedron ($\mathbf{O_h}$), the cube ($\mathbf{O_h}$), the dodecahedron ($\mathbf{I_h}$), and the icosahedron ($\mathbf{I_h}$) (see Section 4-3.b). In the octahedron four edges meet in one vertex and in the icosahedron five, so these figures cannot be constructed from tetracoordinate carbon plus hydrogen; however, the all-carbon compound buckminsterfullerene (C_{60}, see below) has the shape of a truncated icosahedron. Hydrocarbons $(CH)_n$ corresponding to the other three Platonic solids have been synthesized: cubane $(CH)_8$, dodecahedrane

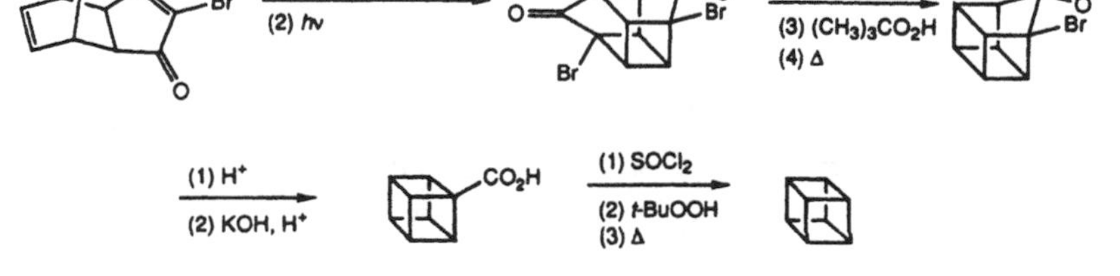

Figure 11.97. Synthesis of cubane.

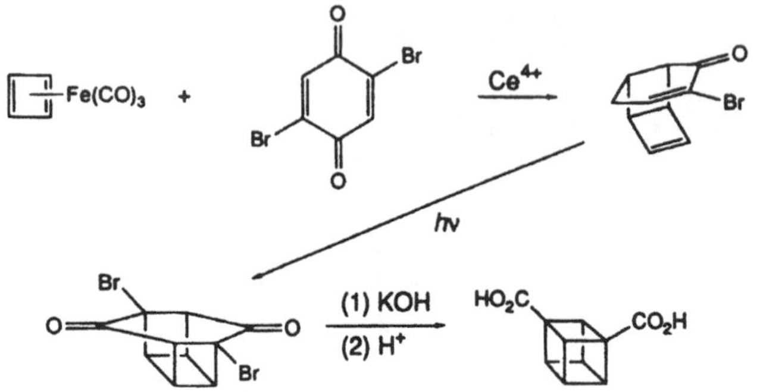

Figure 11.98. Alternative synthesis of a cubane.

$(CH)_{20}$, and tetrahedrane $(CH)_4$, though the last one was synthesized only as a tetra-*tert*-butyl derivative.

Cubane was synthesized in 1964[430] (see also refs. 431 and 432). The synthesis is schematized in Figure 11.97; the essential features are photochemical dimerization $(2 + 2)$ of a brominated dicyclopentadiene, ring contraction of five- to four-membered rings by Favorskii rearrangement, and free radical decarboxylation. An alternative synthesis,[433] shown in Figure 11.98, is interesting not only because it is quite short, but also because the starting material is a cyclobutadiene complex.

Cubane, as expected, displays a single ^1H resonance (at 4.04 ppm)[434] and a single ^{13}C resonance (at 47.3 ppm).[435] The structure as determined by electron diffraction[436] shows a bond length (157.5 pm, 1.575 Å) longer than that in cyclobutane. Earlier X-ray data[437] indicated a C–C–H angle of 123°–127° suggesting a high degree of s character in the C–H bond concomitant with the high degree of p character in the C–C bonds imposed by the 90° bond angle. Surprisingly, the ^{13}C–H coupling constant of 153.8 Hz[435] suggests only 30.8% s character in the C–H bond if one accepts the relation[438] $\%s = 0.2J^1_{13_{C-H}}$. Perhaps in a structure as highly strained as cubane this relation must be considered suspect. Based on heat of combustion data,[439] the strain energy in cubane was evaluated[351] to be 166 kcal mol^{-1} (695 kJ mol^{-1}). A somewhat smaller value, (154.7 kcal mol^{-1}, 647 kJ mol^{-1}) has also been computed.[440] It may or may not be coincidental that these values are close to six times the strain energy in cyclobutane (Table 11.4). (For a review of cubane chemistry, see ref. 432).

Tetra-*tert*-butyltetrahedrane (Fig. 11.99) was synthesized in 1978 by Maier's group.[441] The rather straightforward synthetic scheme shown in Figure 11.99 conceals both the limited availability of the starting material and the enormous amount of work

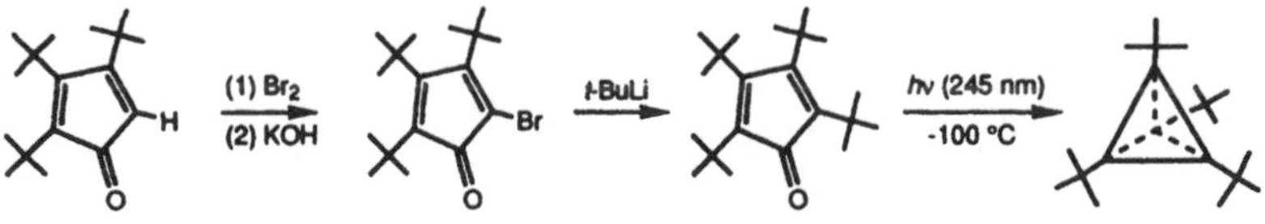

Figure 11.99. Synthesis of tetra-*tert*-butyltetrahedrane.

required in establishing the proper conditions for each step. The material, characterized by X-ray crystallograph py[442] and spectrally[443] (e.g., ^{1}H NMR, $\delta = 1.18$ ppm; ^{13}C NMR, $\delta = 9.27, 27.16, 31.78$ ppm; the high-field tetrahedrane carbon, similar to a cyclopropane one, is characteristic; interestingly, the primary and quaternary carbon atoms of the *tert*-butyl group are reversed, the latter being at higher field) is quite stable: It melts at 135°C with decomposition and is thermally converted to tetra-*tert*-butylcyclobutadiene, which can be photochemically reconverted to the tetrahedrane. The activation energy for the interconversion is remarkably high: $\Delta H^{\ddagger} = 25.5$ kcal mol^{-1} (106.7 kJ mol^{-1}) and $\Delta S^{\ddagger} = -10.3$ cal mol^{-1} K^{-1} (−43.1 J mol^{-1} K^{-1}). The strain in tetrahedrane is very large, having been estimated at 130–150 kcal mol^{-1} (544–628 kJ mol^{-1}).[24,440] All attempts to synthesize tetrahedrane itself have failed for reasons that have been explained elsewhere.[24,90,444]

The last of the accessible Platonic solid shaped hydrocarbons to be synthesized was dodecahedrane (Fig. 11.100).[445,446] (The original synthesis comprises 23 steps and is therefore not reproduced here.) The structure of the compound as determined by X-ray diffraction[447] shows the expected $\mathbf{I_h}$ symmetry, nearly normal C–C bond distances (153.5–154.1 pm, 1.535–1.541 Å) and a C–C–C bond angle of about 108°. It will be interesting to assess the strain in dodecahedrane, because it is one of the few molecules where different force fields yield totally different strain estimates around 40 kcal mol^{-1} (167 kJ mol^{-1}), by the Engler–Schleyer force field[351] but about twice as much by the Allinger et al., force field.[350]

Numerous other interesting caged hydrocarbons have been synthesized[24,448] but we shall mention only one more here: adamantane (Fig. 11.101), not only because of its high symmetry ($\mathbf{T_d}$) and its relation to the diamond lattice, but also because, unlike the hydrocarbons so far discussed in this section, it is a highly stable molecule and constitutes the product of thermodynamically controlled isomerization of many other $C_{10}H_{16}$ hydrocarbons.

Adamantane (for reviews see refs. 24, 449, and 450) was first isolated in 1933 from Czechoslovakian petroleum[451]; its structure was correctly assigned at that time and confirmed some years later by a rational, if lengthy and low-yielding synthesis.[452] The substance remained a rarity, however, until Schleyer discovered[453] that both hydrogenated *exo*- and *endo*-bicyclopentadienes on treatment with aluminum chloride gave adamantane in about 20% yield. Other $C_{10}H_{16}$ hydrocarbons isomerize more cleanly; for example, twistane (Fig. 4.10) is isomerized to adamantane rapidly in nearly 100% yield on treatment with aluminum chloride.[454]

Although a model of adamantane appears strain-free, the molecule has about 7.6 kcal mol^{-1} (31.8 kJ mol^{-1}) strain. This is much more than the strain in cyclohexane.

Figure 11.100. Dodecahedrane.

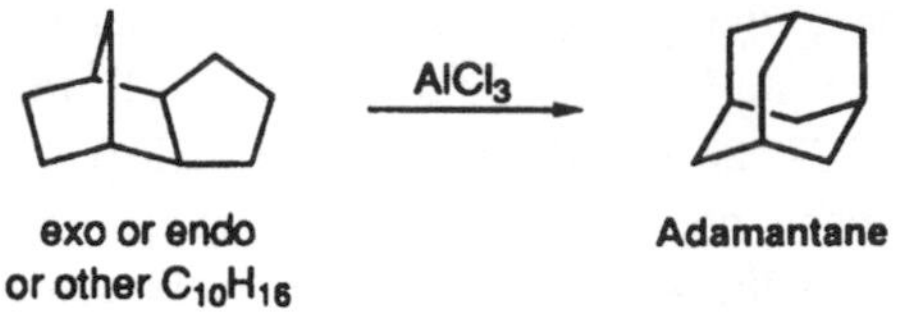

Figure 11.101.   Adamantane synthesis.

In adamantane, the framework structure constrains the bond angles to be nearly perfectly tetrahedral. However, as we have seen before, a C–CH$_2$–C bond angle does not want to be exactly tetrahedral; the optimum (in propane) is 112.5° and in the slightly strained cyclohexane the angle is 111.5°. It is the further reduction to 109.5° which causes the additional strain in adamantane, thus demonstrating once again that molecular models can be misleading.

Finally, we mention the carbon alleotrope C$_{60}$, which was originally prepared by vaporization of graphite (by a focused pulse laser) into a high-density helium flow. This experiment[455,456] produces a number of large carbon clusters of which C$_{60}$ was discovered, by mass spectrometry, to be by far the most abundant. The authors postulated that the molecule C$_{60}$ is a closed-shell polygon with 60 vertices and 32 faces of which 20 are hexagonal (aromatic) and 12 are pentagonal (Fig. 11.102). Because of the similarity of the structure with a geodesic dome designed by the architect Buckminster Fuller, the compound has been called "buckminsterfullerene"; it has also been called "footballene," which is the Chemical Abstracts name, or "soccerballene." Subsequent to its initial detection in the mass spectrometer, buckminsterfullerene was prepared and isolated by evaporating graphite electrodes into an atmosphere of helium or argon at 50–100 mbar (38–75 mm Hg) followed by extraction of the sootlike product with benzene.[457] This procedure produces mostly C$_{60}$, a mustard-colored crystalline solid, along with the more deeply colored C$_{70}$ (of similar structure) in about 5:1 ratio. Buckminsterfullerene (C$_{60}$) is purified by chromatography[458]; or by selective complexation with *p-tert*-butylcalix[8]arene (cf. Fig. 11.85).[459] Although X-ray structure determination of C$_{60}$ is complicated by orientational disorder caused by a "rolling" of the near-spherical molecules in the crystal relative to each other,[460] X-ray structure analysis of the derivative C$_{60}$(OsO$_4$).4-*tert*-butylpyridine confirms the truncated icosahedral (**I$_h$**) structure of the parent molecule.[461] The bond lengths are 139 pm (1.39 Å) for the

Figure 11.102.   Buckminsterfullerene.

six-six-membered-ring fusion and 143 pm (1.43 Å) for the six-five-membered-ring fusion.

The discovery of a new alleotrope of carbon has engendered a great deal of interest in buckminsterfullerene and the higher fullerenes (such as C_{70}). A number of reviews[462–467] are available and, in 1992, an entire issue of the journal *Accounts of Chemical Research* (Vol. 25, No. 3) was devoted to the topic.

REFERENCES

1. Cahn, R. S., Ingold, Sir C., and Prelog, V. *Angew. Chem. Int. Ed. Engl.* **1966**, *5*, 385.

2. Cross, L. C. and Klyne, W., collators. *Pure Appl. Chem.* **1976**, *45*, 11.

3. Anet, F. A. L. *Tetrahedron Lett.* **1990**, *31*, 2125.

4. Wislicenus, J. *Ber. Dtsch. Chem. Ges.* **1901**, *34*, 2565.

5. McCoy, L. L. and Nachtigall, G. W. *J. Am. Chem. Soc.* **1963**, *85*, 1321.

6. Bjerrum, N. *Z. Phys. Chem.* **1923**, *106*, 219.

7. Perkin, W. H. *J. Chem. Soc.* **1894**, *65*, 572.

8. Baeyer, A. *Justus Liebigs Ann. Chem.* **1890**, *258*, 145.

9. Bloomfield, J. J. and Fuchs, R. *J. Chem. Soc. B* **1970**, 363.

10. Inoue, Y., Kurosawa, K., Nakanishi, K., and Obara, H. *J. Chem. Soc.* **1965**, 3339.

11. Sicher, J., Šipoš, F., and Jonáš, J. *Coll. Czech. Chem. Commun.* **1961**, *26*, 262.

12. Delben, F. and Crescenzi, V. *J. Solution Chem.* **1978**, *7*, 597.

13. Kuhn, L. P. *J. Am. Chem. Soc.* **1952**, *74*, 2492; **1954**, *76*, 4323.

14. Barnier, J.-P. and Conia, J.-M. *Bull. Soc. Chim. Fr.* **1976**, 281.

15. Monti, J. P., Faure, R., and Vincent, E. J. *Org. Magn. Reson.* **1975**, *7*, 637.

16. Chukovskaya, E. Ts., Dostovalova, V. I., Kamyshova, A. A., and Freidlina, R. Kh. *Bull. Acad. Sci. USSR Div. Chem. Soc.* **1981**, *30*, 801 (Engl. transl. p. 1470).

17. Eliel, E. L. and Pietrusiewicz, K. M. *Org. Magn. Reson.* **1980**, *13*, 193.

18. (a) Christl, M., Reich, H.-J., and Roberts, J. D. *J. Am. Chem. Soc.* **1971**, *83*, 3463. (b) Christl, M. and Roberts, J. D. *J. Org. Chem.* **1972**, *37*, 3443.

19. Dalling, D. K. and Grant, D. M. (a) *J. Am. Chem. Soc.* **1967**, *89*, 6612; (b) *ibid.* **1972**, *94*, 5318.

20. Goering, H. L. and Serres, C. *J. Am. Chem. Soc.* **1952**, *74*, 5908.

21. Baeyer, A. von *Ber. Dtsch. Chem. Ges.* **1885**, *18*, 2277.

22. Ramsay, O. B. "Molecular Models in the Early Development of Stereochemistry," in Ramsay, O.B., ed., *van't Hoff–Le Bel Centennial*, ACS Symposium Series 12, American Chemical Society, Washington, DC, 1975, p. 74.

23. Liebman, J. F. and Greenberg, A. *Chem. Rev.* **1976**, *76*, 311.

24. Greenberg, A. and Liebman, J. F. *Strained Organic Molecules,* Academic Press, New York, 1978.

25. Burkert, U. and Allinger, N. L. *Molecular Mechanics,* ACS Monograph 171, American Chemical Society, Washington, DC, 1982.

26. Wiberg, K. B. *Angew. Chem. Int. Ed. Engl.* **1986**, *25*, 312.

27. Liebman, J. F. and Greenberg, A. *Molecular Structure and Energetics*, Vol. 2, VCH, Deerfield Park, FL, 1987.

28. Schleyer, P. v. R., Williams, J. E., and Blanchard, K. R. *J. Am. Chem. Soc.* **1970**, *92*, 2377.

29. *TRC Thermodynamic Tables, Hydrocarbons*, Vol. VII, Thermodynamic Research Center, The Texas A&M University System, College Station, TX, 1991.

30. Chickos, J. S., Hesse, D. G., Panshin, S. Y., Rogers, D. W., Saunders, M., Uffer, P. M., and Liebman, J. F. *J. Org. Chem.* **1992**, *57*, 1897.

31. Sachse, H. *Ber. Dtsch. Chem. Ges.* **1890**, *23*, 1363; *Z. Phys. Chem.* **1892**, *10*, 203.

32. Brown, H. C., Fletcher, R. S., and Johannesen, R. B. *J. Am. Chem. Soc.* **1951**, *73*, 212 (footnote 21 there).

33. Prelog, V. "Bedeutung der vielgliederigen Ringverbindungen für die theoretische organische Chemie," in Todd, A. R., ed., *Perspectives in Organic Chemistry*, Interscience, New York, 1956, p. 96.

34. Prelog, V. *Bull. Soc. Chim. Fr.* **1960**, 1433.

35. Sicher, J. *Prog. Stereochem.* **1962**, *3*, 202.

36. Bloomfield, J. J., Owsley, D. C., and Nelke, J. M. *Org. React.* **1976**, *23*, 259.

37. Ruzicka, L., Brugger, W., Pfeiffer, M., Schinz, H., and Stoll, M. *Helv. Chim. Acta* **1926**, *9*, 499.

38. Ruggli, P. *Justus Liebigs Ann. Chem.* **1912**, *392*, 92.

39. Ziegler, K. "Methoden zur Herstellung und Umwandlung grosser Ringsysteme," in Houben-Weyl, *Methoden der Organischen Chemie*, 4th ed., Vol. 4, pt. 2, Thieme Verlag, Stuttgart, Germany, 1955, p. 729.

40. Molander, G. A. *Acc. Chem. Res.* **1998**, *31*, 603; Yet, L. *Chem. Rev.* **2000**, *100*, 2963.

41. Nicolaou, K. C. *Tetrahedron* **1977**, *33*, 683; Trost, B. M. and Verhoeven, T. R. *J. Am. Chem. Soc.* **1980**, *102*, 4743.

42. Abraham, R. J., Banks, H. D., Eliel, E. L., Hofer, O., and Kaloustian, M. K. *J. Am. Chem. Soc.* **1972**, *94*, 1913.

43. Illuminati, G. and Mandolini, L. *Acc. Chem. Res.* **1981**, *14*, 95.

44. Hammond, G. S. "Steric Effects on Equilibrated Systems," in Newman, M. S., ed., *Steric Effect in Organic Chemistry*, Wiley, New York, 1956, p. 425.

45. Nilsson, H. and Smith, L. *Z. Phys. Chem.* **1933**, *166A*, 136.

46. Tenud, L., Farooq, S., Seibl, J., and Eschenmoser, A. *Helv. Chim. Acta* **1970**, *53*, 2059.

47. (a) Baldwin, J. E. *J. Chem. Soc. Chem. Commun.* **1976**, *734*, 738. (b) Baldwin, J. E., Cutting, J., Dupont, W., Kruse, L., Silbermann, L., and Thomas, R. C. *J. Chem. Soc. Chem. Commun.* **1976**, 736. See also Johnson, C. D. *Acc. Chem. Res.* **1993**, *26*, 476.

48. Baldwin, J. "Rules for Ring Closure," Ciba Foundation Symposium 53, *Further Perspectives on Organic Chemistry*, Elsevier, New York, 1978, p. 85.

49. Bürgi, H. B., Dunitz, J. D., Lehn, J. M., and Wipff, G. *Tetrahedron* **1974**, *30*, 1563.

50. Baldwin, J. E. and Reiss, J. A. *J. Chem. Soc. Chem. Commun.* **1977**, 77.

51. Baldwin, J. E., Thomas, R. C., Kruse, L. I., and Silbermann, L. *J. Org. Chem.* **1977**, *42*, 3846.

52. Baldwin, J. E. and Kruse, L. I. *J. Chem. Soc. Chem. Commun.* **1977**, 233.

53. Baldwin, J. E. and Lusch, M. J. *Tetrahedron* **1982**, *38*, 2939.

54. Mohr, E. *J. Prakt. Chem.* **1918**, 98 [2], 315.

55. Aschan, O. *Chemie der Alicyclischen Verbindungen*, Vieweg & Sohn, Braunschweig, Germany, 1905, p. 329.

56. Shriner, R. L., Adams, R., and Marvel, C. S. "Stereochemistry," in Gilman, H., ed., *Organic Chemistry*, Wiley, New York, 1938, p. 238.

57. Jensen, F. R. and Bushweller, C. H. (a) *J. Am. Chem. Soc.* **1969**, *91*, 3223; (b) *ibid.* **1969**, *91*, 5774.

58. Barton, D. H. R. *Experientia* **1950**, *6*, 316; reprinted in *Top. Stereochem.* **1971**, *6*, 1.

59. Eliel, E. L., Allinger, N. L., Angyal, S. J., and Morrison, G. A. *Conformational Analysis*, Wiley, New York, 1965; reprinted by American Chemical Society, Washington, DC, 1981.

60. Russell, C. A. "The Origins of Conformational Analysis," in Ramsay, O. B., ed., *van't Hoff–Le Bel Centennial*, ACS Symposium Series 12, American Chemical Society, Washington, DC, 1975, p. 159.

61. Eliel, E. L. *J. Chem. Educ.* **1975**, *52*, 762.

62. Hassel, O. *Tidsskr. Kjemi Bergvesen Met.* **1943**, [*3*] 5, 32; Engl. transl. Hedberg, K. *Top. Stereochem.* **1971**, *6*, 11.

63. Davis, M. and Hassel, O. *Acta Chem. Scand.* **1963**, *17*, 1181.

64. Alekseev, N. V. and Kitaigorodskii, A. I. *J. Struct. Chem. USSR* **1963**, *4*, 163 (Engl. transl. p. 145); Geise, H. J., Buys, H. R., and Mijlhoff, F. C. *J. Mol. Struct.* **1971**, *9*, 447.

65. See also Dommen, J., Brupbacher, T., Grassi, G., and Bauder, A. *J. Am. Chem. Soc.* **1990**, *112*, 953 and refs. cited therein.

66. Dunitz, J. D. *J. Chem. Educ.* **1970**, *47*, 488.

67. Romers, C., Altona, C., Buys, H. R., and Havinga, E. *Top. Stereochem.* **1969**, *4*, 39.

68. Barton, D. H. R., Hassel, O., Pitzer, K. S., and Prelog, V. *Nature (London)* **1953**, *172*, 1096; *Science*, **1954**, *119*, 49.

69. Anet, F. A. L. and Anet, R. "Conformational Processes in Rings," in Jackman, L. M. and Cotton, F. A., eds., *Dynamic Nuclear Magnetic Resonance Spectroscopy*, Academic Press, New York, 1975, p. 543.

70. Sandström, J. *Dynamic NMR Spectroscopy*, Academic Press, New York, 1982.

71. (a) Aydin, R. and Günther, H. *Angew. Chem. Int. Ed. Engl.* **1981**, *20*, 985; (b) Höfner, D., Lesko, S. A., and Binsch, G. *Org. Magn. Reson.* **1978**, *11*, 179.

72. van de Graaf, B., Baas, J. M. A., and van Veen, A. *Rec. J. R. Neth. Chem. Soc.* **1980**, *99*, 175.

73. Anderson, J. E. *Top. Curr. Chem.* **1974**, *45*, 139.

74. Kellie, G. M. and Riddell, F. G. *Top. Stereochem.* **1974**, *8*, 225.

75. Allinger, N. L. *J. Am. Chem. Soc.* **1977**, *99*, 8127.

76. Jensen, F. R. and Bushweller, C. H. *Adv. Alicycl. Chem.* **1971**, *3*, 139.

77. Larnaudie, M. *J. Phys. Radium* **1954**, *15*, 650.

78. Berlin, A. J. and Jensen, F. R. *Chem. Ind. (London)* **1960**, 998.

79. Reeves, L. W. and Strømme, K. O. *Can. J. Chem.* **1960**, *38*, 1241.

80. Fishman, A. I., Remizov, A. B., and Stolov, A. A. *Dokl. Akad Nauk SSSR Phys. Chem.* **1981**, *260*, 683 (Engl. transl. p. 868).

81. Anet, F. A. L., Bradley, C, H., and Buchanan, G. W. *J. Am. Chem. Soc.* **1971**, *93*, 258.

82. Winstein, S. and Holness, N. J. *J. Am. Chem. Soc.* **1955**, *77*, 5562.

83. Eliel, E. L. *Chem. Ind (London)* **1959**, 568.

84. Marshall, J. L. *Carbon–Carbon and Carbon–Proton NMR Couplings: Applications to Organic Stereochemistry and Conformational Analysis*, Verlag Chemie, Deerfield Park, FL, 1983.

85. Anteunis, M. J. O. "Some Results and Limitations in Conformational Studies of Six-Membered Heterocycles," in Chiordoglu, G., ed., *Conformational Analysis, Scope and Present Limitations*, Academic Press, New York, 1971, p. 32.

86. Eliel, E. L. and Ro, R. S. (a) *J. Am. Chem. Soc.* **1957**, *79*, 5992; (b) *ibid.* **1957**, *79*, 5995.

87. Hirsch, J. A. *Top. Stereochem.* **1967**, *1*, 199.

88. Bushweller, C. H. "Stereodynamics of Cyclohexane and Substituted Cyclohexanes. Substituent A Values," in Juaristi, E., ed., *Conformational Behavior of Six-Membered Rings,* VCH, New York, 1995, p. 25.

89. Schneider, H.-J. and Hoppen, V. *J. Org. Chem.* **1978**, *43*, 3866.

90. Eliel, E. L. and Wilen, S. H. *Stereochemistry of Organic Compounds*, Wiley, New York, 1994.

91. Allinger, N. L. and Tribble, M. T. *Tetrahedron Lett.* **1971**, 3259.

92. Booth, H. and Everett, J. R. (a) *J. Chem. Soc. Chem. Commun.* **1976**, *278*; (b) *J. Chem. Soc. Perkin 2* **1980**, 255.

93. Wiberg, K. B., Hammer, J. D., Castejon, H., Bailey, W. F., DeLeon, E. L., and Jarret, R. M. *J. Org. Chem.* **1999**, *64*, 2085.

94. Eliel, E. L. and Kandasamy, D. *J. Org. Chem.* **1976**, *44*, 3899.

95. Allinger, N. L., Szkrybalo, W., and Van-Catledge, F. A. *J. Org. Chem.* **1968**, *33*, 784.

96. Manoharan, M. and Eliel, E. L. (a) *Tetrahedron Lett.* **1983**, *24*, 453; (b) *J. Am. Chem. Soc.* **1984**, *106*, 367.

97. Booth H. and Grindley, T. B. *J. Chem. Soc. Chem. Commun.* **1983**, 1013.

98. Leonard, J. E., Hammond, G. S., and Simmons, H. E. *J. Am. Chem. Soc.* **1975**, *97*, 5052.

99. Allinger, N. L., Miller, M. A., Van-Catledge, F. A., and Hirsch, S. A. *J. Am. Chem. Soc.* **1967**, *83*, 4345.

100. Eliel, E. L. and Enanoza, R. M. *J. Am. Chem. Soc.* **1972**, *94*, 8072.

101. Jordan, E. A. and Thorne, M. P. *Tetrahedron* **1986**, *42*, 93.

102. Carr, C. A., Robinson, M. J. T., and Tchen, C. D. A. *Tetrahedron Lett.* **1987**, *28*, 897.

103. Corey, E. J. and Feiner, N. F. *J. Org. Chem.* **1980**, *45*, 765.

104. Eliel, E. L. and Haubenstock, H. *J. Org. Chem.* **1961**, *26*, 3504.

105. Hudlicky, T. and Cebulak, M. *Cyclitols and Their Derivatives*, VCH, New York, 1993.

106. Angyal, S. J. *Q. Rev. Chem. Soc.* **1957**, *11*, 112.

107. Allinger, N. *J. Am. Chem. Soc.* **1957**, *79*, 3443.

108. Kellie, G. M. and Riddell, F. G. *J. Chem. Soc. Perkin 2* **1975**, 740.

109. van Arkel, A. E. *Rec. Trav. Chim.* **1932**, *51*, 1081; **1933**, *52*, 1013; **1934**, *53*, 246.

110. Edward, J. T., Farrell, P. G., Kirchnerova, J., Halle, J.-C., and Schaal, R. *Can. J. Chem.* **1976**, *54*, 1899.

111. Lemieux, R. U., Kullnig, R. K., Bernstein, H. J., and Schneider, W. G. *J. Am. Chem. Soc.* **1958**, *80*, 6098.

112. McConnell, H. M. *J. Chem. Phys.* **1957**, *27*, 226.

113. (a) Eliel, E. L., Gianni, M. H., Williams, T. H., and Stothers, J. B. *Tetrahedron Lett.* **1962**, 741. (b) Eliel, E. L. and Gianni, M. H. *Tetrahedron Lett.* **1962**, 97.

114. Stothers, J. B. *Carbon-13 NMR Spectroscopy*, Academic Press, New York, 1972.

115. Duddeck, H. *Top. Stereochem.* **1986**, *16*, 219.

116. Morin, F. G. and Grant, D. M. "Use of Carbon-13 Nuclear Magnetic Resonance in the Conformational Analysis of Hydroaromatic Compounds," in Rabideau, P. W., ed., *Conformational Analysis of Cyclohexenes, Cyclohexadienes, and Related Hydroaromatic Compounds*, VCH, New York, 1989, p. 127.

117. Rabideau, P. W., ed., *Conformational Analysis of Cyclohexenes, Cyclohexadienes and Related Hydroaromatic Compounds*, VCH, New York, 1989.

118. Vierhapper, F. W. and Willer, R. L. *Org. Magn. Reson.* **1977**, *9*, 13.

119. Eliel, E. L., Bailey, W. F., Kopp, L. D., Willer, R. L., Grant, D. M., Bertrand, R., Christensen, K. A., Dalling, D. K., Duch, M. W., Wenkert, E., Schell, F. M., and Cochran, D. W. *J. Am. Chem. Soc.* **1975**, *97*, 322.

120. Eliel, E. L., Haubenstock, H., and Acharya, R. V. *J. Am. Chem. Soc.* **1961**, *83*, 2351.

121. Allinger, N. L., Freiberg, L. A., and Hu, S.-E. *J. Am. Chem. Soc.* **1962**, *84*, 2836.

122. Richer, J. C., Pilato, L. A., and Eliel, E. L. *Chem. Ind. (London)* **1961**, 2007.

123. Schreiber, J. and Eschenmoser, A. *Helv. Chim. Acta* **1955**, *38*, 1529.

124. Deslongchamps, P. *Stereoelectronic Effects in Organic Chemistry*, Pergamon Press, New York, 1983.

125. Saunders, W. H. and Cockerill A. F. *Mechanisms of Elimination Reactions*, Wiley, New York, 1973.

126. Grob, C. A. and Schiess, P. W. *Angew. Chem. Int. Ed. Engl.* **1967**, *6*, 1. Grob, C. A. *Angew. Chem. Int. Ed. Engl.* **1969**, *8*, 535.

127. Curtin, D. Y. and Harder R. J. *J. Am. Chem. Soc.* **1960**, *82*, 2357.

128. Anet, F. A. L. "Conformational Analysis of Cyclohexenes," in ref. 117, p. 1.

129. Scharpen L. H., Wollrab, J. E., and Ames, D. P. *J. Chem. Phys.* **1968**, *49*, 2368.

130. Ogata, T. and Kozima, K. *Bull. Chem. Soc. Jpn.* **1969**, *42*, 1263.

131. Chiang, J. F. and Bauer, S. H. (a) *J. Am. Chem. Soc.* **1969**, *91*, 1898; (b) *ibid.* **1970**, *92*, 1614.

132. Geise, H. J. and Buys, H. R. *Recl. Trav. Chim. Pays-Bas* **1970**, *89*, 1147.

133. Auf der Heyde, W. and Lüttke, W. *Chem. Ber.* **1978**, *111*, 2384.

134. Anet, F. A. L. and Haq, M. Z. *J. Am. Chem. Soc.* **1965**, *87*, 3147.

135. Burke, L. *Theor. Chim. Acta* **1985**, *68*, 101.

136. Dashevsky, V. G. and Lugovsky, A. A. *J. Mol. Struct.* **1972**, *12*, 39.

137. Bucourt, R. *Top. Stereochem.* **1974**, *8*, 159.

138. Anet, F. A. L. and Yavari, I. *Tetrahedron* **1978**, *34*, 2879.

139. Lipkowitz, K. B. "Application of Empirical Force-Field Calculations to the Conformational Analysis of Cyclohexenes, Cyclohexadienes and Hydroaromatics," in ref, 117, p. 209.

140. Lambert, J. B., Clikeman, R. R., Taba, K. M., Marko, D. E., Bosch, R. J., and Xue, L. *Acc. Chem. Res.* **1987**, *20*, 454.

141. Zefirov, N. S., Samoshin, V. V., and Akhmetova, G. M. *J. Org. Chem. USSR* **1985**, *21*, 244 (Engl. transl. p. 203).

142. Senda, Y. and Imaizumi, S. *Tetrahedron* **1974**, *30*, 3813.

143. Lessard, J., Tan P. V. M., Martino, R., and Saunders, J. K. *Can. J Chem.* **1977**, *55*, 1015, 1017.

144. Lu, K. C., Chiang, R. L., and Chiang, J. F. *J Mol. Struct.* **1980**, *64*, 229.

145. Sakashita, K. *Nippon Kaguku Zasshi* **1960**, *81*, 49; *Chem. Abstr.* **1960**, *54*, 12015b.

146. Valls, J. *Bull. Soc. Chim. Fr.* **1961**, 432. Valls, J. and Toromanoff, E. *Bull Soc. Chim. Fr.* **1961**, 758.

147. Barili, P. L., Belluci, G., Marioni, F., Morelli, I., and Scartoni, V. *J. Org. Chem.* **1972**, *37*, 4353.

148. Fahey, R. C. *Top. Stereochem.* **1968**, *3*, 237.

149. Ottar, B. *Acta Chem. Scand.* **1947**, *1*, 263.

150. Naumov, V. A. and Bezzubov, V. M. *J. Struct Chem. USSR* **1967**, *8*, 530 (Engl. ttransl. p. 466).

151. Wohl, R. A. *Chimia* **1974**, *28*, 1.

152. Fürst, A. and Plattner, P. A. *Abstract Papers of the 12th International Congress of Pure and Applied Chemistry*, New York, 1951, p. 409.

153. Lamaty, G., Tapiero, C., and Wylde, R. *Bull. Soc. Chim. Fr.* **1966**, 4010.

154. Rabideau P. W. and Sygula, A. "Conformational Analysis of 1,3-Cyclohexadienes and Related Hydroaromatics," in ref. 117, p. 65.

155. Traetteberg, M. *Acta Chem. Scand.* **1968**, *22*, 2305.

156. Oberhammer, H. and Bauer, S. H. *J. Am. Chem. Soc.* **1969**, *91*, 10.

157. Rabideau, P. W. "Conformational Analysis of 1,4-Cyclohexadienes and Related Hydroaromatics," in ref. 117, p. 89.

158. Carreira, L. A., Carter, R. O., and Durig, J. R. *J. Chem. Phys.* **1973**, *59*, 812.

159. Raber, D. J., Hardee, L. E., Rabideau, P. W., and Lipkowitz, K. B. *J. Am. Chem. Soc.* **1982**, *104*, 2843.

160. Jeffrey, G. A. *J. Am. Chem. Soc.* **1988**, *110*, 7218.

161. Rabideau, P. W. *Acc. Chem. Res.* **1978**, *11*, 141.

162. Lambert, J. B. "Conformational Analyses of Six-Membered Carbocyclic Rings with Exocyclic Double Bonds," in ref. 117, p. 47.

163. Dillen, J. and Geise, H. J. (a) *J. Mol. Struct.* **1980**, *69*, 137; (b) *J. Chem. Phys.* **1979**, *70*, 425.

164. Allinger, N. L., Tribble, M. T., and Miller, M. A. *Tetrahedron* **1972**, *28*, 1173.

165. Huet, J., Maroni-Barnaud, Y., Anh, N. T., and Seyden-Penne, J. *Tetrahedron Lett.* **1976**, 159.

166. Alonso, J. *J. Mol. Struct.* **1981**, *73*, 63.

167. Lambert, J. B., Carhart, R. E., and Corfield, P. W. R. *J. Am. Chem. Soc.* **1969**, *91*, 3567.

168. Anet, F. A. L., Chmurny, G. N., and Krane, J. *J. Am. Chem. Soc.* **1973**, *95*, 4423.

169. Ibrahim, M. R. *J. Phys. Org. Chem.* **1990**, *3*, 443.

170. Ashby, E. C. and Laemmle, J. T. *Chem. Rev.* **1975**, *75*, 521.

171. Boone, J. R. and Ashby, E. C. *Top. Stereochem.* **1979**, *11*, 53.

172. Wigfield, D. C. *Tetrahedron* **1979**, *35*, 449.

173. Wheeler, O. H. and Zabicky, J. C. *Can. J. Chem.* **1958**, *36*, 656.

174. (a) Brown, H. C. and Krishnamurthy, S. *J. Am. Chem. Soc.* **1972**, *94*, 7159. (b) Krishnamurthy, S. and Brown, H. C. *J. Am. Chem. Soc.* **1976**, *98*, 3383.

175. Eliel, E. L., Martin, R. J. L., and Nasipuri, D. *Org. Syn.* **1967**, *47*, 16; Coll. Vol. 5, p. 175.

176. Reetz, M. T. and Stanchev, S. *J. Chem. Soc. Chem. Commun.* **1993**, 328; see also Reetz, M. T., Harmat, N., and Mahrwald, R. *Angew. Chem. Int. Ed. Engl.* **1992**, *31*, 342.

177. Dauben, W. G., Fonken, G. J., and Noyce, D. S. *J. Am. Chem. Soc.* **1956**, *78*, 2579.

178. Bürgi, H. B., Dunitz, J. D., and Shefter, E. *J. Am. Chem. Soc.* **1973**, *95*, 5065.

179. Došen-Mićović, L., Li, S., and Allinger, N. L. *J. Phys. Org. Chem.* **1991**, *4*, 467.

180. Jensen, F. R. and Beck, B. H. *J. Am. Chem. Soc.* **1968**, *90*, 1066.

181. Gerig, J. T. *J. Am. Chem. Soc.* **1968**, *90*, 1065. Gerig, J. T. and Rimmerman, R. A. *J. Am. Chem. Soc.* **1970**, *92*, 1219.

182. Bernard, M., Canuel, L., and St.-Jacques, M. *J. Am. Chem. Soc.* **1974**, *96*, 2929.

183. Yursha, I. A., Kabo, G. Ya., and Andreevskii, D. N. *Neftekhimiya* **1974**, *14*, 688; *Chem. Abstr.* **1975**, *82*, 57332g.

184. Peerebom, M., van de Graaf, B., and Baas, J. M. A. *Rec. J. R. Neth. Chem. Soc.* **1982**, *101*, 336.

185. Lambert, J. B. and Clikeman, R. R. *J. Am. Chem. Soc.* **1976**, *98*, 4203.

186. (a) Johnson, F. *Chem. Rev.* **1968**, *68*, 375. (b) Johnson, F. and Whitehead, A. *Tetrahedron Lett.* **1964**, 3825.

187. Corey, E. J. and Enders, D. *Tetrahedron Lett.* **1976**, 3.

188. Fraser, R. R. and Dhawan, K. L. *J. Chem. Soc. Chem. Commun.* **1976**, 674. Fraser, R. R., Banville, J., and Dhawan, K. L. *J. Am. Chem. Soc.* **1978**, *100*, 7999.

189. Maruoka, K., Miyazaki, T., Ando, M., Matsumura, Y., Sakane, S., Hattori, K., and Yamamoto, H. *J. Am. Chem. Soc.* **1983**, *105*, 2831.

190. Riddell, F. G. *The Conformational Analysis of Heterocyclic Compounds*, Academic Press, New York, 1980.

191. Armarego, W. L. F. *Stereochemistry of Heterocyclic Compounds*, Parts 1 and 2, Wiley-Interscience, New York, 1977.

192. Lambert, J. B. and Featherman, S. I. *Chem. Rev.* **1975**, *75*, 611.

193. Delpuech, J. J. "Six-Membered Rings," in Lambert, J. B. and Takeuchi, Y., eds., *Cyclic Organic Stereodynamics*, VCH, New York, 1992, p. 169.

194. Lambert, J. B., Mixan, C. E., and Johnson, D. H. *J. Am. Chem. Soc.* **1973**, *95*, 4634.

195. Lambert, J. B., Keske, R. G., Carhart, R. E., and Jovanovich, A. P. *J. Am. Chem. Soc.* **1967**, *89*, 3761.

196. Eliel, E. L. *Acc. Chem. Res.* **1970**, *3*, 1.

197. Bushweller, C. H., Golini, J., Rao, G. U., and O'Neil, J. W. *J. Am. Chem. Soc.* **1970**, *92*, 3055.

198. Pihlaja, K. and Nikander, H. *Acta Chem. Scand.* **1977**, *B31*, 265.

199. Eliel, E. L., Hargrave, K. D., Pietrusiewicz, K. M., and Manoharan, M. *J. Am. Chem. Soc.* **1982**, *104*, 3635.

200. (a) Eliel, E. L. and Willer, R. L. *J. Am. Chem. Soc.* **1977**, *99*, 1936. (b) Willer, R. L. and Eliel, E. L. *J. Am. Chem. Soc.* **1977**, *99*, 1925.

201. Anet, F. A. L. and Yavari, I. *J. Am. Chem. Soc.* **1977**, *99*, 2794.

202. Vierhapper, F. W. and Eliel, E. L. *J. Org. Chem.* **1979**, *44*, 1081.

203. Eliel, E. L., Kandasamy, D., Yen, C.-Y., and Hargrave, K. D. *J. Am. Chem. Soc.* **1980**, *102*, 3698.

204. Crowley, P. J., Robinson, M. J. T., and Ward, M. G. *Tetrahedron* **1977**, *33*, 915.

205. Eliel, E. L. and Pietrusiewicz, K. M. *Top. C-13 NMR Spetrosc.* **1979**, *3*, 171.

206. Eliel, E. L. *Angew. Chem. Int. Ed. Engl.* **1972**, *11*, 739.

207. Gittins, V. M., Wynn-Jones, E., and White, R. F. M. "Ring Inversion in Some Six-Membered Heterocyclic Compounds," in Orville-Thomas, W. E., ed., *Internal Rotations in Molecules*, Wiley, New York, 1974, p. 425.

208. Eliel, E. L. and Knoeber, M. C. *J. Am. Chem. Soc.* **1966**, *88*, 5347; **1968**, *90*, 3444.

209. Jones, A. J., Eliel, E. L., Grant, D. M., Knoeber, M. C., and Bailey, W. F. *J. Am. Chem. Soc.* **1971**, *93*, 4772.

210. Riddell, F. G. and Robinson, M. J. T. *Tetrahedron* **1967**, *23*, 3417.

211. Kopp, L. D. "Conformational Analysis of Nitrogen Containing Heterocycles," Ph.D. Dissertation, University of Notre Dame, Notre Dame, IN. 1973; *Diss. Abstr. Int. B* **1973**, *34*, 1425.

212. Bernath, G., Fülöp, F., Kálmán, A., Argay, G., Sohár, P., and Pelczer, I. *Tetrahedron* **1984**, *40*, 3587.

213. Lambert, J. B. and Keske, R. G. *J. Org. Chem.* **1966**, *31*, 3429.

214. Kirby, A. J. *The Anomeric Effect and Related Stereoelectronic Effects at Oxygen*, Springer, New York, 1983.

215. Juaristi, E. and Cuevas, G. *The Anomeric Effect*, CRC Press, Boca Raton, FL, 1995.

216. Graczyk, P. P. and Mikołajczyk, M. *Top. Stereochem.* **1994**, *21*, 159.

217. Thatcher, G. R. J., ed., *The Anomeric Effect and Related Stereoelectronic Effects*, ACS Symposium Series 539, American Chemical Society, Washington, DC, 1993.

218. Stoddart, J. F. *Stereochemistry of Carbohydrates*, Wiley-Interscience, New York, 1971.

219. Franck, R. W. *Tetrahedron* **1983**, *39*, 3251.

220. (a) de Hoog, A. J., Buys, H. R., Altona, C., and Havinga, E. *Tetrahedron* **1969**, *25*, 3365. (b) de Hoog, A. J. and Havinga, E. *Rec. Trav. Chim. Pays-Bas* **1970**, *89*, 972.

221. Edward, J. T. *Chem. Ind (London)* **1955**, 1102.

222. Lemieux, R. U. and Chu, N. J. *Abstract Papers American Chemical Society 133rd Meeting*, 1958, p. 31N.

223. Lemieux, R. U. "Rearrangements and Isomerizations in Carbohydrate Chemistry," in de Mayo, P., ed. Wiley, New York, 1964. Lemieux, R. U. *Pure Appl. Chem.* **1971**, *25*, 527.

224. Lemieux, R. U., Koto, S., and Voisin, D. "The Exo-Anomeric Effect," in Szarek, W. A. and Horton, D., eds., *Anomeric Effect, Origin and Consequences*, ACS Symposium Series 87, American Chemical Society, Washington, DC, 1979, p. 17.

225. Wolfe, S., Whangbo, M.-H., and Mitchell, D. J. *Carbohyd. Res.* **1979**, *69*, 1.

226. Praly, J.-P. and Lemieux, R. U. *Can. J Chem.* **1987**, *65*, 213.

227. Box, V. G. S. *Heterocycles* **1990**, *31*, 1151.

228. Jeffrey, G. A. "The Structural Properties of the Anomeric Center in Pyranoses and Pyranosides," in Szarek, W. A. and Horton, D., eds., *Anomeric Effect, Origin and Consequences*, ACS Symposium Series 87, American Chemical Society, Washington, DC, 1979, p. 50.

229. Pinto, B. M., Johnston, B. D., and Nagelkerke, R. *J. Org. Chem.* **1988**, *53*, 5668.

230. Juaristi, E., Flores-Vela, A., Labastida, V., and Ordoñez, M. *J. Phys. Org. Chem.* **1989**, *2*, 349.

231. Liebman, J. F. and Greenberg, A. *Chem. Rev.* **1989**, *89*, 1215.

232. Cremer, D. and Gauss, J. *J. Am. Chem. Soc.* **1986**, *108*, 7467.

233. Wiberg, K. B., Bonneville, G., and Dempsey, R. *Isr. J Chem.* **1983**, *23*, 85.

234. Wiberg, K. B. and Fenoglio, R. A. *J. Am. Chem. Soc.* **1968**, *90*, 3395.

235. Greenberg, A. and Harris, J. *J. Chem. Educ.* **1982**, *59*, 539.

236. de Boer, T. J. *Chimia* **1977**, *31*, 483.

237. Breslow, R., Pecoraro, J., and Sugimoto, T. *Org. Syn.* **1977**, *57*, 41; Coll. Vol. VI, 361 (1988).

238. Breslow, R. (a) *Pure Appl. Chem.* **1971**, *28*, 111; (b) *Acc. Chem. Res.* **1973**, *6*, 393.

239. Walborsky, H. M. and Impastato, F. J. *J. Am. Chem. Soc.* **1959**, *81*, 5835.

240. Moriarty, R. M. *Top. Stereochem.* **1974**, *8*, 270.

241. Egawa, T., Fukuyama, T., Yamamoto, S., Takabayashi, F., Kambara, H., Ueda, T., and Kuchitsu, K. *J. Chem. Phys.* **1987**, *86*, 6018.

242. Block, E., Corey, E. R., Penn, R. E., Renken, T. L., Sherwin, P. F., Bock, H., Hirabayashi, T., Mohmand, S., and Solouki, B. *J. Am. Chem. Soc.* **1982**, *104*, 3119.

243. Durig, J. R., Little, T. S., and Lee, M. J. *J. Raman Spectrosc.* **1989**, *20*, 757.

244. Kozina, M. P., Mastryukov, V. S., and Mil'vitskaya, E. M. *Russ. Chem. Rev.* **1982**, *51*, 1337 (Engl. transl. p. 765).

245. Hargittai, I. and Hargittai, M. *Stereochemical Applications of Gas-Phase Electron Diffraction*, VCH, New York, 1988.

246. Altona, C., Geise, H. J., and Romers, C. *Tetrahedron* **1968**, *24*, 13.

247. Duax, W. L., Weeks, C. M., and Rohrer, D. C. *Top. Stereochem.* **1976**, *9*, 271.

248. Altona, C. and Sundaralingam, M. *J. Am. Chem. Soc.* **1972**, *94*, 8205.

249. Altona, C. *NATO Adv. Study Inst. Ser. A* **1982**, *45*, 161.

250. Haasnoot, C. A. G., de Leeuw, F. A. A. M., de Leeuw, H. P. M., and Altona, C. *Biopolymers* **1981**, *20*, 1211.

251. Pitner, T. P., Edwards, W. B., Bassfield, R. L., and Whidby, J. F. *J. Am. Chem. Soc.* **1978**, *100*, 246.

252. Aston, J. G., Schumann, S. C., Fink, H. L., and Doty, P. M. *J. Am. Chem. Soc.* **1941**, *63*, 2029.

253. Kilpatrick, J. E., Pitzer, K. S., and Spitzer, R. *J. Am. Chem. Soc.* **1947**, *69*, 2483.

254. Fuchs, B. *Top. Stereochem.* **1978**, *10*, 1.

255. Adams, W. J., Geise, H. J., and Bartell, L. S. *J. Am. Chem. Soc.* **1970**, *92*, 5013.

256. Hilderbrandt, R. L. and Shen, Q. *J. Phys. Chem.* **1982**, *86*, 587.

257. Hendrickson, J. B. *J. Am. Chem. Soc.* **1961**, *83*, 4537.

258. Cox, J. D. and Pilcher G. *Thermochemistry of Organic and Organometallic Compounds*, Academic Press, New York, 1970.

259. Allinger, N. L., Dodziuk, H., Rogers, D. W., and Naik, S. N. *Tetrahedron* **1982**, *38*, 1593.

260. Prelog, V. and Kobelt, M. *Helv. Chim. Acta* **1949**, *32*, 1187.

261. Brown, H. C. and Ichikawa, K. *Tetrahedron* **1957**, *1*, 221.

262. Dale, J. *Top. Stereochem.* **1976**, *9*, 199.

263. Anet, F. A. L. "Medium-Sized Oxygen Heterocycles," in ref. 264, p. 35.

264. Glass, R. S., ed., *Conformational Analysis of Medium-Sized Heterocycles*, VCH, New York, 1988.

265. Stoddart, J. F. and Szarek, W. A. *Can. J. Chem.* **1968**, *46*, 3061.

266. Tochtermann, W. *Top. Curr. Chem.* **1970**, *15*, 378.

267. Hendrickson, J. B. *J. Am. Chem. Soc.* **1967**, *89*, 7036, 7043, 7047.

268. Dale, J. *Stereochemistry and Conformational Analysis*, Verlag Chemie, New York, 1978, p. 194.

269. Ermolaeva, L. I., Mastryukov, V. S., Allinger, N. L., and Almenningen A. *J. Mol. Struct.* **1989**, *196*, 151.

270. Ménard, D. and St.-Jacques, M. *Tetrahedron* **1983**, *39*, 1041.

271. Désilets, S. and St.-Jacques, M. (a) *J. Am. Chem. Soc.* **1987**, *109*, 1641; (b) *Tetrahedron* **1988**, *44*, 7027.

272. Traetteberg, M. *J. Am. Chem. Soc.* **1964**, *86*, 4265.

273. Butcher, S. S. *J. Chem. Phys.* **1965**, *42*, 1833.

274. Anet, F. A. L. *J. Am. Chem. Soc.* **1964**, *86*, 458.

275. Jensen, F. R. and Smith, L. A. *J. Am. Chem. Soc.* **1964**, *86*, 956.

276. Anet, F. A. L. *Top. Curr. Chem.* **1974**, *45*, 169. See also ref. 263, p. 51.

277. Anet, F. A. L. and Basus, V. J. *J. Am. Chem. Soc.* **1973**, *95*, 4424.

278. Pakes, P. W., Rounds, T. C., and Strauss, H. L. *J. Phys. Chem.* **1981**, *85*, 2469, 2476.

279. Dorofeeva, O. V., Mastryukov, V. S., Allinger, N. L., and Almenningen, A. *J. Phys. Chem.* (a) **1985**, *89*, 252; (b) **1990**, *94*, 8044; (c) Dorofeeva, O. V., Mastryukov, V. S., Siam, K., Ewbank J. D., Allinger, N. L., and Schaefer, L. *J. Strukt. Chem. (USSR)* **1990**, *31*, 167 (Engl. transl. p. 153).

280. Paquette, L. A. *Tetrahedron* **1975**, *31*, 2855; *Acc. Chem. Res.* **1993**, *26*, 57.

281. Bastiansen, O., Hedberg, L., and Hedberg. K. *J. Chem. Phys.* **1957**, *27*, 1311.

282. Anet, F. A. L., Bourn, A. J. R., and Lin, Y. S. *J. Am. Chem. Soc.* **1964**, *86*, 3576.

283. Dunitz, J. D. *Pure Appl. Chem.* **1971**, *25*, 495.

284. Anet, F. A. L. and Wagner, J. J. *J. Am. Chem. Soc.* **1971**, *93*, 5266.

285. Bryan, R. F. and Dunitz, J. D. *Helv. Chim. Acta* **1960**, *43*, 1.

286. Anet, F. A. L. and Krane, J. *Isr. J. Chem.* **1980**, *20*, 72.

287. Dunitz, J. D., Eser, H., Bixon, M., and Lifson, S. *Helv. Chim. Acta* **1967**, *50*, 1572.

288. Hilderbrandt, R. L., Wieser, J. D., and Montgomery, L. K *J. Am. Chem. Soc.* **1973**, *95*, 8598.

289. Anet, F. A. L., Cheng, A. K., and Wagner, J. J. *J. Am. Chem. Soc.* **1972**, *94*, 9250.

290. Anet, F. A. L. and Rawdah, T. N. (a) *J. Am. Chem. Soc.* **1978**, *100*, 7810; (b) *ibid.* **1978**, *100*, 7166.

291. Dunitz, J. D. and Shearer, H. M. M. *Helv. Chim. Acta* **1960**, *43*, 18.

292. Atavin, E. G., Mastryukov, V. S., Allinger, N. L., Almenningen, A., and Sep, R. *J. Mol. Struct.* **1989**, *212*, 87.

293. Schneider, H.-J. and Thomas, F. *Tetrahedron* **1976**, *32*, 2005.

294. Anet, F. A. L. and Cheng, A. K. *J. Am. Chem. Soc.* **1975**, *97*, 2420.

295. Rubin, B. H., Williamson, M., Takeshita, M., Menger, F. M., Anet, F. A. L., Bacon, B., and Allinger, N. L. *J. Am. Chem. Soc.* **1984**, *106*, 2088.

296. Gallagher, M. J. "Cyclic Compounds: Conformation and Stereochemistry," in Verkade, J. G. and Quin, L. D., eds., *Phosphorus-31 NMR Spectroscopy and Stereochemical Analysis*, VCH, New York, 1987, p. 297.

297. Olah, G., ed., *Cage Hydrocarbons*, Wiley, New York, 1990.

298. Marchand, A. P. "Policyclic Cage Molecules: Useful Intermediates in Organic Synthesis and an Emerging Class of Substrates for Mechanistic Studies," in ref. 299, p. 1.

299. Osawa E. and Yonemitsu, G., eds. *Carbocyclic Cage Compounds*, VCH, New York, 1992.

300. Naemura, K. "High-Symmetry Chiral Cage-Shaped Molecules," in ref. 299, p. 61.

301. Cox, K. W., Harmony, M. D., Nelson, G., and Wiberg. K. B. *J. Chem. Phys.* **1969**, *50*, 1976.

302. Kirmse, W. and Hase, C. *Angew. Chem. Int. Ed. Engl.* **1968**, *7*, 891.

303. Wiberg, K. B. and de Meijere, A. *Tetrahedron Lett.* **1969**, 519.

304. Paukstelis, J. V. and Kao, J.-l. *J. Am. Chem. Soc.* **1972**, *94*, 4783.

305. Svyatkin, V. A., Ioffe, A. I., and Nefedov, O. M. *Bull. Acad. Sci. USSR Div. Chem. Sci.* **1988**, *37*, 78 (Engl. transl. p. 69).

306. Pirkle, W. H. and Lunsford, W. B. *J. Am. Chem. Soc.* **1972**, *94*, 7201.

307. Wiberg, K. B., Lupton, E. C., Wasserman, D. J., de Meijere, A., and Kass, S. R. *J. Am. Chem. Soc.* **1984**, *106*, 1740.

308. Mann, G. *Z. Chem.* **1966**, *6*, 106.

309. Meinwald, J., Tufariello, J. J., and Hurst, J. J. *J. Org. Chem.* **1964**, *29*, 2914.

310. Spelbos, A., Mijlhoff, F. C., Bakker, W. H., Baden R., and Van den Eden, L. *J. Mol. Struct.* **1977**, *38*, 155.

311. Wiberg, K. B. and Wendoloski, J. J. *J. Am. Chem. Soc.* **1982**, *104*, 5679.

312. Wiberg, K. B., Burgmaier, G. J., and Warner, P. *J. Am. Chem. Soc.* **1971**, *93*, 246.

313. Szeimies, G., Harnisch, J., and Baumgartel, O. *J. Am. Chem. Soc.* **1977**, *99*, 5183.

314. Wagner, H.-U., Szeimies, G., Chandrasekhar, J., Schleyer, P. v. R., Pople, J. A., and Binkley, J. S. *J. Am. Chem. Soc.* **1978**, *100*, 1210.

315. a) Wilberg, K. B., Bonneville, G., and Dempsey, R. *Isr. J. Chem.* **1983**, *23*, 85. b) Wiberg, K. B. and Bonneville, G. *Tetrahedron Lett.* **1982**, *23*, 5385.

316. Harnisch, J., Baumgartel, O., Szeimies, G., van Meersche, M., Germain, G., and Declercq, J.-P. *J. Am. Chem. Soc.* **1979**, *101*, 3370.

317. Szeimies-Seebach, U. and Szeimies, G. *J. Am. Chem. Soc.* **1978**, *100*, 3966.

318. Finke, H. L., McCullough, J. P., Messerly, J. F., Osborn, A., and Douslin, D. R. *J. Chem. Thermodyn.* **1972**, *4*, 477.

319. Van den Enden, L. and Geise, H. J. *J. Mol. Struct.* **1981**, *74*, 309.

320. Allinger, N. L. and Coke, J. L. (a) *J. Am. Chem. Soc.* **1960**, *82*, 2553; (b) *ibid.* **1959**, *81*, 4080.

321. Barrett, J. W. and Linstead, R. P. *J. Chem. Soc.* **1936**, 611.

322. Chang, S.-J., McNally, D., Shary-Tehrany, S., Hickey, M. J., and Boyd, R. H. *J. Am. Chem. Soc.* **1970**, *92*, 3109.

323. Owen, L. N. and Peto, A. G. *J. Chem. Soc.* **1955**, 2383.

324. Barkworth, P. M. R. and Crabb, T. A. *Org. Magn. Reson.* **1981**, *17*, 260.

325. Hückel, W. *Justus Liebigs Ann. Chem.* **1925**, *444*, 1.

326. Van den Enden, L., Geise, H. J., and Spelbos, A. *J. Mol. Struct.* **1978**, *44*, 177.

327. Schucker, R. C. *J. Chem. Eng. Data* **1981**, *26*, 239.

328. Speros, D. M. and Rossini, F. D. *J. Phys. Chem.* **1960**, *64*, 1723.

329. Dalling, D. K., Grant, D. M., and Johnson, L. F. *J. Am. Chem. Soc.* **1971**, *93*, 3678.

330. Mann, B. E. *J. Magn. Reson.* **1976**, *21*, 17.

331. Linstead, R. P., Doering, W. E., Davis, S. B., Levine, P., and Whetstone, R. R. (a) *J. Am Chem. Soc.* **1942**, *64*, 1985, 1991, 2003, 2006, 2009, 2014. (b) Linstead, R. P. and Whetstone, R. R. *J. Chem. Soc.* **1950**, 1428.

332. Eliel, E. L. *Stereochemistry of Carbon Compounds*, McGraw-Hill, New York, 1962.

333. Johnson, W. S. *J. Am. Chem. Soc.* **1953**, *75*, 1498.

334. Allinger, N. L., Honig, H., Burkert, U., Asolnai, L., and Huttner, G. *Tetrahedron* **1984**, *40*, 3449.

335. Allinger, N. L., Gorden, B. J., Tyminski, I. J., and Wuesthoff, M. T. *J. Org. Chem.* **1971**, *36*, 739.

336. Honig, H. and Allinger, N. L. *J. Org. Chem.* **1985**, *50*, 4630.

337. Allinger, N. L. and Wuesthoff, M. T. *J. Org. Chem.* **1971**, *36*, 2051.

338. Grossel, M. C., Cheetham, A. K., and Newsam, J. M. *Tetrahedron Lett.* **1978**, 5229.

339. Rabideau, P. W., Wetzel, D. M., and Paschal, J. W. *J. Org. Chem.* **1982**, *47*, 3993.

340. Ferrier, W. G. and Iball, J. *J. Chem. Ind. (London)* **1954**, 1296.

341. Herbstein, F. H., Kapon, M., and Reisner, G. M. *Acta Crystallogr. B* **1986**, *42*, 181.

342. Fieser, L. F. and Fieser, M. *Steroids*, Reinhold, New York, 1959.

343. Klyne, W. and Buckingham, J. *Atlas of Stereochemistry*, 2nd ed., Vol. 1, Chapman and Hall, London, 1978, p. 121.

344. Fawcett, J. K. and Trotter, J. *J. Chem. Soc. B* **1966**, 174.

345. Wiberg, K. B. and Connor, D. S. *J. Am. Chem. Soc.* **1966**, *88*, 4437.

346. a) Almenningen, A., Andersen, B., and Nyhus, B. A. *Acta Chem. Scand.* **1971**, *25*, 1217. b) See also Levin, M. D., Kaszynski, P., and Michl, J. *Chem. Rev.* **2000**, *100*, 169.

347. Applequist, D. E., Renken, T. L., and Wheeler, J. W. *J. Org. Chem.* **1982**, *47*, 4985.

348. Newton, M. D. and Schulman J. M. *J. Am. Chem. Soc.* **1972**, *94*, 773.

349. Wiberg, K. B. "Experimental Thermochemistry," in ref. 27, p. 151.

350. Allinger, N. L., Tribble, M. T., Miller, M. A., and Wertz, D. H. *J. Am. Chem. Soc.* **1971**, *93*, 1637.

351. Engler, E. M., Andose, J. D., and Schleyer, P. v. R. *J. Am. Chem. Soc.* **1973**, *95*, 8005.

352. Chiang, J. F., Wilcox, C. F., and Bauer, S. H. *J. Am. Chem. Soc.* **1968**, *90*, 3149.

353. Dallinga, G. and Toneman, L. H. *Recl. Trav. Chim. Pays-Bas* **1968**, *87*, 795.

354. Yokozeki, A. and Kuchitsu, K. *Bull. Chem. Soc. Jpn.* **1971**, *44*, 2356.

355. Ermer, O. and Dunitz, J. D. *Helv. Chim. Acta* **1969**, *52*, 1861.

356. Yokozeki, A., Kuchitsu., K., and Morino, Y. *Bull. Chem. Soc. Jpn.* **1970**, *43*, 2017.

357. Zimmerman, H. E. and Paufler, R. M. *J. Am. Chem. Soc.* **1960**, *82*, 1514.

358. Yamamoto, S., Nakata, M., Fukuyama, T., Kuchitsu, K., Hasselman, D., and Ermer, O. *J. Phys. Chem.* **1982**, *86*, 529.

359. Zefirov, N. S. and Palyulin, V. *Top. Stereochem.* **1991**, *20*, 171.

360. Laszlo, I. *Recl. Trav. Chim. Pays-Bas* **1965**, *84*, 251.

361. Brown, W. A. C., Martin, J., and Sim, G. A. *J. Chem. Soc.* **1965**, 1844.

362. Mastryukov, V. A., Popik, M. V., Dorofeeva, O. V., Golubinskii, A. V., Vilkov, L. V., Belikova, N. A., and Allinger, N. L. *J. Am. Chem. Soc.* **1981**, *103*, 1333.

363. Jaime, C., Ōsawa, E., Takeuchi, Y., and Camps, P. *J. Org. Chem.* **1983**, *48*, 4514.

364. Skancke, P. N. *THEOCHEM* **1987**, *36*, 11.

365. Warner, P. M. and Peacock, S. *J. Comput. Chem.* **1982**, *3*, 417.

366. Bredt, J. *Justus Liebigs Ann. Chem.* **1924**, *437*, 1.

367. Fawcett, F. S. *Chem. Rev.* **1950**, *47*, 219.

368. Warner, P. M. *Chem. Rev.* **1989**, *89*, 1067.

369. Marshall, J. A. and Faubl, H. *J. Am. Chem. Soc.* **1967**, *89*, 5965.

370. Wiseman, J. R. *J. Am. Chem. Soc.* **1967**, *89*, 5966.

371. Becker, K. B. and Pfluger, R. W. *Tetrahedron Lett.* **1979**, 3713.

372. Sellers, S. F., Klebach, T. C., Hollowood, F., Jones, M., and Schleyer, P. v. R. *J. Am. Chem. Soc.* **1982**, *104*, 5492.

373. Alder, R. W. *Acc. Chem. Res.* **1983**, *16*, 321; *Tetrahedron* **1990**, *46*, 683.

374. Lehn, J.-M. *Acc. Chem. Res.* **1978**, *11*, 49; *Pure Appl. Chem.* **1978**, *50*, 871; *Pure Appl. Chem.* **1980**, *52*, 2303.

375. Jolley, S. T., Bradshaw, J. S., and Izatt, R. M. *J. Heterocycl. Chem.* **1982**, *19*, 3.

376. Hayward, R. C. *Chem. Soc. Rev.* **1983**, *12*, 285.

377. Parker, D. *Adv. Inorg. Chem.* **1983**, *27*, 1.

378. Dietrich, B. "Cryptate Complexes," in Atwood, J. L., Davies, J. E. D., and MacNicol, D. D., eds., *Inclusion Compounds*, Vol. 2, Academic Press, New York, 1984, p. 337.

379. Weber, E. and Vögtle F. *Top. Curr. Chem.* **1981**, *98*, 1.

380. Gokel, G. W. *Crown Ethers and Cryptands*, The Royal Society of Chemistry, London, 1991.

381. Cram, D. J. and Trueblood, K. N. *Top. Curr. Chem.* **1981**, *98*, 43.

382. Cram, D. J. *From Design to Discovery*, American Chemical Society, Washington, DC, 1990. Cram, D. J. and Cram, J. M. *Container Molecules and their Guests*, Monographs in Supramolecular Chemistry, No. 4, Royal Society of Chemistry, Cambridge, UK, 1994.

383. Tanner, M. E., Knobler, C. B., and Cram, D. J. *J. Am. Chem. Soc.* **1990**, *112*, 1659.

384. Gutsche, C. D., Dhawan, B., No, K. H., and Muthukrishnan, R. *J. Am. Chem. Soc.* **1981**, *103*, 3782.

385. Gutsche, C. D. *Calixarenes*, Monographs in Supramolecular Chemistry, No. 1, Royal Society of Chemistry, Cambridge, UK, 1989; *Calixarenes Revisited, ibid.*, No. 6, 1998.

386. Ungaro, R. and Pochini A. "Flexible and Preorganized Molecular Receptors based on Calixarenes," in ref. 387, p. 57.

387. Schneider, H.-J. and Dürr, H., eds. *Frontiers in Supramolecular Organic Chemistry and Photochemistry*, VCH, New York, 1991.

388. Vicens, J. and Böhmer, V., eds., *Calixarenes: A Versatile Class of Macrocyclic Compounds*, Kluwer, Boston, 1991; Böhmer, V. *Angew Chem. Int. Ed. Engl.* **1995**, *34*, 713.

389. Lehn, J.-M. *Science*, **1993**, *260*, 1762.

390. Weber, E., ed. *Top. Curr. Chem.* **1993**, Vol. 165.

391. Altman, J., Babad, E., Itzchaki, J., and Ginsburg, D. *Tetrahedron Suppl. 8, Part 1* **1966**, 279.

392. Wiberg, K. B. and Walker, F. H. *J. Am. Chem. Soc.* **1982**, *104*, 5239. See also Wiberg, K. B., Dailey, W. P., Walker, F. H., Waddell, S. T., Crocker, L. S., and Newton, M. *J. Am. Chem. Soc.* **1985**, *107*, 7247; Hedberg, L. and Hedberg, K. *J. Am. Chem. Soc.* **1985**, *107*, 7257.

393. Ginsburg, D. (a) *Acc. Chem. Res.* **1972**, *5*, 249; (b) *Propellanes, Structure and Reactions*, Verlag Chemie, Weinheim, Germany, 1975; (c) *Top. Curr. Chem.* **1987**, *137*, 1.

394. Wiberg, K. B. *Chem. Rev.* **1989**, *89*, 975.

395. Tobe, Y. "Propellanes," in ref 299, p. 125.

396. Schill, G. *Catenanes, Rotaxanes and Knots*, Academic Press, New York, 1971.

397. Walba, D. M. *Tetrahedron* **1985**, *41*, 3161.

398. Dietrich-Buchecker, C. O. and Sauvage, J.-P. *Chem. Rev.* **1987**, *87*, 795; Sauvage, J.-P. and Dietrich-Buchecker, C., eds. *Molecular Catenanes, Rotaxanes and Knots*, Wiley-VCH, New York, 1999.

399. Sauvage, J.-P., ed. *New J. Chem.* **1993**, *17*, No. 10/11.

400. Amabilino, D. B. and Stoddart, J. F. *Chem. Rev.* **1995**, *95*, 2725. Fyfe, M. C. T. and Stoddart, J. F. *Acc. Chem. Res.* **1997**, *30*, 393.

401. Wasserman, E. *J. Am. Chem. Soc.* **1960**, *82*, 4433.

402. Schill, G. *Nachr. Chem. Tech.* **1967**, *15*, 149.

403. Reuter, C., Mohry, A., Sobanski, A., and Vögtle, E. *Chem. Eur. J.* **2000**, *6*, 1674. (Pretzelanes are catenanes in which the two interlocking rings are covalently linked by a bridge. The resulting molecular shape resembles that of a pretzel.)

404. Chambron, J.-C., Dietrich-Buchecker, C., and Sauvage, J.-P. *Top. Curr. Chem.* **1993**, *165*, 131. *Id.* "From Classical Chirality to Topological Chiral Knots" in Semlyen, J. A., ed. *Large Ring Molecules*, Wiley, New York, 1996.

405. Wasserman, E. *Sci. Am.* **1962**, *207* [5], p. 94.

406. Harrison, I. T. *J. Chem. Soc. Perkin 1* **1974**, 301.

407. Schill, G., Beckmann W., Schweickert, N., and Fritz, H. *Chem. Ber.* **1986**, *119*, 2647.

408. Harrison, I. T. and Harrison, S. *J. Am. Chem. Soc.* **1967**, *89*, 5723.

409. Merrifield, R. B. *Science* **1965**, *150*, 178.

410. Agam, G., Gravier, D., and Zilkha, A. *J. Am. Chem. Soc.* **1976**, *98*, 5206.

411. Ogino, H. *J. Am. Chem. Soc.* **1981**, *103*, 1303.

412. Ashton, P. R., Philp. D., Spencer, N., and Stoddart, J. F. *J. Chem. Soc. Chem. Commun.* **1991**, 1677.

413. Raymo, F. M. and Stoddart, J. F. *Chem. Rev.* **1999**, *99*, 1643.

414. Agam, G. and Zilkha, A. *J. Am. Chem. Soc.* **1976**, *98*, 2514.

415. Schill, G. and Zollenkopf, H. *Justus Liebigs Ann. Chem.* **1969**, *721*, 53.

416. Schill, G. and Lüttringhaus, A. *Angew. Chem. Int. Ed. Engl.* **1964**, *3*, 546.

417. Sauvage, J.-P. and Weiss, J. *J. Am. Chem. Soc.* **1985**, *107*, 6108.

418. Dietrich-Buchecker, C. O. and Sauvage, J.-P. *Angew. Chem. Int. Ed. Engl.* **1989**, *28*, 189.

419. Ashton, P. R., Goodnow, T. T., Kaifer, A., Reddington, M. V., Slawin, A. M. Z., Spencer N., Stoddart, J. F., Vicent, C., and Williams, D. J. *Angew. Chem. Int. Ed. Engl.* **1989**, *28*, 1396.

420. Stoddart, J. F. "Template Directed Synthesis of New Organic Materials," in ref. 387, p. 251.

421. Anelli, P. L., Spencer, N., and Stoddart, J. F. *J. Am. Chem. Soc.* **1991**, *113*, 5131.

422. Amabilino, D. B. and Stoddart, J. F. *Pure Appl. Chem.* **1993**, *65*, 2351.

423. Ashton, P. R., Baxter, I., Cantrill, S. J., Fyfe, M. C. T., Glink, P. T., Stoddart, J. F., White, A. J. P., and Williams, D. J. *Angew. Chem. Int. Ed. Engl.* **1998**, *37*, 1294.

424. Balzani, V., Credi, A., Raymo, F. M., and Stoddart, J. F. *Angew. Chem. Int. Ed. Engl.* **2000**, *39*, 3348; Ashton, P. R. and others, *Chem. Eur. J.* **2000**, *6*, 3558.

425. Gibson, H. W. "Rotaxanes," in Semlyen, J. A., ed., *Large Ring Molecules*, Wiley, New York, 1996.

426. Schmieder, R., Hübner, G., Seel, C., and Vögtle, F. *Angew. Chem. Int. Ed. Engl.* **1999**, *38*, 3528.

427. Sauvage, J.-P. *Acc. Chem. Res.* **1998**, *31*, 611.

428. (a) Walba, D. M., Richards, R., Sherwood, S. P., and Haltiwanger, R. C. *J. Am. Chem. Soc.* **1981**, *103*, 6213. (b) Walba, D. M., Richards, R. M., and Haltiwanger, R. C. *J. Am. Chem. Soc.* **1982**, *104*, 3219; Walba, D. M., Homan, T. C., Richards, R. M., and Haltiwanger, R. C. *New J. Chem.* **1993**, *17*, 661.

429. Dietrich-Buchecker, C., Rapenne, G., Sauvage, J.-P., De Cian, A., and Fischer, J. *Chem Eur. J.* **1999**, *5*, 1432.

430. Eaton, P. E. and Cole, T. W. *J. Am. Chem. Soc.* **1964**, *86*, 3157.

431. Griffin, G. W. and Marchand, A. P. *Chem. Rev.* **1989**, *89*, 997.

432. Higuchi, H. and Ueda, I. "Recent Developments in the Chemistry of Cubane," in ref. 299, p. 217.

433. Barborak, J. C., Watts, L., and Pettit, R. *J. Am. Chem. Soc.* **1966**, *88*, 1328.

434. Edward, J. T., Farrell, P. G., and Langford, G. E. *J. Am. Chem. Soc.* **1976**, *98*, 3075.

435. Della, E. W., Hine, P. T., and Patney, H. K. *J. Org. Chem.* **1977**, *42*, 2940.

436. Almenningen, A., Jonvik, T., Martin, H. D., and Urbanek, T. *J. Mol. Struct.* **1985**, *128*, 239.

437. Fleischer, E. B. *J. Am. Chem. Soc.* **1964**, *86*, 3889.

438. Muller, N. and Pritchard, D. E. *J. Chem. Phys.* **1959**, *31*, 1471.

439. Kybett, B. D., Carroll, S., Natalis, P., Bonnell, D. W., Margrave, J. L., and Franklin, J. L. *J. Am. Chem. Soc.* **1966**, *88*, 626.

440. Wiberg, K. B., Bader, R. F. W., and Lau, C. D. H. *J. Am. Chem. Soc.* **1987**, *109*, 985, 1001.

441. Maier, G., Pfriem, S., Schäfer, U., and Matusch, R. *Angew. Chem. Int. Ed. Engl.* **1978**, *17*, 520. Maier, G., Pfriem, S., Schäfer, U., Malsch, K.-D., and Matusch, R. *Chem. Ber.* **1981**, *114*, 3965.

442. Irngartinger, H., Goldmann, A., Jahn, R., Nixdorf, M., Rodewald, H., Maier, G., Malsch, K.-D., and Emrich, R. *Angew. Chem. Int Ed. Engl.* **1984**, *23*, 993.

443. Maier, G., Pfriem, S., Malsch, K.-D., Kalinowski, H.-O., and Dehnike, K. *Chem. Ber.* **1981**, *114*, 3988.

444. Zefirov, N. S., Koz'min, A. S., and Abramenkov, A. V. *Russ. Chem. Rev.* **1978**, *47*, 289 (Engl. transl. p. 163).

445. Ternansky, R. J., Balogh, D. W., and Paquette, L. A. *J. Am. Chem. Soc.* **1982**, *104*, 4503. Paquette, L. A., Ternansky, R. J., Balogh, D. W., and Kentgen, G. *J. Am. Chem. Soc.* **1983**, *105*, 5446.

446. Paquette, L. A. *Chem. Rev.* **1989**, *89*, 1051.

447. Gallucci, J. C., Doecke, C. W., and Paquette, L. A. *J. Am. Chem. Soc.* **1986**, *108*, 1343.

448. Scott, L. T. and Jones, M. *Chem. Rev.* **1972**, *72*, 181.

449. Fort, R. C. *Adamantane, The Chemistry of Diamond Molecules*, Marcel Dekker, New York, 1976.

450. Ganter, C. "A New Approach to Adamantane Rearrangements," in ref 299, p. 293.

451. Landa, S. and Macháček, V. *Coll. Czech. Chem. Commun.* **1933**, *5*, 1.

452. Prelog, V. and Seiwerth, R. *Ber. Dtsch. Chem. Ges.* **1941**, *74B*, 1644, 1769.

453. Schleyer, P. *J. Am. Chem. Soc.* **1957**, *79*, 3292.

454. Whitlock, H. W. and Siefken, M. W. *J. Am. Chem. Soc.* **1968**, *90*, 4929.

455. Kroto, H. W., Heath, J. R., O'Brien, S. C., Curl, R. F., and Smalley, R. E. *Nature (London)* **1985**, *318*, 162.

456. Curl, R. F. and Smalley, R. E. *Science*, **1988**, *242*, 1017. Kroto, H. W. *Science* **1988**, *242*, 1139.

457. Krätschmer, W., Lamb, L. D., Fostiropoulos, K., and Huffman, D. R. *Nature (London)* **1990**, *347*, 354.

458. Taylor, R., Hare, J. P., Abdul-Sada, A. K., and Kroto H. W. *J. Chem. Soc. Chem. Commun.* **1990**, 1423.

459. Atwood, J. L., Koutsantonis, G. A., and Raston, C. L. *Nature (London)* **1994**, *368*, 229.

460. Hawkins, J. M., Lewis, T. A., Loren, S. D., Meyer, A., Heath, J. R., Saykally, R. J., and Hollander, F. J. *J. Chem. Soc. Chem. Commun.* **1991**, 775.

461. Hawkins, J. M., Meyer, A., Lewis, T. A., Loren, S., and Hollander, F. J. *Science* **1991**, *252*, 312.

462. Kroto, H. W., Allaf, A. W., and Balm, S. P. *Chem. Rev.* **1991**, *91*, 1213.

463. Kroto, H. W. and Walton, D. R. M. "Postfullerene Organic Chemistry," in ref 299, p. 91.

464. Hammond, G. S. and Kuck, V. J., eds., *Fullerenes: Synthesis, Properties and Chemistry of Large Carbon Clusters*, ACS Symposium Series 481, American Chemical Society, Washington, DC, 1992.

465. Billups, W. E. and Ciufolini, M. A., eds., *Buckminsterfullerenes*, VCH, New York, 1993.

466. Kroto, H. W., Fischer, J. E., and Cox, D. E., eds., *The Fullerenes*, Elsevier, New York, 1993. Hirsch, A., ed. *Fullerenes and Related Structures*, Springer, New York, 1998.

467. Rabideau, P. W. and Sygula, A. *Acc. Chem. Res.* **1996**, *29*, 235.

12

CHIROPTICAL PROPERTIES

12-1. INTRODUCTION

"Chiroptical properties" of chiral substances arise from their nondestructive interaction with anisotropic radiation (polarized light), properties that can differentiate between the two enantiomers of a chiral compound. The term, whose use was introduced by Lord Kelvin in 1894 (Section 1-3),[1-3] encompasses the classical spectroscopic qualitative and quantitative manifestations of chirality: optical activity and optical rotatory dispersion (ORD), the change of optical rotation with wavelength. Widespread application of another chiroptical technique, circular dichroism (CD), is more recent. Vibrational CD and its counterpart in Raman spectroscopy are relatively new chiroptical techniques, and so is the emission of circularly polarized light, circular polarization of emission (CPE).

The foregoing techniques, which are discussed in detail in this chapter, are mainly concerned with "natural optical activity," that is, with inherent properties of nonracemic samples of chiral substances. Properties resulting from optical activity induced in achiral substances or in racemic chiral samples by magnetic and electric fields are not considered to be chiroptical phenomena according to some authorities. These phenomena are not treated in this book.

Application of the chiroptical techniques to structural analysis is elaborated in Sections 12-4 to 12-6 of this chapter. It will become evident that the application of these techniques is in some cases dependent on the presence of a chromophore in the substance being analyzed; the correlation between the specific chiroptical technique, its underlying principle, and the need for a chromophore is outlined in Table 12.1.

TABLE 12.1. Chiroptical Techniques

Technique	Principle	Accessible Chromophore
Polarimetry and optical rotary dispersion (ORD)	Refraction	No
Circular dichroism (CD)	Absorption	Yes

12-2. OPTICAL ACTIVITY AND ANISOTROPIC REFRACTION

a. Origin and Theory

Optical activity (or optical rotatory power) results from the *refraction* of right and left circularly polarized light (cpl) to different extents by chiral molecules.[4] The source of the rotation, and hence of ORD as well, is birefringence, that is, unequal slowing down of right (R) and left (L) circularly polarized light ($n_R \neq n_L$, where n is the index of refraction) as the light passes through the sample. In contrast, CD is the consequence of the difference in *absorption* of right and left cpl ($\varepsilon_R \neq \varepsilon_L$, where ε is the molar absorption coefficient).

Let us now consider what happens when a beam of monochromatic polarized radiation passes through a nonracemic sample of a chiral substance. Light is electromagnetic radiation. It has associated with it time-dependent electric and magnetic fields. In ordinary radiation, the electric field associated with the light waves oscillates in all directions perpendicular to the direction of propagation (Fig. 12.1a). Such radiation is called isotropic (or unpolarized). In contrast, if the radiation is filtered so as to remove all oscillations other than in one direction, say, in the x,z-plane (Fig. 12.1b), then the light is anisotropic and said to be linearly polarized or, less rigorously, plane polarized.[5]

The relation of the electric field vector **E** and the magnetic field vector **H** of a linearly polarized light beam to the direction of propagation at a given time is shown in Fig. 12.2a. The two fields oscillate at right angles to one another and in phase. A different view (Fig. 12.2b) shows only the magnitude and direction of the electric field vector as a function of time t and at a given distance $z = z_0$ from the light source.[6] Both relations are cosine functions described by Eq. 12.1:

$$E = E_0 \cos(2\pi \nu t - 2\pi z/\lambda) = E_0 \cos \omega(t - z/c_0) \qquad (12.1)$$

where ν is the light frequency, $\lambda = c_0/\nu$ is its wavelength (c_0 is the speed of light in vacuum), E_0 is the maximum amplitude of the wave, and $\omega = 2\pi \nu$.

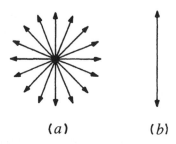

(a)   *(b)*

Figure 12.1. (*a*) Isotropic and (*b*) anisotropic (linearly polarized) light beams (electric field only) viewed along the z axis toward the light source. [Adapted with permission from Solomons, T. W. G. *Organic Chemistry*. Copyright © 1978 John Wiley & Sons, Inc., New York, pp. 244–245.]

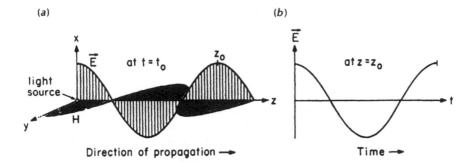

Figure 12.2. Linearly polarized light (*a*, at a given time and *b*, at a given place.) [(*a*) Reprinted with permission from Brewster, J. H. *Top Stereochem.* **1961**, *2*, 1. Copyright © 1967 John Wiley & Sons, Inc. and (*b*) adapted with permission from Snatzke, G. *Chem. Unserer Zeit.* **1981**, *15*, 78.]

Circularly polarized light may be described by examining the movement of the electric field vector only. The tip of the electric field vector **E** follows a helical path along the surface of a cylinder that is aligned with the axis of propagation of the light; it is helpful in this connection to think of the helix as being pushed out of the light source in the direction of propagation *but not rotated out.* Figure 12.3*a* defines a right circularly polarized ray viewed toward the light source and **E** is seen to have traveled toward the observer in a clockwise fashion. Time dependent measurement of the angle is opposite in sense since the observer encounters **E** in the order 6, 5, 4, 3, 2, 1, 0 (Fig. 12.3*b*).[6–8]

The definition of the sense of cpl may be a source of some confusion. Helical motion is a combination of translation and rotation occurring at the same time. It is a fact that a static *P* helix (right handed)—as in a screw—is of the same sense when viewed from either end, "front to back" and, after reversal, "back to front." This fact is of some significance in the designation of absolute configuration of molecules having helical symmetry (cf. Section 13-7). It becomes necessary to stipulate from which end the helix is being viewed the moment the two components of helical motion (translation and rotation) are "decoupled." As the above description points out and as indicated in Figure 12.3, analysis of cpl requires that (a) the light ray be viewed *toward* the light source only, and that (b) since the ray is not static, a distinction must be made between an instantaneous view ($t = t_0$, equivalent to the observer traveling with the light wave, Fig. 12.3*a*), and a view at a fixed point ($z = z_0$, equivalent to the observer looking at a slit through which the light source emerges some time after the light is "turned on," Fig. 12.3*b*). These are alternative and complementary, not contradictory, definitions of the sense of cpl.

In addition, when left and right cpl envisaged by drawing (electric field) vectors in the *xy* plane (as in Figs. 12.4 and 12.6), it is essential to stipulate the direction in which one is viewing the vectors. This requirement is akin to one needed when observing a transparent clock. In Figure 12.4, when viewed *toward* the light, the --- vector is defined as left cpl. In Figure 12.6, the same convention applies.

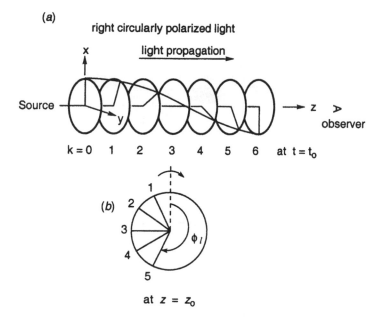

Figure 12.3. Definition of right cpl. (*a*) At $t = t_0$, the electric field vector describes a right-handed helix as viewed toward the light source (z increases as $z = k\ell/12$, $k = 0, 1, 2, \ldots$). [Adapted from Harada, N. and Nakanishi, K. *Circular Dichroic Spectroscopy. Exciton Splitting in Organic Stereochemistry*, University Science Books, Mill Valley, CA, p. 439.] (*b*) Right cpl viewed at $z = z_0$ as a function of time with light traveling toward the observer. [Adapted with modification from Snatzke, G. *Chem. Unserer Zeit.* **1981**, *15*, 78.]

Linearly polarized light may be envisaged (represented or conceptualized) mathematically and graphically as a combination of left and right (hence oppositely) coherent rotating beams of cpl. In an isotropic medium, the two components travel at the same velocity, hence in-phase, but in opposite senses (Fig. 12.4). The resultant vector sum, which exhibits the properties of linearly polarized light, is shown as

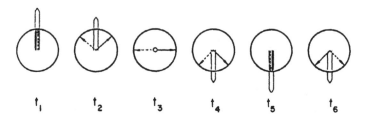

Figure 12.4. Addition of left (---) and right (—) cpl (of equal frequency, wavelength, and intensity) yields linearly polarized light. Time dependence ($t_1 \rightarrow t_6$) of the view along the light from a given point $z = z_0$; only the electric field vector is shown. [Adapted with permission from Snatzke, G. *Chem. Unserer Zeit.* **1981**, *15*, 78.]

(a)

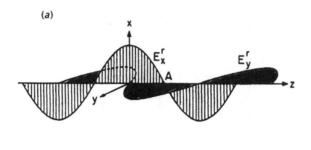

(b)

Figure 12.5. (*a*) Right cpl ray (only electric fields are shown) from one-quarter wave retardation (see also Fig. 12.2). (*b*) The instantaneous electric field is one-quarter wavelength out of phase. [Reprinted with permission from Brewster, J. H. *Top. Stereochem.* **1967**, *2*, 1. Copyright © 1967 John Wiley & Sons, Inc.]

travelling in the *x,z* plane. The amplitude of the vector sum that decreases and increases as shown in Figure 12.2*a* is double that of each cpl beam.

Let us now consider what happens when linearly polarized light (equivalent to opposite cpl beams of equal intensity) is passed through a biaxial crystal (Fig. 12.5), where linearly polarized monochromatic light is subject to one-quarter wave retardation. Then consider what happens when linearly polarized light is passed through a sample of a chiral compound containing unequal amounts of the two enantiomers. Let us assume for the moment that the frequency of the radiation is in a region of the spectrum that is free of absorption bands.

One of the enantiomers interacts with, say, the right cpl beam in such fashion that, within the sample, its velocity is different from that of the corresponding left cpl beam. Both beams are slowed down relative to their (equal) velocities prior to entrance into the sample but to a different extent (anisotropic refraction; note that the refractive index n for any medium equals c_0/c, where c is the velocity of light in that medium and c_0 is the velocity of light in a vacuum), hence the interaction of the two chiral rays with the chiral sample must have been different. One might say that their interaction was diastereomeric in nature.

If the enantiomer in excess slows down the left cpl beam more than the right beam ($c_L < c_R$, hence $n_L > n_R$), then the sample is defined as being dextrorotatory. In a time-dependent view toward the light source, the linearly polarized light resulting from addition of the two cpl beams appears clockwise rotated relative to the incident beam; α is positive (Fig. 12.6). The difference in light velocity corresponding to a difference in refractive index is given by the Fresnel equation, Eqs. 12.2 or 12.3:

$$\alpha = (n_L - n_R)\pi\ell/\lambda_0 \qquad \text{(in rad)} \qquad (12.2)$$

or

$$\alpha = (n_L - n_R)1800\ell/\lambda_0 \qquad \text{(in deg)} \qquad (12.3)$$

where n_L and n_R are the indexes of refraction of the left and right cpl beams in the medium, ℓ is the path length [in centimeters (Eq. 12.2); in decimeters (Eq. 12.3)], and λ_0 is the wavelength in vacuum of the light beam (in centimeters); since $360° = 2\pi$ rads and $1 \text{ dm} = 10 \text{ cm}$, Eqs. 12.2 and 12.3 are equivalent.[9] When $n_R \neq n_L$, the medium is said to be *circularly birefringent* and to exhibit *optical activity*. As an example of the magnitude of the refractive index difference, consider that at 589 nm (D line of sodium), optically active 2-butanol exhibits a rotation of $\alpha = 11.2°$ without solvent at 20°C (path length 1 dm). We calculate (12.3), $n_L - n_R = \Delta n = 11.2$ (589 × 10^{-7})/1800(1) = 3.66 × 10^{-7}. This very small number represents $\Delta n/n = (3.66 \times 10^{-7}/1.3954) \times 100$ or 2.63 × 10^{-5}% of the value of the isotropic refractive index of 2-butanol.

A classical and qualitative interpretation of how polarization affects the passage of light through a sample all of whose molecules have the same chirality sense is given in the Feynman lectures.[10] Assume the sample to consist of molecules in the shape of a helix. [The use of helical molecules in this analysis is in keeping with the view that chiral molecules in general may be envisaged as molecular helices.[7,11] Suppose that light falls on such a molecule with the beam linearly polarized along the long axis of the molecule. Although the effect of the beam is independent of the orientation of the molecules, for the sake of simplicity, we have chosen the x direction as that both of the polarization and of the long axis of the helix (Fig. 12.7).

The electric field E (cf. Fig. 12.2) exerts a force on charges (essentially on the electrons only since its effect on the heavier atomic nuclei is negligible) in the helical molecules. The up and down motion of the charges generates a current in the direction of the polarization (the x axis). The current in turn generates an electric field that is

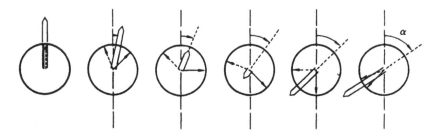

Figure 12.6. The origin of optical activity. Rotation of linealry polarized light by superposition of left (---) and right (—) cpl. Time-dependent view toward the light source from a given point $z = z_0$ (left to right). As shown, the rotation is positive (dextrorotation). [Adapted with permission from Snatzke, G. *Chem. Unserer Zeit.* **1981**, *15*, 78.]

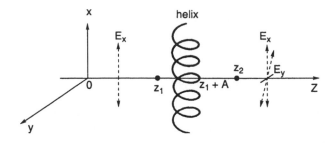

Figure 12.7. Interaction of a beam of light linearly polarized in the x direction with a chiral molecule.

polarized in the same direction as that of the impinging radiation. Thus the radiation is absorbed and reemitted.

In addition, in a three-dimensional molecule, such as the illustrated helix, electrons driven by the field E_x are constrained to move also in the y direction. Much of the induced electric field generated by the current moving in the y direction produces no radiation since the field arising from current moving in the $+y$ direction is canceled by that from current moving in the $-y$ direction on the opposite side of the helix. However, as the light propagates along the z axis, the fields E_y resulting from the transverse electron motion in the molecule, do not travel together since they are separated by the cross-sectional distance A corresponding to the distance across the spiral. The delay equals A/c (in seconds), where c is the speed of light in the medium. This delay results in a phase difference of $\pi + \nu A/c$, where ν is the frequency of light, such that the E_y fields do not exactly cancel each other.

The upshot of the foregoing analysis is that although the impinging radiation is entirely polarized in the x direction, the emerging radiation has a small component polarized in the y direction. The resultant net polarization is tilted away (rotated) from the x plane in a sense ultimately determined by the handedness of the interacting molecules. That is the origin of the optical rotation.

It is evident that replacement of the enantiomerically homogeneous sample in the above experiment by one containing an equal number of R and S molecules of a given structure, that is, by a racemate, would lead to zero rotation. Each molecule individually rotates the plane of polarization from that of the incident radiation. However, the net rotation is zero since the number of molecules rotating the plane in one sense equals that rotating the plane to the same extent but in the other direction. Intermediate situations occur for mixtures of chiral molecules containing an excess of one enantiomer. The quantitative treatment of the optical rotation of such mixtures is taken up in Section 12-5.a.

How does the foregoing analysis treat the interaction of *achiral* molecules with polarized radiation? Linearly polarized radiation encounters molecules in chiral conformations or chiral vibrational states such that its plane of polarization is rotated. Virtually all formally achiral molecules are subject to such interaction. However, bulk samples *in the liquid or gaseous states* possess molecules in every conceivable conformation and vibrational state such that the rotation generated by a molecule in one of these conformations or states is canceled by another having the mirror-image conformation or state (see Fig. 2.7 for an example), thus leading to a net zero rotation just as in the case of a racemate. Mislow and Bickart[12] called the statistical cancellation of local chiral effects *stochastic achirality*.

This interpretation of the optical inactivity of achiral molecules is supported in part by the finding that optical activity in the solid state is observed even with certain achiral compounds: Those such as quartz (SiO_2) that crystallize in enantiomorphic space groups,[130] and those that crystallize in chiral conformations, for example, *meso*-tartaric acid.[136]

b. Optical Rotatory Dispersion

The measurement of specific rotation, [α] (Section 1-3), as a function of wavelength is called *optical rotatory dispersion* (ORD). In the absence of significant absorption by the analyte, one observes monotonic changes in [α] as a function of wavelength in accord with the Fresnel equation (Eqs. 12.2 and 12.3). More specifically, the absolute value of the rotation increases as the wavelength decreases, that is, as the wavelength tends toward the UV (Fig. 12.8).

It has been known for a long time that the monotonic increase in rotation gives way to an anomaly (see below) in the vicinity of an electronic absorption band. It is therefore standard procedure to compare the isotropic (ordinary UV/vis) absorption spectrum with the ORD curve (the combined monotonic and anomalous change in optical rotation with wavelength) even though the latter is a measure of refraction and

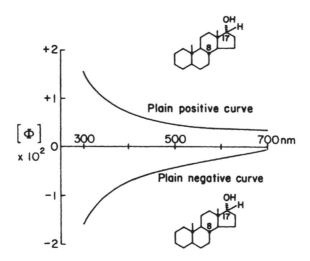

Figure 12.8. Optical rotatory dispersion in transparent regions of the spectrum. [Crabbé, P. In Snatzke, G., ed., *Optical Rotatory Dispersion and Circular Dichroism in Organic Chemistry*, p. 2. Copyright © 1967 Heyden & Son. Adapted with permission of John Wiley & Sons, Ltd.]

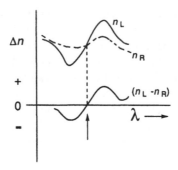

Figure 12.9. Dependence of Δn on wavelength. Anomalous dispersion.

not of absorption. Figure 12.9 illustrates the fact that anomalous ORD arises from the superposition of two anomalies, namely, those in the indexes of refraction of left and right circularly polarized light beams passing through the sample.

Typical "simple" ORD curves, such as that of D-camphor (Fig. 12.10), are similar in shape to the sum of the refractive index curves ($n_L - n_R$) shown in Figure 12.9, consistent with the relationship between α and Δn (Eq. 12.3), and the optical null occurs at precisely the same wavelength. The crossover point closely corresponds to

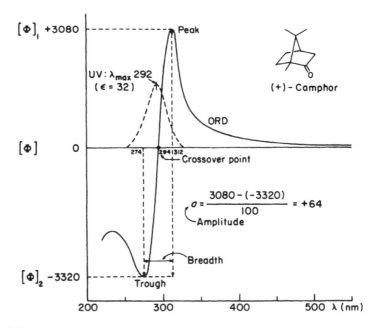

Figure 12.10. Anomalous ORD curve of (1R,4R)-(+)-camphor (–) exhibiting a single positive CE. Nomenclature of ORD curves; the crossover point at 294 nm is an "optical null," [Φ] = 0. The isotropic UV spectrum of camphor (---) is superposed on the ORD curve. [Adapted with permission from Crabbé, P. *ORD and CD in Chemistry and Biochemistry,* Academic Press, Orlando, FL, 1972, p. 6.]

the ε_{max} of the UV spectrum provided that the latter does not reflect a superposition of several close lying electronic transitions.

In contrast to a plain curve (Fig. 12.8), an "anomalous" ORD curve exhibits both a maximum and a minimum, and a point of inflection. (On occasion, one or more of these features may be hidden.) The anomaly is called the Cotton effect (CE).[14] (A simple ORD curve is one that shows a single CE.) Figure 12.10 shows a single CE curve exhibiting also a change in sign, and illustrates terms that are often used to describe curves without reproducing them. The curve is called *positive* when the rotation first increases as the wavelength decreases; conversely, it is called *negative* when the rotation magnitude first decreases when going towards shorter wavelengths.[15] The molar amplitude of the curve is given by the following relation (Eq. 12.4):

$$a = \frac{|[\Phi]_1| + |[\Phi]_2|}{100} \tag{12.4}$$

in which $|[\Phi]_1|$ and $|[\Phi]_2|$ are the absolute values of the molar rotations at the first and second extrema ("peak and trough," respectively). The wavelength difference between two extrema is called the breadth; the breadth can vary strongly with the wavelength at which the CE occurs.

The wavelength of the CE coincides with the λ_{max} (UV/vis) of an electronic transition (Fig. 12.10); that is, the optical null $[\Phi] = 0$ at 294 nm lies close to, but not precisely at, the same wavelength as the UV λ_{max} (292 nm). The UV spectral band shape is not strictly symmetrical, nor is the shape of the ORD curve. Both curves are influenced slightly by transitions lying at shorter wavelengths. This influence is sufficient to cause the difference cited. In addition, $[\Phi] = 0$ does not lie precisely at the midpoint between peak and trough either in intensity or in wavelength.

Figure 12.11*a* illustrates a less symmetrical single Cotton effect ORD curve in which the sign of the first CE encountered as the wavelength is scanned toward the UV is opposite to that of $[\alpha]_D$. Figure 12.11 implies that at least one other CE is to be found at shorter wavelengths. The shape of the ORD curve observed is a consequence of superposition of the CE shown and of the "background" curve shown in Figure 12.11*b*; only the plain part of the negative curve is shown (---) because the corresponding CE is at too short a wavelength to be experimentally manifested.

The sign of the ORD curve reflects the configuration of the chromophore, or of the stereogenic centers that perturb the chromophore, even in the presence of other stereocenters. For a simple example that illustrates the powerful advantage of ORD over the specific rotation, see Figure 12.8. A similar advantage applies to CD (see below). The nearly mirror-image, plain, ORD curves in Figure 12.8 are generated from two diastereomeric sterols whose configuration differs only at the carbinol stereocenter. The configurational relationship between the two stereoisomers can be inferred only from the ORD curve, whereas it could not have been gleaned safely from the fact that the two isomers have oppositely signed $[\alpha]_D$ values below 700 nm. Both ORD and CD serve as the principal spectroscopic probes of absolute configuration of stereocenters in chiral molecules and, in a complementary sense, also of conformation.

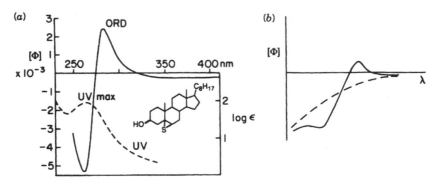

Figure 12.11. (*a*) Single Cotton effect ORD curve: positive CE with negative rotation in the visible (UV $\lambda_{max} = 264$ nm). [Adapted with permission from Djerassi, C. *Proc. Chem. Soc. London* **1964**, 315. Copyright © Royal Society of Chemistry, Science Park, Milton Road, Cambridge CB4 4WF, UK.]. (*b*) Shape of an ORD curve that stems from superposition of a positive CE (–) near 264 nm and of a negative (background) CE (--) lying at shorter wavelength. [Adapted with permission from Snatzke, G. *Chem. Unserer Zeit.* **1981**, *15*, 78.]

12-3. CIRCULAR DICHROISM AND ANISOTROPIC ABSORPTION

In addition to the anisotropic refraction of polarized light by chiral matter (circular birefringence), a second chiroptical phenomenon is observed in *nontransparent* regions of the spectrum, namely, CD. The latter phenomenon reflects the anisotropic *absorption* of cpl by chiral samples containing an excess of one enantiomer. Anisotropic absorption, which is also a CE, takes place only in spectral regions in which absorption bands are found in the isotropic UV or visible electronic spectrum (Fig. 12.12).

What is the origin of this absorption? An electronic (or vibrational) transition associated with a chirotopic chromophore in a chiral molecule causes right and left cpl to be absorbed differentially. Provided that the sample contains an excess of one enantiomer, the intensities of the two cpl beams are no longer equal in exiting the sample; the absorbances $A_L \neq A_R$ and $\Delta A = A_L - A_R$ are measures of the CD. If the molar concentrations are known, then, since $\Delta A = \Delta \varepsilon c \ell$, where c is the concentration in moles per liter (mol L^{-1}) and ℓ is the path length in centimeters (cm), we may write $\varepsilon_L - \varepsilon_R = \Delta \varepsilon$, where ε_L and ε_R are the molar absorption coefficients for left and right cpl, respectively, at the absorption wavelength.

The sign of $\Delta \varepsilon$ defines the sign of the CD (Fig. 12.12*a*, e.g., illustrates a positive CD curve). The signs of the CD curve and that of the corresponding ORD curve in the region of an anomaly are the same (rule of Natanson and Bruhat).[16] This correspondence is easily perceived if only one transition is present in a given wavelength range (Fig. 12.12*b*). At a given wavelength, both phenomena, ORD and CD, reflect the interaction of polarized light with the same chirotopic chromophore.

Since the absorbance of left and right cpl by the sample is unequal, $A_L \neq A_R$, and $A = \log(I_0/I)$ (I_0 is the intensity of the impinging light while I is that of the transmitted light), the two cpl components are now of unequal intensity ($I_L \neq I_R$). During its

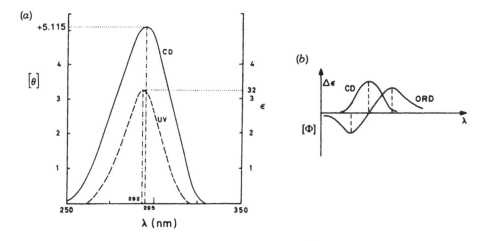

Figure 12.12. (*a*) UV (electronic absorption, EA) and CD (positive CE) spectra of (1*R*,4*R*)-(+)-camphor. [Adapted with permission from Crabbé, P. *ORD and CD in Chemistry and Biochemistry*, Academic Press, Orlando, FL, 1972, p. 6.]. (*b*) CD and ORD spectra describing the positive CE of a single electronic (isolated) transition. [Adapted with permission from Snatzke, G. *Chem. Unserer Zeit.* **1981**, *15*, 78.]

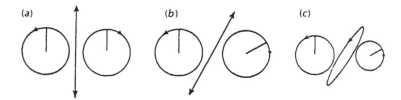

Figure 12.13. Elliptically polarized light. (*a*) Equal velocities of transmission give *no rotation*; (*b*) unequal velocities of transmission give *rotation*; (*c*) unequal velocities and unequal absorptions give *rotation* and elliptical polarization. [Reprinted with permission from Lowry, T. M. *Optical Rotatory Power*, Dover, New York, 1964, p. 152.]

passage through the sample in a region where absorption takes place, the incident linearly polarized light is converted into elliptically polarized light, that is, the resultant electric field vector traces an elliptical path. This conversion is schematically represented in Figure 12.13.

Elliptically polarized light as defined in Figure 12.14 is the most general form of polarized light; linear and circular polarization are special cases of elliptical polarization. The eccentricity of the ellipse [$(a - b)/a$] is 1 for linearly polarized light ($b = 0$), and 0 (zero) for circularly polarized light ($a = b$). The major axis of the ellipse traces the angle of rotation α and the ellipticity ψ is defined by $\tan \psi = b/a$, where b and a are the minor and major axes, respectively, of the ellipse that characterizes the elliptically polarized light (Fig. 12.14).

By analogy with rotation α, one may define a specific ellipticity [ψ] and a molar ellipticity [θ] (Eqs. 12.5 and 12.6, respectively):

$$[\psi] = \frac{\psi}{c\ell} \qquad \text{in } 10^{-1} \text{ deg cm}^2 \text{ g}^{-1} \qquad (12.5)$$

$$[\theta] = \frac{[\psi]M}{100} \qquad \text{in } 10 \text{ deg cm}^2 \text{ mol}^{-1} \qquad (12.6)$$

where the symbols c, ℓ, and M have the same meanings as they do in the definitions of $[\alpha]$ and $[\Phi]$ (Section 1-3).

When the ellipticity is small, which is a common occurrence, $\tan \psi \cong \psi$ and the latter is proportional to ΔA (Eqs. 12.7 and 12.8).

$$\psi = 32.982 \ \Delta A = 32.982 \ \Delta\varepsilon \, c' \ell \qquad (12.7)$$

$$[\theta] = 3298.2 \ \Delta\varepsilon \qquad (12.8)$$

In Eq. 12.7, the units of c' are moles per liter (mol L^{-1}) and those of ℓ' are centimeters (cm).

Ellipticities and molar ellipticities are dependent on conditions of measurement. Consequently, the temperature, wavelength, and concentration of the sample should always be specified. While it is possible to measure the ellipticity ψ directly (ellipsometry), this is difficult to do in practice and, due to the incorporation of a photoelastic modulator to generate the cpl (cf. Section 12-2.a), all current commercial CD spectrometers measure ΔA even if they are calibrated in ellipticity (see Eq. 12.7). For descriptions of CD instrumentation, see Crabbé and Parker,[17] Lambert et al.,[18] and Mason.[19]

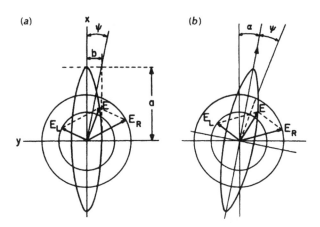

Figure 12.14. Elliptically polarized light (a) in a region where $\alpha = 0°$ and (b) in a region where α = positive viewed toward the light source. Electric field vectors $E_R > E_L$ are both smaller than E_0 (incident cpl); the resultant vector **E** traces an elliptical path. The ellipticity angle ψ is given by the geometric construction: arc tangent of minor axis b/major axis a, where $a = E_R + E_L$ and $b = E_R - E_L$. Since by definition, $\Delta\varepsilon = \varepsilon_L - \varepsilon_R$, ψ is positive if $\varepsilon_L > \varepsilon_R$. [Adapted from Velluz, L., Legrand. M., and Grosjean, M. *Optical Circular Dichroism*, Verlag Chemie, Weinheim, 1965, pp. 22–23.]

The correspondence of CD maxima with the wavelength of anomalous ORD crossover and the correspondence in sign of the two phenomena in a given sample suggests that it may be possible to calculate a CD curve from the ORD spectrum, and vice versa. This is, in fact, possible through application of the Kronig–Kramers theorem, a general relationship between absorption and refraction.[20–23]

Less rigorous, but still useful, expressions (Eqs. 12.9 and 12.10) have been proposed that relate the molar amplitude a of an ORD curve (Fig 12.10) to the intensity of the CD curve, $\Delta\varepsilon$[24]:

$$a = 40.28 \, \Delta\varepsilon \qquad (12.9)$$

and similarly, in view of Eq. 12.8, the ORD molar amplitude is related to the molar ellipticity $[\theta]$ of the CD spectrum (Fig. 12.12):

$$a = 0.0122 \, [\theta] \qquad (12.10)$$

These equations were derived for the $n-\pi^*$ carbonyl transition; they should be used with caution with other chromophores.[25]

Since the two types of measurement, ORD and CD, would appear to provide complementary information, it is fair to inquire into the need for both. Commercial ORD instruments became readily available beginning about 1955 while the first commercial CD spectrometer did not make its appearance until 1960. In the interval since 1960, both types of measurements have been reported, often for the same substance.

Circular dichroism spectra are inherently simpler to interpret; the key point is that CD is nonzero only in the vicinity of an EA band (Fig. 12.15). Thus a typical CD spectrum is less cluttered, bands are better separated, and comparison of anisotropic absorption bands with EA is more straightforward than with ORD. In Figure 12.15, the rotation observed (plain curve) in the region between 350 and 600 nm is a background or skeletal effect due to CEs below 250 nm. Even the sign of the CE at about 290 nm is nearly masked by the strong skeleton effect. This is not a problem with CD. Therefore, where a choice is possible, CD measurements are preferred over ORD. Consistent with the disappearance of commercial ORD spectrometers from the market, CD has by now essentially replaced ORD as the main chiroptical technique in the study of chiral substances.[26]

However, since ORD of optically active compounds is observable over the entire wavelength range (Fig. 12.15), this technique can provide information about CEs that are outside the range of commercial CD spectrometers. Plain ORD curves, especially in regions where $[\theta] = 0$, can reveal the configuration of a chromophore as illustrated in Figure 12.8. For the same reason, ORD may also be useful in the measurement of chiroptical properties of compounds having low optical activity[18] or when the sign of an ORD curve is opposite to that of a small $[\alpha]_D$ as in Fig. 12.11a.

Configurational assignments based on comparisons of specific rotations measured at single wavelengths far removed from CEs can be equivocal since it is not possible to tell if a dispersion (ORD) curve changes sign at lower wavelength. As an example, whereas the $[\alpha]_D$ values of the meta and para isomers of α-(iodophenoxy)propionic acids are positive, that of the ortho isomer is negative. The configuration of the (−)-o-

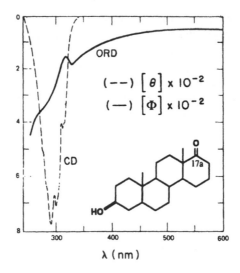

Figure 12.15. The CD and ORD curves of a simple hydroxyketone. The chirotopic chromophore responsible for the CE near 290 nm is the carbonyl group at position 17a. The shoulders in the CD band are due to vibrational fine structure. [Adapted with permission from Crabbé, P. and Parker, A. C. In Weissberger, A. and Rossiter, B. W., eds., *Physical Methods of Chemistry,* Part IIIC, *Techniques of Chemistry,* Vol. 1, p. 209. Copyright © 1972 John Wiley & Sons, Inc.]

isomer is shown to be the same as those of the (+)-*m*- and (+)-*p*-isomers, since the shapes and signs of the three ORD curves are alike [the ORD curve of the (−)-*o*-isomer crosses the zero rotation axis below 350 nm].[27]

12-4. APPLICATIONS OF OPTICAL ROTARY DISPERSION AND CIRCULAR DICHROISM

a. Determination of Configuration and Conformation: Theory

The principal application of CD and ORD is in the assignment of configuration or conformation.[28,29] In principle, the theories of optical activity should permit one to calculate *ab initio* the magnitude and sign of a CE from the constitution, the relative and absolute configuration of stereocenters, and the conformation of a given compound. In practice, however, we are not at the point where the aforementioned *ab initio* calculations are possible for any but small molecules.

The following qualitative treatment is intended to be just sufficient to show how the sign of a CE of a specific transition may be related to the absolute configuration of the molecule giving rise to the CD spectrum. With this background, empirical or semiempirical generalizations called sector and helicity rules may be rationalized and applied. Sector rules are widely used in the assignment of configuration by inspection of CD spectra of homologous and analogous compounds bearing the identical functional group.

Typical anomalous ORD and electronic CD arise in the same regions of the spectrum in which isotropic radiation excites electrons formally associated with atoms, groups, or moieties in achiral *or* chiral molecules (e.g., C=O, –CH=CH–C=O, C_6H_5). When polarized radiation is used, similar excitation in the identical spectral regions give rise to an absorption whose intensity differs for left and right cpl. Because of this relationship between isotropic and anisotropic absorption (cf. Section 12-3), theoretical treatments of chiroptical effects are little concerned with the prediction of wavelengths at which CEs are observed.

In general, any two of the three structural components, constitution, configuration, and conformation, must be known if one is to infer the third from chiroptical spectra (CD or ORD). All three structural components must be known if one is to predict ORD/CD. Hence, as a rule, for chiral molecules possessing torsional degrees of freedom, it is not possible to deduce information about both configuration and conformation simultaneously from chiroptical data. In this connection, the Octant rule (see below) has been called a "one-way rule," that is, one for which prediction of the sign of a CE *from* a known configuration is possible, but for which the converse, assignment of configuration from the CE without further structural information is not.[30]

Let us digress for a moment to recall what is required for electromagnetic radiation, whether isotropic or anisotropic, to be absorbed by matter: (a) the energy $h\nu$ of the impinging photon must correspond to the energy difference ΔE between the pertinent ground- and excited-state orbitals, and (b) the excitation (electronic) must be associated with the migration of charge thus generating a momentary electric dipole that is usually called the *electric transition moment* μ; a similar argument pertains to vibrational spectroscopy. The electric transition moment is related to the area of the UV absorption band by Eq. 12.11:

$$D = \mu^2 = 9.188 \times 10^{-39} \int (\epsilon/\lambda)d\lambda \qquad (12.11)$$

where D is called the dipole strength. In practice, the integrated area under the band may be approximated by measuring ϵ_{max}, λ_{max}, and $\Delta\lambda$ (the width of the band at $\epsilon_{max}/2$) if the band has a Gaussian shape (Eq. 12.12):

$$\epsilon_{max}\Delta\lambda/\lambda_{max} \approx \int (\epsilon/\lambda) \Delta\lambda \qquad (12.12)$$

Thus a measurable "substance property" (absorption intensity) is related to a theoretically calculable "molecular property," namely, the transition moment.

While a precise calculation of μ is beyond the purview of this book, its magnitude and direction may be estimated by means of a qualitative molecular orbital (MO) method or "recipe" described by Snatzke.[6,31,32] Circular dichroism requires not only that charge displacement take place during an electronic transition, but charge rotation as well; that is, the excited electron must be constrained to rotate or to be excited along a helical path.

b. Classification of Chromophores

The chromophores that are analyzed by means of CD measurements naturally fall into two broad classes as proposed by Moscowitz[22] on the basis of symmetry considerations (Chapters 4 and 5)[6,19,33]:

1. Chromophores that are *inherently achiral* by symmetry, such as the carbonyl and carboxyl groups, the ordinary C=C double bonds (alkenes), and the sulfoxide moiety. Each of these, when considered without substituents, contains at least one mirror plane. Chiral molecules containing inherently achiral (symmetric) chromophores exhibit CEs as a consequence of "chiral perturbations" arising in the chromophore during the electronic excitation of the latter. These perturbations are exerted by substituents located in the vicinity of the chromophore or by the molecular skeleton itself.

   In the preceding statement we have purposely used the language that one finds in most descriptions of such chromophores in the literature. However, as has already been pointed out, since *all* points in a chiral molecule are in a locally chiral environment, the notion of an inherently achiral moiety in a chiral molecule is fiction. It might then seem that the proposed bipartite classification of chromophores is invalid. In fact, the classification has an experimental basis (see below); and it may also be retained as a matter of convenience.

2. Chromophores that are *inherently chiral*. This type of chromophore includes compounds, such as the helicenes, in which the entire molecule acts as a single chromophore. Other examples are disulfides, biaryls, enones, cyclic 1,3-dienes (e.g., Chapter 11), and strained (twisted) alkenes (for the latter, cf. Section 9-1.d). In all of these, the chirality is built into the chromophore. The rotational strengths R of inherently chiral chromophores tend to be very large (see Table 12.2).

When two or more chromophores (identical or not) of either class are in close physical proximity (and have very large values of μ, corresponding to ε of several thousand) yet do not have overlapping orbitals, a third type of CE is observed. For the typical case involving just two such chromophores, two CD bands are observed having opposite signs. This phenomenon is called exciton chirality or exciton splitting. The deduction of absolute configurations from such CD spectra is particularly straightforward (cf. Section 12-4.d).[8]

Transitions of inherently chiral chromophores may also be classified on the basis of symmetry and group theoretical considerations. Examination of character tables allow one to identify those point groups having representations that permit electromagnetic transitions having finite rotational strength.[35]

The foregoing classification, while based on symmetry considerations, is directly related to the strength of the observed electronic transition. Inherently achiral transitions tend to be weak; they have low "rotational strengths." Inherently chiral transitions tend to have high rotational strengths, that is, to produce strong CEs. It has been suggested that high optical activity, for example, an ORD molar amplitude a of

TABLE 12.2. Electronic Transition Magnitudes[a]

	Wavelength			g Number[b]	
	$\dfrac{\lambda}{(nm)}$	UV ε	CD $\Delta\varepsilon$	$g = \dfrac{\Delta\varepsilon}{\varepsilon}\,(\times 10^3)$	Transition
3-Methylcyclohexanone	298	16	+0.48	30	$n-\pi^*$
	185	1200	+1.0	0.8	$n-\sigma^*\,(3s)$
(−)-β-Pinene	200	1.08×10^4	−17.1	2	$\pi_x - \pi_x^*$
	181	0.9×10^4	+17.0	2	$\pi_x - \pi_y^*$
(+)-Hexahelicene (3, pg. 113)	325	2.8×10^4	+196	7.0	$\pi - \pi^{*c}$
	244	4.8×10^4	−216	7.7	$\pi - \pi^{*c}$
(+) ... NH$_2$ / NH$_2$	247	7×10^4	−245	3	$\pi - \pi^*$
	231	6×10^4	+135	2	couplet

[a]Comparison of isotropic (UV) and anisotropic (CD) electronic transition magnitudes in selected chiral molecules. Adapted with permission from Mason. S. F. *Molecular Optical Activity and the Chiral Discriminations*, Cambridge University Press, Cambridge, UK, 1982, p. 49. The data on (+)$_{589}$-hexahelicene (in CHCl$_3$) is taken from Newman, M. S., Darlak, R. S., and Tsai, L. *J. Am. Chem. Soc.* **1967**, *89*, 6191 with $\Delta\varepsilon = [\theta]/3300$.
[b]The dimensionless ratio of the circular dichroic to isotropic absorbance, previously called anisotropy or dissymmetry factor.[34]
[c]The transitions for hexahelicene are both presumed to be of the $\pi - \pi^*$ type.[81]

the order of 100,000, constitutes strong evidence for the presence of an inherently chiral chromophore in the molecule.[4] Circular dichroic spectra exhibiting exciton splitting have the highest amplitudes of the three types of transitions.

It should by now be apparent that rotational strengths, and the more easily observed measures of ORD and CD magnitude (amplitude *a*, $\Delta\varepsilon$, and molar ellipticity [θ]) of individual transitions, have diagnostic value. It is instructive then to compare the magnitudes of measured CD bands of some specific compounds (Table 12.2). The table clearly reveals that there is a significant gradation in CD intensity ($\Delta\varepsilon$) with differences of the order of 10^2 between an inherently achiral chromophore (the ketone

C=O group) and the inherently chiral chromophore comprising the entire skeleton of the helicene. The CD couplet exhibited by the diaminobinaphthyl is attributed to exciton splitting.

Table 12.2 also reveals that all the isotropic transitions are strong and allowed with the exception of the n–π^* carbonyl band of the 3-methylcyclohexanone where μ must be close to zero. All the other transitions are said to be electric dipole allowed. For chromophores giving rise to CD, the magnitude of $\mathbf{m}$ is of considerable importance.

The disulfide (–S–S–) chromophore has absorption bands in the vicinity of 300 and 250 nm. It has been known for a long time, on the basis of X-ray data and of theoretical considerations, that in the absence of geometrical constraints, this structural element [torsional barrier 5–15 kcal mol^{-1} (21–63 kJ mol^{-1})] preferentially adopts a skewed conformation with $\omega \cong 90°$. It was pointed out by Beychok[36] that useful stereochemical information might be extracted from the CD spectra of chiral compounds containing the disulfide chromophore but that evidence relating the "screw sense" of the –S–S– moiety with such spectra was not yet available. Such data were soon provided by Carmack and Neubert,[37] who investigated the CD spectra of compound **2** (Fig. 12.16). On the assumption that the dithiane ring would preferentially adopt a chair conformation, the absolute configuration of compound **2** prepared from (S,S)-$trans$-1,2-cyclohexanedicarboxylic acid **B** would be S,S as shown in **A**. In **A**, the dithiane is required to adopt an M configuration (left-handed helix **C** as viewed along the –S–S– bond); the alternative P configuration of –S–S– would involve a twist–boat conformation in the six-membered ring.

The longest wavelength absorption band of **2** at 290 nm is associated with a CD band having n–σ^* character. A negative CD band ([θ] – 16,700) was found to correspond to the M configuration of the disulfide. A second absorption band at 240 nm (CD band at 241 nm), was found to have the opposite sign, that is, a positive CD band was found at 241 nm for the disulfide M configuration. These findings were incorporated into an empirical disulfide helicity rule.[35,38] Theoretical calculations that

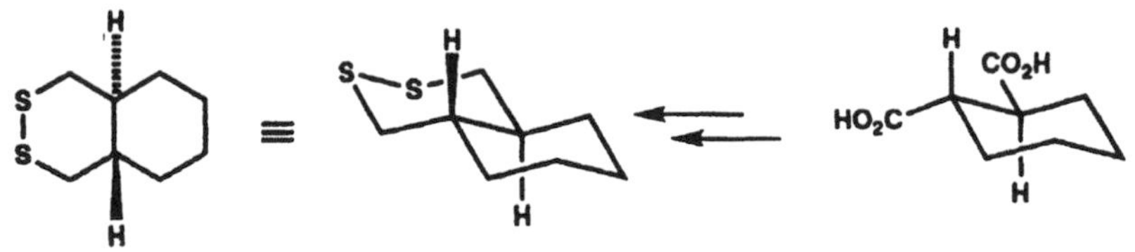

Figure 12.16. Conformation of ($9S$,$10S$)-(–)-$trans$-2,3-dithiadecalin.

treat the −S−S− moiety as an inherently chiral chromophore have shown that this helicity rule holds only for dihedral angles $\omega < 90°$. At $\omega = 0°$, the CE vanishes and at $\omega > 90°$, the signs are inverted.[39]

c. Sector and Helicity Rules

Introduction. Semiempirical rules correlating absolute stereochemistry (conformation or configuration) with chiroptical properties, especially with CD spectra and, in particular, the signs of individual CEs, are of two types: (a) sector rules and (b) helicity (or chirality) rules. The former type pertains to inherently achiral chromophores only while the latter are applicable to inherently chiral ones.[9,40]

When chirotopic atoms or groups are present (even in nearly symmetrical molecules), these groups are able to perturb the electronic transitions of symmetric chromophores sufficiently to generate chiroptical properties. The designation "Sector rule" stems from the division of 3D space surrounding such symmetrical chromophores into sectors by nodal or symmetry planes as well as by nodal surfaces. In fact, some authors use the term "symmetry rules" to stress the connection.[41] Such rules are designed to assess the contributions of perturbing groups to the sign of the CE according to their location in one or another sector surrounding the chromophore. In order to understand how these rules may be applied, a brief excursion into the history of the theories of optical activity is necessary since it turns out that symmetry rules are related to (one might say, designed to be consistent with) the several "mechanisms" for the generation of rotational strength that have been proposed and developed over the years.[19,35]

Numerous sector rules have been described for inherently achiral chromophores, carbonyl and C=C (alkenes) in pairticular. Extensive reviews and summaries of these are given in Crabbé,[42] Legrand and Rougier,[38] Snatzke and Snatzke,[43] and Kirk.[44] These are examined in some detail in this section.

Circular dichroism of compounds containing inherently chiral chromophores has also given rise to empirical generalizations (helicity rules) relating the signs of CEs with the sense of helicity of part of a molecule. Cotton effects of helicenes, enones, and peptides, among others, are very large. This effect is characteristic of the presence of both magnetic and electric transition moments. First-order determination of the sign of the CE and even of its magnitude directly from theory is more readily feasible for inherently chiral chromophores than is true for inherently achiral ones.

Saturated Ketones: The Octant Rule. The investigation of chiroptical properties of chiral ketones began around 1954 in the laboratory of Djerassi at Wayne State University when instrumental developments permitted the construction of the first easily usable spectropolarimeter reaching into the UV (to 300 nm). The experimental studies fortunately coincided with the growing understanding of conformational effects that followed the seminal contributions of Barton.[45] The initial ORD measurements were made on anancomeric (see p. 442) steroidal ketones.[46,47] Steroids as a class had been thoroughly studied by then and numerous samples of well-characterized compounds were available for the study. Moreover, the rigid steroid backbone was chosen purposely to minimize conformational ambiguities. In addition to the more obvious structural effects, such as the location of the carbonyl group, this permitted the ready assessment of the configurational and conformational effects of numerous substituent groups on the CE of the carbonyl chromophore.

The carbonyl chromophore had been chosen as the "test case" because of its ubiquity in the domain of organic chemistry.[20,48] Use of the carbonyl group was an inspired "stroke of luck" in view of two factors: Its UV absorption wavelength ($n-\pi^*$ transition) is in a readily accessible spectral region (λ_{max}, ca. 300 nm) and its absorption band is so far removed from the next absorption band of higher energy (ca. 190 nm) that there is no danger from overlap of the two nor of confusing the nature of the transition under study.[49] Moreover, the weakness of the $n-\pi^*$ transition avoids experimental difficulties in measuring the rotation right through the absorption band. This measurement was essential since one of the key differences between these studies and the few earlier ones was the emphasis on CE wavelengths, band shapes, and signs.

The very first generalization established as a result of the ORD studies of the Djerassi school was the *axial haloketone rule*.[20,50] Just as IR and UV spectral measurements are sensitive to the conformation of halogen substituents in six-membered rings (cf. Section 11-4.d), so too are the ORD parameters: CE intensities and wavelengths.[20] Based on measurements carried out on steroidal ketones, it was observed that equatorial α-halogen (or acetoxy) substituents on either side of the carbonyl group have little effect on the ORD parameters; the sign of the CE at about 300 nm is unchanged in these cases relative to that which is found for the unsubstituted ketone. However, in addition to causing bathochromic shifts, the corresponding axial substituents (a) cause an increase in the amplitude of the CE, and (b) may invert the sign of the CE according to the configuration of the stereogenic atom bearing the halogen atom; this finding was later expanded to include SR, SO$_2$R, NR$_2$, and other substituents.[51] The sign of the CE depends on the constitution, the conformation, and the configuration of the haloketone in the vicinity of the carbonyl group. Knowledge of any two of the foregoing factors and of the sign of the CE would suffice to reveal the third.

It was suggested that, in order to predict the sign of the CE, the carbonyl group should be viewed along the O=C bond in the direction of the ring with the carbonyl carbon placed at the "head" of the chair (Fig. 12.17). If the axial α-halogen atom is found to the right of the observer, then a positive CE is found; conversely, if the axial halogen atom is on the left, a negative CE obtains. Axial fluorine substituents have been shown to have effects opposite to those of the other halogens: The *S* enantiomer (X = F) in Figure 12.17 would exhibit a negative CE.

Applications of the rule include (a) the determination of configuration, e.g., by correlation of CE signs with those of analogous compounds having independently established configurations, assuming that the other two factors, constitution and conformation, are known or can be ascertained, and (b) determination of the constitution of a compound. The determination of the configuration of an 11-bromoketone is illustrated in Fig. 12.18, **B**; for an example of a configurational

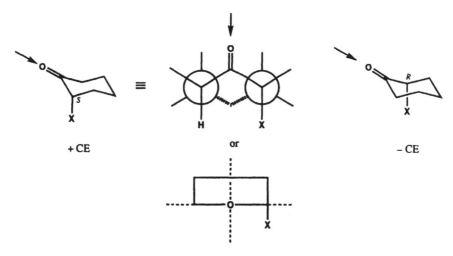

Figure 12.17. Axial haloketone rule.

assignment in the 1-decalone series, see Djerassi and Staunton[52] and Eliel.[53] Application (b) is illustrated by the establishment of the location of a halogen atom from the sign of the CE following halogenation of a nonracemic sample of a chiral ketone (Fig. 12.18, **A**; the axial orientation of the halogen—as independently established by UV and/or IR—is implied).

The rule also permits demonstration of conformational mobility as a function of solvent polarity. On chlorination of (R)-(+)-3-methylcyclohexanone, a crystalline 2-chloro-5-methyl derivative is isolated that exhibits a negative CE in octane. With the constitution of the product and the axial position of the halogen having been independently established, the negative CE is consistent only with trans stereochemistry (Fig. 12.19, **A** and **B**). However, a positive CE is observed in methanol. The latter finding is interpreted to mean that the conformational equilibrium has shifted in favor of the higher dipole diequatorial conformer in the presence of a

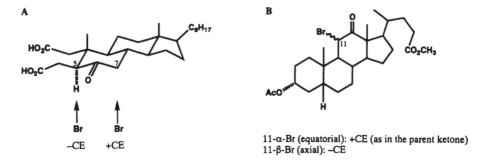

Figure 12.18. Applications of the axial haloketone rule. (**A**) Position of the halogen substituent. The bromo derivative exhibited a negative CE, hence substitution occurred at C(5). (**B**) determination of the absolute configuration of the 11-bromo substituent in an 11-bromo-12-ketosteroid (ref. 20, p. 123).

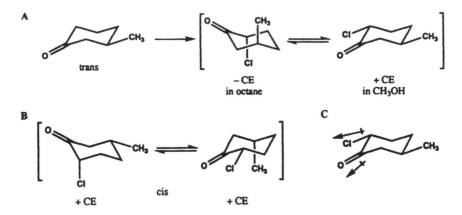

Figure 12.19. Conformational mobility in 2-chloro-5-methylcyclohexanones. (**A**) trans isomer; (**B**) the conformational equilibrium of the cis isomer is shown for comparison; (**C**) dipole repulsion in the trans isomer.

more polar solvent. Presumably, this is due to reduced repulsion between the diequatorially disposed carbonyl and chlorine dipoles in the presence of a solvent of high dielectric moment (Fig. 12.19, **C**) (cf. Chapter 11).[54]

The axial haloketone rule is also applicable to cyclohexanones that exist in a boat form. Of the two constitutionally related 2-bromo-2-methylcholestan-3-ones (Fig. 12.20), the 2β-bromo-2α-methyl isomer **A** exhibits a strong positive CE as anticipated; the 2α-bromo-2β-methyl diastereomer (shown spectroscopically to have its halogen in an axial orientation) unexpectedly exhibits a negative CE. The results were explained by assuming that in the latter isomer, ring A assumes a boat conformation (shown in Fig. 12.20, **B**).[55] The destabilization of the chair form **B′** is

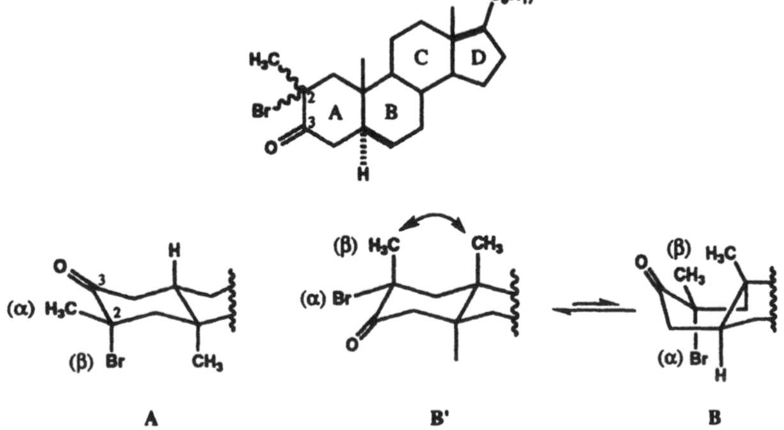

Figure 12.20. Demonstration of a boat form from chiroptical data. In A, the steroid A ring has been inverted for ease of comparison with Fig. 12.17.

ascribed to unfavorable axial 1,3-dimethyl interactions and to the unfavorable equatorial orientation of the bromine adjacent to the carbonyl group.

The above discussion of the axial haloketone rule was presented in some detail for two reasons: (a) The clarity of the results beautifully illustrate the power of chiroptical techniques, and (b) the data on which it was based led fairly rapidly to the first of the sector rules, namely, the octant rule.

The *octant rule* is an empirical generalization that relates the sign of the CE of the carbonyl chromophore, as measured at about 300 nm in saturated cyclic ketones, to the configuration of chiral centers present in the vicinity of the chromophore.[56] Its most common application is the deduction of configurational or of conformational information from CEs.[26,44,49]

In spite of being empirical, the rule is firmly based on theoretical studies of the $n-\pi^*$ transition that is responsible for the aforementioned CE observed in chiral ketones. Accordingly, the symmetry properties of this transition provide a framework around which the influence of stereocenters, both near and at some remove from the symmetrical chromophore, may be assessed. A set of Cartesian coordinates is drawn right through a carbonyl group with the origin at the midpoint of the C=O bond and the z axis collinear with the bond as shown in Figure 12.21. The carbonyl group is bisected by x,z and y,z planes that are nodal and symmetry planes of the MOs involved in the electronic transition. The x,y plane (both in character and in location) is merely an approximation to the nodal surfaces associated with (or orbitally determined by) the π^* virtual orbital. Substituent groups attached to the carbonyl carbon atom lie in the y,z plane facing away from the observer.

The coordinate system divides space around the carbonyl group into eight sectors or octants. The principal premise of the rule is that an atom anywhere in the vicinity of the carbonyl group [say, at point $P(x, y, z)$] makes a contribution to, and ultimately determines the sign of, the $n-\pi^*$ CE, according to its location, the sign being determined by the simple product xyz of its coordinates. For example, the contribution

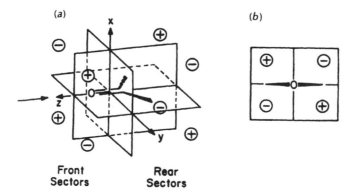

Front Sectors   Rear Sectors

Figure 12.21. Octant rule for saturated ketones. (*a*) Signs of the sectors in a left-handed Cartesian coordinate system; (*b*) projection of the rear (−z hemisphere) sectors. [Adapted with permission from Snatzke, G. *Chem. Unserer Zeit.* **1982**, *16*, 160.]

of an atom located in a region corresponding to the lower right rear sector (Fig. 12.21*b*) whose coordinates are −*x*, +*y*, −*z*, is positive in a left-handed coordinate system (see discussion below) and, therefore, would predict a positive CE. The same atoms located in the mirror-image lower left rear sector would induce a negative CE.

A useful mnemonic for learning the signs of the octant rule sectors is that the *upper, front, right* sector is positive. Since sector signs alternate in all directions, the signs of all remaining sectors are easily deduced from that of the "UPFRont" (upper, positive, front, right) sector.

The octant rule was first applied to cyclohexanones whose geometry (bond lengths and angles) were well known and whose conformations were fixed by the ring fusions exhibited in steroids. The cyclohexane skeleton is disposed within the coordinate system as shown in Figure 12.22*a*, with the α- and α′-carbon atoms in the *yz* plane, and the remaining carbon atoms projecting upward.

Carbon-4 lies in the *xz* plane. Since, as a first approximation, substituents located on or close to nodal planes make no contribution to the CE (in fact, chiral adamantanones bearing methyl groups lying in symmetry planes do exhibit weak CEs,[51] we see that only groups attached to C(2), C(3), C(5), and/or C(6) can make such contributions. Moreover, since equatorial substituents at C(2) and C(6) lie in the *yz* plane, they have little, if any effect, on the CE. Contributions occurring in more than one ring are additive, and a semiquantitative assessment of such contributions is sometimes possible (see below). Incidentally, it should be evident that axial haloketone rule is but a special case of the Octant rule (compare Fig. 12.17 with Figs. 12.21 and 12.22).

Carbon atoms, halogen atoms other than fluorine, and sulfur make contributions as predicted above. However, the halogens (again with the exception of fluorine) dominate the CE, completely overriding contributions by alkyl groups, as a consequence of their much higher atomic refraction.[58] In this connection, it may be noted that fluorine has the lowest atomic refraction of all the halogens: its refraction is *less* than that of hydrogen. Nitrogen- and oxygen-containing groups may exhibit

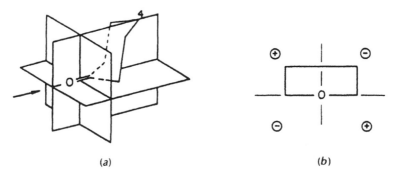

(a)   (b)

Figure 12.22. (*a*) Stereoprojection of the cyclohexanone ring (chair form) in the octant diagram; (*b*) projection of cyclohexanone bonds. View facing the carbonyl oxygen with signs of rear octants. [Reprinted with permission from Snatzke. G. *Angew. Chem. Int. Ed. Engl.* **1968**, 7, 14.]

either octant or antioctant (inverse) behavior. The $-N(CH_3)_3^+$ group, for example, exhibits antioctant behavior.

An effect due to a particular substituent is called *consignate* (more accurately sector consignate) if the sign of its contribution to the CD or ORD intensity is the same as the sign of the product of the Cartesian coordinates for the sector in which the substituent is located. A contrary relationship between the sign of the contribution to the CD or ORD and the sign of the sector is termed *dissignate* (or antioctant), for example. fluorine which makes contributions inverse to those of the other halogens.[59] Naturally, con- or dissignate behavior is entirely dependent on the choice of signs of the coordinates, that is, on the coordinate system convention adopted (see above).

Contributions by hydrogen have usually been ignored. This contribution is simply the result of cancellation of effects by hydrogen atoms appearing in oppositely signed sectors. In addition, the atomic refraction of H, hence, the inherent magnitude of its contribution is small. More recent studies have shown that, in some circumstances, contributions particularly from C_α–H make dominant octant–dissignate contributions.[60]

The effect of deuterium as a ketone substituent has also been studied.[61–64] Deuterium makes a significant dissignate contribution such that the configuration of deuterium substituted ketones may be determined. It has been established that the rotatory strengths of transitions due to conformers in conformationally flexible systems [e.g., (R)-(2-$^2H_1$)- and (S)-(3-$^2H_1$)-cyclohexanones] are additive. These results require the assumption that axial and equatorial conformers of these ketones are present in nearly equal amounts; steric isotope effects on the equilibrium are below the experimental error limit (see also Chapter 11).[63]

The compound (+)-3-methylcyclohexanone is known to exhibit a positive CE. Octant rule projections for both axial and equatorial conformers of this compound are shown in Figure 12.23 along with the corresponding configurations at C(3). The C(3) and C(5) carbon atoms make equal and opposite contributions to the CE. Only the positive contribution of the methyl group is unmatched by a negative one. It is evident from the illustration that the preferred equatorial conformation is predicted by the

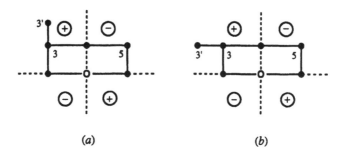

(a)   (b)

Figure 12.23. Octant rule projections for (+)-3-methylcyclohexanone (rear sectors). (*a*) Projection for the axial conformer (*S* configuration); (*b*) projection for the equatorial conformer (*R* configuration). [Reprinted with permission from Charney, E. *The Molecular Basis of Optical Activity. Optical Rotatory Dispersion and Circular Dichroism*, p. 176. Copyright © 1979 John Wiley & Sons, Inc.]

Octant rule if the R configuration obtains (as is indeed known). Arguments based on the additivity of rotatory strengths exclude the possibility that twist conformations play a major role in the conformational equilibrium.[65] Conversely, if the preferred conformation (equatorial) had been established independently (e.g., by NMR, cf. Chapter 11), the configuration would have been deduced as being R (Fig. 12.23).

In the case of compounds with reduced conformational mobility, such as 2-methylcyclopentanone, application of the octant rule suffices to reveal the absolute configuration. The compound (−)-2-methylcyclopentanone exhibits a sizeable negative CE, $[\theta]_{306}$−4786 (dioxane); accordingly, the methyl substituent must project into the lower left rear octant (projection analogous to that in Fig. 12.22b) and the configuration assigned as 2R.[66] An alternative interpretation of the data holds that it is not the contribution of the methyl group that gives rise to the relatively large CE, but rather the helicity of the relatively rigid twisted cyclopentanone skeleton in that conformation that has the methyl group locked in the quasiequatorial position.[44]

The octant rule projections for the three steroidal ketones shown in Figure 12.24 along with the relevant chiroptical data illustrate the fact that semiquantitative assessment of CE magnitudes is often possible: The larger the number of carbon atoms, groups, and/or rings in a given sector, uncompensated by like contributions in sectors of opposite sign, the greater the CE magnitude. The larger positive CE observed for 5α-cholestan-2-one **4** relative to that observed for 5α-cholestan-3-one **5** (Fig. 12.24) is evident, even predicted, from the projections: In the latter, only the methylene C(6) and C(7) atoms found in the positive upper left rear sector make significant contributions to the CE. The angular methyl groups lie in the x,z plane, as

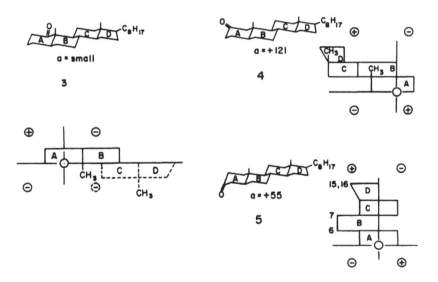

Figure 12.24. Semiquantitative assessment of CE magnitudes. Octant rule projection for isomeric 1-, 2-, and 3-cholestanones (**3–5**, respectively) and experimentally observed CE amplitudes (see Figs. 12.25 and 12.26, respectively). The projection outlined with dashed lines is that in a front octant. [Adapted with permission from Snatzke, G. *Angew. Chem. Int. Ed. Engl.* **1968**, 7, 14.]

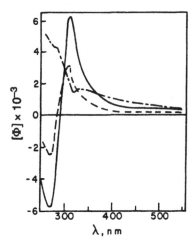

Figure 12.25. The ORD spectra of 5α-cholestan-1-one **3** (–··–), -2-one **4** (–), and -3-one **5** (--) (methanol solution). [Adapted from Djerassi, C. *Optical Rotatory Dispersion*, McGraw-Hill, New York, 1960, p. 42.]

do many of the carbon atoms of the C(17) side chain. Neither these nor the remote C(15) and C(16) of **5** make appreciable contributions to the CE, which is less than one-half as large as that of **4**.

The 5α-cholestan-1-one **3** exhibits a negative CE that is much more easily revealed by the CD spectrum (Fig. 12.26) than by the ORD spectrum (Fig. 12.25). The ORD spectrum is more difficult to interpret due to the presence of strong background

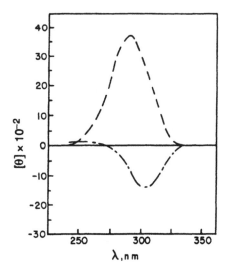

Figure 12.26. The CD spectra of 5α-cholestan-1-one **3** (– · –) and -3-one **5** (--) (methanol solution). [Reprinted with permission from Djerassi, C., Records, R., Bunnenberg, E., Mislow, K., and Moscowitz, A. *J. Am. Chem. Soc.* **1962**, *84*, 4552. Copyright © 1962 American Chemical Society.]

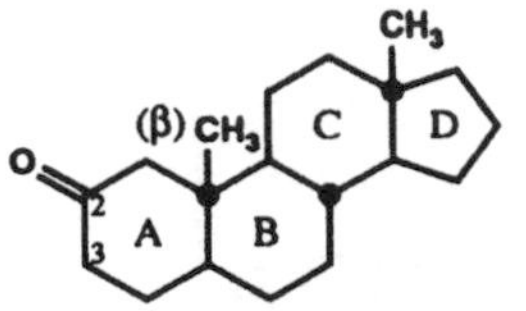

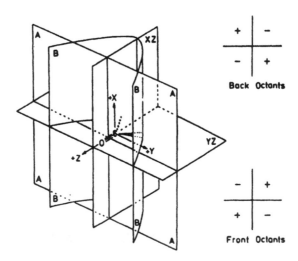

	$\delta\Delta\varepsilon$
Ring B	+ 1.3
Ring C	– 0.15
Ring D	+ 0.35
β-CH₃	+ 0.5
(Calcd.)	+ 2.0
Found	+ 1.78

Figure 12.27. Numerical contributions to the CE magnitude.[67] In this computation, ring D is treated as a six-membered ring.

rotation caused by CEs lying at shorter wavelengths. The latter are not evident in the CD spectrum. The negative CE may be due to uncompensated contributions principally from the front sectors as well as a smaller contribution from rear sectors; the front sectors are indicated in Fig. 12.24 by means of dashed lines.

Figure 12.27 illustrates the utility of empirically determined numerical contributions or increments ($\delta\Delta\varepsilon$) to CE magnitudes.[67] A cautionary note is in order, however. Such contributions are likely to be accurate only within narrowly defined structural domains. For applications, for example, in the decalone system, see Kirk and Klyne[68]; for a summary, see Kirk.[44]

Many theoretical studies have helped to define the limits of the octant rule. One such theoretical study by Bouman and Lightner[69] has revealed the curved geometry of the third nodal surface of the octant rule (Fig. 12.28). In this model, surface B takes

Figure 12.28. Third nodal surface B of the octant rule as proposed by Bournan and Lightner.[69] Sign change regions computed for the carbonyl n–π^* Cotton effect. [Reprinted with permission from Lightner, D. A., Chang, T. C., Hefelfinger, D. T., Jackman, D. E., Wijekoon, W. M. D., and Givens, III, J. W. *J. Am. Chem. Soc.* **1985,** *107,* 7499. Copyright © 1985 American Chemical Society.]

the place of plane A of the original octant rule. As earlier pointed out (see above), the A (x,y) plane is the only one of the three surfaces delineating the sectors that is not determined by the ground state of the carbonyl group. The computed regions of sign change have accounted for a number of results previously characterized as being dissignate.[70,71] From the foregoing, it will be evident that earlier "exceptions" to the octant rule were responsible for later minor modifications.

The interpretation of CD spectra of compounds containing β-axial substituents in cyclohexanones has proven to be particularly troublesome.[72,73] The contribution of axial β-methyl groups to the (negative) CE is relatively weak and strongly solvent dependent, for example, in the 2-adamantanone derivative **6** shown in Figure 12.29. This contribution was interpreted as being dissignate.[74-76.] Part of the difficulty may result from the fact that the substituents lie close to the curved nodal surface of the octant rule, hence the substituents may actually project into a front sector so that its contribution is really consignate.[69,71] Alternatively, the dissignate behavior has been interpreted as being due to a specific through-space interaction between orbitals of the β substituent and the carbonyl group.[31] Another interpretation is given by Mason.[19]

Computer programs have been developed that permit prediction of the sign, as well as reasonably good relative intensities, of the CD associated with the $n-\pi^*$ transition of rigid and of conformationally flexible ketones, that is, computer modeling of the octant rule.[77] For conformationally flexible molecules, the approach relies on molecular mechanics calculations and searches for a set of low-energy conformations. Intensities are parametrized by means of a distance factor formula derived from the concept that the magnitude of the perturbation by a substituent group is inversely proportional to the distance between the chromophore and the substituent;[31] in addition, intensities are affected by the distance between the nodal plane and the substituent.

Other Sector Rules.   In addition to the octant rule, other sector rules applicable to inherently achiral chromophores have been described. Readers are referred to the most complete summaries of sector and helicity rules available as of this writing, those of Legrand and Rougier[38]; the book by Ciardelli and Salvadori (see especially the summary in Chapter 1)[33]; and that by Crabbé.[42] A compact but useful list may be found in the review by Snatzke and Snatzke.[43]

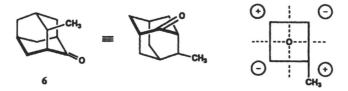

6

Figure 12.29. Dissignate behavior of axial 4β-methyl-2-adamantanone **6**.[74] At right, the signs of the octant projections are those of the rear sectors.

Helicity Rules.   The analysis of CEs has been carried out in terms of the two types of chromophores, inherently achiral and inherently chiral, discussed earlier (p. 550). However, as already mentioned, this distinction is only a convenient fiction.[78] We should then not be surprised to find that some molecular systems have been successfully treated both as containing inherently achiral and inherently chiral chromophores.

Let us consider an example to make this point clear. Cyclopentanones and certain cyclohexanones, constrained into rigid twisted and hence chiral conformations, exhibit CEs of intermediate magnitude. As a first approximation, the sign of the CE is given by the octant rule as if the chromophore were still inherently achiral (Fig. 12.30; see also Section 12-4.c), though this may be fortuitous.[44]

Generalizations regarding CEs of inherently chiral chromophores are called helicity (and sometimes chirality) rules (Section 12-4.b). Typical examples of such chromophores are unsaturated ketones, dienes, skewed (twisted) alkenes, disulfides (these have already been discussed), and helicenes.

Often, the interaction between two inherently achiral chromophores present in one molecule suffices to generate the high rotational strength that is the hallmark of inherently chiral chromophores. This type of composite (inherently chiral) chromophore was first observed in chiral β,γ-unsaturated ketones in which the intramolecular C=O and C=C groups are disposed in a twisted array.[80]

Shortly thereafter, a β,γ-unsaturated ketone helicity rule (Fig. 12.31) for transitions of the $n-\pi^*$ type (ca. 300–310 nm) was described.[81] The rule effectively states that homoconjugated C=C–CH_2–C=O arrays disposed in conformations represented by Figure 12.31a exhibit strongly negative CEs at about 300 nm. These strong CEs are limited to conformations with dihedral angles (ω) between 100° and 120°. Correspondingly large isotropic absorption intensities are also found, especially in nonpolar solvents.

The homoconjugated enone rule was shown to be consignate with the ketone Octant rule with the C=C (considered as a substituent) dominating the sign of the CE (generalized octant rule) (Fig. 12.31c).[80]

The helicity rule is also applicable to homoconjugated aldehydes and acid derivatives. In 1961 on the basis of this rule, dehydronorcamphor **7** and bicyclooctenone **8** (Fig. 12.32) were conjectured to have intense positive CEs. The accuracy of this prediction was confirmed in 1962.[82] The absolute configuration of the

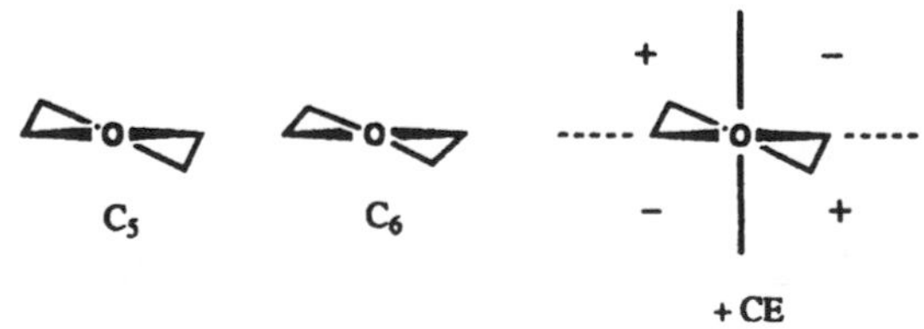

Figure 12.30. Application of the octant rule to skewed cyclanones. The chirality sense shown gives rise to a positive CE. [Adapted with permission from Kirk, D. N. *Tetrahedron* **1986**, *42*, 777. Copyright © 1986 Pergamon Press, Ltd, Headington Hill Hall, Oxford, OX3 0BW, UK.]

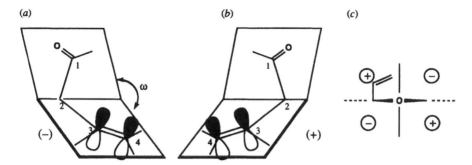

Figure 12.31. (*a, b*) β,γ-Unsaturated (homoconjugated) ketone helicity rule showing the signs of the CEs generated at about 300 nm. [Adapted with permission from Moscowitz, A., Mislow, K., Glass, M. A. W., and Djerassi, C. *J. Am. Chem. Soc.* **1962**, *84*, 1945. Copyright © 1962 American Chemical Society.]. (*c*) Generalized octant rule (the signs are those of the rear sectors).

pheromone stegobinone **9** (derived from the bread beetle) was deduced from the negative sign of the strong CE at 290 nm.[83,84] The absolute configuration at C(7) may be inferred if it is recalled that conformations in which C–C bonds synperiplanar to C=O are energetically preferred to the corresponding (C=O)–C–H conformations (cf. p. 390); this places the C(7) methyl group rather than the corresponding methine hydrogen in the groove of the helicity rule diagram (Fig. 12.32, right).

A relatively simple semiempirical quantitative chirality rule for β,γ-enones has shown that the sign and the magnitude of the CE are directly related to the angle between the C=O and C=C bonds.[85]

The D-glucal **10** (Fig. 12.33*a*) is a cyclic enol ether that incorporates a nonplanar C=C–O moiety locked into a chiral conformation. Figure 12.33*b* illustrates the case with a negative torsion angle for C(5)–O–C(1)=C(2).

Assuming that the relative configurations of the OH and CH$_2$OH groups are known from NMR or other results, the absolute configurations of the chiral centers at C(3), C(4), and C(5) (Fig. 12.33*a*) follow from conformational analysis: L-glucal in this conformation requires all substituents to be axial while D-glucal **10** would have them all equatorial.[31]

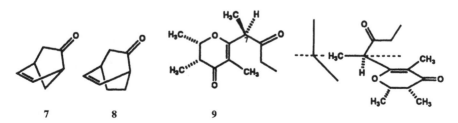

Figure 12.32. Homoconjugated ketones. At right, absolute configuration of C(7) of stegobinone **9** as inferred from the preferred conformation.

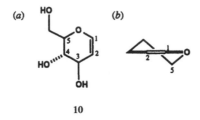

10

Figure 12.33. (*a*) Structure of D-glucal **10**; (*b*) twisted conformation with M^- (left-handed) helicity.

Conjugated 1,3-dienes absorb in the 220–280-nm range (longest wavelength or π_2–π_3^*) according to substitution and ring size, if any. When the diene is skewed, an inherently chiral chromophore is present. The corresponding absolute configuration is correlated with the sign of the CE by means of the diene helicity rule shown in Figure 12.34.[86,87] Calculations show that *P* helicity is correlated with a positive CE and *M* helicity gives rise to a negative CE.[35] Exceptions to the rule are found especially when the double bonds are not homoannular.[88]

α-Phellandrene **11** (Fig. 12.34) behaves according to this rule. At room temperature and above, a negative CE is found; however, as the temperature is reduced, the magnitude of the CE decreases, becomes zero, and eventually (down to −168°C!) becomes positive. The two conformations consistent with these results are shown in Figure 12.34*b* [the configuration at C(5) as shown in the figure is *S*] with the thermodynamically preferred conformation present at low temperature bearing a pseudoequatorial isopropyl group.[89] An interesting point is that it is the conformation with the pseudoaxial isopropyl group that dominates the CE at room temperature; this explanation was advanced by Burgstahler and co-workers in 1961, and confirmed by

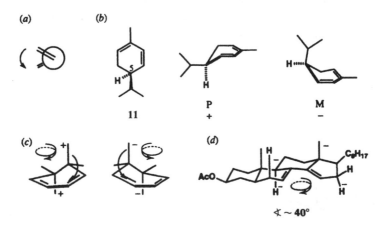

Figure 12.34. (*a*) Diene helicity rule; (*b*) *P* and *M* absolute configurations of (+)-α-phellandrene **11**; (*c*) reinforcing contributions of allylic pseudoaxial bonds to the CE; (*d*) opposing diene and allylic pseudoaxial contributions. In the example, the CE is negative. [Adapted with permission from Burgstahler. A. W. and Backhurst. R. C. *J. Am. Chem. Soc.* **1970**, *92*, 7601. Copyright © 1970 American Chemical Society.]

Snatzke and co-workers by means of the just mentioned low-temperature measurements.[89,90] The greater stability of the pseudoaxial conformation at room temperature may be due to a higher entropy associated with the freer rotation of the pseudoaxial (relative to that of the pseudoequatorial) isopropyl group; the pseudoequatorial isopropyl group in the other conformation may be constrained by the flanking ring hydrogen atoms, and therefore be unable to freely rotate.

d. Exciton Chirality

When two chromophores are in close spatial proximity to one another and so disposed that a chiral array results, interaction (dynamic coupling) between the individual chromophores gives rise to distinctive CE couplets, often called exciton coupling, in the CD spectrum from which the configuration of the chiral array may be easily deduced. The term exciton (coined by Davydov in connection with his molecular exciton theory)[91] applies to the nondegenerate excited states of a polychromophoric system.

In 1969, Harada and Nakanishi described a method for determining the configuration of glycols from the sign of the CE of the strong $\pi-\pi^*$ transition of the glycol dibenzoate derivatives.[92] Although considered by them to be an extension of the benzoate sector rule (see above), the new method has little in common with this sector rule (other than the chromophore). The high amplitude of the CEs of such dibenzoates arise mainly from dipole–dipole interaction between the electric transition moments of the two identical but nonoverlapping benzoate chromophores. An example is given in Figure 12.35.

The CD spectrum of compound **12** displays the typical bisignate couplet that characterizes exciton splitting. The two CEs are centered at 307 nm near the UV absorption maximum [310.5 nm, ε 49,000 (EtOH)]. The longer wavelength CE of (−)-**12** is negative while the shorter wavelength CE of the couplet is positive. This relationship between the two CEs constitutes "negative chirality" as defined by Harada and Nakanishi[92] and consequently, since the relative configuration of the stereocenters at C(1) and C(6) is known to be 1,5-*cis*-5,6-*trans* (by absence of C_2 symmetry as deduced from the ^{1}H NMR spectrum that shows two different sets of *p*-dimethylaminobenzoate peaks as well as two different methine peaks),[93] the configuration of **12** is 1*S*,5*S*,6*R*.

Harada and Nakanishi called this type of analysis the *exciton chirality* method.[8,94] The method itself is historically based on the coupled oscillator method (see above) as extended especially by Moffitt and Tinoco, who applied it to the analysis of chiroptical properties of biopolymers (Section 12-4.f).[19] In 1962, Mason deduced the absolute configuration of the alkaloid calycanthine (Fig. 12.36, **13**) by application of the coupled oscillator method.[19,95] This determination represents the first application of the exciton chirality method to nonpolymeric systems.

The very large magnitude of the characteristic split CEs observed has led to the classification of systems exhibiting exciton splitting as a third type of chromophore, neither inherently chiral nor inherently achiral. Alternatively, dibenzoates, biaryls,

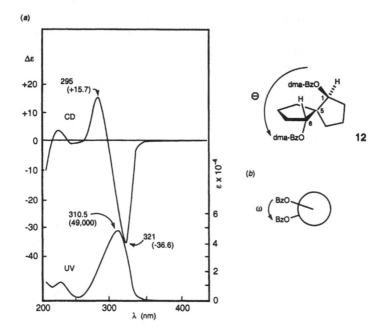

Figure 12.35. (*a*) The UV and CD spectra of the bis(*p*-dimethylaminobenzoate) (dma-BzO) ester of (−)-spiro[4,4]nonane-1,6-diol **12** in ethanol; (*b*) dibenzoate chirality rule. [Adapted from Harada, N. and Nakanishi, K. *Circular Dichroic Spectroscopy, Exciton Splitting in Organic Stereochemistry.* University Science Books, Mill Valley, CA, 1983.

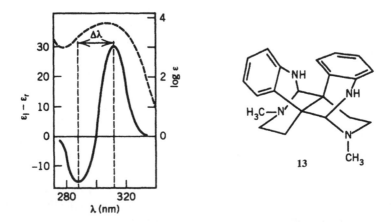

Figure 12.36. The UV (upper) and CD (lower) spectra of calycanthine **13** in ethanol. Exciton splitting. Example of positive chirality. Δλ represents the "Davydov splitting." [Adapted with permission from Mason, S. F. *Proc. Chem. Soc. London* **1962**, 362. Copyright © Royal Society of Chemistry, Science Park, Milton Road, Cambridge CB4 4WF, UK.]

and other polychromophoric systems exhibiting exciton splitting have been considered to be inherently chiral chromophores by Mason[19] and by Snatzke[40] in spite of the lack of electron delocalization between the individual chromophoric units.

The exciton chirality method is readily applicable to a wide variety of systems consisting of two nonconjugated (or homoconjugated) chromophores. These range from the highly symmetrical (e.g., 1,1'-binaphthyl; Fig. 12.37, **14**)[19] to those in which the two chromophores are very different from one another, for example, allyl benzoates (see above), and the p-chlorobenzoate derivatives of cinchonine and cinchonidine (**19** and **21**, respectively, in Fig. 7.9a.[94] Similarly, configurations of 1° amines and some α-amino acids and amino sugars may be predicted by analysis of the CD spectra of their salicylideneamino derivatives (salicylideneamino chirality rule).[96] Intramolecularly hydrogen-bonded Schiff's base derivatives prepared, for example, by reaction of an arylalkylamine with salicylaldehyde (Fig. 12.37, **15**) give rise to CEs attributed to exciton coupling between the salicylideneamino and phenyl chromophores (but no bisignate signals, presumably because the $\pi-\pi^*$ transitions of the phenyl chromophores are at too short a wavelength). For a thorough survey of applications of the exciton chirality method, see the book by Harada and Nakanishi.[8]

Systems consisting of three chromophores, for example, triptycenes as well as tribenzoates of sugars and steroids are also amenable to analysis by the exciton chirality method.[8] The cyclotriveratrylenes (e.g., **16**; Fig. 12.37, $R_1 = OCH_3$, $R_2 = OH$) give rise to split CD bands from which the configuration may be deduced by application of exciton theory. The CD is the extremely sensitive to the nature and position of the substituents, for example, acetylation of the OH groups of **16** gives rise to **17** in which the handedness of the CD couplets is reversed.[97] Even the case of **18** with $R_1 = D$, $R_2 = H$ (Fig. 12.37) gives rise to a measurable and interpretable CD spectrum.[97b]

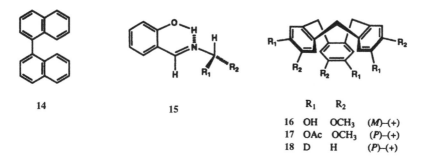

	R_1	R_2	
16	OH	OCH₃	(M)–(+)
17	OAc	OCH₃	(P)–(+)
18	D	H	(P)–(+)

Figure 12.37. Compounds whose configurations have been determined by the exciton chirality method: 1,1'-binaphthyl **14**; a primary amine $R_1R_2CHNH_2$ as its salicylideneamino derivative **15** (as enolimine tautomer); cyclotribenzylenes **16**, **17**, and **18** with $R_1 \neq R_2$.

e. Other Applications: Induced ORD and CD

Introduction.   In addition to "structural analyses" described in the preceding sections (and below), CD measurements, in particular, have been applied also to other analytical problems, which are illustrated below. For reviews, see Barrett,[96] Scopes,[26] Legrand and Rougier,[38] Snatzke and Snatzke,[98] and Purdie and Brittain.[29] Qualitative and quantitative analysis of the plant growth regulator abscisic acid (Fig. 12.38, **19**), in plant extracts has been described by measurement of a specific CE in preference to polarimetric measurements at a single (or a few) wavelength.[99]

Chiral reaction byproducts, particularly when present in very small amounts, are more easily identified and monitored in nonracemic samples by CD than by many other methods, for example, conjugated unsaturated ketones ($\lambda_{max} \approx 340$ nm) in samples of the corresponding saturated ketones ($\lambda_{max} \approx 300$ nm). A similar application of CD is the identification of diene **20** in **21** (Fig. 12.38) isolated from frankincense.[100] The critical micelle concentration of a chiral detergent was determined by means of CD.[101] A CD detector has been applied to analyses and chromatographic resolutions by high-performance liquid chromatography (HPLC) to permit direct determination of the enantiomeric purity.[102,103]

Numerous other applications involve the induction of ORD or CD either in achiral or chiral compounds under the influence of temperature, different solvents or environments, or under the influence of a variety of reagents. Optical activity can be induced in racemates by cpl via the photointerconversion of the enantiomers, as well

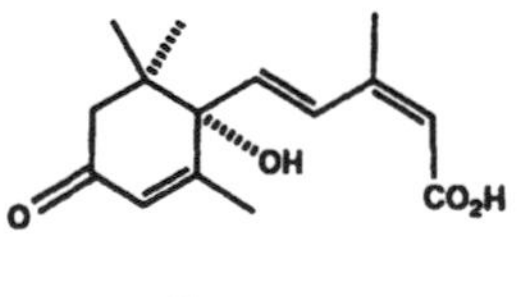

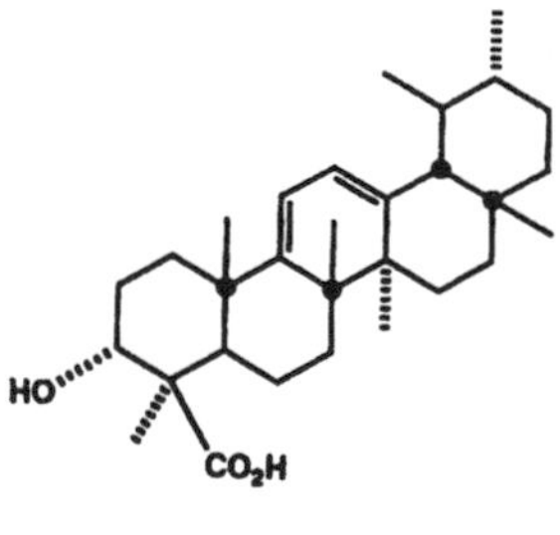

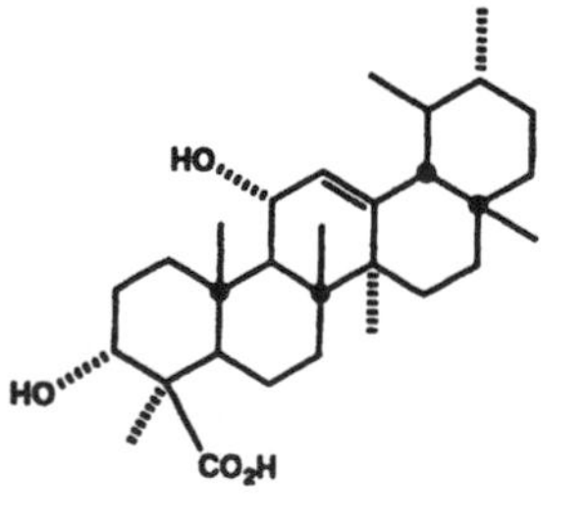

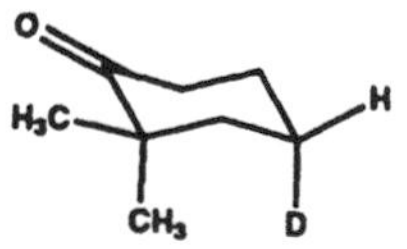

Figure 12.38. Structures **19–22**.

as by the preferential photodestruction of one enantiomer of a racemate.[104,105] Chiroptical properties induced in achiral compounds are often called "extrinsic" to distinguish them from those (*intrinsic* properties) of chiral compounds.

Changes in the CD spectra of adducts of guanosine with carcinogenic hydrocarbons induced by pH conveniently permit determination of the pK_a values that reflect the point of attachment of the hydrocarbon on the base.[106]

Optical rotary dispersion is a sensitive reporter of solvent-induced changes in the position of equilibrium in conformationally mobile systems. In a study aimed at assessing specific solvent effects, the conformational equilibrium of (+)-*trans*-2-chloro-5-methylcyclohexanone (cf. Section 12-2.c) was reexamined in 28 solvents; $[\Phi]_{330}$ values ranged from 680 in dimethyl sulfoxide (DMSO) to 8.3 in cyclohexane.[107]

Temperature dependence of the CD is generally interpreted as being caused by equilibration between at least two chiral species.[108] The method is sufficiently sensitive to estimate the conformational energy of deuterium versus hydrogen. In (4S)-2,2-dimethyl-4-deuteriocyclohexanone (Fig. 12.38, **22**), the gem-dimethyl groups serve as "chiral probes" for the deuterium that resides in the octant rule x,z symmetry plane. Changes in the rotational strength of the carbonyl $n-\pi^*$ transition as a function of temperature can be related to the conformational equilibrium, leading to the conclusion that axial D is preferred over H with $\Delta H° = -3.3$ cal mol^{-1} (-13.8 J mol^{-1}).[65]

A well-known and dramatic example of induced CD involves the achiral metabolite bilirubin **32** (Fig. 12.39a), a major constituent of bile, that is produced in humans by degradation of heme and is responsible for the pigmentation evident in jaundice. Bilirubin can exhibit induced CD when simply dissolved in α-methylbenzylamine,[109,110] when examined in the presence of cyclodextrins,[111] in the presence of albumins,[112] on complexation with cinchona alkaloids in CH_2Cl_2 solution,[113] or in covalent derivatives, such as the N-acetyl-L-cysteine adducts to the C(18) vinyl group.[114] The nature of the inherently dissymmetric chromophore exhibited by bilirubin and its chiral as well as achiral analogues was first recognized and studied by Moscowitz et al.[115] Intramolecular hydrogen bonding permits a helical conformation to be easily achieved. Except in covalent derivatives, the induced CD results from an asymmetric transformation of the two interconverting enantiomeric conformations of bilirubin (Fig. 12.39b).

The induction of CD in a racemic helical polymer on dissolution in a nonracemic chiral solvent has been described in Section 7-2.d.

Cottonogenic Substituents. In order to overcome the inability of CD and ORD spectrometers to record Cotton effects in the UV region below about 210 nm, so-called transparent functional groups have been derivatized so that they would reveal themselves through new chromophores that could be studied in easily accessible wavelength regions.[20,26,48,116] To emphasize the generation of CEs, such derivatives have been dubbed *cottonogenic* (or chromophoric).[116] For example, simple alcohols do not exhibit CD bands down to 200 nm; and, while they do exhibit plain ORD curves, deductions of configuration from such plain curves are either unfeasible or

(a)

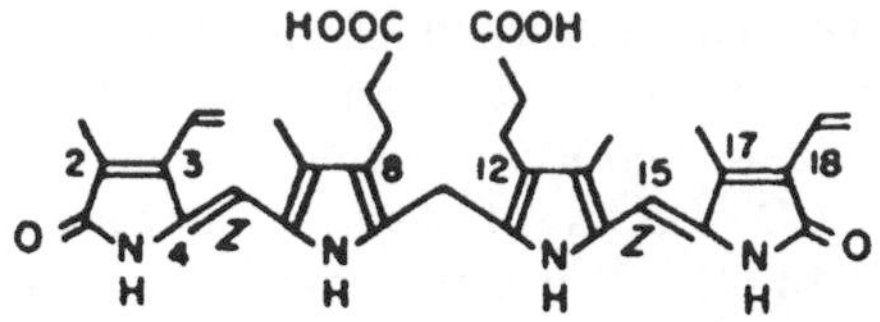

23

(b)

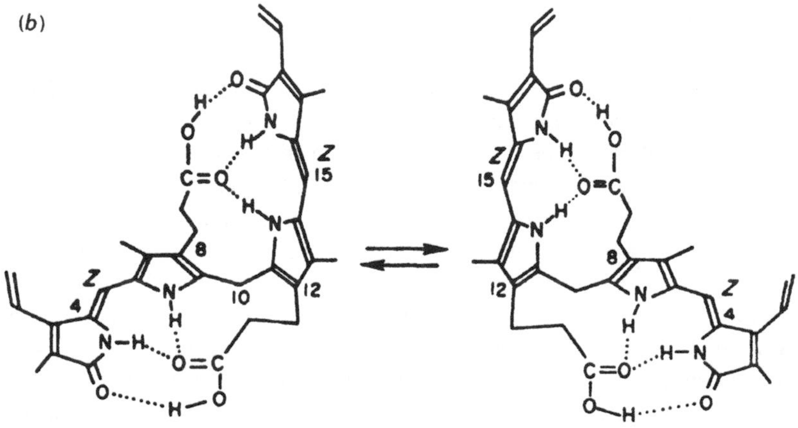

Figure 12.39. (*a*) Structure of (4Z,15Z)-bilirubin **23**; (*b*) interconverting intramolecularly hydrogen-bonded enantiomeric conformers of **23**. [Reproduced with permission from Lightner, D. A., Gawronski, J. K., and Wijekoon, W. M. D. *J. Am. Chem. Soc.* **1987**, *109*, 6354. Copyright © 1987 American Chemical Society.]

unsafe.[42] Similarly, deductions about configuration were difficult to make for many simple aliphatic and alicyclic amines. These functional groups were converted, for example, to nitrites or to hemiphthalates in the case of alcohols and to phthalimides or to isothiocyanates in the case of amines. Thus, for example, 2-amino-1-butanol exhibits three CEs from 200 to 350 nm when derivatized as the isothiocyanate of the methyl carbonate ester (Fig. 12.40, **24**).[117]

Extensive lists of such derivatives and discussion of the configurational correlations made possible by examination of their CEs have been given by Crabbé,[118,119] Sjöberg,[116] and Crabbé and Parker.[17] The principal types of derivatives that have been studied are, for alcohols, esters [in the early literature, acetates, xanthates, and nitrites (–O–N=O), and more recently, benzoates and cinnamates]; for amines, dithiocarbamates, *N*-salicylidene and *N*-phthaloyl derivatives, dimedone (enamine) derivatives, *N*-nitroso and *N*-chloro derivatives; for carboxylic acids, acylthioureas, and thionamides.

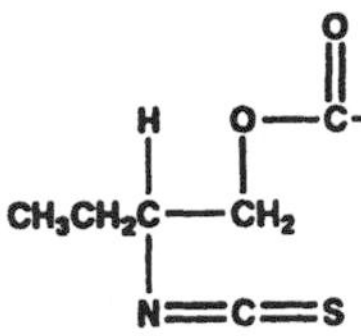

Figure 12.40. Cottonogenic derivative of 2-amino-1-butanol.

Organometallic derivatives have figured prominently in studies of ORD and CD. Cuprammonium solutions of glycols and of amino alcohols exhibit CEs in the vicinity of 600 nm in the metallic $d-d$ transition region as a result of 1:1 in situ complex formation between the copper and a bidentate ligand. This finding has been called the Cupra A effect. The sign of this CE has been correlated with the sense of chirality particularly in the carbohydrate series.[120,121] For example, (2S)-diol (Fig. 12.41, **25**) exhibits two CEs when dissolved in Cupra A solution: $[\theta]_{540} - 50$ and $[\theta]_{320} + 370$ both of these being associated with a complex of structure **A** (Fig. 12.41) having an absolute configuration with a negative chirality sense.[122,123] The method may be applicable to α-hydroxy acids as well.[124]

The NMR lanthanide shift reagents, such as Pr(DPM)$_3$ (DPM = dipivaloylmethane) and the nickel complex Ni(acac)$_2$ (acac = 2,5-hexanedione), induce long wavelength CE's originating in the transitions of the metallic element. These CEs may be empirically linked to the configurations of vicinal glycols and amino alcohols.[125]

Molybdenum, rhodium, and ruthenium form acylate (e.g., acetate) complexes of the general formula [Met$_2$(O$_2$CR)$_4$; Fig. 12.42, **A**] whose ligands may be exchanged in situ with or add typically nonabsorbing chiral compounds, for example, carboxylic acids, diols, amino alcohols, peptides, and nucleosides. The chiral complexes exhibit several CEs, the most reliable for determination of absolute configuration being those in the 300–400-nm range. Since the conformation of the ligands is restricted in the complex, the sign of the CE may be used for the empirical assignment of configuration of the complexing ligands (mostly bidentate) molecules in a reliable way.[6,126,127]

Secondary alcohols, alkenes, epoxides, and ethers rapidly form in situ complexes by binding to the axial position of [Rh$_2$(O$_2$CCF$_3$)$_4$] (Fig. 12.42, **A**). The CD spectra of the complexes exhibit up to five CEs in the 300–600-nm range. In the case of 2°

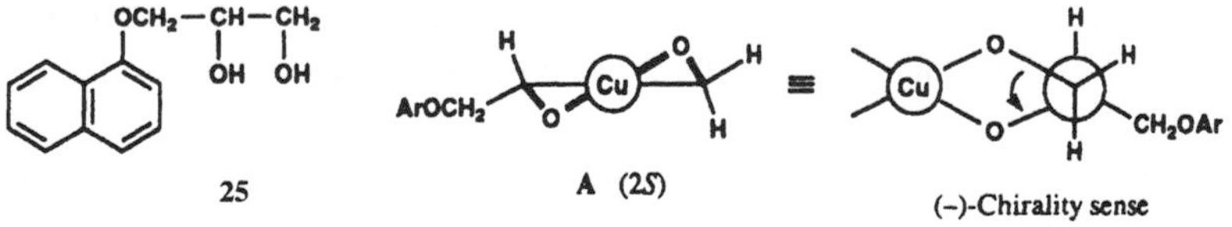

Figure 12.41. Cupra A complex of (2S)-diol **25** in the λ conformation with the aryloxymethylene substituent in the equatorial position. [Reproduced with permission from Nelson, W. L., Wennerstrom, J. E., and Sankar, S. R. *J. Org. Chem.* **1977**, *42*, 1006. Copyright © 1977 American Chemical Society.]

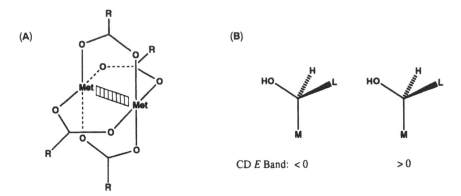

Figure 12.42. (A) Acylate complexes of type [Met$_2$(O$_2$CR)$_4$], Met = Mo^{2+}, Rh^{2+}, Ru^{2+}, Ru^{3+}, (R = CH$_3$, CF$_3$, C$_3$H$_7$); (B) relation of absolute configuration of 2° alcohol to the sign of the CE at about 350 nm (*E* band) generated by complex formation with [Rh$_2$(O$_2$CCF$_3$)$_4$]; M and L are medium and large groups, respectively. [Adapted with permission from Gerards, M. and Snatzke, G. *Tetrahedron: Asymmetry* **1990,** *1,* 221. Copyright © 1990 Elsevier Science Ltd, The Boulevard, Langford Lane, Kidlington, OX5 1GB, UK.]

alcohols, the configuration can be established from the sign of the CE at about 350 nm (*E* band) by means of empirical rules (Fig. 12.42, **B**).[128]

Liquid Crystal Induced Optical Activity and CD. "Liquid crystals" is a general term that describes intermediate phases (or mesophases) existing between the solid crystalline state and the isotropic liquid state. On warming, liquid crystals are characterized by the stepwise occurrence of reversible transitions between the solid and several possible mesophases. Many compounds possessing rodlike or, less commonly disclike, molecular structures exhibit such behavior. When such compounds are heated, the regular 3D organization of the solid gives way to a two-dimensional (2D) layered structure wherein the rods are aligned with their long axes orthogonal or tilted to the layer planes (smectic phases). On further warming, the smectic phase gives way to a turbid nematic phase in which the rods remain oriented parallel along their long axes (thread like) but the layers disappear. In the case of chiral molecules, the smectic phase on warming forms a twisted nematic (cholesteric) phase that retains local nematic packing. However, the long axes of molecules in individual layers are rotated about an axis that is perpendicular to the layer planes formed into a stack. Cholesteric phases thus have helical structure. The three types of mesophases, smectic, nematic, and cholesteric, all of which are visibly turbid, together are called liquid crystals. On further warming, both nematic and cholesteric phases give way to isotropic liquids.[129]

Cholesteric liquid crystals, formed exclusively from chiral molecules (e.g., cholesteryl benzoate), are characterized by the helix pitch *p* of the molecule (the pitch is the periodicity of twist; it is given by the distance corresponding to a 360° turn of the long axis) and by the handedness of the helix.[130] Cholesteric mesophases exhibit extremely high molar rotations (e.g., in excess of 50,000).

Cholesteric phases can apparently induce CD in *achiral* solute molecules (liquid crystal induced circular dichroism, LCICD). The cholesteric liquid crystal does not behave just as a chiral solvent; it can be shown that the induction is caused by the macroscopic helical structure of the mesophase.[131] The CD spectra of anthracene and even substituted benzenes dissolved, for example, in a mixture of cholesteryl nonanoate and cholesteryl chloride, exhibit CEs in the absorption bands of the solute.[132]

When nonracemic *chiral* solutes are dissolved in achiral nematic phases, the latter are transformed into cholesteric phases. This observation, made by Friedel in 1922,[133] was ignored for nearly 50 years. Small amounts of virtually any chiral substance, including those that do not form mesophases by themselves, can effect this transformation. One of the most widely used thermotropic liquid crystals (those whose order is altered on heating) for induction of optical activity or CD is *N*-(4-methoxybenzylidene)-4′-butylaniline (MBBA) **26**, Figure 12.43.

The induction of cholesteric phases may be observed with a polarizing microscope in the visible range, and by infrared ORD.[134] At low concentrations, the induced pitch (*p*, in μm; methods for the measurement of the pitch are described in Solladié and Zimmerman)[130] is inversely proportional to the molar concentration of the solute *c* and to its enantiomeric purity *r* (Eq. 12.13). It has been pointed out that while the sign

$$1/p = \beta_M r c \tag{12.13}$$

(handedness) and the magnitude of the helical twisting power β_M may characterize a chiral compound just as well as the specific rotation [α], the origin of the two properties is quite different. The rotation is a measure of the interaction of light with a chromophore in the molecule, while β is dependent on the nature of, and the extent of interaction between, solute and solvent molecules.[135]

As a consequence of their very high rotatory power, induced cholesteric mesophases may serve to detect or to amplify very small optical activities, for example, in compounds whose chirality is due solely to isotopic substitution,[136] or to observe the rotatory power in a very small amount of a compound.[137]

In general, it can be said that close matching of the structure and conformation of the chiral solutes and the nematic director yields the largest twisting power β_M and the best correlation of the configuration of the solute molecules with the sign of β_M. High twisting power is a sign of strong solute–solvent interaction.[130,138]

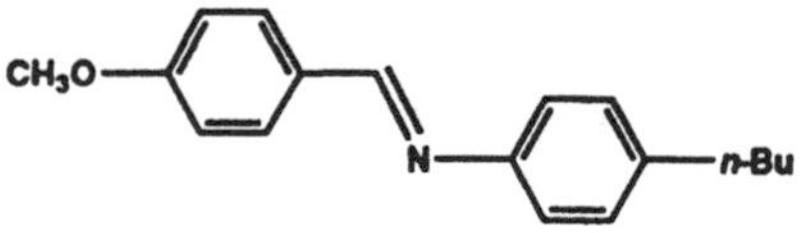

26

Figure 12.43. *N*-(4-Methoxybenzylidene)-4′-butylaniline (MBBA).

f. Circular Dichroism of Chiral Polymers

The use of CD spectroscopy, and to a lesser extent ORD, is especially important in the study of macromolecules because it is one of the few spectroscopic techniques that is able to probe the helical secondary structure that characterizes such molecules. In contrast to X-ray diffraction techniques, CD and ORD provide information about the conformations of large molecules in solution. As a consequence, applications of CD to the analysis of biopolymers, in particular, exceed those to virtually any other class of compounds. Numerous books, chapters, and review articles have summarized the many applications of chiroptical properties to the study of biopolymer conformations.[33,35,139–144] This section provides but a brief overview of the chiroptical properties of biopolymers and of synthetic polymers, a very active research area endowed with a large literature.

As we have already seen, CD is very sensitive to the interaction between neighboring chromophores. In polypeptides and in polynucleotides, for example, interaction between adjacent amide groups and adjacent aromatic nuclei, respectively, during light absorption accounts for most of the intensity of the CD bands.

Since the configurations of the constituent amino acids and nucleotides are known, the CD principally provides information about the conformations (i.e., the secondary structure) of the biopolymers that are built up from these chiral monomers. The key structural elements that are responsible for the wavelengths and intensities of CD bands are the relative orientation of the neighboring chromophores and the distance between them. Since the magnitude of CD bands that arise from interaction between chromophores falls off rapidly as the chromophores are further removed from one another, in polychromophoric systems it is only adjacent absorbing groups that need be considered in first-order analyses of CD spectra (the amplitude of exciton splitting depends approximately on the inverse square of the distance between the chromophores). These chromophores have a greater effect on CD spectra than does the primary structure, that is, the number, kind, and location of stereocenters in the backbone and side chains of chiral macromolecules.[143,145]

Biopolymers. Let us first consider the case of polypeptides and proteins. As a consequence of the conformational mobility of single bonds flanking the chiral center of amino acids, linear polypeptides spontaneously assume the principal conformations in solution shown in Figure 12.44.

These conformations are stabilized especially by *intra*molecular hydrogen bonds in the case of α helices (typically right handed in the case of polymers of L-amino acids) and by *intra*molecular as well as *inter*molecular hydrogen bonds in the case of β forms. When such conformations are not able to form, for example in polymers of a single amino acid having repeating R groups bearing like charges, such as poly(L-glutamate) (at pH 8) or poly(L-lysine) (at pH 7), the polymer molecules assume flexible and partially disordered arrangements called random coils. When the charges are neutralized, for example, at pH 4.5 for poly(L-glutamic acid) (Fig. 12.45a) and at pH 12 for poly(L-lysine) (Fig. 12.45b), conformations approximating α helices form spontaneously.

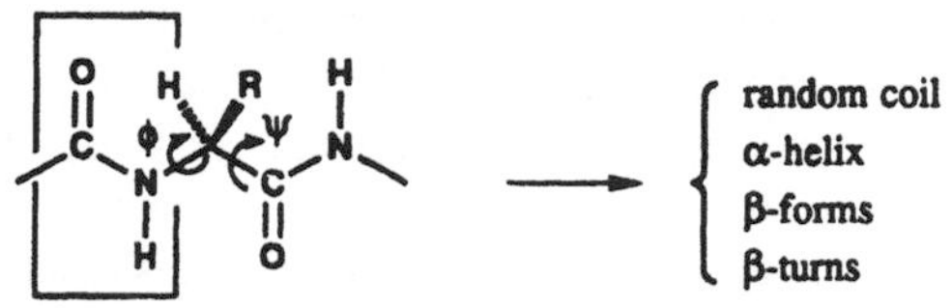

Figure 12.44. Organization of polypeptides into their principal conformational forms. The atoms within the rectangle constitute a rigid planar unit.

The CD of polypeptides in the random coil conformation does not approximate the sum of the monomer (dipeptide) contributions to the CD. This finding reflects the fact that polypeptides are partially organized even in the random coil form.[139] The term random coil is therefore a misnomer. Designations, such as unordered (or aperiodic) form(s), are to be preferred.

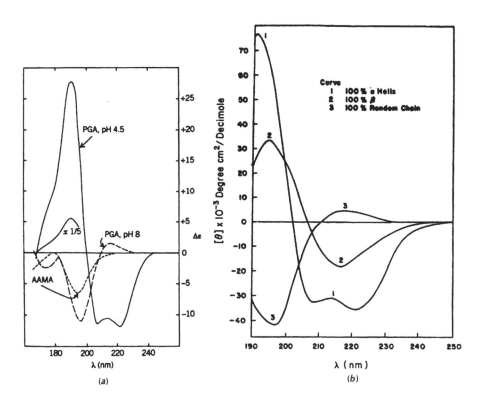

Figure 12.45. (a) CD of poly(L-glutamic acid) (–) and poly(L-glutamate) (--): α helix (PGA, pH 4.5) and random coil (PGA, pH 8). CD of N-acetyl-L-alanine-N′-methylamide (AAMA,----). [Reproduced with permission from Johnson. W. C., Jr., and Tinoco, I., Jr. *J. Am. Chem. Soc.* **1972**, *94*, 4389. Copyright © 1972 American Chemical Society]. (b) CD of poly(L-lysine): (1) α helix; (2) β form: (3) random coil. [Reproduced with permission from Greenfield, N. and Fasman, G. D. *Biochemistry* **1969**, *8*, 4108. Copyright © 1969 American Chemical Society.]

The organization of peptides into stereoregular forms begins when the number of peptide units in the polymer reaches a range of 5–12.[146] As the polypeptide molecules organize themselves by extending the chains and by hydrogen bonding intermolecularly (between NH and C=O groups) with like chains either in parallel or antiparallel ways (β forms), the CD bands increase in intensity and change in sign, the CD spectrum becoming virtually enantiomeric with that of the random coil form (Fig. 12.45*b*).

When intramolecular hydrogen bonding prevails, coiling of the polymer chain into an approximate α helix takes place. Under such conditions, there is a dramatic increase in the intensity of the CE in the vicinity of 190 nm. The peptide backbone itself having become helical, absorption of radiation by numerous properly oriented like chromophores is enhanced giving rise to large CEs by the interaction of the chromophores with near neighbors (exciton coupling; see below and Section 12-4.d). The CD spectroscopy allows the presence of the α helices in proteins to be inferred with greater certainty than that of any other secondary structural feature.[147]

The CD spectrum of poly(L-alanine), which is a fairly typical homopolypeptide, in the α helix form is shown in Figure 12.46. The experimental (boldface) CD spectrum exhibits three distinctive extrema, at 191, 207, and 221 nm. The spectrum can be

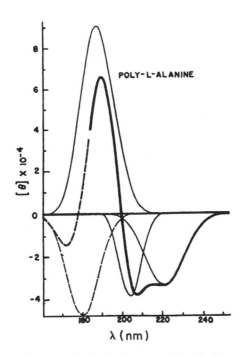

Figure 12.46. Resolved CD spectrum of the pure helical form of poly(L-alanine) in trifluoroethanol–trifluoroacetic acid (98.5:1.5 v/v). The bold faced curve represents the experimental data. The 180-nm negative CD band (––) is inferred to facilitate and improve the curve resolution. [Reproduced with permission from Quadrifoglio, F. and Urry, D. W. *J. Am. Chem. Soc.* **1968**, *90*, 2755. Copyright © 1968 American Chemical Society.]

deconvoluted into component (theoretical) Gaussian curves the addition of which approximates the experimental spectrum. This analysis reveals the presence of the negative exciton couplet due to splitting of the $\pi–\pi^*$ transition as found in right-hand α helixes of L-polypeptides. The long wavelength negative band at 221 nm is due to the carbonyl $n–\pi^*$ chromophore. In the absence of solvent interactions or of strongly absorbing (chromophoric) side chains, as in phenylalanine, the CD spectra of many polypeptides are essentially identical well into the vacuum UV region (to ca. 140 nm).

A second major stereochemical feature distinguishable in CD spectra is the so-called β sheet conformation. The latter, exemplified by poly(L-lysine) (at pH 11.1), consists of adjacent molecules aligned side by side either in parallel or in antiparallel senses. Beta sheets exhibit a negative band at about 216 nm, a positive band between 195 and 200 nm, and a negative band near 175 nm (Fig. 12.45b, curve 2). The CD spectral features of β forms are subject to more variation than are those of α helix forms. At lower pH, poly(L-lysine) (pH 5.7) becomes unordered, and, although strong CD bands are still evident especially in the vicinity of 200 nm, these are of limited usefulness in conformational analysis (Fig. 12.45b, curve 3).[145]

Interest in the measurement of chiroptical properties increased beginning in the late 1950s when it became evident that ORD (and later CD) spectra could provide measures of the extent of α helix and β forms present in polypeptides and proteins in solution.[148]

Quantitative analysis of the secondary structure of proteins[149] is generally based on treatment of the CD spectrum as a linear combination of ellipticities at specified wavelengths contributed by each conformational form.[150,151] Such analyses have been refined to include additional conformations that are present in proteins, such as β turns (structural domains in which peptide chains reverse direction).[152] Equations of the form

$$[\theta]_\lambda = f_H[\theta]_H + f_\beta[\theta]_\beta + f_t[\theta]_t + f_R[\theta]_R \qquad (12.14)$$

for such analyses have now been incorporated in the data handling software of contemporary CD spectrometers. In this equation, $[\theta]_\lambda$ is the mean residue ellipticity at a given wavelength, that is, the ellipticity of the macromolecule per peptide unit. The other $[\theta]$ terms are the corresponding ellipticities for the α-helical H, β-form β, β-turn t, and unordered R forms, respectively. Simultaneous solution of the equation with ellipticity data measured at several wavelengths leads to the f values, f being the fraction of each conformational form present in the protein.

Originally, Fasman used the CD of synthetic homopolymers [e.g., poly(L-lysine)] that may be prepared in each of the three distinct conformational forms: α helix, β pleated, and random coil, as reference spectra for such analyses. In spite of numerous assumptions inherent in this quantitative treatment, for example, lack of interference from side-chain chromophores,[153] estimates of protein composition calculated in this way are in good agreement with fractional conformations determined by X-ray diffraction.[145,151] Quantitative analysis of this type has been extended into the 205–165-nm wavelength range.[154]

In an alternative approach developed by Saxena and Wetlaufer,[155] reference spectra for the several conformational forms were computed from the CD spectra of proteins (lysozyme, myoglobin, and ribonuclease were used in the original treatment) whose 3D structures, including the fraction of each conformational domain present, were established by X-ray diffraction (Fig. 12.47).

The latter approach overcomes two limitations inherent in that of Fasman, namely, (a) reference homopolypeptides used in the analysis are of much greater molecular weight than are the α helix and β form regions of typical globular proteins, and (b) CD spectra of globular proteins incorporate contributions due to the interaction between α-helical and β-form fragments within a given protein. Use of proteins to generate the basis spectra, of necessity, compensates for differences in the lengths of the conformational forms present in proteins, and to some extent, also for the interaction between segments that have different forms.[145]

Other macromolecular conformations of polypeptides and proteins having characteristic CD spectra that have been examined closely are the two helical froms (one right handed and the other left handed) of poly(L-proline), and the double and triple helical forms of the proteins myosin (muscle protein) and collagen, respectively. Polypeptides of proline and its derivatives, in which hydrogen bonding is precluded,

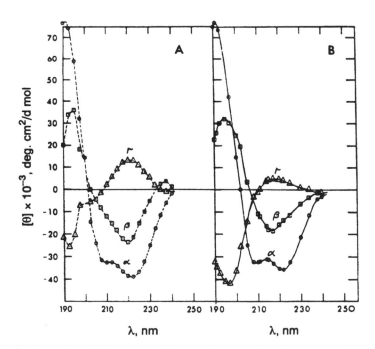

Figure 12.47. Comparison of CD spectra of three conformational modes computed (A) from X-ray diffraction data and CD spectra of lysozyme,. myoglobin and ribonucleases, and (B) from CD spectra of the pure conformational forms of poly(L-lysine) (*r* = random coil). [Reproduced with permission from Saxena. V. P. and Wetlaufer. D. B. *Proc. Natl. Acad. Sci. USA* **1971**, *66*, 969.]

form stable conformational forms as a result solely of restricted rotation about the polypeptide backbone.[139]

Although nucleic acid and polynucleotide CD spectra, for example, those of transfer-ribonucleic acid (tRNA) and deoxyribonucleic acid (DNA), have been analyzed in ways similar to those described for proteins, the process is much more complicated. These polymers form triple, double, as well as single helices because of differences among the strongly absorbing monomer components and because the heterocyclic bases that are present strongly bind to one another. Not only the nature of the bases present (there are at least four distinct chromophores: adenine, guanine, cytosine, and uracil or thymine) but also their sequences significantly affects the DC. In spite of these complications, some impressive computations of CD spectra of polynucleotides have been achieved.[145]

Figure 12.48 illustrates the CD spectrum of the polynucleotide polyadenylic acid (poly A, a polymer of adenosine 5′-phosphoric acid) in the region in which the planar but chirotopic 6-aminopurine chromophore absorbs. The sign of the CD between 260 and 280 nm is inverted and its ellipticity is increased nearly 10-fold relative to that of the monomer (adenylic acid, Fig. 12.48). Most of the CD intensity can be accounted for by interaction between adjacent (nearest neighbor) bases that are oriented (rotated along the polymer backbone) so that they form a stack. Much of the increased CD intensity is lost on denaturation.[153]

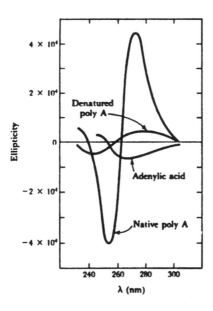

Figure 12.48. The CD spectra of adenylic acid, native polyadenylic acid (poly A), and denatured poly A. [Reprinted with permission from Freifelder, D. *Physical Biochemistry*, p. 467. Copyright © 1976 W. H. Freeman and Company, New York.]

Even in systems as simple as dinucleoside phosphates (containing a chromophore dimer), for example, adenosyl adenosine (ApA), the ORD revealed stacking of the bases and a conformation that obviously is the beginning of a right hand helix.[156]

A hexanucleotide composed of L-deoxyribose in lieu of the naturally occurring D-deoxyribose (the only chiral structural component of DNA) has been synthesized. This DNA oligomer exhibits a CD spectrum that is the mirror image of the synthetic (natural) D hexamer. The CD spectrum is known to reflect principally the conformation of the hexanucleotide; it exhibits the same sign inversion at 295 nm signaling a conformational transition (in the case of the L hexamer) from left-handed to right-handed double helical conformation under high salt conditions (the inverse of that of the D hexamer). It is inferred that L-DNA and D-DNA must possess the same conformation and dynamic conformational properties, except for the sense of chirality.[157]

Circular dichroism spectra provide information about prosthetic groups, that is, defined structural moieties, such as hemes, that are attached to proteins. Analysis of prosthetic groups requires that they absorb in spectral regions that do not interfere with polypeptide backbone absorption bands. For example, reduction of cytochrome and of hemoglobin leads to distinctive ellipticity changes (in ferro- vs. ferri- and in oxy- vs. deoxy-, respectively) in the 240–360-nm region where the heme moiety absorbs.

Heme–heme interaction due to stacking of the large and relatively flat prosthetic groups in proteins is revealed by CD as is the interaction between heteroaromatic moieties in nucleotides. Information about the conformation of macromolecules in biomembranes is also derived from CD spectra. The bulk of the purple membrane in halobacteria is composed of the protein bacteriorhodopsin. The CD spectrum is especially sensitive to the physical state of the membrane, that is, whether its suspension has or has not been sonicated or whether the membrane has been dissolved. Application of correction factors to the CD ellipticities makes it possible to obtain reliable estimates of the amount of α helix present in the membrane.[144]

Another very useful application of CD is the detection of *changes* in conformation, for example, on denaturation (visible changes associated with the unfolding of a peptide chain or its reorganization into a new conformation), on reaction of a chiral polymer with a chemical agent (including the solvent and the pH), or on binding of a substrate, inhibitor, or coenzyme to an enzyme.

Synthetic Polymers.   While they contain stereocenters, the stereoregular (isotactic and syndiotactic) diastereomeric forms of vinyl polymers, such as polypropylene, $\{CH_2—CH(CH_3)\}_n$, are considered to be achiral. Several explanations for this view have been advanced. While this subject cannot be examined in detail here, the explanations involve considerations of symmetry elements (the molecules of stereoregular polymers may, in some cases, be considered to be centrosymmetric), and/or the concept of cryptochirality (Section 12-5.a).[158,159]

Poly(methylene-1,3-cyclopentane), an isotactic polymer whose main chain incorporates rings with trans-oriented attachment atoms, is chiral (Fig. 12.49). Optically active samples have been obtained by enantioselective cyclopolymerization of 1,5-hexadiene. The magnitude of its rotation, $[\Phi]_D^{20}$ 22.8 (c = 7.8, CHCl$_3$) per

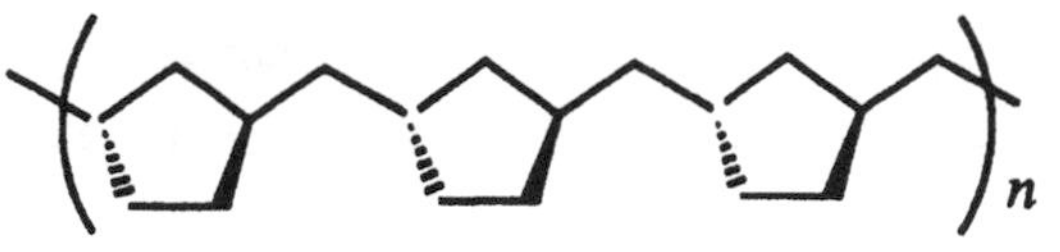

Figure 12.49. Isotactic poly(methylene-1,3-cyclopentane). The nonracemic polymer produced actually contains only about 68% trans rings.[160]

monomer unit, relative to that of optically active *trans*-1,3-dimethylcyclopentane, $[\Phi]_D^{20}$ 3.1, suggests that part of the rotation magnitude of the polymer is conformational in origin.[160]

In contrast, atactic vinyl homopolymers, even when prepared from achiral monomers, are chiral. Based on statistical considerations, samples of such polymers containing the shorter chains (degree of polymerization, $n < 60$) are "conventionally racemic" (the probability of finding equal numbers of enantiomeric pairs is high) while those having longer chains, that is, having a high degree of polymerization ($n > 70$), would be expected to be cryptochiral, that is, optically inactive as a result of intermolecular compensation of many different enantiopure diastereomers whose rotations are expected to average to zero.[161]

Most stereoregular macromolecules assume helical conformation in the solid (crystal) state[159,162] and chiroptical properties can be, and have been, measured in the solid state even for isotactic polypropylene.[159] However, in solution, stereoregular polymers behave essentially as random coils.[163] In solution such polymers do not usually exhibit chiroptical properties associated with the presence of helices: formation of the M and P helical conformations is equally probable and the helices are unstable (such systems are stochastically achiral; see Section 12-2.a). Stereoregular vinyl polymers exemplify the dictum that the application of the term chiral to any system or molecular model "depends on the conditions of measurement".[12]

Only when the side chain R in $\{CH_2-CHR\}_n$ is very bulky does restricted rotation about CC single bonds of the polymer backbone stabilize helical conformations yielding conformationally rigid polymers even in solution. In any event, the observation of CD depends not only on the presence of an excess of one stable enantiomeric polymer conformation but also on the absorption wavelength of its chromophores. Two types of conformationally rigid synthetic chiral polymers (atropisomeric polymers)[164] that are devoid of chiral side-chain groups yet exhibit chiroptical properties are illustrated in Figure 12.50.

Polymerization of (S)-(+)-*sec*-butyl isocyanide $CH_3CH_2CH(CH_3)N{=}C$ generates an optically active polymer (Fig. 12.50, **A**, R = *sec*-butyl). Simple molecular modeling considerations [assuming that the steric fit in this poly (S)-(+)-*sec*-butylimino-methylene is the controlling parameter in determining the screw sense of the polymer helix] led to the conclusion that levorotatory fractions $[\alpha]_D < 0$ are composed mostly of P helixes. Calculation of the sign of the CD band in the vicinity of 300 nm led to the same conclusion.[165,166]

An analogous polymer (with degree of polymerization, $n \cong 20$, hence actually a large oligomer) devoid of chirotopic atoms in the side chain,

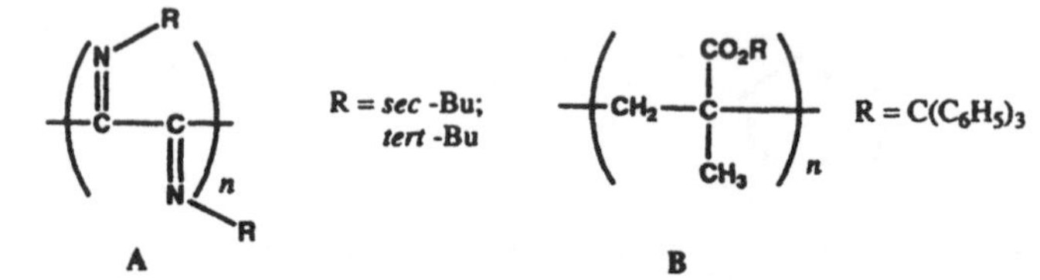

Figure 12.50. Conformationally stable helical polymers. (A) poly(*sec*- and *tert*-butyliminomethylene); (B) poly(triphenylmethyl methacrylate).

poly(*tert*-butyliminomethylene) (Fig. 12.50, A, R = *tert*-butyl), could be resolved chromatographically using the optically active *sec*-butyl analogue (see above) as an enantioselective stationary phase. The apparent sole stereochemical element of this remarkable macromolecule is its helicity (though this is an obvious oversimplification since the nitrogen atoms in the imino side chains are stereogenic, alkyl groups having syn-anti relationships about the >C=N bond).[167] If one assumes that "parallel screws have a smoother mutual fit," then optically active fractions of the *tert*-butyl polymer consist mainly of helical molecules having a like helical sense. Comparison of the experimental CD spectrum with the calculated spectrum (see above) leads to the conclusion that fractions with $[\alpha]_D < 0$ consist mainly of helical molecules having a *P* helix sense.[165] The same conclusion is reached by analysis of the polymerization mechanism.[168]

Stable helical conformations have been inferred also for the homopolymer of triphenylmethyl methacrylate (Fig. 12.50, **B**). When obtained by polymerization of triphenylmethyl methacrylate in the presence of (–)-sparteine–butyllithium, one of the helical forms predominates. Applications of the highly levorotatory polymer as an enantioselective stationary phase in HPLC are described in Chapter 7. Hydrolysis of polymer **B** and methylation of the resulting poly(methacrylic acid) yields optically inactive isotactic poly(methyl methacrylate).[169] It thus appears that methacrylates with less bulky pendant ester groups are unable to maintain stable helical conformations.[170,171]

Polyisocyanates, $-\{NR_2-CO\}_n$, readily form helical conformations both in the solid state and in solution.[172] Since these helical forms readily undergo enantiomerization, they do not exhibit chiroptical properties. As has already been pointed out (Section 7-2.d), when dissolved in (*R*)-2-chlorobutane, racemic poly(*n*-hexyl isocyanate) undergoes an asymmetric transformation of the first kind, that is, the equilibrium between the two enantiomers is displaced and the polymer exhibits a positive CD at about 250 nm (a spectral region in which the chiral solvent is transparent) attributed to an excess of right handed helix.[173]

A large optical rotation is exhibited by poly[(*R*)-1-deuterio-*n*-hexyl isocyanate] $[\alpha]_D^{25}$ –367 (CHCl₃). It has been suggested that the helical sense, that is, the ratio of enantiomeric polymer helixes in this polymer, is strongly influenced by a cooperative conformational equilibrium isotope effect due to a difference in energy of α-D versus α-H involving many deuterium atoms.[174–176]

Vinyl polymers containing chiral substituents may, of course, be prepared in optically active forms, as may polymers containing heteroatoms, for example, polypropylene oxide, $-\!(CH_2\!-\!CH(CH_3)\!-\!O)_n$, and/or double bonds in the polymer backbone.[177] Space limitations preclude a detailed exposition of the chiroptical properties of such polymers. For a summary of such properties, see Farina.[159]

It is only since the 1980s that CD studies have permitted the observation of CEs in the vacuum UV region of hydrocarbon polymers devoid of aromatic groups. The CD spectra of films of poly-(S)-4-methyl-1-hexene and poly-(R)-3,7-dimethyl-1-octene exhibit a CD band at 158 nm that is ascribed to conformations containing helical segments having a common helix sense.[178]

12-5. APPLICATIONS OF OPTICAL ACTIVITY

a. Polarimetry

The actual measurement of optical activity may be carried out with either manual or photoelectric polarimeters. Manual polarimeters have changed relatively little since the first instruments were developed some 140 years ago.[16] Photoelectric polarimeters, the type nowadays commonly found in research laboratories, have greatly reduced the tedium formerly associated with the measurement of optical rotation with manual instruments. Moreover, photoelectric polarimeters are much more accurate and sensitive, permitting the rapid and meaningful recording of quite small absolute rotation values α to about $\pm 0.002°$ and, consequently, the use of smaller samples. Polarimeters fitted with microcells may even serve advantageously as detectors in HPLC resolutions.[179–182] A laser-based polarimetric HPLC detector has been shown to be sensitive to as little as 12 ng of sample.[183] For the advantages of the use of CD detectors in HPLC, see Salvadori et al.[184] and Mannschreck.[180]

The laser polarimetric detector has been adapted to HPLC analysis not only of optically active samples but also, in a different way, as a universal detector for achiral, that is, optically inactive, substances. In this technique, termed "indirect polarimetry," the mobile phase is optically active, containing, for example, (–)-2-methyl-1-butanol or (+)-limonene, and the detector output due to the optically active solvent is zeroed. Under these conditions, any optically inactive fraction passing through the detector cell is sensed since the concentration of the optically active solvent is thereby reduced. The response of the detector is universal, like that of a refractive index detector, but is more sensitive than the latter.[185–187] The simultaneous measurement of absorbance and optical rotation during the liquid chromatographic resolution of chiral substances on enantioselective stationary phases make possible the determination of the enantiomer composition in spite of extensive peak overlap.[102,179,188] For a brief discussion of polarimetry and its instrumentation, see the review by Lyle and Lyle,[189] for a more extensive treatment, see Heller and Curmé.[190]

The measurement of optical activity has traditionally been the method of choice to establish the nonracemic character of a sample of a chiral compound and, when quantitatively expressed as a ratio $[\alpha]/[\alpha]_{max}$, of its enantiomeric composition (optical purity). In contemporary practice, chiroptical measurements have to a large extent been replaced by NMR and by chromatographic analyses for the purpose of determining enantiomeric compositions (cf. Chapter 6). Nevertheless, the use of $[\alpha]$ for this and other purposes continues. The reasons are that the measurement is easy to carry out and one may wish to compare experimental values of $[\alpha]$ with those in the literature. While substantial collections of optical rotation data exist, for example those in various handbooks and chemical supplier catalogs, it should not be assumed that values of $[\alpha]$ provided are those of enantiomerically pure compounds. A consistent set of specific rotation data for amino acids including temperature coefficients has been compiled by Itoh.[191]

Optical activity has been used (a) to determine if a given unknown substance is chiral or achiral; (b) to ascertain the enantiomeric composition of chiral samples, either qualitatively or quantitatively; (c) to study equilibria; the mutarotation or change in rotation of equilibrating stereoisomers as a function of time is one such phenomenon.[53,192]; and (d) to study reaction mechanisms. Other chiroptical techniques, namely, ORD and CD, have increasingly replaced polarimetry in these applications, especially in the past 20 years. For reviews of applications of polarimetry, see Lowry,[16] Eliel,[53] Legrand and Rougier,[38] and Purdie and Brittain.[29]

Polarimetric methods remain useful for quality control in pharmacology and food-related industries[193,194]; there are also numerous applications in forensic, clinical, pharmaceutical, and agricultural chemistry.[29] The percentage of sucrose in commercial samples is still being determined by polarimetry (saccharimetry); in the trade this is called "direct polarization." The cost of raw sugar is based on the results of the polarimetric analysis; if the analyte solution is dark, the raw sugar is first clarified by precipitation of the dark side products with basic lead acetate. An example of application (d) (see above) is the methanolysis of the tosylate of (R)-(+)-$C_6H_5CH_2SCH_2CH(CH_3)CH_2OH$ that leads to a partially racemized methyl ether. The intervention of a cyclic (symmetrical, and hence achiral) intermediate, via neighboring group participation, was inferred.[195]

The magnitude of rotation α, in degrees, fundamentally depends on the number of molecules of the sample being traversed by the linearly polarized light as well as on their nature, hence optical activity is not a colligative property. Values of α are affected by many variables, among which are wavelength, solvent, concentration, temperature, and presence of soluble impurities. It must also be mentioned that large molecules, such as proteins, may spontaneously orient themselves in solution, and consequently no longer be isotropic. The measurement of the rotatory power of such substances may then be complicated by the occurrence of linear dichroism.[190]

As already pointed out in Section 1-3, rotation magnitudes are usually normalized to a quantity called the *specific rotation* $[\alpha]$ that was introduced by Biot in 1835,[16,196] Eq. 12.15,

$$[\alpha] = \alpha/\ell\rho = \alpha/\ell c \qquad (12.15)$$

where ℓ is the length of the cell in decimeters, ρ (for undiluted liquids) is the density in grams per milliliter (g mL^{-1}), and c is the concentration also in grams per milliliter. The units of $[\alpha]$ are 10^{-1} deg cm^2 g^{-1} (see also Eq. 1.1 and Section 1-3).

Comparison of specific rotations of homologues, and of organic compounds generally, is more significant if a modified Biot equation is used in which the quantity called the *molar rotation* $[\Phi]$ depends on the number of moles of substance traversed by the linearly polarized light, Eq. 12.16:

$$[\Phi] = [\alpha]M/100 \qquad (12.16)$$

where M is the molecular weight. The units of $[\Phi]$ are 10 deg cm^2 mol^{-1} (see also Eq. 1.2).[9]

The cumulative effect of the above-mentioned variables on $[\alpha]$ or $[\Phi]$ is potentially very large. A practical consequence is that precise reproduction of published rotation values, from laboratory to laboratory, or even from day to day in the same laboratory, is difficult to achieve.[189] This sensitivity to numerous variables[197] and the absence of major tabulations of critically evaluated absolute rotation data is responsible for the decreasing reliance on optical activity as a measure of enantiomeric composition.

The sign of rotation is often the only experimental criterion for the specification of configuration. It is important to stress how frequently and how easily this property may change for a given substance, for example, (*R*)-2-hydroxy-1,1′-binaphthyl has $[\alpha]_D^{20}$ +4.77 ($c = 0.86$, tetrahydrofuran (THF)), +13.0 ($c = 1.12$, THF), and $[\alpha]_D^{20}$ −5.2 ($c = 1.03$, CH$_3$OH).[198] Even tartaric acid, one of our configurational standards, does not exhibit an invariant sense of rotation: $[\Phi]_{578}$ −12.9 (24°C), −0.9 (57°C), and +10.8 (94°C) (all $c = 10$, dioxane); +21.3 (24.7°C, H$_2$O), +6.6 (24°C, EtOH), +0.3 [25.3°C, *N,N*-dimethylformamide (DMF)], −12.9 (24°C, dioxane), and −14 (25.2°C, Et$_2$O) (all $c = 10$) [both sets of data measured on (*R,R*)-tartaric acid].[199] When the sign of rotation shows a strong solvent, concentration, wavelength, or temperature dependence, the association of such sign with a given configurational descriptor is arbitrary. This serves to emphasize the crucial importance of specifying and recording the precise experimental conditions of measurement of optical rotations and of chiroptical properties in general. In particular, confusion can arise when the sign of rotation is related to a given configuration and the solvent is not specified.

Occasionally, specific rotations of samples are very small. When that situation arises, especially in resolutions or in stereoselective syntheses in which strongly rotating reagents are used, exceptional care must be taken to insure that the rotation of the product is not spurious. A small amount of impurity having a large $[\alpha]$ may overwhelm (or at least seriously falsify) the rotation of a sample having a small $[\alpha]$.[200,201] Achiral contaminants, particularly solvents, will also affect the optical activity of a sample.[189,197]

Traces of achiral compounds, including solvent residues, normally would be expected to reduce $[\alpha]$ (by dilution of the sample) and hence to artificially lower the optical activity of nonracemic samples (but not the enantiomeric purity of the chiral solute). However, the converse may also be observed, for example, the optical activity of 1-phenylethanol at 589 nm is increased when acetophenone, a possible contaminant, is present in the sample.[202] The enhanced optical activity of the alcohol comes about because chiroptical properties are *induced* in the achiral ketone by the alcohol; in the example, the induction is superimposed on and swamps the typical and opposite dilution effect (cf. Section 12-4.e).

The case of low rotation warrants further comment. There are two situations in which no optical rotation is observed with enantiomerically enriched samples: (a) the experimental device used (by implication this includes the eye) is of insufficient sensitivity, and (b) the specific conditions of measurement are such that α is, in fact, accidentally equal to zero.

In the first situation, the measurement threshold is such that there is no clear signal (rotation) distinguishable from instrumental noise. The condition is one of *operational null*. Progressive dilution of a solution of an optically active compound eventually leads to a sample that is no longer palpably optically active when the operational null threshold is crossed. Such a sample no longer reveals its enantiomeric excess; the sample is said to be *cryptochiral*.[12]

Notable examples of enantiomerically enriched compounds that are cryptochiral as a consequence of inherently low optical rotation magnitude are shown in Figure 12.51. The cryptochirality condition may conceivably be lifted by measuring a different chiroptical property, for example, vibrational circular dichroism (VCD).

The second type of cryptochirality arises when the measurement of rotation accidentally takes place in the vicinity of a change in sign (see below and above). For an example involving a change in concentration of dimethyl α-methylsuccinate, $CH_3O_2CCH(CH_3)CH_2CO_2CH_3$, see Berner and Leonardsen.[203] At a certain

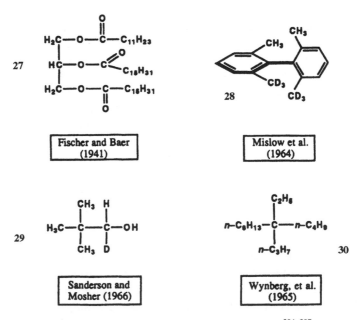

Figure 12.51. Compounds illustrating cryptochirality.[204–207]

concentration, the measured rotation is necessarily zero (crossover point) and the sample is then accidentally cryptochiral. Note that a distinction between stochastic achirality (cf. Section 12-2.a) and cryptochirality cannot be made unless the former be lifted by a change in measuring device or the latter by a change in conditions, the latter being easier to achieve.

The dependence of optical rotation on the wavelength of the light, ORD, has been discussed in Sections 12-2.b and 12-4.

Effect of Temperature.   The effect of temperature on chiroptical properties may be ascribed to the following phenomena[38]: (a) changes in density of the solute and/or the solvent that alter the number of molecules being observed; (b) changes in the population of vibrational and rotational energy levels of the chiral solute; (c) displacement of solute–solvent equilibria; (d) displacement of conformational equilibria; and (e) aggregation and microcrystallization of the chiral solute (cf. enantiomer discrimination, Section 6-2).

In general, $[\alpha]$ changes 1–2% per degree Celsius, but larger changes (up to 10% per degree Celsius) are not unknown, for example, $[\alpha]_D$ of aspartic acid, $HO_2CCH(NH_2)CH_2CO_2H$, in water ($c = 0.5\%$) is 4.4 at 20°C, 0 at 75°C, and −1.86 at 90°C. The change in sign at 75°C (temperature of cryptochirality, see above) is noteworthy.[208]

Strong dependence of the optical rotation on temperature may be found even among hydrocarbons, for example, 3-phenyl-1-butene whose neat rotation, $[\alpha]_D^{22} - 5.91$, for the enantiomerically pure R enantiomer increases linearly 0.18°/°C from 16°C to 29°C. Here, it is likely that the temperature exerts a strong conformational bias.[209]

Effect of Solvent.   The "nonspecific" influence of solvent on the specific rotation may be corrected by calculation of a quantity called the specific rotivity Ω' that includes the refractive index of the solvent n_s [190]:

$$\Omega' = [3\alpha]/(n_s^2 + 2) \qquad (12.17)$$

Several examples of dramatic changes in the angle of rotation as a function of solvent have been given above. Many instances of changes in sign of $[\alpha]$ have also been recorded for amphoteric substances, such as the amino acids, as the pH is changed.[208] An exceptional example of the effect of solvent on $[\alpha]$ is given in Figure 12.52. Given examples such as these, it is disconcerting how frequently the mention of the solvent is omitted from experimental descriptions of the optical rotation.

Care in choosing the solvent to be used in the measurement of $[\alpha]$ is necessary in view of the several specific types of interaction that are possible between solute and solvent. In general, one recognizes the intervention of hydrogen bonds when oxygen-containing solutes, such as carboxylic acids, aldehydes and ketones, and alcohols, are dissolved in hydroxylic solvents; in some cases reactions, such as hemiacetal formation, may occur. In addition, dipole–dipole interactions and changes

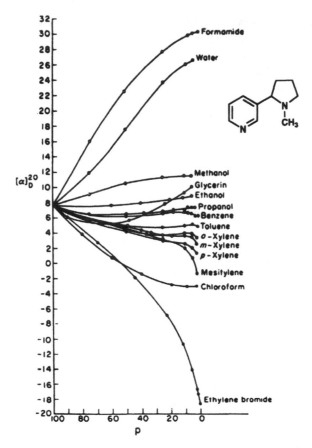

Figure 12.52. Specific rotation of nicotine in various solvents (p = concentration of solute in grams per 100 grams of solution). At p = 100, the "bulk" rotation $[\alpha]_{100}$ should be a constant, as observed, and at p = 0, $[\alpha]$ should tend to the intrinsic rotation $\{\alpha\}$ (p. 592). [Adapted with permission from Winther, C. *Z. Phys. Chem.* **1907**, *60*, 621.]

in conformer populations are important sources of solvent-induced variations in rotation magnitude.[189]

The effect of intermolecular solute association of polar solutes on $[\alpha]$ in nonpolar solvents has already been pointed out (see above). Solute–solute association effects may be leveled out or suppressed in polar solvents by competition with (concentration-independent) solute–solvent association. Polar solvents, such as ethanol, may break up solute–solute association leading to a smaller concentration dependence of $[\alpha]$, as is found with nicotine (Fig. 12.52). Such findings illustrate the desirability of using methanol or ethanol as a solvent in polarimetry.

In some instances, hydrogen bonding is known to be responsible for changes in $[\alpha]$ with concentration and/or solvent. Compounds **31** and **33** (Fig. 12.53) exhibit a remarkable solvent dependence of the sign of $[\alpha]_D$ for the *RR/SS* (syn) **31** diastereomer that is not found in the case of the *RS/SR* (anti) diastereomer **33**. The sign of $[\alpha]_D^{20}$ of

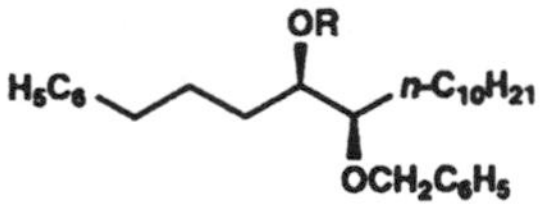

31 (R = H) *R,R* shown
32 (R = Ac)

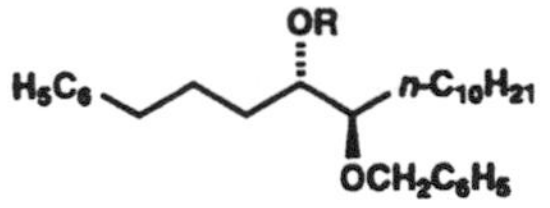

33 (R = H) *S,R* shown

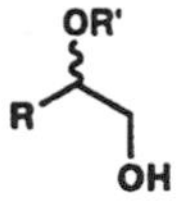

34

Figure 12.53. Structures **31–34**.

(4*R*,5*R*)-**31** is (+) in methanol and (−) in chloroform. This difference has been ascribed to a conformational change: The predominant methanol-solvated (OH/OCH$_2$C$_6$H$_5$) anti conformer gives way to a (OH/OCH$_2$C$_6$H$_5$) gauche intramolecularly hydrogen-bonded conformation in chloroform. Such sign reversal is not seen in the benzyl ether–acetate derivative **32**, in the corresponding diol (the latter appears to prefer the gauche conformation regardless of solvent) or in the diol acetonide. Reversal of the sign of [α]$_D$ would seem to be precluded in the predominant zigzag (all-anti) conformation of the molecular skeleton. A similar sign reversal was observed in a series of 2-alkoxy alcohols **34** (Fig. 12.53) presumably for the reasons advanced above. The free diol (*S*)-1,2-dodecanediol exhibits [α]$_D^{20}$ −10.1 (EtOH) but +0.9 (CHCl$_3$). This suggests that here, too, the intramolecularly hydrogen-bonded conformer prevails in CHCl$_3$.[210]

A particularly clear-cut example of a conformational equilibrium that is responsible for changes in chiroptical properties over the range of 210–350 nm is shown by ketone (+)-**35** (Fig. 12.54). As the solvent is changed from cyclohexane to acetonitrile or methanol, the effect of increasing solvent polarity on the dipole–dipole repulsion (as well as solvation effects) between the adjacent permanent dipoles (C=O and C–Br) causes a conformational change: The bromine changes from axial to equatorial and significant changes in the CD ensue.[211] Striking changes in [α]$_D$ of propylene oxide, including sign reversal, are observed as the solvent is changed from benzene to water. In this instance, we are cautioned against ascribing the effect of solvent directly to conformational changes.[212]

Effect of Concentration.   Equation 12.15 (p. 586) suggests that specific rotation should be independent of concentration. It is not hard to find evidence that this

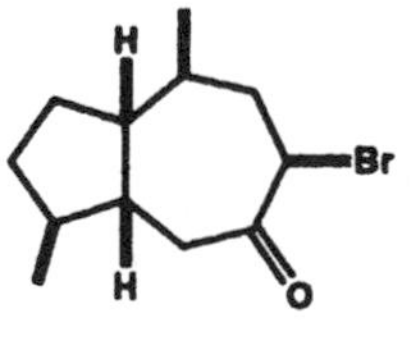

35

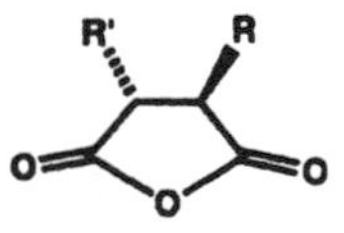

36

Figure 12.54. Structures **35** and **36**.

constancy holds only over very narrow concentration ranges and, in some solvents, not at all (the example of nicotine is found in Fig. 12.52; other examples may be found in the book by Lowry).[16]

As early as 1838, Biot suggested that the specific rotation followed a linear relationship, such as that of Eq. 12.18,

$$[\alpha] = a + bc \tag{12.18}$$

where a and b are constants and c is the concentration.[16] Constant a has been equated with a new quantity called the "intrinsic rotation" $\{\alpha\}$, a true constant corresponding to the specific rotation in a given solvent at infinite dilution; $[\alpha]_{c\to 0} = \{\alpha\}$.[190] Obviously, $\{\alpha\}$ can only be calculated since experimentally, as the concentration is reduced to zero the rotation must vanish.

The intrinsic rotation is the specific rotation for a system free of solute–solute interactions. However, solute–solvent interactions are maximized in $\{\alpha\}$, which can differ greatly from solvent to solvent, for example, for nicotine (Fig. 12.52). Since, obviously $\alpha = 0°$ at 0% solute, the values at very low concentrations must be extrapolated. Conversely, as the concentration of solute increases, solute–solute interactions become dominant and the effect of the solvent eventually vanishes: $[\alpha]_{c\to 100} = [\alpha]_{neat}$ tends to a constant value that is identical for all solvents.

A recent report on the specific rotation of (S)-2-phenylpropanal, $CH_3CH(C_6H_5)CH=O$, makes it clear that even at relatively low concentrations in benzene ($c = 1–4$), changes of the order of 1–2% are found in $[\alpha]_D$ as the concentration is doubled (see Table 12.3).[213] The accurate determination of optical purities is thus seen to be dependent on the careful measurement of rotations as well as on comparison of the resulting $[\alpha]$ values with reference $[\alpha]$ values measured in the *same* solvent, at the *same* temperature, and at the *same* concentration.[213] The reader is also reminded that $[\alpha]_D$ reflects, but is not necessarily linearly related to, the enantiomeric composition (Horeau effect; see Section 6-5.c).[214]

TABLE 12.3. Influence of Dilution on $[\alpha]_D^{25}$ for (S)-2-Phenylpropanal[a,b]

Concentration (g/100 mL^{-1})[c]	$[\alpha]_D^{25}$	$[\alpha]_D^{21}$
Neat	161.8	166.6
46.43	177.9	182.2
18.57	190.5	195.4
9.29	196.6	201.9
7.43	202.7	207.9
3.72	205.8	211.3
1.49	209.1	214.7

[a]Reprinted with permission from Consiglio, G., Pino, P., Flowers, L. I. and Pittman, C. U., Jr. *J. Chem. Soc. Chem. Commun.* **1983**, 612. Copyright © Royal Society of Chemistry, Science Park, Milton Road, Cambridge CB4 4WF, UK.
[b]Optical purity 68%.
[c]Benzene solution.

b. Empirical Rules and Correlations: Calculation of Optical Rotation

Ever since simple curiosity about optical activity gave way to its application, efforts have been made to calculate the magnitude and sign of the optical rotation in relation to structure and configuration.

One of the very oldest correlations between structure and rotatory power is that of Walden who observed that the molar rotations of diastereomeric salts in dilute solution are additive properties of the constituent ions.[13,215] Arithmetic manipulation of the molar rotations of diastereomeric salts, such as those obtained in a resolution, may thus permit one to estimate the enantiomeric purity achieved during a resolution mediated by these salts.

Analogously, molar rotations of inclusion compounds would be expected to be additive properties of the host and guest molar rotations. In both cases, additivity of rotations would not necessarily obtain when strong intermolecular interaction takes place.

In the case of covalent compounds, early correlation attempts also made use of the concept that the rotations of compounds containing several chiral centers might be calculated by adding rotation contributions from each of these centers. This concept, incorporated in van't Hoff's empirical "Principle of Optical Superposition," that individual chiral centers in a chiral compound make independent contributions to the molar rotation,[216,217] is still successfully being applied in very limited contexts.

The relative configurations of the diastereomeric (R)-O-methylmandelate esters of **37** (Fig. 12.55) were assigned by application of the van't Hoff principle. Contributions to the specific rotations from the octalin portions of the ester molecules were estimated from the rotations of (+)-dihydromevinolin **38** and lactone **39** (Fig. 12.56) to be approximately 100 (148.6 – 48.8) while that of the (R)-O-methylmandelate portion

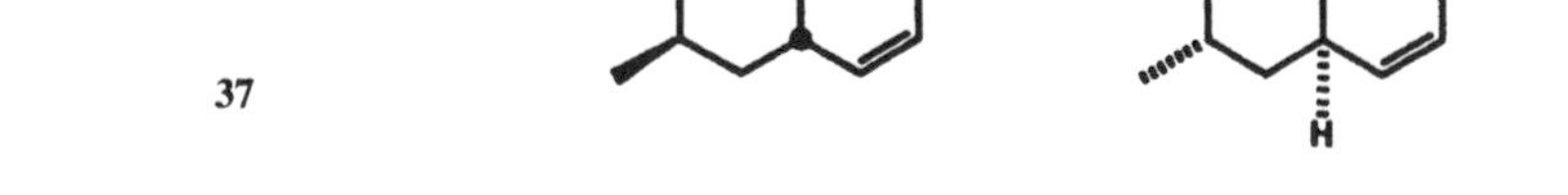

$[\alpha]_D$ –19.9 $[\alpha]_D$ +130

A + B = +130   A = +55 (due to the mandelate)
A – B = –20   B = +75 (due to the octalin)

Figure 12.55. Application of van't Hoff s "principle of optical superposition." Simultaneous solution of the two equations at left gives the values of **A** and **B** shown at right. Assignment of the relative configurations depends on the correct attribution of rotations **A** and **B** (see discussion in the text).[219]

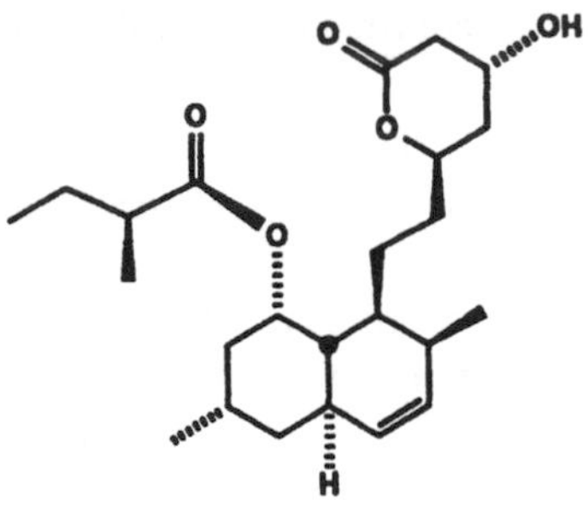

38   **39**

$[\alpha]_D + 148.6$ $[\alpha]_D + 48.8$

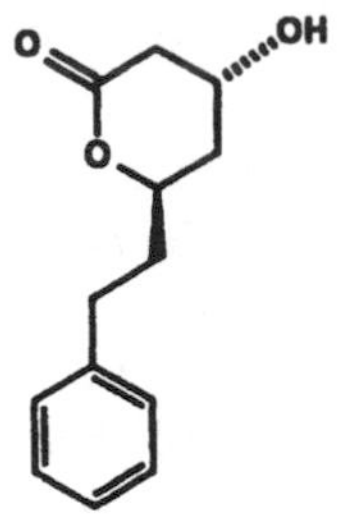

Figure 12.56. Rotations of (+)-dihydromevinolin **38** and of lactone **39**.

was independently known to be strong and positive [(*S*)-(−)-methyl O-methylmandelate] has $[\alpha]_D^{25}$ −124.[218] Accordingly, the configurational assignments were ±75 for the octalin portion of the esters and +55 for the O-methylmandelate (respectively, **B** and **A** in Fig. 12.55).[219] These assignments were confirmed by chemical correlations.

> van't Hoff's principle is likely to be invalid when the stereocenters contributing to the molar rotation are close together ("vicinal action" limitation). However, when the stereocenters are separated by several saturated atoms, as in the above example, the principle holds reasonably well.[53]

Along with van't Hoff s "principle of optical superposition," another empirical rule, Freudenberg's "rule of shift" (also termed Rotational Displacement Rule),[220] is considered quite reliable for establishing the relative configuration of pairs of compounds (by examining the sign and magnitude of molar rotation changes as these pairs are subjected to like chemical changes, e.g., derivatization). Illustrations are given by Freudenberg,[220] Lowry,[221] Eliel,[53] Barrett,[96] and Potapov.[222] For applications to carbohydrates, see Eliel et al.[223] For other empirical rules relating the signs of optical rotations of certain functional group types (allylic alcohols, amino acids, lactones, nucleosides, sugars) to their configurations, see Snatzke.[224]

Empirical and semiempirical treatments for the prediction of the magnitude of optical rotations from structural formulas have been developed that are based on polarizability and one-electron theories developed beginning in the 1930s (for a summary, see Charney).[35] These theories have also had as a goal the prediction of configuration from the sign of the experimentally measured rotatory power at a single wavelength. A new semiempirical theory that has been applied to the calculation of the optical activity of the saccharides is illustrative of recent developments in this area.[225,226] Ab initio calculations of chiroptical properties is presently limited to small molecules, such as *trans*-1,2-dimethylcyclopropane (see Section 12-4.a).[227] While the examination of the several theories of optical rotatory power is not possible in this

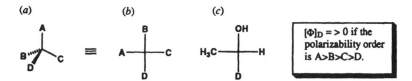

Figure 12.57. (*a*) Prediction of the sign of rotation in a system exhibiting atomic asymmetry; (*b*) Fischer projection; and (*c*) application to ethanol-1-*d*.

book, nor even a description of all of the treatments alluded to above, we will illustrate in detail at least the most successful of the empirical ones that is also easily accessible to organic chemists, namely, that of Brewster.[7,228–230] Brewster's model fits within the framework of the coupled oscillator theory of optical activity (Section 12-4.c). It is still being applied some three decades after its appearance, albeit in modified form.

The reader will recall that optical activity arises because nonracemic samples of chiral compounds are circularly birefringent (Section 12-2). Since the refractive index is related to the polarizability of atoms and groups in molecules, that is, the sensitivity of these molecular constituents to deformation by electrical fields, as well as to their relative positions, it should not surprise us that the rotatory power of chiral molecules also should be related to polarizabilities.[7]

Brewster's original formulation empirically factored the optical rotatory power into two components, as had earlier been proposed by Whiffen[231]; contributions to the rotation result from differences in the polarizability of atoms attached to asymmetric atoms (this was called atomic asymmetry), and from conformational dissymmetry (see below).[228] Nowadays we might attribute these contributions to local chirality and chiral conformations, respectively.

Since both atomic and conformational dissymmetry describe chiral screw patterns of polarizability, both contribute to the molar rotation in conformationally flexible molecules. Although there is no simple way of assessing its magnitude (see below), the contribution of the atomic asymmetry component is small especially when the polarizabilities of two of the attachment atoms are equal or nearly so.[232]

Compounds whose molecules are adequately described by the model in Fig. 12.57*b*, where the substituents A–D are atoms or small groups having average cylindrical or conical symmetry (for average symmetry, see Section 4-4), and whose absolute configuration is depicted by Figure 12.57*a*, are dextrorotatory when the polarizability order of the attached atoms is A > B > C > D. The polarizability order is given by the atomic refractions. In the case of attachment atoms in C≡C and C=C, atomic refractions are calculated by taking one-half the value of the group refraction (the latter are labeled[*]; a slightly more complicated apportionment is used in the case of CN, C_6H_5, and CO_2H): I (13.954) > Br (8.741) > SH (7.729) > Cl (5.844) > C≡C (3.580 = 7.159[*]/2) > CN (3.580; 5.459[*]) > C=C (3.379 = 6.757[*]/2) > C_6H_5 (3.379 = 6.757[*]/2) > CO_2H (3.379; 4.680[*]) > CH_3 (2.591) > NH_2, (2.382) > OH (1.518) > H

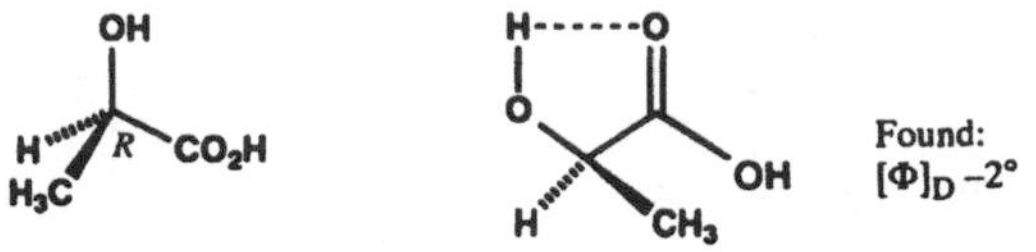

Found:
[Φ]$_D$ –2°

(+) Predicted   (–) Predicted

Figure 12.58. Influence of intramolecular hydrogen bonding on the molar rotation of lactic acid. The observed rotation of lactic acid is [Φ]$_D$–2 (c 1.24, water), however, the sodium salt is dextrorotatory.

(1.028) > D (1.004) > F (0.81). For example, on the basis of the rule, (*R*)-ethanol-1-*d* is predicted to be dextrorotatory at 589 nm ([Φ]$_D$ > 0). This is in accord with experiment.[233]

This polarizability order is insufficient, however, in determining the rotation since it turns out that polarizability is affected by the nature of the attachment atoms; for example, NH_2 and OH must be ranked ahead of groups whose attachment atom is carbon (C≡C, CN, C_6H_5, C=C, CO_2H, CH_3) when they are alpha to a phenyl group.[53] Moreover, if two of the groups can interact intramolecularly (e.g., by hydrogen bonding) there is a conformational dissymmetry contribution to the rotation (see below). If the atomic asymmetry and the conformational dissymmetry components (the latter was originally called conformational asymmetry) predict the same sense of rotation, then Figure 12.57 still gives the configuration accurately. However, if the two components are predicted to have oppositely signed rotations, then the model may lead to ambiguous results. Lactic acid is an example of the latter situation (Fig. 12.58).

An interesting empirical approach to the contribution of conformational dissymmetry to the sign and approximate magnitude of optical rotation has been developed by Brewster.[228–230]

The configuration of allenes and of alkylidenecycloalkanes (cf. Chapter 13) is predicted by Lowe's rule,[234] at least for structures in which substituents do not contribute to conformational dissymmetry (Fig. 12.59) and their molar rotations are calculable by means of Brewster's helix model.[7]

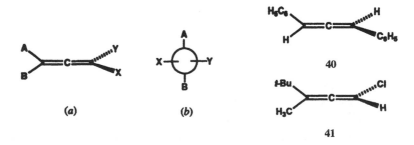

(*a*)   (*b*)   40   41

Figure 12.59. Absolute configuration of allenes (*a*) by Lowe's rule. The model shown in (*a*), corresponding to (*b*) in Newman notation, is dextrorotatory at 589 nm if the polarizability order is X > Y, A > B (right-hand helix) and levorotatory if the polarizability order of X and Y is reversed.

12-6. VIBRATIONAL OPTICAL ACTIVITY

Vibrational optical rotatory dispersion, vibrational circular dichroism (VCD) and Raman optical activity (ROA) are all manifestations of the interaction of polarized IR radiation with chiral substances. The three spectroscopic techniques together are aspects of vibrational optical activity (VOA).[19,235,236]

Few infrared ORD measurements have been reported.[237,238] On the other hand, VCD data and, to a lesser extent, ROA data are being reported with increasing frequency. As of 1992, VCD measurements are still being carried out on home-built dispersive or Fourier Transform IR (FTIR) spectrometers modified for VCD.[235,239]

Vibrational optical activity arises from vibrational transitions in the electronic ground state of chiral molecules. Vibrational circular dichroism spectra are characterized by numerous usually well-resolved bands (CEs) consistent with the

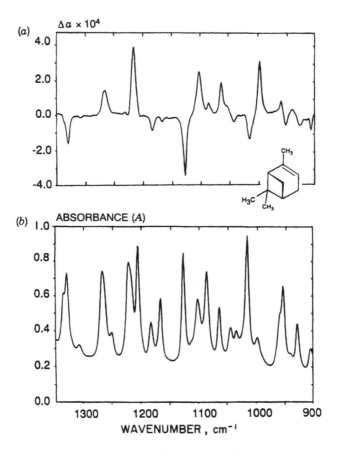

Figure 12.60. Spectra of (–)-α-pinene in the mid-IR range: (a) VCD and (b) isotropic IR. Ordinate units are absorbance (A) for IR, and the difference between absorbance for left and right circularly polarized radiation, $\Delta A = A_L - A_R$ for VCD. [Personal communication from L. A. Nafie, 1992. Reproduced by permission.]

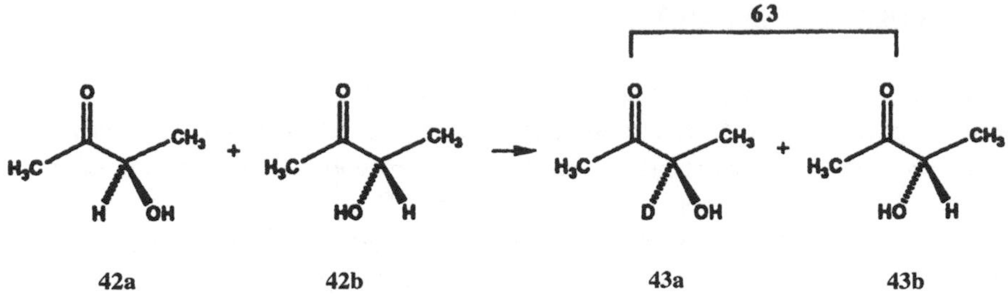

Figure 12.61. Proton exchange in acetoin.

presence of many dissymmetric vibrational motions that absorb in the IR region of the spectrum; but not all IR bands find their counterpart in VCD spectra. Within a given spectral region, the latter are generally simpler than IR spectra.[240] A typical VCD spectrum is shown in Figure 12.60. Just as in electronic CD, VCD spectra are often recorded with abscissa units of $\Delta\varepsilon$.

The advantage of VCD relative to electronic CD is strikingly illustrated by the proton exchange of racemic acetoin, $CH_3COCH(OH)CH_3$ (Fig. 12.61, **42**) catalyzed by the enzyme acetolactate decarboxylase. In D_2O, proton exchange at the stereogenic center is observed to stop at 50% incorporation. The product of the exchange reaction (Fig. 12.61) exhibits only a very weak negative CE at 279 nm (associated with the $n-\pi^*$ carbonyl transition) relative to the strong negative CD typical of a chiral ketone, such as that observed in (R)-$(-)$-acetoin.[42a] This finding is due to the fact that the reaction product is a 50:50 mixture of two compounds having nearly equal and opposite CDs.

Examination of the ν_{C-D} region by VCD, on the other hand, reveals several negative CEs from 2050 to 2200 cm^{-1}. The product mixture is clearly not racemic, contary to what might be inferred from the electronic CD spectrum [a 50:50 mixture of (S)-acetoin **42b** and (R)-acetoin-d **43a**, would be expected to have nearly no rotation at any wavelength; however, since the IR spectra of **42** and **43** differ, they can be distinguished in VCD spectra]. Also, the H–D exchange must be stereoselective: only (R)-acetoin undergoes exchange to give acetoin-d **43a** with retention of configuration (**43a** is the so-called isotopomer of **42b**) and the same product exhibits positive VCD in the ν_{C-H} region of the spectrum due to unreacted (S)-acetoin **42b** with an intensity about one-half that of pure **42a**.[241]

REFERENCES

1. Kelvin, Lord (W. Thompson). *The Second Robert Boyle Lecture* in J. Oxford Univ. Junior Scientific Club, **1894**, [18], 25.

2. Prelog, V. *Proc. K. Ned. Akad. Wet.* **1968**, *B71*, 108.

3. Weiss, U. *Experientia* **1968**, *24*, 1088.

4. Mislow, K. *Introduction to Stereochemistry*, Benjamin, New York, 1965.

5. Lambert, J. B., Shurvell, H. F., Lightner, D. A., and Cooks, R. G. *Introduction to Organic Spectroscopy*, Macmillan, New York, 1987.

6. Snatzke, G. *Chem. Unserer Zeit* **1981**, *15*, 78.

7. Brewster, J. H. *Top. Stereochem.* **1967**, *2*, 1.

8. Harada, N. and Nakanisi, K. *Circular Dichroic Spectroscopy–Exciton Coupling in Organic Stereochemistry*, University Science Books, Mill Valley, CA, 1983.

9. IUPAC. "Basic Terminology of Stereochemistry (III.1)," *Pure App. Chem.* **1996**, *68*, 2193.

10. Feynman, R. P., Leighton R. B., and Sands, M. *The Feynman Lectures on Physics*, Vol. I, Addison-Wesley, Reading, MA, 1963.

11. Izumi, Y. and Tai, A. *Stereo-Differentiating Reactions*, Kodansha, Tokyo, and Academic, New York, 1977.

12. Mislow, K. and Bickart, P. *Isr. J. Chem.* **1976**, *15*, 1.

13. Jacques, J., Collet, A., and Wilen, S. H. *Enantiomers, Racemates and Resolutions*, Wiley, New York, 1981.

14. Cotton, A. *Competes Rendus* **1895**, *120*, 989, 1044; *Ann. Chim. Phys.* **1896**, *8* [7], 347. Bootsma, G. A. and Schoone, J. C. *Acta Crystallogr.* **1967**, *22*, 522.

15. Djerassi, C. and Klyne, W. *Proc. Chem. Soc. London* **1957**, 55.

16. Lowry, T. M. *Optical Rotatory Power*, Dover, New York, 1964; this is a reprint of the book originally published in 1935 by Longmans, Green and Co., London.

17. Crabbé, P. and Parker, A. C. "Optical Rotatory Dispersion and Circular Dichroism," in Weissberger, A. and Rossiter, B . W., eds., *Physical Methods of Chemistry*, Part IIIC, *Techniques of Chemistry*, Vol. 1, Wiley, New York, 1972, Chap. 3.

18. Lambert, J. B., Shurvell, H. F., Verbit, L., Cooks, R. G., and Stout, G. H. *Organic Structural Analysis*, Part 3, Macmillan, New York, 1976.

19. Mason, S. F. *Molecular Optical Activity and the Chiral Discriminations*, Cambridge University Press, Cambridge, UK, 1982.

20. Djerassi, C. *Optical Rotatory Dispersion: Applications to Organic Chemistry*, McGraw-Hill, New York, 1960.

21. Moffitt, W. and Moscowitz. A. *J. Chem. Phys.* **1959**, *30*, 648.

22. Moscowitz, A. *Tetrahedron* **1961**, *13*, 48.

23. Emeis, C. A., Ooosterhoff, L. J., and De Vries, G. *Proc. R. Soc. London A* **1967**, *297*, 54.

24. Mislow, K. *Ann. N.Y. Acad. Sci.* **1962**, *93*, 459, and references cited therein.

25. Crabbé, P. "An Introduction to Optical Rotatory Dispersion and Circular Dichrois in Organic Chemistry," in Snatzke, G., ed., *Optical Rotatory Dispersion and Circular Dichroism in Organic Chemistry*, Heyden, London, 1967, Chap. 1.

26. Scopes, P. M. *Fortschr. Chem. Org. Naturst.* **1975**, *32*, 167.

27. Sjöberg, B., through Djerassi, C. *Optical Rotatory Dispersion,* McGraw-Hill, New York, 1960, p. 236.

28. Nakanishi, K., Berova, N., and Woody, R. W., eds., *Circular Dichroism: Principles and Applications*, Second ed., Wiley-VCH, New York, 2000, and chapters therein.

29. Purdie, N. and Brittain, H. G., eds. *Analytical and Applications of Circular Dichroism*, Elsevier, Amsterdam, 1994, and chapters therein.

30. Dugundiji, J., Marquarding, D., and Ugi, I. *Chem. Scr.* **1976**, *9*, 74.

31. Snatzke, G. *Angew. Chem. Int. Ed. Engl.* **1979**, *18*, 363.

32. Snatzke, G. *Pure Appl. Chem.* **1979**, *51*, 769.

33. Ciardelli, F. and Salvadori, P., eds. *Fundamental Aspects and Recent Developments in Optical Rotatory Dispersion and Circular Dichroism*, Heyden, London, 1973.

34. Kuhn, W. *Trans. Faraday Soc.* **1930**, *26*, 293.

35. Charney, E. *The Molecular Basis of Optical Activity. Optical Rotatory Dispersion and Circular Dichroism*, Wiley, New York, 1979.

36. Beychok, S. *Science* **1966**, *154*, 1288.

37. Carmack, M. and Neubert, L. A. *J. Am. Chem. Soc.* **1967**, *89*, 7134.

38. Legrand, M. and Rougier, M. J. "Application of the Optical Activity to Stereochemical Determinations," in Kagan, J. B., ed., *Stereochemistry, Fundamentals and Methods*, Vol. 2, Thieme, Stuttgart, Germany, 1977, p. 33.

39. Linderberg, J. and Michl, J. *J. Am. Chem. Soc.* **1970**, *92*, 2619.

40. Snatzke, G. "Chiroptical Properties of Organic Compounds: Chirality and Sector Rules," in Mason, S. F., ed., *Optical Activity and Chiral Discrimination*, Reidel, Dordrecht, The Netherlands, 1979, p. 25.

41. Schellman, J. A. *Acc. Chem. Res.* **1968**, *1*, 144.

42. Crabbé, P. *ORD and CD in Chemistry and Biochemistry: An Introduction*, Academic Press, New York, 1972.

43. Snatzke, G. and Snatzke, F. "Chiroptische Methoden," in Kienitz, H., Bock, R., Fresenius, W., Huber, W., and Tolg, G., eds., *Analytiker-Taschenbuch*, Vol. 1, Springer, Berlin, 1980, p. 217.

44. Kirk, D. N. *Tetrahedron* **1986**, *42*, 777.

45. Barton, D. H. R. *Experientia* **1950**, *6*, 316.

46. Djerassi, C., Foltz, E. W., and Lippman, A. E. *J. Am. Chem. Soc.* **1955**, *77*, 4354.

47. Djerassi, C. *Steroids Made It Possible*, American Chemical Society, Washington, DC, 1990.

48. Djerassi, C. *Proc. Chem. Soc. London* **1964**, 314.

49. Klyne, W. and Kirk, D. N. "The Carbonyl Chromophore: Saturated Ketones," in Ciardelli, F. and Salvadori, P., eds., *Fundamental Aspects and Recent Developments in Optical Rotatory Dispersion and Circular Dichroism*, Heyden, London, 1973, Chap. 3.1.

50. Djerassi, C., and Klyne, W. *J. Am. Chem. Soc.* **1957**, *79*, 1506.

51. Djerassi, C., Osiecki, J., Riniker, R., and Riniker, B. *J. Am. Chem. Soc.* **1958**, *80*, 1216.

52. Djerassi, C. and Staunton, J. *J. Am. Chem. Soc.* **1961**, *83*, 736.

53. Eliel, E. L. *Stereochemistry of Carbon Compounds*, McGraw-Hill, New York, 1962.

54. Djerassi, C., Geller, L. E., and Eisenbraun, E. J. *J. Org. Chem.* **1960**, *25*, 1.

55. Djerassi, C., Finch, N., and Mauli, R. *J. Am. Chem. Soc.* **1959**, *81*, 4997.

56. Moffitt, W., Woodward, R. B., Moscowitz, A., Klyne, W., and Djerassi, C. *J. Am. Chem. Soc.* **1961**, *83*, 4013.

57. Lightner, D. A. and Toan, V. V. *J. Chem. Soc. Chem. Commun.* **1987**, 210.

58. Djerassi, C. *Tetrahedron* **1961**, *13*, 13.

59. Klyne, W. and Kirk, D. N. *Tetrahedron Lett.* **1973**, 1483.

60. Kirk, D. N. *J. Chem. Soc. Perkin 1* **1976**, 2171.

61. Lightner, D. A., Chang, T. C., and Horwitz, J. *Tetrahedron Lett.* **1977**, 3019; *ibid.* **1978**, 696.

62. Numan, H. and Wynberg, H. *J. Org. Chem.* **1978**, *43*, 2232.

63. Sundararaman, P. and Djerassi, C. *Tetrahedron Lett.* **1978**, 2457; *ibid.* **1979**, 4120.

64. Sundararaman, P., Barth, G., and Djerassi, C. *J. Org. Chem.* **1980**, *45*, 5231.

65. Lee, S.-F., Barth, G., and Djerassi, C. *J. Am. Chem. Soc.* **1980**, *102*, 4784. Barth, G. and Djerassi, C. *Tetrahedron* **1981**, *37*, 4123.

66. Partridge, J. J., Chadha, N. K., and Uskokovic, M. R. *J. Am. Chem. Soc.* **1973**, *95*, 532.

67. Ripperger, H. *Z. Chem.* **1977**, *17*, 250.

68. Kirk, D. N. and Klyne, W. *J. Chem. Soc. Perkin 1* **1974**, 1076.

69. Bouman, T. D. and Lightner, D. A. *J. Am. Chem. Soc.* **1976**, *98*, 3145.

70. Lightner, D. A., Crist, B. V., Kalyanam, N., May, L. M., and Jackman, D. E. *J. Org. Chem.* **1985**, *50*, 3507.

71. Lightner, D. A., Chang, T. C., Hefelfinger, D. T., Jackman, D. E., Wijekoon, W. M. D., and Givens, J. W. III, *J. Am. Chem. Soc.* **1985**, *107*, 7499.

72. Gorthey, L. A., Samori, B., Fuganti, C., and Grasselli, P. *J. Am. Chem. Soc.* **1981**, *103*, 471.

73. Rodgers, S. L., Kalyanam, N., and Lightner, D. A. *J. Chem. Soc. Chem. Commun.* **1982**, 1040.

74. Snatzke, G., Ehrig, B., and Klein, H. *Tetrahedron* **1969**, *25*, 5601.

75. Jacobs, J. J. C. and Havinga, E. *Tetrahedron* **1972**, *28*, 135.

76. Lightner, D. A. and Wijekoon, W. M. D. *J. Org. Chem.* **1982**, *47*, 306.

77. Wilson, S. R. and Cui, W. *J. Org. Chem.* **1989**, *54*, 6047.

78. Deutsche, C. W., Lightner, D. A., Woody, R. W., and Moscowitz, A. *Ann. Rev. Phys. Chem.* **1969**, *20*, 407. See also ref. 79.

79. Mislow, K. and Siegel, J. *J. Am. Chem. Soc.* **1984**, *106*, 3319.

80. Mislow, K., Glass, M. A. W., Moscowitz, A., and Djerassi, C. *J. Am. Chem. Soc.* **1961**, *83*, 2771.

81. Moscowitz, A. *Adv. Chem. Phys.* **1962**, *4*, 67.

82. Bunnenberg, E., Djerassi, C., Mislow, K., and Moscowitz, A. *J. Am. Chem. Soc.* **1962**, *84*, 2823.

83. Hoffmann, R. W., Ladner, W., Steinbach, K., Massa, W., Schmidt, R., and Snatzke, G. *Chem. Ber.* **1981**, *114*, 2786.

84. Snatzke, G. *Chem. Unserer Zeit* **1982**, *16*, 160.

85. Schippers, P. H. and Dekkers, H. P. J. M. *J. Am. Chem. Soc.* **1983**, *105*, 79.

86. Moscowitz, A., Charney, E., Weiss, U., and Ziffer, H. *J. Am. Chem. Soc.* **1961**, *83*, 4661.

87. Weiss, U., Ziffer H., and Charney, E. *Tetrahedron* **1965**, *21*, 3105.

88. Koolstra, R. B., Jacobs, H. J. C., and Dekkers, H. P. J. M. *Croat. Chem. Acta* **1989**, *62*, 115.

89. Snatzke, G., Kovats, E., and Ohloff, G. *Tetrahedron Lett.* **1966**, 4551.

90. Burghstahler, A. W., Ziffer, H., and Weiss, U. *J. Am. Chem. Soc.* **1961**, *83*, 4660.

91. Davydov, A. S. *Theory of Molecular Excitons*, Kasha, M. and Oppenheimer, M., Jr. (translators), McGraw-Hill, New York, 1962.

92. Harada, N. and Nakanishi, K. *J. Am. Chem. Soc.* **1969**, *91*, 3989.

93. Harada, N., Ochiai, N., Takada, K., and Uda, H. *J. Chem. Soc. Chem. Commun.* **1977**, 495.

94. Harada, N. and Nakanishi, K. *Acc. Chem. Res.* **1972**, *5*, 257.

95. Mason, S. F. *Proc. Chem. Soc. London* **1962**, 362.

96. Barrett, G. C. "Applications of Optical Rotatory Dispersion and Circular Dichroism," in Bentley, K. W. and Kirby, G. W., eds., *Techniques of Chemistry*, Vol. IV, 2nd ed., Part I, Wiley, New York, 1972, Chap. 8. Smith, H. E. *Chem. Rev.* **1983**, *83*, 359.

97. Collet, A. and Gottarelli, G. *J. Am. Chem. Soc.* **1981**, *103*, 204. Snatzke, F. and Snatzke, G. *Fresenius' Z Anal. Chem.* **1977**, *285*, 97.

98. Canceill, J., Collet, A., Gabard, J., Gottarelli, G., and Spada, G. P. *J. Am. Chem. Soc.* **1985**, *107*, 1299. Dorffling, K. *Naturwissenschaften* **1967**, *54*, 23.

99. Cornforth, J. W., Milborrow, B. V., and Ryback, G. *Nature (London)* **1966**, *210*, 627.

100. Snatzke, G. and Vertesy, L. *Monatsh. Chem.* **1967**, *98*, 121.

101. de Weerd, R. J. E. M., van Hal, H. M. P. J., and Buck, H. M. *J. Org. Chem.* **1984**, *49*, 3413.

102. Drake, A. F., Gould, J. M., and Mason, S. F. *J. Chromatogr.* **1980**, *202*, 239.

103. Salvadori, P., Rosini, C., and Bertucci, C. *J. Org. Chem.* **1984**, *49*, 5050.

104. Zandomeneghi, M., Cavazza, M., and Pietra, F. *J. Am. Chem. Soc.* **1984**, *106*, 7261.

105. Cavazza, M., Festa, D., Veracini, C. A., and Zandomeneghei, M. *Chirality* **1991**, *3*, 257.

106. Nakanishi, K., Kasai, H., Cho, H., Harvey, R. G., Jeffrey, A. M., Jennette, K. W., and Weinstein, I. B. *J. Am. Chem. Soc.* **1977**, *99*, 258.

107. Menger, F. M. and Boyer, B. *J. Org. Chem.* **1984**, *49*, 1826.

108. Moscowitz, A., Wellman, K., and Djerassi, C. *J. Am. Chem. Soc.* **1963**, *85*, 3515.

109. Blauer, G. *Isr. J. Chem.* **1983**, *23*, 201.

110. Lightner, D. A., Gawronski, J. K., and Wijekoon, W. M. D. *J. Am. Chem. Soc.* **1987** *109*, 6354.

111. Lightner, D. A., Gawronski, J. K., and Gawronska, K. *J. Am. Chem. Soc.* **1985**, *107*, 2456.

112. Lightner, D. A., Reisinger, M., and Landen, G. L. *J. Biol. Chem.* **1986**, *261*, 6034.

113. Lightner, D. A., Reisinger, M., and Wijekoon, W. M. D. *J. Org. Chem.* **1987**, *52*, 5391.

114. Lightner, D. A., McDonagh, A. F., Wijekoon, W. M. D., and Reisinger, M. *Tetrahedron Lett.* **1988**, *29*, 3507.

115. Moscowitz, A., Krueger, W. C., Kay, I. T., Skewes, G., and Bruckenstein, S. *Proc. Natl. Acad. Sci. USA* **1964**, *52*, 1190.

116. Sjöberg, B. "Optical Rotatory Dispersion and Circular Dichroism of Chromophoric Derivatives of Transparent Compounds," in Snatzke, G., ed., *Optical Rotatory Dispersion and Circular Dichroism in Organic Chemistry*, Heyden, London, 1967, Chap. 11.

117. Halpern, B., Patton, W., and Crabbé P. *J. Chem. Soc. (B)* **1969**, 1143.

118. Crabbé, P. *Optical Rotatory Dispersion and Circular Dichroism in Organic Chemistry*, Holden-Day, San Francisco, 1967; see especially Chap. 11.

119. Crabbé, P. *Applications de la Dispersion Rotatooire Optique et du Dichorisme Circulaire Optique en Chimie Organique*, Gauthers Villars, Paris, 1968.

120. Reeves, R. E. *Adv. Carbohydrate Chem.* **1951**, *6*, 107.

121. Reeves, R. E. "Optical Rotations in Cuprammonium Solutions for Configurational and Conformational Studies," in Whistler, R. L., ed., *Methods in Carbohydrate Chemistry*, Vol. V, Academic, New York, 1965, p. 203.

122. Bukhari, S. T. K., Guthrie, R. D., Scott, A. I., and Wrixon, A. D. *Tetrahedron* **1970**, *26*, 3653.

123. Nelson, W. L., Wennerstrom, J. E., and Sankar, S. R. *J. Org. Chem.* **1977**, *42*, 1006.

124. Nelson, W. L. and Bartels, M. J. *J. Org. Chem.* **1982**, *47*, 1574.

125. Dillon, J. and Nakanishi, K. *J. Am. Chem. Soc.* **1975**, *97*, 5409, 5417.

126. Frelek, J., Konowal, A., Piotrowski, G., Snatzke, G., and Wagner, U. "Absolute Configuration of Natural Products from Circular Dichroism," in Atta-ur-Rahman and Le Quesne, P. W., eds., *New Trends in Natural Products Chemistry 1986*; *Studies in Organic Chemistry*, Vol. 26, Elsevier, Amsterdam, The Netherlands, 1986, p. 477.

127. Frelek, J., Majer, Z., Perkowska, A., Snatzke, G., Vlahov, I., and Wagner, U. *Pure Appl. Chem.* **1985**, *57*, 441.

128. Garards, M. and Snatzke, G. *Tetrahedron: Asymmetry* **1990**, *1*, 221.

129. Brown, G. H. and Crooker, P. P. *Chem. Eng. News* **1983**, *61* [Jan. 31] 24.

130. Solladié, G. and Zimmerman, R. G. *Angew. Chem. Int. Ed. Engl.* **1984**, *23*, 348.

131. Saeva, F. D. and Wysocki, J. J. *J. Am. Chem. Soc.* **1971**, *93*, 5928.

132. Saeva, F. D., Sharpe, P. E., and Olin, G. R. *J. Am. Chem. Soc.* **1973**, *95*, 7656, 7660.

133. Friedel, G. *Ann. Phys. (Paris)* **1922**, *18* [9], 273.

134. Korte, E. H., Schrader, B., and Bualek, S. *J. Chem. Res. Synopsis* **1978**, 236; *Miniprint* **1978**, 3001.

135. Gottarelli, G., Spada, G. P., and Solladié, G. *Nouv. J. Chim.* **1986**, *10*, 691.

136. Gottarelli, G., Samorí, B., Fuganti, C., and Grasselli, P. *J. Am. Chem. Soc.* **1981**, *103*, 471.

137. Gaubert, P. *C. R. Hebd. Seances Acad. Sci.* **1939**, *208*, 43.

138. Solladié, G. and Gottarelli, G. *Tetrahedron* **1987**, *43*, 1425.

139. Goodman, M., Verdini, A. S., Choi, N. S., and Masuda, Y. *Top. Stereochem.* **1970**, *5*, 69.

140. Scheraga, H. A. *Chem. Rev.* **1971**, *71*, 195.

141. Ciardelli, F., et al., "Circular Dichroism and Optical Rotatory Dispersion in Polymer Structure Analysis," in Hummel, D. O., ed., *Proceedings of the Fifth European Symposium on Polymer Spectroscopy, 1978*, Verlag Chemie, Weinheim, Germany, 1979, pp. 181–216. Jirgensons, B. *Optical Activity of Proteins and Other Macromolecules,* 2nd ed., Springer, New York, 1973.

142. Sélégny, E., ed. *Optically Active Polymers,* Reidel, Dordrecht, The Netherlands, 1979.

143. Johnson, W. C., Jr. "Circular Dichroism and Its Empirical Application to Biopolymers," in Glick, D., ed., *Methods of Biochemical Analysis*, Vol. 31, Wiley, New York, 1973, p. 61.

144. Urry, D. W. "Absorption, Circular Dichroism and Optical Rotatory Dispersion of Polypeptides, Proteins, Prosthetic Groups, and Biomembranes," in Neuberger, A. and

Van Deenen, L. L. M., eds., *Modern Physical Methods in Biochemistry*, Part A, *New Comprehensive Biochemistry*, Vol. 11A, Elsevier, Amsterdam, The Netherlands, 1985, Chap. 4.

145. Cantor, C. R. and Schimmel, P. R. *Biophysical Chemistry Part II: Techniques for the Study of Biological Structure and Function*, Freeman, San Francisco, 1980, Chap. 8. Woody, R. W. "Circular Dichroism of Peptides," in Udenfriend, S. and Meienhofer, J., eds., *The Peptides*, Academic Press, New York, Vol. 7, Hruby, V. J., ed., Chap. 2.

146. Blout, E. R. and Stryer, L. *Proc. Natl. Acad. Sci. USA* **1959**, *45*, 1591.

147. Regan, L. and DeGrado, W. F. *Science* **1988**, *241*, 976.

148. Doty, P. *Sci. Am.* **1957**, *197* [Sept.], 173.

149. Yang, J. T., Wu C.-S., and Martinez, H. M. "Calculation of Protein Conformation from Circular Dichroism," in Hirs, C. H. W. and Timasheff, S. N., eds., *Methods in Enzymology*, Vol. 130, Academic Press, New York, 1986, p. 208.

150. Greenfield, N., Davidison, B., and Fasman, G. D. *Biochemistry* **1967**, *6*, 1630.

151. Greenfield, N. and Fasman, G. D. *Biochemistry* **1969**, *8*, 4108.

152. Chang, T. C., Wu, C.-S. C., and Yang, J. T. *Anal. Biochem.* **1978**, *91*, 13.

153. Freifelder, D. *Physical Biochemistry: Applications to Biochemistry and Molecular Biology*, Freeman, San Francisco, 1976, Chap. 16.

154. Brahms, S. and Brahms, J. *J. Mol. Biol.* **1980**, *138*, 149.

155. Saxena, V. P. and Wetlaufer, D. B. *Proc. Natl. Acad. Sci. USA* **1971**, *68*, 969.

156. Tinoco, I., Jr. "Circular Dichroism of Polymers: Theory and Practice," in Selégny, E., ed., *Optically Active Polymers*, Reidel, Dordrecht, The Netherlands, 1979, p. 1.

157. Urata, H., Shinohara, K., Ogura, E., Ueda, Y., and Akagi, M. *J. Am. Chem. Soc.* **1991**, *113*, 8174.

158. Goodman, M. *Top. Stereochem.* **1967**, *2*, 73.

159. Farina, M. *Top. Stereochem.* **1987**, *17*, 1.

160. Coates, G. W. and Waymouth, R. M. *J. Am. Chem. Soc.* **1991**, *113*, 6270.

161. Green, M. M. and Garetz, B. A. *Tetrahedron Lett.* **1984**, *25*, 2831.

162. Vogl, O. and Jaycox, G. D. *Polymer* **1987**, *28*, 2179.

163. Bovey, F. A. *Chain Structure and Conformation of Macromolecules*, Academic Press, New York, 1982, Chap. 7.

164. Nolte, R. J. M. and Drenth, W. "Atropisomeric Polymers," in Fontanille, M. and Guyot, A., eds., *Recent Advances in Mechanistic and Synthetic Aspects of Polymerization*, Reidel, Dordrecht, The Netherlands, 1987, p. 451ff.

165. van Beijnen, A. J. M., Nolte, R. J. M., Drenth, W., and Hezemans, A. M. F. *Tetrahedron* **1976**, *32*, 2017.

166. Drenth, W. and Nolte, R. J. M. *Acc. Chem. Res.* **1979**, *12*, 30.

167. Green, M. M., Gross, R. A., Schilling, F. C., Zero, K., and Crosby, C., III. *Macromolecules* **1988**, *21*, 1839.

168. Kamer, P. C. J., Nolte, R. J. M., and Drenth, W. *J. Chem. Soc. Chem. Commun.* **1986**, 1789.

169. Okamoto, Y., Suzuki, K., Ohta, K., Hatada, K., and Yuki, H. *J. Am. Chem. Soc.* **1979**, *101*, 4763.

170. Cram, D. J. and Sogah, D. Y. *J. Am. Chem. Soc.* **1985**, *107*, 8301.

171. Okamoto, Y., Nakano, T., and Hatada, K. *Polym. J. (Tokyo)* **1989**, *21*, 199.

172. Bur, A. J. and Fetters, L. J. *Chem. Rev.* **1976**, *76*, 727.

173. Khatri, C. A., Andreola, C., Peterson, N. C., and Green, M. M. 204th American Chemical Society National Meeting, Washington, DC, August 1992, Polymer Division Preprints.

174. Green M. M., Park, J.-W., Sato, T., Lifson, S., Selinger, R. L. B., and Selinger *Angew. Chem. Int. Ed.* **1999**, *38*, 3138.

175. Lifson, S., Andreola, C., Peterson, N. C., and Green, M. M. *J. Am. Chem. Soc.* **1989**, *111*, 8550.

176. Green, M. M., Lifson, S., and Teramoto, A. *Chirality* **1991**, *3*, 285. See also Green, M. M. "A Model for How Polymers Amplify Chirality," Chapter 17 in ref. 28.

177. Pino, P. "Optical Rotatory Dispersion and Circular Dichorism in Conformational Analysis of Synthetic High Polymers," in Ciardelli, F. and Salvadori, P., eds., *Fundamental Aspects and Recent Developments in Optical Rotatory Dispersion and Circular Dichroism*, Heyden, London, 1973, Chap. 4.4.

178. Ciardelli, F. and Salvadori, P. *Pure Appl. Chem.* **1985**, *57*, 931.

179. Mannschreck, A., Eiglsperger, A., and Stühler, G. *Chem. Ber.* **1982**, *115*, 1568.

180. Mannschreck, A. *Chirality* **1992**, *4*, 163.

181. Pirkle, W. H., Sowin, T. J., Salvadori, P., and Rosini, C. *J. Org. Chem.* **1988**, *53*, 826.

182. Lloyd, D. K. and Goodall, D. M. *Chirality*, **1989**, *1*, 251.

183. Yeung, E. S., Steenhoek, L. E., Woodruff, S. D., and Kuo, J. C. *Anal. Chem.* **1980**, *52*, 1399.

184. Salvadori, P., Bertucci, C., and Rosini, C. *Chirality* **1991**, *3*, 376.

185. Bobbitt, D. R. and Yeung, E. S. *Anal. Chem.* **1984**, *56*, 1577.

186. Bobbitt, D. R. and Yeung, E. S. *Anal. Chem.* **1985**, *57*, 271.

187. Yeung, E. S. *Acc. Chem. Res.* **1989**, *22*, 125.

188. Mannschreck, A., Mintas, M., Becher, G., and Stühler, G. *Angew. Chem. Int. Ed. Engl.* **1980**, *19*, 469.

189. Lyle, G. G. and Lyle, R. E. "Polarimetry," in Morrison, J. D., ed. *Asymmetric Synthesis*, Vol. 1, Academic Press, New York, 1983, Chap. 2.

190. Heller, W. and Curmé, H. G. "Optical Rotation—Experimental Techniques and Physical Optics," in Weissberger, A. and Rossiter, B. W., eds., *Physical Methods of Chemistry*, Part IIIC, *Techniques of Chemistry*, Vol. I, Weissberger, A., ed., Wiley, New York, 1972, Chap. 2.

191. Itoh, T. "Quality of Amino Acids" in Kaneko, T., Izumi, Y., Chibata, I., and Itoh, T., eds., *Synthetic Production and Utilization of Amino Acids*, Kodansha, Tokyo and Wiley, New York, 1974, Chap. 5.

192. Arjona, O., Pérez-Ossorio, R., Pérez-Rubalcaba, A., Plumet, J., and Santesmases, M. J. *J. Org. Chem.* **1984**, *49*, 2624.

193. Lowman, D. W. *J. Am. Soc. Sugar Beet Technol.* **1979**, *20*, 233.

194. Chafetz, L. *Pharm. Technol.* **1991**, *15*, 52.

195. Eliel, E. L. and Knox, D. E. *J. Am. Soc. Chem. Soc.* **1985**, *107*, 2946.

196. Biot, J. B. *Mém. Acad. R. Sci. Inst. France* **1835**, *13*, 116 through Kuhn, W. "Theorie und Grundgesetze der optischen Aktivität," in Freudenberg, K., ed., *Stereochemie*, Franz Deuticke, Leipzig and Vienna, 1933, p. 318.

197. Schurig, V. *Kontake (Darmstadt)* **1985**, 54.

198. Kabuto, K., Yasuhara, F., and Yamaguchi, S. *Bull. Chem. Soc. Jpn.* **1983**, *56*, 1263.

199. Hargreaves, M. K. and Richardson, P. J. *J. Chem. Soc.* **1957**, 2260.

200. Baldwin, J. E., Hackler, R. E., and Scott, R. M. *J. Chem. Soc. D* **1969**, 1415.

201. Goldberg, S. I., Bailey, W. D., and McGregor, M. L. *J. Org. Chem.* **1971**, *36*, 761. See especially Note 16.

202. Yamaguchi, S. and Mosher, H. S. *J. Org. Chem.* **1973**, *38*, 1870.

203. Berner, E. and Leonardsen, R. *Justus Liebigs Ann. Chem.* **1939**, *538*, 1.

204. Fischer, H. O. L. and Baer, E. *Chem. Rev.* **1941**, *29*, 287.

205. Mislow, K., Graeve, R., Gordon, A. J., and Wahl, G. H., Jr. *J. Am. Chem. Soc.* **1964**, *86*, 1733.

206. Sanderson, W. A. and Mosher, H. S. *J. Am. Chem. Soc.* **1966**, *88*, 4185.

207. Wynberg, H., Numan, H., and Dekkers, H. P. J. M. *J. Am. Chem. Soc.* **1977**, *99*, 3870.

208. Greenstein, J. P. and Winitz, M. *Chemistry of the Amino Acids*, Wiley, New York, 1961.

209. Cross, G. A. and Kellogg, R. M. *J. Chem. Soc. Chem. Commun.* **1987**, 1746.

210. Ko, K.-Y. and Eliel, E. L. *J. Org. Chem.* **1986**, *51*, 5353.

211. Kuriyama, K., Iwata, T., Moriyama, M., Ishikawa, M., Minato, H., and Takeda, K. *J. Chem. Soc. C* **1967**, 420.

212. Kumata, Y., Furukawa, J., and Fueno, T. *Bull. Chem. Soc. Jpn.* **1970**, *43*, 3920.

213. Consiglio, G., Pino., P., Flowers, L. I., and Pittman, C. U., Jr. *J. Chem. Soc. Chem. Commun.* **1983**, 612.

214. Horeau, A. and Guetté, J.-P. *Tetrahedron* **1974**, *30*, 1923.

215. Walden, P. *Z. Phys. Chem.* **1894**, *15*, 196.

216. van't Hoff, J. H. *Die Lagerung der Atome im Raume,* 2nd ed., Vieweg, Brunswick, Germany, 1894, p. 119. See also, Guye, P. A. and Gautier, M. *C. R. Hebd. Séances Acad. Sci.* **1894**, *119*, 740. Kuhn, W. "Theorie und Grundgesetze der optischen Aktivität," in Freudenberg, K., ed., *Stereochemie,* Franz Deuticke, Leipzig, Germany, 1933, p. 317.

217. Kondru, R. K., Beratan, D. N., Friestad, G. K., Smith, A. B., and Wipf, P. *Org. Lett.* **2000**, *2*, 1509.

218. Barth, G., Voelter, W., Mosher, H. S., Bunnenberg, E., and Djerassi, C. *J. Am. Chem. Soc.* **1970**, *92*, 875.

219. Hecker, S. J. and Heathcock, C. H. *J. Am. Chem. Soc.* **1986**, *108*, 4586.

220. Freudenberg, K., Todd, J., and Seidler, R. *Justus Liebigs Ann. Chem.* **1933**, *501*, 199.

221. Freduenberg, K. "Konfigurative Zusammenhänge optisch aktiver Verbindungen," in Freudenberg, K., ed., *Stereochemie,* Franz Deuticke, Leipzig and Vienna, 1933, p. 662.

222. Potapov, V. M. *Stereochemistry*, Mir, Moscow, 1979.

223. Eliel, E. L., Allinger, N. L., Angyal, S. J., and Morrison, G. A. *Conformational Analysis*, Wiley, New York, 1965, pp. 381–394.

224. Snatzke, G. "Application of Circular Dichorism, Optical Rotatory Dispersion and Polarimetry in Organic Stereochemistry," in Korte, F., ed., *Methodicum Chimicum*, Vol. 1A, Academic Press, New York, and Thieme, Stuttgart, Germany, 1974, Chap. 5.7.

225. Stevens, E. S. and Sathyanarayana, B. K. *Carbohydr. Res.* **1987**, *166*, 181.

226. Sathyanarayana, B. K. and Stevens, E. S. *J. Org. Chem.* **1987**, *52*, 3170.

227. Bohan, S. and Bouman, T. D. *J. Am. Chem. Soc.* **1986**, *108*, 3261.

228. Brewster, J. H. *J. Am. Chem. Soc.* **1959**, *81*, 5475.

229. Brewster, J. H. *J. Am. Chem. Soc.* **1959**, *81*, 5483, 5493.

230. Brewster, J. H. *Tetrahedron* **1961**, *13*, 106.

231. Whiffen, D. H. *Chem. Ind. (London)* **1956**, 964.

232. Boter, H. L. *Recl. Trav. Chim. Pays-Bas* **1968**, *87*, 957.

233. Klyne, W. and Buckingham, J. *Atlas of Stereochemistry,* 2nd ed., Vol. I, Oxford University Press, New York, 1978, p. 211.

234. Lowe, G. *Chem. Commun.* **1965**, 411.

235. Nafie, L. A. Personal communications to S.H.W.

236. Freedman, T. B., Paterlini, M. G., Lee, N.-S., Nafie, L. A., Schwab, J. M., and Ray, T. *J. Am. Chem. Soc.* **1987**, *109*, 4727.

237. Korte, E. H. and Schrader, B. *Messtechnik (Braunschwig)* **1973**, *81*, 371.

238. Korte, E. H. *Appl. Spectrosc.* **1978**, *32*, 568.

239. Stinson, S. C. *Chem. Eng. News,* **1985**, *63* [Nov. 11], 21.

240. Nafie, L. A., Keiderling, T. A., and Stephens, P. J. *J. Am. Chem. Soc.* **1976**, *98*, 2715.

241. Drake, A. F., Siligardi, G., Crout, D. H. G., and Rathbone, D. L. *J. Chem. Soc. Chem. Commun.* **1987**, 1834.

13

CHIRALITY IN MOLECULES DEVOID OF CHIRAL CENTERS

13-1. INTRODUCTION AND NOMENCLATURE

In Chapter 1 it was pointed out that a necessary and sufficient condition for a molecule to be chiral is that it not be superposable with its mirror image. The presence of a (single, configurationally stable) chiral center in the molecule is a sufficient condition for the existence of chirality but not a necessary one. In this chapter we shall turn our attention to chiral molecules devoid of chiral centers. We shall include some types of molecules (certain spiranes and metallocenes) in which, for nomenclatural purposes, a chiral center may be defined to exist[1] (cf. Fig. 13.4) even though these molecules are closely akin to others in which no chiral centers can be discerned.

Classes of molecules to be discussed here[2,3] are allenes; cumulenes with even numbers of double bonds (cf. Chapter 9 for cumulenes with odd numbers of double bonds); alkylidenecycloalkanes; spiranes; the so-called atropisomers (biphenyls and similar compounds in which chirality is due to restricted rotation about a single bond); helicenes, propellerlike structures; and molecules, such as cyclophanes, chiral *trans*-cycloalkenes, ansa compounds, and arene-metal complexes including metallocenes, which are said to contain a "plane of chirality."

Allenes, alkylidenecycloalkanes, biphenyls, and so on, are said to possess a "chiral axis."[4] If we stretch a tetrahedron along its S_4 axis, it is desymmetrized to a framework of $\mathbf{D}_{2d}$ symmetry (Fig. 13.1). With proper substitution, the long axis of this framework constitutes the chiral axis. Because of the intrinsically lower symmetry of the

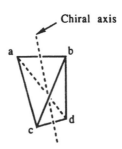

Figure 13.1. Chiral axis.

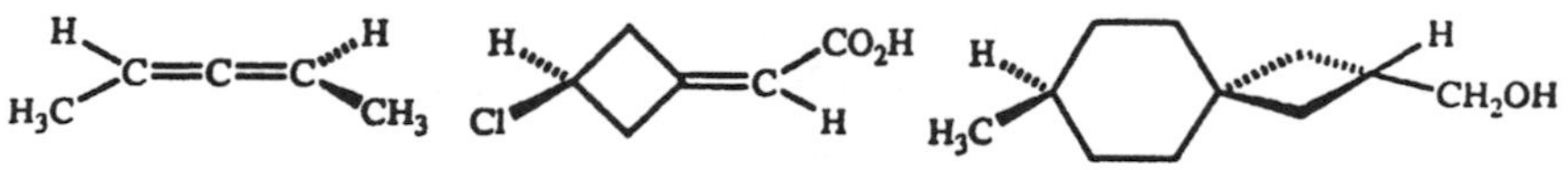

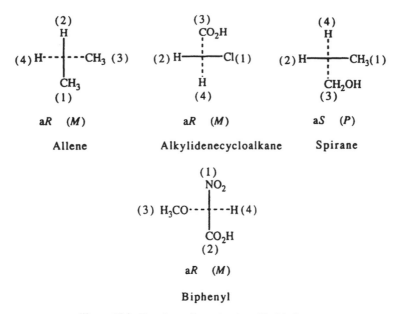

Allene Alkylidenecycloalkane Spirane

Biphenyl

Figure 13.2. Molecules with chiral axes.

framework shown in Figure 13.1 compared to a tetrahedron, it no longer takes four different substituents to make the framework chiral: A necessary and sufficient condition for chirality is that a ≠ b and c ≠ d. Thus, even when a = c and/or b = d, the framework retains chirality, for example, in abC=C=Cab (see below).

To specify the sense of chirality (i.e., configuration) of a molecule possessing a chiral axis (*axial chirality*, examples are shown in Fig. 13.2) an additional sequence rule is needed: Near groups precede far groups. The application of this rule to the molecules shown in Figure 13.2 is shown in Figure 13.3. In all cases the molecules in

a*R* (*M*) a*R* (*M*) a*S* (*P*)

Allene Alkylidenecycloalkane Spirane

a*R* (*M*)

Biphenyl

Figure 13.3. Descriptors for molecules with chiral axes.

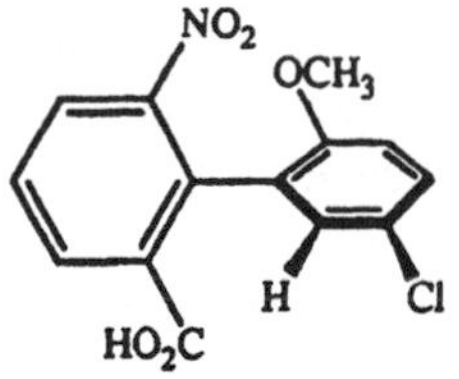

Figure 13.2 are viewed from the left. However, the reader should take note that the same configurational descriptor results when the molecules are viewed from the right, so no specification in this regard is needed. In the case of biphenyl it is important to note that the ring substituents are to be explored from the center on outward, regardless of the rule given above. Thus, in the biphenyl in Figure 13.2, in the right ring the sequence is $C—OCH_3 > C—H$; the chlorine atom is too far out to matter, a decision being made before it is reached in the outward exploration. The fiducial atoms (i.e., those that determine the configurational symbol, cf. p. 421) are the same when the molecule is viewed from the right. The descriptors aR and aS are sometimes used to distinguish axial chirality from other types, but the use of the a prefix is optional.

Molecules with chiral axes may alternatively be viewed as helices (in this respect they resemble the helicenes to be discussed below) and their configuration may be denoted as P or M, in a manner similar to that of conformational isomers (Chapter 10).[5] For this designation, only the ligands of highest priority in front and in the back of the framework are considered (ligands 1 and 3 in Fig. 13.3). If the turn from the priority front ligand 1 to the priority rear ligand 3 is clockwise, the configuration is P; if counterclockwise it is M. Thus three of the four structures in Figures 13.2 and 13.3 are aR (chiral axis nomenclature) or M (helix nomenclature); the spirane is aS or P. (The correspondence of aR with M and aS with P is general.)

Figure 13.4 shows molecules with chiral planes. The definition of a chiral plane is less simple and clear-cut than that of a chiral center or axis. It is a plane that contains as many of the atoms of the molecule as possible, but not all; in fact, the chirality is due (and solely due) to the fact that at least one ligand (usually more) is *not* contained

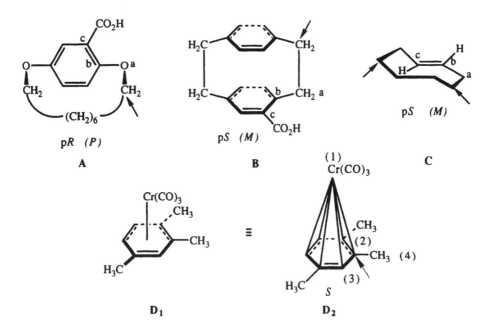

Figure 13.4. Molecules with chiral planes.

in the chiral plane. Thus the chiral plane of the "ansa compound" **A** (in which the alicyclic ring is too small for the aromatic one to swivel through) is the plane of the benzene ring (the same is formally true of the arenechromium tricarbonyl compound **D**); in the paracyclophane **B**, the more highly substituted benzene ring (bottom) is considered the chiral plane and in *trans*-cyclooctene **C** the chiral plane is that of the double bond. To find the descriptor for planar chiral molecules one views the chiral plane from the out-of-plane atom closest to the plane (if there are two or more candidates, one chooses the one closest to the atom of higher precedence according to the sequence rules, cf. Section 5-2). This atom, sometimes called the "pilot atom," is marked with an arrow in Figure 13.4 (for compound **C** there are two equivalent such atoms). Then, if the adjacent three atoms a, b, and c (again chosen by precedence if there is a choice) describe a clockwise array in the chiral plane, the configuration is p*R*, if the array is counterclockwise, the descriptor is p*S*. (The prefix "p" may be used to signal planar chirality.)

Compound **D**, although it would also appear to have a chiral plane, is conventionally treated as having chiral centers by replacing the η_6 π bond by six σ single bonds, as shown in structure **D$_2$**. The (central) chirality is now determined for the atom of highest precedence (the ring carbon marked by an arrow) and the descriptor is thus found to be S^1 (see also ref. 6 and ref. 7a, p. 222).

Planar chirality, like axial chirality, may alternatively be looked at as a type of helicity.[5] To determine the sense of the helix one uses the pilot atom plus atoms a, b, and c specified as above. It is then seen (Fig. 13.4) that p*R* compounds correspond to *P* and p*S* corresponds to *M*, opposite to the correlation in axial chirality (see above).

13-2. ALLENES

a. Historical Overview and Natural Occurrence

It was already pointed out by van't Hoff[8] that an appropriately substituted allene should exist in two enantiomeric forms. A simple case is shown in Figure 13.5, **A**; a necessary and sufficient condition for such an allene to be chiral is that a ≠ b. The reason for the dissymmetry is that the groups a and b at one end of the system lie in a plane at right angles to those at the other end. If the doubly bonded carbon atoms are viewed as tetrahedra joined edge to edge, a view that was originally proposed by van't Hoff (see also Chapter 9), the noncoplanarity of the two sets of groups follows directly

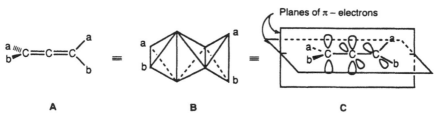

Figure 13.5. Dissymmetric allene.

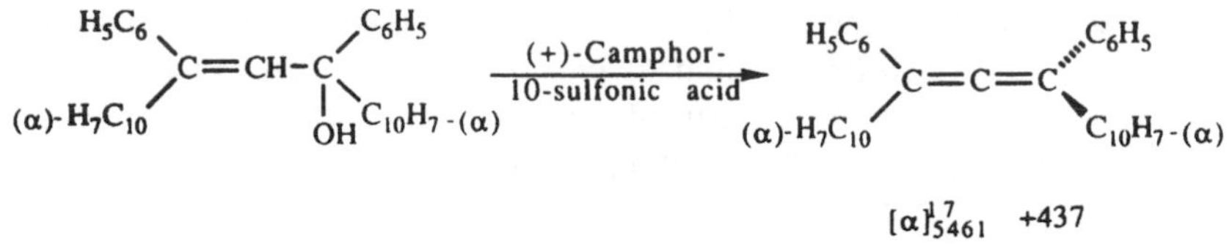

Figure 13.6. Asymmetric synthesis of optically active allene.

from the geometry of the system (Fig. 13.5, **B**). If, on the other hand, one views a double bond as being made up of pairs of σ and π electrons, orbital considerations indicate that the two planes of the π bonds attached to the central carbon atom must be orthogonal, and since the a and b groups attached to the trigonal carbon lie in a plane at right angles to the plane of the adjacent π bond, their planes are orthogonal to each other (Fig. 13.5, **C**).

The experimental realization of van't Hoff's prediction proved to be quite difficult, and 60 years elapsed before the first optically active allene was obtained in the laboratory.[9] The route chosen was one of asymmetric synthesis: Dehydration of 1,3-diphenyl-1,3-α-naphthyl-2-propen-1-ol with (+)-camphor-10-sulfonic acid gave (+)-1,3-diphenyl-1,3-di-α-naphthylallene (Fig. 13.6) in slight preponderance over its enantiomer [enantiomer excess (ee) ca. 5%]. Fortunately, the optically active allene forms a conglomerate (cf. Chapter 6) and the pure enantiomer could be separated from the racemate by fractional crystallization without excessive difficulty. The material has the high specific rotation $[\alpha]_{546}^{17} + 437$ (benzene), $[\alpha]_D^{20} + 351$ (cyclohexane). Use of (−)-camphor-10-sulfonic acid gave the enantiomer of $[\alpha]_{546}^{17} + 438$ (benzene).

In 1952, it was recognized that optically active allenes also occur in nature. In that year Celmer and Solomons[10] established the structure of the antibiotic mycomycin, a fungal metabolite, to be that of a chiral allene: $HC \equiv C - C \equiv C - CH = C = CH - CH = CH - CH = CH - CH_2CO_2H$. Since then a number of other chiral allenes have been found in nature (for tabulations see refs. 11–14.)

In recent years. numerous optically active allenes have been obtained in a variety of ways (resolution, transformation of chiral precursors, and enantioselective synthesis) and only the highlights of allene stereochemistry can be presented here. Fortunately, several detailed reviews are available.[3,11-15]

b. Synthesis of Optically Active Allenes

In addition to classical resolution[13] and the method shown in Fig. 13.6 there are a number of schemes of synthesizing chiral allenes from precursors having chiral centers. An example is the reductive rearrangement of tetrahydropyranyl ethers of chiral acetylenic alcohols shown in Figure 13.7,[16] which proceeds in 75–100% optical yield, depending on R (which contained an alcohol function in all cases). The addition is trans, that is, hydride approaches the acetylene from the side opposite to that of the —OTHP (THP = tetrahydropyran) leaving group. A conceptually similar scheme involves the transformation of the methanesulfonate of a chiral acetylenic alcohol to

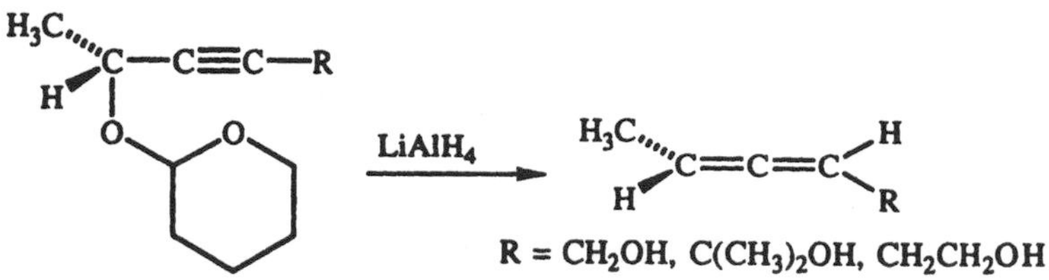

Figure 13.7. Reductive rearrangement of chiral acetylenic carbinol to chiral allene.

a chiral allenic halide by means of lithium copper halides[17] (for related reactions, see refs. 18 and 19).

An orthoester Claisen rearrangement, which also proceeds highly stereospecifically, has been used by Mori's group[20] in the preparation of an intermediate X (Fig. 13.8) toward the synthesis of the sex pheromone produced by the male dried-bean beetle, (E)-$CH_3(CH_2)_7CH=C=CH-CH=CHCO_2CH_3$. This reaction proceeds by suprafacial attack, that is, the enol ether on the bottom side of the molecule (as drawn) attaches itself to the acetylenic triple bond from the same side.

c. Determination of Configuration and Enantiomeric Purity of Allenes

Experimental Methods. As was discussed in Section 5-3.a, the classical absolute method of determining configuration is by anomalous X-ray scattering (Bijvoet method). There seems to be only one application of this method to allenes, namely, the elucidation of the relative and absolute configuration of the allenic ketone shown in Figure 13.9 isolated from ant-repellant secretions of a species of grasshopper.[21] This compound contains both chiral centers and allenic chirality and although it has been correlated with other naturally occurring chiral allenes, the chiral allenic moiety has not been cut out. However, neither the Bijvoet method nor correlative X-ray methods have yet been applied to the determination of absolute configuration of a chiral allene devoid of other chiral elements (chiral centers).

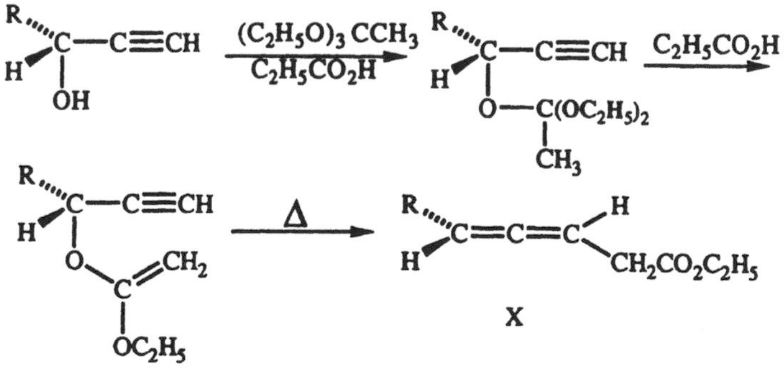

Figure 13.8. Orthoester Claisen rearrangement of chiral acetylenic carbinol to chiral allene.

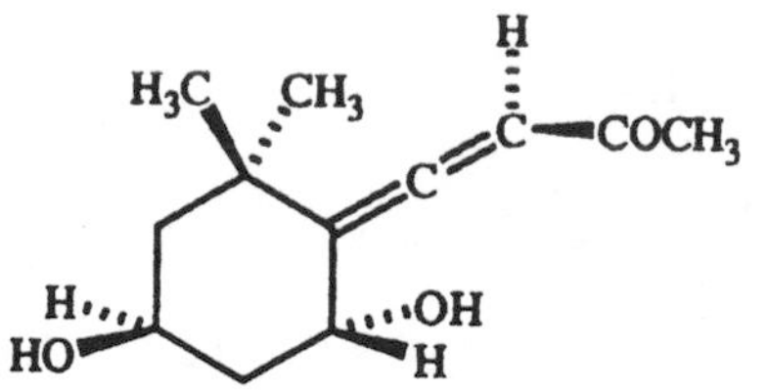

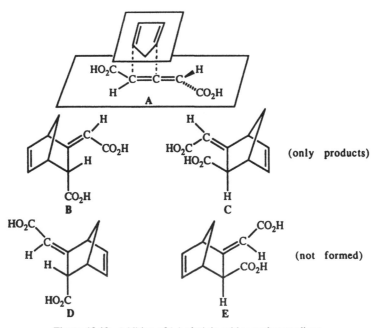

Figure 13.9. Allene whose configuration has been elucidated by X-ray structure analysis.

There are, however, a number of indirect methods, notably mechanistic correlation of chiral allenes with molecules possessing chiral centers of known configuration, methods based on interpretation of optical rotatory dispersion–circular dichroism (ORD/CD) spectra (cf. Chapter 12) and theoretical methods. Only a few examples can be presented; detailed discussions may be found elsewhere[3,11,13] (see also refs. 7 and 22).

A correlation of configuration, based on the Diels–Alder reaction of the so-called glutinic acid (pentadienoic acid **A**) with cyclopentadiene,[23] is shown in Figure 13.10. If the (−) acid has the aR configuration (cf. Fig. 13.10, **A**), there are four diastereomers (**B–E**) that can be formed in the addition, two (**B** and **C**) by approach from the less hindered side (out-of-plane hydrogen, see **A**) and two (**D** and **E**) by approach from the more hindered side (out-of-plane carboxyl). If the (−) acid has the aS configuration, the enantiomers of **B–E** would result. (It is to be noted that the two double bonds of glutinic acid are symmetry equivalent being interconverted by a C_2 operation.)

Figure 13.10. Addition of (−)-glutinic acid to cyclopentadiene.

According to the Alder–Stein rules,[24,25] the endo addition product **B** should be the major or exclusive one, the mode of addition being as shown in **A**, where the double bonds of the cyclopentadiene have maximum overlap with the C=O π bond of the in-plane carboxylate group (see also ref. 2, p. 295). However, many exceptions to the Alder–Stein endo addition rule are now known.

In fact, only two (separable) products (shown to be **B** and **C**; see below) were formed in the addition. Both had proximate carboxyl groups, as demonstrated by cyclic anhydride formation upon treatment with acetic anhydride; this excludes **D** and **E**. The endo configuration of the carboxyl group in **B** was proved by iodolactone formation, whereas the exo configuration of the carboxyl group in **C** (which formed no iodolactone) was demonstrated by hydrogenation to the known cis dicarboxylic acid **F** as shown in Figure 13.11.

With the relative configuration of **B**, and therefore also of **C**, thus firmly established, it remained to elucidate their absolute configuration. This feat was accomplished (cf. Fig. 13.11) by selectively hydrogenating the ring double bond, esterifying, ozonizing the exocyclic double bond, hydrolyzing, and decarboxylating. In this way it was found that (+)-**B** formed from (−)-glutinic acid gave (+)-norcamphor known to have the 1S configuration as shown in Figure 13.11. Diastereomer (−)-**C**, also formed from (−)-glutinic acid, similarly gave (−)-norcamphor, identified as the (−)-2,4-dinitrophenylhydrazone, as expected. The configuration of (−)-glutinic acid is thus aR, as shown in Figure 13.10, **A**.

A Priori Methods.   The Lowe–Brewster rules[26–28] are summarized in Figure 13.12. If ligand A is more polarizable than B and C than D, the molecule as shown is levorotatory at the sodium D line. (−)-Glutinic-acid (Fig. 13.10) is shown in the proper perspective in Figure 13.12b. It is clear that the correct prediction is made in this case (and a number of others) if reasonable assumptions for relative polarizabilities are made.

A case where the rules fail is 1,2-cyclononadiene; the (+) isomer has been shown to be R,[29] whereas the Lowe–Brewster rules predict it to be S. The failure has been explained[29]; the correct configuration can be derived from the CD spectrum.[30]

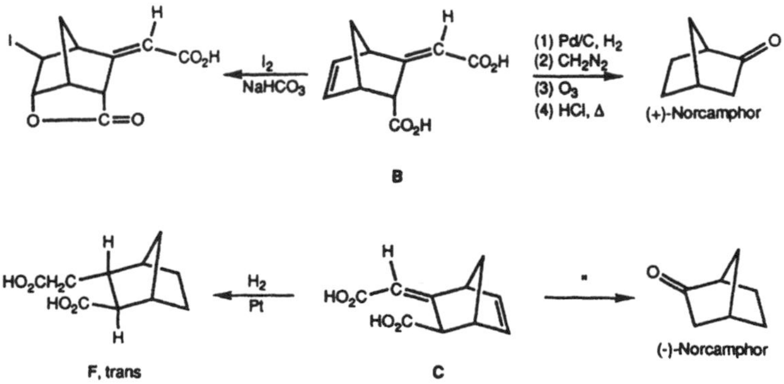

Figure 13.11. Determination of relative and absolute configuration of **B** (Fig. 13.10).

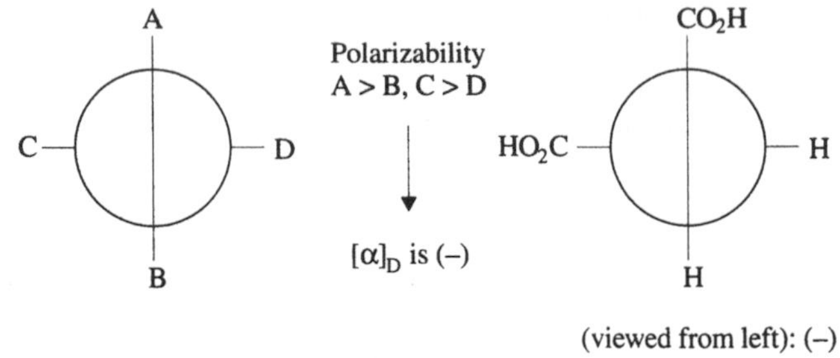

Figure 13.12. (a) Lowe–Brewster rule (also called "Lowe's rule") and (b) its application to (–)-glutinic acid.

Vibrational circular dichroism (VCD) has been used in the configurational correlations of allenes.[31] Especially useful is the VCD of the C=C=C stretch near 1950 cm^{-1}: positive VCD corresponds to the S configuration of the allene.

Enantiomeric purity of allenes may be determined by the usual methods (use of chiral shift reagents in NMR, chromatography or NMR analysis of derivatives with chiral reagents, chromatography on chiral stationary phases, and so on; cf. Section 6-5). However, the application of such methods to allenes so far seems to have been limited.[13]

d. Cyclic Allenes, Cumulenes, and Ketene Imines

1,2-Cyclonoadiene is the smallest stable cyclic-allene to be isolated[32,33] (see also pp. 642, 644) although 1,2-cyclooctadiene has been detected spectroscopically[34] and both it and 1,2-cycloheptadiene can be trapped as platinum complexes[35,36]; trapping has succeeded even with the very unstable 1,2-cyclohexadiene[37,38] cf. also refs. 39 and 40. The cyclic diallene dodeca-1,2,7,8-tetraene-5,11-dione (Fig. 13.13) has been obtained both as the meso isomer **A** and in optically active form **B**,[41] whereas only the meso isomer has been synthesized in the case of lower homologue deca-1,2,6,7-tetraene.[33,42] Of particular interest is the synthesis of the doubly bridged allene **C** (Fig. 13.13) obtained through a dihalocyclopropane synthesis both in racemic and (by use of a butyllithium-(–)-sparteine combination) in optically active form.[43,44]

According to van't Hoff (cf. Fig. 13.5), any properly substituted cumulene with an even number of double bonds, RR'C=C(=C=C)$_n$=CRR' should be chiral. For

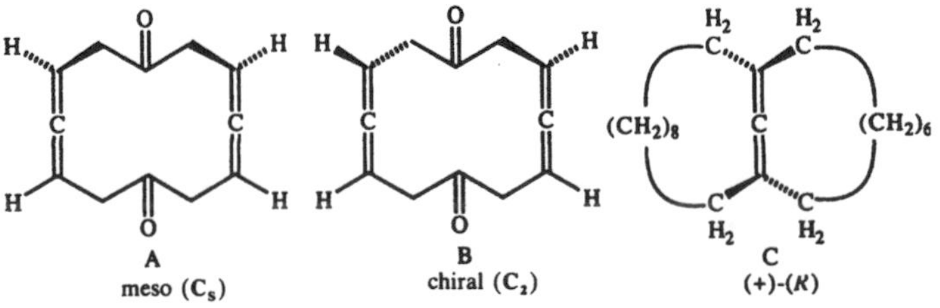

Figure 13.13. Cyclic bis-allenes and a doubly bridged allene.

allene $n = 0$; cumulenes of this type with $n \geq 2$ seem to be unknown (for cumulenes with odd numbers of double bonds, see Section 9-1.c). In any case such cumulenes, if obtainable, would probably not be configurationally stable, since the barrier to rotation around the C=C double bonds in a cumulene decreases with increasing number of double bonds.[45] Although tetraaryl substituted pentatetraenes were first synthesized in 1964,[46] resolution of $(CH_3)_3C(C_6H_5)C{=}C{=}C{=}C{=}C(C_6H_5)C(CH_3)_3$ (ref. 47) was achieved[45] only in 1977 through chromatography with an enantioselective stationary phase. The optical activity of material purified by crystallization at $-80°C$ was $[\alpha]_D^{22} \pm 336$ (both enantiomers were obtained) and the material racemized in a few hours in n-nonane solution at $-15°C$ with an activation barrier $\Delta G^{\ddagger} = 114.8$ kJ mol^{-1} (27.4 kcal mol^{-1}). This barrier is slightly less than the cis–trans interconversion barrier of the next lower homologue [triene, 122.5 kJ mol^{-1} (29.3 kcal mol^{-1}) in chlorobenzene] but considerably above that of the next higher one [pentaene, 86.9 kJ mol^{-1} (20.8 kcal mol^{-1}) in nitrobenzene]. It is, of course, much lower than the racemization barrier in an allene RCH=C=CHR [R = CH$_3$ or (CH$_3$)$_3$C, 46–47 kcal mol^{-1} (192–197 kJ mol^{-1})],[48] which, in turn, is appreciably lower than the 62.2 kcal mol^{-1} (260 kJ mol^{-1}) cis–trans interconversion barrier in *trans*-2-butene (see Section 9-1.a).

Among the nitrogen analogues of allenes, the ketene imines RR′C=C=NR″ and their quaternary immonium salts, RR′C=C=$\overset{+}{N}$R″R‴, and the carbodiimides RN=C=NR are of particular interest. Racemization barriers in both cases were determined indirectly, by observing the coalescence of diastereotopic groups in R [R = (CH$_3$)$_2$CH— or C$_6$H$_5$CH$_2$C(CH$_3$)$_2$—]. For the ketene imine,[49] barriers generally varied between 37 and 63 kJ mol^{-1} (8.8 and 15.1 kcal mol^{-1}); these species are clearly not resolvable. The barrier in (CH$_3$)$_2$CHN=C=NCH(CH$_3$)$_2$, 6.7 ± 0.2 kcal mol^{-1} (28.0 ± 0.8 kJ mol^{-1}) is even lower[50]; however, diferrocenylcarbodiimide has been resolved and its absolute configuration determined.[51] The barriers in ketene immonium salts[52] are too high to be measured by NMR (> 115 kJ mol^{-1}, 27.5 kcal mol^{-1}) and these compounds should be resolvable.

13.3. ALKYLIDENECYCLOALKANES

Although optically active alkylidenecycloalkanes, in contrast to allenes, probably do not occur in nature, their resolution in the laboratory preceded that of allenes by over a quarter of a century. The resolution of 4-methylcyclohexylideneacetic acid (Fig. 13.14, **A**) was already reported in 1909.[53] The absolute configuration of this compound was assigned by Gerlach[54] through correlation with (2R)-(–)-isoborneol of known configuration. The key intermediate in this correlation is cis-(R)-(+)-4-methyl-cyclohexylacetic-α-d acid **C**, obtained from the resolved α-deuterated analogue of 4-methylcyclohexylideneacetic acid, **B**, by catalytic hydrogenation (with Pd/BaSO$_4$), which proceeds cleanly (i.e., without atom or bond migration) and is thus assumed to be syn. The two diastereomeric products **C** and **D** (Fig. 13.14) were separated and their relative configurations (cis or trans) ascertained by comparison with known samples; their optical rotations $[\alpha]_{546}$, +0.44 and +0.65, respectively, were appreciable. To

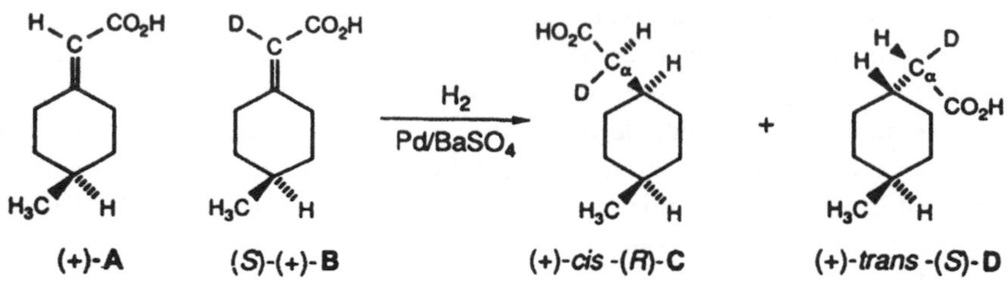

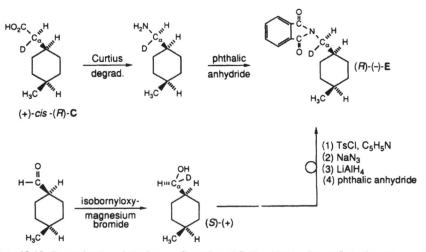

Figure 13.14. Correlation of the configuration of (+)-4-methylcyclohexylideneacetic acid. The configurational descriptors in the products relate to Cα.

determine the absolute configuration of (+)-**C**, it was converted to the corresponding amine by Curtius degradation, which is known to proceed with retention of configuration (cf. Section 5-5.f; see also ref. 2). The (−) amine obtained was then synthesized by an alternative route involving an asymmetric reduction with isobornyloxymagnesium bromide made from (2R)-(−)-isoborneol (Fig. 13.15). It is known from earlier work[55] that the isobornyloxymagnesium bromide reduction of RCHO leads to (S)-RCHDOH. Since the conversion of the alcohol to the amine involves a single inversion (in the tosylate to azide step), the (−)-cis-phthalimide (Fig. 13.15, **E**) is R and its precursor acid (+)-**C** is also R. Reference to Figure 13.14 then indicates that (+)-**B** must be S or P, and since chiral compounds bearing hydrogen and deuterium in the same position generally differ little in rotation (cf. ref. 55b) (+)-**A** is assigned the S or P configuration also. The same assignment is obtained through the trans isomer (+)-**D** which, by further degradation, is assigned the S configuration (Fig. 13.14).[54]

In the above correlation, an alkylidenecycloalkane is converted to a compound with a chiral center of known (or demonstrable) configuration by a reaction (catalytic

Figure 13.15. Determination of absolute configuration of **C** (Fig. 13.14). The configurational descriptors refer to Cα.

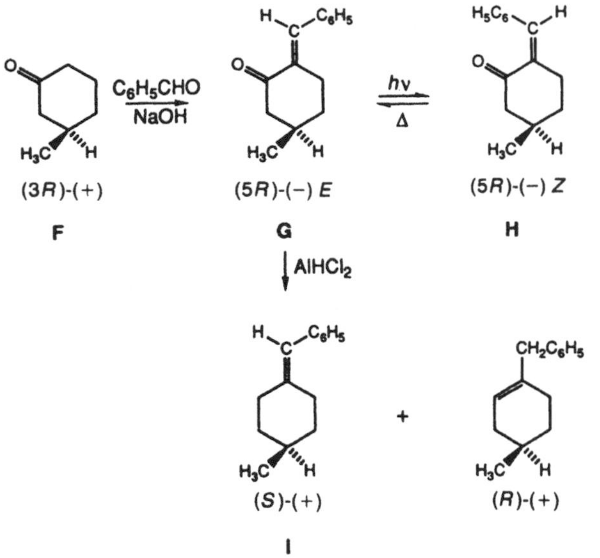

Figure 13.16. Assignment of configuration of (+)-1-benzylidene-4- methylcyclohexane.

hydrogenation) of known mechanism. The opposite course, conversion of a compound with central chirality to a chiral alkylidenecyclohexane, was taken by Brewster and Privett[56] as shown in Figure 13.16. The compound (R)-(+)-3-methylcyclohexanone (**F**) of known configuration was condensed with benzaldehyde to give one of the corresponding (5R)-(−)-2-benzylidene derivatives (**G**). Photochemical isomerization led to a thermally less stable compound (**H**) with a lower λ_{max} and ε_{max} than that of the original isomer. It was concluded that the product of condensation was the E isomer (**G**) and the less stable photoproduct the Z isomer (**H**, Fig. 13.16). Removal of the carbonyl group from the E isomer gave as one of two products (+)-1-benzylidene-4-methylcyclohexane (**I**) whose configuration is thus shown to be S or P (Fig. 13.16). The Lowe–Brewster rules (Fig. 13.12) evidently apply to the alkylidenecyclohexanes shown in Figures 13.14 and 13.16 (assumed polarizabilities $CH_3 > H$, $C_6H_5 > H$, and $CO_2H > H$) but the proviso is added[56] that the rotation in the visible must not be controlled by a near-UV Cotton effect (CE) for this to be so.

The Gerlach and Brewster–Privett assignments of compounds **A** and **I** have been correlated by Walborsky's group,[57,58] who also established, by correlation, the configuration of a number of additional alkylidenecycloalkanes.

Alkylidenecycloalkanes with substituents in the 2 and 3 positions (cf. refs. 58 and 59) display central rather than axial chirality, the carbon atoms in positions 2 or 3 being stereogenic centers. In addition, there is cis–trans isomerism. However, apparent axial chirality may be found in cis-3,5-disubstituted alkylidenecyclohexanes and similar structures (Fig. 13.17, **A**), and a nitrogen analogue of this type (Fig. 13.17, **B**) has, in fact, been synthesized and resolved.[60a] The configuration of this compound

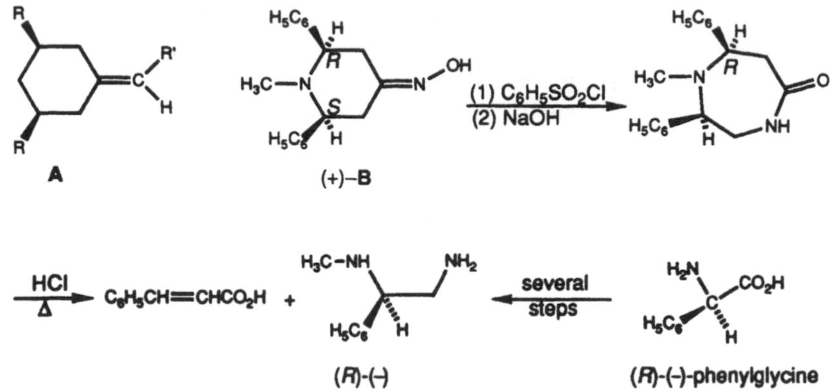

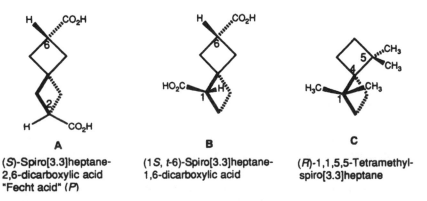

Figure 13.17. Absolute configuration of (+)-1-methyl-2,6-diphenyl-4-piperidone oxime **B**.

has been elucidated[60b] as shown in Fig. 13.17. Beckmann degradation (cf. Fig. 9.20) followed by β elimination and hydrolysis gave (*R*)-(−)-2-methylamino-2-phenylethylamine, whose configuration was established by correlation with the known (*R*)-(−)-phenylglycine (Fig. 13.17). The configuration of the (+) oxime is therefore *Z* (equivalent to *syn-R*).

13-4. SPIRANES

The name "spirane," from the Latin *spira* meaning twist or whorl, implies that spiranes (cf. Fig. 13.2) are not planar; it is their nonplanarity that gives rise to their chirality.

Among the chiral spiranes (Fig. 13.18) one may discern three types: **A**, which definitely displays axial chirality similar to that of allenes and alkylidenecycloalkanes (see above); **B**, which, like corresponding alkylidenecycloalkanes (see above), displays central rather than axial chirality; and **C**, which conceptually would appear

	A	B	C

A

(*S*)-Spiro[3.3]heptane-
2,6-dicarboxylic acid
"Fecht acid" (*P*)

B

(1*S*, *t*-6)-Spiro[3.3]heptane-
1,6-dicarboxylic acid

C

(*R*)-1,1,5,5-Tetramethyl-
spiro[3.3]heptane

Figure 13.18. Types of spiranes.

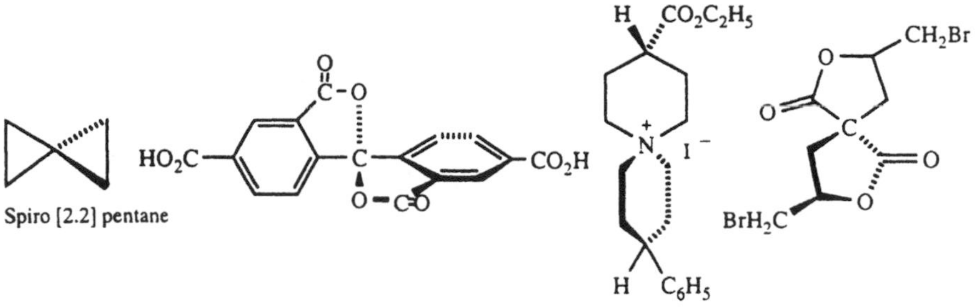

[Figure 13.19 structures: Spiro [2.2] pentane (A); a spirodicarboxylic acid structure (B); a quaternary nitrogen spiro compound with H, CO₂C₂H₅, I⁻, N⁺, BrH₂C, H, C₆H₅ labels (C); a spiro lactone structure with CH₂Br and BrH₂C labels (D)]

A   **B**   **C**   **D**

Figure 13.19. Examples of spiranes.

to display axial chirality but, for purposes of nomenclature, is considered to have a chiral center.[1] Compound **A** is described as indicated in Figure 13.3, the descriptor is aS or P. Compound **B** has four stereoisomers (2 pairs of enantiomers); C(1) is a chiral center, whereas C(6) displays cis–trans isomerism and the stereoisomer shown is 1S,6-trans. To name **C** one arbitrarily gives one ring preference over the other; the more substituted branch in that ring then has priority 1 and the less substituted has priority 3, whereas the corresponding priorities in the arbitrarily less favored ring are 2 and 4. The spiro center C(4) is considered a chiral center and the configuration is then 4R.

The most strained saturated spirane, spiro[2.2]pentane (Fig. 13.19, **A**), was apparently first synthesized in 1896,[61] although it was not then recognized as such.[62] Its strain of 65 kcal mol^{-1} (272 kJ mol^{-1}) is only about 10 kcal mol^{-1} (42 kJ mol^{-1}) greater than that of two isolated cyclopropane rings[63,64] (for an interpretation, see ref. 65). Chirality in spiranes, first recognized by Aschan,[66] was demonstrated in 1920[67] by resolution of a spirodicarboxylic acid (Fig. 13.19, **B**). This compound is of type **C** in Figure 13.18; the central carbon atom can be described as a chiral center. However, a compound of type **A** in Figure 13.18 was resolved 5 years later[68]; it is shown in Figure 13.19, **C**. It is of interest that the spiro center of compound **C** is a quaternary nitrogen rather than a carbon atom. Compound **D** in Figure 13.19 is also interesting; it has a spiro center and two conventional chiral centers; contemplation of models (cf. Fig. 13.20) indicates the existence of three diastereomeric racemates, which have, in fact, been isolated.[69]

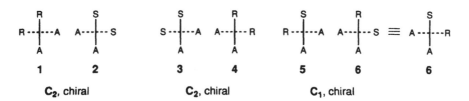

1   **2**   **3**   **4**   **5**   **6**   **6**

C_2, chiral C_2, chiral C_1, chiral

Figure 13.20. Combination of spirane and conventional chiral centers.

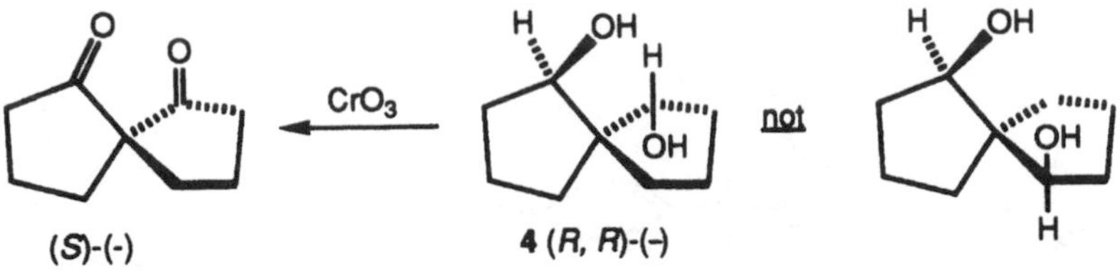

Figure 13.21. Configuration of spiro [4.41 nonane-1,6-dione. Intramolecular hydrogen bonding is possible in isomer **1** and in its enantiomer (**2**, Fig. 13.20).

Several assignments of absolute configuration of spiranes of type **C** in Figure 13.18 have been accomplished (see also ref. 7b, p. 106; ref. 22 pp. 150–151), only one of which will be discussed here. It is concerned with the dione shown in Figure 13.21 and is based on that of its diol precursor shown in the same figure. The relative configuration of this diol had been established by the absence of intramolecular hydrogen bonding and by reductive correlation with a single one of the two diastereomeric mono-ols. The latter result demands C_2 symmetry, that is, structures **5** and **6** (Fig. 13.20) are excluded. The former result (absence of intramolecular hydrogen bonding) excludes **1** (Fig. 13.20) and its enantiomer **2**. Finally, the absolute configuration of the diol was established by Horeau's rule (cf. ref. 71 and p. 96), to be *R,R* (Fig. 13.21). Oxidation of this diol then must lead to the (*S*)-dione (Fig. 13.21), chiral center naming) which, experimentally, was found to be levorotatory.

13-5. BIPHENYLS AND ATROPISOMERISM

a. Introduction

In the examples given so far, the chiral axis is sustained (and the screw or helical sense of the molecule maintained) either by the "stiffness" (high barrier to rotation) of a double bond (allenes) or by the molecular framework as a whole (spiranes) or by a combination of the two (alkylidenecycloalkanes). We now come to molecules with a chiral axis whose helical sense is maintained through hindered rotation about single

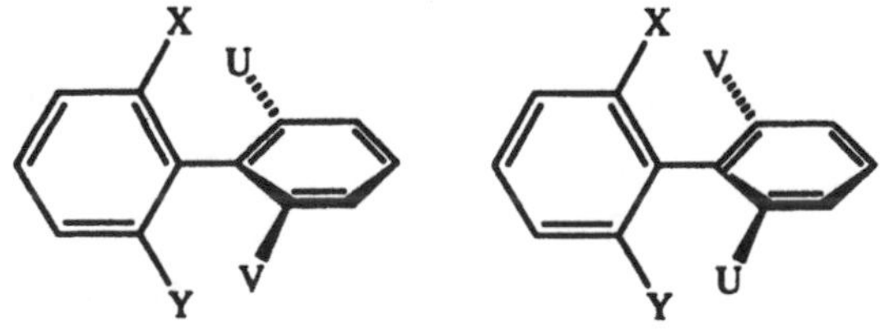

Figure 13.22. Enantiomeric chiral biphenyls.

bonds, the hindrance in general being due to steric congestion. The classical examples of such molecules are the biphenyls (or biaryls in general) shown in Figure 13.22. If $X \neq Y$ and $U \neq W$ and, moreover, the steric interaction of X—U, X—V, and/or Y—V, Y—U is large enough to make the planar conformation an energy maximum, two nonplanar, axially chiral enantiomers (Fig. 13.22) exist. If the interconversion through the planar conformation is slow enough they may, under suitable circumstances, be isolated (resolved). This type of enantiomerism was first discovered by Christie and Kenner[72] in the case of 6,6'-dinitro-2,2'-diphenic acid (Fig. 13.22, $X = U = CO_2H$; Y $= V = NO_2$), which they were able to resolve. It was later called[73] "atropisomerism" (from Greek *a* meaning not, and *tropos* meaning turn).

Reference to Chapter 10 suggests that atropisomerism is a type of conformational (rotational) isomerism in which the conformational isomers or conformers can be isolated. It is immediately obvious that the term suffers from all the problems discussed in Sections 2-4 and 3-1.b: How slow must the interconversion of the enantiomers be (i.e., how long is their half-life) before one speaks of atropisomerism? At what temperature is this measurement to be made? Does atropisomerism still exist when isolation of stereoisomers becomes difficult or impossible but their existence can be revealed by NMR (or other spectral) study, and so on. Ōki[74] arbitrarily defined the condition for the existence of atropisomerism as one where the isomers can be isolated and have a half-life $t_{1/2}$ of at least 1000s (16.7 min). This value still does not define the required free energy barrier, which evidently now depends on temperature; it is 22.3 kcal mol^{-1} (93.3 kJ mol^{-1}) at 300 K, 26.2 kcal mol^{-1} (109.6 kJ mol^{-1}) at 350 K, and 14.7 kcal mol^{-1} (61.5 kJ mol^{-1}) at 200 K. Though this definition is entirely arbitrary, it is convenient and quite essential if the concept of atropisomerism is to be maintained at all.

b. Biphenyls and Other Atropisomers of the sp^2–sp^2 Single-Bond Type

General Aspects.   Atropisomers are numerous in number and type and only a very brief treatment can be given here. Biphenyl isomerism has been extensively discussed earlier,[2,3,75,76] especially in regard to the structural attributes needed to "restrict" rotation. Half-lives of racemization of numerous biphenyls have been determined with the following general findings:

1. Most tetra-ortho substituted biphenyls (Fig. 13.22, U, V, X, Y $\neq$ H) are resolvable and quite stable to racemization unless at least two of the groups are fluorine or methoxy. A nonresolvable, tetra-ortho substituted biphenyl is shown in Figure 13.23, **A**.[75] It should be noted that although the condition $U \neq V$ and $X \neq Y$ is not fulfilled by this molecule, the perpendicular conformation lacks a plane of symmetry because of the meta substituents ($Cl \neq CO_2H$).

2. Tri-ortho substituted biphenyls are readily racemized (short $t_{1/2}$ values) when at least one of the groups is small (CH_3O or F), otherwise racemization tends to be slow (but is possible, generally at elevated temperatures).

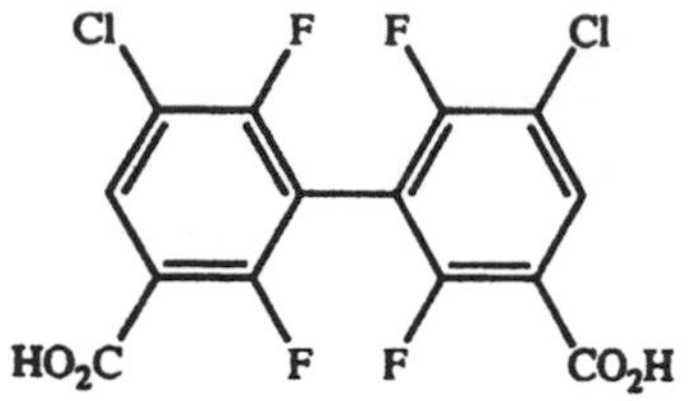

Figure 13.23. A tetra-*o*-fluorosubstituted biphenyl **A** and 1,1′-binaphthyl **B**.

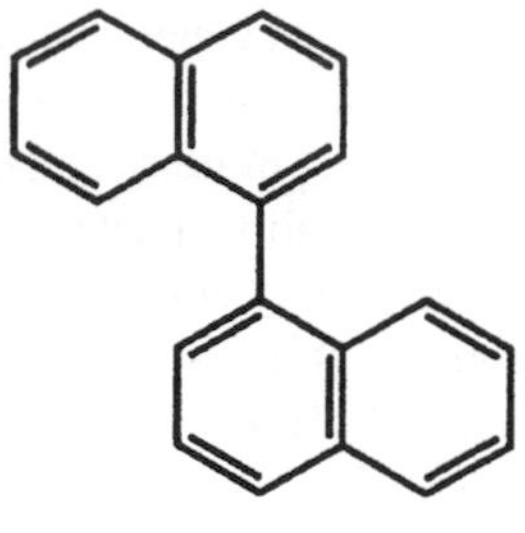

3. Di-ortho substituted biphenyls are generally resolvable only if the substituents are large. An interesting example is 1,1′-binaphthyl (Fig. 13.23, **B**), originally obtained optically active by deamination of the resolved 4,4′-diamino derivative.[77] The racemate exists in two crystalline modifications, a racemic compound, mp 145°C and a conglomerate (cf. Section 6-3), mp 158°C. The latter is easily resolved either spontaneously or by seeding of the melt or solution[78]; above the melting point the enantiomers are in rapid equilibrium ($t_{1/2} \approx 0.5$ s at 160°C; $\Delta G^{\ddagger} = 23.5$ kcal mol^{-1} (98.3 kJ mol^{-1}).

4. Mono-ortho substituted biphenyls are, in general, not resolvable although the (+)-camphorsulfonate of the arsonium salt shown in Figure 13.24 shows mutarotation (cf. p. 477) suggesting that an interconversion of diastereomers occurs in solution because the two diastereomers are not equally stable (asymmetric transformation of the first kind, cf. Section 7-2.d).

5. Substituents, in the meta position tend to increase racemization barriers by what is known as a "buttressing effect," that is, by preventing the outward bending of an ortho substituent, which would otherwise occur in the transition state (coplanar conformation) for racemization. (This outward bending allows the ortho substituents to slip past each other more readily by energy minimization of the activated complex (cf. Section 2-6).

6. The apparent size of substituents (as gauged by racemization rates of differently ortho-substituted biphenyls) is $I > Br \gg CH_3 > Cl > NO_2 > CO_2H \gg OCH_3 > F > H$. This order roughly parallels van der Waals radii ($I > Br > C > Cl > N > O > F > H$; in polyatomic groups allowance must be made for the outer substituents) and is quite different from that of the ΔG^0 values in cyclohexanes (axial-equatorial equilibrium, Table 11.6). In contrast to

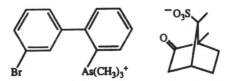

Figure 13.24. Mutarotating mono-orthosubstituted biphenyl.

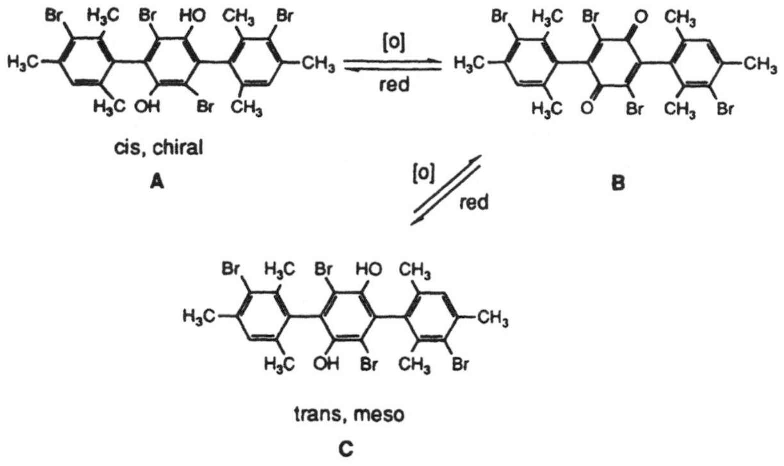

Figure 13.25. cis–trans Isomerism in terphenyls and diphenylquinones.

synaxial substituents in cyclohexanes, ortho substituents on the two rings in biphenyls point at each other, so their interaction should increase with increasing van der Waals (and bond) radii.

7. Activation barriers to racemization can be calculated quite closely by molecular mechanics (cf. Section 2-6); in fact, calculation of barriers of this type constitutes the first application of what is now called the molecular mechanics or force field method.[79] Semi-empirical methods have also been applied to the calculation of barrier height, with some success.[80]

8. Diastereomers are found not only in biphenyls with chiral substituents but also in terphenyls. The compound shown in Figure 13.25, **A** is an example; the cis isomer has been resolved, whereas the trans isomer, which was separated from the cis, cannot be resolved because it has a center of symmetry.[81]

   Oxidation of the optically active cis hydroquinone **A** gave an optically active quinone **B**, which was reduced back to optically active **A**. This finding showed that atropisomerism is possible in structures other than biphenyls. Additional cases of resolvable atropisomers are shown in Figure 13.26.[82–84] Examples of thioamides were discussed earlier (Section 9-1.e; see also ref. 74).

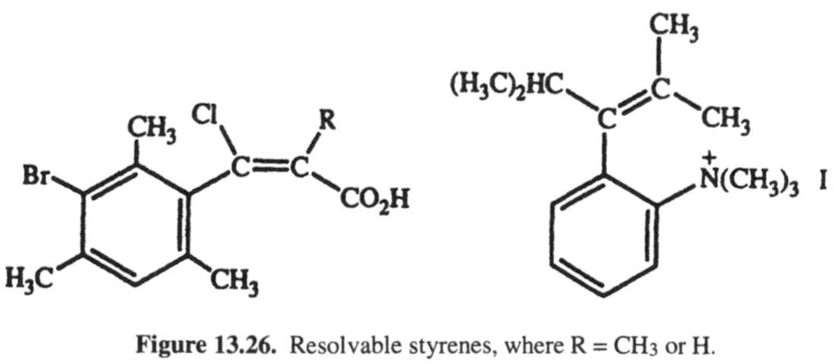

Figure 13.26. Resolvable styrenes, where R = CH₃ or H.

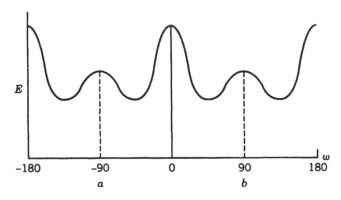

Figure 13.27. Energy profile for rotation in biphenyls.

The energy profile in unbridged biphenyls is more complex than the discussion so far might imply. Conceptually, one might want to think of a "hindered" biphenyl (Fig. 13.22) as being constituted of two orthogonal phenyl rings; the condition that $X \neq Y$ and $U \neq V$ serves to abolish a plane of symmetry that would otherwise exist in that conformation. In fact, however, in solution the rings in biphenyls are neither coplanar nor orthogonal in the lowest energy conformation. The tendency for nonplanarity, which is imposed by the steric demands of the ortho substituents, is opposed by π-electron overlap, which produces maximum stabilization when the rings are coplanar. Even biphenyl itself is nonplanar in the ground state, the inter-ring torsion angle being 44° in the vapor phase[85] though the rings are coplanar in the crystal phase, perhaps because of packing forces.[86] The result is a compromise with the interplanar angle in biphenyls varying from 42° to 90°.[87,88] The energy profile for rotation in ortho substituted biphenyls is shown in Figure 13.27; the maxima (barriers to rotation) occur at 0° and 180° and the regions to the two sides of the 0° maximum correspond to the two enantiomers which may or may not be isolable, depending on the energy difference between the minimum and the lower maximum (0° or ±180°). The curves to the right and left of the ω = 0° maximum are mirror images of each other, but these parts of the curves for each enantiomer, in turn, display a secondary maximum at ±90° with minima near ±44° (and ±136°) on both sides. The maximum at 90° is due to the complete absence of resonance (π-orbital overlap) stabilization of the biphenyl system at this angle. Nonetheless the curve may be quite flat in the ±90° region and in some cases of large steric repulsion between ortho substituents (cf. ref. 87) there may actually be a minimum in energy at 90° instead of a low maximum. For biphenyl itself and biphenyls lacking ortho substituents, the barriers at 0° and 90° are of comparable magnitude, in the range of 1.4–2.0 kcal mol^{-1} (6–8 kJ mol^{-1}).

The strong conjugation band in the UV spectra of biphenyl, o-methylbiphenyl and some ortho, ortho′ disubstituted biphenyls in the 240–250-nm region, which disappears upon more extensive ortho substitution or with more bulky ortho substituents,[89–91] is an indication of residual conjugation (π-orbital overlap); that is,

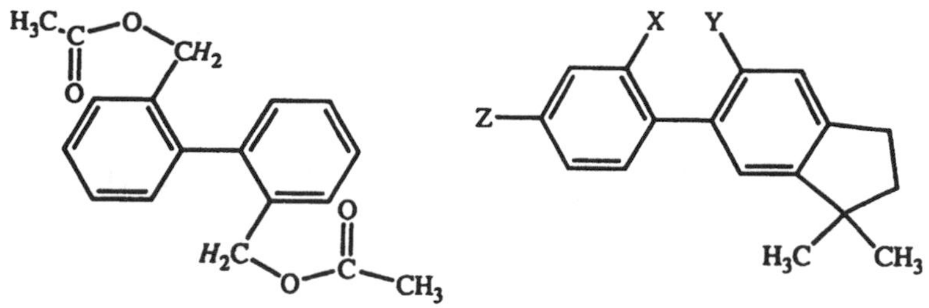

Figure 13.28. Biphenyls in which the rotational barrier has been determined by NMR.

the interplanar angle must be less than 90° in these molecules. In o,o'-dichlorobiphenyl, the torsion angle is near 70° with the two chlorine substituents being syn to each other,[92,93] perhaps because of attractive van der Waals forces.

Nonplanarity can sometimes be demonstrated by NMR spectroscopy even when the 0° barrier is too low for stereoisomers to be isolated. Thus o,o'-diacetoxymethylbiphenyl (Fig. 13.28, **A**) shows an $(AB)_2$ system for the italicized protons at room temperature, whereas at 127°C the pattern collapses to a single line.[94] The CH_2 protons would be enantiotopic (cf. Chapter 8) if the biphenyl system were planar (the biphenyl plane, in that case, would be a symmetry plane relating the two hydrogen atoms), but since the system is nonplanar at room temperature, these protons become diastereotopic and anisochronous (cf. Section 8-4.d). On heating, topomerization (p. 328) occurs and the protons coalesce. From the coalescence pattern, an energy barrier to biphenyl rotation of 13 kcal mol^{-1} (54.4 kJ mol^{-1}) was estimated; this barrier is evidently too low for resolution to be possible. Clearly, this method can be used to predict if a system might reasonably be resolved (and if so, how easily) before resolution is attempted. It does not even require optically active substances though it does require the presence of NMR active diastereotopic nuclei.

Configuration of Biphenyls and Binaphthyls.   Early assignments of configuration of biphenyls and binaphthyls[7,22] were based on correlations with optically active compounds possessing chiral centers of known configuration. Such correlations (e.g., refs. 95 and 96) involved mechanistic arguments including the application of Prelog's rule (Section 5-5.g).[97] The following example is illustrative: Meerwein-Ponndorf reduction of racemic ketone **A** (Fig. 13.29) with (S)-$(+)$-pinacolyl alcohol or (S)-$(+)$-octanol (**B**) was interrupted short of completion (kinetic resolution, cf. Section 7-5). The two faces of the cyclic ketone are homotopic (by virtue of the existence of a C_2 axis in **A**), but the faces in **A** and its enantiomer are (externally) enantiotopic, and therefore their respective interactions with (S)-$(+)$-**B** give rise to diastereomeric transition states. The more favorable interaction (from model considerations) is with (R)-**A**, which is therefore reduced [to (R)-**C**] more rapidly than is (S)-**A** to (S)-**C**. When the reaction is interrupted short of completion, the predominant product must therefore be (R)-**C** and the predominant ketone left-over is (S)-**A**. Since the product

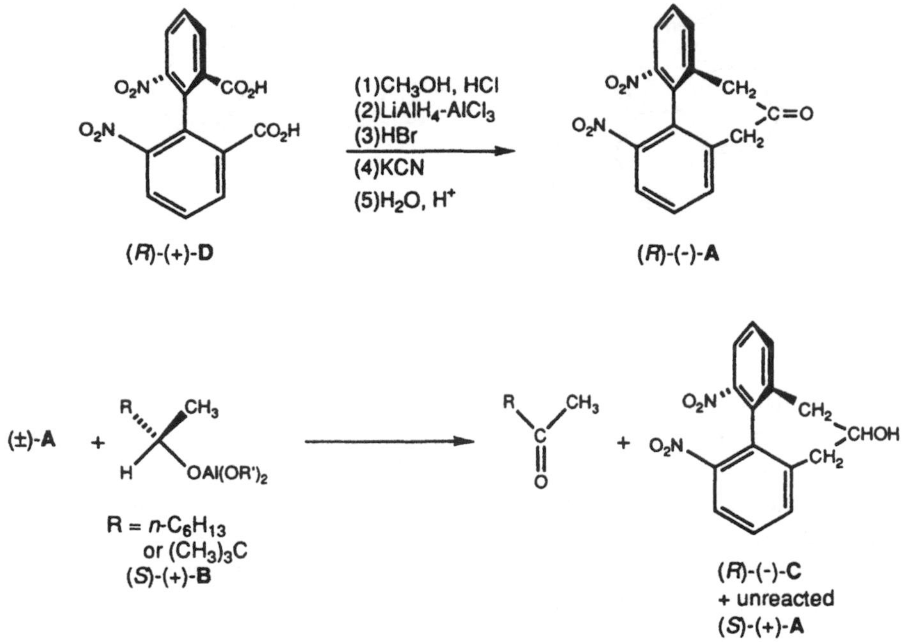

Figure 13.29. Configuration determination of o,o'-dicarboxy-o,o'-dinitrobiphenyl.

alcohol is levorotatory and the left-over ketone is dextrorotatory, it follows (provided the model is correct) that $(-)$-**C** is R and $(+)$-**A** is S; consistency was checked by reduction of $(+)$-**A** to $(+)$-**C** by a symmetric reducing agent, $Al(Oi\text{-}Pr)_3$. Since (S)-$(+)$-**A** is obtained by cyclization of $(-)$-6,6'-dinitro-2,2'-diphenic acid **D**, $(-)$-**D** is also shown to be S. A number of other biphenyls were configurationally connected with $(+)$- or $(-)$-**D** by correlative methods.

Meanwhile, these biphenyl configurations have been confirmed by the Bijvoet X-ray method (Section 5-3.a) of a cobalt complex of 2,2'-diamino-6,6'-dimethylbiphenyl.[99] The $(+)$ ligand has the R configuration; this is in accord with the above-described configurational assignment of (R)-$(+)$-**D** (Fig. 13.29), which can readily be converted chemically into the dextrorotatory 2,2'-diamino-6,6'-dimethyl analogue.[98] The Bijvoet method has also been applied to $(+)$-2,2'-dihydroxy-1,1'-binaphthyl-3,3'-dicarboxylic acid[100]; the crystal contained a molecule of bromobenzene of solvation which provided the desired heavy atom. The $(+)$ isomer is R and it was chemically correlated with a number of other binaphthyls, including (S)-$(+)$-1,1'-binaphthyl itself and the (S)-$(-)$-1,1'-binaphthyl-2,2'- dicarboxylic acid whose absolute configuration had been earlier determined.[95b]

Atropisomerism about an sp^2–sp^2 single bond should, in principle, be possible in appropriately substituted butadienes (Fig. 13.30, **A**) provided the substituents R and R' are large enough. In fact, fulgenic acid (Fig. 13.30, **B**) was resolved as early as 1957[101]; however, it racemized completely in 20 min at room temperature. More

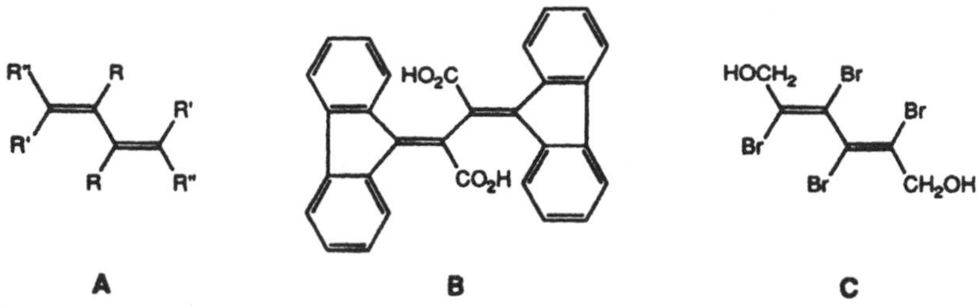

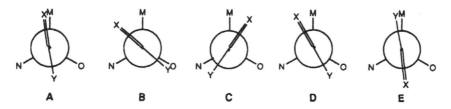

Figure 13.30. Resolvable 1,3-butadienes.

recently, barriers to rotation in a variety of hindered butadienes have been measured by NMR spectroscopy (e.g., ref. 102). Compound **C** (Fig. 13.30) has been resolved both by classical methods[103] and by chromatography on the optically active stationary phase, triacetylcellulose.[104] The racemization barrier in this diene is only 23.7 kcal mol^{-1} (99.2 kJ mol^{-1}) and so it racemizes rapidly in solution on standing.

c. Atropisomerism About sp^2–sp^3 Single Bonds

Whereas atropisomerism about sp^2–sp^2 single bonds involves either a twofold or a fourfold barrier (Fig. 13.27), atropisomerism about an sp^2–sp^3 single bond is either threefold or sixfold, as implied in Figure 13.31. If X is small, the situation will be similar to that in propene (Section 10-2.a); that is, conformer **A** (and corresponding conformers with X eclipsing N or O) will correspond to an energy minimum (since double bonds tend to be eclipsed with single bonds), and conformer **B** and similar conformers in which Y eclipses M or N (such as **E**) will correspond to energy maxima; thus the barrier is threefold. But if X is large and the barrier is dominated by steric interactions, conformers **C** and **D** and corresponding conformers in which X flanks N or O on either side will correspond to energy minima and **A, B** and **E** (plus three other eclipsed conformations) in which X or Y eclipses M, N, or O will correspond to energy maxima and the barrier will be sixfold.

In practice, the situation tends to be less complex, because if X=C–Y is part of an aromatic system, the C=X and C–Y bond orders are equal (or nearly equal) and conformers **A** and **E** have nearly identical torsional interactions. Assuming that X and

Figure 13.31. sp^2–sp^3 barrier. Only A, B, and E are eclipsed conformations.

Y are of moderate size and M is appreciably smaller than N and O, only conformers **A** and **E** need be considered, plus the lower of the two barriers between them, and the situation then formally resembles that in biphenyls (Fig. 13.27) with two minima and two maxima. However, since conformers **A** and **E** are diastereomeric (not enantiomeric) sp^2–sp^3 atropisomerism involves separation and equilibration of diastereomers, not resolution and racemization of enantiomers.

One might think that increasing the size of M, as well as N, O, X, or Y in Figure 13.31 would increase the activation energy for rotation, but this is not necessarily so. For example the **A** $\rightleftharpoons$ **E** interconversion in Figure 13.31 involves passage of X by N and Y by O (or vice versa) but does not require either X or Y to pass by M. Thus if M is made larger (but not so large as to make **A** or **E** energy maxima) the main effect may be to raise the ground-state energy of **A** and/or **E**, and thereby to *lower* the activation energy, assuming that the transition state energy level is largely unaffected (cf. Fig. 11.30). This does, in fact, happen.

The first instance of an sp^2–sp^3 barrier was reported by Chandross and Sheley,[105] who found nonequivalent ortho methyl groups in 9-mesitylfluorene (Fig. 13.32, **A**) at all temperatures studied. A much lower barrier was found in the 9-chloro compound **B**.

Later studies[106] indicated the barriers shown in Table 13.1 (mesityl column). The magnitude of the barriers is **A** > **C** > **B** (Fig. 13.32), in other words, as predicted above, the smallest 9-substituent leads to the highest barrier, and vice versa. Replacing the mesityl group by 2,6-dimethoxyphenyl, on the other hand, lowered the barrier considerably (Table 13.1, 2,6-dimethoxyphenyl column); in other words, diminishing the size of X and Y (Fig. 13.31) lowers the barrier.

High barriers can be observed in compounds of the type shown in Figure 13.33, where atropisomers can actually be isolated and their interconversion studied by classical kinetic methods.

d. Atropisomerism About sp^3–sp^3 Bonds

Barriers to rotation in ethanes have been discussed in Chapter 10. In principle, if these barriers are made high enough, and if the structures are appropriately desymmetrized, atropisomers should be isolable. The framework within which atropisomerism of this type has been most successfully demonstrated contains the triptycene

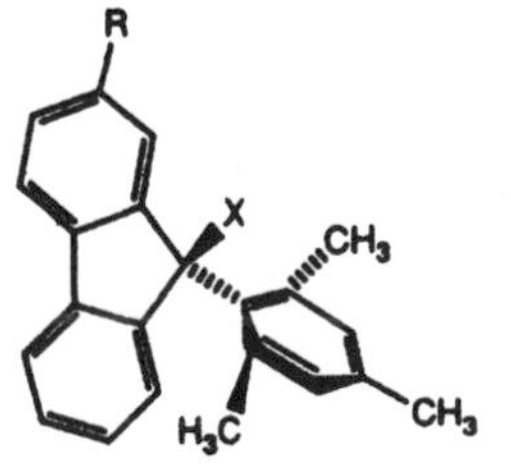

A, X = H, R = H
B, X = Cl, R = H
C, X = OH, R = H
D, X = Cl, R = CH(CH₃)₂

Figure 13.32. 9-Mesitylfluorenes.

TABLE 13.1. Barriers to Rotation about the Aryl-to-Fluorenyl Bond in 9-Arylfluorenes

		ΔG^0			
		Mesityl		2,6-Dimethoxyphenyl	
Compound	9-Substituent	(kcal mol^{-1})	(kJ mol^{-1})[a]	(kcal mol^{-1})	(kJ mol^{-1})[a]
A	H	>25[b]	104[b] (> 190)	20.6	86.2 (145)
C	OH	20.2	84.5 (145)	14.4	60.2 (24)
B	Cl	16.2	67.8 (66)	9.2	38.5 (−81)

[a]Numbers given in parentheses are coalescence temperatures (°C) at which the free energy of activation for rotation was obtained. From ref. 106.
[b]The actual value is probably about 27 kcal mol^{-1} (113 kJ mol^{-1}); see text.

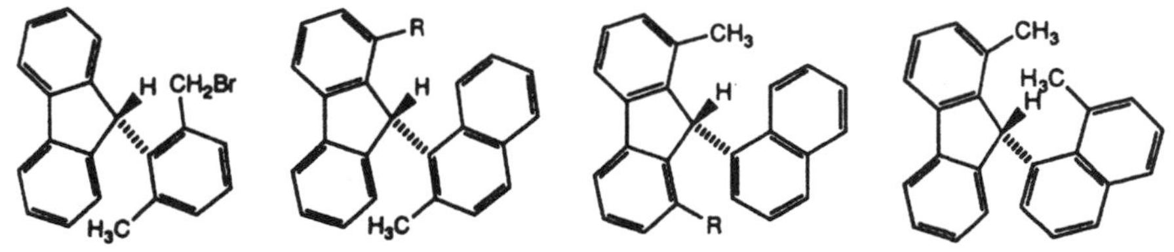

A [107]
27.1 kcal mol^{-1}
113.4 kJ mol^{-1}

B, R = CH$_3$ [108]
33.3 kcal mol^{-1}
139.3 kJ mol^{-1}

C, R = H [109]
29.9 kcal mol^{-1}

D, R = H [110]
21.4 kcal mol^{-1}
89.5 kJ vol^{-1}

E. R = CH$_3$ [111a]
20.6 kcal mol^{-1}

F (sc)* [111b]
23.2 kcal mol^{-1}
97.0 kJ mol^{-1} and
barriers therein

Figure 13.33. Structures of 9-Arylfluorenes and Barriers therein.

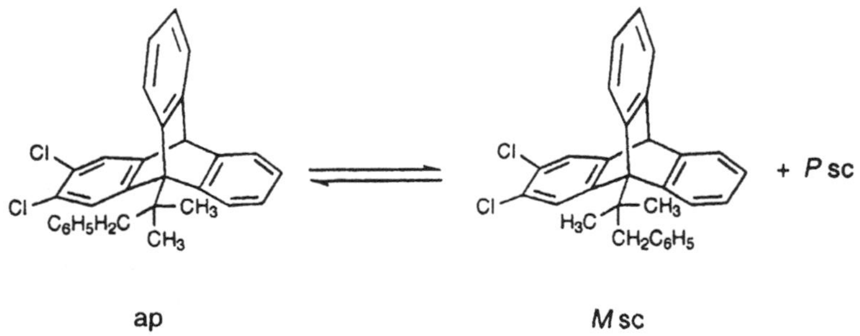

ap   *M* sc

Figure 13.34. Atropisomerism in the 3,4-dichlorotriptycene system.

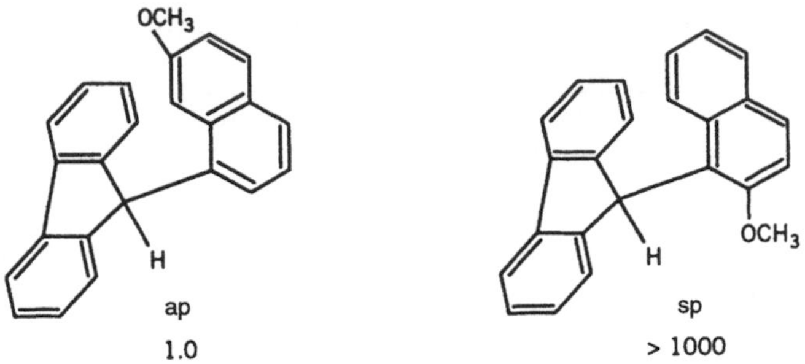

Figure 13.35. Relative rates of lithiation of *ap*- and *sc*-9(2-methoxy-1-naphthyl)fluorene. [Reprinted with permission from Ōki, M. *Recent Advances in Atropisomerism Topics in Stereochemistry*, Vol. 14, Wiley, New York, p. 72. Copyright ©1983 John Wiley and Sons. Inc.]

(tribenzobicyclo[2.2.2]octatriene, Fig. 13.34 or dibenzobicyclo-[2.2.2]octatriene (not shown) structures. Much of this work has been pioneered by Ōki and co-workers and has been reviewed by him.[74,112]

The triptycene systems shown in Figure 13.34 were prepared by in situ addition of dichlorobenzyne to an appropriately substituted anthracene (*ap*) and benzyne to an appropriately substituted dichloroanthracene (*sp*), respectively.[113] Equilibration to a statistical (1:2) mixture proceeded on heating with an activation energy (E_a) of 36.6 kcal mol^{-1} (153 kJ mol^{-1}). A number of similar systems, with different substituents on the aromatic rings and different functional groups in the aliphatic Cabc portion have been examined.[74]

Just as interesting as the stereoselective synthesis of one or other of two atropisomers (diastereomers) of the 9-alkyltrypticene type is their differential reactivity. The clearest examples here involve the sp^3–sp^2 type; thus (cf. Fig. 13.35) in replacement of the acidic H(9) in the fluorene moiety by lithium (with butyllithium) the *sp* isomer reacts over 1000 times as fast as the *ap*, presumably because the methoxy group is available for chelation in the former but not in the latter.[114] The anion appears to have the expected *sp* configuration, since on treatment with water it gives the *sp* starting material.

13-6. MOLECULAR PROPELLERS

This section deals with particular kind of atropisomerism involving "molecular propellers"[115,116] because of their analogy with the (chiral) propellers (two-, three-, or more bladed) of airplanes or boats. Molecules of this type consist of two or more subunits (the blades) radiating from a central axis of rotation (propeller axis), which may be a single atom or a combination of atoms (e.g., a C_2-ethanoid or C_2-ethenoid or C_2-benzenoid unit). Each blade must be twisted in the same sense. If the blades are identical in structure, this may lead to symmetry as high as D_n, but the term "molecular propeller" is not confined to cases where the subunits (blades) are identical.

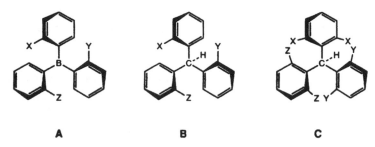

Figure 13.36. Three-bladed propeller molecules.

As in an airplane propeller, helicity can be imposed by having planar blades (subunits), which are all tilted in the same sense or by having truly helical blades, all of the same sense of helicity. The latter case has apparently not yet been realized on the molecular scale.

A straightforward, if perhaps not simple example of a three-bladed propeller is the tri-ortho-substituted triarylboron shown in Figure 13.36, **A**. If the three aryl rings are not coplanar, the molecule is chiral. If we assume, for the moment, that the rings are perpendicular to the plane defined by the boron and the three attached carbon atoms of the aryl rings, there are four diastereomeric arrangements, one with all substituents (X, Y and Z) on the same side of the plane and three with one atom (X, Y, or Z) on one side, and the other two on the other side. Since each arrangement exists in two enantiomeric forms, there are four racemic pairs. If we now change the assumption to one where the rings are not orthogonal to the boron plane but tilted, all in the same direction, with respect to it, then the system acquires helicity with either a right-handed or a left-handed pitch and the number of stereoisomers doubles to 8 racemic pairs. Finally, if the central atom is also made chiral, as in the triarylmethane (Fig. 13.36, **B**), the number of stereoisomers is doubled again, to 16 racemic pairs. This number will be reduced if either two or all three of the rings are identical (X = Y or X = Y = Z) or if the rings have local C_2 axes, as in Fig. 13.36, **C**). The number of racemic pairs in these various cases[115] is shown in Table 13.2.

TABLE 13.2. Number of Racemic Pairs for ArAr'Ar''Z and ArAr'Ar''ZX Systems

System	Number of Identical Rings	Number of Rings with C_2 Axes			
		0	1	2	3
ArAr'Ar''Z	0	8	4	2	1
	2	4	3	1	1
	3	2	a	a	1
ArAr'Ar''ZX	0	16	8	4	2
	2	8	4	2	1
	3	4	a	a	1

aThe system is achiral and there are no diastereomers.

The interconversion of stereoisomers has been considered in terms of "flip mechanisms"[117] by Gust and Mislow.[118] A "flip" is defined as a passage of one or more rings through the plane perpendicular to that of the central atom and its three neighbors ("reference plane"; cf. 13.36, **A**). In compounds **B** and **C** in Figure 13.36, the "flip" carries a ring through a plane containing the central atom, the attached atom of the ring in question, and the singular atom, which is hydrogen in cases **B** and **C**. One can then discern four kinds of mechanisms, depicted in Figure 13.37, called the zero-ring, one-ring, two-ring, or three-ring flip. (The ring or rings that do not flip pass through the reference plane in the transition state.) Each flip mechanism reverses the helicity and, for molecules of the type shown in Figure 13.36, **A** and **B** (i.e., without local C_2 axes) leads to a different diastereomer. For each stereoisomer, eight single-step isomerization paths are possible (one zero-ring, one three-ring, three one-ring, and three two-ring flips) and for the 16 isomers of **A** there are $(16 \times 8)/2$ or 64 interconversion paths possible.[115] The same is true for **B**, since the "flips" do not affect the chiral center, but in **B** only diastereomer interconversion is possible by flipping, whereas in **A** flipping will eventually interconvert enantiomers.

We shall mention here only two examples from several investigations of molecular propellers of the type shown in Figure 13.36 pioneered by Mislow and co-workers. One concerns the tris-1-(2-methylnaphthyl)borane system shown in Figure 13.38, **A**.[119] According to Table 13.2, this system (Ar_3B) should exist as two racemic pairs, one with C_3 symmetry and one with C_1. Low-temperature ${}^1H$ NMR spectroscopy in fact discloses two diastereomers; but at 85°C the spectrum coalesces to that of a single species. Spectral study of the thermodynamics and kinetics of the system shows that the symmetrical isomer (all methyl groups on the same side) is of lower enthalpy

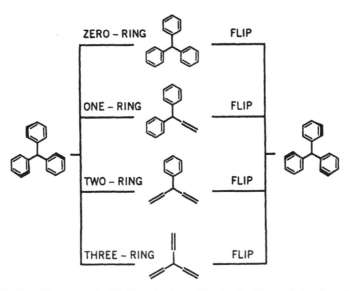

Figure 13.37. Transition states for "flip" mechanisms. [Reprinted with permission from Gust, D. and Mislow, K. *J. Am. Chem. Soc.* **1973**, *95*, 1535. Copyright ©1973 American Chemical Society.]

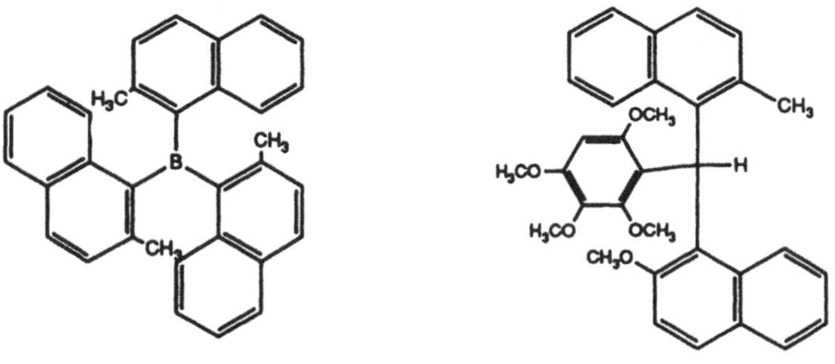

A   **B**

Figure 13.38. Molecular propellers.

($\Delta H^0 = 0.61$ kcal mol^{-1}, 2.55 kJ mol^{-1}) but also, as one might expect from its symmetry number of 3 (cf. Chapter 4), of lower entropy (3.1 cal deg mol^{-1}, 13.0 J deg^{-1} mol^{-1}). The barrier to diastereomer interconversion is quite low [15.9 or 16.2 kcal mol^{-1} (66.5 or 67.8 kJ mol^{-1}), depending on the starting isomer]. The C_1 isomer can enantiomerize without diastereoisomerization with a barrier of 14.6 kcal mol^{-1} (61.1 kJ mol^{-1}). A detailed analysis of the four flip mechanisms and the kind of NMR coalescences they would produce[119] leads to the conclusion that only the one- and two-ring flip mechanisms are compatible with the data; on steric grounds, the two-ring flip mechanism is more likely.

Molecule **B**, Figure 13.38[120] belongs to the general case **B**, Figure 13.36 and, according to Table 13.2 should have 16 racemic pairs. Indeed, at $-40°$C this compound shows a multitude of signals in the ^{1}H NMR. As the temperature is raised, however, many of these signals coalesce, and at 87°C only two sets of signals in a nearly 1:1 ratio remain; evidently two diastereoisomers are quite stable and not readily interconverted. Indeed two crystalline isomers (albeit not quite pure) can be isolated from the material; they interconvert (on the laboratory time scale) on heating to 122° (with an activation energy of 30.4–30.6 kcal mol^{-1} (127.2–128.0 kJ mol^{-1}) from one side or the other. Analysis of the flip mechanisms discloses that the two-ring flip will *not* interconvert all isomers but will leave two separate families of "residual stereoisomers" (cf. Section 3-1.b). These two families are interconverted one into the other only at much higher temperatures, presumably by a one-ring flip mechanism. [Molecular mechanics calculations show the three-ring and zero-ring flips (Fig. 13.37) to have even higher activation energies.] Evidently the rotation of the rings at 87°C, while quite free according to the two-ring flip mechanism, which interconverts 8 of the 16 racemic pairs, is yet not entirely unrestricted, or else there would be no residual diastereoisomerism. The rotation of the three rings is said to be "correlated."[120] Unlike in the case of biphenyls (cf. Fig. 13.27) the two residual stereoisomers; (or stereoisomer sets) can*not* be differentiated by the torsion angles of any *one* of the three rings because in each set members will be found that have the same torsion angles for a given aryl ring. Thus it is the *relationship* between the torsion angles of all three aryl rings, not the individual torsion angles, that defines and differentiates the two residual stereoisomers, hence the term "correlated rotation." [The Prelog–Klyne nomenclature (cf. Table 2.2) cannot be applied in this case.]

13-7. HELICENES

In the chiral molecules described so far, helicity was due to some form of restricted rotation about a chiral axis, due to a high bond order, a rigid framework, or steric factors, combined with an appropriate substitution pattern. In this section we consider molecules whose helicity is inherent in the molecular framework.[121–124]

In 1947, Newman and Hussey[125] succeeded in resolving 4,5,8-trimethyl-phenanthrene-1-acetic acid (Fig. 13.39, **A**) and correctly ascribed its optical activity to nonplanarity enforced by the crowding of the 4,5-methyl substituents. The molecule was of low optical stability and racemized in a matter of minutes, but much more stable molecules of this type (Fig. 13.39, **B**) were synthesized and resolved subsequently.[126] This research culminated in the synthesis of the first optically active helicene, hexahelicene (Fig. 13.39, **C**).[127] Resolution was effected by complexation with α-(2,4,5,7-tetranitro-9-fluorenylideneaminooxy)propionic acid TAPA, structure **89**, Fig. 7.22; for details, see p. 232). The material has the remarkable rotation $[\alpha]_D^{24} - 3640$ (CHCl$_3$) and begins to racemize only at the melting point of 266°C. Contemplation of models shows that the molecule is helical and that passage through a planar transition state for racemization would meet with extraordinary steric difficulty. That the molecule racemizes at all $[\Delta G_{300}^{\ddagger} = 36.2$ kcal mol^{-1} (151.5 kJ mol^{-1}); $t_{1/2} = 13.4$ min at 221.7°C] must mean that the transition state is in fact not planar but that the two ends of the helix slip across the mean plane of the molecule one after the other.[122,124] As one passes from hexahelicene to [9]helicene, the activation free energy to racemization increases only modestly to 43.5 kcal mol^{-1} (182 kJ mol^{-1}).[128a]

Higher carbon helicenes, up to [14]helicene,[129] as well as a number of heterohelicenes, especially those containing thiophene units (cf. ref. 130) up to 15 rings[131] have meanwhile been synthesized, as have double helicenes (Fig. 13.40). Compounds **A** and **B** are [10]helicenes with and without a reversal of helicity at the center and may be considered as two superposed hexahelicenes. Depending on whether the two halves have opposite (P, M) or identical (P, P or M, M) twists, the

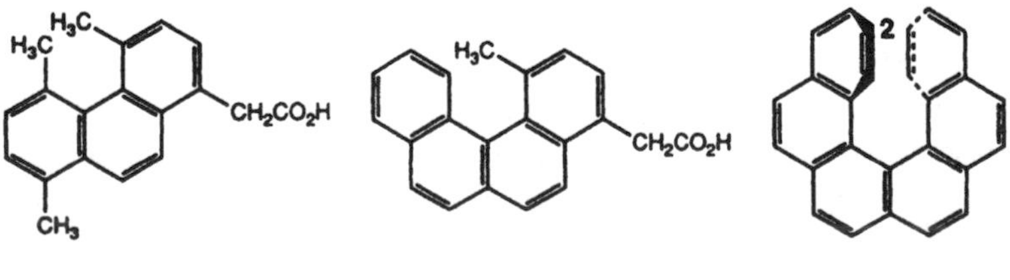

Figure 13.39. Chirality due to "molecular overcrowding."

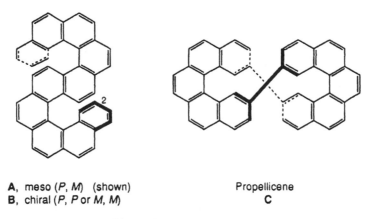

A, meso (*P, M*) (shown)   Propellicene
B, chiral (*P, P* or *M, M*)   **C**

Figure 13.40. Double helicenes.

compound will have a center of symmetry (meso isomeor, **A**) or a C_2 axis (active isomer or racemic pair **B**).[132] The compounds were made photochemically from stilbene precursors and separated chromatographically. As might be expected from the analogy with hexahelicenes, the two diastereomers are interconverted on heating above 320°C. Compound **C**, "propellicene" or bi-2,13-pentahelicenylene[133] results from the fusion of two pentahelicenes (though it is synthesized in a different way) and has **D₂** symmetry. It is both a helicene and a molecular propeller.

The layered nature of the helicene molecule expresses itself in the ¹H NMR spectrum. Thus H(2) in hexahelicene (Fig. 13.39, **C**) lies in the shielding region of the aromatic ring at the other end of the helix and its chemical shift ($\delta = 6.65$) is about 1 ppm upfield of the normal phenanthrene resonance; in [13]helicene this proton is shifted to 5.82 ppm. In compound **A**, Fig. 13.40, the H(2) resonates at 6.40 ppm, slightly upfield from the helicene value, but in isomer **B** it is found at 7.12 ppm, perhaps because the two end rings, while shielded by the middle ones, deshield each other (cf. ref. 122, pp. 99, 101).

X-ray crystallographic data of hexahelicene[134] and of higher helicenes (cf. ref. 122, p. 113) leave no doubt as to the helical nature of these molecules. In hexahelicene, the interplanar angle of the terminal rings is 58.5°. Hexahelicene forms ostensibly chiral crystals and thus appears to be a conglomerate (cf. Section 6-3). However, single crystals of hexahelicene give solutions with only a 2% enantiomeric excess. This disappointing result has been ascribed[135] to "lamellar twinning" of alternate layers of *P*- and *M*-hexahelicenes; the layers are 10–30 μm thick. Nonetheless, once hexahelicene is resolved to the extent of about 20% ee, enantiomerically pure crystals can then be obtained from the solution. Similar problems are not encountered with [7]-, [8]-, and [9]helicene whose conglomerates can be separated by repeated recrystallization after initial seeding with hand-picked crystals.[128b]

As already mentioned for hexahelicene, the specific rotation of helicenes is remarkably high reaching a value of $[\alpha]_D^{25}$ 9620 in [13]helicene.[136] As one proceeds to higher helicenes, $[\alpha]_D^{25}$ increases, but with decreasing increments (cf. ref. 122, p. 91). Because of the very high rotation, combined with the fact that they are readily

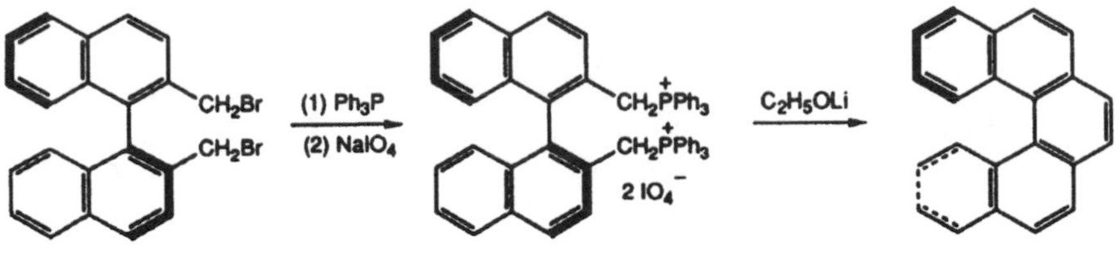

(S)-(−)   (P)-(+)

Figure 13.41. Configurational correlation of pentahelicene with binaphthyl.

synthesized photochemically from stilbene precursors in the presence of oxidants (stilbene → phenanthrene synthesis), helicenes have been prime targets for asymmetric synthesis with circularly polarized light, a technique that had not been noted for success with other types of molecules but did succeed with helicenes[137,138] (see also ref. 139). Even though optical yields were very small (rarely more than 1%), the high specific rotation of the products left no doubt about the accomplishment of a photochemical asymmetric synthesis. The extent of asymmetric synthesis appears to reach a maximum with octahelicene, drops rapidly with the nona and deca compounds and is nil for [11]–[13]helicenes. Left and right circularly polarized light appropriately yield products of opposite rotation. The possibility that optical activation might be due to preferential asymmetric photochemical destruction (a known process[140]) of one of the helicene enantiomers was excluded by control experiments.

The absolute configuration of (−)-hexahelicene has been determined as *M* (as shown in Fig. 13.39, **C**) by a Bijvoet X-ray structure determination of the (−)-2-bromo derivative which was then chemically converted to (−)-hexahelicene.[141] The absolute X-ray method had earlier been applied to a heterohelicene[142] and, in both cases, is in agreement, with the best available calculations (hexahelicene: ref. 143, based on CD spectra; heterohelicene: ref. 144) though not with earlier ones, using less refined methodology.[145]

A very straightforward chemical correlation of (+)-pentahelicene with (S)-(−)-2,2′-bisbromomethyl-1,1′-binaphthyl of known configuration (Section 13-5.b)[146,147] is shown in Figure 13.41; the configuration of (+)-pentahelicene is *P*.

13-8. MOLECULES WITH PLANAR CHIRALITY

a. Introduction

Among molecules with planar chirality,[148] examples of which have been shown in Figure 13.4, the cyclophanes are the most important. Other examples are bridged annulenes, *trans*-cyclooctene, and related molecules, which may alternatively be considered to possess axial chirality (cf. ref. 3) and metallocenes and other metal complexes of arenes.

b. Cyclophanes

The topic of cyclophanes is very extensive[149–153]; only a short presentation of the stereochemical aspects of these molecules can be given here.

The chirality of cyclophanes was discovered by Lüttringhaus and Gralheer[154] in compounds of type **A** in Fig. 13.42, which, at the time, were called "ansa compounds" (from Latin *ansa* meaning handle) but would now be called 1,n-dioxa[n]para-cyclophanes. The first compound to be resolved was **A**, $n = 12$, X = Br; the two bulky ortho substituents prevent the benzene ring from swiveling through the larger ring formed by the dioxamethylene chain, thus this is a form of atropisomerism. When the bulk of X is reduced by reduction of **A**, X = Br to **A**, $n = 12$, X = H the product becomes nonresolvable. Evidently, rotation of the phenyl ring is now rapid enough for racemization to occur on a time scale faster than that of the experiment. However, when $n = 10$, X = H, compound **A** can be resolved by the classical salt formation with alkaloids (Section 7-3.a) and racemizes only extremely slowly even at 200°C. The compound with $n = 11$, X = H[155] is intermediate in enantiomer stability; it racemizes at 82.5°C with $t_{1/2} = 30.5$ h [$E_a = 28.4$ kcal mol^{-1} (119 kJ mol^{-1})]. The [10]paracyclophanecarboxylic acid (Fig. 13.42, **B**, $n = 10$), in which the bridge is purely carbocyclic, has also been prepared and resolved.[156] The [n]paracyclophane with the smallest bridge chemically stable at room temperature is [6]paracyclophane[157a] and its carboxylic acid derivative (Fig. 13.42, **B**, $n = 6$).[157b] X-ray structure determination of the latter shows the substituted carbon atoms of the benzene ring, C(1) and C(4), to be bent out of the plane of the other four by about 20°; the attached benzylic carbon atoms of the bridge are bent up further by about 20° (relative to the plane of the three nearest aromatic ring carbon atoms) and the bond angles in the bridge are distorted to an average of 126.5°.

The configuration of the levorotatory acid **A** in Figure 13.42 ($n = 11$, X = H) has been determined by correlation with a compound of known configuration possessing a chiral center, as shown in Figure 13.43.[158] The key assumption, which appears eminently reasonable, is that catalytic hydrogenation of the aromatic ring occurs from the side opposite the bridge and that the product **B** is stereochemically stable at the

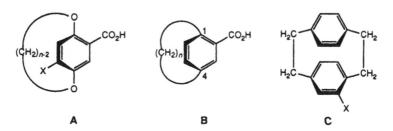

Figure 13.42. Chiral [n]cyclophanes.

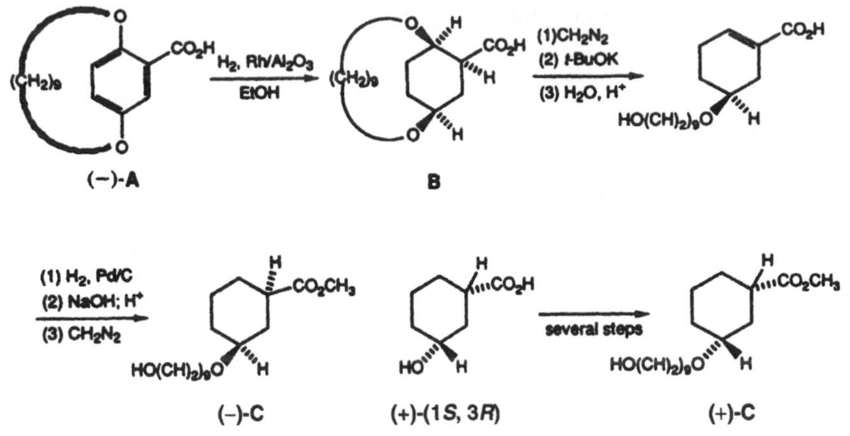

Figure 13.43. Configurational correlation of (−)-1,11-dioxa[11] cyclophanecarboxylic acid.

ether linkages. Esterification, beta-elimination, hydrogenation, followed by equilibration (to the cis isomer of the carboxylic acid), and reesterification yields (−)-**C**, which served as a relay. The dextrorotatory enantiomer, (+)-**C**, was prepared in several steps from (+)-cis-3-hydroxycyclohexanecarboxylic acid, which is known to be R at the carbinol carbon.[7,22] Hence (−)-**C** is 3S and **B** has the configuration shown in Figure 13.43; it follows that the bridge in (−)-**A** is forward when the CO_2H group is on the right. The configuration of (−)-**A** is therefore S or M.

The smallest [m.n]paracyclophane, [2.2]paracyclophane (Fig. 13.42, **C**, X = H) was first obtained accidentally as a by-product of the pyrolytic polymerization of xylene to poly-p-xylylene.[159] Systematic work on this system is largely due to Cram and co-workers.[160, 161] Since the benzene rings cannot freely turn in this structure, the corresponding acid (Fig. 13.42, **C**, X = CO_2H) is chiral and has been resolved[162]; the absolute configuration of the levorotatory enantiomer has been determined as (−)-(R) or (−)-(P) by X-ray diffraction (Bijvoet method).[163]

c. *trans*-Cycloalkenes

When the saturated segment of a *trans*-cycloalkene (Fig. 13.44) is sufficiently short, it is forced out of the plane of the olefinic moiety, which is normally a symmetry plane

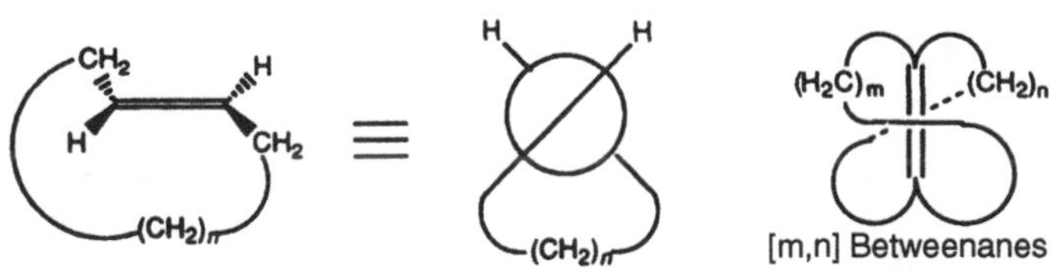

Figure 13.44. *trans*-Cycloalkenes.

of the molecule. The *trans*-cycloalkene then becomes chiral (cf. refs. 44 and 148), its chirality being of the planar type.

That (*E*)-cyclononene is, in principle, chiral was recognized early[164]; however, the two enantiomers of this compound are rapidly interconverted and it was not until 10 years later that the configurationally much more stable (*E*)-cyclooctene was resolved through conversion to and separation of diastereomeric platinum(IV) complexes, *trans*-$C_8H_{14}.PtCl_2(R^*NH_2)$, where R = (+)-$C_6H_5CH_2CH(CH_3)$- (ref. 165).

The configuration of (−)-(*E*)-cyclooctene was shown[166] to be R or P by oxidative (OsO$_4$) conversion to (*S*,*S*)-(+)-1,2-cyclooctanediol whose configuration, in turn, was demonstrated by transformation to its dimethyl ether, which was synthesized independently from (*R*,*R*)-(+)-tartaric acid of known configuration (Fig. 13.45).

The R or P configuration of (−)-(*E*)-cyclooctene was confirmed by X-ray crystallography of the *trans*-$PtCl_2[(+)$-$H_2NCH(CH_3)C_6H_5]$ complex that correlates the configuration of the levorotatory (*E*)-alkene with that of (*R*)-(+)-$C_6H_5CH(CH_3)NH_2$ (cf. Section 5-5.a); an absolute determination of the configuration of the same complex by the Bijvoet method (Section 5-3.a) confirmed the R assignment to the (−)-alkene (as well as the R configuration of (+)-α-methylbenzylamine)[167] (see ref. 7a, p. 216 and ref. 44 for additional correlations). Initial theoretical predictions of the configuration of (−)-cyclooctene based on the twist of the alkene moiety[168] gave the wrong answer, but later predictions[169] do agree with experiment.

In larger *trans*-cycloalkenes, swiveling of the double bond through the polymethylene bridge becomes more facile and the racemization barrier (E_a) drops accordingly: (*E*)-cyclooctene, 35.6 kcal mol^{-1} (149 kJ mol^{-1})[170]; (*E*)-cyclononene, 20 kcal mol^{-1} (83.7 kJ mol^{-1}), measured at low temperature[171]; (*E*)-cyclodecene, 10.7 kcal mol^{-1} (44.8 kJ mol^{-1}), measured by dynamic NMR.[172]

(*E*)-Cycloheptene is surprisingly stable; it can be obtained at −78°C by methyl benzoate sensitized photoisomerization of the cis isomer; the barrier to thermal isomerization to the latter is $\Delta H^{\ddagger} = 17$ kcal mol^{-1} (71 kJ mol^{-1}) and the substance persists for several minutes at 1°C[173,174] (see also ref. 175 for a review on *trans*-cycloalkenes).

An interesting class of cyclic polyunsaturated compounds is represented by the cyclic enediynes containing the −C≡C−CH=CH−C≡C− moiety. Such molecules,

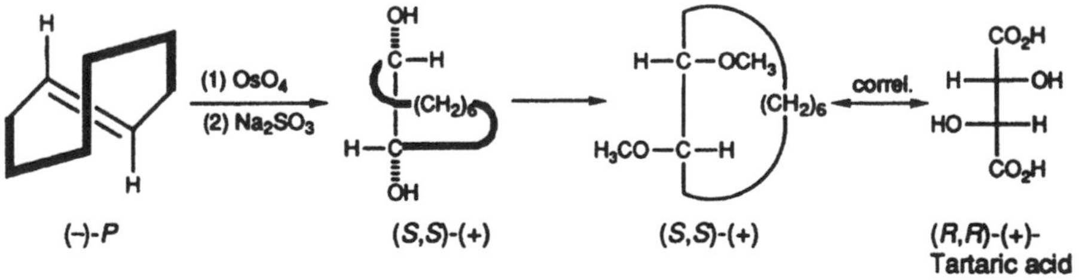

| (−)-P | (*S,S*)-(+) | (*S,S*)-(+) | (*R,R*)-(+)- Tartaric acid |

Figure 13.45. Configurational correlation of (−)-(*E*)-cyclooctene with (+)-tartaric acid. The osmium tetroxide oxidation takes place from the rear (*Si*, *Si*) face, the front face being screened by the hexamethylene bridge.

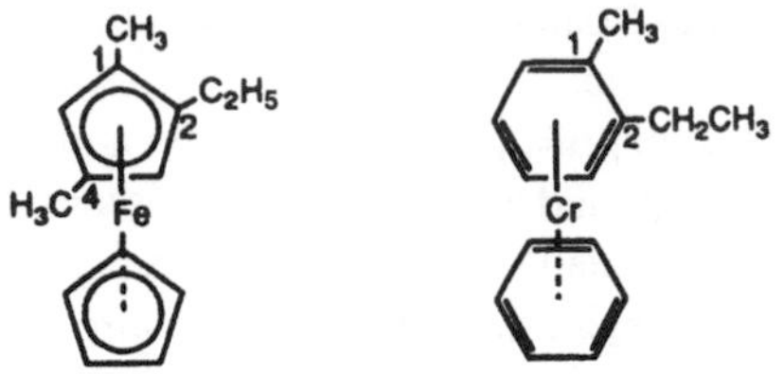

Figure 13.46. Chiral metallocenes.

which include the natural product calicheamicin, readily cyclize thermally in an exothermic process to reactive benzene 1,4-diradicals in a so-called Bergman cyclization.[176] The diradicals, in turn, promote DNA cleavage, a fact that has generated interest in their enediyne precursors.[177]

The so-called "betweenenanes" (Fig. 13.44) are trans-bridged *trans*-cycloalkenes in which the second bridge impedes swiveling at least when the rings are small; such compounds should thus be resolvable and optically stable. Indeed, Nakazaki et al.[178] synthesized [8.8]betweenenane (Fig. 13.44, $m = n = 8$) in optically active form by photochemical asymmetric isomerization of a Z precursor (ketone) in a chiral solvent and later resolved the racemate by chromatography; both procedures gave material of relatively low enantiomeric purity. Marshall and Flynn[179] later obtained nearly enantiomerically pure [10,10]betweenenane (Fig. 13.44, $m = n = 10$) by a reaction sequence involving Sharpless oxidation; since the steric course of the Sharpless oxidation is known, the R configuration could be assigned to (+)-[10,10]-betweenenane.

d. Metallocenes and Related Compounds

Metallocenes (metal "sandwich compounds") and other aryl metal complexes display chirality when the arene is properly substituted (Fig. 13.46). The parent compounds may display average symmetry as high as D_{nh} (e.g., D_{5h} for ferrocene; cf. Fig. 4.18 and discussion on p. 56) since the molecules generally pivot rapidly around the arene–metal axis. Yet an appropriate set of substituents on one of the aromatic rings will destroy all symmetry planes and lead to chiral structures (Fig. 13.46; see also Fig. 13.4, **D**). Although, conceptually, this is presumably a case of planar chirality, it is, for purposes of nomenclature, treated in terms of chiral centers (cf. Section 13.1).

A detailed discussion of metallocenes is outside the scope of this book and the reader is referred to two reviews by Schlögl[6,180]

REFERENCES

1. Cahn, R. S., Ingold, Sir C., and Prelog, V. *Angew. Chem. Int. Ed. Engl.* **1966**, *5*, 385.

2. Eliel, E. L. *Stereochemistry of Carbon Compounds,* McGraw-Hill, New York, 1962.

3. Krow, G. *Top. Stereochem.* **1970**, *5*, 31.

4. Cahn, R. S., Ingold, C. K., and Prelog, V. *Experientia* **1956**, *12*, 81.

5. Prelog, V. and Helmchen, G. *Angew. Chem. Int. Ed. Engl.* **1982**, *21*, 567.

6. Schlögl, K. *Top. Stereochem.* **1967**, *1*, 39.

7. Klyne, W. and Buckingham, J. *Atlas of Stereochemistry,* 2nd. ed. (a) Vol. 1, (b) Vol. 2, Chapman and Hall, London, 1978.

8. van't Hoff, J. H. *La Chimie dans l'Espace,* Bazendijk, Rotterdam, The Netherlands, 1875, p. 29.

9. Maitland, P. and Mills, W. H. *Nature (London)* **1935**, *135*, 994; *J. Chem. Soc.* **1936**, 987.

10. Celmer, W. D. and Solomons, I. A. *J. Am. Chem. Soc.* **1952**, *74*, 1870, 2245; **1953**, *75*, 1372.

11. Rossi, R. and Diversi, P. *Synthesis* **1973**, 25.

12. Murray, M. "Methoden zur Herstellung und Urnwandlung von Allenen bezw. Kumulenen," in Houben-Weyl, 4th ed., Vol V/2a, Thieme, Stuttgart, Germany, 1977.

13. Runge, W. "Stereochemistry of Allenes," in Landor, S. R., ed., *The Chemistry of the Allenes,* Vol. 3, Academic Press, New York, 1982.

14. Landor, S. R. "Naturally Occurring Allenes," in Landor, S. R., ed., *The Chemistry of the Allenes,* Vol. 3, Academic Press, New York, 1982, p. 679.

15. Runge, W. "Chirality and Chiroptical Properties," in Patai, S., ed., *The Chemistry of Ketenes, Allenes and Related Compounds,* part 1, Wiley, New York, 1980, pp. 45, 99.

16. Olsson, L.-I. and Claesson, A. *Acta Chem. Scand.* **1977**, *B31*, 614.

17. Elsevier, C. J., Meijer, J., Tadema, G., Stehouwer, P. M., Bos, H. J. T., and Vermeer, P. *J. Org. Chem.* **1982**, *47*, 2194.

18. Elsevier, C. J., Stehouwer, P. M., Westmijze, H., and Vermeer, P. *J. Org. Chem.* **1983**, *48*, 1103.

19. Hayashi, T., Okamoto, Y., and Kumada, M. *Tetrahedron Lett.* **1983**, *24*, 807.

20. Mori, K., Nukada, T., and Ebata, T. *Tetrahedron* **1981**, *37*, 1343.

21. DeVille, T. E., Hursthouse, M. B., Russell, S. W., and Weedon, B. C. L. *J. Chem. Soc. D Chem. Commun.* **1969**, 754, 1311. Hlubucek, J. R., Hora, J., Russell, S. W., Toube, T. P., and Weedon, B. C. L. *J. Chem. Soc. Perkin I* **1974**, 848.

22. Buckingham, J. and Hill, R. A. *Atlas of Stereochemistry,* 2nd ed., Supplement, Chapman and Hall, New York, 1986.

23. Agosta, W. C. *J. Am. Chem. Soc.* **1964**, *86*, 2638.

24. Alder, K. and Stein, G. *Justus Liebigs Ann. Chem.* **1934**, *514*, 1.

25. Alder, K. and Stein, G. *Angew. Chem.* **1937**, *50*, 510.

26. Lowe, G. *Chem. Commun.* **1965**, 411.

27. Brewster, J. H. *Top. Stereochem.* **1967**, *2*, 1.

28. Crabbé, P., Velarde, E., Anderson, H. W., Clark, S. D., Moore, W. R., Drake, A. F., and Mason, S. F. *J. Chem. Soc. Chem. Commun.* **1971**, 1261.

29. Moore, W. R., Anderson, H. W., Clark, S. D., and Ozretich, M. *J. Am. Chem. Soc.* **1971**, *93*, 4932.

30. Rauk, A., Drake, A. F., and Mason, S. F. *J. Am. Chem. Soc.* **1979**, *101*, 2284.

31. Narayanan, U., Keiderling, T. A., Elsevier, C. J., Vermeer, P., and Runge, W. *J. Am. Chem. Soc.* **1987**, *110*, 4133.

32. Moore, W. R. and Bertelson, R. C. *J. Org. Chem.* **1962**, *27*, 4182.

33. Skattebøl, L. *Acta Chem. Scand.* **1963**, *17*, 1683.

34. Marquis, E. T. and Gardner, P. D. *Tetrahedron Lett.* **1966**, 2793.

35. Visser, J. P. and Ramakers, J. E. *J. Chem. Soc. Chem. Commun.* **1972**, 178.

36. Balci, M. and Jones, W. M. *J. Am. Chem. Soc.* **1980**, *102*, 7608.

37. Wiitig, G. and Fritze, P. *Justus Liebigs Ann. Chem.* **1968**, *711*, 82.

38. Wentrup, C., Gross, G., Maquestiau, A., and Flammang, R. *Angew. Chem. Int. Ed. Engl.* **1983**, *22*, 542.

39. Greenberg, A. and Liebman, J. F. *Strained Organic Molecules,* Academic Press, New York, 1978, p. 127.

40. Johnson, R. P. "Structural Limitations in Cyclic Alkenes, Alkynes and Cumulenes," in Liebman, J. F. and Greenberg, A., eds., *Molecular Structure and Energetics,* VCH, Weinheim, Germany, Vol. 3, 1986, p. 85. Johnson, R. P. *Chem. Rev.* **1989**, *89*, 1111.

41. Garratt, P. J., Nicolaou, K. C., and Sondheimer, F. *J. Am. Chem. Soc.* **1973**, *95*, 4582.

42. Nakazaki, M. *Top. Stereochem.* **1984**, *15*, 199.

43. Nakazaki, M., Yamamoto, K., Maeda, M., Sato, O., and Tsutsui, T. *J. Org. Chem.* **1982**, *47*, 1435.

44. Nakazaki, M., Yamamoto, K., and Naemura, K. *Top. Curr. Chem.* **1984**, *125*, 1.

45. Bertsch, K. and Jochims, J. C. *Tetrahedron Lett.* **1977**, 4379.

46. Kuhn, R., Fischer, H., and Fischer, H. *Chem. Ber.* **1964**, *97*, 1760.

47. Karich, G. and Jochims, J. C. *Chem. Ber.* **1977**, *110*, 2680.

48. Roth, W. R., Ruf, G., and Ford, P. W. *Chem. Ber.* **1974**, *107*, 48.

49. Lambrecht, J., Gambke, B., von Seyerl, J., Huttner, G., Kollmannsberger-von Nell, G., Herzberger, S., and Jochims, J. C. *Chem. Ber.* **1981**, *114*, 3751. Jochims, J. C., Lambrecht, J., Burkert, U., Zsolnai, L., and Huttner, G. *Tetrahedron* **1984**, *40*, 893.

50. Anet, F. A. L., Jochims, J. C., and Bradley, C. *J. Am. Chem. Soc.* **1970**, *92*, 2557.

51. Schlögl, K. and Mechtler, H. *Angew. Chem Int. Ed. Engl.* **1966**, *5*, 596.

52. Lambrecht, J., Zsolnai, L., Hutter, G., and Jochims, J. C. *Chem. Ber.* **1982**, *115*, 172.

53. Perkin, W. H., Pope, W. J., and Wallach, O. *Justus Liebigs Ann. Chem.* **1909**, *371*, 180; *J. Chem. Soc.* **1909**, *95*, 1789.

54. Gerlach, H. *Helv. Chem. Acta* **1966**, *49*, 1291.

55. a) Streitwieser, A., Wolfe, J. R., and Schaeffer, W. D. *Tetrahedron* **1959**, *6*, 338. b) See also Arigoni, D. and Eliel, E. L. *Top. Stereochem.* **1969**, *4*, 127.

56. Brewster, J. H. and Privett, J. E. *J. Am. Chem. Soc.* **1966**, *88*, 1419.

57. Walborsky, H. M. and Banks, R. B. *J. Org. Chem.* **1981**, *46*, 5074. Walborsky, H. M., Banks, R. B., Banks, M. L. A., and Duraisamy, M. *Organometalllics* **1982**, *1*, 667.

58. Duraisamy, M. and Walborsky, H. M. *J. Am. Chem. Soc.* **1983**, *105*, 3252.

59. Hanessian, S., Delorme, D., Beaudoin, S., and Leblanc, Y. *J. Am. Chem. Soc.* **1984**, *106*, 5754.

60. (a) Lyle, R. E. and Lyle, G. G. *J. Org. Chem.* **1959**, *24*, 1679. (b) Lyle, G. G. and Pelosi, E. T. *J. Am. Chem. Soc.* **1966**, *88*, 5276.

61. Gustavson, G. *J. Prakt. Chem.* **1896**, *54*[2], 97.

62. Applequist, D. E., Fanta, G. F., and Henrikson, B. W. *J. Org. Chem.* **1958**, *23*, 1715.

63. Humphrey, G. L. and Spitzer, R. *J. Chem. Phys* **1950**, *18*, 902.

64. Fraser, F. M. and Prosen E. J. *J. Res. Natl. Bur. Stand.* **1955**, *54*, 143.

65. Bernett, W. A. *J. Chem. Educ.* **1967**, *44*, 17.

66. Aschan, O. *Ber. Dtsch. Chem. Ges.* **1902**, *35*, 3396.

67. Mills, W. H. and Nodder, C. R. *J. Chem. Soc.* **1920**, *117*, 1407.

68. Mills, W. H. and Warren, E. H. *J. Chem. Soc.* **1925**, *127*, 2507.

69. Leuchs, H. and Gieseler, E. *Ber. Dtsch. Chem. Ges.* **1912**, *45*, 2114.

70. Gerlach, H. *Helv. Chim. Acta* **1968**, *51*, 1587.

71. Horeau, A. "Determination of the Configuration of Secondary Alcohols by Partial Resolution," in Kagan, H. B., ed., *Stereochemistry, Fundamentals and Methods,* Vol. 3, Thieme, Stuttgart, Germany, 1977, p. 51.

72. Christie, G. H. and Kenner, J. H. *J. Chem. Soc.* **1922**, *121*, 614.

73. Kuhn, R. "Molekulare Asymmetrie," in Freudenberg, K., ed., *Stereochemie,* Deutike, Leipzig, Germany, 1933, p. 803.

74. Ōki, M. *Top. Stereochem.* **1983**, *14*, 1.

75. Adams, R. and Yuan, R. C. *Chem. Rev.* **1933**, *12*, 261.

76. Shriner, R. L., Adams, R., and Marvel, C. S. "Stereoisomerism," in Gilman, H., ed., *Organic Chemistry,* 2nd ed., Vol. 1, Wiley, New York, 1943, p. 343.

77. Cooke, A. S. and Harris, M. M. *J. Chem. Soc.* **1963**, 2365.

78. Wilson, K. R. and Pincock, R. E. *J. Am. Chem. Soc.* **1975**, *97*, 1474.

79. Westheimer, F. H. and Mayer, J. E. *J. Chem. Phys.* **1946**, *14*, 733. Westheimer, F. H. "Calculation of the Magnitude of Steric Effects," in Newman, M. S., ed., *Steric Effects in Organic Chemistry,* Wiley, New York, 1956, p. 523.

80. Kranz, M., Clark, T., and Schleyer, P. v. R. *J. Org. Chem.* **1993**, *58*, 3317.

81. Knauf, A. E., Shildneck, R. R., and Adams, R. *J. Am. Chem. Soc.* **1934**, *56*, 2109.

82. Mills, W. H. and Dazeley, G. H. *J. Chem. Soc.* **1939**, 460.

83. Adams, R. and Miller, M. W. *J. Am. Chem. Soc.* **1940**, *62*, 53.

84. Adams, R., Anderson, A. W., and Miller, M. W. *J. Am. Chem. Soc.* **1941**, *63*, 1589.

85. Almenningen, A., Bastiansen, O., Fernholt, L., Cyvin, B. N., Cyvin, S. J., and Samdal, S. *J. Mol. Struct.* **1985**, *128*, 59.

86. Brock, C. P. *Mol. Cryst. Liq. Cryst.* **1979**, *52*, 163. Brock, C. P. and Minton, R. *J. Am. Chem. Soc.* **1989**, *111*, 4586.

87. Ingraham., L. L. "Steric Effects on Certain Physical Properties," in Newman, M. S., ed., *Steric Effects in Organic Chemistry,* Wiley, New York, 1956, p. 479.

88. Bastiansen, O. and Samdal, S. *J. Mol. Struct.* **1985**, *128*, 115.

89. Pickett, L. W., Walter, G. F., and France, H. *J. Am. Chem. Soc.* **1936**, *58*, 2296.

90. O'Shaughnessy, M. T. and Rodebush, W. H. *J. Am. Chem. Soc.* **1940**, *62*, 2906.

91. Hall, D. M. *Prog. Stereochem.* **1969**, *4*, 1.

92. Bastiansen O. *Acta Chem. Scand.* **1950**, *4*, 926.

93. Rømming, C., Seip, S. M., and Aanesen Øymo, I.-M. *Acta Chem. Scand.* **1975**, *A28*, 507.

94. Meyer, W. L. and Meyer, R. B. *J. Am. Chem. Soc.* **1963**, *85*, 2170.

95. (a) Newman, P., Rutkin, P., and Mislow, K. *J. Am. Chem. Soc.* **1958**, *80*, 465. (b) Mislow, K. and McGinn, F. A. *J. Am. Chem. Soc.* **1958**, *80*, 6036.

96. Berson, J. A. and Greenbaum, M. A. *J. Am. Chem. Soc.* **1958**, *80*, 445.

97. Mislow, K., Prelog, V., and Scherrer, H. *Helv. Chim. Acta* **1958**, *41*, 1410.

98. Mislow, K. *Angew. Chem.* **1958**, *70*, 683.

99. Pignolet, L. H., Taylor, R. P., and Horrocks, W. DeW. *Chem. Commun.* **1968**, 1443.

100. Akimoto, H., Shiori , T., Iitaka, Y., and Yamada, S. *Tetrahedron Lett.* **1968**, 97.

101. Goldschmidt, S., Riedle, R., and Reichardt, A. *Justus Liebigs Ann. Chem.* **1957**, *604*, 121.

102. Köbrich, G., Mannschreck, A., Misra, R. A., Rissmann, G., Rosner, M., and Zundorf, W. *Chem. Ber.* **1972**, *105*, 3794. Mannschreck, A., Jonas, V., Bödecker, H.-O., Elbe, H.-L., and Köbrich, G. *Tetrahedron Lett.* **1974**, 2153. Becher, G. and Mannschreck, A. *Chem. Ber.* **1983**, *116*, 264.

103. Rosner, M. and Köbrich, G. *Angew. Chem. Int. Ed. Engl.* **1974**, *13*, 741.

104. Becher, G. and Mannschreck, A. *Chem. Ber.* **1981**, *114*, 2365.

105. Chandross, E. A. and Sheley, C. F. *J. Am. Chem. Soc.,* **1968**, *90*, 4345.

106. Rieker, A. and Kessler, H. *Tetrahedron Lett.* **1969**, 1227.

107. Nakamura, M. and Ōki, M. *Tetrahedron Lett.* **1974**, 505.

108. Ford, W. T., Thompson, T. B., Snoble, K. A. J., and Timko, J. M. *J. Am. Chem. Soc.* **1975**, *97*, 95.

109. Siddall, T. H. and Stewart, W. E. *J. Org. Chem.* **1969**, *34*, 233.

110. Kajigaeshi, S., Fujisaki, S., Kadoya, N., Kondo, M., and Ueda, K. *Nippon Kagaku Kaishi* **1979**, 239; *Chem. Abstr.* **1979**, *90*, 203375g.

111. Mori, T. and Ōki, M. (a) unpublished results cited in ref. 74, p. 38. (b) *Bull. Chem. Soc. Jpn.* **1981**, *54*, 1199.

112. Ōki, M. *The Chemistry of Rotational Isomers,* Springer, New York, 1993.

113. Yamamoto, G. and Ōki, M. *Bull. Chem. Soc. Jpn.* **1975**, *48*, 3686.

114. Nakamura, M. and Ōki, M. *Chem. Lett.* **1975**, 671.

115. Mislow, K., Gust, D., Finocchiaro, P., and Boettcher, R. J. *Top. Curr. Chem.* **1974**, *47*, 1.

116. Mislow K. *Acc. Chem. Res.* **1976**, *9*, 26; *Chemtracts-Org. Chem.* **1989**, *2*, 151.

117. Kurland, R. J., Schuster, I. I., and Colter, A. K. *J. Am. Chem. Soc.* **1965**, *87*, 2279.

118. Gust, D. and Mislow, K. *J. Am. Chem. Soc.* **1973**, *95*, 1535.

119. Blount, J. F., Finocchiaro, P., Gust, D., and Mislow, K. *J. Am. Chem. Soc.* **1973**, *95*, 7019.

120. Finocchirao, P., Gust, D., and Mislow, K. *J. Am. Chem. Soc.* **1974**, *96*, 3198, 3205.

121. Martin, R. H. *Angew. Chem. Int. Ed Engl.* **1974**, *13*, 649.

122. Laarhoven, W. H. and Prinsen, W. J. C. *Top. Curr. Chem.* **1984**, *125*, 63.

123. Meurer, K. P. and Vögtle, F. *Top. Curr. Chem.* **1985**, *127*, 1.

124. Oremek, G., Seiffert, U., and Janecka, A. *Chem.-Ztg.* **1987**, *111*, 69.

125. Newman, M. S. and Hussey, A. S. *J. Am. Chem. Soc.* **1947**, *69*, 3023.

126. Newman, M. S. and Wise, R. M. *J. Am. Chem. Soc.* **1956**, *78*, 450.

127. Newman, M. S. and Lednicer, D. *J. Am. Chem. Soc.* **1956**, *78*, 4765.

128. Martin, R. H. and Marchant, M.-J. (a) *Tetrahedron* **1974**, *30*, 347; (b) *ibid.* **1974**, *30*, 343.

129. Martin, R. H. and Baes, M. *Tetrahedron* **1975**, *31*, 2135.

130. Wynberg, H. *Acc. Chem. Res.* **1971**, *4*, 65.

131. Yamada, K.-I., Ogashiwa, S., Tanaka, H., Nakagawa, H., and Kawazura, H. *Chem. Lett.* **1981**, 343.

132. Laarhoven, W. H. and Cuppen, Th. H. J. M. *Recl. Trav. Chim. Pays-Bas* **1973**, *92*, 553.

133. Thulin, B. and Wennerström, O. *Acta Chem. Scand.* 1976, B30, 688.

134. DeRango, C., Tsoucaris, G., Declerq, J. P., Germain, G., and Putzeys, J. P. *Cryst. Struct. Commun.* **1973**, *2*, 189.

135. Green, B. S. and Knossow, M. *Science* **1981**, *214*, 795.

136. Martin, R. H. and Libert, V. *J. Chem. Res.* **1980**, *Synopsis*, 130; *Miniprint*, 1940.

137. Kagan, H., Moradpour, A., Nicoud, J. F., Balavoine, G., Martin, R. H., and Cosyn, J. P. *Terahedron Lett.* **1971**, 2479. Moradpour, A., Nicoud, J. F., Balavoine, G., Kagan, H., and Tsoucaris, G. *J. Am. Chem. Soc.* **1971**, *93*, 2353. Moradpour, A., Kagan, H., Baes, M., Morren, G., and Martin, R. H. *Tetrahedron* **1975**, *31*, 2139.

138. Bernstein, W. J., Calvin, M., and Buchardt O. *Tetrahedron Lett.* **1972**, 2195; *J. Am. Chem Soc.* **1972**, *94*, 494; *ibid.* **1973**, *95*, 527.

139. Buchardt, O. *Angew. Chem. Int. Ed. Engl.* **1974**, *13*, 179.

140. Kuhn, W. and Knopf, E. *Z. Phys. Chem.* **1930**, *7B*, 292.

141. Lightner, D. A., Hefelfinger, D. T., Powers, T. W., Frank, G. W., and Trueblood, K. N. *J. Am. Chem. Soc.* **1972**, *94*, 3492.

142. Groen, M. B., Stulen, G., Visser, G. J., and Wynberg, H. *J. Am. Chem. Soc.* **1970**, *92*, 7218.

143. Hug, W. and Wagnière, G. *Tetrahedron* **1972**, *28*, 1241.

144. Groen, M. B. and Wynberg, H. *J. Am. Chem. Soc.* **1971**, *93*, 2968.

145. Moscowitz, A. J. *Tetrahedron* **1961**, *13*, 48.

146. Bestmann, H. J. and Both, W. *Angew. Chem. Int. Ed. Engl.* **1972**, *11*, 296.

147. Mazaleyrat, J.-P. and Welvart, Z. *Nouv. J. Chem.* **1983**, *7*, 491.

148. Schlögl, K. *Top. Curr. Chem.* **1984**, *125*, 27.

149. Vögtle, F. and Neuman, P. *Top. Curr. Chem.* **1974**, *48*, 67. Vögtle, F. and Hohner, G. *Top. Curr. Chem.* **1978**, *74*, 1. Vögtle, F., ed. *Cyclophanes I, II, Topics in Current Chemistry*, Vols. 113 and 115, Springer, New York, 1983. Vögtle, F. *Cyclophane Chemistry: Synthesis, Structure and Reactions*, Wiley, New York, 1993.

150. Boekelheide, V. *Acc. Chem. Res.* **1980**, *13*, 65.

151. Keehn, P. M. and Rosenfeld, S. M., eds. *Cyclophanes*, Academic Press, New York, Vols. 1 and 2, 1983.

152. Kiggen, W. and Vögtle F. "More than Twofold Bridged Phanes," in Izatt, R. M. and Christensen, J. J., eds., *Synthesis of Macrocycles*, Wiley, New York, 1987, p. 309.

153. Diederich, F. *Cyclophanes*, Monographs in Supramolecular Chemistry, No. 2, The Royal Society of Chemistry, London, 1991.

154. Lüttringhaus, A. and Gralheer, H. *Justus Liebigs Ann. Chem.* **1942**, *550*, 67; *ibid* **1947**, *557*, 108, 112.

155. Lüttringhaus, A. and Eyring, G. *Justus Liebigs Ann. Chem.* **1957**, *604*, 111.

156. Blomquist, A. T., Stahl, R. E., Meinwald, Y. C., and Smith, B. H. *J. Org. Chem.* **1961**, *26*, 1687.

157. a) Kane, V. V., Wolf, A. D., and Jones, M. *J. Am. Chem. Soc.* **1974**, *96*, 2643. b) Tobe, Y. and others *J. Am. Chem. Soc.* **1983**, *105*, 1376.

158. Schwartz, L. H. and Bathija, B. L. *J. Am. Chem. Soc.* **1976**, *98*, 5344.

159. Brown, C. J. and Farthing A. C. *Nature (London)* **1949**, *164*, 915. Brown, C. J. *J. Chem. Soc.* **1953**, 3265. Farthing, A. C. *J. Chem. Soc.* **1953**, 3261.

160. Cram, D. J. and Steinberg H. *J. Am. Chem. Soc.* **1951**, *73*, 5691.

161. Cram, D. J., Hornby, R. B., Truesdale, E. A., Reich, H. J., Delton, M. H., and Cram, J. M. *Tetrahedron* **1974**, *30*, 1757.

162. Cram, D. J. and Allinger, N. L. *J. Am. Chem. Soc.* **1955**, *77*, 6289.

163. Frank, G. W. Unpublished result cited in Tribout, J., Martin, R. H., Doyle, M., and Wynberg, H. *Tetrahedron Lett.* **1972**, 2839 (footnote b, p. 2842).

164. Blomquist, A. T., Liu, L. H., and Bohrer, J. C. *J. Am. Chem. Soc.* **1952**, *74*, 3643.

165. Cope, A. C., Ganellin, C. R., and Johnson, H. W. *J. Am. Chem. Soc.* **1962**, *84*, 3191.

166. Cope, A. C. and Mehta, A. S. *J. Am. Chem. Soc.* **1964**, *86*, 5626.

167. Manor, P. C., Shoemaker, D. P., and Parkes, A. S. *J. Am. Chem. Soc.* **1970**, *92*, 5260.

168. Moscowitz, A. and Mislow, K. *J. Am. Chem. Soc.* **1962**, *84*, 4605.

169. Levin, C. C. and Hoffmann, R. *J. Am. Chem. Soc.* **1972**, *94*, 3446.

170. Cope, A. C. and Pawson, B. A. *J. Am. Chem. Soc.* **1965**, *87*, 3649.

171. Cope, A. C., Banholzer, K., Keller, H., Pawson, B. A., Whang, J. J., and Winkler, H. J. S. *J. Am. Chem. Soc.* **1965**, *87*, 3644.

172. Binsch, G. and Roberts, J. D. *J. Am. Chem. Soc.* **1965**, *87*, 5157.

173. Inoue, Y., Ueoka, T., Kuroda, T., and Hakushi, T. *J. Chem. Soc. Perkin 2* **1983**, 983. Inoue, Y., Ueoka, T., and Hakushi, T. *J. Chem. Soc. Perkin 2* **1984**, 2053.

174. Squillacote, M., Bergman, A., and De Felippis, J. *Tetrahedron Lett.* **1989**, *30*, 6805.

175. Warner, P. M. *Chem. Rev.* **1989**, *89*, 1067.

176. Jones, R. R. and Bergman, R. G. *J. Am. Chem. Soc.* **1972**, *94*, 660.

177. Nicolaou, K. C. and Smith, A. L. *Acc. Chem. Res.* **1992**, *25*, 497.

178. Nakazaki, M., Yamamoto, K., and Maeda, M. *J. Org. Chem.* **1980**, *45*, 3229.

179. Marshall, J. A. and Flynn, K. E. *J. Am. Chem. Soc.* **1983**, *105*, 3360.

180. Schlögl, K. *Pure Appl. Chem.* **1970**, *23*, 413.

INDEX

aa Conformer, 386
$A^{(1-2)}$ strain, 470–472
$A^{(1-3)}$ strain, 470, 471
A value, 442
Ab initio calculations, 21, 380, 464, 501
  of chiroptical properties, 594
Abiotic theories, 139
Absolute configuration, 67, 77, 80, 84, 89
Absolute rotation, 143
Absorption edge (X-rays), 76
Abzymes, 270
Accidental degeneracy, 59
Accidental isochrony, 321
Acetaldehyde, 312, 389
  enzymatic reduction of, 330–332
Acetaldehyde-1-d, 94, 331
Acetamide, 392
Acetic acid, 307, 309, 310
Acetoin, 598
2-Acetonaphthone, triplet energy of, 369
Acetone, 308, 309
  preferred conformation of, 390
  triplet energy of, 369
Acetophenone oxime, 341
Acetylcholine, 386
Acetylmandelic acid, 218, 219, 222
Acetylphenylalanine methyl ester, 272
N-Acetylphenylglycine, 244

N-Acetylvaline $tert$-butyl ester, 128
Achiral, diastereomers, 304
  ligand, 310
  molecules, 540
  shift reagents, 151, 155
α_1-Acid glycoprotein (α_1-AGP), 172
Acid halides, conformation of, 390
Acid quenching, 414
Acid strength, in determination of E-Z
      configuration, 356
Acids, kinetic resolution of, 263
cis-Aconitic acid, 305
Acrolein, 394
Acrylic acid, 394
Acrylyl chloride, 394
Activation enthalpy of ring closure, 432, 433
Activation entropy of ring closure, 432, 433
Active-site models, 270
Acyclic diastereomers, physical and
      spectral properties of, 398ff
Acyl halides, racemization of, 281
Acyl oxygen bond, 391, 392
Acylases, 271
Acylate complexes [$Met_2(O_2CR)_4$], CD of, 573
Acyloin reaction, 432
Adamantane, 515, 516
  strain energy in, 516
  symmetry of, 57

Adamantanone, 563
Adamantanones, chiral, 558
4-(1-Adamantyl)homoadamant-3-ene, 504
Addition criterion for heterotopicity,
    307–314
Addition,
    of metal hydrides to cyclohexanones,
        467, 468
    of organometallics to cyclohexanones,
        467–469
Adenine dinucleotide, *see NAD⁺*
Adenosine triphosphate (ATP), 271, 305
Adenosyladenosine, 582
Adrenaline, see Epinephrine
Adsorbed additives, 77–79
ag Conformer, 386
Aggregation, 589
α₁-AGP, *see* α₁-Acid glycoprotein
Agreement factor, *see* R factor
Agrobacterium radiobacter, 208
Alanine, 20, 198, 244
    dimerization of, 52, 53
    N-carboxyanhydride, 141
L-Alanyl-L-proline, 348
Alcohols, kinetic resolution of, 262
Aldehydes, conformation of, 389, 390
    symmetry of, 52
Alder-Stein rules, 615
Aldohexoses, 40, 476, 477
Aldol condensation, adaptation to kinetic
        resolutions, 264
    intramolecular, 436
Aldopentose, 40, 43
Aldopyranoses, 477
Aldose sugars, 39–41
Aldotetroses, 40, 41
Alkalase™, *see* Subilopeptidase
Alkaloids, 220, 266
Alkanes, higher, 380–383
Alkene triplet formation, 367
Alkene-consuming reactions, 352
Alkene-forming reactions, 352
Alkenes, 339ff.
    chemical equilibration of, 366ff.
    chiral, determination of enantiomeric
        purity of, 159
    diastereomerism of, 4, 32, 33
        disubstituted, 358, 363
        kinetic resolution of, 265

racemization of, 280
skewed, 564
steric crowding in, 343, 344
symmetrical, 360
thermodynamically controlled
    syntheses of, 369
trisubstituted, 358, 359, 361, 363
twisted, 344, 345
2-Alkoxyoxanes, 478
2-Alkoxythianes, 478
Alkyl groups, conformational energies of, 447
Alkyl halides, resolution of, 236
Alkyl oxygen bond, 391
N-Alkylamides, conformational energies of,
    392
Alkylbenzenes, 395–397
N-Alkyl-D-glucamines, 212, 217
Alkylidenecycloalkanes, 608, 617–620
    configuration of, 596
Alkylidenephosphorane, 351, 352
1-Alkyl-4-methylcyclohexanes, 446
β-Alkylstyrenes, configuration of, 397
2-Alkylthiooxanes, 474
2-Alkylthiothianes, 478
9-Alkyltriptycenes, 632
Alkynes, addition to, 350
Allenes, 51, 159, 304, 608, 611ff
    bridged, 642
    configuration of, 596
    cyclic, 616
    doubly bridged, 616, 642
    kinetic resolution of, 265
    optically active, 612, 613
    stereoisomerism in, 342
Allenic alcohols, 235
N,N,N-Allylethylmethylanilinium iodide,
    206
Allylic alcohols, epoxidation of, 261
    kinetic resolution of, 262
Allylic strain, *see* A^(1.3) strain
α,β-effects (NMR), 456–458
Alternating axis of symmetry, 3, 36, 45–47,
    49, 52, 53, 310, 314, 315
Ambrosia beetle, 134
    aggregation pheromone of, 272
Amides, 392, 393
    barriers in, 65, 347
    bond order in, 65
Amines, inversion of, 30, 33–35, 329

kinetic resolution of, 263
D-Amino acid oxidase, 135
α-Amino acids, 133
  analysis of, 172
  configurational notation of, 75
  dating of, 287
  racemization of, 285
D-Amino acids, 134, 135
L-Amino acids, 134
β-Amino alcohols, kinetic resolution of, 265
3-Amino-1,4-benzodiazepin-2-one, 242
3-*endo*-Aminoborneol, 214
2-Amino-l-butanol, 216
  Cottonogenic derivatives of, 572
α-Aminocaproic acid, 78, 79
ε-Aminocaproic acid, 78, 79
2-Amino-2-deoxyglucopyranoses, 222
2-Amino-1,2-diphenylethanols, 213
1-Amino-2-methoxymethylpyrrolidine, 224
3-Aminomethylpinane, 213
2-Amino-l-phenyl-1,3-propanediol, 213
2-Amino-l-propanol, TFA derivative, 168
Aminostratigraphy, 287
Ammonia, symmetry of, 54
Ammonium malate, 202
Amphetamine, 211, 223
Amplitude (ORD), 535, 543
Analysis, of amino acids, 176
Anancomeric, 442, 494
Anderson-Shapiro reagent, 149
Angle bending, 25
Angle of pucker, 481, 492
Angle of rotation, 5–7
Angle strain, 429–431, 467, 480, 482
Anharmonicity, 19
Anisochronous nuclei, 318, 319, 323–326
Anisochronous protons, in stereochemical
  analysis, 320, 321
Anisochrony, 124, 156, 157, 318, 319, 325
  origin of, 323–326
  self-induced, 125
Anisole, conformation of, 396
Anisometric aggregates, 103
Anisotropic absorption, 544
Anisotropic radiation, 534
Anisotropic refraction, 535, 538, 544
  vibration, 18
Anisotropy factor (or ratio), *see* g number
p-Anisyl α-methylbenzyl ketone, 207

Anomalous ORD curve, 542, 543
Anomeric center, 417
Anomeric effect, 386–388, 476, 478
  generalized, 386–388
  in second- and third-row elements, 480
Anomers, 477
Anosmia, 134
Ansa compounds, 608, 611, 639
Anthracene, 55
α-l-(9-Anthrytl)-2,2,2-trifluoroethoxyacetic
  acid, 148
Anti, 14, 22. *See also* Antiperiplanar
  conformer, destabilization of, 384
  isomer, 14
Antiaromatic 8π-electron transition, 489
Antiaromatic species, 481
Antibodies, 179
Antibody proteins, application to kinetic
  resolution, 270
Anticlinal, 16
Antimer, *see* Enantiomer
Antiperiplanar, 16
  addition, 369, 370
  attack, 464
  conformation, 378, 394, 398
  elimination, 350, 351, 369, 370, 461
  orbitals, 461–463
Antipodes, *see* Optical antipodes
Arabinose, 40, 87, 88
Arenechromium tricarbonyl, 610, 611
Arenes, 159
Arginine, 217, 220, 332
Aromatic solvent induced shifts (ASIS), 151
9-Arylfluorenes, barriers in, 630, 631
Arylglycines, enzymatic resolution of, 271
ASIS, *see* Aromatic solvent induced shifts
Asparagine, 133, 198
Aspartame, 133, 175
Aspartase, 333–335
Aspartic acid, 287, 333–335
  rotation of, 589
Aspartic-3-*d* acid, 334
N-Aspartylphenylalanine methyl ester, *see*
  Aspartame
Association, effect on specific rotation, 6
Asymmetric, 49
  atom, 32
  destruction, 50. *See also* Kinetic
    resolution

Asymmetric (*continued*)
  photodestruction, 140
  photolysis, 140
  synthesis, 49, 312. *See* also
    Enantioselective synthesis and
    configuration assignment, 95–99
  in chromatographic resolutions, 208
  transformation
    of cyanohydrin, 243
    of diastereomers, 240
    of the first kind, 205, 240
    of racemates, 204
    of the second kind, 241
Asymmetry, 46
"At a glance" nomenclature, 81
Atactic vinyl homopolymers, 583
ATEA reagent, 158
Atomic asymmetry, 595, 596
Atomic coordinates, 8
Atomic refractions, 595
Atrolactic acid, 97, 215
Atropisomeric binaphthols, 171
Atropisomeric polymers, 583
Atropisomerism, 622ff
  about *sp2-sp3* single bonds, 629–631
  about *sp3-sp3* bonds, 630–632
Atropisomers, 66, 608, 622
Automerization, *see* Topomerization
Autoracemization, 283
Auwers-Skita rule, *see under* von Auwers
Auxin, *see* Plant growth stimulant
Average symmetry, 52, 56, 58, 309, 450
Average symmetry plane, 52
Averaging heterotopic nuclei, 325, 326
Axial, 34
  atoms, 34, 438
  chirality, 317, 608, 609
  haloketone rule, 554, 558
  ligands, 442–447
  methylcyclohexane interaction, 495
  preference, of substituents, in saturated
    heterocycles, 476
  prochirality, 317
  protons, 454, 455
Axis of chirality, *see* Axial chirality and
    Chiral axis
Axis of prostereoisomerism, 304
Axis of rotation, 3

Axis of symmetry, 45, 48, 49, 52–57, 61,
    88, 89, 309, 310
Azide displacement, 93
Azlactone, 285

B. coagulans, 272
Bacillus sp. LP-75, 273
Background curve (ORD), 543
Backside attack, stereoelectronic,
    requirement of, 461
Bacteriorhodopsin, 582
Baeyer strain, *see* Angle strain
Baeyer-Villiger rearrangement, 94
Baldwin's rules, 434–436
Ball-and-stick models, 25
Bandwidth (NMR), 454
Barbiturate enantiomers, 136
Barrelene, 502, 503
Barrier, in biphenyls, 626
  to E-Z Isomerization about C=N and
    N=N bonds, 346, 347
  height of, 380
  height of, from microwave spectrum, 399
  for interconversion of cis-trans isomers,
    339, 342–344
  to nitrogen inversion, in piperidine, 411
  to ring flip, 634, 635
  to ring inversion, in *cis*-decalin, 495
  to rotation, *see* Rotational barrier
Beckmann rearrangement, 94, 352
Beilstein nomenclature, 81
Bending force constants, 22
Benzaldehyde, conformation of, 394
Benzene, oxidation of, 349
Benzene-d_6, 318
Benzenechromium tricarbonyl, 54
Benzoate sector rule, 567
Benzocycloheptene, 487
Benzoin, 245
N-Benzoylphenylglycine, 244
Benzyl α-^{12}C benzyl-α-^{13}C sulfoxide, 60
Benzyl *p*-tolyl sulfone-, ^{16}O,^{18}O, 60, 109,
    110
2-(*N*-Benzylamino)-l-butanol, 230
cis-N-Benzyl-2-(hydroxymethyl)cyclo-
    hexylamine, 214
1-Benzylidene-4-methylcyclohexane, 619
N-Benzyloxycarbonylalanine (Z-Ala), 214,
    215

Benzylquininium chloride, 253
Benzyne, 632
β-effect, *see* α-effect
Betweenenanes, 640, 642
Bi-2,13-pentahelicenylene, 637
Biaryl chromophore, 550
Biaryls, 622ff
  racemization of, 279, 281
Biassed model compound, 442
Bicyclic ligand, in CIP system, 70, 71
Bicyclooctenone, Cotton effect of, 564
Bicyclopentadienes, conversion to
  adamantane, 515, 516
Bicyclo[1.1.0]butane, 492
  strain in, 492
Bicyclo[1.1.0]but-1(3)-ene, 493
Bicyclo[1.1.1]pentane, 501
Bicyclo[2.1.0]pent-1(4)-ene, 493
Bicyclo[2.1.0]pentane, 492
trans-Bicyclo[2.2.0]hexane, 493
Bicyclo[2.2.0]hex-1(4)ene, 493
Bicyclo[2.2.2]octane, 502, 504
Bicyclo[2.2.2]octane-2,6-dione, 503
Bicyclo[2.2.2]octatriene, 502
Bicyclo[3.1.0]hex-1(5)-ene, 493
trans-Bicyclo[3.2.0]heptane, 492
Bicyclo[3.3.0]nonane, 494
  hetero analogues of, 494
Bicyclo[3.3.1]nonane, 502
Bicyclo[3.3.1]-1(2)-nonene, 504
trans-Bicyclo[4.1.0]heptane, 492
Bicyclo[4.2.0]octanes, 492
trans-Bicyclo[5.1.0]octane, 492
Bicyclo[n.1.0]alkanes, strain in, 492
Bidentate ligand, in CIP system, 70
Bijvoet method, 76, 77
  application to allene, 613
  application to biaryls, 628
Bijvoet pairs, 76
Bilirubin, induced CD in
  α-methylbenzylamine, 571
Bimolecular elimination reaction, *see* E2
  reaction
Bimolecular nucleophilic substitution
  reaction, *see* S_N2 reaction
1,1'-Binaphthalene-2,2'-diol, *see*
  2,2'-Dihydroxy-1,1'-binaphthyl
1,1'-Binaphthyl, 107, 202, 624
  configuration of, 569, 628

exciton chirality in, 569
racemization of, 279
1,1'-Binaphthyl-2,2'-dicarboxylic acid, 628
1,1'-Binaphthyl-2,2'-diyl hydrogen
  phosphate, 86, 148, 176, 218, 220,
  243
Binaphthylphosphoric acid, *see* 1,1'-
  Binaphthyl-2,2'-diyl hydrogen
  phosphate
Binaphthyls, configuration of, 627, 628
Binary phase diagram, 107, 200, 229
Binuclear lanthanide shift reagents, 159
Biodiscrimination, 132, 137
Biomembranes, 582
Biopolymers, chiroptical properties of, 567
  conformation of, 576
  secondary structure of, 579
Biot's law, 1, 5, 586. *See also* Specific
  rotation
Biotic theories, 138
Biparaphenylene-34-crown-10, 508
Biphenyl-2,2',6,6'-tetracarboxylic acid,
  α-methylbenzylamides of, 628
Biphenyls, 30, 31, 49, 51, 53, 55, 65, 622ff
  configuration of, 627, 628
  di-ortho substituted, 624
  half-lives of racemization of, 623
  mono-ortho substituted, 624
  tetra-ortho substituted, 30, 31, 344, 623
  tri-ortho substituted, 623
  UV spectra of, 626, 627
2,2'-Bisbromomethyl-1,1'-binaphthyl, 638
1,4-Bis-O-(4-chlorophenyl)-L-threitol, 223
4,4'-Bispyridine units, 508, 510
Bis-allenes, cyclic, 616
Bisecting conformation, 388, 389
Bjerrum's law, 356, 425
Boat, 439, 486–489
Boat conformation, in perhydroanthracene,
  497
Boat-boat, 488, 489
Boat-chair, 488
Boat-chair-boat, 490, 491
Boat-chair-chair, 491
Boat/twist-boat family, in cycloheptane,
  486, 487
Boiling point, 379, 398
  in determination of E-Z configuration,
  353–355

Boiling point (*continued*)
  of diastereomers, 451, 453
Boltzmann distribution (of conformers), 448
Bond, definition of, 11, 12
Bond angle strain, 21, 22
Bond angles, 8, 9
  in alkenes, 340
  in cyclohexane, 437
  in cyclopentane, 482
  enlargement of, 382, 383
  in ethene, 340
  in isobutylene, 340
  in propene, 340
Bond compression strain, 21, 22
Bond dipoles, 24
Bond distance, *see* Bond lengths
Bond lengthening, 387, 388
Bond lengths, 8, 9, 19
  anomalies in, 387, 388
Bond order, 65, 66
Bond rotation, 10, 65, 376
Bond shortening, 388
Bond strain, 429
Bond stretching, 21–23
Bonded atoms, 11, 12
Bonded distance, 11
Bonded OH, 12, 400, 401
Bonding-antibonding orbital interaction, 378
Boranes, kinetic resolution of, 265
Borneol, 162
endo-Bornyl-1,2,4-triazoline-3,5-dione, 226
endo-Bornylamine, 214, 225
Boron trifluoride, 56
Bovine serum albumin (BSA), 172
Branched hydrocarbons, 382, 383
Branched molecules, 9
Breadth (ORD), 543
Bredt's rule, 503, 504
Brewster's helix model, 596
Brewster's model, 595
Bridged allenes, 642
Bridged rings, 501–504
2-Bromobutane, 34–36, 65, 312
  dehydrohalogenation of, 352, 353
2-Bromo-2-butenes, 358
1-Bromo-1-chloro-2-ethyl-2-
        methylcyclopropane, 422
Bromochlorofluoromethane, 13, 30, 32, 49
Bromochloromethane, 310, 311

Bromocyclobutane, 304, 481
Bromocyclohexane, 481
1-Bromo-1,2-dichloroethane, 340, 341
Bromodichloromethane, 307
2-Bromohexahelicene, 638
Bromolactic acid, 87
2-Bromo-2-methylcholestan-3-ones, 556
p-Bromomethyltrityl chloride, 508
α-(2-Bromophenoxy)-propionic acid, 136
2-Bromo-4-phenylcyclohexanols, 462, 463
2-Bromopropanoic acid, 272
Bromostilbenes, configuration of, 351
Bromosuccinic acid, 90
Brucine, 211, 216, 220, 230, 235, 243
BSA, *see* Bovine serum albumin
Buckminsterfullerene (C_{60}), 58, 516, 517
Bullvalene, 13
Bürgi-Dunitz trajectory, 435, 469
1,3-Butadiene, 54, 393, 394
Butadienes, hindered, 628, 629
Butane, 378–380, 382
  anti conformer of, 65, 378, 379, 486
  gauche conformer of, 65, 378, 379, 486
  gauche-anti enthalpy difference in, 378, 379
  gauche-anti free energy difference in,
        379, 380
Butane conformers, enthalpy difference
        between, 65, 378, 379
Butane-gauche interaction, 378, 495
chiral-2,3-Butanediol, 148, 161, 162, 223
2,3-Butanediols, IR spectra of, 400, 401
1-Butanol, 272
2-Butanol, 234
2-Butanol-3-*d*, 94, 95
2-Butanone, 390
2-Butanone-3-*d*, 95
Butatriene, 342
1-Butene, 389
1-Butene conformers, barrier to
        interconversion of, 389
2-Butene, 352, 353, 410, 617
  barrier in, 339, 366
  isomerism in, 66
4-*tert*-Butoxycyclopent-2-one, 225
Buttressing effect, 624
Buttressing problems, 450
Butyl (*R*)-(+)-2-bromopropionate, 272
tert-Butyl hydroperoxide, 98
tert-Butylacetaldehyde, 390

tert-Butylbenzene, 395
p-tert-butylcalix[8]arene, 506, 516
sec-Butylcyclobutanes, 45–48
tert-Butylcyclohexane, 441, 475
4-*tert*-Butylcyclohexanecarboxylicacids,
    453, 454
4-*tert*-Butylcyclohexanol, 95, 441, 442, 456
4-*tert*-Butylcyclohexanone, 95, 468
4-*tert*-Butylcyclohexene, 464
4-*tert*-Butylcyclohexene oxide, reduction
    of, 466, 467
  ring opening of, 467
4-*tert*-Butylcyclohexyl phenyl sulfide, 93,
    455, 456
4-*tert*-Butylcyclohexyl *p*-toluenesulfonates,
    93, 455, 456
  bimolecular elimination of, 461
4-*tert*-Butylcyclohexylamine, 453, 454
4-*tert*-Butyl-1,1-difluorocyclohexane, 313
2-Butylene oxide, 321, 322
n-Butylethyl-*n*-hexyl-*n*-propylmethane, 60
Butyllithium (–)-sparteine combination,
    584, 616
N-Butyloxycarbonyl-L-phenylglycine, 221
4-*tert*-Butylpyridine, 416

C_{60}, *see* Buckminsterfullerene
C_{70}, 516
$CaCl_2$ alcoholates, in stereoisomer
    separation, 249
Cahn-Ingold-Prelog system, 14, 67ff. *See
    also* CIP system
Calicheamicin, 642
Calixarenes, 506
Calorimetric entropy, 377
Calycanthine, absolute configuration of,
    567, 568
CAMP, *see* Cyclic adenosine
    monophosphate
ω-Camphanic acid, 148, 221, 222
Camphor, 2, 115, 502–504
  CD of, 545
  dimers of, 278
  nucleophilic addition to, 469
  ORD of, 542
D-Camphorquinone, 263, 264
Camphor-10-sulfonic acid, 165, 211, 219,
    225, 242, 612
Camphor-10-sulfonyl chloride, 263

Catidida cylindracea, 270, 272
Capto-dative alkene, 343, 345
Carbamalactic acid, 219
Carbamates, 236
N-Carbamoyl-D-amino acid
    amidohydrolase, 208
Carbamoylphenylglycine, 208
Carbinol proton, 455, 456
Carbon atoms, sp2 hybridized, 539
Carbon dioxide, symmetry of, 56
Carbon monoxide, symmetry of, 54
Carbon-13 NMR spectra, 318–322
  of diastereomers, 404–406
  of substituted cyclohexanes, 455–458
Carbon-13 NMR spectroscopy, 441
  and conformational equilibria, 440
  in determination of E-Z configuration,
    360, 361
Carbon-carbon sigma bond overlap, 384
Carbonyl-bis-(L-valine ethyl ester), 167
Carboxylic acids, 160
  enzymatic resolution of, 271
o-Carboxyphenyl methyl sulfoxide, 69
Carcerand, 504
cis-Caronaldehyde, 222
Carotene, 369
Cartesian coordinates, 8, 9
Carvone, 133
Catalin models, 26
Catalyzed equilibration of cis-trans isomers,
    367, 368
Catalyzed isomerization, general
    mechanism of, 367, 368
Catenanes, 505ff
  directed synthesis of, 508
  statistical synthesis of, 507
  template synthesis of, 508
[2]Catenanes, 508
Cavitates, 233
C–C σ bond strength, 339
C=C bond length, 339
C=C bond with single bond character, 343
C=C stretching vibration, 356, 357
CD, 451, 534, 535, 544. *See also* Circular
    dichroism
  of chiral polymers, 576
  of cyclohexane derivatives, 451
  induced, 570
  induced, temperature dependence of, 571

CD (*continued*)
  induction of, 205
CDA, *see* Chiral derivatizing agent
CE, *see* Cotton effect
Cellulose triacetate (TAC), microcrystalline,
    172, 239, 344
Center of chirality, *see* Chiral center
Center of prostereoisomerism, 304
Center of stereoisomerism, *see* Stereogenic
    center
Center of symmetry, 3, 42, 45–47, 51, 52,
    61, 310, 314, 315
CFPA, *see* α-Cyano-α-fluorophenylacetic
    acid
CH_2-O bond length in anomeric effect, 387,
    388
CH_3-C-H bond angle, 382, 383
CH_3-lone pair interactions, 472, 474
CH_3/CH_3 barrier, 380
CH_3/CH_3 eclipsing, 378, 380
CH_3/H eclipsing, 378, 380
CH_3/H synaxial interactions, 472
Chair, 486–488
Chair conformer of cyclohexane, 436–440,
    485
  symmetry of, 55
  inversion of, 319, 439
Chair inversion barrier, in cyclohexanone,
    457
Chair-chair, 488, 494
Chair-to-twist activation enthalpy, 431
Chair-twist energy differences in
    diheteracyclohexanes, 474
Chair/twist-chair family, in cycloheptane,
    486
Charge interaction factors, 393
CHD=CHD, barrier in, 339
CHDBrCHDBr, 371
Chemical Abstracts nomenclature, 81
Chemical correlation of configuration, 84ff
Chemical-shift equivalent nuclei, *see*
    Isochronous nuclei
Chemical shifts, 456
  averaging of, 401, 441
  in determination of E-Z configuration,
    358ff
Chiral, 3, 4, 14
  adjuvant, 95

axis, 304, 305, 608. *See also* Axial
    chirality
axis nomenclature, 609
center, 32, 39, 44, 72, 79, 303, 313
crown ether, 131
crystals, 541
derivatizing agent (CDA), 147, 148
detergent, critical micelle concentration
    of, 570
deuterium compounds, 60, 87, 95
ionophore, 178
lanthanide shift reagents, *see* Chiral shift
    reagents
media, 36, 319
mobile phases, 164
molecules, devoid of chiral centers, 608
perturbations, 550
plane, 305, 610, 611. *See also* Planar
    chirality
protein selectors, 172
shift reagent (CSR), 157, 312, 319
solvating agent (CSA), 132, 163
solvents, 319
space groups, 110
Chiral and meso stereoisomers, distinction
    between, 320, 321
Chiral, racemic, 4
Chiralcel™ HPLC columns, 173
Chirality, 4, 46, 394
  due to isotopic composition, 60
  rule, 67
  sense, *see* Sense of chirality
Chirasil-Val™, 167, 168
Chiro-inositols, 453
Chiroptical methods, 145
Chiroptical properties, 534, 591
  ab initio calculations of, 594
  of chiral ketones, 553
  of compounds with low optical activity,
    547
  of cyclohexane derivatives, 451
  induced in achiral compounds, 571
Chiroptical techniques, 534
Chloramphenicol, 204, 213, 220
Chloroacetic acid, 307
Chloroacetylene, 54
Chloroallene, 310
2-Chlorobutane, 205
3-Chloro-2-butanol, 236

2-Chloro-2-butenoic acids, 358
Chlorocholine, 386
Chlorocyclohexane, 10, 34, 36, 437
Chlorocyclopentane, 484
Chloroethene, 51, 304
α-Chloroethylbenzene, 60
Chloroform, symmetry of, 54
Chlorohydrins, rates of ring closure of, 434
1-Chloro-2-iodoethylene, 354, 355
2-Chloro-5-methylcyclohexanone, 555, 571
3-Chloro-1,2-propanediol, resolution of, 263
3-(3-Chlorophenyl)-3-hydroxypropanoic
        acid, 115
2-Chloropropanoic acid, 151, 152
Chlorosuccinic acid, 90
m-Chlorotoluene, 51
Cholecystokinin, 242
5α-Cholestan-1-one, Cotton effect of, 560
5α-Cholestan-2-one, Cotton effect of, 560
5α-Cholestan-3-one, 318
    Cotton effect of, 560
Cholesteric behavior, 200
Cholesteric liquid crystals, 574
Cholesteric phase, 574
Cholesterol, 131
Cholic acid, 235, 236
Chromatographic column, enantioselective,
        165
Chromatographic resolution, 227
    preparative, 160, 236
Chromatography, of diastereomers, 161
    dynamic enantioselective, 328
    racemization during, 283
Chromophores, classification of, 550
    inherently achiral, 550
    inherently chiral, 550
Chrompack™, 167, 168
Chrysanthemic acid, 162
Cinchona alkaloids, 172
Cinchonidine, 212
    p-chlorobenzoate derivative of, 569
Cinchonine, 212
    p-chlorobenzoate derivative of, 569
Cinchotoxine, 210, 262
CIP sequence, 74
CIP system, 67ff. *See also*
        Cahn-Ingold-Prelog system
Circular birefringence, 544
Circular dichroism, 37, 77, 97. *See also* CD

Circular dichroism spectra, 547
Circularly birefringent medium, 539
Circularly polarized light, 535, 536, 545
cis, 341
Cis conformer, 388, 389
Cis-trans equilibria, position of, 362ff
Cis-trans interconversion, directed, 368–371
Cis-trans isomerism, 72, 73, 340
    about C=N bonds, 346, 347
    about N=N bonds, 346
Cis-trans isomerization, 362ff
    about C=N bond, 346, 347
    photochemical, 368
    by reversible 1,4 addition, 367
Cis-trans isomers, catalyzed equilibration
        of, 367, 368
    ΔH° between, 362–365
    determination of configuration of, 348ff
Cisoid ring juncture, 497
Citraconic acid, 335
Citral, determination of E-Z configuration, 362
Citramalate, 127
Citric acid, 305, 306
Citric acid cycle, 306
Clathrate inclusion complexes of TOT, 206
Clathrate inclusion compounds, 233
Clathrates, 283
Clathration, diastereomer separation by, 253
Closed-shell repulsion, 23
CMP, *see* Cyclic cytidine monophosphate
Coalescence, 326, 627
Coalescence temperature, 326–328, 402,
        440
    relation to barrier, 402, 403
Cocaine, 137, 138
Coenzymes, 271
Cofactors, 271
Collinear atoms, 9
Column chromatography of diastereomers, 161
Common rings, 430
Compression effect, 361
Computer modeling, 26
Concentration effect in polarimetry, 587,
        591, 592
Conductometric titration, 414
Configuration, 9, 13–15, 17, 30, 39, 65ff,
        320, 321
    absolute, 65–67, 75–79
    relative, 65–67, 79, 84ff, 423ff

Configuration assignment, 84ff
Configuration assignment (*continued*)
   by carbon-13 NMR, 455
   by CD/ORD, 548
   by NMR, 156, 454–458
Configurational correlation, 84ff
Configurational descriptors, 67ff
Configurational isomers, 10, 66
Configurational notation, 67ff
Conformation, 9, 10, 15–17, 26
   assignment by CD/ORD, 548
   of carbon chain, 404
Conformation population difference,
   325–327
Conformational Analysis in acyclic systems,
   377
Conformational biassing, 441, 442
Conformational composition, 404
Conformational dissymmetry model, 595
Conformational energy, 442–447
   additivity of, 446, 450
   in substituted cyclohexenes, 463, 464
Conformational energy difference, 412
   dependence on solvent, 383
   in polar molecules, 383
Conformational entropy, 432
   loss of, 432
Conformational equilibria in di- and
   trisubstituted cyclohexenes, 466
Conformational equilibrium constant, 440
Conformational isomers, *see* Conformers
Conformational locking, 441
Conformational population, temperature
   dependence of, 489, 490
Conformational rule, 379, 451
Conformational topomers, 319, 320, 438
Conformationally heterogeneous system,
   rate constant for, 409
Conformationally mobile systems, ORD in,
   571
Conformations, families of, 485
Conformer composition from dipole
   moment, 399
Conformer equilibration, *see*
   Conformational energy
Conformers, 10, 14, 16, 35, 66, 376
   entropy differences between, 440
Confrontation analysis, 89
Conglomerate, 106, 198, 202

Conglomerate crystals, manual sorting of,
   198
Conglomerates, frequency of formation of,
   107, 108, 110
   identification of, 200
Conical symmetry, 54
Conjugated dienes, 393, 394
Connectivity, *see* Constitution
Connectivity matrix, 11
Consecutive reactions, in kinetic
   resolutions, 261
Consignate, 559
Constitution, 10–13, 17, 30, 66
   as inferred from ORD/CD, 549
Constitutional formula, 11, 12
Constitutional heterotopicity, 315, 316
Constitutional isomerism, 10, 11
Constitutional isomers, 10, 11, 41, 315
Constitutionally heterotopic faces, 315
Constitutionally heterotopic ligands, 315
Contact method of Kofler, 200
Corannulenes, 56
Corey-lactone II, 254
Coronand, 504
Coronene, 56
Correlation of chiral with prochiral center,
   322–324
COSY, *see* Proton correlated spectroscopy
Cotton effect (CE), 543
   of haloketones, in relation to structure,
   554
   magnitude of, 551, 560, 562
   of chiral ketones, 553
Cottonogenic rhodium complexes, 573
Cottonogenic substituents, 571
Coulomb's law, 478
Coulombic interaction of dipoles, 384
Coulombic term, 23
Counterpoise method, 446, 450
Coupe du roi, 52
Coupled oscillator method, 567
Coupling constants, 401, 404, 441, 454
   averaging of, 401
   from ^{13}C satellite spectrum, 360
   in determination of E-Z configuration,
   358, 360, 361
   and electronegativity, 404
Coupling constants, $^{19}F/^{19}F$ and E-Z
   configuration, 358, 360

^{1}H/^{13}C, 441
^{1}H/^{19}F, and E-Z configuration, 360
^{1}H/^{1}H, 403, 454
  and E-Z configuration, 358, 360
Courtauld models, 26
Covalent Pirkle CSPS, 169
CPE, *see* Circular polarization of emission
C-P-K models, 26
Cram's rule, 98
Crotonic acid, configuration of, 349, 357
Crown ether stationary phases, 272
Crown ethers, chiral, 233
Crown conformation, 488, 489
CRR, *see* Chiral relaxation reagents
[2.2.2]Cryptand, 505
Cryptochiral, 60, 107, 109
Cryptochiral sample, 124, 200
Cryptochirality, 582, 583, 588, 589
Cryptone, racemization of, 281
Crystal axis, 78
Crystal growth, 78
Crystal habit, 2
Crystal morphology, 2, 84
Crystal shape, 109
Crystal space group, 200
Crystallization, "differentiated," 201
  localization of, 201
  from the melt, 254
  preferential, 201ff
  Crystallization-induced asymmetric
    transformation, 206, 241
Crystallographic space groups, 111
Crystals, chiral, 541
CSA, *see* Chiral solvating agent
C$_2$-symmetric CDAS, 149
Cubane, 57
Cubic symmetry, 57
Cumulenes, isomerism in, 342, 343
Cupra A effect, 573
Cuprammonium solutions of aminoalcohols,
    Cotton effect of, 573
  of glycols, Cotton effect of, 573
Curtin-Hammett kinetics, 411, 413
Curtin-Hammett principle, 407ff
Curtius rearrangement, 94, 618
α-Cyano-α-fluorophenylacetic acid
    (CFPA), 148, 150
Cyanohydrins, 235
  from cyclohexanones, 467

Cyclic intermediates, 371
  five-membered, 371
  three-membered, 371
Cyclic ligand, in CIP system, 70–72
Cycloalkane-1,2-dicarboxylic acids,
    dissociation constants of, 426
Cycloalkanes, ring strain in, 429–431
(*E*)-Cycloalkenes, 503, 504, 610, 611
Cyclobutadiene complex, 514
Cyclobutane, 9, 56, 422, 423, 481, 492
  conformational equilibrium in
    substituted, 481, 482
  entropy of ring closure, 432
  strain in, 429–431, 481, 514
  symmetry of, 56
Cyclobutane-1,2-dicarboxylic acids, 426,
    482
Cyclobutane-1,2-diol, 426, 427
Cyclobutane-1,3-diols, 313
Cyclobutanes, 1,2-disubstituted, 423, 482
  1,3-disubstituted, 423, 482
  monosubstituted, 481
Cyclobutanol, 313, 316
Cyclobutanone, 310, 312, 467
Cyclobutene, 482
Cyclobutyl bromide, *see* Bromocyclobutane
Cyclodecane, 430, 431, 456, 489–491
(*E*)-Cyclodecene, 641
Cyclodextrins (CDs), 233
  derivatized, 234
  methylated, 168
α-Cyclodextrin (cyclohexaamylose), 50
β-Cyclodextrin (cycloheptaamylose), 153,
    167, 233, 508
γ-Cyclodextrin (cyclooctaamylose), 153
Cyclodimerization, 368
Cyclododecane, 430, 431, 491
  monosubstituted, 491
Cyclododeca-1,2,7,8-tetraene, 616, 617
Cycloheptaamylose, *see* β-Cyclodextrin
1,2-Cycloheptadiene, 616
Cycloheptane, 430, 431
  barrier in, 486, 487
  conformation of, 486, 487
Cycloheptane-1,2-diol, 426, 427
Cycloheptatriene, 487
Cycloheptene, 487
(*E*)-Cycloheptene, 641
Cyclohexaamylose, *see* α-Cyclodextrin

Cyclohexadecane, 491
1,3-Cyclohexadiene, 466
Cyclohexane, 429–431, 436–439, 485
    derivatives, physical properties of, 451ff
    inversion in, 437–440
    inversion barrier in, 438, 439, 474
    planar, 59
    symmetry of, 59
    vicinal coupling constants in, 438, 454
Cyclohexane ring, flattening of, 437, 438
Cyclohexane p-toluenesulfonates, reaction
        with thiophenolate, 456, 461
Cyclohexane-d_{11}, 327, 438
Cyclohexane-1,2-dicarboxylic acids, 426,
        482, 552
trans-1,2-Cyclohexanedicarboxylic
        anhydride, 222
Cyclohexane-1,2-diols, 427
Cyclohexane-1,3-diols, 252
Cyclohexane-1,4-diols, 252
Cyclohexanes, conformation and reactivity
        in, 457ff
    di- and polysubstituted, 447ff
    monosubstituted, 439ff
Cyclohexanol, conformational equilibrium
        in, 442, 445
Cyclohexanols, oxidation rates of, 460
    separation of, 248
Cyclohexanone, 466–468
    addition to, 467–469
    chair inversion barrier in, 467
    HCN addition to, 467
    reduction of, with sodium borohydride,
        467
    strain in, 467
    twist form of, 467
Cyclohexanones, addition of nucleophiles
        to, 467–469
    reaction of, with organometallics, 468
    reduction of, 467, 468
Cyclohexene, 463–466
    epoxidation of, 466
    inversion barrier in, 464
    torsion angles in, 463
Cyclohexene oxide, 466, 467
Cyclohexenes, 3-substituted, 464
    4-substituted, 464, 465
Cyclohexyl bromide, conformers of, 440
Cyclohexyl-d_8 methyl-d_3 ether, 454

Cyclohexyl ring, in CIP system, 72
Cyclohexyl p-toluenesulfonate, bimolecular
        elimination of, 461
Cyclohexyl tosylate, see Cyclohexyl
        p-toluenesulfonate
1,2-Cyclononadiene, 610
Cyclononane, 429–431, 489, 490
Cyclononatetraenyl anion, 56
(E)-Cyclononene, 640, 641
Cyclooctaamylose, see γ-Cyclodextrin
1,2-Cyclooctadiene, 616
Cyclooctadienes, 489
Cyclooctane, 430, 431, 485–489
Cyclooctane-1,2-diol, 426, 427, 641
Cyclooctatetraene, 489
Cyclooctatetraenes, 225
    racemization of, 280
Cyclooctatetraenyl dianion, 56
Cyclooctatriene, 489
Cyclooctene, 355
(E)-Cyclooctene, 354, 355, 504, 640, 641
    configuration of, 641
trans-Cyclooctene, see (E)-Cyclooctene
Cyclopentadiene, strain in, 485
Cyclopentane, 482–485
Cyclopentane-1,2-dicarboxylic acid, 426
Cyclopentane-1,2-diols, 41, 42, 426, 427
Cyclopentanes, 1,2-disubstituted, 484, 485
    monosubstituted, 484
Cyclopentanone, 467, 482, 483
    addition to, 485
Cyclopentanone skeleton, twisted, 560
Cyclopentene, strain in, 405
Cyclopeptides, 53
Cyclophanes, 608, 639, 640
Cyclopropane, 421, 422, 429–431, 480, 492
    conjugation with, 394, 395
    derivatives of, 421, 422
    electronic stabilization of, 480
    internuclear and interorbital angles in, 480
    strain in, 480
    symmetry of, 66
Cyclopropane-1,2-dicarboxylic acids, 421,
        424–426
Cyclopropanecarboxaldehyde, 394, 395
Cyclopropanes, 1,2-disubstituted, 421, 422
    1,3-disubstituted, 422
    enantiomerization of, 282
    stereoisomerism in, 421, 422

substituted, 422
tetrasubstituted, 422
Cyclopropanone, 480, 481
Cyclopropene, strain in, 480
Cyclopropenium cation, 480
Cyclopropylbenzene, 394, 395
Cyclotetradecane, 491
Cyclotridecane, 491
Cylindrical symmetry, 56
Cytochrome, 582

2D graph, 11, 13
D,L, 88
2D NMR, 17, 19, 20, 361, 362
2D NOESY experiment, 19, 20, 362
Dansylated amino acids, 175
analysis of, 175
Dating, of fossil samples, 287
Debromination, iodide-promoted, 410
Debye-Scherrer diagrams, *see* X-ray
powder diagram
cis-Decalin, 248, 494, 495
substituted, 496
trans-Decalin, 248, 494, 495
substituted, 496
Decalins, heat of isomerization of, 495
heats of combustion of, 495
1-Decalones, 74, 75, 241
Decane-1,10-diol, 507
Degenerate isomerization, *see*
Topomerization
Degenerate structure, 13
Dehydroabietylamine, 211, 212, 263
Dehydronorcamphor, Cotton effect of, 564
Delocalization energy, 24
Deltamethrin, 137
Denaturation, 581
Density, 379, 398
in determination of E-Z configuration,
353–355
of diastereoisomers, 451, 453
and racemate type, 110
Deoxyephedrine, 211, 215
Deoxyribonucleic acid (DNA), CD of, 581
L-Deoxyribose, 582
Deracemization, 245
Descriptors, *see also* CIP system and
Configurational descriptors
aR and aS, 610

of heterotopic ligands, assignment of,
322–324
pR and pS, 611
for stereoheterotopic ligands or faces,
316, 317
Determinate mechanisms, 139, 141
Determination of enantiomer composition,
see Enantiomeric composition
Deuterium, conformational energy of, 571
Deuterium compounds, chiral, 87, 98
Deuterium substituted ketones,
configuration of, 559
Dextrorotatory sample, 539
Dextrorotation, 5
Diagenetic temperature, 287
Dialkyl peroxides, barrier in, 397, 398
Diamagnetic anisotropy, 455
2,2'-Diamino-1,1'-binaphthyl, 202, 551
1,10-Diaminodecane, 508
2,2'-Diamino-6,6'-dimethylbiphenyl, 628
1,12-Diaminododecane, 508
Diamond lattice conformation, 491
Dianin's compound, 254
Diastereodifferentiation, 329
Diastereomer discrimination, 103
in the solid state, 130
Diastereomer excess, 143
Diastereomer mixture, phase diagram for,
229
Diastereomer-mediated resolutions, types
of, 215
Diastereomeric complexes, 231, 321
Diastereomeric salts, dissociation of, 231
p and n nomenclature for, 215
solubility differences of, 228
Diastereomeric "solvates", 125
Diastereomeric sulfoxides, 34
Diastereomers, 31–33, 39ff, 67, 310, 315,
316
achiral, 304
addition compounds of, 230
carbon-13 spectra of, 404–406
covalent, 217, 222, 227, 236
distinction of, by dipole moment, 351,
354, 355
nomenclature of, 79–84
separation, methods for, 246
separation of, by chromatography, 161,
250

Diastereomers (*continued*)
  by distillation, 231, 247
  solid solutions of, 230
Diastereoselective stationary phases, 161
Diastereoselective synthesis, 95–98, 350
Diastereoselectivity, 329
Diastereotopic, 315
  atoms, *see* diastereotopic ligands
  faces, 307, 310, 312–314, 329
  fluorine atoms, 319, 320
  hydrogen atoms, 352
  ligands, 307, 310, 312–315, 318–320,
    323–326, 329
  methyl groups, 319, 362
  nuclei, 312, 318, 320, 323
  protons, 319–321
  relationships, 154
  sets of hydrogen atoms, in cyclohexane,
    438
  solvates, 155
Diastereotopicity, in biphenyls, 627
Diatomic molecule, 8
Diaxial addition, 464
trans-Diaxial elimination, 461
1,3-Diazane system, N-substituted, 475, 476
Diazodicarboxylic acid, 202
Dibasic solute model, 156
Dibenzenechromium, 55–56
Dibenzobicyclo[2.2.2]octatriene,
    atropisomerism in, 632
Dibenzoylcyclobutadiene, 55
Dibenzophosphole, 351, 352
Dibenzoyltartaric acid, 218, 219, 263
Dibenzylmethylamine, inversion of, 329
2,3-Dibromobutanes, 236
  debromination of, 410
2,3-Dibromocholestanes, 500
1,3-Dibromocyclobutanes, 482
trans-1,2-Dibromocyclohexane, 24, 61
trans-1,4-Dibromocyclohexane, 24, 61
1,2-Dibromo-1,2-diphenylethanes, 351
1,2-Dibromoethane, 308, 384, 385
Dibromo[2.2]paracyclophane, 52
1,2-Dibromopropane, 236
Di-*tert*-butyl peroxide, 398
Di-*sec*-butylcyclobutane, 46, 47
1,2-Di-*tert*-butylethylene, 364, 368
Dibutyltin(IV) oxide, 126
1,2-Dicarbomethoxycyclopentanes, 485

1,3-Dichloroallene, 49
Dichloroanthracene, 632
1,4-Dichlorobenzene, 55
meso-2,3-dichlorobutane, 52
1,3-Dichlorocyclobutanes, 33, 42
2,3-Dichlorocyclopropane-1-carboxylic
    acid, 422
1,4-Dichloro-2,5-dibromobenzene, 54
1,2-Dichloroethane, 10, 13, 14, 16, 30, 31,
    36, 42, 49, 61
  anti-gauche enthalpy difference in, 384, 385
1,2-Dichloroethenes, 42, 54, 304, 357, 366
Dichloromethane, 54
2,4-Dichloropentanes, ^{13}C spectra of, 405,
    406
2,4-Dichlorophenoxypropanoic acid, 262
1,2-Dichloropropene, 357
3,5-Dichlorosalicylaldehyde, 242
Dichotomous tree, in determination of point
    groups, 58
Dielectric constant, 24, 385, 478
  effective, 24, 425
Diels-Alder reaction, 614
  application to kinetic resolutions, 265
Diene chromophore, 550, 564, 566
Diene helicity rule, 566
1,3-Dienes, Cotton effect of, 566
  resolution of, 225
Diethyl fumarate, 367
Diethyl maleate, 367
Diethyl tartrate, 223, 235
p-Diethylbenzene, 396
1,3-Diethylbenzene, conformation of, 396
Diferrocenylcarbodiimide, 617
Differential scanning calorimetry (DSC),
    111, 112
Differential thermal analysis (DTA), 111
Diffraction pattern, 17–19
  centrosymmetric, 76
1,1-Difluorocyclohexane, 319, 320, 326, 327
1,2-Difluoroethane, conformation of,
    383–385
2,3-Dihalobutanes, NMR spectra of,
    404–405
1,2-Dihaloethanes, 383–385
  dipole moment of, 383, 399
1,2-Dihaloethenes, 363, 365, 366
2,3-Dihalo-4-methylpentanes, dipole
    moments of, 303

2,4-Dihalopentanes, NMR spectra of, 406
Dihedral angle, *see* Torsion angle
Dihedral point group, 50, 51, 55, 56, 61
Dihedral symmetry, 50
9,10-Dihydroanthracene, 466, 499
  9-substituted, 499
1,4-Dihydrobenzene, 466, 499
Dihydrogen, symmetry of, 56
Dihydromevinolin, 593
1,4-Dihydronaphthalene, 499
Dihydroquinine, 124
2,2′-Dihydroxy-1,1′-binaphthyl, 128, 129,
  154, 176, 273
2,2′-Dihydroxy-1,1′-binaphthyl-3,3′-
  dicarboxylic acid, 628
L-3-(3,4-Dihydroxyphenyl)alanine
  (L-DOPA), 136
1,3-Diiodocyclobutane, 482
Diisopropyl tartrate, 130
1,3-Dichlorocyclobutanes, 33, 42
2,3-Dichlorocyclopropane-l-carboxylic
  acid, 422
1,4-Dichloro-2,5-dibromobenzene, 54
1,2-Dichloroethane, 10, 13, 14, 16, 30, 31,
  36, 42, 49, 61
  anti-gauche enthalpy difference in, 384,
  385
1,2-Dichloroethenes, 42, 54, 304, 357, 366
Dichloromethane, 54
2,4-Dichloropentanes, ^{13}C spectra of, 405, 406
2,4-Dichlorophenoxypropanoic acid, 262
1,2-Dichloropropene, 357
3,5-Dichlorosalicylaldehyde, 242
Dichotomous tree, in determination of point
  groups, 58
Dielectric constant, 24, 385, 478
  effective, 24, 425
Diels-Alder reaction, 614
Diene chromophore, 550, 564, 566
Diene helicity rule, 566
1,3-Dienes, Cotton effect of, 566
  resolution of, 225
Diethyl fumarate, 367
Diethyl maleate, 367
Diethyl tartrate, 223, 235
p-Diethylbenzene, 396
1,3-Diethylbenzene, conformation of, 396
Diferrocenylcarbodiimide, 617

Differential scanning calorimetry (DSC),
  111, 112
Differential thermal analysis (DTA), 111
Diffraction pattern, 17–19
  centrosymmetric, 76
1,1-Difluorocyclohexane, 319, 320, 326, 327
1,2-Difluoroethane, conformation of,
  383–385
2,3-Dihalobutanes, NMR spectra of,
  404–405
1,2-Dihaloethanes, 383–385
  dipole moment of, 383, 399
1,2-Dihaloethenes, 363, 365, 366
2,3-Dihalo-4-methylpentanes, dipole
  moments of, 303
2,4-Dihalopentanes, NMR spectra of, 406
Dihedral angle, *see* Torsion angle
Dihedral point group, 50, 51, 55, 56, 61
Dihedral symmetry, 50
9,10-Dihydroanthracene, 466, 499
  9-substituted, 499
1,4-Dihydrobenzene, 466, 499
Dihydrogen, symmetry of, 56
Dihydromevinolin, 593
1,4-Dihydronaphthalene, 499
Dihydroquinine, 124
2,2′-Dihydroxy-1,1′-binaphthyl, 128, 129,
  154, 176, 273
2,2′-Dihydroxy-1,1′-binaphthyl-3,3′-
  dicarboxylic acid, 628
L-3-(3,4-Dihydroxyphenyl)alanine
  (L-DOPA), 136
1,3-Diiodocyclobutane, 482
Diisopropyl tartrate, 130
(−)-Diisopropylidene-2-keto-L-gulonic acid,
  220
N,N-Diisopropyltartramide, 165
Diketopiperazines, 52, 285
Dilactyldiamide, 202
Dilocular hosts, 275
Dimerization, of enantiomers, 52, 53
  of racemates, 53
p-Dimethoxybenzene, conformation of, 396,
  397
1,2-Dimethoxyethane, 385
Dimethoxymethane, 386
  anomeric effect in, 386–388

Dimethyl
2,3-bis(2,2,2-trifluoroethyl)succinate, 248
Dimethyl ether, barrier to methyl rotation in, 397, 398
constitution of, 11
Dimethyl α-methylsuccinate, 588
Dimethyl peroxide, 397, 398
Dimethyl sulfoxide, 319
Dimethyl tartrate, 114, 129
β,β-Dimethylacrylic acid, 362
β,β-Dimethylallyl-d_2 alcohol, 333
Dimethylamine, barrier to methyl rotation in, 397, 398
7,7-Dimethylbicyclo[4.1.0]hept-1(6)-ene, 493
2,2′-Dimethyl-6,6′-biphenyldicarboxylic acid, 230
2,3-Dimethylbutane, 382, 383
3,3-Dimethyl-1,2-butanediol, 266
3,3-Dimethyl-2-butanol, 150
Dimethylcycloalkanes, ^{13}C resonances of, 427, 428
trans-1,2-Dimethylcyclohexane, 448
cis-1,2-Dimethylcyclohexane, 16, 17, 30, 448, 450
^{13}C NMR of, 457
trans-1,3-Dimethylcyclohexane, synthesis of, 470, 471
cis-1,4-Dimethylcyclohexane, 16
trans-1,4-Dimethylcyclohexane, 16
1,2-Dimethylcyclohexanes, 448–450, 493–495
physical properties of, 453
1,3-Dimethylcyclohexanes, 447–450
physical properties of, 453
1,4-Dimethylcyclohexanes, 16, 17, 448
physical properties of, 453
trans-2,6-Dimethylcyclohexanone, synthesis of, 470–471
trans-1,3-Dimethylcyclopentane, rotation of, 583
2,5-Dimethylcyclopentane-1,1-dicarboxylic acids, 424, 425
trans-1,2-Dimethylcyclopropane, 594
2,2-Dimethylcyclopropanecarboxylic acid, 421
2,2-Dimethyl-4-deuteriocyclohexanone, 571
1′-Dimethylferrocene-3-carboxylic acid, 85

N,N-Dimethylformamide, 362
2,3-Dimethyloxirane, 321, 322
N,S-Dimethyl-S-phenylsulfoximine, 224
N,3-Dimethylpiperidinium salt, 474
2,3-Dimethylsuccinic acids, 123
N-(3,5-Dinitrobenzoyl)leucine, 238
N-(3,5-Dinitrobenzoyl)leucine butyl thioester, 246
N-(3,5-Dinitrobenzoyl)phenylglycine, 238
6,6′-Dinitro-2,2′-diphenic acid, 31, 623, 628
1,2-Diols, enantiomeric purity, determination of, 126
infrared stretching frequency in, 426, 427
1,10-Dioxa[1 1]cyclophanecarboxylic acid, 640
1,3-Dioxane, 472, 474, 475
1,4-Dioxane, 476
1,2,3-Dioxaphosphorinane oxides, 149
2,6-Dioxaspiro[4]nonane, 208
Dioxastannotanes, 126
1,3-Dioxolane, 209
[(DIPAMP)Rh]$^+$, 266, 268
Dipeptides, 306
1,3-Diphenylallene, 596
2,2-Diphenylaziridine, chlorination of, 132
2,2-Diphenyl-N-chloroaziridine, racemization of, 281
4,4-Diphenylcyclohexanone, structure of, 467
1,2-Diphenylcyclopentanes, 485
1,3-Diphenyl-1,3-di-α-naphthylallene, 612
1,3-Diphenyl-1,3-α-naphthyl-2-propen-l-ol, 612
Di-(3-pinanyl)borane triflate, 264
Dipolar molecules, 22, 24
Dipole, local, 61
Dipole interaction, 24, 383, 476
Dipole moment, 19, 61, 62, 351, 357, 399, 441, 442, 478
in determination of E-Z configuration, 353–355
of diastereomers, 482
of 1,2-dihaloethanes, 383–385, 399
measurement of, 99
Dipole rule, 354, 355, 453
Dipole-dipole repulsion, 384, 387
Direct method (X-ray), 18
Directed conversion of trans- to cis-alkene, 368–371

Disjuncture, in CIP system, 71–72
Disparlure, 133, 134
Dispersion force, 23
Dissignate, 559, 563
Dissignate contribution of deuterium, 559
Dissymmetric facets, 2
Dissymmetry, 46
Dissymetry factor, *see g* number
Distomer, 135
1,2-Disubstituted alkenes, determination of
    configuration by physical methods,
    353ff
2,3-Disubstituted butanes, ^{13}C NMR spectra
    of, 404–406
Disubstituted cyclohexanes, 447ff
Disulfide chromophore, 550, 552, 564
Disulfide helicity rule, 552
1,3-Dithiacyclobutane, 481
Dithiahexahelicene, 206
1,3-Dithiane, 472, 474, 475
1,3-Dithiolane-*S*-oxides, 226
Di-*p*-toluyltartaric acid, 218, 219
DNMR, *see* Dynamic nuclear magnetic
    resonance
DNS-amino acids, 175
Dodecaborane dianion, 58
Dodecahedrane, 58, 515
Dodecahedron, 57
Dodecane-1,2-diol, intramolecular hydrogen
    bonding in, 591
L-DOPA, *see*
    3-(3,4-Dihydroxyphenyl)alanine
Double bond, eclipsed, 388–390
    partial, 65, 343
Double quantum coherence spectroscopy,
    361
Double-bond barrier, 339
Double-bond isomerization, 281
Double-bond-no-bond resonance, 387
Double-solubility rule, 117
Dreiding models, 26
Drimanoic acid, 162
Drug receptors, 306
DSC, *see* Differential scanning calorimetry
DTA, *see* Differential thermal analysis
Duplication method, 127, 152, 256
Dynamic coupling, 567
Dynamic disorder, 18

Dynamic enantioselective chromatography,
    see Chromatography, dynamic
Dynamic nuclear magnetic resonance
    (DNMR), 326–329
Dynamic stereochemistry, 1

E (entgegen), 340, 341
Eclipsed conformation, 79, 378, 388–390,
    398
Eclipsing of hydrogen atoms with lone
    pairs, 397
Eclipsing strain, 488, 489
EDC, see Enantiomer distribution constants
Electric field vector, 536, 537
Electric transition moment, 549
Electrical center of gravity, 455
Electroosmosis, 175
Electron correlation, 21
Electron density, 11
Electron diffraction, 4, 17, 19, 34, 437, 463,
    482, 502, 503, 514
    of cyclodecanes, 491
    of decalin, 494
    in determination of E-Z configuration,
        353, 354
    and molecular mechanics, 491
Electron spectra, of stereoisomers, 121
Electron-donating group, 357
Electron-withdrawing group, 357
Electronegativity, 62
Electronic absorption, 545
Electronic transition, 544
Electrophilic addition, 350
    to alkenes, stereoelectronic factors in, 463
    to alkynes, 350
    to cyclohexenes, 464, 466
Electrophoresis, in capillary columns, 175
Electrostatic attraction, 386, 392
Electrostatic energy, 22–24
Electrostatic interaction, 22–24, 383, 386
Electrostatic repulsion, 24, 387
Electrostatic term, 23, 478
E2 eliminations, 350, 351, 410, 461
Elimination with KOH, 551
Elimination with sodium thiophenolate, 351
Ellipsometry, 546
Elliptically polarized light, 545
Ellipticity, 545
Ellipticity data, 579

Empirical force field method, *see* force field
Enamines, A$^{(1,3)}$ strain in, 470, 471
Enantiodifferentiating reactions, 308
Enantioenriched, 4
Enantiomer, 3, 14, 30–32, 36ff, 41, 42. *See also* Enantiomers
Enantiomer composition, 104, 142, 143
Enantiomer composition, determination by chromatographic methods, 161ff
   by enzymatic methods, 176
   by kinetic methods, 177
   by NMR methods, 147ff
   potentiometric, 178
   by radiochemical techniques, 277
Enantiomer discrimination, 103
Enantiomer distribution constants (EDC), 274
Enantiomer excess (ee), 143. *See also* Enantiomer Composition
Enantiomer labeling, 168
Enantiomer mixtures, chromatography of, on achiral stationary phases, 128
   crystallization of, 112, 118
   density of, 110
   distillation of, 105, 106
   electronic spectra of, 121
   infrared spectra of, 120
   mass spectrometry of, 129
   melting point of, 111
   NMR spectra of, in liquid state, 124
   NMR spectra of, in solid state, 122
   solubility of, 115
   sublimation of, 119
   vapor pressure of, 119
   X-ray spectra of, 123
Enantiomeric composition, 586
Enantiomeric enrichment, amplification processes for, 141
   of carboxylic acids, 255, 256
   by crystallization, 253–257
   by distillation, 105
   by "duplication," 256
   by kinetic resolution, 257
   by sublimation, 119
Enantiomeric homogeneity in nature, 138
Enantiomeric molecules, 312
Enantiomeric purity, 143
   of alcohols and amines, determination of, 150
   of carboxylic acids, determination of, 151

   of ketones, determination of, 151
   of sulfoxides, determination of, 153
Enantiomerically enriched, *see* Enantioenriched
Enantiomerically pure, 108. *See* Enantiopure
Enantiomerization, 205, 209
Enantiomers, 3, 310, 315, 316
   odor difference of, 133
   taste difference of, 133
   unequal rate of crystallization of, 203
Enantiomorph, 2
Enantiomorphic crystals, *see* Enantiomorphous crystals
Enantiomorphous crystal classes, 109
Enantiomorphous crystals, 2, 76, 108, 198
Enantiopac™, 172
Enantiopure, 4, 143
Enantioselective protein HPLC stationary phases, 172
Enantioselective reaction, 330
Enantioselective stationary phases, 165, 169
Enantioselective supporting electrolytes, 175
Enantioselective synthesis, 95, 330. *See also* Asymmetric Synthesis
Enantioselective transport, 176
Enantiotopic, 315
   faces, 307, 310–312, 315–317, 330, 331
   ligands, 307, 310–312, 315–317, 330, 331
     assignment of descriptor to, 322
   methyl protons, 319
   nuclei, 319
   protons, 320, 321
   relationships, 154
endo, 502
Endo ring closures, 434
Endo-dig reactions, 436
n-Endo-tet reactions, 435
Endo-tet ring closure, 435
Endo-trig reactions, 435
Endocyclic alkylations of ketone enolates, 436
Endocyclic torsion angle, 493
Energy barrier, 14, 15, 377, 378
Energy contour diagram of pentane, 382
Energy minima, 377–379
Energy minimization, 22, 380
Energy minimum, 21, 22
Energy optimization method, 22
Enone chromophore, 550

Ensemble average, 398, 401
Ensemble averaged NMR spectrum, 404, 406
Ensemble averaged dipole moment, 399
Enthalpy difference, between diastereomers, 363–365
  between stilbene isomers, 364, 365
Enthalpy of activation, 327
Entrainment, 203
Entropies of ring closure, 432
Entropy difference, in decalins, 495
  between stilbene isomers, 363
Entropy of activation, 327
Entropy of mixing, 379, 380, 439, 448
  in enantiomers, 379, 380
  in decalins, 495
Entropy of symmetry, 62, 379, 438, 448
  in decalins, 495
Envelope conformation of cyclopentane, 482–484
Enzymatic dehydration, 305
Enzymatic dehydrogenation, 331
Enzymatic oxidation, 95
  of ethanol, 330–333
Enzymatic reactions, 305, 312, 322, 330–334
Enzymatic reduction, 94
  of acetaldehyde, 330–332
Enzymatic resolution, 208–266
  mathematical relationships in, 269
Enzyme preparations, commercial, 269
Enzyme-catalyzed reactions, 230–232
Enzymes, 5, 305–307
  immobilization of, 271
Ephedrine, 130, 178, 211, 212, 215, 217, 220, 225, 228, 245
Ephedrinium mandelate, 130
Ephedrinium salts, 178
Epichlorohydrin, 263
Epimerization, 240, 277
  catalyzed, 242
Epimers, 277
Epoxidation, 98
  of alkenes, enantioselective, 98
  of allytic alcohols, 261
Epoxide ring opening, 466, 467
  stereoelectronic factors in, 463
2,3-Epoxybutane, 94, 95
Equatorial atoms, 34, 438

conformer, 34
protons, 454
Equilibration, acid-catalyzed, 475
  of alkenes, 362–364, 366ff
Equilibrium mixture, of cis-trans alkenes, 366
Erythro-threo nomenclature, 82
Erythro, 82
Erythrose, 40, 41, 81, 86, 87
Esters, conformation of, 390, 391
Ethanal oxime, 340
Ethanal-1-d, 331
Ethane, 15
  barrier in, 376–378
  potential energy of, 377
  skew form of, 51
  symmetry of, 55
Ethanol, 316
  constitution of, 11
  enzymatic oxidation of, 330–332
Ethanol-1-d, 86, 87, 94, 95, 316, 331, 332
  oxidation of, 330–332
Ethyl acetoacetate, 12
Ethyl formate, 390, 391
Ethyl methyl ether, 386
  barrier to methyl rotation in, 397, 398
Ethyl methyl sulfide, 386
  barrier to methyl rotation in, 397, 398
Ethylbenzene-α-d, 60, 72, 73
Ethylcyclohexane, conformational equilibrium in, 447
Ethylcyclohexanecarboxylates, saponification of, 459–460
Ethylene, 308, 388
  symmetry of, 55
Ethylene-d_2, 371
Ethylene glycol, 386
  IR spectrum of, 400
Ethylene glycol monomethyl ether, 386
Ethylenes, tetrasubstituted, 341
Ethylmethylamine, barrier to methyl rotation in, 397, 398
Ethylmethylbenzylamine, 30
Ethylmethylpropylamine, 15
α-Ethyl-α-methylsuccinic acid, 146, 147
p-Ethylstyrene, 397
Ethyne, symmetry of, 56
Ethynyl group, in CIP system, 70, 74
Eu(dcm)$_3$, 158

Eu(dpm)$_3$, 151, 158
Eu(fod)$_3$, 151, 158
Eu(t-cam)$_3$, 157
Eu(thd)$_3$, 158
EuCl$_3$, 127
Eudismic ratio, 135
Eutectic behavior, 90
Eutomer, 135
Exchange of heterotopic nuclei, 326. *See
    also* Site Exchange
Excimer emission, 121
Excited singlet states, 368, 369
Exciton chirality method, 77, 86, 550,
    567–569
Exciton coupling, *see* Exciton chirality
    method
Exciton splitting, *see* Exciton chirality
    method
Exo, 502
Exo ring closures, 434
Exo-anomeric effect, 478
Exo-dig reactions, 436
Exo-endo nomenclature, 502
n-exo-tet reactions, 435
Exo-tet ring closure, 435
Exo-trig reactions, 436
Exocyclic and endocyclic double bonds,
    relative stability of, 469, 480, 482,
    485
Exocyclic torsion angle, 493, 494
External diastereotopicity, 321
Externally diastereotopic nuclei, 321
Extinction coefficients in IR, 442
Eyring equation, 327
E-Z isomerization, *see* Cis-trans
    isomerization
E-Z nomenclature, 340, 341

Face of a molecule, 317, 318
Factorization, 9
Fecht acid, 620
Fenchone, 104
Fenfluramine, camphorate salt of, 230
Ferrocene, 55, 56
Ferrocenyl cation, 318
Ferrocenylmethylcarbenium ion, 318
Fischer projection, 37–42, 45, 68, 75, 79,
    80, 82, 96, 97
Fisher-Hirschfelder-Taylor models, 26

Five-membered rings, 482, 485
Flash chromatography, stereoisomer
    separation by, 250
Flat region in potential energy curve, 390
Flip mechanisms, 634
Fluid state, 10
Fluorinated haloethanes, 401
Fluorochlorobromomethane, *see*
    Bromochlorofluoromethane
Fluorocholine, 386
2-Fluorocitric acids, separation of
    diastereomers of, 249
Fluoroethanes, 401
Fluxion, 13
Fluxional isomers, *see* Valence bond isomers
Flying wedge formula, 79, 80
Footballene, see Buckminsterfullerene
Force field calculations, 21–24, 434, 469,
    490, 497
  in perhydrophenanthrenes, 497
Force field parametrization, 22–24
Formamide, 392
Fossil samples, dating of, 287
Four-membered rings, 481, 482
Fourfold alternating axis of symmetry, 47,
    53
Fragmentation, concerted, stereoelectronic
    requirement for, 462
Framework-type models, 26
Franck-Condon principle, 396
Frankincense, 570
Free energy difference, between alkene
    isomers, 362–365
  of transition states, 412, 413
Free energy of activation, formula for, 327
"Free" rotation, 376
Freudenberg's rule of shift, 594
Fulgenic acid, 628, 629
Fulvene, 342–343
Fumarase, 333, 334
Fumaric acid, 333–335, 349, 356, 357
  isomerization of, 366
Furanose sugars, 476, 482
Furfural, 51
  rotational barrier in, 328
2-Furylcarbinols, 265
2-(2-Furyl)ethanol, 272
Fused rings, 492ff

g number, 551
γ Effect (NMR), 360, 361, 404, 456–458
Gas chromatography (GC), 161
  in determination of configuration, 97
  retention volume of E-Z isomers, 354
  in separation of diastereomers, 161–163
  in separation of enantiomers, 165ff
Gas phase, structure in, 21
Gauche, 14, 16
Gauche conformer, 14, 378–383, 388–390,
    393, 394
Gauche effect, 384
  attractive, 406
Gauche-anti enthalpy difference in butane,
    378–380
  in pentane, 380–382
GC, see Gas chromatography
Gd(dcm)₃, 158
Gearing, 55
Gem dialkyl effect, see Thorpe-Ingold effect
Generalized anomeric effect, see Anomeric
    effect
Geometric isomerism, see Cis-trans
    isomerism
gg Conformer, 381, 382, 386
Glucal, Cotton effect of, 565
Glucose, anomers of, 476–478
D-Glucose, 40, 80, 81, 88, 476, 477
α-D-Glucose, 477
L-Glucose, 134, 221
Glutamic acid, 224
Glutinic acid, 614, 615
Glyceraldehyde, 39, 40, 67, 69, 70, 87
  configuration of, 87
Glyceric acid, 87
Glycerokinase, 305, 306
Glycerol, 433
Glycerol 1-phosphate, 305, 306
Glycolic-d acid, 86
Glycols, determination of enantiomer
    composition of, 159
Glycosides, axial, 478, 479
  bond distances in, 478
  equatorial, 479
Godfrey models, 26
Gossypol, 223, 225, 236
Gramicidin S, 135
Grant parameters, 456–458
Graphical formulas, 11

Grignard reagent, kinetic resolution of, 266
Ground-state compression, 459–461
Group, 48
Group theory, 48, 49
Guanosine, adducts of, with carcinogenic
    hydrocarbons, 571
Guest, 274
Guest-host complexation, 275
Gulose, 40, 88
Gypsy moth, 134
Gyrochiral molecules, see High symmetry
    chiral molecules

H/D coupling constant, 360
H/CH₃ steric repulsion, 391
H/H eclipsing, 377, 378, 389
Half-chair conformation, 463
  of cyclohexane, 438, 439
  of cyclopentane, 482, 483
2-Halobutanes, 234
2-Halocyclohexanones, conformational
    equilibria in, 556
Haloethanols, 386
Halogen compounds, 159
Halogens, steric interference of, 365
Halohydrin cyclization, 434
α-Haloketones, ORD and CD of, 554–557
Halonium ion, 371
2-Halooctanes, 92
1-Halopropanes, 386
β-Halopropionitriles, 385
Halothane, 168
Handedness, see Chirality
Hapten, 179, 270
Hard organic bases, ee analysis of, 158
Haworth formula, 477
Heat of combustion, 363
  of alkenes, 363, 364
  of cycloalkanes, 429–431
  differences in, effect of phase on, 363,
    364
  of stilbenes, 364
Heat of formation, 21
  of alkenes, 363, 364
  of cycloalkanes, 430
Heat of hydrogenation, for cis-trans
    isomers, 364, 365
Heat of isomerization, of alkenes, 364
Heat of mixing, of enantiomers, 104

Heat of solution, of amino acids, 116
  of crystalline racemates, 104
Heat of sublimation, 105, 119
Heat of vaporization, 364, 379
Heavy atom, 76, 84
Heavy atom method (X-ray), 18
Helical conformation, 385
Helical pitch, 131
Helical polymer, induction of, in chiral
     solvent, 571
Helical polymers, 142
Helical structure, of cholesteric phases, 574
[7]Helicene, 637
[8]Helicene, 637, 638
[9]Helicene, 637, 638
  racemization of, 279
[10]Helicene, 636, 638
[11]Helicene, racemization of, 280
[13]Helicene, 638
[14]Helicene, 636
Helicenes, 551, 564, 636–638
  double, 637
  racemization of, 279, 336
Helices, 610
α-Helices, 20
Helicity rules, 553, 563, 564
Helicity, 633
Helix, 536, 539
α-Helix, 576, 577–579
Helix nomenclature, 610
Heme-heme interaction, 582
Hemiacetals, cyclic, 476
Hemihedral crystals, 2, 3
Hemihedral faces (facets), 2, 3, 198
Hemihedry, 2, 109
Hemoglobin, 582
Hendrickson-Wiberg-Allinger approach, 25
Heptaric acid, 43
Herbicides, chiral, 137
Heteracyclobutanes, 481
Heteracyclohexane, inversion barriers in,
     772, 773
Heteracyclopentanes, 484
Heterochiral, 52, 56, 91, 102, 103
  interactions, 103, 125
  molecules, 4
Heterocycles, saturated, conformational
     analysis of, 472ff
Heterofacial, 90, 91, 94

Heterohelicenes, 636
Heterotopic, 303, 315
  by external comparison, 315
  faces, 307, 310, 313, 314, 317, 320
  by internal comparison, 315
  ligands, 307, 310, 312, 314, 316, 317,
     320, 322
  nuclei, averaging of, 325
  site exchange of, 326–328
Heterotopicity, 316, 317, 320, 425
  and NMR, 318, 319, 326–328
Hexachlorobenzene, 56
Hexafluoroacetone phenylimine, 346, 347
Hexahelicene, 113, 140, 175, 232, 551, 636,
     637
(–)-Hexahelicene, absolute configuration of,
     638
Hexahydrobenzoic acid, 437
Hexahydroxybenzene, 55
Hexaisopropylbenzene, 55
Hexane-2,5-diols, 249
3-Hexanol-2-d, 72, 73
Hexapentaenes, 342
Hexaric acids, 43, 88
Hexasubstituted cyclohexanes, 451
2-Hexenes, 357
3-Hexenes, IR spectra of, 357
Hexoses, 134
L-Hexoses, 134
Hexyl isocyanate, 142
High dilution, 432
High performance liquid chromatography,
     see HPLC
Histidine, 332
Histidine hydrochloride, 198, 202
History of stereochemistry, 2ff
HIV-1 protease, enantiomers of, 135
Hofmann bromamide rearrangements, 94
Hog kidney acylase, 176
Holding group, conformational, 441, 442
Homochiral, 52, 66, 91, 102, 103
  crystals, 201
  interactions, 103, 125
  molecules, 4
  dimerization of, 52, 53
Homoconjugated aldehydes, CD of, 564
Homofacial, 90, 91, 94
Homogeneous, see Neat
Homomeric, 31, 315

Homomethionine, 220
Homomorphic faces, 315
Homomorphic ligands, 303, 306, 307, 312, 315
Homotopic, 315
  faces, 307, 309, 315
  ligands, 307–309, 312, 315, 321
  protons, 320
Horeau effect, 145
Horeau's method, 96, 97
Horeau's rule, 96, 97, 622
Horse liver alcohol dehydrogenase (HLAD), 270
Hot beams, rapid cooling of, 391
HPLC, with chiral solvents, 164
  chiral stationary phases in, 170
  in determination of configuration, 97
  in separation of diastereomers, 163–169
  in separation of enantiomers, 169ff
Huggins electronegativity, 404
Hydantoins, 208, 285
Hydrazine, conformation of, 397, 398
Hydrindane, 493, 494
Hydrobenzoin, 113, 203, 377
Hydroboration, 265
Hydrocarbon polymers, chiroptical properties, 585
Hydrogen bonding, 12, 233, 425
  and infrared spectroscopy, 400, 401
  intermolecular, 12
  intramolecular, 386, 400, 401, 406, 578
  in vicinal diols, 426, 427
Hydrogen bridge, *see* Hydrogen bonding
Hydrogen chloride, symmetry of, 54
Hydrogen peroxide, conformation of, 397
Hydrogen-bond forming solvents, 442, 445
Hydrolases, 271
Hydropyrimidine hydrase, 208
α-Hydroxy acids, 176
  configuration of, 97
1-Hydroxybenzotriazole, 285
2-Hydroxy-1,1′-binaphthyl, rotation of, 587
cis-3-Hydroxycyclohexanecarboxylic acid, 429, 462, 640
  lactonization of, 462
1-Hydroxy-*trans*-decalin, 495, 496
2-Hydroxy-2-phenylcyclopropanecarboxylic acid, 421, 422
p-Hydroxyphenylglycine, 217, 220

5-(*p*-Hydroxyphenyl)hydantoin, 208
2-Hydroxypinane-3-one, 221
Hydroxyproline, 482
5-Hydroxytryptophan, 335
Hyoscyamine, 137
Hyperconjugation, 368, 389
Hyperconjugative interaction, 384
Hyperconjugative resonance, 388

Ibuprofen, 138
Icosahedron, 57, 58, 513, 516
Identity operation, 46, 48
Immobilization of enzymes, 271
Immunogens, 179
Improper axis of symmetry, *see* Alternating axis of symmetry
Improper operation, 46, 59
Inclusion chromatography, 172
  model for, 173
Inclusion compounds, diastereomer separation by formation of, 252
Indenes, analysis of, 177
Induced cholesteric mesophases, 575
Infrared absorption coefficients, 400
Infrared spectra, 4, 356, 357, 400, 401
  and conformational equilibria, 391, 400
  of enantiomers and racemates, 201
  of stereoisomers, 120
Infrared spectroscopy, and intramolecular hydrogen bonding, 401
  time scale in, 399
Inherently achiral chromophores, 551, 553
  interaction between, 564
Inherently chiral chromophores, 550, 551, 553, 564
myo-Inositol, 221
Inositols, 451
"In-out'" isomer, 504
Insect pheromones, 134
Interatomic distances, 26
Interconversion rate, 34
Intermolecular hydrogen bonding, *see* Hydrogen bonding
Internal coordinates, 8, 9
Internal diastereotopicity, 321
Internally diastereotopic nuclei, 321
Internally enantiotopic nuclei, 321
Internuclear distance, 11, 19, 23. *See also* bond length

Intraannular torsion angle 437, 438. *See also* endocyclic torsion angles
Intersystem crossing, 344
Intramolecular bond distance, 25
Intramolecular hydrogen bonding, *see* Hydrogen bonding
Intramolecular interaction, 26
Intrinsic rotation, 592
Intrinsic shift non-equivalence, 325
Inversion, 34, 93. *See also* Walden inversion
  in amines, 30, 33–35
  at nitrogen, *see* Nitrogen
Inversion barrier, in cyclohexane, 438, 439, 474
  in heteracyclohexanes, 472, 474
  in methylenecyclohexane, 469
  in oxane, 472, 474
  in piperidine, 472, 474
Inversion of configuration, 90–93
Inversion path, in *cis*-decalin, 495
Iodonium ion, 371
α-(Iodophenoxy)propionic acids, 547
Ionic Pirkle type CSP, 171
Ionization constants of cyclane-1,2-dicarboxylic acids, 425, 426
Ionol isomers, photochemical conversion of, 369
Ionophores, 275
IR, *see* Infrared
IR stretching frequency, in 1,2-diols, 427
Isoborneol, 503, 618
Isobornyloxymagnesium bromide, 618
Isobutane, barrier in, 380
Isobutyraldehyde, 433
Isochronous nuclei, 318, 319
Isochronous protons, 321
Isochrony, accidental, 318
Isolability criterion, 10, 34–36, 66
Isolability, of isomers, 66
Isolable species, 10
Isoleucine hydrobromide, 77
Isomers, 10
  cis-trans, *see* Cis-trans isomers
Isometric structures, 41, 42
Isometry, 41
Isopropyl alcohol, 310, 311
cis- and *trans*-3-Isopropylcyclohexanols, separation of, 248
α-Isopropyl-α-methylsuccinic acid, 147

Isoserine, 87
Isotactic polypropylene, 583
Isotactic vinyl polymers, 582
Isotopic composition, chirality due to, 60
Isotopic substitution, effect on VCD, 598
Isotopomer, 598
Isotropic refractive index, 539
Isotropic transitions, 552
IUPAC numbering, 41

Japanese beetle pheromone, 134

Karplus relationship, 360, 361, 403, 404, 438, 454
Kekulé models, 430
Kendrew Skeletal Molecular models, 26
Ketene imines, 617
Ketene immonium salts, 617
α-Ketoester, 97
α-Ketoglutaric acid, 305, 306
Ketones, kinetic resolution of, 263
  racemization of, 284
  unsaturated, 564
Kinetic analysis, 414
Kinetic quenching, 408, 411
Kinetic resolution, 257ff
  abiotic catalysis for, 265, 267
  catalyzed by antibody protein, 270
  in configuration assignment, 96, 97
  consecutive reactions in, 261
  mathematical relationships in, 259
  of α-methylbenzyl alcohol, 262
Klyne-Prelog convention, 16, 631
Kofler method, 200
Krebs cycle, 305
Kronig-Kramers theorem, 547

l (for like), 81
L, *see* D
l,u (like, unlike) nomenclature, 81
Lactate dehydrogenase, 176
Lactic acid, 6, 37, 38, 45, 68, 69, 303, 332
  configuration of, 86, 87
  rotation of, 37
(R)-Lactic acid (D), 37
Lactones, 162
  large-ring, 432
Lactonitrile, 310, 312

Lactose, as enantioselective stationary phase, 237
LAD, *see* Liver alcohol dehydrogenase
Lamellar twinning, 637
Lanthanide induced shifts (LIS), 157
Lanthanide shift method, 117
Lanthanide shift reagents (LSRs), 157, 318
as inducers of Cotton effect, 573
Large rings, 430, 432
conformation of, 485ff
Lasalocid A, 229
Laser polarimetric detector, 585
Lattice inclusion compounds, 234
N-Lauroylvaline *tert*-butylamide, 128
Least motion, principle of, 461
LEC, *see* Ligand-exchange chromatography
Leishmania brazilensis panamensis, 37
Lennard-Jones potential, 23
Leucine, 141, 198
Levorotation, 5
Libration, 383
Ligand complementation, 70
Ligand-exchange chromatography (LEC), 232
Like, *see* lk
Limonene, 133, 585
Line shape analysis, 327, 328
Linear dichroism, 586
Linear molecules, 9, 54, 56
Linear polarization, 536
Linear substituents, 445, 446
Linearly polarized light, 1, 536, 545
Lipases, use in organic solvents, 272
Lipodex™, 167, 168
Liquid chromatographic resolution of chiral substances, 585
Liquid crystal induced optical activity, 574
Liquid crystals, 131
Liquid phase, structure in, 21
LIS, *see* Lanthanide induced shifts
Lithium (*S,S*)-α,α′-dimethyldibenzylamide, 245
Lithium trisiamylborohydride, 468
Liver alcohol dehydrogenase (LAD), 330–333
lk (like), 97
Ln(facam)₃, 158, 159
Ln(hfbc)₃, 158, 159
Ln(hfc)₃, *see* Ln(hfbc)₃

Ln(*t*-cam)₃, 158
Lobeline, 218
Local minimum, 24
London force, 23, 442
Lone electron pairs, 475
Lossen rearrangements, 94
Low rotational barriers in alkenes, 342ff
Low-temperature NMR, 441
Lowe's rule, *see* Lowe-Brewster rules
Lowe-Brewster rules, 596, 616
LSR, *see* Lanthanide shift reagent
L-Lysine, 134
hydrochloride, 78, 79
Lysozyme, CD of, 580
Lyxose, 40, 87, 88

M, 610
Macrolides, 432
Macromolecules, CD of, 576
Macroscopic sample, chirality of, 4
Magnetic dipole, 455
Magnetic field vector, 535
"Main axis," *see* Principal axis
Main chain of molecule, 82
Malate, 127. *See also* Malic acid
Malathion, 137, 138
Maleic acid, 247, 335, 349, 356, 357
Malic acid, 87, 91, 219, 333, 334
Malic-3-*d* acid, 333, 334
Malvalic acid, 480
Mandelate racemase, 284
Mandelic acid, 114, 120, 130, 131, 164, 215, 218–221, 227, 239, 245
racemization of, 284
Mannose, 40, 88, 476
Marckwald principle, 214
Mass resolved excitation spectroscopy, *see* MRES
Mass spectrometry, 129
in determination of configuration by Horeau's method, 97
Matrix isolation technique, 393
Matrix-isolation IR spectroscopy, 391
McConnell equation, 455
Mechanical model, 25, 26, 45
Medium rings, 430
conformation of, 485ff
Meerwein-Ponndorf reduction, 627
Melting points, of chiral compounds, 111

Menthol, 133, 162, 214, 218, 245

Menthone, 225
 racemic, 227

Menthoxyacetic acid, 147, 148, 164

Menthoxychloroformate, 162

Menthyl chlorocarbonate, 221

Menthyl 3-(2-hydroxybenzoyl)propionate, 244

10-Mercaptoisoborneol, 226

Merrifield synthesis of rotaxane, 507

Mesaconic acid, 335

9-Mesitylfluorene, 630, 631

Meso and chiral stereoisomers, distinction between, 320–322

Meso diacetates,

Meso isomers, 421
 separation from *rac*- isomers, 248, 249

Mesophases, 574

Metallocenes, 608, 642

Metastable forms, 107

Metastable racemic compounds, 202

Methane, planar, 4
 symmetry of, 57

Methanol, barrier in, 397, 398
 symmetry of, 52

N-(4-Methoxybenzylidene)-4′-butylaniline, 575

1-Methoxybicyclo[2.2.2]oct-5-ene-2-carboxylic acids, 250, 251

4-Methoxycyclohexanols, 252

β-Methoxylactic acid, 69

α-Methoxy-α-methyl-(pentafluorophenyl)-acetic acid (MMPA), 148

2-(6-Methoxy-2-naphthyl)propionic acid, *see* Naproxen

3-Methoxyprolines, 249

α-Methoxy-α-(trifluoromethyl)benzyl isocyanate, 148

α-Methoxy-α-(trifluoromethyl)phenylacetic acid (Mosher's acid, MTPA), 148, 149, 218
 microbial resolution of, 272

Methyl *N*-acetyl-1-pyrenylalaninate, 121

Methyl acrylate, 394

Methyl formate, 390–394

Methyl glucoside, 478

Methyl groups, conformational energy of, 445, 446

conformational equilibrium of, in heteranes, 472–474

Methyl 5-hydroxy-3-cyclohexene-carboxylate, 283

Methyl mercaptan, 12

Methyl O-methylmandelate, 594

Methyl α-methylbenzyl ketone, 313, 314

Methyl L-phenylalaninate, 236

Methyl phenylglycinate, 243, 274

Methyl phenylglycinate.HPF$_6$, 276

Methyl substitution parameters (NMR), 456, 457

Methyl 2,4,5-tetradecatrienoate, 613

N-Methylacetamide, structure of, 392

4β-Methyl-2-adamantanone, Cotton effect of, 563

Methylamine, barrier in, 397, 398

2-Methylamino-2-phenylethylamine, 620

1-Methyl-2-arylpyrrolidines, quaternization of, 413, 414

N-Methylbenzamide, 353

α-Methylbenzyl alcohol, 240. *See also* Phenylmethylcarbinol

α-Methylbenzyl chloride, 93
 racemization of, 483

α-Methylbenzyl 3,5-dinitrobenzoate, 118

α-Methylbenzyl isocyanate, 148, 151, 221, 223

α-Methylbenzyl methacrylate, 266

α-Methylbenzylamine, 104, 131, 151, 153, 154, 156, 162, 211, 213, 216, 228, 641
 racemization of, 281

α-Methylbenzylammonium cinnamate, 108, 203
 DSC of, 112
 hexafluorophosphate, 274
 hydrogen sulfate, 211
 salts, 275

α-Methylbenzylsemioxamazide, 223

2-Methyl-l-butanol, 585

1-Methyl-7-*tert*-butylcycloheptatriene, 487

1-Methyl-*trans*-4-*tert*-butylcyclo-hexanecarboxylic acid, 250, 251

2-Methyl-5-*tert*-butyl-1,3-dioxane, 478

N-Methyl-4-*tert*-butylpiperidine, N-oxidation of, 411, 412

3-Methylcyclohexanol, 427, 429

2-Methylcyclohexanone, 263

3-Methylcyclohexanone, 551, 552, 555, 559

2-Methylcyclohexanone
N,N-dimethylhydrazone, 470, 471

4-Methylcyclohexylacetic-α-d acid, 618

4-Methylcyclohexylideneacetic acid, 617, 618

Methylcyclopentane, 482, 483

2-Methylcyclopentanone, 560

Methylcyclopropane, 394

Methylcyclopropene, 480

Methylcysteine sulfoxide, 85

2-Methyl-trans-decalin, 495

1-Methyl-2,6-diphenyl-4-piperidone oxime, 620

Methylene chloride, 307–309

Methylenecyclobutane, 482

Methylenecyclohexane, 469, 485

Methylenecyclohexane-l-methylcyclohexene equilibrium, 469

Methylenecyclopentane, 485
strain in, 485

Methylenecyclopentane -1-
methylcyclopentene equilibrium, 485

Methylenecyclopropane, 480

α-Methyl-(p-ethylbenzyl) hydrogen phthatate, 121

2-Methyl-2-ethylsuccinic acid, 7

N-Methylformamide, structure of, 392, 393

α-Methylfumaric acid, 335

1-Methylindene, 266

3-Methylindene, 266

α-Methylmaleic acid, 335

(R)-O-Methylmandelate, ester of
(S)-(+)-propanol-1,1,1-d₃, 322, 323

O-Methylmandelic acid, 148, 153, 218, 219, 222

O-Methylmandeloyl chloride, 147

2-Methyl-1-methylenecyclohexane, 469
confomational equilibrium in, 469
inversion barrier in, 469

p-Methyl-trans-β-methylstyrene, 397

Methyloxanes, 472, 474

2-Methyl-l-pentene-2-methyl-2-pentene equilibrium, 469

2-Methylphenylacetaldehyde, see
2-Phenylpropanal

2-Methyl-2-phenylbutanedioic acid, 219

α-Methyl-β-phenylethylamine, 211

N-Methyl-1-phenylethylamine, 34, 35

1-Methyl-5-phenyl-5-propylbarbituric acid, 136

Methylphenylpropylphosphine, 15, 69

α-Methyl-α-phenylsuccinic anhydride, 221

Methylpiperidines, 472, 474

N-Methylpiperidinium salt, 474

2-Methylproline, 286

α-Methylstyrene, 397

10-Methyl-TFAE, 169

Methylthianes, 472, 474

Mevalonolactone, 159

Michaelis constant, 269

Microorganisms, in kinetic resolutions, 270

Microscale determination of configuration, 96, 97

Microwave spectroscopy, 17, 19, 399, 463
in determination of E-Z configuration, 353, 354

Migrating group, 94

Migration of chiral group, 94

Mirror image relationship, 31

Mirror images, 2, 3, 48

Mixing of enantiomers, 104, 105

MMP, 24

MMPA, see α-Methoxy-α-methyl-(α-
pentafluorophenyl)acetic acid

Möbius strip, 505, 513

Molar ellipticity, 546

Molar rotation, 145, 587

Molar volume, see Molecular volume

Molecular energy, a priori calculation of, 21, 22

Molecular formula, 30

Molecular imprinting, 173, 174

Molecular mechanics, 21–25, 380, 384, 385, 447, 464, 497

Molecular modeling, 26

Molecular models, 25, 26, 430

Molecular overcrowding, 636

Molecular propellers, 632ff

Molecular shuttle, 510, 511

Molecular spectroscopy, 45

Molecular structure, a priori calculation of, 20ff

Molecular vibration, 10

Molecular volume, 354, 379

"Molecule," meaning of, 10

Molecules with polar substituents, 383ff, 399

Moments of inertia, 19, 379
Monochromatic polarized radiation, 535
Monoclonal antibodies, 270
Morphine, 135, 136, 211
Mosher's acid, *see* α-Methoxy-α-(trifluoro-
   methyl)phenylacetic acid
MTPA, *see* α-Methoxy-α-(trifluoromethyl)-
   phenylacetic acid
MRES, 396, 397
Mutarotation, 277, 477, 586
Mycomycin, 612
Myeloperoxidase, 135
Myoglobin, CD of, 580
Myosin, 580

n, 215
NAD$^+$, 176, 330–333
NADH, 330–333
Naphthalene, 55
N-(2-Naphthyl)-α-amino acids, 170
1-(1-Naphthyl)ethyl isocyanate, 223
1-(1-Naphthyl)ethylamine (NEA), 154, 213
1-(1-Naphthyl)ethylammonium
   phenylacetate, 211
N-(1-Naphthyl)leucine, 238
1-(1-Naphthyl)propanoic acid, 114
Naproxen, 165, 221, 222
   ethylamine salt, 208
Naproxen methyl ester, 207
NEA, *see* 1-(1-Naphthyl)ethylamine
Neat, 6
Neighboring group participation, in amide
   hydrolysis, 217
   stereoelectronic requirements of, 462
Nematic phase, 574
   NMR in, 17
   transformation into cholesteric phase, 575
Neopentane, 434
Neopentyl-1-*d* alcohol, 60
Network arguments, 98
Neutron diffraction, 17, 19
Newman projection, 79, 282
Nicotinamide adenine dinucteotide, *see*
   NAD$^+$, NADH
Nicotine, 128, 135, 136, 482
   rotation of, 590
Nitrogen, stereochemistry at, 34, 35
Nitrogen inversion, 35
o-Nitrophenyl octyl ether, 275

threo-l-(*p*-Nitrophenyl)-2-amino-1,3-propane
   diol, 204
Nitrous acid deamination, 433
NMR, 306, 307, 318ff
   spectral nonequivance, 155
   spectral properties of cyclohexane
      derivatives, 455
NMR spectra, in determination of E-Z
      configuration, 353, 358ff
   of diastereomers, 146
   of diastereotopic nuclei, 312
NMR spectroscopy, 305, 401ff
   in chiral media, 153, 312
   ^{1}H and conformational equilibria, 440
   time scale in, 399, 401
NOE, *see* Nuclear Overhauser effect,
NOESY, *see* Nuclear Overhauser effect
Nomenclature, cis-trans, in cyclanes, 421,
      422
   local systems of, 423
α,β-Nomenclature, in decalins, 495, 496
   in steroids, 495, 500
Nonbonded atoms, 11, 12, 19
Nonbonded distance, 11
Nonbonded energy, 22, 23, 430
Nonbonded interaction, 22, 23, 25, 103, 390
Nonbonded potential, 23
Noncentrosymmetric crystal classes, 109
Nonconservation of parity, 140
Nonequivalence, in NMR, 306, 307
   intrinsic, 325, 326
Nonlinear n-atomic molecule, 9
Nonplanar alkenes, 340, 342–346
Nonplanar ring, bond angles in, 430
Nonplanarity, in biphenyls, 626
Nonracemic, 4, 108, 143
Nonrigid symmetry, 59
5-Nonyl tartrate, 178
Nootkatone, 133
2,3-Norbornadione-^{16}O,^{18}O, 60
Norbornane, 501, 502
2-Norbornanone, 469, 502, 503
   nucleophilic additions to, 468, 502, 503
Norcamphor, 502, 504, 615. *See also*
      2-Norbornanone
   nucleophilic addition to, 468
Norcaradiene, 487
Normal coordinate analysis, 17, 19

n-π* carbonyl transition, 547, 552, 554, 557, 563, 579
n-σ* orbital overlap, 392, 393
n-σ* transition, 552
Nuclear magnetic resonance, *see* NMR. *See also* individual nuclei
Nuclear Overhauser effect, 17, 19, 20, 361, 362
Nucleophiles, additions of, to cyclic ketones, 502, 503
Nucleophilic additions, to alkynes, 350
Nucleophilic displacement reactions, 19–94
Nucleotides, 482
Null rotation, 7

Observational technique, 10
Octahedron, 57, 513
2-Octanol, 92, 152
Octant rule, 549, 557–563
  exceptions to, 562
2-Octyl acetate, 92
2-Octyl *p*-toluenesulfonate, 92
2-Octyl tosylate, *see* 2-Octyl *p*-toluenesulfonate
Olefinic proton, chemical shift of, 358, 359
Olfactory properties of enantiomers, 134
One-ring flip mechanism, 634
Operational null, 588
Optical activity, 59, 60, 109
  due to isotopic composition, 60
  induced, 534
Optical antipode, *see* Enantiomer
Optical null, 542
Optical purity, 143, 586
Optical resolution, see Resolution
Optical rotation, 1, 2, 5–7, 37, 39, 59–61, 67, 398, 399
  calculation of, 593
  semiempirical theory of, 594
  spurious, 587
Optical rotatory dispersion, 37, 77, 398, 399, 534, 541, 544. *See also* ORD
  of succinic-*d* acid, 333
Optical rotatory power, 535
Optical superposition, principle of, 594
ORD, 547, 548. *See also* Optical rotatory dispersion and CD
  of cyclohexane derivatives, 453
  induced, 470

sign of, 543
ORD curves, anomalous, 542
  crossing the zero rotation axis, 548
  plain, 541
  simple, 542
Order of a group, 48–58, 62, 63
Organoiron reagents, 468
Origin of optically active molecules in nature, 138
Orthoester Claisen rearrangement, 613
Oscillator theory of optical activity, 595
Overlap interaction, 378
2-Oxahydrindane, 494
Oxaloacetic acid, 305, 306
Oxane, inversion barrier in, 472, 473
Oxapadol, 239, 240
Oxaphosphetanes, 351, 352
Oxazolidine-2-selone, 151
Oxazolinones, 285
Oxetane, 481
Oxidation, enzymatic, 331
Oxidoreductases, 271
Oxidoreductions, 270
Oxime ethers, 346
Oximes, 346
  Beckmann rearrangement of, 94, 352, 353
  configurational assignment of, 352, 353
5-Oxo-2-tetrahydrofurancarboxylic acid, 223
Oxycarbenium ion intermediate, 436

p, 215
P, 610
Paclobutrazol, 137, 138, 207
[2.2]Paracyclophane, 640
[6]Paracyclophane, 639
[10]Paracyclophanecarboxylic acid, 639
Paracyclophanes, 304, 305, 611, 639, 640
Parity principle, 139
Partial double bond, *see* Double bond, partial
Pasteur's salt, 198, 202
Pauli's exclusion principle, 378
P cepacia, 273
PEA, see α-Methylbenzylamine
Peak broadening, 137
1,3-Pentadiene, 49
Pentadienoic acid (glutinic acid), 614
Pentahelicene, 266

Pentahydroxypimelic acid, 84
Pentane, 380–382
2,4-Pentanediol, 163, 263
Pentatetraenes, 617
Pentoses, 40
Peptidases, 306
Peptide chain, 20
Peptides, conformation of, 392, 393
  2D NMR in, 19, 20
Perchlorotriphenylamine, 239
  racemization of, 281
Perhydroanthracenes, 496–498
Perhydrophenanthrenes, 496, 497
Perhydrotriphenylene, 51
Permethrin, 230
cis-Permethrinic acid, 230, 231
Permutations, 38
Perspective formulas, 80
Pesticides, chiral, 137
Pfeiffer's rule, 137
pH dependence of rotation, 6
Phantom (duplicate) atom, 70
Phantom ligands, 314
Pharmacological implications, of
  stereoisomerism, 135
Phase behavior, of conglomerates, 90
  of racemic compounds, 90
Phase change (X-rays), 76
Phase diagram, 90
  binary, 107
  of conglomerates, 107, 111
  melting point, 111
  of solid solutions of enantiomers, 115
  ternary, 117, 118
Phase information (X-rays), 18
α-Phellandrene, Cotton effect of, 566
α-Phenethylamine, see
  α-Methylbenzylamine
α-Phenethyl chloride, see α-Methylbenzyl
  chloride
α-Phenethyl methyl sulfide, 313, 314
α-Phenethyl methyl sulfoxides, 313, 314
Phenyl group, conformational energy of, 446
  in CIP system, 72
Phenyl rings, orthogonal, 626
Phenyl p-tolyl sulfoxide, racemization of,
  281
Phenylalanine amide, 217
3-Phenylalaninol, 217

1-Phenyl-1-butanol, 113
3-Phenyl-1-butene, rotation of, 589
3-Phenylbutenoic acid, 341
α-Phenylbutyric anhydride, 96, 263
β-Phenylbutyrolactone, 163, 164
Phenylcyclohexane, 25, 446
3-Phenylcyclopentanecarboxaldehyde,
  462, 463
2-Phenyl-1,3-dioxin-4-ones, 239
1-Phenyl-1,2-ethanediol, 127
α-Phenylethanesulfonic acid, 221
1-Phenylethanol, see α-Methylbenzyl
  alcohol and Phenylmethylcarbinol
α-Phenylglycinate esters, 243
α-Phenylglycine, 113, 208, 217, 227, 274,
  620
  sulfate, 202
2-Phenylglycinol, 163, 164, 217, 237
5-Phenylhydantoin, 208
Phenylmagnesium bromide, addition to
  pyruvate, 97
Phenylmethylcarbinol, rotation of, 587. See
  also α-Methylbenzyl alcohol
1-Phenylmethylidene-4-methylcyclohexane,
  619
Phenylpentadeuteriophenylcarbinol, 60
2-Phenylpropanal, rotation of, 592
2-Phenylpropanoic acid, 237
1-Phenyl-2,2,2-trifluoroethanol, 319
Phloroglucinol, 54, 55
Phosphorus-containing rings, 491
Photochemical asymmetric synthesis, 140,
  638
Photochemical isomerization, of cis-trans
  isomers, 366ff
Photoelastic modulator, 546
Photosensitization, 367
Photostationary state, 366, 368, 369
Physical methods, in determination of E-Z
  configuration, 353ff
Physical properties, of alkenes, 353–356
  of cyclane diastereomers, 425–428
  of cyclohexane derivatives, 453ff
  of stereoisomers, 37, 41–43
π bond, 339
π bonds, orthogonal, 342
π electrons, delocalization of, 343, 344
  in C=N, delocalization of, 346
Pig pancreatic lipase (PPL), 272

Pilot atom, 611
α-Pinene, 254, 256, 265
  racemic, 278
Pinenylamines, 213
π-Orbital overlap, 626
Piperidine, ring inversion barrier in, 473
Pirkle covalent CSPs, 171
Pirkle enantioselective stationary Phases, 169–171, 238
Pitzer potential. *See* Torsional Potential
Pitzer strain. *See* Eclipsing strain, Torsional strain
Pivalaldehyde, 286
N-Pivaloylphenylalanine dimethylamide, racemization of, 284
pKa Values of acids, 442
Plain curves (ORD), 541, 543, 547
Planar chirality, 317, 611, 638–642
Planar methane, 4
Planar prochirality, 317
Plane polarized light, 1, 2, 5–7, 59, 60. *See also* Linearly polarized light
Plane of symmetry, 3, 45–47, 51, 53–57, 61, 88, 89, 310, 312, 314, 315.
Plant growth stimulant (auxin), 136
Plant growth regulators, 137
Platonic solids, 57, 58
β Pleated sheet, *see* β Sheet
p orbitals, lateral overlap of, 339
p-π orbital overlap, 619
Podand, 505, 510
Podophyllotoxin, 223
Point group, *see* Symmetry point group
Point of inversion, *see* Center of symmetry
Polar axis, 124
Polar crystal, 77–78
Polar medium, 385
Polarimeters, photoelectric, 585
Polarimetric detector, laser, 585
Polarimetric measurement, 67
Polarimetry, 5ff, 534, 585–592
  concentration effect in, 587, 591, 592
  solvent effect in, 587, 589, 592
  temperature effect in, 587, 589
Polarizability, 23
Polarizability order, 595
Polarizability theories, 594
Polarization, 399
Polarized light, 1, 5, 534

interaction with chiral molecules, 539, 540
Polyadenylic acid, CD of, 581
Poly-L-alanine, CD spectrum of, 578
Polyamide stationary phases, 172
Poly[(*S*)-(+)-*sec*-butyl isocyanide], 583
Poly(*sec*-butyliaminomethylene), 584
Poly(*tert*-butyliminomethylene), 584
Polydentate ligand, in CIP system, 70, 71
Poly[(*R*)-3,7-dimethyl-l-octene], 585
Poly(D-glutamate), 134
Poly(L-glutamate), 576, 577
Poly(L-glutamic acid), CD spectrum of, 577
Poly(1-hexyl isocyanate), 584
  enantiomerization of, 205
Poly[(*R*)-l-hexyl-l-*d* isocyanate], 584
Polyisocyanates, 205, 584
Poly(DL-leucine), 141
Poly(L-lysine), 576, 579
  CD spectrum of, 577
Polymerization reactions, in kinetic resolutions, 266
Polymers, CD of, 576
Poly(methylene-1,3-cyclopentane), 582, 583
Poly[(*S*)-4-methyl-l-hexene], 585
Polymorphic forms, 207
Polymorphism, 107
  of enantiomers, 108
Polynucleotides, CD of, 576, 581
Polyoxyethylene, 585
Polyoxymethylene, 386
Polypeptides, 12, 393, 577
  α-helix in, 579
  CD of, 576
  Poly(L-proline), 580
Poly(propylene oxide), 584
Poly(triphenylmethyl methacrylate) (PTrMA), 141, 175, 239, 584
Potassium mandelate, 131
Potato starch, as enantioselective stationary phase, 237
Potential energy curves, shallow (flat region in), 390
p-π orbital overlap, 391, 392
PPL, *see* Pig pancreatic lipase
Pr(tpip)₃, 160
Preferential crystallization, 199, 201, 204
  efficiency of, 204
  of hydrobenzoin, successive cycles in, 203

Preferential crystallization (*continued*)
  stirring rate in, 204
Prelog's rule, 97, 627
Prelog-Klyne nomenclature, 16, 631, 632
Prévost reaction, 371
Principal axis, 50, 55
Priority (CIP), 68–70
Prochiral axis, 303, 304, 317
Prochiral center, 303, 304, 316, 322
Prochiral faces, 303, 304
Prochiral plane, 303, 304, 314, 317
Prochirality, 303, 304, 316
Pseudoasymmetry, 43
Product ratio, and ground state
  conformation, 412, 413
Projection formulas, 79, 80
Proline, 217, 224, 286, 482
Proline-containing polypeptides, cis-trans
  isomerism in, 348
Propane, 9, 378
1,2-Propanediol, 126
1,2-Propanediol di-*p*-toluenesulfonate, 254
1-Propanol, 386
2-Propanol, 322–323
2-Propanol-1,1,1-d_3, 72, 73, 322, 323
[1.1.1]Propellane, 505
Propellanes, 505
Propellicene, 637
Propene, 313
Propenes, barrier to methyl rotation in, 388, 389
Propenoic acid, 394
Propenoic acids, pKa values of, 356
Proper axis of symmetry, *see* Axis of
  symmetry; Symmetry axis
Proper operation, 46, 59
Propionaldehyde, 390
Propionic acid, 303
Propoxyphene, 136
Propranolol, 165, 179
  hydrochloride, 153
n-Propylammonium 2-phenoxypropionate,
  255
2-Propylcyclohexanone, 151
Propylene, conformation of, 388, 389. *See
  also* Propene
Propylene oxide, 266
Propylene sulfide, 266
Pro-R, 316, 320, 322

Pro-S, 316, 320, 322
Prostaglandins, 482
Prostereogenic, 304, 313
  assembly, 316
  axis, 304
  centers, 304
  elements, 304
Prostereogenicity, 314
Prostereoisomerism, 304, 305, 320, 335. *See
  also* Prochirality
  in biochemical reactions, 335
Prosthetic groups, 582
Proteins, 576
  dating of, 287
  2D NMR in, 20
  globular, 580
Proton correlated spectroscopy (COSY), 19,
  20
Protonation of nitrogenous bases, effect on
  conformational equilibrium, 446, 447
Protons, antiperiplanar to electronegative
  atoms (NMR), 454
Pseudoasymmetric center, 43, 83, 84
Pseudoaxial positions, 463
Pseudoaxial substituent in
  9,10-dihydroanthracene, 499
Pseudochirality, *see* Pseudoasymmetry
Pseudoephedrine, 131
Pseudoequatorial positions, 463
Pseudomonas sp., 273
Pseudoracemates, 106
Pseudorotation, 483–485, 487
Pseudorotation circuit, 484
Pseudoscalar properties, 37
PTRMA, *see* Poly(triphenylmethyl
  methacrylate)
Puckering, 493
  amplitude, 484
  in cyclobutane, 481
  in cyclopentane, 484
  in rings, 430
Purification by duplication, 256
Push-pull alkene, 343, 366
Push-pull ethylenes, twisted ground states
  in, 345
Pyramidal triligant atoms, 68
Pyranose sugars, 40, 476–478
Pyrethrins, 480
Pyridine-d_3, 318

1-(4-Pyridyl)ethanol, 263
Pyrolytic elimination, 350
Pyrrolidine enamine, 470, 471
Pyrrolizidine alkaloid biosynthesis, 335
Pyruvic acid, 303, 304, 332

Quantum mechanical treatment of
    stereochemistry, 1
Quantum mechanics, in calculation of
    structure, 20, 21
Quartz, 2
Quasi-enantiomeric selector, 172
Quasi-racemate, 85, 90
  method of configuration assignment, 90
Quinidine, 85, 172, 212, 241, 266
Quinine, 153, 154, 172, 212
Quinone, oxidation of, 349
Quinotoxine, 210, 212

r (descriptor), definition, 84
r (reference) nomenclature, 421, 422
R (*rectus*), definition, 68
R factor, 18
Racemase enzymes, 244
Racemate, 102
Racemates, labile, 205
  nature of, 106
Racemic, 4, 142
Racemic compound, 106
  conversion to a conglomerate, 198
  stability of, 110
Racemic mixture, 106
Racemic modification, *see* Racemate
Racemization, 242, 277. *See also* individual
    compounds racemized
  via achiral intermediates, 281
  of amines, 281
  of amino acids, 285
  aldehydes in, 286
  mechanisms for, 285
  via carbanions, 283
  via carbocations, 282
  in chromatography, 283
  half-life of, 278
  by pyramidal inversion, 281
  processes, 278
  thermal methods for, 279
Radicals, in catalyzed equilibration, 367
Radioimmunoassay (RIA), 179
Raman optical activity (ROA), 597

Raman spectra, 4
  of 1,2-dihaloethanes, 383
Raman spectroscopic studies, of 1,3-
    butadiene, 393
Raman spectroscopy, 19, 396
Random coil, 576
Rate constants, averaging of, 409
Rate of interconversion of acyclic
    conformers, 398, 399
Re, 317, 320
Reaction products and rates and
    conformational composition, 407ff
Reaction rate, use of, in configuration
    assignment, 560
Rearrangement reactions, stereoelectronic
    requirements for, 462, 463
Reciprocal resolutions, 211, 238
Reciprocal salt pairs, 215
Rectangular conformation, 490, 491
Reduction, enzymatic, 331
Reentrant angles, 423
Refinement (X-ray), 18
Reflection plane, *see* Plane of symmetry
Refractive index, 379, 398
  in determination of E-Z configuration,
    353–355
  of diastereomers, 453
Relative configuration, 65–67, 79ff, 423ff
Relaxation, in NOE, 362
Replacement criterion, *see* Substitution
    criterion
Residual enantiomers, 35
Residual isomers, 66
Residual species, 36
Residual stereoisomerism, 34–36
Residual stereoisomers, 34, 635
Resolution, 197
  of alkenes and arenes, 231, 265
  chromatographic, 227
    on enantioselective stationary phases, 237
    preparative, 236
  by complex formation, 231ff
  via diastereomers, 209ff
  by distillation, 231, 247
  by inclusion, 232ff
  kinetic, 257ff
    catalytic, 265ff
    enzymatic, 268ff
    by Sharpless protocol, 261
    in stoichiometric reactions, 262ff

Resolution (*continued*)
  kinetic control in, 197
  Marckwald, 211, 215
  mutual, 215
  by preferential crystallization, 201ff
  principles and practice of, 227ff
  reciprocal, 211, 215
  spontaneous, *see* Spontaneous resolution
  strategy in, 253
  thermodynamic control in, 197
  types of, 215
Resolution index (RI), 204
Resolvability criterion, 424
Resolving agents, for acids and lactones,
      217, 218
  for alcohols, diols, thiols, dithiols, and
      phenols, 220–222
  for aldehydes and ketones, 223–225
  for amino acids, 220
  for bases, 218, 219
  desirable characteristics of, 210
  desirable structural features of, 216
  lists of, 216
  miscellaneous, 225–227
  resolving ability of, 228
  synthetic, 211
  use in nonstoichiometric amount, 230
Resolving machine, 277
Resolvosil™, 172
Retention of configuration, 90, 91
Rh(BINAP)(OCH₃)₂+, 265
Rh[CHIRAPHOS]₂+Cl-, 265
RIA, *see* Radioimmunoassay
Ribonolactone, 221
Ribonuclease, CD spectra of, 580
Ribose, 40, 87
Rigid models, 10
Rigid rotation, 31
Ring closure, activation enthalpy for, 432
  activation entropy for, 432
  rates of, 432
Ring compounds, substituted, determination
      of configuration of, 423ff
Ring constraint, 437
Ring formation, stereoelectronic
      requirements of, 462
Ring inversion, 319, 326, 438
  in cyclobutane, barrier to, 481
  in cyclohexane, barrier to, 438
  in higher cycloalkanes, 485

Ring reversal, *see* Ring inversion
Ring strain in cycloalkanes, 429–431
ROA, *see* Raman optical activity
Rotamer, 445
Rotation, 46. *See also* Optical rotation
  activation energy for, in alkenes, 343, 344
  dependence on pH, 6
  about single bonds, 376, 377
  time scale of, 401
Rotation-reflection axis, 47. *See also*
      Alternating axis of symmetry
Rotation-reflection operations, 47
Rotational barrier, 342, 377, 378
  in acetaldehyde, 390
  of alkenes, 339, 340
  in amides, 65, 392
  in propionaldehyde, 390
Rotational entropy, 62, 379
Rotational strength, 550, 559
Rotaxanes, 505ff
  directed synthesis of, 508
  statistical synthesis of, 507
  template synthesis of, 508, 510
[Ru(BINAP)]OAc₂, 265
Ruggli principle, *see* High dilution
Rule of shift, *see* Freudenberg's rule of shift

s, definition, 84
S (*sinister*), definition, 67
S parameter, 228
Saccharides, CD of, 569
Saccharimetry, 586
Sachse-Mohr theory, 436, 437, 495
Saddle point, 381, 394
Salicylideneamino chirality rule, 569
Salt formation, effect of, on conformational
      equilibrium, 446, 447
Sampling of racemates, 109
Sarcolactic acid, 37
Sarin, 137
Sawhorse formula, 80
Scacchi's salt, 110
Scalar properties, 37, 41, 59
Scattering density (X-ray), 17
Scattering pattern (X-ray), 17
Scattering power, for neutrons, 19
Schmidt rearrangements, 94
Schrodinger equation, 21
s-cis Conformer, 393, 394
Screw patterns of polarizability, 595

Secondary alcohols, configuration of, 96, 97
Sector rules, 548, 553, 563, 564
Seignette's salt, 198
Selectand, 166
Selector, 166
L-Selectride™, 468
Selenocholine, 386
Self-association, 125, 393
Semiempirical quantum mechanical
    methods, 21
Sense of chirality, 4, 39, 66, 609
Sensitizer triplet energies, 369
Separation, of diastereomers, by trituration,
    248
Sequence rules, 68ff
Serum reactions, 178
Seven-membered rings, 344, 345
    conformation of, 485–487
Sharpless epoxidation, 98
Sharpless reagent, 261, 262, 265
Sharpless' Rule, 98
β Sheet, 20, 576, 579, 580
Shielding effect of C–C bond, 455
Shielding, angular dependence of, 455, 456
Shift reagents, 157
Si, 317, 320
Single bond, 65
Singlet excited state, 368, 369
Site exchange, 326–328
    rate of, 402
Sixfold barrier, 395, 629
Size of substituents, in biphenyls, 624, 625
Skewed alkenes, 564
Small rings, 430
    formation of, 433, 434
Small-ring ketones, 467
Smectic phase, 574
S_N2 reaction, 91–93, 370, 461
S_N2 rule, 92, 93
Snatzke MO "recipe," 549
Soccerballene, see Buckminsterfullerene
Sodium ammonium tartrate, 2, 3, 110, 198
Sodium deoxycholate, 176, 270
Sodium lactate, 127
Sodium rubidium tartrate, 77
Soft organic bases, ee analysis of, 159
Solenopsin, 470, 471
Solid solution, formation of, 230
Solid state IR spectra, 200

Solid state ^{13}C NMR spectra, of
    enantiomers, 122
Solubility, of enantiomers, 116
Solubility ratio, 118
Solute-solute association, 590
Solute-solvent association, 590
Solvation, 6, 21, 393, 453
    effect of, on conformational equilibrium,
        446, 447
    of polar conformers, 384
    stabilization by, 24
Solvation energy, 21, 24, 383
Solvation term, 478
Solvent, importance of in optical rotation, 6
Solvent dependence, of anomeric effect, 387
    of enthalpy differences, in
        1,2-dihaloethanes, 385
Solvent effect, on NMR anisochrony, 155
    in polarimetry, 587, 589, 592
Soman, 137
Space filling models, 25, 26
Space group, 109, 123, 201
Sparteine, 141, 175, 235, 266
(–)-Sparteine-butyllithium, 266, 584
Specific ellipticity, 549
Specific rotation, 5–7, 145, 586
    of lactic acid, 37
Specific rotivity, 589
Spectral anisochrony, 150
Spectroscopic entropy, 377
Spectroscopic methods, in configuration
    assignment, 85
Spectroscopic observability, 65
Spectroscopic transitions in MRES, 396
Spherand, 504, 505
Spherical symmetry, 58
Spiranes, 51, 608, 620–622
    chiral, 620–622
    application of Lowe's rule to, 596
(–)-Spiro[4.4]nonane-1,6-diol
    bis(p-dimethylaminobenzoate), 568
Spiro[2.2]pentane, 621
Spiro[3.3]hepta-1,5-diene, 231
Spiro[3.3]heptane-1,6-dicarboxylic acid, 620
Spiro[3.3]heptane-2,6-dicarboxylic acid, see
    Fecht acid
Spiro[4.4]nonane-1,6-dione, configuration
    of, 622
Splitting (NMR), see Coupling constants
Spontaneous resolution, 198, 201, 207

Spontaneous resolution (*continued*)
total, 204, 205
Squalene, biosynthesis of, 335
Squalene 2,3-epoxide, 272
Square conformation, 491
Stabilization of gauche conformer by
hydrogen bonding, 386
Staggered conformation, 79, 80, 323, 377,
378, 380
Static disorder, 18
Static stereochemistry, 1, 8
Statistical treatment, of olefinic proton
NMR shifts, 358
Statistics, in conformational equilibria, 381,
384, 385
"Steepest descent," path of, 22
Stegobinone, Cotton effect of, 565
Sterculic acid, 480
"Stereochemical ballast," 135
Stereochemical course of replacement, 90ff
Stereochemical memory effect, 331
Stereochemically cryptic reactions, 388
Stereochemistry, dynamic, 1
history of, 1
static, 1
Stereoconvergence, 349, 350
Stereoelective polymerization, 141, 260
Stereoelectronic control, 460–462
Stereoelectronic effect, 386, 407, 460–462
Stereogenic axis, 33
Stereogenic center, 32, 39, 43, 83, 93, 95,
303
in cyclanes, 422
Stereoheterotopic, 315, 316. *See also*
heterotopic
faces, 316, 329
ligands, 316, 329
Stereoisomer discrimination, 103
effect on reaction rates, 127
in odor perception, 133
Stereoisomerism, pharmacological
implications of, 135
Stereoisomers, 10, 315
Stereoregular polymers, 582
Stereoselective, definition of, 350
Stereoselective synthesis, 303, 305, 329
Stereoselectivity factor, biochemical, 268
Stereospecific, definition of, 350, 410
Stereospecific synthesis, directed, 350
Steric approach control, 468, 469

Steric assistance, 459, 460
Steric effects, 458–461
Steric factors, 407
in amides, 393
in gauche conformers, 380, 386
Steric hindrance, 458–460
Steric inhibition of resonance, 356
Steric interactions, 22–24
Steric repulsion, 378, 393
Steroidal 11-bromoketone, 554
Steroidal ketones, ORD of, 553
Steroids, 73, 74, 317, 318, 482, 499, 500
conformational factors in, 500, 501
Stevens rearrangement, 94
cis-Stilbene, 363
trans-Stilbene, 363, 364
Stilbenes, addition of bromine to, 351
barrier in, 344
configuration of, 351
equilibration of, 363
heats of combustion of, 364
heats of hydrogenation, difference of, 365
isomerization of, 367
Stirring rate, in preferential crystallization,
204
Stochastic achirality, 541
Straight-chain hydrocarbons, heat of
formation of, 430
Strain, 429–431
$A^{(1,3)}$, *See* $A^{(1,3)}$ strain
Strain energy, 22, 430, 431
Strain parameters, 22
Strained alkenes, 363, 364
Strainless rings, 430
s-trans conformation, 393, 394
Stretching frequencies, 10
Structure, 8ff, 30
a priori calculation of, 20ff
determination of, 17ff
tabulation of, 20
Structure amplitude (X-ray), 18
Structure factor (X-ray), 17
Strychnine, 212
Stuart-Briegleb models, 26
Styrene, conformation of, 397
rotational barrier in, 394
Subilopeptidase A (Alcalase™), 272
Substituent effects on ^{13}C NMR shifts,
456–458

Substitution criterion for heterotopicity, 307, 308, 310–313
Substitution shifts in ^{13}C NMR, 456–458
Substitution-addition criteria for heterotopicity, 308–314
Succinic-*d* acid, 333
Succinonitrile, 385
Sucrose, 2, 134
Sugars, configurational notation of, 40, 75
Sulcatol, 134, 272
N-Sulfinylcarbamates, ene reaction of, 265
Sulfoxides, diastereomeric, 85, 314
Sulfoximine reagents, 265
Sulfur hexafluoride, 57
Sulfur pentafluorochloride, SClF$_5$, 54
Superposability, 31, 36
Superposable structure, 31, 48, 54
Superposition, 46
Supersonic jet nozzle, 396
Symbolic addition, 18
Symmetry, 45ff
  cylindrical, 56
Symmetry arguments, 86–89
Symmetry axis, *see* Axis of symmetry
Symmetry criterion for heterotopicity, 309, 311, 312, 314
Symmetry element of the first kind, 310, 314
  of the second kind, 310, 314
Symmetry elements, 45ff
Symmetry group, order of, *see* Order
Symmetry number, 62, 63, 379
Symmetry operation of the first kind, 46
  of the second kind, 46
Symmetry operations, 48, 49
Symmetry operators, 48ff
Symmetry plane, *see also* Plane of symmetry
Symmetry point group, 48ff. *See also* Point group
  of decalins, 495
  order of, *see* Order
Symmetry properties, 84, 88
Symmetry rules, 553
Symmetry-based methods, of configuration assignment, 424, 425
Syn, *see* Synperiplanar
Syn addition, 369
Syn conformation, 384
Syn elimination, 369
Syn hydroxylations, 371
Syn, anti prefixes, 341

Syn-anti nomenclature, 83
Synaxial hydrogen atoms, 442, 472, 475
Synaxial interaction energies, 451, 452
Synaxial methylene interaction, 497
Synaxial substituents, 496
Synclinal, 16
Synclinal conformation, 16, 389
Syndiotactic vinyl polymers, 582
Synperiplanar, 16
Synperiplanar conformation, 16, 394
Synperiplanar orbitals, 461

Tabun, 137
TAC, *see* Cellulose triacetate, microcrystalline
Tailor-made additives, 77–79
TAPA, *see* α-(2,4,5,7-Tetranitro-9-fluorenylideneaminooxy)propionic acid
Tartaric acid, 32, 42, 43, 49, 77, 86, 104, 122, 198, 211, 218, 219, 243, 308–311, 641
Tartaric-2-*d* acid, 308, 311
Tartaric acid enantiomers, heat of mixing of, 104
Tautomerism, 13
Temperature dependence of chemical shifts, 402
Temperature effect in polarimetry, 587, 589
Temperature factors (X-ray), 18
Template method, 508, 509
Ten-membered ring, 490, 491
Terleucine, 220
Ternary phase diagram, 116
Terphenyls, 32, 33, 625
Tetra-*tert*-butylcyclobutadiene, 514, 515
Tetra-*sec*-butylcyclobutane, 46–48
1,1,2,2-Tetra-*tert*-butylethane, 46
Tetra-*tert*-butylethylene, 345, 346
Tetra-*tert*-butyltetrahedrane, 57, 514, 515
Tetrahedrane, 57, 514, 515
Tetrahedron, 3, 4, 37, 38, 45, 57
Tetrahydrofuran-2-carboxylic acid, 231
Tetrahydro-5-oxofurancarboxylic acid, 223
Tetrahydroxyadipic acids, 43, 84
Tetramesitylethylene, 175
1,1,5,5-Tetramethylspiro[3.3]heptane, 620, 621
3,3,6,6-Tetramethyl-1,2,4,5-tetrathiane, 472

α-(2,4,5,7-Tetranitro-9-fluorenylidene-aminooxy)propionic acid (TAPA), 636

1,3,5,7-Tetraoxacyclooctane, 489

Tetrapinanyldiborane, 256, 264

Tetraatomic molecule, 8

Tetroses, 39–41, 86, 87

TFAE, *see* 2,2,2-Trifluoro-1-(9-anthryl)ethanol

TFNE, *see* 2,2,2-Trifluoro-l-(1-naphthyl)ethanol

TFPE, *see* 2,2,2-Trifluoro-l-phenylethanol

Thermal ellipsoids, 18

Thermal energy, 21

Thermal isomerization, of cis-trans isomers, 366

Thermal motion (X-ray), 18

Thermodynamic vs. kinetic control, 432, 433

Thermodynamic parameters, in substituted cyclohexanes, 446

Thiaheterohelicenes,

Thiane sulfonium salt, 476

Thiane sulfoxide, 476

Thietane, 481

Thioamides, barrier in, 65, 346, 347, 625
  bond order, 65
  partial double bonds in, 65, 346

Thorpe-Ingold effect, 433, 434

Threading of rotaxanes, by chemical affinity, 508

Three-bladed propeller, 633–635

Three-membered rings, 480

Three-point contact, 306, 307

Three-point interaction models, 137, 169, 170, 227

Threefold barrier, 629

threo, 82. *See also* Erythro

Threonine, 20, 104
  enantiomers, heat of mixing of, 104

Threose, 40, 41, 81, 86, 87

Through-space effect, 362, 425

Time scale, in NMR spectroscopy, 34, 399, 401

Titanium tetraalkoxides, 98

TLC, *see* Thin-layer chromatography

Toluene, 395

p-Toluenesulfonate group, conformational energy of, 461

Topomerization, 328, 346, 439, 627

Topomers, 328

Torsion, 10, 25

Torsion angle, 8, 9, 14–16, 22, 30, 340, 360, 376, 377, 381, 635
  in 1,3-butadiene, 394
  in butane, 380
  in cyclohexane, 437, 438
  in cyclohexenes, 463
  in cyclopentane, 484
  endocyclic, 437, 493
  exocyclic, 493, 494
  notation, 16, 485
  in peroxides, 397, 398
  sign convention of, 16, 377

Torsion axis, 32

Torsional degrees of freedom, 549

Torsional energy, 378, 380

Torsional potential, 377

Torsional strain, 22, 429, 432, 481, 482

Torsional vibration, 383

Tosylate group, *see* *p*-Toluenesulfonate

TOT, *see* tri-*o*-Thymotide

trans, 341, 422

trans-cis interconversion, 366ff

Transannular strain, 430, 501

Transfer-ribonucleic acid (tRNA), CD of, 581

Transferases, 271

Transformations, between racemic forms, 107

Transition moment, 549

Transition states, planar, 278

Translational entropy, 379
  loss of, 432

Transoid ring juncture, 497

Transport, across membranes, 276

Tree-graph exploration, 68, 69

Trefoil knot, 505–509
  template synthesis of, 508, 509

Tri-*o*-thymotide (TOT), 123, 206, 234
  racemization of, 279

Triacetylcellulose (TAC), 629. *See also* Cellulose acetate, monocrystalline

Triage, 198

Trial structure, 18, 22

Triarylboron, 633

Triarylmethanes, 633

Triatomic molecule, 8

Triazolidinediones, 225

Tribenzoates, of steroids, analysis by exciton chirality method, 569 of sugars, analysis by exciton chirality method, 569

Tricarboxylic acid cycle, *see* Citric acid cycle

1,3,5-Trichlorobenzene, 56, 62

Trichlorocrotonic acid, 349

1,1,1-Trichloroethane, 54

1,1,1-Trideuterio-2-propanol, 60, 322, 323

N-Trifluoroacetyl-α-amino acid isopropyl esters, 166

N-Trifluoroacetyl-L-isoleucine dodecyl ester, 166

N-Trifluoroacetylproline isopropyl ester, 168

2,2,2-Trifluoro-1-(9-anthryl)ethanol (TFAE), 153, 154, 156, 164, 169, 238, 513

2,2,2-Trifluoro-1-(1-naphthyl)ethanol (TFNE), 154

2,2,2-Trifluoro-1-phenylethanol (TFPE), 153, 154, 156

Triglycerides, 124

Trihydroxyglutaric acids, 43, 83, 84, 87

3,7,11-Trimethylcyclododeca-1,5,9-triene, 49, 50

3,3,5-Trimethylcyclohexanone, reaction with nucleophiles, 468

Trimethylene oxide, 481

Trimethylene sulfide, 481

4,5,8-Trimethylphenanthrene-1-acetic acid, 636

Triphenylene, 56

Triplet biradical, resonance-delocalized, 344

Triplet excitation, 367

Tripodal ligand, 14, 15

Triptycenes, analysis by exciton chirality method, 569
    atropisomerism in, 630–632

Tris-1-(2-methylnaphthyl)borane, 634

Trisubstituted alkenes, configuration of, 354, 358–362

Tröger's base, 153, 154, 175, 238, 240, 243

Trolox™, 162

Trolox™ methyl ether, 162

Tropylium cation, 56

Trouton's rule, 453

Tryptophan, 198

Turnover number, 269

β-Turns, 577, 579

Twelve-membered ring, 491

Twist conformations, 438, 439, 466, 560

Twist form, 34

Twist structure in perhydrophenanthrene, 496, 497

Twist-boat, 486–489

Twist-boat family, 485, 487

Twist-boat-boat, 489, 490

Twist-boat-chair, 487–489, 491

Twist-boat-chair-chair, 491

Twist-chair, 486–488, 491

Twist-chair-chair, 488

Twist-to-chair activation energy, 439

Twistane, 51, 515

Twisted alkenes, 344–346

Twisted conformations of alkenes, 344, 368

Two-ring flip mechanism, 634, 635

Tyrosine, decarboxylation of, 335

o-Tyrosine, 220

Tyrosine hydrazide, 220, 222

u (for unlike), 81

ul, 97

Umbrella motion, in tertiary amines, 10, 30, 33

10-Undecenyl (*R*)-*N*-(1-naphthyl)alaninate, 246

Unit cell, 18

Uranocene, 55, 56

Urazoles, 225

Urea, chiral crystals, resolution in, 234

Urey-Bradley nonbonded energy, 22

α,β-Unsaturated carbonyl compounds, cis-trans isomerization of, 367

α,β-Unsaturated ketone, kinetic resolution of, 265

β,γ-Unsaturated ketone helicity rule, 564

α,β-Unsaturated ketones, s-cis, s-trans, 357, 394

UV absorption, of 1,3-butadiene, 393

UV-vis spectroscopy, time scale in, 399

V_1 potential, 384

V_2 potential, 384

V_3 potential, 377, 384

Valence bond tautomerism, 13

Valency angle, 376. *See also* Bond angles

Valine, assignment of diastereotopic methyl groups in, 323, 324

Valine-4-^{13}C, 72, 73

van Arkel rule, *see* Dipole rule

van der Waals compression, 429, 430
    effect on proton chemical shift, 455

van der Waals distance, 23, 378
van der Waals interaction, 23, 458
  attractive, 391, 392, 442, 458
  repulsive, 378, 380, 382, 384, 458
van der Waals radius, 23, 442, 624
van der Waals strain, 430
van't Hoff isochore, 400
VCD, see Vibrational circular dichroism
Vibrational circular dichroism (VCD), 534,
  597
Vibrational entropy, 379, 439
Vibrational frequencies, 19
Vibrational optical activity (VOA), 597, 598
Vibrational optical rotatory dispersion, 597
Vibrational spectra, 4
  in determination of E-Z configuration,
    353, 356, 357
Vibrational spectroscopy, 19, 65. See also
    Infrared and Raman spectroscopy
Vicinal action, 594
Vicinal coupling constants, 360, 361
  in cyclohexanes, 454
  $^1H/^1H$, as a function of torsion angle, 403,
    404
Vicinal nonadditivity of conformational
    energies, 450
Vicinal substituents, 450
Vincadifformine, racemization of, 280
Vinyl group, conformational energy of, 446
2-Vinyltetrahydropyran, 231
Vitamin K_3 2,3-epoxide, 253
VOA, see Vibrational optical activity
von Auwers-Skita rule, 379, 453

Walden inversion, 91, 92. See also
    Inversion of configuration
Walsh orbitals, 394, 395
Water molecule, symmetry of, 54, 62
Wedge notation, 485, 486

Wieland-Miescher ketone, 128, 235, 236,
  254
Winstein-Holness equation, 407–410, 441,
  461
Winstein-Holness kinetics, 410–413
Winstein-Holness rate constant, 409
Wittig reaction, 350–352
Wittig reagents, triarylphosphine derived,
  351
Wittig rearrangement, kinetic resolution in,
  266

X-ray correlation method, 85, 86
X-ray crystallography, 75–77
X-ray diffraction, 4, 17–19, 85
  and configuration assignment, 75–77, 85,
    86
  of cyclodecanes, 490, 491
  in determination of E-Z configuration,
    353
  and force field calculations, 491
X-ray powder diagrams, 124
X-ray powder diffraction spectra, 124, 200
X-ray scattering, anomalous, 76
X-ray structure analysis (configuration), 76,
    77, 85, 86
o-Xylene, 395
Xylose, 40, 87
p-Xylyl dibromide, 508, 510

YADH, see Yeast alcohol dehydrogenase
Yeast, 271
Yeast alcohol dehydrogenase, 330–332
Yeast lipase, 272

Z (zusammen), 340, 341
Zero-point energy, 21
Zigzag formulas, 79, 80, 83
Zinc blende (ZnS), 76
Zwitterionic transition state, 343